Lecture Notes in Physics

The Editorial Policy for Proceedings

The series Lecture Notes in Physics reports new developments in physical research and teaching – quickly, informally, and at a high level. The proceedings to be considered for publication in this series should be limited to only a few areas of research, and these should be closely related to each other. The contributions should be of a high standard and should avoid lengthy redraftings of papers already published or about to be published elsewhere. As a whole, the proceedings should aim for a balanced presentation of the theme of the conference including a description of the techniques used and enough motivation for a broad readership. It should not be assumed that the published proceedings must reflect the conference in its entirety. (A listing or abstracts of papers presented at the meeting but not included in the proceedings could be added as an appendix.)

When applying for publication in the series Lecture Notes in Physics the volume's editor(s) should submit sufficient material to enable the series editors and their referees to make a fairly accurate evaluation (e.g. a complete list of speakers and titles of papers to be presented and abstracts). If, based on this information, the proceedings are (tentatively) accepted, the volume's editor(s), whose name(s) will appear on the title pages, should select the papers suitable for publication and have them refereed (as for a journal) when appropriate. As a rule discussions will not be accepted. The series editors and Springer-Verlag will normally not interfere with the detailed editing except in fairly obvious cases or on technical matters.

Final acceptance is expressed by the series editor in charge, in consultation with Springer-Verlag only after receiving the complete manuscript. It might help to send a copy of the authors' manuscripts in advance to the editor in charge to discuss possible revisions with him. As a general rule, the series editor will confirm his tentative acceptance if the final manuscript corresponds to the original concept discussed, if the quality of the contribution meets the requirements of the series, and if the final size of the manuscript does not greatly exceed the number of pages originally agreed upon.

The manuscript should be forwarded to Springer-Verlag shortly after the meeting. In cases of extreme delay (more than six months after the conference) the series editors will check once more the timeliness of the papers. Therefore, the volume's editor(s) should establish strict deadlines, or collect the articles during the conference and have them revised on the spot. If a delay is unavoidable, one should encourage the authors to update their contributions if appropriate. The editors of proceedings are strongly advised to inform contributors about these points at an early stage.

The final manuscript should contain a table of contents and an informative introduction accessible also to readers not particularly familiar with the topic of the conference. The contributions should be in English. The volume's editor(s) should check the contributions for the correct use of language. At Springer-Verlag only the prefaces will be checked by a copy-editor for language and style. Grave linguistic or technical shortcomings may lead to the rejection of contributions by the series editors.

A conference report should not exceed a total of 500 pages. Keeping the size within this bound should be achieved by a stricter selection of articles and not by imposing an upper limit to the length of the individual papers.

Editors receive jointly 30 complimentary copies of their book. They are entitled to purchase further copies of their book at a reduced rate. As a rule no reprints of individual contributions can be supplied. No royalty is paid on Lecture Notes in Physics volumes. Commitment to publish is made by letter of interest rather than by signing a formal contract. Springer-Verlag secures the copyright for each volume.

The Production Process

The books are hardbound, and quality paper appropriate to the needs of the authors is used. Publication time is about ten weeks. More than twenty years of experience guarantee authors the best possible service. To reach the goal of rapid publication at a low price the technique of photographic reproduction from a camera-ready manuscript was chosen. This process shifts the main responsibility for the technical quality considerably from the publisher to the authors. We therefore urge all authors and editors of proceedings to observe very carefully the essentials for the preparation of camera-ready manuscripts, which we will supply on request. This applies especially to the quality of figures and halftones submitted for publication. In addition, it might be useful to look at some of the volumes already published.

As a special service, we offer free of charge LATEX and TEX macro packages to format the text according to Springer-Verlag's quality requirements. We strongly recommend that you make use of this offer, since the result will be a book of considerably improved technical quality.

To avoid mistakes and time-consuming correspondence during the production period the conference editors should request special instructions from the publisher well before the beginning of the conference. Manuscripts not meeting the technical standard of the series will have to be returned for improvement.

For further information please contact Springer-Verlag, Physics Editorial Department V, Tiergartenstrasse 17, W-6900 Heidelberg, FRG

M. Dienes M. Month S. Turner (Eds.)

Frontiers of Particle Beams: Intensity Limitations

Proceedings of a Topical Course
Held by the Joint US-CERN School
on Particle Accelerators
at Hilton Head Island, South Carolina, USA
7-14 November 1990

Springer-Verlag Berlin Heidelberg GmbH

Editors

M. Dienes
205 E. Harmon, Apt. 908
Las Vegas, NV 89109, USA

M. Month
Brookhaven National Laboratory
Upton, NY 11973, USA
and:
US Particle Accelerator School
Fermilab, P. O. Box 500
Batavia, IL 60510, USA

S. Turner
CERN
CH-1211 Geneva, Switzerland

ISBN 978-3-662-13892-2 ISBN 978-3-540-46797-7 (eBook)
DOI 10.1007/978-3-540-46797-7

Originally published by Springer-Verlag Berlin Heidelberg New York in 1992
Softcover reprint of the hardcover 1st edition 1992

Typesetting: Camera ready by author

58/3140-543210 - Printed on acid-free paper

JOINT US-CERN SCHOOL ON PARTICLE ACCELERATORS

FRONTIERS OF PARTICLE BEAMS: INTENSITY LIMITATIONS

Members of the Organizing Institutions

US Particle Accelerator School
Director: M. Month
M. Paul
S. Winchester
M. Dienes
A. Gonzales

CERN Accelerator School
Head: P.J. Bryant
S. Turner
S. von Wartburg

CEBAF
Director: H. Grunder
Associate Director: C. Leemann

Members of the Programme Committees

J. Bisognano
P. Bryant
A. Chao
J.-P. Delahaye
J. Gareyte
I. Hofmann
S. Holmes
K. Hubner
C. Leemann
H. Mais
M. Month
G. Rees
R. Ruth
J. Seeman
R. Shafer
R. Siemann
W.T. Weng
H. Wiedemann
B. Zotter

Local Organising Committees

C. Kuc
J. Leverenz
M. Paul
S. Turner
S. von Wartburg
S. Winchester

Editors of the Proceedings

M. Dienes
S. Turner
M. Month

Sponsors

US Department of Energy
National Science Foundation
CERN
CEBAF

PREFACE

Since 1985, the CERN Accelerator School (CAS) and the US Particle Accelerator School (USPAS) have jointly organized a series of specialized courses. The first of these, on the topic of nonlinear dynamics and its relationship to accelerator physics, was held in Santa Margherita di Pula, Sardinia. The second was held on South Padre Island, Texas, in 1986 with the title "Frontiers of Particle Beams" and covered the latest concepts in the physics and technology of particle accelerators. Next came the course held in Anacapri, Isola di Capri in 1988. At this point "Frontiers of Particle Beams" was adopted as a series title and this particular course concentrated on the techniques of observing, diagnosing, and correcting the properties of particle beams in accelerators and storage rings. The proceedings of each of these joint schools were published in the *Lecture Notes in Physics* series of Springer-Verlag as volumes 247, 296, and 343 respectively.

The present volume is the proceedings of the latest of these joint schools, held on Hilton Head Island, South Carolina, in 1990. This course dealt with intensity limitations and was centered on a series of lectures which could be divided into the following main categories:

- Self and environmental fields
- Coherent instabilities and their simulation
- Beam-beam interaction
- Other multiparticle effects
- Beam source limitations
- Engineering limitations

The subject of intensity limitation was a natural sequel to the earlier Capri course on observation, diagnosis, and correction of particle beams, which had been directed towards the problems of machine commissioning and had been, at the time, very topical because of the imminent commissioning of several new machines. Since commissioning naturally evolves into a steady struggle to increase intensity and luminosity, the present proceedings are also a natural evolution.

Some 145 participants attended the course, of whom 43 were from Asia and Europe and the rest from North America. All the lecturers were chosen as being at the forefront of the latest developments in this field and came in equal numbers from North America and Europe.

Our host on this occasion was the Continuous Electron Beam Accelerator Facility (CEBAF); on behalf of the CERN and US Accelerator Schools we express our sincere appreciation to its Director and his staff for their support. Special thanks are also due to the CAS Advisory Committee and the USPAS Executive Board for their continuing support, as well as to the Programme and Local Organizing Committees for their dedicated efforts. As usual the lecturers deserve all our praise for accepting a very exacting task and for the serious effort they put into preparing and presenting the various topics as well as writing the chapters presented in this volume. Finally we thank all the participants for their attendance at the course and their lively interest during the lectures and in the ensuing discussions.

Chicago, Illinois
Geneva, Switzerland

January 20, 1992

M. Month
USPAS

S. Turner
CAS

CONTENTS

WELCOME ADDRESS

This Welcome Address is devoted to *Schools* and then to *Accelerators*. What does an accelerator school, such as the one we are embarking upon, hope to achieve? The immediate aim can perhaps be described as an *exposure operation*. With the help of 23 lecturers we are going to review and expose the inner workings of high-intensity particle beams. Newcomers to the field will find it hard to follow in detail because of the high density of information and the short time that we have together. More experienced participants will be able to select topics of interest, but the basic aim is to make you aware of what is known in this field.

The second aim is more durable. The lecturers will write up their talks for the proceedings, which they will try to make as complete but as concise as possible. Several such proceedings have been created by schools like this one and many are devoted to specialist topics within the accelerator field. It is our hope that if you leave for a sabbatical year you will consider it worthwhile to take the proceedings for your speciality with you.

When planning the programme for a school one discovers grey areas which may not be adequately covered in the literature. For example, the most detailed and rigorous derivations of the Vlasov equation are found in plasma physics books. In these derivations, it is usual to invoke Debye shielding and a neutral plasma. Accelerator beams, however, are far from being neutral and their dimensions are often less than the Debye shielding length. If you are interested in the validity of the Vlasov equation when applied to accelerator beams and criteria for judging that validity, you can now refer to one of our earlier school proceedings [1].

One also uncovers work that has never been published. This situation often comes to light in discussions with colleagues who are about to retire. Here again the school can play a useful role by providing the encouragement and opportunity to write up and publish this work, which would otherwise be lost.

A school is also a good framework for preserving the history and traditions of accelerators. Just 10 to 15 years ago, many of the founder members of the accelerator community were still working with us. We are just now at the point where the very early years are slipping beyond living memory. This advancing "curtain" makes it all the more important for a school to record our history and traditions as they develop. The Prize for Achievement in Accelerator Physics and Technology awarded by the U.S. Particle Accelerator School serves this purpose by drawing attention to important personal contributions and main developments.

In connection with the last point, I should now like to pay a small tribute to the founders of our community by recounting the early history of accelerators. This history can be traced from three separate roots. Each root is based on an idea for an acceleration mechanism and all three originated in the twenties.

The first root to be described is generally taken as the principal "history line" since it was the logical consequence of the vigorous physics research programme in progress at the turn of the century. Indeed, particle physics research has always been the driving force behind accelerator development; it is therefore natural also to consider high energy physics as its birthplace. The main events along this history line are listed in Table I. The line is started at the end of the last century to show the natural progression through atomic physics to nuclear physics and the inevitable need for higher-energy and higher-intensity beams of "atomic projectiles" than those provided by natural radioactive sources. In this context, the particle accelerator was a planned development and it fulfilled its goal of performing the first human-controlled splitting of the atom. It was Ernest Rutherford, in the early twenties, who realised this need, but the electrostatic machines then available were far from reaching the necessary voltage and for a few years there was no advance. Suddenly the situation changed, in 1928, when Gurney and Gamov independently predicted tunnelling [2], and it appeared that an energy of 500 keV might just suffice to split the atom. This seemed technologically feasible to Rutherford and he immediately encouraged Cockcroft and Walton to start designing a 500-kV particle accelerator. Four years later, in 1932, they split the lithium atom with 400-keV protons. This was the first fully human-controlled splitting of the atom [3], and it earned them the Nobel Prize in 1951.

Table I Main History Line

1895	Lenard. Electron scattering on gases (Nobel Prize).	< 100-keV electrons. Wimshurst-type machines
1913	Franck and Hertz excite electron shells by electron bombardment.	
1906	Rutherford bombards mica sheet with natural alphas and develops the theory of atomic scattering.	Natural alpha particles of several MeV
1911	Rutherford publishes theory of atomic structure.	
1919	Rutherford induces a nuclear reaction with natural alphas.	
	... Rutherford believes he needs a source of many MeV to continue research on the nucleus. This is far beyond the electrostatic machines then existing, but ...	
1928	Gamov predicts tunnelling and **perhaps 500 keV would suffice.**	
1928	Cockcroft and Walton, encouraged by Rutherford, start designing an 800-kV generator.	
1932	Generator reaches 700 kV and Cockcroft and Walton split lithium atom with only 400-keV protons. They received the Nobel Prize in 1951.	

Figure 1 shows the original apparatus, which is now kept in the Science Museum, London. The top electrode contains the proton source and was held at 400 kV, the intermediate drift tube at 200 kV, and the final drift tube and target at earth potential. This structure can be seen inside the evacuated glass tube in Fig. 1 above the curtained booth in which the experimenter sat while watching the evidence of nuclear disintegrations on a scintillation screen. The voltage generator was at the limit of the in-house technology available to Cockcroft and Walton, and the optimistic design voltage of 800 kV was never reached because of a persistent spark discharge that occurred at just over 700 kV. However, the famous atom-splitting experiment was carried out at 400 kV, well within the capabilities of the apparatus. The **Cockcroft-Walton generator,** as it became known, was widely used for many years afterwards as the input stage (up to 800 kV) for larger accelerators, since it could deliver a high current.

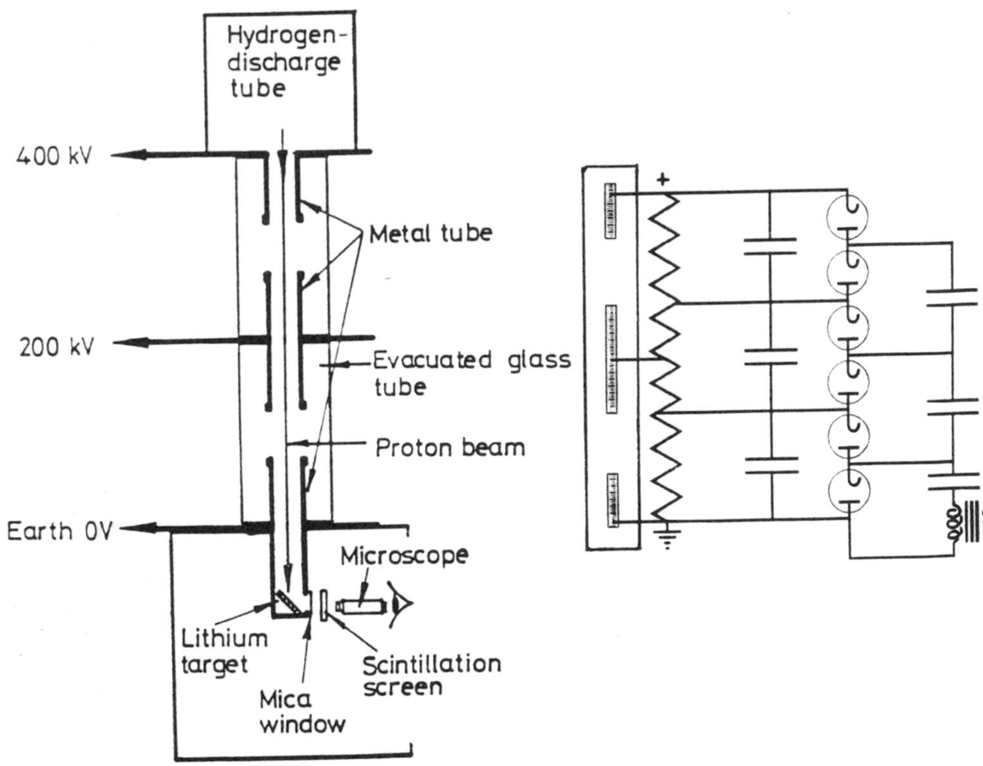

(a) Acceleration column (b) DC generator

Fig. 1 Cockcroft and Walton's apparatus for splitting the lithium nucleus

The direct-voltage accelerators were the first to be exploited for nuclear physics research, but they were limited to the maximum voltage that could be generated in the system (except for the astute double use of the applied voltage in the Tandem). This limitation was too restrictive for the requirements of high energy physics and an alternative was needed.

In fact, an alternative had already been proposed in 1924 by Ising [4] in Sweden. He planned to apply the same voltage to the particle repeatedly, using time-varying fields. This is known as **resonant acceleration**. Ising starts a new history line; its main events are given in Table II. The difference between Cockcroft and Walton's and Ising's acceleration mechanisms depends on whether the fields are static or time-varying. The former corresponds to the Cockcroft-Walton machine and the latter to Ising's resonant acceleration. All present-day high energy accelerators use the latter option of time-varying fields.

Table II The Second History Line

1924	Ising proposes time-varying fields across drift tubes. This is resonant acceleration, which can achieve energies above that given by the highest voltage in the system.
1928	Wideröe demonstrates Ising's principle with a 1-MHz, 25-kV oscillator to make 50-keV potassium ions.
1929	Lawrence, inspired by Wideröe and Ising, conceives the cyclotron.
1931	Livingston demonstrates the cyclotron by accelerating hydrogen ions to 80 keV.
1932	Lawrence's cyclotron produces 1.25-MeV protons, and he also splits the atom just a few weeks after Cockcroft and Walton (Lawrence received the Nobel Prize in 1939).

Ising suggested accelerating particles with a linear series of conducting drift tubes, and Wideröe built a proof-of-principle linear accelerator in 1928 [5]. Alternate drift tubes are connected to the same terminal of an rf generator. The generator frequency is adjusted so that a particle traversing a gap sees an electric field in the direction of its motion and while the particle is inside the drift tube the field reverses so that it is again the direction of motion at the next gap. As the particle gains energy and speed the structure periods must be made longer to maintain synchronism (see Fig. 2).

Clearly, as the velocity increases the drift tubes become inconveniently long unless the frequency can be increased, but at high frequencies the open drift-tube structure is lossy. This problem is overcome by enclosing the structure to form a cavity (in a circular machine) or a series of cavities (in a linear machine), working typically in the MHz range. The underlying principle remains unchanged, but the accelerating structure design has several variants.

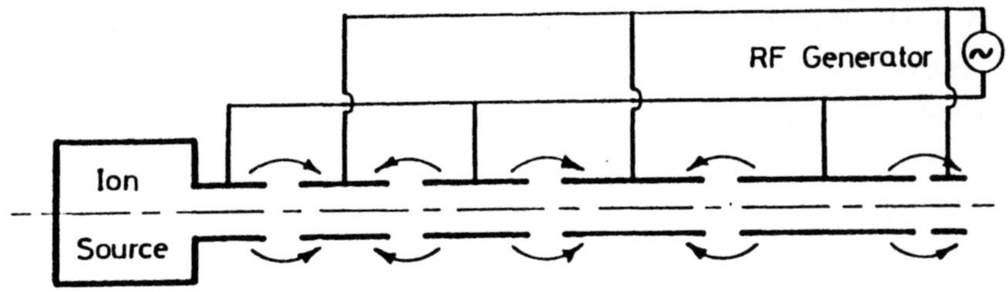

Fig. 2 RF linac

Ising's original idea can be considered as the beginning of the **true accelerator**. Indeed, the next future generation of linear colliders, which will be in the TeV range, will probably still be applying his principle of resonant acceleration, except that the frequency is likely to be in the range of tens of GHz.

Technologically the linear accelerator, or **linac**, was difficult to build and during the 1930s it was pushed into the background by a simpler idea conceived by Ernest Lawrence in 1929 [6], the fixed-frequency cyclotron (see Fig. 3). Lawrence's idea was inspired by a written account of Wideröe's work, and M.S. Livingston demonstrated the principle by accelerating hydrogen ions to 80 keV in 1931. Lawrence's first model worked in 1932 [7]. It was less than a foot in diameter and could accelerate protons to 1.25 MeV. He split the atom only weeks after Cockcroft and Walton. Lawrence received the Nobel Prize in 1939, and by that year the University of California had a 5-ft-diameter cyclotron (the Crocker cyclotron) capable of delivering protons of 20 MeV, twice the energy of the most energetic alpha particles emitted from radioactive sources.

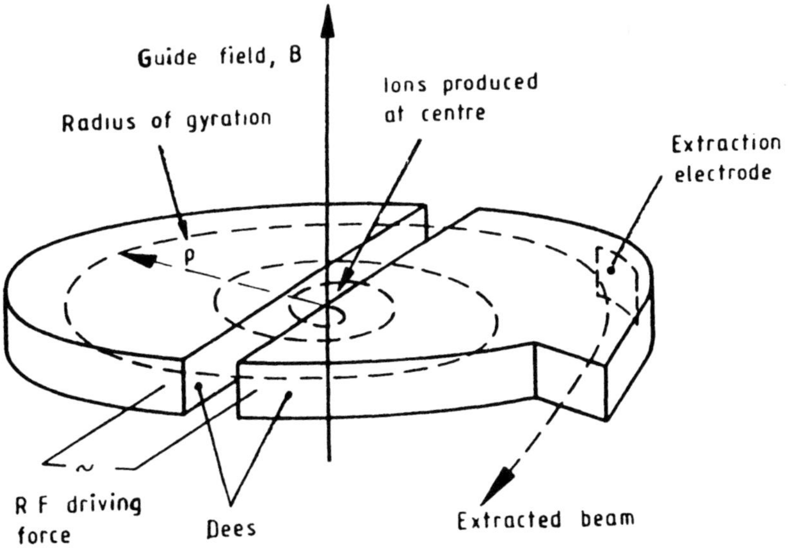

Fig. 3 Schematic cyclotron

In fact two basic geometries can be used to exploit the electric field induced by a time-varying magnetic field (Faraday's law) for acceleration. The first is the basis of Ising's idea. The mechanism can be made more evident by enclosing the gaps in cavities. The fields in a single rf cavity are shown schematically in Fig. 4(a). The azimuthal magnetic field is concentrated towards the outer wall and links the beam. According to Faraday's law, the periodic rise and fall of this magnetic field induces an electric field on the cavity axis, which can be synchronised with the passage of the beam pulse.

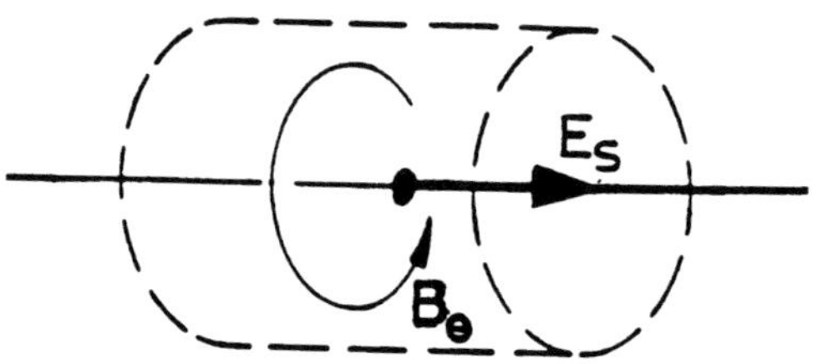

(a) Cavity configuration

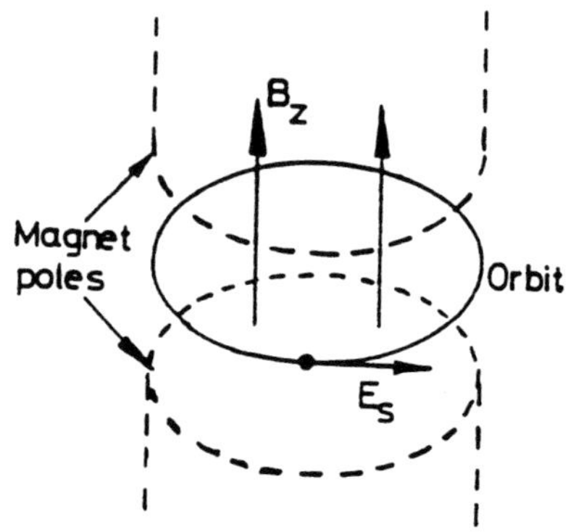

(b) Betatron configuration

Fig. 4 Acceleration configurations

Suppose that this topology is transformed, so that the beam encircles the magnetic field as shown in Fig. 4(b). Wideröe [8, 9] suggested this configuration and the acceleration mechanism, now known as **betatron acceleration**. He called his idea a **strahlung transformator** or **ray transformer**, because the beam effectively formed the secondary winding of a transformer, see Figs. 4(b) and 5. As the flux through the magnet core is increased, it induces an azimuthal emf that drives the charged beam particles to higher and higher energies. The trick is to arrange for the increase in the magnetic field in the vicinity of the beam to correspond to the increase in particle energy, so that the beam stays on the same orbit*. This device, the **betatron,** is insensitive to relativistic effects and is therefore ideal for accelerating electrons. The betatron also has the great advantages of being robust and simple. The one active element is the power converter that drives the large inductive load of the main magnet. The focusing and synchronisation of the beam energy with the field level are both determined by the geometry of the main magnet. As noted in the third history line in Table III, Wideröe put this idea in his laboratory notebook while he was a student but it remained unpublished, only to re-surface many years later when Kerst [10] built the first machine of this type.

---

* Known as the Wideröe condition, or 2-to-1 rule.

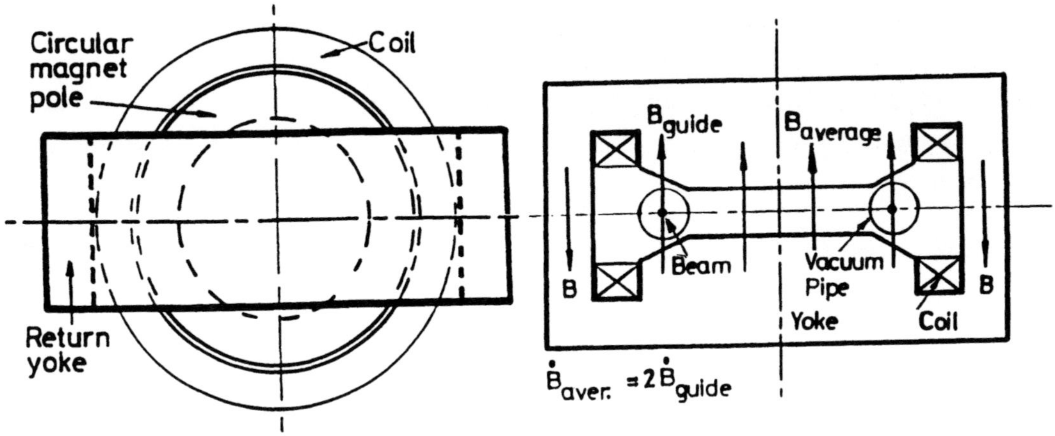

Fig. 5 Strahlung transformator or betatron

Table III The Third History Line

1923	Wideröe, a young Norwegian student, draws in his laboratory notebook the design of the betatron with the well-known 2-to-1 rule. Two years later he adds the condition for radial stability **but does not publish.**
1927	Later, in Aachen, Wideröe makes a model betatron, but it does not work. Discouraged, he changes course and builds the linear accelerator mentioned in Table 2.
1940	Kerst re-invents the betatron and builds the first working machine for 2.2-MeV electrons.
1950	Kerst builds the world's largest betatron of 300 MeV.

Thus originally three acceleration mechanisms were competing at the birth of accelerators in the 1920s. In fact, no new acceleration mechanism entered the picture until the 1960s, when collective acceleration [11] was proposed, and the 1980s, when a series of workshops were specially dedicated to finding new techniques.

The name Wideröe appears in both the second and third history lines, first for building the first proof-of-principle linear accelerator and then for his original ideas for the strahlung transformator. Wideröe worked in accelerators all his life and he is still alive and living in Switzerland.

P. J. Bryant,
Head, CERN Accelerator School

REFERENCES

[1] F. Ruggiero, "Kinetic theory of charged particle beams", in Proc. 3rd Advanced Accelerator Physics Course of CERN Accelerator School, CERN 90-04 (April 1990), 52-66.

[2] R.W. Gurney and E.U. Condon, Nature 122, 439 (1928); G. Gamov, Z. Phys. 51, 204 (1928).

[3] J.D. Crockcroft and E.T.S. Walton, "Experiments with high velocity ions", Proc. Roy. Soc. A 136, 619-30 (1932).

[4] G. Ising, Arkiv för Matematik, Astronomi och Fysik 18, 1-4 (1924).

[5] R. Wideröe, Arch. Elektrotechnik 21, 387-406 (1928).

[6] E.O. Lawrence and N.E. Edlefsen, Science 72, 376-7 (1930).

[7] E.O. Lawrence and M.S. Livingston, "The production of high speed light ions without the use of high voltages", Phys. Rev. 40, 19-35 (April 1932).

[8] W. Paul, "Early days in the development of accelerators", in Proc. Int. Symp. in Honor of Robert R. Wilson, Fermilab, 1979 (Sleepeck Printing Co. Bellwood, IL, 1979), 25-688.

[9] R. Wideröe, "Some memories and dreams from the childhood of particle accelerators", Europhysics News, (2) 9-11 (1984).

[10] D.W. Kerst, "The acceleration of electrons by magnetic induction", Phys. Rev. 60, 47-53 (July 1942).

[11] G.S. James, R.H. Levy, H.A. Bethe and B.T. Fields, "On a new type of accelerator for heavy ions", Phys. Rev. 145, 925 (1966).

INTENSITY LIMITATIONS IN CIRCULAR PARTICLE ACCELERATORS

G. H. Rees

Rutherford Appleton Laboratory, Chilton, Didcot, Oxon, UK

Abstract

First, a brief summary is given of the various intensity limitations that may occur in circular particle accelerators. Then, as an introductory topic for a course on coherent beam instabilities, the validity is discussed of some of the approximations that are used in developing the existing theories.

1. INTRODUCTION

Intensity limitations in particle accelerators may be subdivided into the following categories:

Single-particle effects,
Collective incoherent effects,
Collective coherent effects, and
Practical considerations.

A complete summary of the subject would have to cover the full range of particles, $e^{\pm}$, p, $\bar{p}$, ions, $\mu^{\pm}$ and radioactive ions, and also all the existing types of accelerators, linear accelerators, cyclotrons, synchrotrons, storage rings, etc. This is too large a range of topics to cover in a single fifty-minute talk, so the lecture will concentrate on synchrotrons and storage rings, first listing the various limiting effects and then proceeding to analyse the underlying approximations that are made in some of the existing coherent beam instability theories.

The subject of coherent beam motion has developed progressively since the late nineteen fifties, starting with the observation and analysis of the longitudinal negative mass instability for a coasting beam above transition energy. There followed the discovery of further longitudinal and transverse instabilities for both coasting and bunched beams. Accelerator school lectures tend to follow the historical development, starting with the longitudinal instability of a coasting beam, even though this is not the easiest starting point for the subject. A simpler introduction is via the longitudinal instability of a bunched beam, as the instantaneous coherent synchrotron oscillation frequency is then mainly determined by the accelerator's radio-frequency system and the presence of modulation sideband frequencies is more readily apparent. Bunched beam theory is therefore adopted as the starting point for the main part of the present paper, though it is developed in a slightly different form.

A re-look at the basic theory was prompted by the discovery on RAL's high intensity proton synchrotron, ISIS, of significant discrepancies between observation and theory, both for coasting and bunched beams. In looking again at the theory, it was decided to develop the beam environment interaction in a somewhat modified form to allow for simpler physical interpretation of the coherent effects. Thus, instead of finding the separate response for each sideband frequency in the beam current spectrum, a simple technique has been developed to give a composite modulation response for any pair of modulation sidebands. The advantages of doing this will become apparent as the technique is developed in the lecture. The response is written either in the standard form $G(j\omega,s)$, where s is the Laplace Transform Operator, or as $G(j\omega,ip)$, where a dual complex number algebra is introduced to handle separately the phase shifts of the carrier and modulation frequencies. Antidamping terms may then be readily identified. The response function is applicable to both longitudinal and transverse motion and may be developed further for the special case of coherent synchrobetatron motion. In this case, use is made of a third complex number in the algebra.

For the longitudinal motion of a bunched beam, the usual development of the Vlasov equation is questioned and an alternative development proposed, though a rigorous proof of the suggested modification has not been found. The modified development is introduced to remove two very basic anomalies that exist in the standard treatment. Also, the modification is found to be consistent with the traditional longitudinal envelope equation under space charge. For a more thorough analysis, it seems better to avoid the Vlasov approach and develop directly the basic equations of longitudinal motion. This also appears necessary to allow the correct introduction of Landau Damping terms.

In the case of a coasting beam, the use of the G-function directs attention to a further anomaly; the Keil-Schnell criterion[1] gives the threshold for longitudinal instability in terms of the total ring impedance at a single frequency, whereas the beam interaction function $G(j\omega,ip)$ involves two separate frequencies. This additional anomaly is also addressed in the lecture as are the implications for the theory. Also discussed are aspects of transverse coherent motion and the topic of beam distributions.

Editor's note: This approach is unconventional and lacks a rigorous proof. However, it is applied in these proceedings in the paper 'Some longitudinal dynamics of bunched beams' by S.R. Koscielniak.

2. SINGLE-PARTICLE EFFECTS IN CIRCULAR ACCELERATORS

In general, single particle effects lead only indirectly to intensity limitations. Nevertheless, they constitute some of the major effects that influence the basic design of an accelerator. A list of the most important single-particle effects includes:

- Betatron, synchrobetatron resonances (dynamic aperture, halo etc),
- Closed orbit errors, especially in the presence of sextupoles,
- Electron capture or stripping for the case of ions (change of charge state),
- Multiple scattering in the residual gas,
- Quantum excitation and lifetime, } for $e^{\pm}$ beams,
- Antidamping due to coupling or rf frequency errors, } for $e^{\pm}$ beams,
- Beam-residual gas bremsstrahlung, } for $e^{\pm}$ beams,
- Noise excitation for longitudinal motion,
- Decay lifetime for $\mu^{\pm}$ or radioactive ions.

The first two topics usually involve detailed computations for specified magnet lattices to establish if there is an adequate dynamic aperture and acceptable transverse coupling. This is particularly so for high brightness synchrotron light sources and for accelerators with very long injection filling times where there are enhanced non-linear fields. The closed orbit deviations are not really correctly categorised as a single particle effect for, at high transverse space charge levels, the coherent tune is involved, which is a function of the beam intensity.

The third topic involves ions, including radioactive ions, and is very important for specifying the vacuum and injection requirements of the ion accelerator involved. Relevant published literature in the field is rather sparse apart from a review article of B Franzke in the Proceedings of the 1989 CERN Advanced Accelerator School, Uppsala. This review also covers the more general topic of multiple scattering of particle beams in the residual gas of a circular machine.

There follows the three items listed specifically for $e^{\pm}$ beams, which are well understood. Adequate transverse beam acceptances and rf voltages have to be specified to give the required beam lifetime. Also, the transverse coupling and rf must be controlled and the appropriate vacuum pressure obtained. Ultra high energy protons should also be categorised here.

The last two items are included for completeness. Noise in the frequency or amplitude of the rf system or in the magnetic guide field may all lead to coherent synchrotron motion. Beam control systems usually eliminate the problem, but special care is needed to minimise the frequency noise in the rf system for long-time storage of bunched p or $\bar{p}$ beams. The decay lifetime for $\mu^{\pm}$ and radioactive ions is a fact of life that has to be accommodated.

3. COLLECTIVE INCOHERENT EFFECTS IN CIRCULAR ACCELERATORS

Here, a list of the most important effects includes:

- Betatron resonances together with incoherent transverse space charge,
- Emittance growth due to injection mismatch and space charge,
- Beam-beam interaction for colliding beams,
- Intra-beam scattering, including Touschek effect for $e^{\pm}$,
- Beam-induced gas desorption,
- Beam-induced multipactoring,
- Reduced longitudinal acceptance due to space charge or acceleration,
- Non-linearity in accelerating fields due to beam loading or parasitics,
- Bunch lengthening for $e^{\pm}$, in contrast to bunch widening.

These tend to be more complex than the single particle effects. For example, the crossing of betatron resonances due to space charge tune depression is conceptually simple but computationally very complex. Large 3-D codes are required, involving large numbers of particles, and though such codes have been developed for space charge-dominated beam line transport systems, it is still early days for their application to circular accelerators.[2] The second topic, the growth in emittance due to space charge and mismatched conditions, is known to occur at low energy in linear accelerators, and almost certainly occurs for high space-charge ring injection. The 3-D codes may also be able to simulate such injection processes.

The beam-beam interaction is one of the basic limits for colliding beams in storage rings. Extensive studies have been made of the subject over many years and it is one of the main topics for this school. The space-charge interaction is confined to the vicinity of the interaction regions so the computations per turn are less extensive than those for the case of low energy internal self-fields, but tracking has to continue for many more beam revolutions.

Intra-beam scattering is also a topic at the school; the effects are somewhat different for protons than electrons, where it is known as the Touschek effect. First analysed during the early days of low energy electron storage rings, the Touschek effect involves the elastic scattering of electrons in a bunch that collide due to their different betatron motions. There may result a change in longitudinal momentum larger than that which may be contained by the ring's radio-frequency system, so that a Touschek lifetime may be defined. The intra-beam scattering for protons is more complicated, particularly above the ring transition energy. Then, no equilibrium distribution exists and the coupling between the betatron and synchrotron motions results in a gradual increase for all three bunch dimensions. The effect is important for p-$\bar{p}$ colliders.

Beam-induced gas desorption results from particle collisions with residual gas molecules, producing ions which may subsequently collide with the vacuum chamber walls, releasing further gas atoms. The number of atoms removed from the walls depends on the surface conditions, and the outgassing may reach a level that leads to an exponential pressure blow-up. To raise the threshold level of the effect, it is important to have baked 'hydrocarbon clean' surfaces and adequate pumping speeds. Beam-induced multipactoring arises from the same origins but it is the electrons produced during the ionisation of the gas molecules that cause the effect. The electrons may travel to the vacuum walls during the interval between the passage of the main beam bunches. Multipactoring results if more than one electron is released from the wall per incident electron, so again the choice of the wall surface is important, as is its cleanliness.

Longitudinal space-charge forces introduce longitudinal defocusing at energies below the ring transition energy. The forces may be large for low energy beams, and significant compared to the rf focusing forces; then, it is necessary to increase the rf fields to preserve a specified longitudinal beam acceptance. As the bunch duration is decreased, the defocusing increases, and in fast-cycling synchrotrons there is a point early in the acceleration cycle when the acceptance is a minimum. Careful adjustment of the rf voltage may be needed at this time to compensate for the effect. For very short bunches, there may be severe instantaneous beam loading of the rf cavities, together with the excitation of parasitic resonant modes. Non-linearities may be introduced into the accelerating field waveforms and these may perturb the longitudinal motion. Bunch lengthening for electrons is a consequence of inductive ring impedances providing defocusing forces for $\gamma > \gamma_t$. Experimental results are in agreement with impedance measurements and estimates.

4. COLLECTIVE COHERENT EFFECTS IN CIRCULAR ACCELERATORS

The following is a list of the major collective coherent instability effects:

Longitudinal coherent bunched beam rigid dipole-mode instability,
Longitudinal coherent higher modes (quadrupole etc, also crossing γ_t,)
Coasting beam transverse instability (resistive wall, ions in residual gas),
Coasting beam longitudinal 'microwave' instability,

Transverse bunched beam (head-tail resistive wall or resonant element),
Bunch widening (mode coupling or many even-order mode instabilities),
Coupled bunch instabilities,
Transverse electron-proton instability,
Coherent synchrobetatron instability.

The major part of the talk is on these topics and, as noted, the longitudinal bunched beam cases are considered first. Then, the coasting beam effects are addressed, but with more emphasis placed on the longitudinal than the transverse motion. Transverse coherent bunched beam displacements will be considered only briefly.

Coupled motion between bunches is shown to be the most general coherent bunch motion, with the beam current frequency spectrum at $(l\,h \pm n)\,\omega_o \pm m\,\omega_{s,\beta}$, where l, n and m are integers, h is the rf harmonic number, and ω_o, ω_s, ω_β are, respectively, the angular revolution, synchrotron and betatron frequencies.

The bunch widening observed for $e^{\pm}$ beams is a complex intensity-related effect, with an increase in both the beam momentum spread and the bunch length. There are two possible explanations, one as a coupling mechanism between neighbouring longitudinal higher modes, and one as a turbulent condition where many even order longitudinal modes become simultaneously unstable. The latter appears to be the more likely.

An electron-proton instability may develop for an intense coasting proton beam as a coupled transverse oscillation of the protons with electrons formed by ionisation of the residual gas. The electrons become trapped in the electrostatic potential well of the protons and coupled coherent motion may result. Oscillation frequencies are of order 100 MHz and both dipole and quadrupole mode motions have been observed.

Finally, a development of the G-function, $G_\perp(jn\omega_o, i\omega_s, k\omega_\beta)$ will be considered to obtain the beam-environment interaction for coherent synchrobetatron motion. Here j, i and k are introduced as three independent complex numbers.

5. PRACTICAL INTENSITY LIMIT CONSIDERATIONS

The practicalities for achieving high circulating beam currents are very different in the cases of protons, electrons and ions. For protons, enhanced particle densities are usually obtained in a booster or main synchrotron by the charge exchange of an injected H^- beam in a stripping foil, over many turns. This technique allows considerable flexibility in 'painting' the required beam distributions. For ultra high intensities, the foil scattering, heating and lifetime are major considerations, and recent studies have been directed towards the production of foils with two unsupported edges. Multi-turn injection may then be achieved with a reduced number of proton foil traversals on the subsequent turns.

Other major features of very high intensity proton synchrotrons are the use of:

Rapid cycling magnetic guiding and focusing fields, typically at 50 Hz,
Ceramic vacuum chambers with internal radio-frequency shields,
Modular designs for the remote handling or repair of damaged items, and
Localised longitudinal and transverse beam loss collection and protection units.

For medium and high energy electron machines, high intensities lead to high power in the emitted synchrotron radiation. Heat removal and control of outgassing from the walls then require the use of special vacuum chambers, with enlarged outer chambers located away from the beam region. The high energy γ and X-rays emitted cause air ionisation around the chamber, which leads to corrosion of the adjacent components. Lead cladding of the chambers is adopted to reduce the scale of this effect.

Intense beams of circulating ions have yet to be produced but design studies have been undertaken for the special case of heavy ion fusion storage ring drivers.[3] Typically, 10 GeV storage rings for B_i^+ or U^+ ions were considered, employing conventional multi-turn injection. Very severe practical problems existed for the design of the injection septum magnets.

Finally, the problem of the design of ring rf systems for very heavy beam loading is mentioned. Robinson instability limits for the main rf system (see later) are often exceeded by large factors. One method to raise the limit is to increase the amount of installed rf power and load the cavities, but this is an expensive solution. An alternative is to provide a narrow band width rf feedback loop around the cavity-power amplifier system, centred at the fundamental rf system frequency.

6. INSTABILITY OBSERVATIONS ON THE ISIS PROTON SYNCHROTRON

Initially, a search was made for the coasting beam longitudinal microwave instability at the 70 MeV injection energy with circulating currents of up to 3 A. The main frequencies present were 202.5 MHz and higher multiples, remaining in the beam as the residue of the injector linac bunch structure. Existing theories predict that these components should grow exponentially for currents greater than about 0.1 A for the injector beam momentum spread of $< \pm 2\ 10^{-3}$.

A sensitive high frequency monitor was installed in the ring and its output fed to a spectrum analyser, set at the fixed frequency of 202.5 or 405 or 607.5 MHz. The analog output of the spectrum analyser and the injected beam signal were then observed on adjacent oscilloscope traces. As injection proceeded, the analog signal was seen to grow, but to saturate when ~ 0.5 A had been injected. Subsequently, the signal decayed and, the greater the injected beam, the more rapid was the decay. It was also found that, whatever the level of the injected beam, the analog signal began to decrease whenever injection stopped. Decay rates were a function of the beam size in the ring and the beam momentum spread. The longitudinal beam coupling impedance is dominated by the space-charge term at 70 MeV. Theory predicts that operation is far from stability at the higher injected beam currents and that small resistive impedance components should lead to very rapid growth rates. Thus, the observations are in complete disagreement with theory.

In the transverse plane, the dominant coupling impedance at injection is again due to space charge forces, and a coasting beam transverse instability is predicted due to their combined effect with the resistive wall impedance. This is in fact observed, with the fastest instability growth rate occuring, as expected, for the lowest frequency mode, $(4-Q_v)\omega_o$, with Q_v between 3.7 and 3.9 and ω_o at 2π (0.67) MHz. There is not complete agreement with theory, however, for the growth rate does not scale as $1/\sqrt{4-Q_v}$ over the range of frequencies where the wall thicknesses are greater than the skin depths. When Q_v is reduced to the lower end of its range, the instability is no longer observed, even for beam currents of 3.6 A.

The third instability investigated was that for the bunched beam vertical head-tail motion, which was observed in the 2 to 4 ms interval of the normal 10 ms acceleration cycle. The growth rate was very fast for Q_v values greater than 3.82, leading to total beam loss, but the coherent motion has been almost entirely eliminated by lowering Q_v linearly between 2 and 4 ms from 3.78 to 3.72. In all cases, the resistive wall mode observed has been one where the vertical motion has a single node at the bunch centre, which was not expected from measurements of the frequency, bunch length and the chromaticity. For the actual operating conditions, the form factors, as derived by Sacherer,[4] are small for the case of a single node but a maximum if the vertical motion exhibits 2 nodes along the length of the bunch. The extent of the disagreement with theory will be explored further in the future after sextupoles are installed in the ring to allow adjustment of the chromaticity values.

A final discrepancy between observation and theory has been found for the bunched beam longitudinal quadrupole mode. The ISIS synchrotron operates with rf harmonic number 2 so there are no out-of-phase bunch motions to complicate the observations. Towards the end of the acceleration cycle, it has been found possible to induce quadrupole mode oscillations by adjusting the tuning of the rf cavities. Modulation at twice the synchrotron frequency is observed on the accelerating field waveforms.

The frequency, f, and the bunch length, τ, are then 3 MHz and 100 ns respectively, giving an $f\tau$ product of 0.3. However, theory predicts that the quadrupole mode form factors are zero for all $f\tau$ values less than 0.5. This discrepancy is explained by the modified treatment of longitudinal bunched beam theory given in the lecture.

7. APPROXIMATIONS IN LONGITUDINAL BUNCHED BEAM INSTABILITY THEORY

The traditional approach is via the Vlasov equation, as developed by Sacherer,[4] and used by many authors eg Zotter.[5] A longitudinal density distribution function, $\psi(\phi, \dot{\phi}, t)$, is introduced for the longitudinal phase space, $(\phi, \dot{\phi})$, and the coordinate system is then changed from cartesian to polar:

$$\phi = r \cos \Theta \quad , \quad \dot{\phi}/\Omega = r \sin \Theta$$

$$\psi(r, \Theta, t) = \psi_o(r) + \psi_1(r,\Theta)\, e^{ipt}$$

where Ω is the synchrotron oscillation angular frequency, and ψ_o, ψ are the stationary and perturbed density functions respectively.

Algebraic manipulation of the Vlasov equation, neglecting the products of perturbed quantities, leads directly to the equation:

$$\left(\frac{\partial \Psi_1}{\partial t} - \Omega \frac{\partial \Psi_1}{\partial \Theta}\right) = -\left(\frac{\sin \Theta}{\Omega}\right)\left(\frac{d\Psi_0}{dr}\right)\left(\frac{d^2}{dt^2} + \Omega^2\right)\phi \qquad \ldots.(1)$$

where the RHS may be identified as the coherent excitation. The equation (1) is usually developed by substituting for the term $(d^2/dt^2 + \Omega^2)\phi$, but a physical insight for the motion is quickly lost in the subsequent Bessel function manipulations. Before making the substitution, therefore, it is advisable to consider equation (1) further. Approximations have already been introduced in writing $\dot{\phi}/\Omega = r \sin \Theta$. The reason is that there are coherent excitations at frequencies in addition to those at angular frequency, Ω. These may arise due to external modulation of the accelerating cavities or due to higher-mode frequencies, $m\Omega$. It is not surprising therefore that inspection of equation (1) reveals two anomalous features.

The first anomaly is related to the fact that the LHS of the equation is of first order in ψ_1 whereas the RHS is of second order in ϕ. This indicates that ψ_1 may change more rapidly than ϕ, which it clearly cannot.

The second anomaly is that the RHS contains excitation terms such as $(\sin m\Theta \cos pt)$ so that ψ_1 has solutions of the form $\cos(m\Theta+pt)$ and $\cos(m\Theta-pt)$. Thus, if considered as one-dimensional wave motion, there are solutions in both wave directions. Again this cannot be, for motion in the (r,Θ) space is unidirectional, with one direction below the transition energy and the other direction above. (Above transition, the negative sign on the LHS of equation (1) needs to be changed.)

Further evidence that equation (1) is not strictly correct is the following. Solutions for coherent dipole mode motion may be found by two means, either by solving directly the full equation for ϕ, which is derived in Appendix 1, or by substituting the excitation terms for $(d^2/dt^2 + \Omega^2)\phi$ in (1). If the solutions are compared for the case of a ring containing only resistive coupling impedances, they are seen to be different. The former indicates only a shift in the coherent dipole mode frequency whereas the latter shows there is damping or antidamping.

Despite these problem areas, it is still worth persevering with the Vlasov approach, though a modified development is required to remove the anomalous aspects of equation (1). Intuition suggests that the correct form of the equation should be that of a one-dimensional wave equation, but with excitation in only one direction of the wave. Consider, therefore, applying to both sides of (1) the differential operator

$$\left(\frac{\partial}{\partial t} + \Omega \frac{\partial}{\partial \Theta}\right) \equiv D$$

The LHS assumes the form of the standard 1-D wave equation, while on the RHS, the operator has to act on terms such as $(\sin m\Theta \cos pt)$ or $(\cos m\Theta \cos pt)$. For the former, the RHS contains the term

$$(m \cos m\Theta \cos pt - (p/\Omega) \sin m\Theta \sin pt)$$

which reduces to the required form for unidirectional wave excitation if $p=m\Omega$:

$$m \cos (m\Theta + pt).$$

The analysis is now simplified by introducing a dual (i,k) complex number algebra to represent modulated signals, a technique which is described in Appendix 2. Operating with D on a modulated signal gives

$$D\,(e^{km\Theta}\, e^{ipt}) = (km\Omega + ip)\, e^{km\Theta}\, e^{ipt}$$

$$D\,(e^{km\Theta}\, e^{ipt}) \cong (i + k)\, m\Omega\, e^{km\Theta}\, e^{ipt}$$

So, for the approximation $p = m\Omega$, the Vlasov equation for mode m may be written in the modified complex (i,k) form:

$$\left(\frac{\partial^2\psi_1}{\partial t^2} - \Omega^2\frac{\partial^2\psi_1}{\partial\theta^2}\right) = \left(im + \frac{\partial}{\partial\theta}\right)\left(-(\sin\theta)\left(\frac{d\psi_o}{dr}\right)\left(\frac{d^2}{dt^2} + \Omega^2\right)\phi\right) \quad \ldots(2)$$

Both of the physical anomalies in the equation have now been removed. The LHS and RHS of the equation are each of second order in t and the complex form converts the modulation term on the RHS to the required unidirectional excitation for the LHS. It is probable that there exists a more rigorous and direct derivation of (2) than that given, but such a derivation from the basic Vlasov equation has not been found. The RHS of (2) is now developed, making use of modulation response functions, $G(j\omega, ip)$, to calculate the beam environment interaction. The basis of the G-functions is explained in Appendix 2. Solutions of (2) are more readily obtained than for (1). For mode m=2, the equation leads directly to the traditional envelope equation, and for mode m=1, the equation is not incompatible with the Robinson[6] equation.

8. COMPOSITE MODULATION RESPONSE FUNCTIONS

In the beam current are revolution frequency harmonics and pairs of modulation sidebands, all of which induce fields in the structures that surround the beam. As explained in Appendix 2, the modulation response may be found for each sideband separately or a composite modulation response obtained for any sideband pair. There are many advantages in adopting the latter course; it gives a clearer insight into the source of antidamping terms and it greatly reduces the associated algebra.

The composite response is obtained in either a Laplace Transform form or as a function of two complex numbers, one of which is associated with the carrier frequency and the other with the modulation. It is useful to use both forms. In the simpler cases, equations may be written directly in terms of the operator, s, and then solved to obtain the complex mode frequencies. For other cases, however, such as coupled bunch motion or single sideband feedback systems, separate terms in i appear in the equations in addition to those in ip ($\equiv$ s). Then, it is simpler to use Nyquist stability techniques to investigate the stability of a beam oscillation mode.

Examples of $G(j\omega, s)$ and the equivalent $G(j\omega, ip)$ are given for some longitudinal beam coupling impedances in the Appendix 2. The approximate solution for a parallel resonant structure is

$$G(j\omega, ip) = R / (1 + (2Q/\omega_r)(j\Delta\omega + ip))$$

For a detuned cavity, there is an induced voltage with an in-phase and an out-of-phase component relative to the carrier of the excitation, and each component has its own frequency response characteristic with respect to the modulation.

The beam interaction is given as either the product of the beam current modulation and the modulation response function or as the sum of the products of each sideband current component and the impedance value at that sideband. Separate impedances in a ring may be summed to obtain a total impedance, $Z(j,\omega)$, and so may individual G-functions to give a total value for $G(j\omega,ip)$. The summation for G is straightforward despite the presence of the additional terms in p.

In the case where the individual ring impedances are not known, there remains only the possibility of beam measurements for estimates of the total G-function. For these measurements, beam current modulations may be induced, either by fixed frequency longitudinal excitation at angular frequencies near to $(l\omega_o \pm m\Omega)$ or by excitation with modulations of the form $(\cos l\omega_o t \cos pt)$ or $(\sin l\omega_o t \cos pt)$. Changes may be made in the beam current, the number of bunches, the bunch duration and the modulation frequency until longitudinal beam modulation is observed to vary as a function of the beam current level. Measurements have to be analysed to find an analytic function for G in terms of ω and p. In some cases for protons and ions, additional measurements may be made with coasting beams; beam transfer function measurements lead to values for $Z(\omega-p)$ and $Z(\omega+p)$ for a range of ω and p values.

9. SOLUTIONS FOR THE MODIFIED LONGITUDINAL VLASOV EQUATION

Solutions for the perturbation density ψ_1 are assumed of the form

$$\psi_1 = \sum_{m=1}^{\infty} R_m(r) \cos (m\Theta + pt + \Theta_m)$$

This is a backward wave in (r,Θ) space with Θ_m a constant and operation below transition. The motions for the coherent modes are all independent unless the frequency shifts are large enough to bring two mode frequencies close together.

First, the rigid dipole mode is considered. All the bunches move together in phase and, in the the absence of any incoherent frequency spreads, all the particles have the same coherent phase and energy excursions. Setting m equal to one, then substituting for ψ_1 in equation (2) and equating the real parts leads to

$$(s^2 + \Omega^2)\, R_1(r) \cos (\Theta+pt) = -\,(d\psi_o/dr)\, (s^2 + \Omega^2)\hat{\phi} \cos (\Theta+pt)$$

where $\hat{\phi}$ is the amplitude of the coherent dipole phase motion. It is apparent that the radial mode function, $R_1(r)$, is a scaling factor between ψ_1 and ϕ:

$$R_1(r) = -\,\hat{\phi}\, (d\psi_o/dr)$$

The solution is compatible with the Robinson equation, which remains to be solved in the usual way. A solution making use of the G-function is given in Appendix 3.

Next, the mode m equals two is considered, the so-called quadrupole or envelope mode, and again it is assumed that all the bunches oscillate in phase. The longitudinal excitation terms are derived in Appendix 1, and consideration of these and equation (2) show that the quadrupole mode is excited by voltages of the form $v_s \sin l\phi$. Such voltages have $\cos \Theta$ and $\cos 3\Theta$ terms in their Bessel function expansions, and these, when multiplied with the $\sin \Theta$ term of (2), lead to

$$v_s \sin \Theta \sin l\phi \;=\; 2\, v_s \sin \Theta\, (J_1(lr) \cos \Theta - J_3(lr) \cos 3\Theta) \;=$$

$$v_s\, (J_1(lr) + J_3(lr)) \sin 2\Theta \;=\; 4\, (J_2(lr)/(lr))\, v_s \sin 2\Theta$$

The real excitation term for the RHS of (2) is found, after substituting for $(s^2 + \Omega^2)\phi$ from Appendix 1 and writing in complex form the ($\sin 2\Theta$) dependence of the quadrupole mode:

$$4\, (J_2(lr)/(lr))\, (d\psi_o/dr)\, (\Omega^2/V)\, (2i + \partial/\partial\Theta)\, (-ke^{k2\Theta})\, (\hat{v}_s\, e^{ipt}) \equiv$$

$$8\, (J_2(lr)/(lr))\, (d\psi_o/dr)\, (\Omega^2/V)\, \hat{v}_s \cos (2\Theta + pt)$$

Thus, the quadrupole mode equation may be written.

$$(s^2 + 4\Omega^2)\, R_2(r) \cos (2\Theta + pt) \;=\; -\left(\frac{8J_2(lr)}{(lr)}\right)\left(\frac{d\psi_o}{dr}\right)(s^2 + \Omega^2)\left(\frac{\hat{\phi} \cos (2\Theta + pt)}{\sin l\phi}\right)$$

To obtain the equation as an oscillation of the bunch half phase extent, $\phi_o + \Delta\phi$, the angle Θ is set at zero and an expression found for $\hat{\phi} / \sin l\phi$:

$$\hat{\phi} / \sin l\phi = \Delta\hat{\phi} / \sin l\phi_o.$$

Equating coefficients of s^2, solving for $R_2(r)$ and substituting, gives

$$R_2(r) \;=\; -\, 8\, (J_2(lr)/(lr))\, (d\psi_o/dr)\, (\Delta\hat{\phi} / \sin l\phi_o)$$

$$(s^2 + 4\Omega^2)\, \Delta\hat{\phi}\, e^{ipt} \;=\; -\, (\sin l\phi_o)\, (\Omega^2/V)\, \hat{v}_s\, e^{ipt}$$

where ϕ_o is the equilibrium bunch half-phase extent and the expression for v_s is given in Appendix 1. This equation is equivalent to the traditional envelope[7] equation, as may be seen when the modulation, v_s, is that due only to the longitudinal space charge forces of a beam with bunches of parabolic line density. In this case, the gradient with respect to ϕ of the space charge voltage per turn is

$$\delta = -3\pi (h\,\omega_0 L)\, I / (\phi_0)^3$$

where L is the equivalent negative inductance due to the space charge, and

I is the average bunch current per rf period. Also, if:

Ω_i is the incoherent synchrotron angular frequency under space charge,

Ω_o is the value of Ω_i for zero space charge, and

Ω_2 is the coherent quadrupole mode angular frequency, then:

$$\Omega_i = \Omega_o - \Delta\Omega = \Omega_o(1 - \delta/(2V\cos\phi_s))$$

$$v_1 = (\partial\delta/\partial\phi_o)\,\Delta\phi = 3\delta\,\Delta\phi/\phi_o$$

$$(s^2 + 4\Omega^2)\Delta\phi = -\phi_o\,(\Omega_o{}^2/(V\cos\phi_s))\,(3\delta\,\Delta\phi/\phi_o)$$

$$\Omega_2{}^2 = 4\,\Omega_o{}^2(1 - \delta/(V\cos\phi_s)) + 3\,\Omega_o{}^2\,(\delta/(V\cos\phi_s))$$

$$\Omega_2 = 2\,\Omega_o - \Delta\Omega / 2$$

Thus, the quadrupole mode tune shift under space charge is one half that of the shift in the incoherent frequency. This agrees with the solution for the coherent mode frequency obtained from the traditional envelope equation.

Finally, there are the modes with m greater than two, and these may be analysed in a manner similar to that for the quadrupole mode. The radial mode functions, $R_m(r)$, separate from the equations and thus represent scaling factors for the density, ψ_1, relative to the asymmetric bunch motion (odd order modes) or the motion of the envelope (even order modes). The odd and even order modes separate into two distinct categories, with the excitation for the former and latter having the forms $v_c \cos l\phi$ and $v_s \sin l\phi$, respectively. Also, all the odd order modes have one type of excitation spectra and all the even order modes another; this is discussed in section 10. The general expressions for the radial mode functions are found to be

$$R_m(r) = (-1)^{(m+1)/2}\,(2\,m^2\,J_m(lr)/(lr))\quad(d\psi_o/dr)\,(\hat{\phi} / \cos l\phi_o)\quad \text{for m odd,}$$

$$R_m(r) = (-1)^{(m/2)}\,(2\,m^2\,J_m(lr)/(lr))\,(d\psi_o/dr)\,(\Delta\hat{\phi} / \sin l\phi_o)\quad \text{for m even.}$$

After separating $R_m(r)$, the odd and even order mode equations are in ϕ (see section 10) and $\Delta\phi$ (envelope motion) respectively, and so the next step is to find the ϕ-dependence of the amplitude of v_c and the $\Delta\phi$-dependence of that of v_s. These amplitudes are obtained from the beam current modulations and the G-functions and then substituted in the equations. If the total G-function is of the form for a single resonant structure, the resulting equations are of fourth order. An example of this is the Robinson dipole mode equation, as shown in Appendix 3. For more complex G-functions, higher order equations may result.

10. EXCITATION SPECTRA FOR ODD AND EVEN LONGITUDINAL MODES

First it is necessary to consider the spectrum of a beam with stationary bunch distributions. These are approximately realised with radiating electron beams, after many damping times have elapsed following a disturbance, and with protons and ions, after filamentation has effectively ceased, provided there is also coherent beam stability. The resulting longitudinal distributions are Gaussian for the electrons and are, typically, elliptical two-dimensional densities for the protons and ions. The latter case corresponds to parabolic longitudinal line densities.

The frequency spectrum for both these types of stationary distributions may be found directly by Fourier analysis. As an example, consider a beam of h equidistant bunches, each with parabolic line densities over a bunch duration, 2τ, and containing N particles per bunch. It may be shown that the average and the Fourier harmonics of the beam current, I_b, are

$$I_o = (hN)\, e\, \omega_o / 2\pi$$

$$I_b = I_o + \sum_1^{\infty} I_l \cos(l\phi_o t/\tau) \qquad (\phi_o = h\omega\tau)$$

$$I_l/I_o = 6\,(\sin(l\phi_o) - (l\phi_o)\cos(l\phi_o)) / (l\phi_o)^3$$

$$I_l/I_o = 2 \text{ for } (l\phi_o) \ll 1$$

For short duration bunches, the amplitudes of the harmonics decrease with increasing l until they change sign when $l\phi_o \cong 1.43\,\pi$. Over higher intervals of l, the amplitudes are small, successively negative and positive. Similar spectral characteristics are found for a Gaussian bunch distribution. The spectrum has some widening of the lines if the synchrotron oscillations are not linear and if the bunch distributions are not stationary.

There is a modified spectrum when coherent motion is present, with the appearance of sidebands which are spaced from the main lines at multiples of the synchrotron frequency. The amplitudes of the sidebands are related to the amplitudes of the coherent mode oscillations and may be found, as follows, for small coherent values of ϕ and $\Delta\phi$. In the case of the dipole mode, the entire bunch oscillates in phase about its equilibrium position, and if the coherent motion is small enough so that the bunch shape does not change, the expression for I_b is modified to

$$I_b = I_o + \sum_1^{\infty} I_l\,(\cos l((\phi_o t/\tau) - \hat{\phi}\cos pt))$$

This may be expanded in (cos cos + sin sin) form and then approximated:

$$\Delta I_b = \sum_1^{\infty} I_l \, (a \cos (l\phi_o t/\tau) + b \sin (l\phi_o t/\tau))$$

$$a = (-1 + J_o(l\hat{\phi}) - 2\, J_2(l\hat{\phi}) \cos 2pt + \ldots.)$$

$$b = (2\, (J_1(l\hat{\phi}) \cos pt - J_3(l\hat{\phi}) \cos 3pt + \ldots.)$$

$$\Delta I_b \cong \sum_1^{\infty} (-j\, (l\hat{\phi})\, I_l\, e^{jlh\omega_o t}\, e^{ipt})$$

Thus, the amplitudes of the main sidebands are proportional to both $\hat{\phi}$ and to I_l, and the spectral envelope of $(\Delta I_l/l\hat{\phi})$ has the same shape as that of the stationary distribution. The j term in ΔI_b indicates that there is phase modulation of each Fourier harmonic. In the simplest case, l equals one, and there is phase modulation of the fundamental rf component, which will excite the main accelerating cavities.

All the other modes involve non-rigid bunch motion. The line densities for these cases, $\lambda_m(\phi)$, are found by integrating $(R_m(r) \cos (m\Theta + pt))$ over $\dot{\phi}/\Omega$ at constant ϕ and t. The solutions for the even and odd order modes are found to differ in that they are even and odd functions of ϕ respectively. Thus, the Fourier harmonics at time t have the form $\cos (lh\omega_o t) \cos pt$ for m even and $\sin (lh\omega_o t) \cos pt$ for m odd, the former representing a type of amplitude modulation of each bunch harmonic and the latter a type of phase modulation. There is an approximation involved in finding beam current modulations in this way because $R_m(r)$ contains the term $(d\psi_o/dr)$, which is defined only over the interval $(-\phi_o, \phi_o)$. Thus, values for $\lambda_m(\phi)$ cannot be found for beam excursions outside the interval. The calculated Fourier spectrum is then in error since it is sensitive to the actual bunch extent; the error may be large if the coherent amplitude is not negligible compared to ϕ. (The one case which escapes this limitation is that of a Gaussian bunch shape.) An alternative, simpler approximation is therefore first considered for the case of the next-highest-order coherent mode, the quadrupole oscillation.

For the even order modes, there is envelope motion which is symmetric in ϕ and there is no motion of the bunch centre. Consider therefore, approximating the quadrupole mode by amplitude modulation of the form

$$\Delta I_b = \sum_1^{\infty} \left((\partial I_l/\partial\phi_o)\, \Delta\phi + (\partial^2 I_l/\partial\phi_o^2)\, \Delta\phi^2/2 + \ldots \right)$$

where $\Delta\phi = \Delta\hat{\phi} \cos pt$ and $p \cong 2\,\Omega$. The terms of higher order in $\Delta\phi$ may be neglected and an analytic expression introduced for $(\partial I_l/\partial\phi_o)$ for any specified bunch distribution. For the case of the parabolic line density considered

$$(\partial I_1/\partial\phi_o) = 6\,1\,I_o(-3\sin(1\phi_o) + 3(1\phi_o)\cos(1\phi_o) + (1\phi_o)^2\sin(1\phi_o))\,/(1\phi_o)^4$$

$$(\partial I_1/\partial\phi_o) = -0.4\,(1^2\phi_o)\,I_o \qquad \text{for } (1\phi_o) \ll 1 \qquad \ldots.\ (3)$$

In comparison with the dipole mode, the spectrum is shifted to higher frequency, with the first zero at $1\phi_o = 1.86\pi$ instead of $1\hat{\phi}_o = 1.43\pi$, and the amplitudes of the lower harmonics scale as $(1^2\phi_o\Delta\hat{\phi})$ instead of $(1\hat{\phi})$. The formula explains the observation on the ISIS synchrotron of a quadrupole mode instability when $1\phi_o = 0.3\pi$, whereas the quadrupole mode form factor is usually quoted as zero for $1\phi_o < 0.5\pi$.

Alternatively, using the first approximation discussed, ΔI_1 is

$$\Delta I_1 = I_o \int_{-\pi}^{\pi} \cos 1\phi \left(\int_{-y_1}^{y_1} R_m(r)\cos(m\Theta + pt)\,dy \right) d\phi \qquad \ldots.\ (4)$$

with m=2 for the quadrupole mode, $y = \dot{\phi}/\Omega$ and $y_1 = \hat{y}(\phi)$. The inner integral gives the line density, the outer, the Fourier expansion, while from section 9 it may be seen that $R_m(r)$ is already given in the relevant form for the 1-th Fourier harmonic. Evaluating the integral for $m = 2$ and $(1\phi_o) \ll 1$, it is again found that the amplitude of the lower harmonics scale as $(1^2\phi_o\ \Delta\hat{\phi})$.

Of the two approximations, equation (3) neglects the changing bunch shape during the oscillation and so it underestimates the harmonic amplitudes at the higher frequencies, while equation (4) neglects the changing bunch extent and so is also in error. Equation (3) appears to be adequate for the quadrupole mode, as shown in section 9, where its use leads to agreement with the space-charge envelope equation. Equation (4) is more appropriate for use with the higher even-order modes because of the higher frequency structure that then exists in the bunch line density. If the higher frequency part of the spectrum is the major concern, it is advisable to compare the results of (4) with simulations of the mode for a given distribution in a 1-D longitudinal particle tracking code.

Finally, for the higher odd-order modes ($m = 3,5,7, \ldots$), a modification of (4) may be used, with $\sin 1\phi$ replacing $\cos 1\phi$. In this case, the line density is an odd function of ϕ and the modulation has the form $\sin(1h\omega_o t)\cos pt$. The actual bunch shape is asymmetric, with successive maximum positive and negative phase excursions. The average phase value, $\langle \phi \rangle$, is zero, but a phase detector whose beam signal input is amplified and limited will show the phase modulation at angular frequency $\sim m\Omega$. When m is odd, the $\hat{\phi}$ term in the expression for $R_m(r)$ is to be interpreted as the maximum coherent amplitude of a particle on the bunch periphery.

11. COUPLED BUNCH MOTION

Coupled bunch motion may arise in both the longitudinal and transverse planes. A pattern develops, with phase differences between the individual bunch oscillations, and for M bunches, there are M possible modes of coupled bunch motion. The simplest case is when there are M equidistant bunches, all with equal bunch populations; then, if there is a constant phase shift, $2\pi n/M$, between the coherent motions of neighbouring bunches, the spectrum of the beam current includes the angular frequencies $|(n \pm 1\,M)\,\omega_o + p\,|$. Here 1 and n are integers, n may assume any value in the range 0 to (M-1) and p is the angular mode frequency. The case of n = 0 reverts to all the bunches oscillating in phase, with sideband frequencies at $1M\,\omega_o \pm p$. When $n \neq 0$, the motion may be considered as having carrier frequencies at $1M\,\omega_o$ and sidebands at $1M\,\omega_o \pm (p + n\omega_o)$ for mode n, and at $1M\,\omega_o \pm (p - n\omega_o)$ for mode (M-n).

Longitudinal motion only is considered here, with $p \simeq m\Omega$. A coupled bunch mode is then defined by n and m, with n the coupled bunch number and m the order of the mode. The induced fields for coupled bunch motion are obtained from modified forms of the G-functions, as for example in the following case for a resonant cavity:

$$G\,(j\omega,\,i(p \pm n\omega_o)) = R\,/\,(1 + (2Q/\omega_r)(j\Delta\omega + i(p \pm n\omega_o))$$

If the interaction is with the main rf cavities, the coupled bunch frequencies of most effect are those at $h\omega_o \pm p$, $h\omega_o \pm (p + \omega_o)$ and $h\omega_o \pm (p-\omega_o)$. Analysis for the first pair of sidebands is given in Appendix 3 for the case of no beam control loops; the coherent motion is governed by the Robinson equation. Analysis for the other pairs of sidebands leads to more complex equations, with terms in $(i\omega_o)$ in addition to those in s (=ip). Nyquist techniques are then appropriate to investigate the mode stability. Of most effect above the ring transition energy are the sidebands $1M\,\omega_o \pm (p + n\omega_o)$ and of most effect below, those at $1M\,\omega_o \pm (p-n\omega_o)$.

12. COASTING BEAM LONGITUDINAL INSTABILITIES

The basic equations of motion are the same as in the bunched beam case. However, since there are no external fields, the motion is governed entirely by the effect of the beam induced fields. These commence at injection since there is invariably some initial longitudinal beam structure. The fields subsequently change with time, decreasing if the coherent motion is damped, or increasing if anti-damped. The changing conditions make the motion more difficult to analyse than when the beam is bunched by an external field, large compared to the beam induced perturbations.

In existing theory, the longitudinal perturbation of a coasting beam is assumed to develop as a unidirectional harmonic wave in the angular direction, z/R, where z is the distance along the orbit and R is the mean machine radius. If this picture is compared with the Vlasov treatment for a bunched beam, then it is apparent that the assumptions are different for the directions of the wave perturbations. In the bunched beam case, the direction assumed is the angle θ in the polar coordinates (r, θ) of a local area of phase space over which focusing has been established. The two theories should converge when applied to the self-bunching of a coasting beam but there are different frequency spectra in the two cases and so they are incompatible. Associated with bunching is the appearance of pairs of sideband frequencies, but this feature is neglected in existing coasting beam theory.

The traditional approach is the Vlasov treatment for coasting beams by Neil and Sessler, [8] and this is developed in impedance concepts by Keil and Schnell.[1] In particular, the beam-induced fields are calculated as the product of the current perturbation and the total longitudinal impedance of the ring. Instability thresholds and growth rates are then found as a function of the total ring impedance at a single frequency. From the treatment for bunched beams, it is apparent that estimating the interaction at one frequency alone is insufficient and that $G(j\omega,s)$ has to replace $Z(\omega)$ in the analysis. The main difference between the self-bunching of a coasting beam and the envelope oscillation of a bunched beam is that, for the former, the modulation frequency is continually changing as the bunching develops. This requires that the G-function be applied in the form of a Laplace Transform and not as $G(j\omega,ip)$.

The effects of different impedances are now considered. For a low energy beam, the dominant impedance is that due to the longitudinal space charge, so that $G(j\omega,s) = - L\,(jn\omega_0 + s)$, where $(-L)$ is the equivalent negative inductance due to the space-charge forces. If a low energy beam with longitudinal structure is injected into a ring whose transition energy is less than the beam energy, there will be an induced field due to the first term in G which has the correct phase to cause self-bunching, provided the input perturbation is large enough and the injected beam momentum spread is small enough. As the bunching develops, the induced fields increase progressively and the effect of the second term in G must be included. Alternatively, if the transition energy of the ring is greater than the beam energy, the induced space-charge field becomes defocusing and the initial beam structure begins to migrate away from the regions of high density. This is the effect observed on the ISIS synchrotron. In general, the space charge does not exist in isolation and it is enhanced at high frequency by the effect of an equivalent broad-band impedance, for which it is necessary to estimate an equivalent G-function.

Instability is calculated by Keil-Schnell from $(\Delta p/p)$ values and estimates of coupling impedances, mainly the resistive component of the broad-band impedance and the reactance due to the space charge. The resistive antidamping is difficult to interpret since a pure resistance does not introduce a modulation phase shift. Other terms are involved, as shown from the form of the G-functions developed for bunched beam theory. Attempts have been made to adapt this theory for coasting beams but without success. The alternative is to simulate the motion in a 1-D tracking code based on the beam dynamics and the differential equations for the induced fields.

A general conclusion for coasting beams is that instability should be evaluated from tracking studies and not from the Keil-Schnell criterion. An important parameter, both for debunching or self-bunching, is the size of the initial perturbation relative to the size of the initial beam momentum spread.

13. LANDAU DAMPING DISCUSSION

The underlying basis of Landau Damping in accelerators is that a spread of incoherent oscillation frequencies in a beam aids decoherence; this may be seen from the following. Assume there are N particles, with frequencies spread uniformly over the range Ω_1 to Ω_2, all excited to a phase oscillation amplitude, $\hat{\phi}$; then solve $\int_{\Omega_1}^{\Omega_2} \phi(\Omega)\, d\Omega/(\Omega_2-\Omega_1)$ to show that the averaged coherent motion subsequently decays:

$$\langle\phi\rangle = 2\hat{\phi}\ (\cos\,(\Omega_2 + \Omega_1)t/2)\ (\sin\,(\Omega_2-\Omega_1)t/2)\ /\ (\Omega_2-\Omega_1)t$$

Landau Damping is complicated to estimate for it must be calculated in the presence of coherent excitation, as shown by Hereward[9] and Kohaupt.[10] The starting point is the equation for single-particle motion which, for the longitudinal plane, is given in Appendix 1. The complementary function gives the incoherent motion, while the particular integral is either a function of external fields or of coherent beam excitations. A spread of oscillation frequencies exists because of the non-linearity of the rf focusing, so an approximation is involved in assuming linear motion. The presence of the frequency spread complicates the analysis, and only an outline of how to proceed is given. The first step is the summation of individual particle motions.

Individual particle solutions are found after introducing the G-functions (or some analytic equivalent). Summations are then obtained in terms of the coherent motion, though approximations may be involved and there is the complexity of integrating in the vicinity of a pole, as discussed in reference 9.

In the case of the dipole mode, the coherent phase motion is the averaged phase for all the beam particles, with the incoherent motions averaging to zero. A spread of incoherent angular frequencies is assumed, with a distribution function, $\psi(\Omega)$, over the range $\Omega_1 < \Omega_c < \Omega_2$. An equation may be written for the coherent phase motion ($\phi = \hat{\phi} \cos \Omega_c t$) by referring to the Appendices 1, 2 and 3 and to the analysis in the references 9 and 10. After introducing a total G-function in the separated form $V \Delta\phi_v = G_1(s)\ I_1\ (1\phi)$ and $v = G_2(s)\ I_1\ (1\phi)$, the equation is (ψ in units $1/\omega$)

$$\phi \quad = \quad (s^2 G_1(s) - \Omega_c{}^2 \tan \phi_s\ G_2(s))\ (I_1/V)\ (1\phi) \int_{\Omega_1}^{\Omega_2} \frac{\psi(\Omega)\ d\Omega}{(s^2+\Omega^2)}$$

The integral may be evaluated as explained in reference 9 and the equation then rearranged. There are real and imaginary components for the integral so the final differential equation has terms both in i and s. For the special case of no frequency spread, the equation reduces to the fourth order Robinson equation. For the general case, because of the terms in i and s, use may be made of Nyquist techniques to determine if the Landau Damping produces stability.

A different approach is required for the higher modes because the average value of phase is no longer a measure of the coherent effect and the excitation is a function of both the incoherent and coherent oscillations. As an example, consider the quadrupole mode where there is no motion of the bunch centre but only of the envelope. Root mean square values may be used as a measure of the motion, or alternatively, the parameter

$$\langle|\phi|\rangle \quad = \quad \bar{\phi}_o + \bar{\phi} \quad = \quad \bar{\phi}_o + \Delta\bar{\phi} \cos 2\ \Omega_c t.$$

The single particle phase equation is introduced and averages taken to obtain $\bar{\phi}$. A double integral is required, involving the distribution functions, $\psi(\Omega)$ and $\psi_o(r)$:

$$(s^2 + \Omega^2)\ \phi \quad = \quad -\ \Omega_c{}^2\ (v_s/V)\ \sin 1\phi$$

$$v_s \quad \cong \quad G_3(s)\ (\partial I_1/\partial\bar{\phi}_o)\ (\Delta\bar{\phi} \cos 2\ \Omega_c t)$$

$$\bar{\phi} \quad = \quad \left(\frac{\Omega_c{}^2 G_3(s)}{V}\right) \left(\frac{\partial I_1}{\partial \bar{\phi}_o}\right)\ \bar{\phi} \int_{\Omega_1}^{\Omega_2} \int_{-\phi_o}^{\phi_o} \frac{\psi(\Omega) \left|\ \psi_o(\hat{\phi})\ \sin(1\hat{\phi} \cos(\Omega t+\alpha))\right|\ d\hat{\phi}\ d\Omega}{(s^2 + \Omega^2)}$$

The amplitudes, $\hat{\phi}$, and phase angles, α, refer to values for the individual incoherent motions, and averaging over α is also required. In order to evaluate the Landau Damping, approximations are required, both for this case and for modes $m > 2$.

14. TRANSVERSE COHERENT MOTION

When a beam oscillates coherently in the transverse plane, the carrier frequencies involved are multiples of the revolution frequency as in the longitudinal case, but the modulation frequency is very different because the betatron tune is more than an order of magnitude larger than the synchrotron tune. There is thus a much greater separation for any sideband frequencies. Another difference is that whereas the longitudinal coherent effects are not strongly influenced by the transverse motion, the reverse holds only for a coasting beam. In the case of a bunched beam, the transverse coherent oscillations are linked to the longitudinal motion via the effect of the chromaticity, which strongly influences the oscillation pattern. Usually, it is the incoherent synchrotron motion which is involved but, for very large rings, there is the possibility of coupled coherent synchrobetatron oscillations.

Theory for transverse coherent motion is developed through the concept of a transverse coupling impedance, defined as a function of frequency. A more useful concept is that of a transverse G-function, as will be indicated briefly here. As an example, the low frequency wall impedance of a ring is considered, assumed to be in the form of a circular pipe of radius, b, for which the transverse coupling impedance is found to be

$$Z_{\perp}(\omega) \quad = \quad 2c\ Z(\omega)\ /\ b^2\omega \quad = \quad a\sqrt{\omega}\ (1+j)\ /\ \omega$$

where a is the constant, $(Rc/b^3)\sqrt{2\mu_o/\sigma}$, for which R is the mean machine radius, c the velocity of light, μ_o the permittivity of free space and σ the wall conductivity. The term (1+j) represents the effect of the skin depth resistance and inductance, while the factor ω in the denominator is due to the differential wall currents and their associated magnetic and electric fields, with the former the integral of the latter. To obtain the equivalent transverse G-function, it is necessary to replace the j factor by (j + s/w) and the ω term by $(j\omega + s)$, as is explained in Appendix 2. The skin depth inductance is thus represented in the G-function by a factor not involving j or s, which signifies that it contributes to a coherent tune shift but not to damping or antidamping.

The resistive wall term, on the other hand, retains in the denominator, the factor $(j\omega + s)$, where s is the modulation at an angular frequency $\simeq \omega_o\ Q$ and Q is the transverse betatron tune. Different expressions are required for s in the coasting and bunched beam cases. Since there is only a single oscillation frequency per mode for the former, while there are sideband pairs for the latter, s is written as $(\pm jp)$ and (ip) respectively.

In the coasting beam case, the displacement is of the form cos $(n\omega_o \pm p)t$ and the excitation is then proportional to $(\pm a\sqrt{\omega}$ sin pt / $(n\omega_o \pm p))$, with $\omega = |n\omega_o \pm p|$. This indicates that all the $(n\omega_o + p)$ modes are damped, and that the $(n\omega_o - p)$ modes have an antidamping force when $n\omega_o > p$, that is for $n > Q$. The fastest growth rate for the resistive wall instability occurs when $(n\omega_o - p)$ is least, that is for the lowest frequency mode, with the integer n just greater than Q.

In the bunched beam case, the sideband frequencies for mode n are widely separated so there are different skin depths at the two frequencies. The transverse G-functions for the upper and lower sidebands are then found separately, as is explained in Appendix 2, before being summed. $G_\perp (jn\omega_o, ip)$ is used in place of $Z_1(\omega)$ to give the transverse excitation, and it has the same units, ohms per metre. The resistive wall $G_\perp$ functions are

$$G_\perp \text{ (lower)} = \frac{a(1+ij)\sqrt{\omega_o|n-Q|}}{2\,(jn\omega_o + ip)} \cong \frac{a(i-j)\sqrt{|n-Q|}}{2\,(n-Q)\sqrt{\omega_o}}$$

$$G_\perp \text{ (upper)} = \frac{a(1-ij)\sqrt{\omega_o(n+Q)}}{2\,(jn\omega_o + ip)} \cong -\frac{a(i+j)\sqrt{n+Q}}{2\,(n+Q)\sqrt{\omega_o}}$$

The factor ½ arises because there are two sidebands, the factor i represents excitation of the form $-\cos n\omega_o t \sin pt$ for beam displacements $\cos n\omega_o t \cos pt$ and the term -j represents $\sin n\omega_o t \cos pt$. There is a larger effect for the lower sideband so that, on summation, the term in i is positive and that in j is negative. The positive term in i shows that there is an antidamping force, though it is less than half that in the coasting beam case, for equivalent current components. There remains the term in -j which contributes no force to a particle at the bunch centre but gives a tune shift effect when away from this point. The term is usually neglected but may have an effect for long bunches, or when there are many displacement nodes along the bunch, or for coherent synchrobetatron motion.

Excitation spectra are different from those in the longitudinal plane. There is the higher modulation frequency and there are two contributors to the angular revolution frequency harmonics, $n\omega_o$. Fourier harmonics of ω_o arise because of the standing wave displacement pattern along the length of the bunch and also because of the pattern due to the chromatic phase shifts. Sum and difference frequencies of the two patterns exist and this explains why many oscillation nodes may exist along the bunch and still the lowest resistive wall mode is excited.

Resonant cavity deflecting modes may also be represented by $G_{\perp}$-functions, but an equivalent circuit is required for the mode and also the deflecting force must change sign for positive and negative beam displacements. The four possible excitation terms are represented by 1,j,i and ij, as in the longitudinal plane, with the terms in j representing a $\pi/2$ phase shift for the carrier and the terms in i, a $\pi/2$ phase shift for the modulation. The terms are used in the longitudinal case to identify the phase and amplitude modulation of the rf voltages but, in the transverse plane, the terms in j are usually neglected. If coherent synchrobetatron motion occurs, the situation is more complicated because the excitation fields are the products of the carrier and two modulation frequencies:

$$e^{jn\omega_o t} \quad e^{iQ_s\omega_o t} \quad e^{kQ_\beta\omega_o t} .$$

The use of three complex numbers allows eight types of excitation terms to be identified by the labels 1, j, i, k, ji, jk, ik and ijk. For example, 1 represents $\cos n\omega_o t \cos Q_s\omega_o t \cos Q_\beta\omega_o t$ while (ik) represents $\cos n\omega_o t \sin Q_s\omega_o t \sin Q_\beta\omega_o t$. This is a convenient identifier for selecting the excitation terms of most interest.

Considerations for beam distributions are very similar to those for the longitudinal plane. Gaussian transverse distributions arise naturally for electron beams due to the synchrotron radiation, while for protons and ions, the distributions depend on the injection process. The H^- injection stripping scheme discussed allows some flexibility in the distribution obtained, both in the transverse and longitudinal planes. Generally, attempts are made to reduce the peak densities at the centre of a beam but, for instability studies, it is possible to obtain special distributions. One such possibility is the distribution which is hollow in longitudinal phase space, for which there has been much debate.

15. SUMMARY FOR COLLECTIVE COHERENT EFFECTS

In looking again at longitudinal instability theories, the Robinson equation and the Envelope equation stand up well to re-examination, but the usual development of the Vlasov equation does not, neither for bunched nor for coasting beams. Approximations in the Vlasov approach for a bunched beam result in incompatibility with the Robinson theory and to physical anomalies when using the equation. In particular, the equation predicts phase shifts for the modulations which are in error by $\pi/2$; this may be seen directly for a purely resistive longitudinal impedance which does not phase-shift the modulations of the induced field harmonics relative to those of the beam current, whereas the equation indicates that it does.

A modified development has been found for the Vlasov equation which removes the physical anomalies, is compatible with the Robinson theory and is in agreement with the Space-Charge Envelope equation. It remains, however, to find a more rigorous proof of the modified equation than that given in the text.

Modulation response functions have been introduced to simplify both the calculation of the beam environment interaction and the subsequent solutions of the modified Vlasov equation. It is shown that summations are required of the G-functions for individual ring impedances and that the total G-functions are more useful in evaluating instability than are total impedance values. If impedance values are used, then separate summations must be obtained for all of the sideband frequencies involved in the interaction. The G-functions prove easy to use and lead to simplified derivations of the Robinson equation, the higher mode equations, coupled bunch oscillations and the special effect of rf feedback in the main accelerating system.

Coasting beam longitudinal instabilities are more complex than those for a bunched beam as the modulation frequency changes while the instability develops. The self-bunching leads to the presence of sideband frequencies as in the case of bunched beams and so the G-functions are again relevant. Since this is so, it leads one to question why the thresholds and growth rates predicted by existing theory (Keil-Schnell) are given as a function of the total coupling impedance at a single frequency when interactions at more than one frequency are involved. This may be traced to the fact that the beam interaction is incorrectly estimated in the theory as the product of the beam current modulation and a total Z value (instead of a total G-value). Another feature of importance for the coasting beam case is the initial condition of the beam. The longitudinal structure which results from injection may influence whether or not an instability develops.

Landau Damping may be evaluated for bunched beams via the basic equations of longitudinal motion, though some approximations are required in most cases. It remains an open question on how to introduce the analysis into the Vlasov equation approach for bunched beams.

For coherent motion in the transverse plane, a clearer picture of the beam interaction emerges by making use of transverse G-functions. They show directly the presence of additional coherent excitation terms which are usually not considered. They also show why a purely resistive impedance may cause instability in the transverse plane whereas it does not in the longitudinal. The extension of the G-function concept to the case of coherent synchrobetatron motion may be made by introducing a third complex number. This allows the convenient identification of all eight possible types of excitation terms.

REFERENCES

1. E Keil and W Schnell, Concerning Longitudinal Stability in the ISR, CERN Internal Report, ISR-TH-RF/69-48, (1969).

2. C R Prior, Space Charge Code Development at RAL.

3. L W Jones, Beam Scraping Problems in Storage Rings; The Black Cloud, Proceedings of the Heavy Ion Fusion Workshop, LBL-10301, (Sept 1980), p 294-296.

4. F Sacherer, Bunch Lengthening and Microwave Instability, CERN Internal Reports CERN/PS/BR 77-5 and 6 (1977) and Proc. IEEE Trans. Nucl.Sc., Vol NS-24, No 3 (1977), p 1393-1395.

5. B Zotter, Longitudinal Stability of Bunched Beams, CERN Internal Report, CERN SPS/81-18, (1981).

6. K W Robinson, Stability of Beam in Radiofrequency System, CEAL Internal Report CEAL-1010, (1964).

7. D Neuffer, Longitudinal Motion in High Current Ion Beams - A Self Consistent Phase Space Distribution with an Envelope Equation, Proc. IEEE Trans. Nucl. Sci., Vol NS-26, No 3 (1979), p 3031-3033.

8. V K Neil and A M Sessler, Longitudinal Resistive Instabilities of Intense Coasting Beams in Particle Accelerators, Rev. Sc. Inst. Vol 36, (1965) p 429.

9. H G Hereward, The Elementary Theory of Landau Damping, CERN Report, CERN 65-20 (1965).

10. R D Kohaupt, What is Landau Damping? Plausibilities, Fundamental Thoughts, Theory, DESY Report M-86-02 (1986).

APPENDIX 1. EQUATION OF LONGITUDINAL PHASE MOTION

This is derived from two separate equations, one giving the rate of change of phase of a particle and the other its rate of change of angular momentum. Phase and angular momentum are canonical coordinates, and it is convenient to measure the phase, ϕ, relative to the synchronous phase, ϕ_s, and the angular momentum, $\Delta E/\omega_o$, relative to that of a particle with the synchronous energy ($\Delta E=0$).

$$d\phi/dt = - h^2(\omega\ \partial\omega/\partial E)\ (\Delta E/h\omega_o) + d(\Delta\phi_v)/dt$$

$$\frac{d}{dt}\left(\frac{\Delta E}{h\omega_o}\right) = \left(\frac{e}{2\pi h}\right)\left((V+v)\sin(\phi+\phi_s) - V\sin\phi_s + \sum_1 v_1 \sin 1(\phi+\phi_1+\phi_s)\right)$$

where $\Delta\phi_v$ is the phase modulation of the accelerating voltage, $V \sin h\omega_o t$,

v is the amplitude modulation of V,

v_1 is the voltage modulation at parasitic frequencies, $1h\omega_o/2\pi$, $1=1,..\infty$,

ϕ_1 is the phase angle relative to the main accelerating voltage, and

e is the magnitude of the electronic charge.

The resultant equation, writing s for d/dt and approximating is

$$(s^2 + \Omega^2)\phi = s^2(\Delta\phi_v) - (\Omega^2/V)\left((v(\tan\phi_s + \sin\phi) + \sum_1 (v_s \sin 1\phi + v_c \cos 1\phi)\right)$$

with $v_s = v_1 \cos 1(\phi_1 + \phi_s) / \cos\phi_s$

$v_c = v_1 \sin 1(\phi_1 + \phi_s) / \cos\phi_s$

The later terms are ignored for the dipole mode equation:

$$(s^2 + \Omega^2)\phi = s^2(\Delta\phi_v) - (\Omega^2/V)\ (v \tan\phi_s)$$

For the higher modes, the later terms have to be expanded:

$$\sin 1\phi = \sin 1(r\cos\Theta) = 2\sum_1^\infty (-1)^{n+1} J_{2n-1}(1r)\cos(2n-1)\Theta$$

$$\cos 1\phi = \cos 1(r\cos\Theta) = J_o(1r) + 2\sum_1^\infty (-1)^n J_{2n}(1r)\cos 2n\Theta$$

where $J_n(1r)$ is the Bessel function or order n and argument $(1r)$.

APPENDIX 2. COMPOSITE MODULATION RESPONSE FUNCTIONS

A beam current modulation, ΔI, is assumed of the form cos ωt cos pt. It will be represented in a dual complex number algebra by

$$\Delta I = \Re[e^{j\omega t}\, e^{ipt}]$$

where the real part, $\Re$, of the complex argument is found in two stages. First, one complex number is assumed to be real and the real part of the argument is found; then the first complex number is assumed imaginary and the real part of the remainder found. Either complex number may be assumed to be real initially. The numbers j and i are distinct, so that product terms ji remain as ji and are not equal to -1.

The voltage modulation produced by ΔI in an impedance, Z, may be written

$$\Delta V = \Re[Z(\omega-p)\, e^{j(\omega-p)t} + Z(\omega+p)\, e^{j(\omega+p)t}]/2$$

$$\Delta V = \Re[G(j\omega,ip)\, e^{j\omega t}\, e^{ipt}]$$

Here G is a composite modulation response for both modulation sidebands. Also

$$\Delta V = \Re[(a + bj + ci + dji)\, e^{j\omega t}\, e^{ipt}]$$

$$\Delta V = a \cos wt \cos pt - b \sin \omega t \cos pt - c \cos \omega t \sin pt + d \sin wt \sin pt$$

Thus, the four possible output modulations are identified by the terms 1,j,i and ji. As an example of using the technique, consider an inductance, L, independent of frequency. The value of ΔV due to ΔI is:

$$\Delta V = \Re[L(j\omega + ip)\, e^{j\omega t}\, e^{ipt}]$$

$$\Delta V = L[-\omega \sin \omega t \cos pt - p \cos \omega t \sin pt]$$

Both expressions are obtained by direct differentiation and are completely consistent. Thus, the G-function and its Laplace Transform equivalent are

$$G(j\omega,ip) = L(j\omega + ip)$$

$$G(j\omega,s) = L(j\omega + s)$$

Another example is that of a series resonant circuit:

$$Z = R + j\omega L + 1/j\omega C$$

$$G(j\omega,s) = R + L(j\omega + s) + 1/C(j\omega + s)$$

The composite modulation response has the dimensions of an impedance and the terms in s indicate phase shifts in the modulation of both the in-phase and quadrature components of the output. For a parallel resonance, admittance analysis may be used:

$$1/G = 1/R + 1/L(j\omega + s) + C(j\omega + s)$$

$$G \cong R \,/\, [1 + (2Q/\omega_r)\,(j\Delta\omega + s)]$$

Here, Q is the product $\omega_r RC$,
ω_r is the angular resonant frequency, and
$\Delta\omega$ is the difference in angular frequencies $(\omega - \omega_r)$.

Note that three parameters are needed to define the circuit and that G is valid only if R, L and C are independent of ω, and if $p \ll \omega$. If there is an ω-dependence, G must be found as the sum of the responses for the lower and upper sidebands, calculated separately

$$G = ((1+ji)G_l + (1-ji)G_u) \,/\, 2$$

where G_l and G_u are G values calculated for R, L and C at $(\omega-p)$ and $(\omega+p)$ respectively. This holds since ΔI may be written in the alternative form

$$\Delta I = (\cos(\omega-p)t + \cos(\omega+p)t) \,/\, 2.$$

The contributions of the two sidebands to the G-function are not as one expects intuitively. When G_l and G_u are equal, the quadrature (ji) terms of the two sidebands cancel each other and each sideband makes an equal contribution to G.

More complex G-functions may be found in a way similar to that for the parallel resonant system, but first the equivalent circuit of the structure must be obtained. Thus, coupled cavity systems may be analysed or a cavity with many parasitic resonant modes. Care is needed for the case of very short bunches, where high frequencies are involved, for then it is necessary to separate the instantaneous from the resonant interaction of the beam with cavity-like structures.

The G-values for individual ring impedances may all be summed to obtain a total G-function. The summation is more complex than that for individual impedances because of the presence of the s-terms. It is evident that though the individual G-functions may be found for individual equivalent circuits, the total G-function is not so readily obtained from a total impedance value. In general, there is insufficient information in a total impedance unless it is known throughout the frequency range of interest.

APPENDIX 3. DERIVATION OF THE ROBINSON EQUATION USING THE G-FUNCTION

Refer to section 10 for an expression for the beam current modulation, to Appendix 2 for the G-function of tuned accelerating cavities and to Appendix 1 for the equation of longitudinal phase motion. Introduce appropriate phase angles for ΔI_b, solve for the amplitude and phase modulations of V from $G(j\omega,s)$ and ΔI_b, and introduce these into the phase equation to arrive at the Robinson fourth order equation:

$$\Delta I_b = j I_h \, (\Delta\phi_v - \phi) \, e^{\,j(3\pi/2 - \phi_s)}$$

$$v + j\,V\,\Delta\phi_v = R\,\Delta I_b \,/\, (1 + (2Q/\omega_r)(j\Delta\omega + s))$$

$$(s^2 + \Omega^2)\,\phi = s^2(\Delta\phi_v) - \Omega^2 \tan\phi_s \,(v/V)$$

$$[\,(s^2 + \Omega^2)((\,s + (\omega_r/2Q))^2 + (\Delta\omega)^2) + (D\,\omega_r/2Q)(\Delta\omega\,\Omega^2/\cos\phi_s)\,]\,\phi = 0$$

where $\Delta\omega = h\omega_o - \omega_r$, $D = RI_h / V$, I_h is the h-th harmonic component of beam current and ω_r is the angular resonant frequency of the cavities. All cavities are assumed identical and R is the sum of their shunt resistances. In practice, the equation for the coherent dipole motion is usually of higher order than fourth because it is modified by the effect of the beam control loops.

The Robinson equation indicates instability for two regions of cavity tuning. When the cavity is tuned for reactive beam loading compensation, instability occurs when the ratio of the total beam power to the power dissipated in the cavities and generator impedances, $D \sin\phi_s$, is > 1. Instability also occurs if the cavity is detuned in the direction opposite to that for reactive compensation.

If there is rf feedback around the cavity-amplifier system, the G-function is changed to: $R \,/\, (1 + A\,e^{-(j\Delta\omega + s)T} + (2Q/\omega_r)(j\Delta\omega + s))$, where A and T are the open loop feedback gain and delay respectively. A modified form of the Robinson equation may then be derived, which shows instability for three separate conditions. Two are as for no feedback, except that in the case of reactive beam loading compensation, the ratio $D \sin\phi_s$ may be increased to a factor much greater than unity. If the feedback gain is made too large, the equation also indicates instability at the point when the feedback loop itself becomes unstable.

Wake Fields and Impedances

T. Weiland, R. Wanzenberg
Technische Hochschule, Darmstadt, Germany

1 Introduction

In this lecture the concepts of wake fields and impedances are introduced to describe the electromagnetic interaction of a bunch of charged particles in the accelerator with its environment. The various components of the environment are the vacuum chamber, cavities, bellows, dielectric coated pipes, and all other kinds of obstacles the beam has to pass on its passage through the accelerator.

Since the purpose of this paper is tutorial and not original nor exhaustive, material is freely taken from previous papers and reports when appropriate. The plan of the lecture is to present first some typical examples from accelerator physics where wake field effects are important.

Then, in section 2, the concept of wake potential is introduced. Multipole expansions are studied for structures with cylindrical symmetry. The interdependence of longitudinal and transverse wake forces are examined in the light of the Panofsky–Wenzel theorem. Even for fully 3–D structures results of numerical calculations are presented.

In section 3, the impedance is introduced as the Fourier transform of the wake potential. The coupling of the beam to one mode of a cavity leads to the concept of the loss parameter. The fundamental theorem of beam loading is explained. The connection between the loss parameter and the shunt impedance is established.

In section 4, different methods of actually calculating the wake potentials are introduced. Analytical calculations for a pill-box cavity are presented. Numerical calculations in the time and frequency domain for arbitrarily shaped structures are shown. On the basis of examples it is demonstrated that mode summation methods are only appropriate for long–range wake potential calculations while short–range wakes are only properly calculated by computer codes working in the time domain. In order to avoid the impression that the wake potential computations by codes like TBCI or MAFIA appear to come from a magic black box, the basic methods with Maxwell grid equations are explained.

The effects of wakes and impedances and their importance for accelerators of all types are explained in section 5. Since instabilities are an extensive subject this section can only be a summary and not in the least exhaustive.

As a conclusion, in section 6, strategies in accelerator design are presented. It is demonstrated that increasing the shunt impedance of a cavity will also increase parasitic losses to higher-order modes. Hence an increase of the shunt impedance can spoil the performance of the whole accelerator if higher currents are used.

1.1 Basic Concepts

Consider a point charge q moving in free space at a velocity v close to the speed of light. The electromagnetic field carried by a relativistic point charge is treated in many text books of electrodynamics [1]. The field distribution is shown in Fig. 1. The electromagnetic field is Lorentz contracted into a thin disk perpendicular to the particle's direction of motion, which

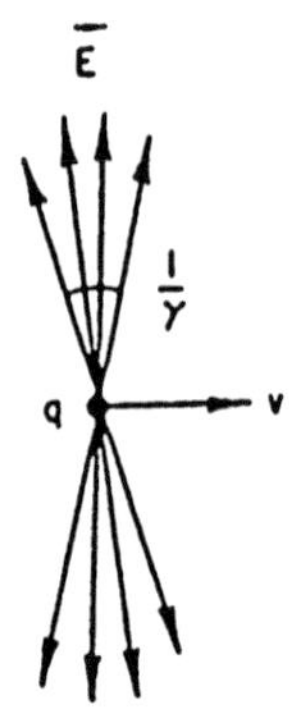

Figure 1: Electromagnetic field carried by a relativistic point charge

we choose to be the z - axis in a cylindrical coordinate system. The opening angle of the field distribution is on the order of $1/\gamma$, where $\gamma = (1-(v/c)^2)^{-1/2}$.

In the ultra–relativistic limit $v \to c$ (or $\gamma \to \infty$) the disk containing the field shrinks into a δ - function distribution. The non–vanishing field components are

$$E_r = \frac{q}{2\pi\,\varepsilon_0 r}\delta(z-c\,t), \qquad H_\varphi = \frac{E_r}{Z_0}, \qquad \text{with } Z_0 = 377\,\Omega$$

Since the electric field $\boldsymbol{E}$ points strictly radially outward from the point charge, all field components are identically zero both ahead and behind the point charge, and hence there are no forces on a test particle either preceding or following the charge q. This is a consequence of the *principle of causality.*

For v slightly less then c this is not strictly true. But if we look at typical charges and energies of $e^+\,e^-$ high energy accelerators

Machine	Charge /nC	Energy /GeV	$\gamma = (1-(v/c)^2)^{-1/2}$
PETRA	40	20	40000
LEP	50	50	100000
SLC	2	50	100000

and compare the space charge force $V_s = e/(4\,\pi\,\epsilon_0\,d^3\,\gamma^2)$ (d the rms distance between two electrons in the bunch) with the accelerating force V_{acc}, it is obvious that in a good approximation any space charge effects can be neglected for sufficiently long bunches. Due to the higher rest mass of protons the argument does not apply to them for the same energy range. Hence for protons or even heavy ions it is necessary to be more cautious.

In the following section we will restrict ourselves to the ultra–relativistic case ($\gamma = \infty$, $v = c$), therefore space-charge forces are neglected.

1.2 Some simple examples

Consider now some typical situations where electromagnetic fields occur behind the point charge q moving with velocity c through a structure.

Let us start with a charge moving through a **cylindrical** pipe on the z-axis. All electric field lines terminate transversely on surface charges on the wall of the pipe. If the pipe is perfectly conducting these image charges move in synchronism with the charge on the axis. Hence there will be no wake fields behind the charge. But the situation is somehow different if the material in the walls has a non–vanishing resistivity. Figure 2 shows the **resistive–wall wake fields**

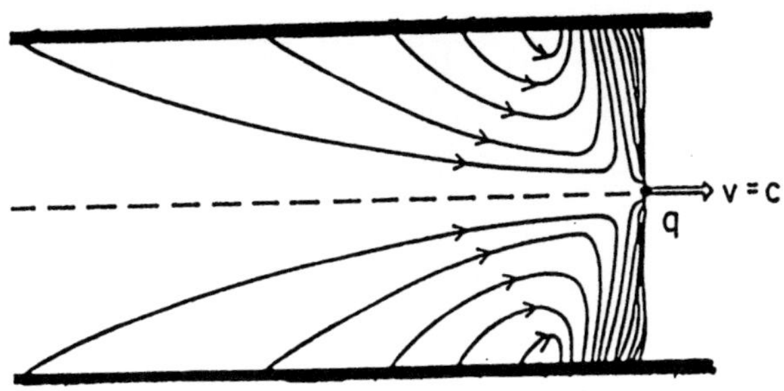

Figure 2: Resistive–wall wake fields in a pipe with materials in the walls with finite conductivity

generated by a point charge q traversing a pipe of which the wall has **finite conductivity**. For a Gaussian bunch ($\sigma = 1$ mm rms bunch length) of a total charge of 10 nC traversing a copper pipe with radius 1 cm, the typical maximum wake field is in the order of 1 kV/m [2].

The next example is a dielectric–coated pipe which could be used as a travelling-wave acceleration section. The phase velocity of the wave is below the velocity of light due to the dielectric coating. At a first glance one might think that the wake fields are significantly smaller than in common iris–loaded structures. But calculations show [3] that this is not the case. For a charge of 10 nC typical wake fields of about 500 kV/m are found with an inner pipe radius of 1 cm, a coating thickness of 3 cm and a dielectric constant of $2 \times \varepsilon_0$. Another problem is the bad thermal conductivity of the dielectric coating which leads to severe heating problems due to electric losses.

The last example of the introductory section is the important case of an **obstacle in a beam pipe**. Figure 3 shows the electric field lines for five successive time steps calculated by the computer code TBCI. For a Gaussian charge distribution (rms length $\sigma = 1$ cm) with a total charge of 10 nC, the wake fields behind the bunch are about 20 MV/m. The pipe has a radius of 1 cm, and the iris radius is 0.2 mm.

From these basic considerations we have learned that for electron accelerators the dominant forces are caused by geometrical changes along the beam pipe. Space-charge effects are negligible for ultra–relativistic particles. Resistive–wall wakes are in most cases negligible. Dielectric–coated structures should always be checked in detail for the specific application.

2 Wake Fields

2.1 Wake Fields in a Resonant Cavity with Beam Pipes

The examples of the Introduction provide a qualitative understanding of wake fields, and how they are generated. Before we proceed to some mathematical descriptions in terms of wake potentials, consider another example, an ultra–relativistic bunch with total charge q_1 traversing a small cavity along the z-axis (see Fig. 4).

The electromagnetic force on any test charge q_2 is given by the Lorentz equation as a function of space and time coordinates:

$$\boldsymbol{F}(\boldsymbol{r},t) = q_2\left(\boldsymbol{E}(\boldsymbol{r},t) + \boldsymbol{v} \times \boldsymbol{B}(\boldsymbol{r},t)\right).$$

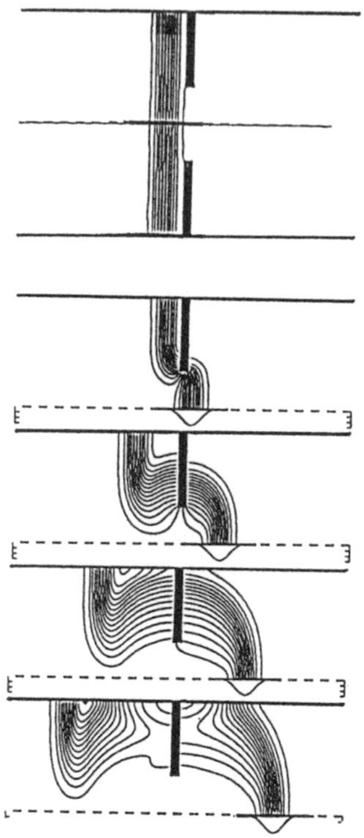

Figure 3: Wake fields generated by a Gaussian charge distribution passing an obstacle in the beam pipe

The fields $\boldsymbol{E}$ and $\boldsymbol{B}$ are generated by the Gaussian charge distribution ρ. They are solutions of the Maxwell equations and have to obey several boundary conditions.

$$\nabla \times \boldsymbol{B} = \mu_0 \boldsymbol{j} + \frac{1}{c^2}\frac{\partial}{\partial t}\boldsymbol{E} \qquad \nabla \cdot \boldsymbol{B} = 0$$

$$\nabla \times \boldsymbol{E} = -\frac{\partial}{\partial t}\boldsymbol{B} \qquad \nabla \cdot \boldsymbol{E} = \frac{1}{\epsilon_0}\rho.$$

In our case the charge distribution is

$$\rho(\boldsymbol{r},t) = q_1\,\lambda_\perp(x,y)\,\lambda(z-ct), \qquad \lambda(s) = \frac{1}{\sigma\sqrt{2\pi}}\exp\left(-\frac{(s-s_0)^2}{2\,\sigma^2}\right)$$

where $\lambda_\perp$ is the transverse distribution and λ is the longitudinal Gaussian distribution (centered around s_0). The corresponding current density is given by

$$\boldsymbol{j}(\boldsymbol{r},t) = c\,\boldsymbol{e}_z\,\rho(\boldsymbol{r},t).$$

A calculation of the electrical fields lines for three subsequent time steps is shown in figure 4. After interaction of the bunch with the cavity there remain electromagnetic fields in the cavity.

The bunch has lost energy into one or, in general, several cavity modes. Consider now a test charge q_2 following the bunch at a distance s with the same velocity $v \approx c$. The Lorentz force is

$$\boldsymbol{F}(s,t) = q_2 \left(\boldsymbol{E}(z = ct - s, t) + c\, \boldsymbol{e}_z \times \boldsymbol{B}(z = ct - s, t) \right).$$

We can distinguish between two regimes, the transient regime due to short-range wake effects and the resonant regime due to long-range ones. For small distances s (inside the bunch) transient wake-field effects are dominant, while for $s \gg \sigma$ (σ the bunch length) the wake field is dominated by resonant effects due to several cavity modes ringing with their specific frequencies.

The net momentum change of the test charge can be calculated as an average over the Lorentz force:

$$\Delta \boldsymbol{p} \sim \int dt\, \boldsymbol{F}(s,t).$$

This leads to the concept of wake potentials.

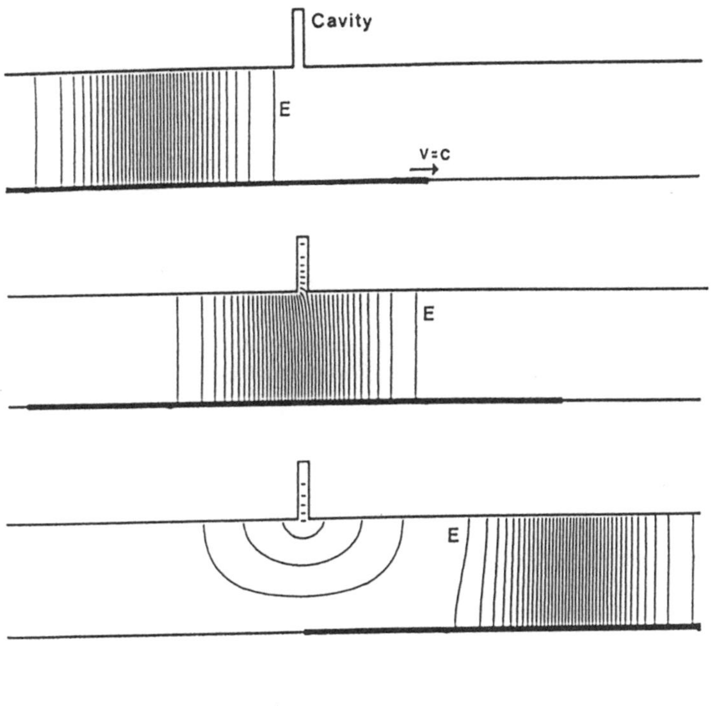

Figure 4: Wake fields generated by a Gaussian bunch traversing a cavity. The electric field is represented by lines of constant $\int dt\, r\, H_\varphi \sim r\, |E|$.

2.2 Basic Definitions

Consider a charge distribution with total charge q_1 traversing a structure with offset r_1 parallel to the z-axis with the speed of light (see Fig. 5). Then the **Wake Potential** is defined as

$$\boldsymbol{W}(r_1, s) = \frac{1}{q_1} \int_{-\infty}^{\infty} dz [\boldsymbol{E}(r_1, z, t) + c\, \boldsymbol{e}_z \times \boldsymbol{B}(r_1, z, t)]_{t=(s+z)/c}.$$

The distance s is measured from the head of the bunch q_1 in the opposite direction to $\boldsymbol{v}$. The change of momentum of a test charge q_2 following behind the bunch at a distance s on the same path is given by:

$$\Delta \boldsymbol{p} = q_1 q_2 \boldsymbol{W}(s)$$

Since $\boldsymbol{e}_z \cdot (\boldsymbol{e}_z \times \boldsymbol{B}) = 0$ the longitudinal component of the wake potential is simply given by

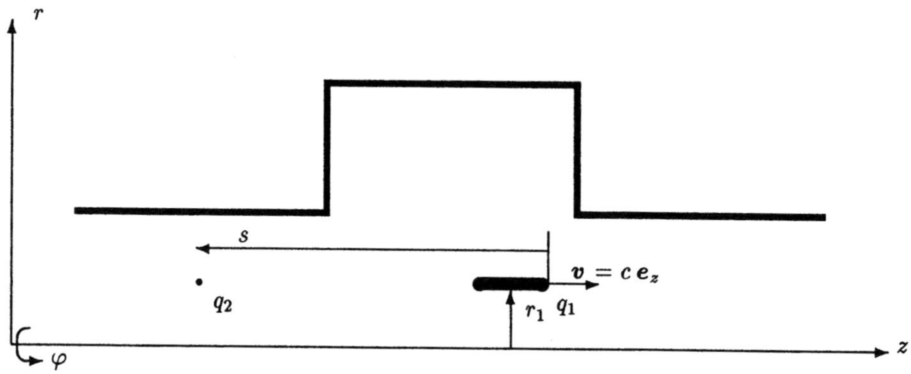

Figure 5: A bunch with total charge q_1 traverses a cavity followed by a test charge q_2

$$W_{||}(r_1, s) = \frac{1}{q_1} \int_{-\infty}^{\infty} dz\, E_z(r_1, z, (s+z)/c).$$

In Fig. 6 the normalized longitudinal component of the wake potential for the above considered example with the small cavity is shown. The dotted line represents the Gaussian charge distribution in the range from $-5\,\sigma$ to $5\,\sigma$. Due to the transient wake field effects the head of the bunch (left hand side of the figure) is decelerated while a test charge at a certain position behind the bunch will be accelerated.

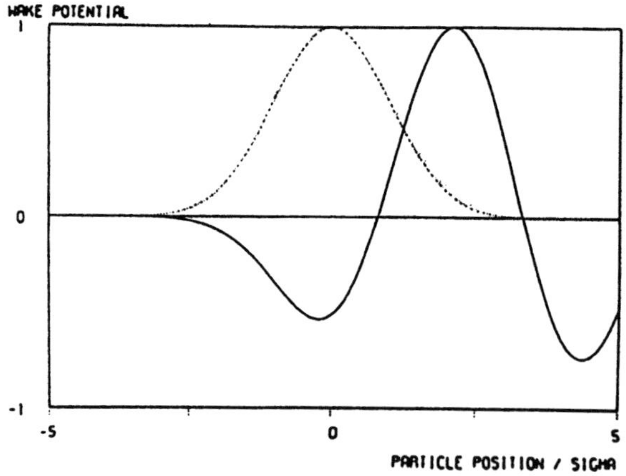

Figure 6: Longitudinal wake potential

2.3 Panofsky–Wenzel Theorem

In principle it is sufficient to know the longitudinal wake potential for all transverse positions (x,y) or (r,φ) since it is then possible to reconstruct the transverse component of the wake potential.

From the Maxwell equation $\nabla \times \boldsymbol{E} = -\frac{\partial}{\partial t}\boldsymbol{B}$ we obtain:

$$\boldsymbol{e}_z \times \frac{\partial}{\partial t}\boldsymbol{B} = \frac{\partial}{\partial z}\boldsymbol{E}_\perp - \nabla_\perp E_z.$$

Since the total derivative of the transverse component of the electric field with respect to z is given by

$$\frac{d}{dz}\boldsymbol{E}_\perp(x,y,z,\frac{z+s}{c}) = \left(\frac{\partial}{\partial z} + \frac{1}{c}\frac{\partial}{\partial t}\right)\boldsymbol{E}_\perp(x,y,z,\frac{z+s}{c}),$$

the derivative of the transverse wake potential with respect to s can be written as

$$\frac{\partial}{\partial s}\boldsymbol{W}_\perp(x,y,s) = \frac{1}{q_1}\int_{-\infty}^{\infty} dz \left((\frac{d}{dz}\boldsymbol{E}_\perp)(\boldsymbol{r},\frac{z+s}{c}) - \nabla_\perp E_z(\boldsymbol{r},\frac{z+s}{c}) \right).$$

If $\boldsymbol{E}_\perp$ vanishes at the boundaries a relation, better known as the Panofsky-Wenzel theorem [4], between the transverse and the longitudinal wake potential is obtained:

$$\frac{\partial}{\partial s}\boldsymbol{W}_\perp(x,y,s) = -\nabla_\perp W_{||}(x,y,s).$$

Integration of the transverse gradient of the longitudinal wake potential provides the transverse wake potential:

$$\boldsymbol{W}_\perp(x,y,s) = -\nabla_\perp \int_{-\infty}^{s} ds' \, W_{||}(x,y,s').$$

We now wish to prove another important property of the longitudinal wake potential, namely that $W_{||}(x,y,s')$ **is a harmonic function of the transverse coordinates:**

$$\nabla_\perp^2 W_{||}(x,y,s) = 0.$$

From the Maxwell equations and the identity $\nabla \times (\nabla \times \boldsymbol{E}) = \nabla(\nabla \cdot \boldsymbol{E}) - \nabla^2 \boldsymbol{E}$ we obtain an equation for $\boldsymbol{E}$

$$\nabla^2 \boldsymbol{E} - \frac{1}{c^2}\frac{\partial^2}{\partial^2 t}\boldsymbol{E} = \frac{1}{\varepsilon_0}\left(\nabla\rho + \frac{1}{c^2}\frac{\partial}{\partial t}\boldsymbol{j}\right).$$

Since ρ is a function of $z - ct$ and $j_z = c\rho$ the following relation holds:

$$\frac{\partial}{\partial t}j_z = -c^2\frac{\partial}{\partial z}\rho.$$

Consequently E_z is a solution of the wave equation, which can be written as

$$\nabla_\perp^2 E_z = \left(\frac{1}{c^2}\frac{\partial^2}{\partial^2 t} - \frac{\partial^2}{\partial^2 z}\right)E_z.$$

For further calculation it is appropriate to use a **Fourier representation of E_z**

$$E_z(x,y,z,t) = \frac{1}{(2\pi)^2}\int_{-\infty}^{\infty} d\omega \int_{-\infty}^{\infty} dk_z e^{i(\omega t - k_z z)}\,\widetilde{E_z}(x,y,k_z,\omega).$$

The wave equation in Fourier space reads

$$\nabla_\perp^2 \widetilde{E_z} = \left(k_z^2 - (\omega/c)^2\right) \widetilde{E_z}.$$

Inserting this relation into the definition of the longitudinal wake potential yields

$$\begin{aligned}
&\nabla_\perp^2 W_{||}(x,y,s) \\
&= \frac{1}{q_1}\int_{-\infty}^{\infty} dz\, \nabla_\perp^2 E_z(x,y,z,(s+z)/c) \\
&= \frac{1}{q_1}\int_{-\infty}^{\infty} dz \frac{1}{2\pi^2}\int_{-\infty}^{\infty} d\omega \int_{-\infty}^{\infty} dk_z \exp(i(\frac{\omega}{c}-k_z)z)\left(k_z^2 - (\frac{\omega}{c})^2\right) \widetilde{E_z}(x,y,k_z,\omega)\exp(i(\frac{\omega}{c})s) \\
&= \frac{1}{q_1}\frac{1}{2\pi}\int_{-\infty}^{\infty} d\omega \int_{-\infty}^{\infty} dk_z \exp(i(\frac{\omega}{c}-k_z)z)\left(k_z^2 - (\frac{\omega}{c})^2\right)\delta(\frac{\omega}{c}-k_z)\,\widetilde{E_z}(x,y,k_z,\omega)\exp(i(\frac{\omega}{c})s) \\
&= 0
\end{aligned}$$

We have used the Fourier representation of the δ - function

$$\delta(\frac{\omega}{c}-k_z) = \frac{1}{2\pi}\int_{-\infty}^{\infty} dz\, \exp(i\,(\omega/c - k_z)\,z).$$

This identity can be interpreted in our case as meaning that only electromagnetic fields synchronous to the motion of bunch contribute to the wake potential. This is strictly only true for ultra–relativistic bunches ($v = c$). In general we have

$$\nabla_\perp^2 W_{||} \sim \frac{1}{\gamma^2} \to 0 \quad \text{for } \gamma \to \infty.$$

So, for sufficiently high energies, $W_{||}$ is a harmonic function in the transverse coordinates for all positions s. This information will help us to evaluate the r dependency of the wake potential in cylindrical symmetric structures in the next subsection.

Even for fully 3-D structures this property of $W_{||}$ leads to a numerical method for calculating the wake potential very efficiently.

2.4 Wake Potential in Cylindrically-Symmetric Structures

Consider now a cylindrically-symmetric acceleration cavity with side tubes of radius a (see Fig. 7). The actual shape in the region $r > a$ is of no importance for the following investigations. Two charges pass the structure from left to right at the speed of light, q_1 at a radius r_1 and q_2 at a radius r_2. We wish to find an expression for the net change in momentum $\Delta \boldsymbol{p}(r_1, r_2, \varphi_1, \varphi_2, s)$ experienced by q_2 due to the wake fields generated by q_1.
Let us start with the simpler situation $r_1 = r_2 = r$, $\varphi_1 = \varphi_2 = \varphi$

$$\Delta p_z(r,\varphi,s) = q_1\, q_2\, W_{||}(r,\varphi,s).$$

The wake potential can be expanded in a multipole series

$$W_{||}(r,\varphi,s) = \mathrm{Re}\left\{\sum_{m=-\infty}^{\infty} e^{i\,m\,\varphi}\, G_m(r,s)\right\}.$$

Since $W_{||}(r,\varphi,s)$ is a harmonic function in r,φ, we have

$$\begin{aligned}
\nabla_\perp^2 W_{||}(r,\varphi,s) &= \left(\frac{1}{r}\frac{\partial}{\partial r} r \frac{\partial}{\partial r} + \frac{1}{r^2}\frac{\partial^2}{\partial \varphi^2}\right) W_{||}(r,\varphi,s) \\
&= \mathrm{Re}\left\{\sum_{m=-\infty}^{\infty} e^{i\,m\,\varphi}\left(\frac{1}{r}\frac{\partial}{\partial r} r \frac{\partial}{\partial r} - \frac{m^2}{r^2}\right) G_m(r,s)\right\} \\
&= 0.
\end{aligned}$$

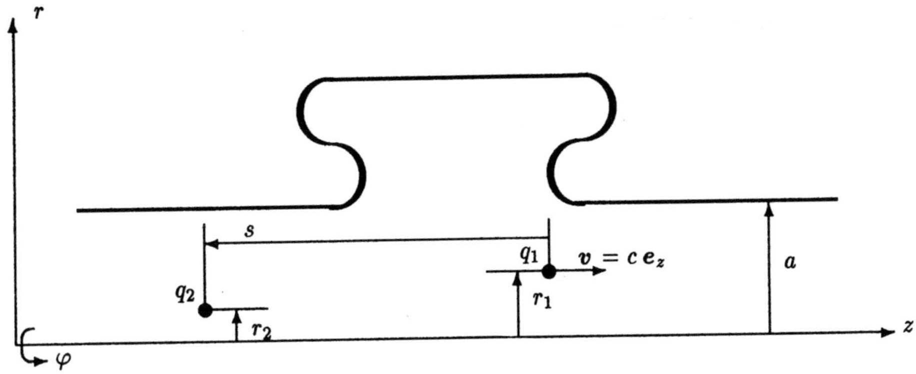

Figure 7: A bunch with total charge q_1 traversing a cavity with offset r_1 followed by a test charge q_2 with offset r_2

So, for all m, the expansion functions $G_m(r,s)$ have to fulfill the Poisson equation:

$$\frac{1}{r}\frac{\partial}{\partial r}\left(r\frac{\partial}{\partial r}G_m(r,s)\right) - \frac{m^2}{r^2}G_m(r,s) = 0.$$

The solutions are

$$\begin{aligned} G_0(r,s) &= U_0(s) + V_0(s)\ln r \\ G_m(r,s) &= U_m(s)\,r^m + V_m(s)\,r^{-m} \qquad \text{for } m > 0. \end{aligned}$$

Keeping only the solutions which are regular at the origin ($r = 0$) the longitudinal wake potential can be written as

$$W_{\|}(r,\varphi,s) = \sum_{m=0}^{\infty} r^m\, u_m(s)\,\cos m\varphi,$$

with expansion functions $u_m(s)$ depending on the details of the given cavity geometry.

The above relation describes the momentum change $\Delta p_z = q_1\, q_2\, W_{\|}$ due to fields that are present before the charge q_2 enters. From conservation of energy it follows that the fields generated by the charge q_1 obtain the energy from the longitudinal momentum, since r_1 and r_2 were defined to be constant. Thus it follows that, for the general case of a charge q_1 at $r = r_1$ generating the fields that act on a second charge q_2 at $r = r_2$, the **longitudinal wake potential is given by:**

$$W_{\|}(r_1,r_2,\varphi_1,\varphi_2,s) = \sum_{m=0}^{\infty} {r_1}^m\, {r_2}^m\, w_m(s)\,\cos m\,(\varphi_2 - \varphi_1).$$

The transverse wake potential is, according to the Panofsky–Wenzel theorem given by

$$\begin{aligned} \boldsymbol{W}_{\perp}(r_1,r_2,\varphi_1,\varphi_2,s) &= -\left(\boldsymbol{e}_r\frac{\partial}{\partial r_2} + \boldsymbol{e}_\varphi\frac{1}{r_2}\frac{\partial}{\partial\varphi_2}\right)\int_{-\infty}^{s} ds'\, W_{\|}(r_1,r_2,\varphi_1,\varphi_2,s') \\ &= \sum_{m=0}^{\infty}\Big\{-\boldsymbol{e}_r\, m\, {r_1}^m\, {r_2}^{m-1}\int_{-\infty}^{s} ds'\, w_m(s')\,\cos m\,(\varphi_2-\varphi_1) \\ &\qquad +\boldsymbol{e}_\varphi\, m\, {r_1}^m\, {r_2}^{m-1}\int_{-\infty}^{s} ds'\, w_m(s')\,\sin m\,(\varphi_2-\varphi_1)\Big\}. \end{aligned}$$

The characteristic wake functions $w_m(s)$ have to be calculated by solving Maxwell's equations for the given geometry and initial conditions:

$$w_m(s) = \frac{\int_{-\infty}^{\infty} dz\, E_{zm}(r_2, \varphi_2, z, (z+s)/c))}{r_1^m\, r_2^m\, \cos m(\varphi_2 - \varphi_1)}.$$

Normally a computer code like TBCI is used to do this job.

Since the dependency on the radial position is now known the integration can be performed for any radial position, especially for the tube radius $r = a$. Since E_z vanished at the metallic tube boundary only the cavity gap contributes to the integral. The explicit knowledge of the wake potential for different radial positions makes it possible to transform an originally infinite integration range to a finite range over the cavity gap.

It should be mentioned that for many practical cases, due to the $(r/a)^m$ dependency, the longitudinal wake potential is dominated by the monopole term, and the transverse ones by the dipole term:

$$\begin{aligned} W_{\|}(r_1, r_2, \varphi_1, \varphi_2, s) &= w_0(s)\, \cos m\,(\varphi_2 - \varphi_1), \\ \boldsymbol{W}_{\perp}(r_1, r_2, \varphi_1, \varphi_2, s) &= r_1 \int_{-\infty}^{s} ds'\, w_1(s') \quad (-\boldsymbol{e}_r\, \cos m\,(\varphi_2 - \varphi_1) \\ &\quad + \boldsymbol{e}_\varphi\, \sin m\,(\varphi_2 - \varphi_1))\,. \end{aligned}$$

2.5 Fully 3-D Structures

While for cylindrically-symmetric structures the dependency of the wake potential on the transverse coordinate is known, in the general situation calculations of the wake potentials can only be made by numerical methods. Since even for the general situation $W_{\|}$ is a harmonic function

$$\nabla_{\perp}^2\, W_{\|}(x, y, s) = 0,$$

it is possible to find $W_{\|}(x, y, s)$ for all possible beam positions (x, y) by a numerical solution of Poisson's equation if one knows the wake potential on the boundary. This indirect method has the advantage that the integration domain for each test beam on the boundary is finite (see Fig. 8).

The transverse wake potential can be calculated from the longitudinal one using the Panofsky–Wenzel theorem. Figure 9 shows the transverse gradient of the longitudinal wake potential in a beam tube.

The boundary values (test beams) on the tube radius are found by a numerical integration of the Maxwell equations (see section 4.2 fields for more details).

3 Impedances

3.1 Definitions

The Fourier transform of the wake potential is called the impedance or coupling–impedance

$$Z_{\|}(x, y, \omega) = \frac{1}{c} \int_{-\infty}^{\infty} ds\, W_{\|}(x, y, s)\, \exp(-i\frac{\omega}{c}\, s).$$

Wake potential and impedance are two descriptions of the same thing, the coupling between the beam and its environment. The wake potential is the time domain description, the impedance

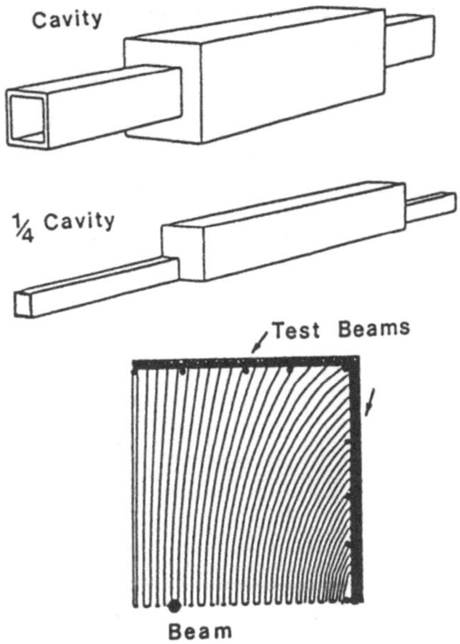

Figure 8: 3-D cavity structure with two symmetry planes. Due to the symmetries it is sufficient to treat only a quarter cavity numerically. The lower picture shows lines of constant longitudinal wake potential.

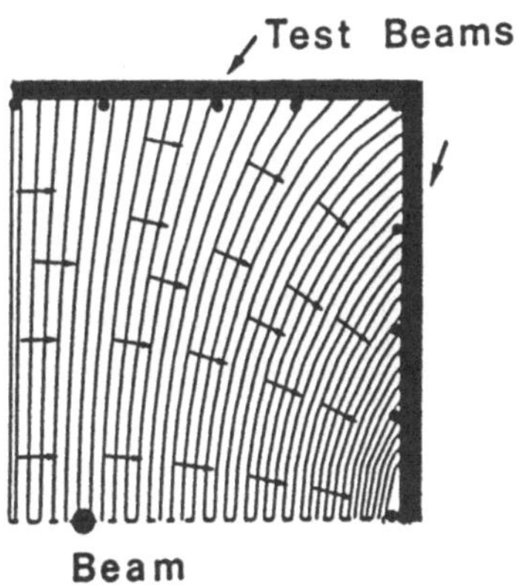

Figure 9: Gradient of the longitudinal wake potential. An integration provides the transverse wake potential according to Panofsky–Wenzel theorem.

the frequency domain description. The reason for the usefulness of the impedance is that it often contains a number of sharply defined frequencies corresponding to the modes of the cavity or the long-range part of the wake potential. Figure 10 shows the real part of a typical impedance for a cavity. Below the cut-off frequency of the beam pipe there is a sharp peak for each cavity mode. Above the cut-off frequency a continuous spectrum of beam pipe modes contribute to the impedance. These are important for the short-range wake-field effects over the bunch, especially for very short bunches.

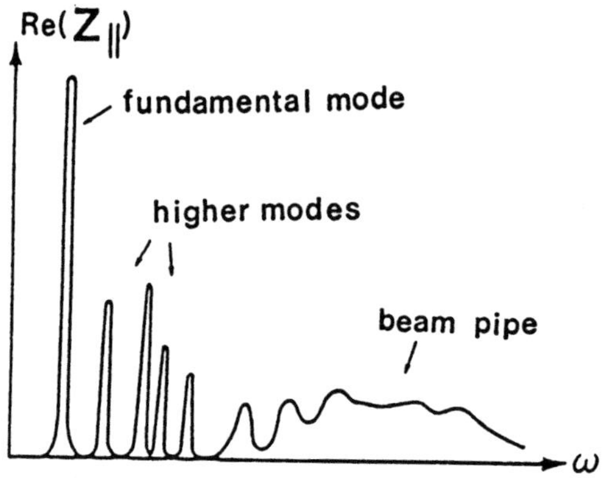

Figure 10: Real part of the impedance for a cavity with side pipes. The sharp peaks correspond to cavity modes while above the cut-off frequency the continuous spectrum of beam pipe modes contributes to the impedance.

For the transverse impedance it is often convenient to use a definition with an extra factor $-i$:

$$Z_\perp(x,y,\omega) = \frac{-i}{c}\int_{-\infty}^{\infty} ds\, W_\perp(x,y,s)\, \exp(-i\frac{\omega}{c}s).$$

The reason is that the Panofsky–Wenzel theorem then reads in the frequency domain

$$\frac{\omega}{c}\, Z_\perp(x,y,\omega) = \nabla_\perp\, Z_\parallel(x,y,\omega).$$

We should mention some properties of the impedance here. First, since the wake potential is a real function the real part of the impedance is an even function of the frequency while the imaginary part is an odd function of it:

$$\mathrm{Re}\left\{Z_\parallel(\omega)\right\} = \mathrm{Re}\left\{Z_\parallel(-\omega)\right\} \qquad \mathrm{Im}\left\{Z_\parallel(\omega)\right\} = -\,\mathrm{Im}\left\{Z_\parallel(-\omega)\right\}.$$

Hence for the wake potential in terms of the impedance we have

$$\begin{aligned} W_\parallel(s) &= \frac{1}{2\pi}\int_{-\infty}^{\infty} d\omega\, Z_\parallel(\omega)\, \exp(i\frac{\omega}{c}s) \\ &= \frac{1}{2\pi}\int_{-\infty}^{\infty} d\omega\, \left(\mathrm{Re}\left\{Z_\parallel(\omega)\right\}\cos(\frac{\omega}{c}s) - \mathrm{Im}\left\{Z_\parallel(\omega)\right\}\sin(\frac{\omega}{c}s)\right). \end{aligned}$$

Furthermore due to **causality** $\mathrm{Re}\left\{Z_\parallel(\omega)\right\}$ and $\mathrm{Im}\left\{Z_\parallel(\omega)\right\}$ are **not** independent of each other. From $W_\parallel(-s) = 0$ for all $s > 0$ it follows that

$$\int_{-\infty}^{\infty} d\omega\, \mathrm{Re}\left\{Z_\parallel(\omega)\right\}\cos(\frac{\omega}{c}s) = -\int_{-\infty}^{\infty} d\omega\, \mathrm{Im}\left\{Z_\parallel(\omega)\right\}\sin(\frac{\omega}{c}s) \qquad \text{for all } s > 0.$$

So only the real (or imaginary) part of the impedance is really needed

$$W_{||}(s) = \frac{1}{\pi} \int_{-\infty}^{\infty} d\omega \, \mathrm{Re}\left\{ Z_{||}(\omega) \right\} \cos(\frac{\omega}{c} s).$$

In the next section the energy losses of the beam to the different modes of a cavity are studied in detail. This will lead to a method for calculating some parts of the impedance.

3.2 Loss Parameters

Consider a closed pill–box cavity (Fig. 11). The electric field is an infinite series of eigen modes of the cavity ringing with the frequencies $\omega_n = 2\pi f_n$

$$\boldsymbol{E}(\boldsymbol{r}, t) = \mathrm{Re}\left\{ \sum_{n=0}^{\infty} \boldsymbol{E}_n(\boldsymbol{r}) \, e^{i \omega_n t} \right\}.$$

We wish to study the coupling between the beam and the various modes. A point charge traversing the cavity with the speed of light ($z = ct$) " sees " the (complex) voltage:

$$V_n = \int_{-\infty}^{\infty} dz \, E_{z,n}(z) \, e^{i \omega_n z/c}.$$

The total field energy stored in mode n is given by

$$U_n = \frac{\epsilon_0}{2} \int d^3 r |\boldsymbol{E}_n(\boldsymbol{r})|^2.$$

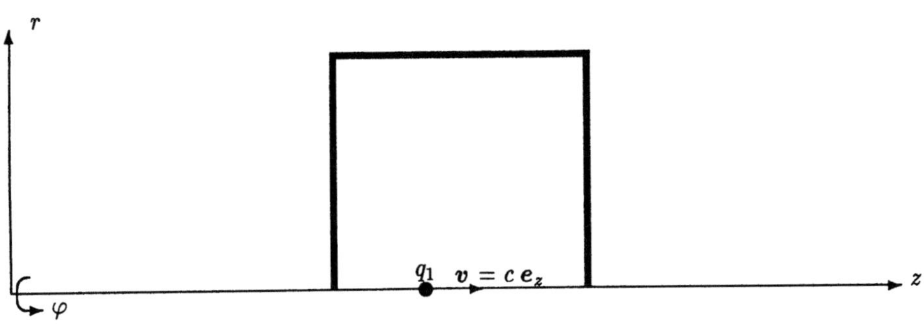

Figure 11: Closed pill–box cavity excited by a point charge q_1

We now define the **loss parameter** k_n

$$k_n = \frac{|V_n|^2}{4 U_n}$$

as a phase and amplitude independent quantity.

The wake potential for $s > 0$ of a point charge is given by (see Ref. [5] or the Appendix)

$$W_{||}^{\delta}(s) = \sum_{n=0}^{\infty} 2 k_n \cos(\omega_n s/c).$$

In the next section we will show that the self wake is only one–half of the wake directly behind the point charge (fundamental theorem of beam loading). So the energy lost by the point charge into the mode n is given by

$$\Delta W_n = {q_1}^2\, k_n.$$

The wake potential of a Gaussian bunch with charge density

$$\rho(\boldsymbol{r},t) = q_1\, \lambda(z-ct), \qquad \lambda(s) = \frac{1}{\sigma\sqrt{2\pi}} \exp\left(-\frac{(s-s_0)^2}{2\,\sigma^2}\right)$$

is obtained by a convolution integral with the point charge wake potential $W_{\|}^{\delta}$

$$W_{\|}(s) = \int_0^{\infty} ds'\, \lambda(s-s')\, W_{\|}^{\delta}(s').$$

For $\sigma \ll s - s_0$ the result of this integral is

$$W_{\|}^{\delta}(s) = \sum_{n=0}^{\infty} 2\, k_n \exp\left(-\frac{1}{2}\left(\frac{\omega_n}{c}\right)^2 \sigma^2\right) \cos(\omega_n\, s/c).$$

It is now possible to define various loss parameters for the coupling of the beam with the different modes of the cavity.

Total loss parameter

$$k_{tot} = \int_{-\infty}^{\infty} ds\, \lambda(s)\, W_{\|}(s)$$

$\Delta W = {q_1}^2\, k_{tot}$ is the total energy lost by the Gaussian bunch. All modes of the cavity and all wave guide modes are included.

Loss to the fundamental mode

$$k_{fu} = k_0 \exp\left(-\frac{1}{2}\left(\frac{\omega_n}{c}\right)^2 \sigma^2\right).$$

This parameter describes the desired coupling between the fundamental cavity mode and the beam.

Loss to higher–order modes

$$k_{hm} = \sum_{n=1}^{N} 2\, k_n \exp\left(-\frac{1}{2}\left(\frac{\omega_n}{c}\right)^2 \sigma^2\right).$$

In addition to the fundamental mode there are some (N) resonant cavity modes.

Loss into the beam pipe

$$k_{tube} = k_{tot} - k_{fu} - k_{hm}.$$

The beam can also lose energy into wave guide modes of the side tubes. Especially very short bunches will excite that kind of mode in the continuous spectrum of the structure.

Total parasitic loss

$$\begin{aligned} k_{par} &= k_{tot} - k_{fu} \\ &= k_{tube} + k_{hm}. \end{aligned}$$

All parasitic effects are summarized in the loss parameter k_{par}.

Figure 12 shows all the above defined loss parameters for a single cell of a PETRA cavity as a function of the bunch length σ. For very short bunches parasitic losses are dominant. Consequently a careful cavity design always requires a closer look at higher–order mode effects and tube radiation losses. For a further discussion on this topic see section 6.

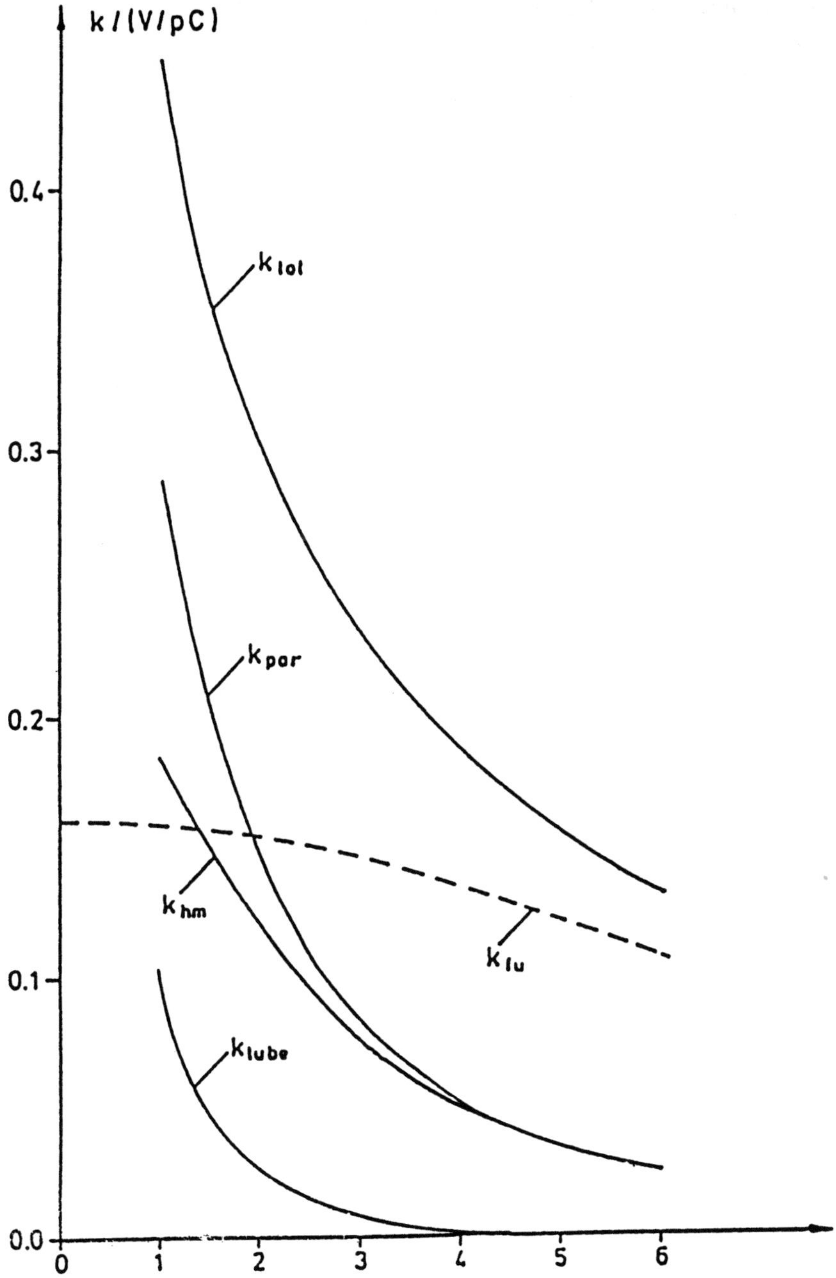

Figure 12: The loss parameters k_{tot} , k_{par} , k_{hm} , k_{tube} and k_{fu} as a function of $\sigma_{r.m.s.}$(Gaussian bunch) for a single cell of a PETRA cavity.

3.3 Fundamental Theorem of Beam Loading

We already mentioned that the self wake is only one–half of the wake directly behind the point charge q_1. In this section the coupling of a point charge to **one** cavity mode ringing with frequency $\omega = 1\pi f$ is studied in detail.

The induced voltage behind the point charge q_1 is

$$V(s) = \hat{V} \cos(\omega s/c) \qquad s > 0,$$

with $\hat{V} \sim q$. The dependency on s is a pure cos function without any extra phase. Otherwise a second point charge q following at an appropriate distance could gain more energy out of the cavity mode than the first point charge lost into that mode. That would clearly lead to a contradiction of the conservation of energy.

Now consider two point charges with the same charge q following each other at a distance $\lambda/2$ ($\lambda = c/f$ the wavelength of the mode). Let V_1 and V_2 be the induced voltages by charge # 1 and # 2 (see Fig. 13).

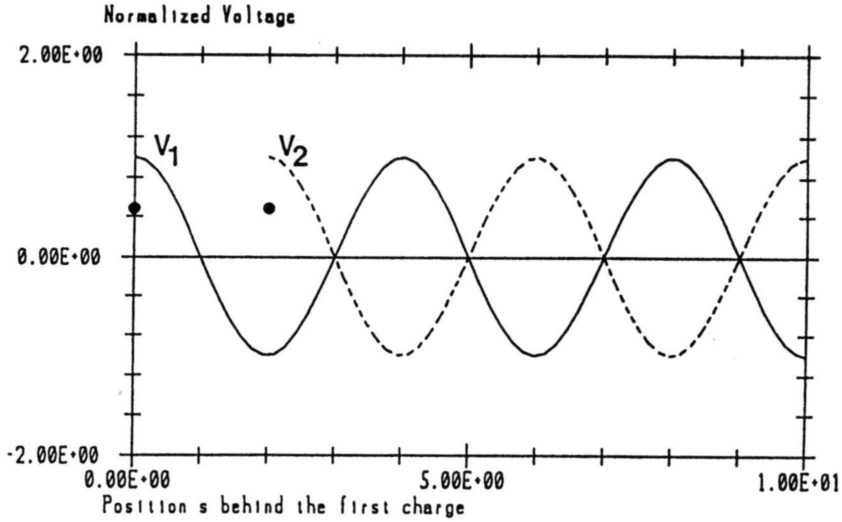

Figure 13: Induced voltages by point charge # 1 and # 2

The point charge # 1 loses the energy ΔW_1 into the cavity mode:

$$\Delta W_1 = k\, q^2 = q\, \eta\, V_1$$

where η is an unknown factor describing the difference between the self-wake and the wake directly behind the charge.

Since the second charge follows at a distance $\lambda/2$ it takes the whole stored energy out of the mode:

$$\begin{aligned} \Delta W_2 &= \Delta W_1 = k\, q^2 \\ &= q\,(V_1 - \eta\, V_2). \end{aligned}$$

This provides the following equation for the unknown factor η

$$q\, \eta\, V_1 = q\,(V_1 - \eta\, V_2).$$

Since $V_1 = V_2$ for identical point charges we obtain

$$\eta = \frac{1}{2}.$$

Consequently the voltage induced by a point charge is given by

$$V(s) = \begin{cases} 2\,q\,k\,\cos(\omega\,s/c) & s > 0 \\ q\,k & s = 0 \\ 0 & s < 0. \end{cases}$$

The energy lost into the cavity mode is

$$\Delta W = q\,V(0) = q^2\,k.$$

3.4 Shunt Impedance and Quality Factor

Various loss parameters were defined to describe the energy lost by the beam to cavity modes. But once excited by a beam such a mode does not ring forever. Due to the finite conductivity of the cavity walls the field amplitudes slowly reduce. The energy in the cavity modes is transformed into heat.

The wall currents are induced in a thin surface layer characterized by the skin depth

$$\delta = \sqrt{\frac{2}{\omega\,\mu\,\kappa}}$$

where $\omega = 2\pi\,f$ is the frequency of the mode, κ the conductivity of the wall material and μ the permeability.

The wall losses can be calculated approximately from the fields obtained for the modes established for a wall with infinite conductivity. If $\boldsymbol{H}_t$ is the magnetic field tangential to the surface the total power dissipated into the walls is given by a surface integral

$$P = \frac{1}{2}\int_{\partial V} d^2r\,\frac{|\boldsymbol{H}_t|^2}{\kappa\,\delta},$$

while the energy stored in the mode is

$$U = \frac{\epsilon_0}{2}\int_V d^3r\,|\boldsymbol{E}|^2.$$

The **quality factor** of a cavity mode is defined as

$$Q = \frac{\omega\,U}{P} = \frac{2\pi}{T}\,\frac{\text{stored energy}}{\text{powerloss}}.$$

Typical values for copper cavities are

$$Q \approx 10000 \qquad f = 3\ \text{GHz}$$
$$Q \approx 35000 \qquad f = 500\ \text{MHz}.$$

The connection with the attenuation constant α for the fields is

$$\alpha = \frac{\omega}{2\,Q}, \qquad \boldsymbol{H},\,\boldsymbol{E} \sim e^{-\alpha\,t}.$$

Since $U \sim \boldsymbol{E}^2 \sim e^{-\alpha\,t}$ we have

$$P = \frac{d}{dt}U = -2\,\alpha\,U.$$

The voltage induced by a point charge q into a mode with frequency ω has to be modified for long distances s

$$V(s) = 2\,q\,k\,\cos(\omega\,s/c)\,\exp(-\frac{\omega}{2\,Q}\,\frac{s}{c}).$$

Although the induced voltage $V(s)$ is completely determined by the parameters k, Q and ω another quantity, the **shunt impedance** R, is commonly used. Formally it can be defined by

$$R = \frac{4\,k\,Q}{\omega} \quad \text{(US)} \quad \text{or} \quad R = \frac{2\,k\,Q}{\omega} \quad \text{(electrical eng.)}.$$

Their meaning will become obvious in the following example. Consider a circular accelerator with one cavity and (say) four identical bunches following each other at a distance T_1 in time (Fig. 14).

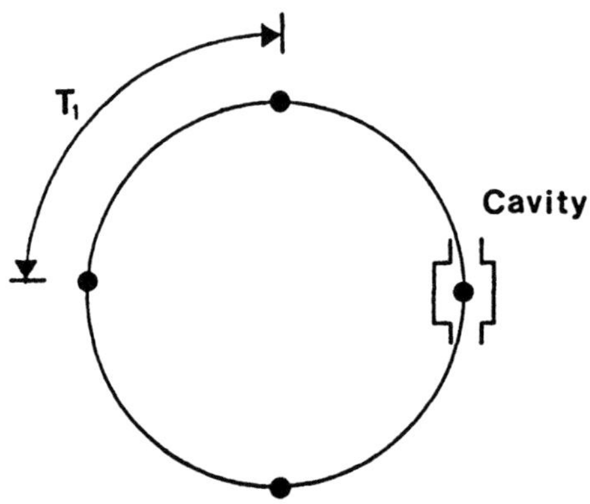

Figure 14: Circular accelerator with one cavity successively traversed by four bunches

The induced voltage in **one cavity** mode after n bunches have traversed the cavity is:

$$\hat{V}(j) = \sum_{j=1}^{n} 2k\,q\,\cos(\omega\,j\,T_1)\exp(-\frac{j\,T_1\,\omega}{2\,Q}).$$

Let us now assume the worst case that all bunches are in phase with the cavity mode

$$\omega\,T_1 = 2\,\pi.$$

In the limit $n \to \infty$ we obtain

$$\begin{aligned}\hat{V}(\infty) &= 2\,k\,q\sum_{j=1}^{\infty}\exp(\frac{\omega\,j\,T_1}{2\,Q}) \\ &= 2\,k\,q\,\frac{1}{1-\exp(-\omega\,T_1/\,2Q)} \\ &\approx 2\,k\,q\,\frac{2\,Q}{\omega\,T_1}.\end{aligned}$$

The long range wake amplitude of **one** mode including losses due to wall currents is

$$\begin{aligned}\hat{V}(\infty) &= \frac{4\,k\,Q}{\omega}\,\frac{q}{T_1} \\ &= R\,I.\end{aligned}$$

This result is of the form voltage equals shunt impedance times current.

Typical values for a 500 MHz copper cavity are

$$R = 20\,\frac{\mathrm{M}\Omega}{\mathrm{m}}$$

(US shunt impedance definition).

4 Analytical and Numerical Calculations of Wake Fields

4.1 Analytical Calculation for a Pill Box

For a closed pill-box cavity all modes can be calculated analytically [6]. Consider a pill box with radius R and gap length g (Fig. 15). The normalized electromagnetic field of the monopole

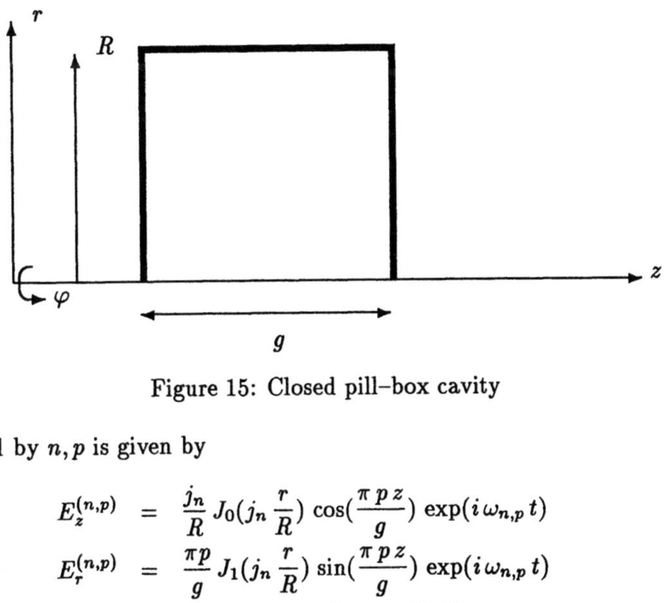

Figure 15: Closed pill–box cavity

modes numbered by n, p is given by

$$\begin{aligned} E_z^{(n,p)} &= \frac{j_n}{R}\, J_0(j_n\,\frac{r}{R})\,\cos(\frac{\pi\,p\,z}{g})\,\exp(i\,\omega_{n,p}\,t) \\ E_r^{(n,p)} &= \frac{\pi p}{g}\, J_1(j_n\,\frac{r}{R})\,\sin(\frac{\pi\,p\,z}{g})\,\exp(i\,\omega_{n,p}\,t) \\ H_\phi^{(n,p)} &= i\,\omega_{n,p}\,\epsilon_0\, J_1(j_n\,\frac{r}{R})\,\cos(\frac{\pi\,p\,z}{g})\,\exp(i\,\omega_{n,p}\,t), \end{aligned}$$

where j_n is the nth zero of the Bessel function $J_0(x)$ and

$$\left(\frac{\omega_{n,p}}{c}\right)^2 = \left(\frac{j_n}{R}\right)^2 + \left(\frac{\pi p}{g}\right)^2.$$

Hence the voltage becomes

$$\begin{aligned} V_{n,p} &= \int_z^g dz\, E_z(r=0, z, t=z/c) \\ &= \frac{i\,\omega_{n,p}\,R}{j_n\,c}\,(1-(-1)^p\exp(i\omega_{n,p}\,g/c)). \end{aligned}$$

The stored energy is given by

$$\begin{aligned} U_{n,p} &= \frac{\mu_0}{2} \int_0^R dr\, r \int_0^{2\pi} d\varphi \int_0^g dz H_\varphi^{(n,p)} \left(H_\varphi^{(n,p)}\right)^* \\ &= \frac{\pi \epsilon_0}{4} \left(\frac{\omega_{n,p}}{c}\right)^2 g\, R^2\, J_1^2(j_n). \end{aligned}$$

The loss parameters can now be calculated from

$$\begin{aligned} k_{n,p} &= \frac{V_{n,p}\, V_{n,p}^*}{4\, U_{n,p}} \\ &= \frac{1}{\pi\, \epsilon_0\, g} \frac{2}{1+\delta_{0,p}} \frac{1-(-1)^p \cos(\omega_{n,p}\, g/c)}{j_n^2\, J_1^2(j_n)} \end{aligned}$$

so that the expression for the point charge (q) monopole wake potential becomes

$$W_{||}^\delta(s) = 2\, q \sum_{n=1}^{\infty} \sum_{p=0}^{\infty} k_{n,p} \cos(\omega_{n,p}\, s/c).$$

The wake potential is the sum of all voltages induced in all modes.

For a Gaussian bunch with charge density

$$\rho(r,t) = q_1\, \lambda(z-ct), \qquad \lambda(s) = \frac{1}{\sigma\sqrt{2\pi}} \exp\left(-\frac{(s-s_0)^2}{2\,\sigma^2}\right)$$

the wake potential is given by

$$W_{||}(s) = \int_0^\infty ds'\, \lambda(s-s')\, W_{||}^\delta(s').$$

Figure 16 shows the wake potential of a Gaussian bunch ($\sigma = 2.5$ cm) due to a pill–box cavity ($R = 5$cm, $g = 10$cm). The dashed line is obtained from an analysis based on 40 modes, while the solid line is calculated with the computer code TBCI with an 11×21 mesh.

There is a good agreement between the two different methods for the long range ($s \gg \sigma$) wake potential. But even for such a simple example as a pill–box cavity it is very hard to compute the wake potential by a mode analysis since many modes are needed. The reason is that for s inside the bunch the charge distribution contributing to the wake potential is cut off due to causality. Hence the Fourier spectrum of the charge distribution contains many (say > 1000) modes. Since an accelerator is not made of closed boxes the mode analysis is not sufficient to calculate the wake potential. The continuous spectrum of wave guide modes in the beam pipes contribute also to the impedance, especially to the short range wake.

Figure 17 shows the real part of the impedance for a PETRA cavity. Above the cut–off frequency f_c the spectrum is continuous.

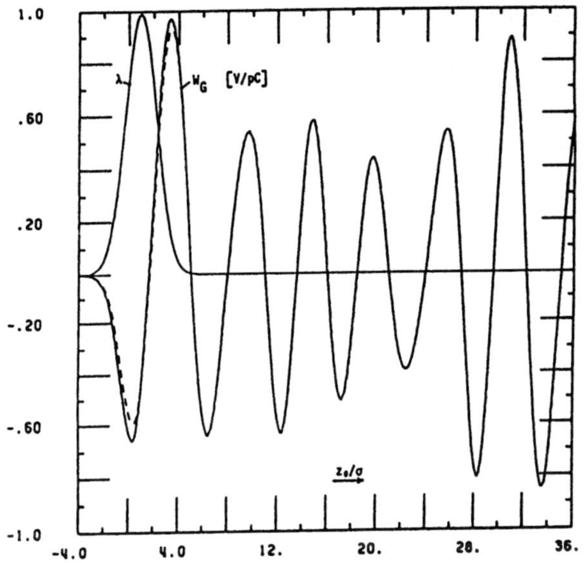

Figure 16: Normalized wake potential of a Gaussian bunch due to a pill–box cavity for $-4\sigma \leq s \leq 36\sigma$. The dashed line shows a mode analysis result (40 modes), while the solid line is obtained with TBCI on an 11×21 mesh.

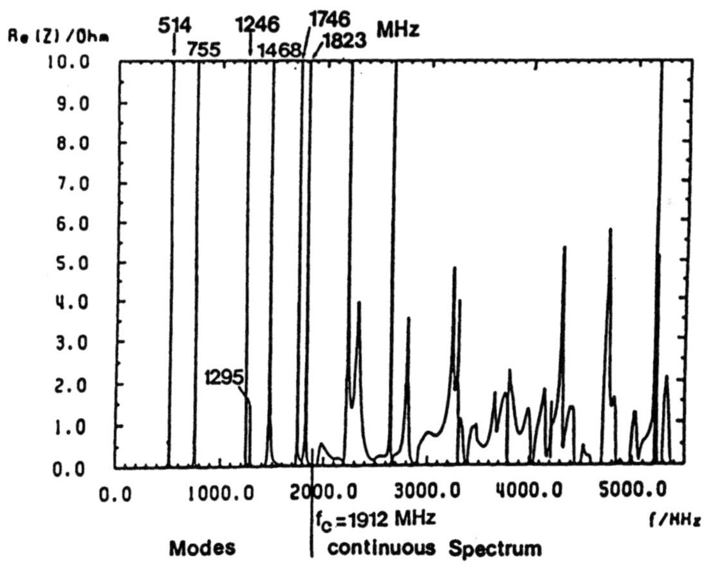

Figure 17: Real part of the impedance of a PETRA cavity

4.2 Numerical Calculations

4.2.1 Grid Maxwell Equations

A numerical approach for the solution of Maxwell's equations is presented. Based on a finite difference lattice the method transforms each of the four Maxwell equations into an equivalent matrix expression that can be subsequently treated by matrix mathematics and suitable numerical methods for solving matrix problems. The algorithm, although derived from integral equations, can be considered to be a special case of finite difference formalisms. A large variety of two and three-dimensional field problems can be solved by computer programs based on this approach: electrostatics and magnetostatics, low-frequency eddy currents in solid and laminated iron cores, high-frequency modes in resonators, waves on dielectric or metallic waveguides, transient fields of antennas and waveguide transitions, transient fields of free-moving bunches of charged particles etc.

Firstly, we will explain the general method. Then we will focus on the calculation of wake potentials, cavity modes and loss parameters.

The Method. In order to avoid specializations of Maxwell's equations prior to numerical solution it is advantageous to solve Maxwell's equations directly, rather than solving a partial differential equation derived from them. Using SI units, **E** and **H** for the electric and magnetic field strength, **D** and **B** for the flux densities and **J** for the current density the equations to solve are:

$$\begin{aligned} \oint_{\partial A} \mathbf{E} \cdot d\mathbf{s} &= -\int\!\!\int_A \frac{\partial \mathbf{B}}{\partial t} \cdot d\mathbf{A}, \\ \oint_{\partial A} \mathbf{H} \cdot d\mathbf{s} &= \int\!\!\int_A \left(\frac{\partial \mathbf{D}}{\partial t} + \mathbf{J} \right) \cdot d\mathbf{A}, \end{aligned}$$

$$\int\!\!\int_V \mathbf{B} \cdot d\mathbf{A} = 0,$$

$$\int\!\!\int_V \left(\frac{\partial \mathbf{D}}{\partial t} + \mathbf{J} \right) \cdot d\mathbf{A} = 0,$$

with the following relations:

$$\mathbf{D} = \epsilon \mathbf{E}, \qquad \mathbf{B} = \mu \mathbf{H}, \qquad \mathbf{J} = \kappa \mathbf{E} + \rho \mathbf{v}.$$

A grid G is defined in the orthogonal coordinate system $\mathbf{r} = \mathbf{r}(u, v, w)$ as

$$\begin{aligned} \mathrm{G} = \{(u_i, v_j, w_k);\ & u_1 < u_i < u_I,\ i = 2,..,I-1; \\ & v_1 < v_j < v_J,\ j = 2,..,J-1; \\ & w_1 < w_k < w_K,\ k = 2,..,K-1\}. \end{aligned}$$

All nodes of G are numbered linearly by

$$\begin{aligned} n &= 1 + (i-1)M_u + (j-1)M_v + (k-1)M_w, \\ n &= 1,..,N\ = IJK. \end{aligned}$$

Usually one uses $M_u = 1$, $M_v = I$, $M_w = IJ$ or any permutation of these three assignments. Figure 18 shows such a three-dimensional grid.

The functions which describe the field distribution in the volume under consideration are replaced by discrete field values in each cell of the grid. Thus one has to allocate components to grid nodes, lines, areas or elementary volumes. We will not choose the obvious allocation of

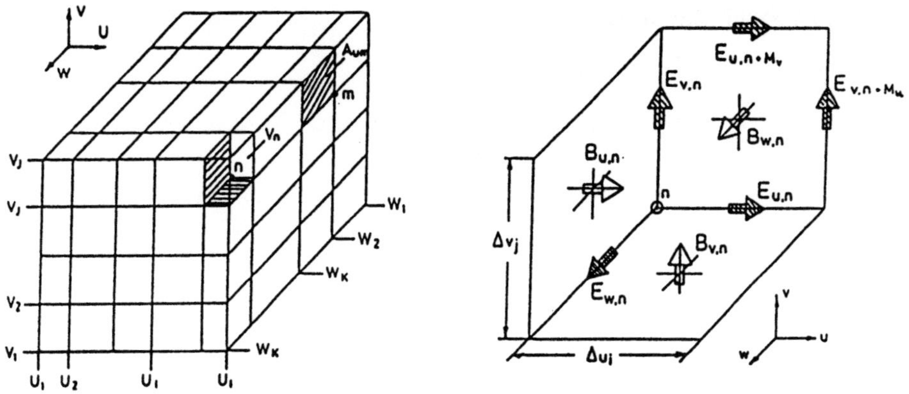

Three–dimensional grid in the orthogonal not necessarily cartesian system (u,v,w).

Allocation of unknown field components in the grid **G** showing the indexing.

Figure 18: Three–dimensional grid used for the numerical calculations

calculating all six field components at the grid nodes. Such an allocation would cause serious problems at surfaces of materials where some of the field components are not continuous. Instead, the components of **E** are placed at the mid points of the edges of the grid cells and the components of **B** at the centre of each face, as shown in Figure 19, extending the Yee Algorithm [7] to more general field problems [8, 9, 10].

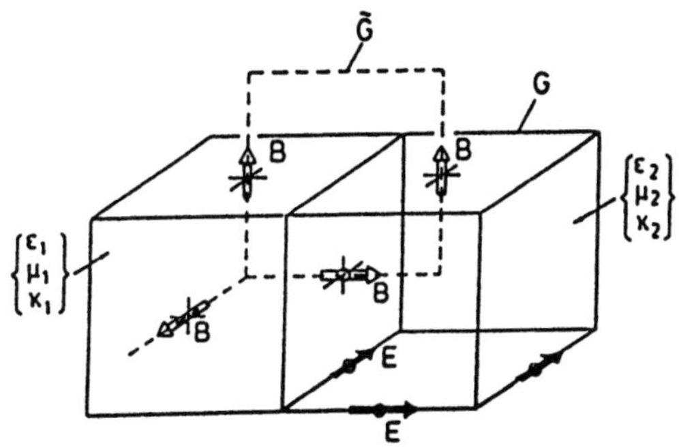

Figure 19: Two neighbouring mesh cells showing the allocation of the electric and magnetic field components in **G** and $\tilde{\mathbf{G}}$

This kind of allocation has the distinct advantage that the transition from one cell to the next involves only continuous components, tangential **E** fields and perpendicular **B** fields (see Fig. 19); thus Maxwell's equations are always satisfied, in this respect, even when different material fillings are involved. We now apply a first-order integration formula in order to approximate the left hand side of the first Maxwell equation and obtain the following algebraic expression replacing the integral along the border of an elementary cell:

$$\begin{aligned}\oint_{\partial A_{w,n}} \mathbf{E}\cdot ds = \; & +(u_{i+1}-u_i)E_{u,n} + (v_{j+1}-v_j)E_{v,n+M_u} \\ & -(u_{i+1}-u_i)E_{u,n+M_v} - (v_{j+1}-v_j)E_{v,n} \\ & +O((u_{i+1}-u_i)^2) + O((v_{j+1}-v_j)^2).\end{aligned}$$

For the right-hand side of that equation, we obtain (again by the lowest-order integration formula)

$$\int\int_{A_{w,n}} \frac{\partial \mathbf{B}}{\partial t} \cdot d\mathbf{A} = \dot{\mathbf{B}}_{w,n}(u_{i+1} - u_i)(v_{j+1} - v_j) + O\Big(((u_{i+1} - u_i)(v_{j+1} - v_j))^2\Big).$$

Equating the two expressions yields a discrete replacement for the first Maxwell equation on each surface of the grid cells and thus on every area composed of mesh–cell surfaces. In order to describe all these equations for all surfaces we introduce a basic discretization matrix with only two bands and with elements taking only the values 0, +1 or -1:

$$\mathbf{P_u} := (\mathbf{P_u})_{m,p} = \begin{cases} -1 & ; p = m \\ +1 & ; p = m + M_u \\ 0 & ; \text{else} \end{cases} \quad {}_{m,p=1,\ldots,N.}$$

Out of such simple matrices we combine the matrix $\mathbf{C}$ that replaces the contour integral operator in the first Maxwell equation for all mesh points:

$$\mathbf{C} = \begin{pmatrix} \mathbf{0} & -\mathbf{P}_w & \mathbf{P}_v \\ \mathbf{P}_w & \mathbf{0} & -\mathbf{P}_u \\ -\mathbf{P}_v & \mathbf{P}_u & \mathbf{0} \end{pmatrix}.$$

All unknown components of the electric field $\mathbf{E}$ are put into a vector of dimension 3N=3(IJK)

$$\mathbf{e} = (E_{u,1}, E_{u,2}, \ldots., E_{w,N})^t$$

and we replace $\mathbf{D}$ by $\mathbf{d}$, $\mathbf{B}$ by $\mathbf{b}$, $\mathbf{H}$ by $\mathbf{h}$ and $\mathbf{J}$ by $\mathbf{j}$ respectively. The topological part of the contour integral (**curl**) is represented by the matrix $\mathbf{C}$. The actual length of the integration path is put into a diagonal matrix:

$$\mathbf{D_s} = \mathbf{Diag}(\Delta u_1, .., \Delta u_N, \Delta v_1, .., \Delta v_N, \Delta w_1, .., \Delta w_N),$$

where

Δu_n, Δv_n and Δw_n are the mesh step sizes. The surfaces of the grid cells compose the diagonal matrix:

$$\mathbf{D_A} = \mathbf{Diag}(A_{u,1}, \ldots A_{u,N}, A_{v,1}, .., A_{v,N}, A_{w,1}, .., A_{w,N}),$$

$$A_{u,n} = \Delta v_n \Delta w_n, \qquad A_{v,n} = \Delta w_n \Delta u_n, \qquad A_{w,n} = \Delta u_n \Delta v_n.$$

Finally we can give a matrix equation using all these definitions and which replaces the first Maxwell equation on all cell surfaces:

$$\boxed{\mathbf{C D_s e} = -\mathbf{D_A} \dot{\mathbf{b}}}\,.$$

The special allocation of field components in the grid generates a second, dual grid $\tilde{\mathbf{G}}$ in which the magnetic flux densities are allocated as are the electric field components in the original grid $\mathbf{G}$, see Fig. 19.

The dual grid is defined by

$$\begin{aligned} \tilde{\mathbf{G}} = \{(\tilde{u}_i, \tilde{v}_j, \tilde{w}_k);\ & \tilde{u}_1 < \tilde{u}_i < \tilde{u}_I,\ i = 2, .., I-1; \\ & \tilde{v}_1 < \tilde{v}_j < \tilde{v}_J,\ j = 2, .., J-1; \\ & \tilde{w}_1 < \tilde{w}_k < \tilde{w}_K,\ k = 2, .., K-1\}, \end{aligned}$$

$$\tilde{u}_i = (u_{i+1} + u_i)/2;\ 1 \le i \le I-1;\ \tilde{u}_I = 0,$$
$$\tilde{v}_j = (v_{j+1} + v_j)/2;\ 1 \le j \le J-1;\ \tilde{v}_J = 0,$$
$$\tilde{w}_k = (w_{k+1} + w_k)/2;\ 1 \le k \le K-1;\ \tilde{w}_K = 0.$$

We can also define matrices that hold the mesh step sizes and cell areas as for the original grid G:

$$\tilde{\mathbf{D}}_{\mathbf{s}} = \mathbf{Diag}(\tilde{\Delta}u_1, .., \tilde{\Delta}u_N, \tilde{\Delta}v_1, .., \tilde{\Delta}v_N, \tilde{\Delta}w_1, .., \tilde{\Delta}w_N),$$
$$\tilde{\mathbf{D}}_{\mathbf{A}} = \mathbf{Diag}(\tilde{A}_{u,1}, .., \tilde{A}_{u,N}, \tilde{A}_{v,1}, .., \tilde{A}_{v,N}, \tilde{A}_{w,1}, .., \tilde{A}_{w,N}),$$
$$\tilde{A}_{u,n} = \tilde{\Delta}v_n\,\tilde{\Delta}w_n, \qquad \tilde{A}_{v,n} = \tilde{\Delta}w_n\,\tilde{\Delta}u_n, \qquad \tilde{A}_{w,n} = \tilde{\Delta}u_n\,\tilde{\Delta}v_n.$$

When solving the second Maxwell equation in a way similar to the first one, we have to face the fact that the magnetic field **H** is not defined and thus has to be calculated from the flux density piecewise along the integration path. Also, the integration over the electric flux density is no longer a simple product of an area with a component but the sum over four parts on each of which the flux has to be determined from the defined **E** and the material constants. Finally we obtain

$$\boxed{\tilde{\mathbf{C}}\,\tilde{\mathbf{D}}_{\mathbf{s}}\,\mathbf{D}_{\mu}^{-1}\,\mathbf{b} = \tilde{\mathbf{D}}_{\mathbf{A}}\,(\dot{\mathbf{d}} + \mathbf{j})}\,.$$

Note that the contour integral operator on the dual grid, $\tilde{\mathbf{C}}$, (**curl**) is related to **C** simply by

$$\tilde{\mathbf{C}} = \mathbf{C}^{\mathbf{t}}.$$

The diagonal matrices which take care of the piecewise different values for the material constants are

$$\mathbf{D}_{\mu}, \quad \tilde{\mathbf{D}}_{\epsilon}, \text{ and } \quad \tilde{\mathbf{D}}_{\kappa}.$$

Note that μ, ϵ and κ may have different properties in the directions u, v and w. Although not originally defined in the grid, one may relate the field strength, flux density and current density, and formally define a magnetic field and an electric flux density by

$$\mathbf{b}{=}\mathbf{D}_{\mu}\mathbf{h} \quad , \quad (\mathbf{B}{=}\mu\mathbf{H}),$$
$$\mathbf{d}{=}\tilde{\mathbf{D}}_{\epsilon}\mathbf{e} \quad , \quad (\mathbf{D} = \epsilon\mathbf{E}),$$
$$\mathbf{j}{=}\tilde{\mathbf{D}}_{\kappa}\mathbf{e} + \mathbf{D}_{\rho}\mathbf{v} \quad , \quad (\mathbf{J} = \kappa\mathbf{E} + \rho\mathbf{v}).$$

In order to complete the transformation of Maxwell's equations to the grid space, we approximate the divergence equation by integrating **B** over surfaces of each mesh cell of **G**. We can then write this equation by defining the discrete **div** operator on **G** as

$$\mathbf{S} = (\mathbf{P_u}|\mathbf{P_v}|\mathbf{P_w})$$

and obtain

$$\boxed{\mathbf{S}\mathbf{D}_{\mathbf{A}}\mathbf{b} = \mathbf{0}}$$

This equation allocates the (non-existent) magnetic charges to nodes of $\tilde{\mathbf{G}}$. The corresponding matrix on the dual grid $\tilde{\mathbf{G}}$ is simply given by

$$\tilde{\mathbf{S}} = (\mathbf{P_u^t}|\mathbf{P_v^t}|\mathbf{P_w^t}).$$

So the continuity equation in the grid space reads as

$$\boxed{\tilde{\mathbf{S}}\tilde{\mathbf{D}}_{\mathbf{A}}(\dot{\mathbf{d}} + \mathbf{j}) = \mathbf{0}}$$

Thus electric charges are defined on nodes of the original grid **G**.

Properties of the grid equations. One of the outstanding features of Maxwell's grid equations is that properties of analytical solutions have their exact analogues in the grid space. The analytical identity **div curl** $\equiv$ **0** in the grid space is

$$\mathbf{div\,curl} \equiv \mathbf{0} \leftrightarrow \tilde{\mathbf{S}}\,\tilde{\mathbf{C}} \equiv \mathbf{S}\,\mathbf{C} \equiv \mathbf{0}.$$

The analytical property that scalar potential fields are curl free is also found in the grid space as

$$\mathbf{curl\,grad} \equiv \mathbf{0} \Leftrightarrow \mathbf{C}^{\mathbf{t}}\,\mathbf{S}^{\mathbf{t}} \equiv \tilde{\mathbf{C}}\,\tilde{\mathbf{S}}^{\mathbf{t}} \equiv \mathbf{0}.$$

These properties not only offer a unique tool to test numerical results for their physical correctness but also avoid the occurrence of incorrect solutions in the calculation of three-dimensional eigenmodes in resonators, or at least their identification [9]. This is quite important since numerical errors in the matrix algorithms may occur and cannot generally be detected. In the field of accelerator physics, one often investigates unknown phenomena by using such codes, so that an incorrect solution could lead to physical misinterpretations.

The Maxwell Grid Equations. Without specifying anything about the shape of materials or the time dependence of the fields, we have obtained a set of matrix equations that approximate Maxwell's equations on a double grid $\{G, \tilde{G}\}$ ($\mathbf{R}^3$ and $\mathbf{R}^{3N}$ represent the physical spaces and $\mathbf{R}_+$ the time)

$$\begin{array}{rcl}
\textbf{Real Space} & \Leftrightarrow & \textbf{Grid Space} \\
\mathbf{R}^3 \otimes \mathbf{R}_+ & & \mathbf{R}^{3\mathbf{N}} \otimes \mathbf{R}_+ \\
\oint_{\partial A} \mathbf{E}\cdot ds = -\int\int_A \frac{\partial \mathbf{B}}{\partial t}\cdot d\mathbf{A} & \Leftrightarrow & \mathbf{C}\mathbf{D_s}\mathbf{e} = -\mathbf{D_A}\dot{\mathbf{b}} \\
\oint_{\partial A} \mathbf{H}\cdot ds = \int\int_A \left(\frac{\partial \mathbf{D}}{\partial t} + \mathbf{J}\right)\cdot d\mathbf{A} & \Leftrightarrow & \tilde{\mathbf{C}}\tilde{\mathbf{D}}_\mathbf{s}\mathbf{h} = \tilde{\mathbf{D}}_\mathbf{A}(\dot{\mathbf{d}} + \mathbf{j}) \\
\int\int_V \mathbf{B}\cdot d\mathbf{A} = 0 & \Leftrightarrow & \mathbf{S}\mathbf{D_A}\mathbf{b} = \mathbf{0} \\
\int\int_V \left(\frac{\partial \mathbf{D}}{\partial t} + \mathbf{J}\right)\cdot d\mathbf{A} = 0 & \Leftrightarrow & \tilde{\mathbf{S}}\tilde{\mathbf{D}}_\mathbf{A}(\dot{\mathbf{d}} + \mathbf{j}) = \mathbf{0} \\
\mathbf{D} = \epsilon\mathbf{E} & \Leftrightarrow & \mathbf{d} = \tilde{\mathbf{D}}_\epsilon\mathbf{e} \\
\mathbf{B} = \mu\mathbf{H} & \Leftrightarrow & \mathbf{b} = \mathbf{D}_\mu\mathbf{h} \\
\mathbf{J} = \kappa\mathbf{E} + \rho\mathbf{v} & \Leftrightarrow & \mathbf{j} = \tilde{\mathbf{D}}_\kappa\mathbf{e} + \mathbf{D}_\rho\mathbf{v} \\
\mathbf{div\,curl} \equiv \mathbf{0} & \Leftrightarrow & \mathbf{S}\,\mathbf{C} = \tilde{\mathbf{S}}\,\tilde{\mathbf{C}} = \mathbf{0} \\
\mathbf{curl\,grad} \equiv \mathbf{0} & \Leftrightarrow & \tilde{\mathbf{C}}^{\mathbf{t}}\,\tilde{\mathbf{S}}^{\mathbf{t}} = \mathbf{C}^{\mathbf{t}}\,\mathbf{S}^{\mathbf{t}} = \mathbf{0}.
\end{array}$$

4.2.2 Short and Long–Range Wakes

In order to solve these equations one has only to perform matrix manipulations and to solve the established matrix problem numerically.

Short–range wake fields can be calculated efficiently in the time domain. For transient fields with a central time difference formula , step size δt, no currents and no losses, one obtains the Yee-algorithm [7]. In the presence of free moving charges the algorithm has more constraints to ensure charge conservation [11]. The upper index n denotes the time in units of δt. These recursive formulae allow the easy calculation of transient fields of antennas, particle beams in accelerators [11] or wave propagation problems even in the lossy case

$$\begin{aligned} \mathbf{b}^{n+1} &= \mathbf{b}^{n} - \Delta t\, \mathbf{D}_A^{-1}\mathbf{C}\mathbf{D}_s\mathbf{e}^{n+0.5}, \\ \mathbf{e}^{n+1.5} &= \mathbf{D}_a\, \mathbf{e}^{n+0.5} + (1-\mathbf{D}_a)\, \tilde{\mathbf{D}}_\kappa^{-1}\tilde{\mathbf{D}}_A^{-1}\tilde{\mathbf{C}}\tilde{\mathbf{D}}_s\mathbf{D}_\mu^{-1}\mathbf{b}^{n+1} - (1-\mathbf{D}_a)\, \tilde{\mathbf{D}}_\kappa^{-1}\mathbf{j}_e^{n+1} \end{aligned}$$

with

$$\mathbf{D}_a = \exp(-\tilde{\mathbf{D}}_\varepsilon^{-1}\mathbf{D}_\kappa\, \Delta t).$$

Detailed setting up of the final matrix problem can be found in the literature [9, 10]. Knowing the electric and magnetic fields for each time step it is possible to obtain the wake potential directly by a numerical integration of

$$\boldsymbol{W}(x,y,s) = \frac{1}{q}\int_{-\infty}^{\infty} dz[\boldsymbol{E}(x,y,z,t) + c\,\boldsymbol{e}_z \times \boldsymbol{B}(x,y,z,t)]_{t=(s+z)/c},$$

if the charge q moves along the z–axis.

As mentioned in section 2.5 it is advantageous to perform the integration for all positions (x,y) at the side tubes because the integration interval is then finite. The longitudinal wake potential for *any* position (x,y) is then obtained as a solution of the equation

$$\nabla_\perp^2 W_{||}(x,y,s) = 0$$

and the transverse one by an application of Panofsky–Wenzel theorem.

The long–range wake potential is dominated by the interaction of the beam with some cavity modes. For arbitrarily-shaped cavities the eigenmodes have to be calculated numerically. The electric fields simultaneously have to obey the following two equations, that can be directly derived from Maxwell's equations

$$\begin{aligned} \operatorname{curl}\frac{1}{\mu}\operatorname{curl}\mathbf{E} &= \omega^2\epsilon\,\mathbf{E} \\ \operatorname{div}\epsilon\,\mathbf{E} &= 0. \end{aligned}$$

The numerical formulation can be derived again by matrix manipulation:

$$\begin{aligned} \tilde{\mathbf{C}}\tilde{\mathbf{D}}_\mathbf{S}\,\mathbf{D}_\mu^{-1}\,\mathbf{D}_\mathbf{A}^{-1}\,\mathbf{C}\,\mathbf{D}_\mathbf{S}\,)\,\mathbf{e} &= \omega^2\,\tilde{\mathbf{D}}_\mathbf{A}\,\tilde{\mathbf{D}}_\epsilon\;\mathbf{e} \\ \tilde{\mathbf{S}}\tilde{\mathbf{D}}_\mathbf{A}\,\tilde{\mathbf{D}}_\epsilon\,\mathbf{e} &= 0. \end{aligned}$$

In order to convert the algorithm to a symmetric form we must perform a diagonal transformation, which physically replaces the field strength by the root of the local energy density (except from a scalar factor):

$$\mathbf{e}' = (\tilde{\mathbf{D}}_\epsilon\,\tilde{\mathbf{D}}_\mathbf{A}\,\mathbf{D}_\mathbf{S}\,)^{1/2}\mathbf{e}.$$

One obtains a 3N x 3N eigenvalue equation, the eigenvalues of which are the resonant frequencies squared [9]:

$$(\tilde{\mathbf{D}}\tilde{\mathbf{C}}\mathbf{D})\,(\tilde{\mathbf{D}}\tilde{\mathbf{C}}\mathbf{D})^t\;\mathbf{e}' = \omega^2\,\mathbf{e}'$$

with the matrices

$$\begin{aligned} \mathbf{D} &= (\mathbf{D}_\mathbf{A}^{-1}\,\tilde{\mathbf{D}}_\mathbf{S}\,\mathbf{D}_\mu^{-1}\,)^{1/2}, \\ \tilde{\mathbf{D}} &= (\tilde{\mathbf{D}}_\mathbf{A}^{-1}\,\mathbf{D}_\mathbf{S}\,\tilde{\mathbf{D}}_\epsilon^{-1}\,)^{1/2}. \end{aligned}$$

The resulting matrix has a multiply–vanishing eigenvalue for $\omega = 0$ as the existence of static solutions cannot be excluded by curl operations only. These solutions can be identified easily by resubstituting the eigenvectors in the div equation. The analytical analogue to this combination of equations is the equation curl curl**E** = grad div**E** − ∇^2**E** and div**E** = 0 thus solving ∇^2**E**. After numerical solution of the cavity modes the loss parameter k can be easily calculated from the voltage

$$V = \int dz E_z(\boldsymbol{r}) \exp(i\omega z/c)$$

and the total energy stored in the specific mode

$$U = \frac{\epsilon_0}{2} \int d^3r |\boldsymbol{E}(\boldsymbol{r})|^2$$

(see also section 3.2).

4.3 Examples

The method presented in the preceding sections to solve the Maxwell grid equations in two and three dimensions is implemented in the CAD–program packet MAFIA [12]. We now present some calculations obtained with the programs TBCI and URMEL, which are now included into the newest release of MAFIA. The acronym MAFIA stands for the solution of **MA**xwell's equations by the **F**inite **I**ntegration **A**lgorithm. A short description of the modules is given in the following table:

Module	Description	Special Features
M	Mesh generator, translates the physical problem into mesh data	Many predefined shapes, interactive remeshing
S	Static solver, electro- and magnetostatic problems	Open boundary, nonlinear magnetic material properties, advanced multigrid solver for very large meshes
R E	Matrix generator, eigenvalue solver for resonator and waveguide problems	Periodic boundary conditions, resonance frequencies (several hundred modes), waveguide modes and propagation parameters
T2 T3	Time domain solver	Losses, open boundary, stimulation by initial field, current, waveguide, incident waves; wake potentials
TS2 TS3	Particle–in–cell code, calculates equation of motion for free-moving charges	Losses, stimulation by initial field, calculation of trajectories, velocities, charge densities etc.
W3	Eddy–current solver, frequency domain	Losses, current stimulation
P	Post processor, displays results and calculates secondary field quantities	1, 2 and 3-D graphics of scalar and vector fields, power loss integrals, field energy, far–field pattern, shunt impedance

Figure 20 shows the interconnections between the various programs of the third release of MAFIA. The first example is a radiofrequency quadrupole (RFQ) that has been calculated with the MAFIA frequency domain module R/E. The device might be tuned for a certain minimum eigenmode or in order to minimize the losses. Three–dimensional calculations are important especially for the end region of such an RFQ. A careful design using the MAFIA mesh generator has been carried out, using about 75,000 unknowns. The lowest eigenmode calculated, 344.78 MHz is in good agreement with measurements and other calculations. Figure 21 displays the power loss which is strongest on the outside wall. The power loss can be calculated by the MAFIA-postprocessor from the field solutions.

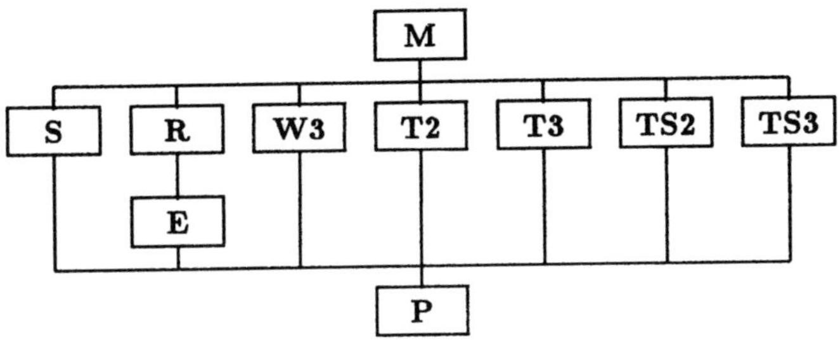

Figure 20: : The MAFIA System with its inter-relationships

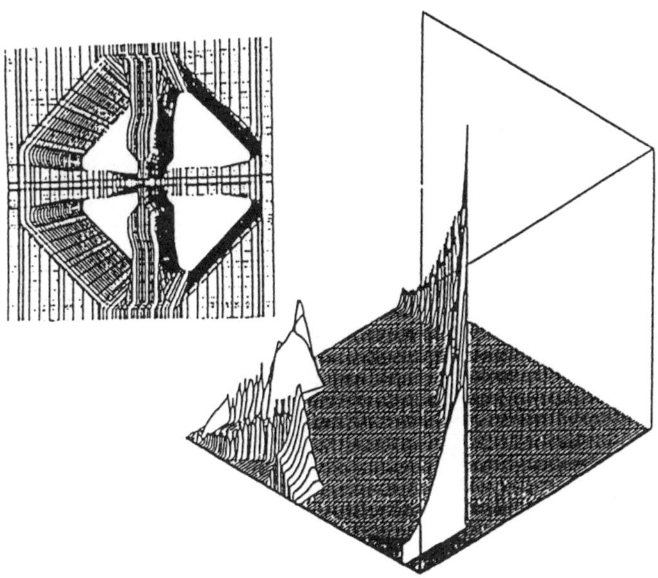

Figure 21: RFQ geometry and power loss from MAFIA

The figure 22 shows the geometry of a kicker magnet calculated for the GSI with MAFIA. With the modules R and E the modes of the structures have been calculated in order to evaluate the loss parameters.

The next two examples show wake field calculations with the computer code TBCI. A Gaussian bunch traverses a bellow structure (Fig. 23) and a PETRA cavity (Fig. 24) respectively.

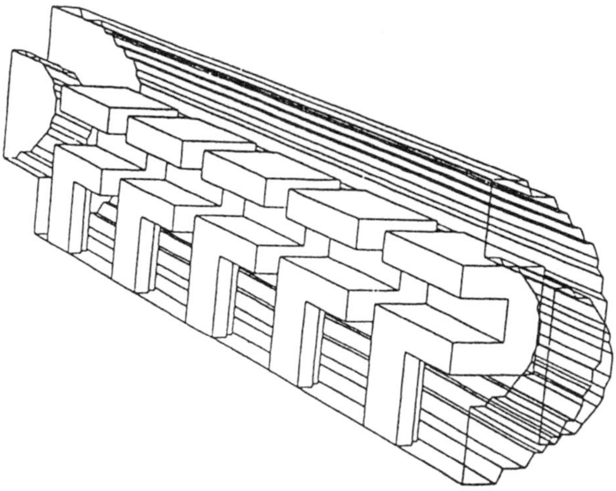

Figure 22: Geometry of a kicker magnet from MAFIA

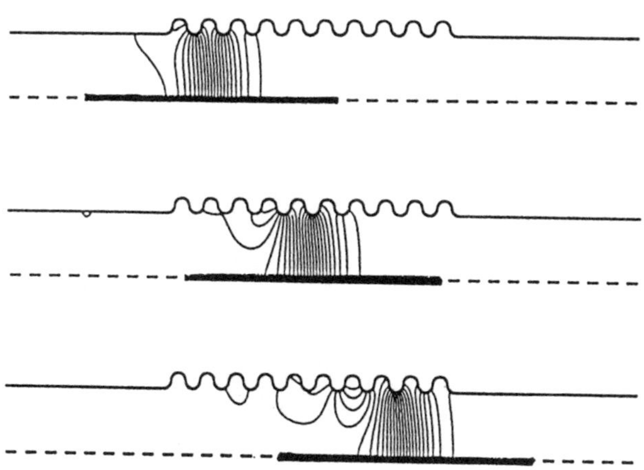

Figure 23: Wake fields in a bellows structure. The bunch is represented by the heavy line and the electric field by lines of constant $\int dt\, r\, H_\varphi \sim r\, |E|$.

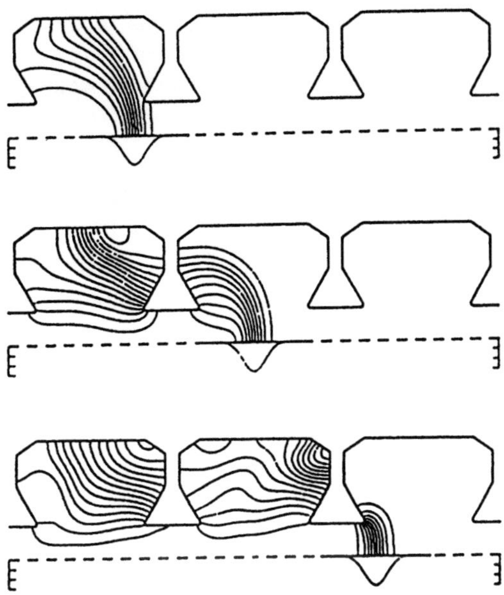

Figure 24: **Wake fields in a PETRA cavity generated by a Gaussian bunch. The electric field is** represented by lines of constant $\int dt\, r\, H_\varphi \sim r\, |E|$.

5 Effects of Wakes and Impedances

5.1 Heating

Due to wake fields the bunch loses energy which heats the traversed structure. We can distinguish between heating effects due to short–range wakes and those due to long–range wakes. The former are single–turn effects in a storage ring. The heating due to the instantaneous energy loss normally causes no problems for structures with a good thermal conductivity. Long–range wake field effects are important if a single mode is driven resonantly by successive turns of the bunches. All effects can be calculated from the loss parameters and we will give an example for each.

Transient effects. Consider a storage ring characterized by the
- revolution time $T_0 = 7.68\,\mu$s,
- number of bunches $N_b = 4$,
- charge per bunch $q_b = 40\,$nC.

The heating of a PETRA cavity (5 cells) due to transient losses can be calculated from the parasitic loss parameter

$$k_{par}(\sigma) = 1.5 \times 10^{12}\,\frac{\mathrm{V}}{\mathrm{As}},$$

for a Gaussian bunch with rms length of $\sigma = 1$ cm. The heating power is given by

$$P = N_b\, q_b^2\, k_{par}\, \frac{1}{T_0} \approx 1250\,\mathrm{W}$$

and causes no problems for a large copper structure like a PETRA cavity.

Resonant effects. We now consider the heating due to the resonant excitation of one mode characterized by the loss parameter k, the revolution frequency ω and the quality factor Q. Note that k and ω are determined by the geometrical shape while Q depends on the material properties. In section 3.4 we calculated the induced voltage if one mode is driven resonantly by the beam. We obtained for the cavity voltage

$$V = R\,I \qquad \text{with}\, R = \frac{4\,k\,Q}{\omega}$$

where R is the shunt impedance (US definition) and I the current in the storage ring.

For $I = 160$ mA ($N_b = 210$ bunches in HERA with $q_b = 16$ nC) and a shunt impedance of $R = 1\,\text{M}\Omega$ one obtains a heating power of

$$P = R\,I^2 \approx 25\,\text{kW},$$

if in the worst case all bunches are in phase with the mode.

All accelerator components, and not only cavities, have a finite shunt impedance. For high beam currents (e.g. B–factories with $I \approx 1$ A) even small shunt impedances will generate high heating powers if such a mode is driven resonantly.

5.2 Instabilities

As already mentioned in the introduction this section can only give a brief overview of instabilities. For a more detailed treatment we refer to the lecture by A. Hofmann in these proceedings, and the literature.

We can again distinguish between short–range wake field effects within one bunch and long–range wake field effects from bunch to bunch. Furthermore longitudinal and transverse effects on the beam can be considered. In this section we will focus on single–bunch effects, especially on bunch lengthening. The goal is to explain how the wake potential can be included into tracking calculations leading to a quantitative prediction of bunch lengthening.

The electromagnetic interaction between a bunch and its environment can be described by a Green's function, the δ–function wake potential. This potential gives the energy loss of a witness charge behind a leading point charge after the passage through the entire structure as a function of their distance apart. Such δ–function wake potentials cannot be calculated exactly for a real structure but an approximation can be obtained by mode analysis with additional analytical extentions.

However, it is not necessary to know the δ–function wake potential. An approximate Green's function can be calculated by means of the computer code TBCI. Instead of a leading point charge, a leading short Gaussian bunch (super particle) is used. This is a good approximation if the rms lengths of the super particles are small compared to the total rms bunch length. (For a detailed analysis of the errors see [13].)

The longitudinal equations of motion are

$$\begin{aligned} \dot{\tau} &= \alpha\,\frac{\epsilon}{E_0} \\ \dot{\epsilon} &= e\,V_{rf}\,\sin(\omega_{rf}\,\tau) \end{aligned}$$

where τ is the particle position in time with origin at zero of the rf voltage, ϵ is the particle energy deviation from the nominal energy E_0, and V_{rf} is the rf voltage and α the momentum compaction.

Longitudinal wake field effects can be simply included:

$$\dot{\epsilon} = e\,V_{rf}\,\sin(\omega_{rf}\,\tau) - e\,w(\tau)$$

where $w(\tau)$ is the wake potential at position τ which can be computed by summing all contributions of the particles in front of τ:

$$w(\tau) = \frac{q}{N} \sum_{i=1,\tau_i>\tau}^{N} w_s(\tau_i - \tau)$$

where q is the total charge in the bunch, N the number of super particles, and w_s is the wake potential of one super particle calculated numerically by TBCI. For practical tracking calculations the super particles can be filled into bins to speed up evaluation of the sums [13].

Radiation losses and quantum excitations can be also included into the tracking equations. Figure 25 shows the predicted and measured bunch length as a function of single–bunch current. The good agreement of the calculations with the measurements demonstrates the usefulness of the method.

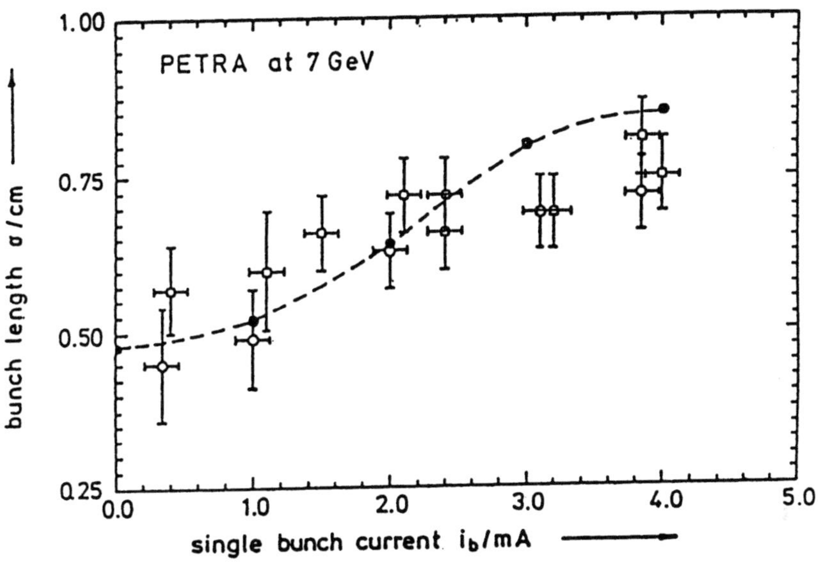

Figure 25: Bunch length as a function of the single–bunch current at an energy of 7 GeV in PETRA. The dashed line represents the result of tracking calculations.

The transverse wake-field effects can also be studied by tracking calculations. Again super particles can be used to calculate the wake potential. Every super particle is kicked by the leading particles. While different approaches agreed well with each other there was a factor 3 to 4 difference between the simulated and measured results. The reason was that the transverse impedance of separator tanks, bellows and vacuum chamber joints outside the cavities were not included in the calculations.

A rough comparison of the transverse wake effects of a cavity, a bellows and a pipe junction (Fig. 26) shows that all accelerator components are important rather than only the rf cavities.

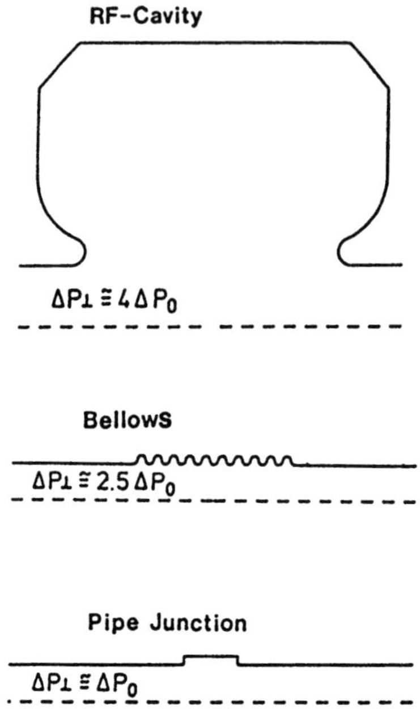

Figure 26: Comparison of the transverse wake field effects of a cavity, a bellows and a pipe junction.

6 Strategies in Accelerator Design

6.1 Design Procedure – an Overview

Since one important reason for current limitations in accelerators is instabilities caused by wake fields, all accelerator components should be carefully analysed at the design stage with respect to their contribution to the impedance.

It took over ten years to develop tools to compute impedances very precisely and include the numerical results into instability theory. Today rather sophisticated computer codes are available for calculating the impedance of the various types of accelerator components such as cavities, bellows, vacuum chambers, etc. Heating and instability effects can be evaluated very precisely from these numerical results. One can trust the computer calculations and it is no longer necessary to use rather crude models.

While we can calculate impedances very well, we can not avoid impedances due to accelerating cavities. But by a careful design it is possible to reduce wake field effects significantly.

A good strategy for the design of an accelerator is

- Minimize parasitic effects of unavoidable components (e.g. cavities).
- Avoidable impedances should be less (equal) than the unavoidable impedances.
- Optimize the overall performance of the accelerator.

In Fig. 27 an overview of the design procedure for any accelerator component is given. To skip steps of the procedure is certainly a much faster way. But in order to avoid current limitations due to impedances it is always better to go through all steps of the design procedure.

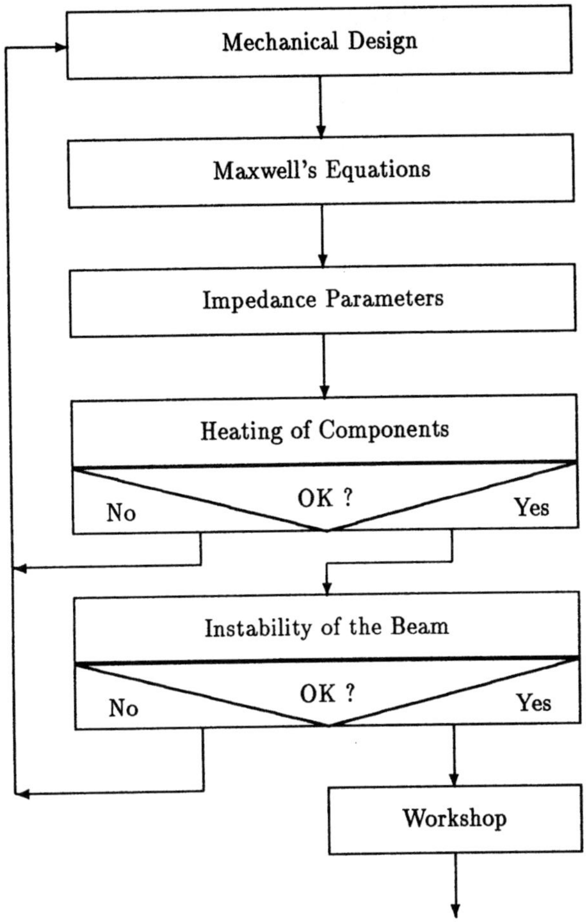

Figure 27: Design procedure for any accelerator component

6.2 Parasitic Losses versus Shunt Impedance

In this section we will give one example of how to minimize parasitic effects of cavities. The accelerating voltage as a function of klystron power P and shunt impedance R (we use here the definition from electrical engineering) is given by

$$V_{acc} = \sqrt{2\,R\,P}.$$

In order to save power the shunt impedance can be optimized. Figure 28 shows three examples of typical cavity shapes with different shunt impedances.

The nose–cone structure gives the highest shunt impedance. But this type of cavity has also the highest parasitic impedance. While a little klystron power is saved, one loses the ability to accelerate higher currents. Even a simple iris–loaded structure has a lower parasitic impedance than the nose–cone structure. Parasitic effects can be further reduced with the third structure shown in Fig. 28. This kind of shape is now widely used for superconducting cavities. It is possible to design a single-mode cavity based on this kind of shape. In this cavity no higher-order modes exist. All parasitic effects are due to the continuous part of the impedance. For a quantitative comparison of the various loss parameters we refer to the literature [14]. The conclusion of this section is that it is better to optimize the cavity with respect to parasitic effects than with respect to the shunt impedance.

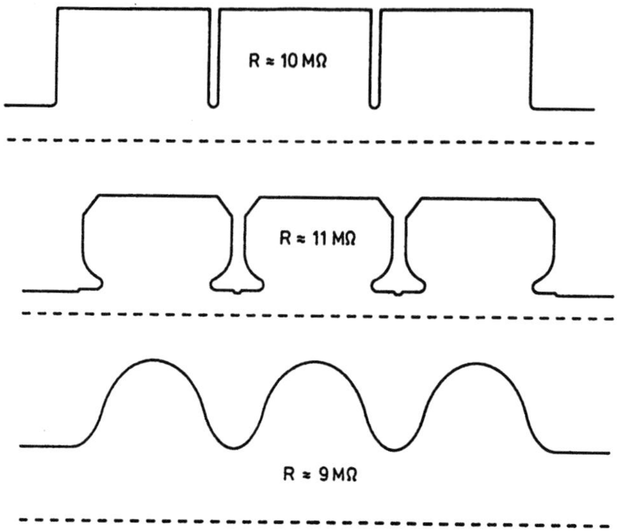

Figure 28: Three differently shaped 500 MHz cavities with their shunt impedances

References

[1] J.D. Jackson: *Classical Electrodynamics*, 2nd edition, Wiley, New York, 1975

[2] A. Piwinski, *Longitudinal and Transverse Wake Fields in Flat Vacuum Chambers*, DESY 84/097, Oct. 1984

[3] K. Steinigke, *Calculation of the Propagation Characteristics of a Circular Waveguide Excentrically Loaded with a Dielectric Rod*, Frequenz 44 (1990) 1

[4] W.K.H. Panofsky, W.A. Wenzel, *Some Considerations Concerning the Transverse Deflection of Charged Particles in Radio-Frequency Fields*, Rev. Sci. Instrum. 27 (1956), 947

[5] K.L.F. Bane, P.B. Wilson, T. Weiland, *Wake Fields and Wake Field Acceleration* in Physics of High Energy Particle Accelerators (BNL/SUNY Summer School, 1983) AIP Conf. Proc. No 127 (1984), 876, ed. M. Month et al.

[6] T. Weiland, B. Zotter, *Wake Potentials of a Relativistic Current in a Cavity*, Part. Acc. 11,143-151 (1981)

[7] K.S. Yee, *Numerical Solution of Initial Boundary Value Problems Involving Maxwell's Equations in Isotropic Media.* IEEE, AP-14, 1966, pp.302-307

[8] T. Weiland: *A Discretization Method for the Solution of Maxwell's Equations for 6-Component Fields.* AEÜ, Electron. and Commun., 31(1977),pp.116-120

[9] T. Weiland, *On the Numerical Solution of Maxwell's Equations and Applications in the Field of Accelerator Physics.*, Particle Accelerators 15 (1984), pp.245-292 and references therein

[10] T. Weiland, *On the Unique Solution of Maxwellian Eigenvalue Problems in Three Dimensions,* Particle Accelerators 17 (1985), pp.227-242

[11] T. Weiland, *Transient Electromagnetic Fields Excited by Bunches of Charged Particles in Cavities of Arbitrary Shape.* Proceedings of the 11th International Conference on High Energy Accelerators, Geneva 1980, Birkhauser Verlag, Basel, pp. 570

[12] MAFIA code group: M. Bartsch, M. Dehler, M. Dohlus, F. Ebeling, P. Hahne, R. Klatt, F. Krawczyk, Zhang M., T. Pröpper, D. Schmitt, P. Schütt, T. Weiland, H. Wolter; Technische Hochschule Darmstadt, Fachbereich 18, Fachgebiet Theorie Elektromagnetischer Felder, Schloßgartenstr. 8, D-6100 Darmstadt, Germany
M. Marx, S.G. Wipf; Deutsches Elektronen-Synchrotron DESY, Notkestr. 85, D-2000 Hamburg 52, Germany
B. Steffen; Kernforschungsanlage Jülich KFA, D-5170 Jülich, Germany

[13] T. Weiland, *On the Quantitative Prediction of Bunch Lengthening in High Energy Electron Storage Rings*, DESY 81–088 (1981)

[14] R. Klatt, T. Weiland, *Comparison of Six Accelerating Cavities With Respect to Collective Effects*, DESY M-84-06 (1984)

[15] R. Wanzenberg, *Longitudinale Strahldynamik im Resonanten Wake Field Beschleuniger*, Thesis, Universität Hamburg 1990, and DESY M-90-08 Aug. 1990

Acknowledgement

We would like to thank M. Dohlus who wrote an article for the MAFIA code group from which the section about the grid Maxwell equations has been adopted for this chapter.

Appendix : Wake potential and impedance, mode expansion

The goal of the appendix is to prove that the longitudinal wake potential of a point charge traversing any closed resonator structure along the z-axis with the velocity of light can be written as

$$W_{||}(s) = \sum_n 2\, k_n \cos(\omega_n\, s/c),$$

for all $s > 0$. The k_n are the loss parameters of the structure. The calculations are first made in the frequency domain [15]. Then the wake potential is obtained by a Fourier transformation.

The electric and magnetic fields can be written in terms of the vector $\boldsymbol{A}$ and the scalar potential Φ

$$\boldsymbol{E} = -\frac{\partial}{\partial t}\boldsymbol{A} - \nabla\Phi \qquad \boldsymbol{B} = \nabla \times \boldsymbol{A}$$

and are equivalent to the Maxwell equations

$$\begin{aligned} -\nabla^2 \boldsymbol{A} + \frac{1}{c^2}\frac{\partial^2}{\partial t^2}\boldsymbol{A} &= \mu_0 \boldsymbol{j} - \frac{1}{c^2}\frac{\partial}{\partial t}\nabla\Phi \\ -\nabla^2\Phi &= \frac{1}{\epsilon_0}\rho. \end{aligned}$$

The Coulomb gauge was used for the potentials:

$$\nabla \cdot \boldsymbol{A} = 0.$$

The vector potential $\boldsymbol{A}$ can be expanded into the modes $\boldsymbol{a}_n$ of the closed resonator structure.

$$\boldsymbol{A}(\boldsymbol{r},t) = \sum_n q_n(t)\,\boldsymbol{a}_n(\boldsymbol{r})$$

where the $\boldsymbol{a}_n$ have to fulfill the equation

$$-\nabla^2 \boldsymbol{a}_n(\boldsymbol{r}) = \left(\frac{\omega_n}{c}\right)^2 \boldsymbol{a}_n(\boldsymbol{r})$$

and the ω_n are the eigen frequencies of the structure.

Since the a_n form a complete orthogonal system (U_n is a normalization factor)

$$\frac{\epsilon_0}{2}\int d^3r\,\boldsymbol{a}_n^*(\boldsymbol{r})\,\boldsymbol{a}_m(\boldsymbol{r}) = U_n\,\delta_{nm}\,,$$

the following relation for the expansion coefficients q_n holds:

$$(\omega_n^2\, q_n(t) + \frac{d^2}{dt^2}q_n(t))\,U_n = \frac{1}{2}\int d^3r\;\boldsymbol{a}_n^*(\boldsymbol{r})\,\boldsymbol{j}(\boldsymbol{r},t) - \frac{\epsilon_0}{2}\frac{\partial}{\partial t}\int d^3r\;\boldsymbol{a}_n^*(\boldsymbol{r})\,\nabla\Phi(\boldsymbol{r},t).$$

For the Coulomb gauge $\nabla\,\boldsymbol{a}_n = 0$ we have $\boldsymbol{a}_n^*\,\nabla\Phi = \nabla(\boldsymbol{a}_n^*\Phi)$. Hence, according to Gauss' theorem, the second integral can be transformed into a surface integral which vanishes due to the boundary condition for Φ.

For the Fourier transform $\widetilde{q_n}(\omega) = \int_{-\infty}^{\infty} dt\, q_n(t)\exp(-i\omega t)$ of the expansion coefficient we obtain:

$$\widetilde{q_n}(\omega) = \frac{1}{2\,U_n}\frac{-1}{\omega^2 - {\omega_n}^2}\int d^3r\,\boldsymbol{a}_n^*\,\widetilde{\boldsymbol{j}}(\boldsymbol{r},\omega),$$

while the Fourier transform of the vector potential is

$$\widetilde{\boldsymbol{A}}(\boldsymbol{r},\omega) = \sum_n \widetilde{q_n}(\omega)\boldsymbol{a}_n(\boldsymbol{r})$$

and the scalar potential can also be expanded into a complete orthogonal system:

$$\Phi(\boldsymbol{r},t) = \sum_n r_n(t)\,\phi_n(\boldsymbol{r}).$$

Let ϕ_n be the eigenvectors of

$$-\nabla^2\phi_n(\boldsymbol{r}) = \left(\frac{\Omega_n}{c}\right)^2 \phi_n(\boldsymbol{r})\,,$$

which fulfill the boundary condition $[\phi_n]_{\partial V} = 0$.

We choose the T_n in such a way that we have

$$\frac{\epsilon_0}{2}\left(\frac{\Omega_n}{c}\right)^2 \int d^3r\,\phi_n^*(\boldsymbol{r})\,\phi_m(\boldsymbol{r}) = T_n\,\delta_{nm}.$$

For the expansion coefficients r_n for Φ we obtain

$$r_n(t) = \frac{1}{2T_n} \int d^3r\, \phi_n^*(\boldsymbol{r})\, \rho(\boldsymbol{r},t).$$

Calculation of the impedance
The longitudinal impedance is given by

$$Z(x,y,s) = \frac{1}{q} \int_{-\infty}^{\infty} dz\, \widetilde{E}_z(\boldsymbol{r},\omega)\, \exp(i\omega\, z/c).$$

The electric field $\widetilde{\boldsymbol{E}}(\boldsymbol{r},\omega) = i\omega \widetilde{\boldsymbol{A}}(\boldsymbol{r},\omega) - \nabla\widetilde{\Phi}(\boldsymbol{r},\omega)$ is excited by a point charge q moving parallel to the z–axis.

$$\begin{aligned} \rho(\boldsymbol{r},t) &= q\delta(z-ct)\,\delta(x-x_0)\,\delta(y-y_0), \\ \boldsymbol{j}(\boldsymbol{r},t) &= c\,\boldsymbol{e}_z\, \rho(\boldsymbol{r},t). \end{aligned}$$

The Fourier transforms of ρ and $\boldsymbol{j}$ are

$$\begin{aligned} \widetilde{\rho}(\boldsymbol{r},\omega) &= \frac{q}{c} \exp(-i\omega z/c)\,\delta(x-x_0)\,\delta(y-y_0), \\ \widetilde{j}_z(\boldsymbol{r},\omega) &= q \exp(-i\omega z/c)\,\delta(x-x_0)\,\delta(y-y_0). \end{aligned}$$

Hence, the Fourier transform of the electric field is given by

$$\begin{aligned} \widetilde{\boldsymbol{E}}(\boldsymbol{r},\omega) &= q\sum_n \frac{-i\omega}{\omega^2-{\omega_n}^2} \boldsymbol{a}_n(\boldsymbol{r}) \frac{1}{2U_n} \int_{-\infty}^{\infty} d\acute{z}\, a_{nz}^*(x_0,y_0,\acute{z}) \exp(-i\omega\acute{z}/c) \\ &\quad -\frac{q}{c} \sum_n \nabla\phi_n(\boldsymbol{r}) \frac{1}{2T_n} \int_{-\infty}^{\infty} d\acute{z}\phi_n^*(x_0,y_0,\acute{z}) \exp(-i\omega\acute{z}/c). \end{aligned}$$

We define the complex voltages:

$$\begin{aligned} V_n(x,y,\omega) &= \int_{-\infty}^{\infty} dz\, a_{zn}(x,y,z) \exp(i\omega z/c) \\ v_n(x,y,\omega) &= \int_{-\infty}^{\infty} dz \left(\frac{\partial}{\partial z}\phi_n\right)(x,y,z) \exp(i\omega z/c). \end{aligned}$$

Now the longitudinal impedance can be calculated:

$$\begin{aligned} Z_0(x,y,\omega) &= \sum_n \frac{-i\omega}{\omega^2-{\omega_n}^2} \frac{1}{2U_n} V_n(x,y,\omega)\, V_n^*(x_0,y_0,\omega) \\ &\quad + \sum_n \frac{i}{\omega} \frac{1}{2T_n} v_n(x,y,\omega)\, v_n^*(x_0,y_0,\omega). \end{aligned}$$

Since ϕ_n vanished at the boundaries of the structure the following relations could be used:

$$\frac{1}{c}\int_{-\infty}^{\infty} dz\, \phi_n^*(x_0,y_0,z) \exp(-i\omega z/c) = \frac{-i}{\omega}\int_{-\infty}^{\infty} dz \left(\frac{\partial}{\partial z}\phi_n^*\right)(x_0,y_0,z) \exp(-i\omega z/c).$$

The longitudinal **wake potential** is obtained by a Fourier transformation of the impedance:

$$W_{||}(x,y,s) = \frac{1}{2\pi} \int_C d\omega\, Z(x,y,\omega) \exp(i\omega\, s/c).$$

Due to causality we have to choose the retarded solution. The contour C is shown in Figure 29. For $s > 0$ we can close the integration path in the upper half plane and for $s < 0$ in the

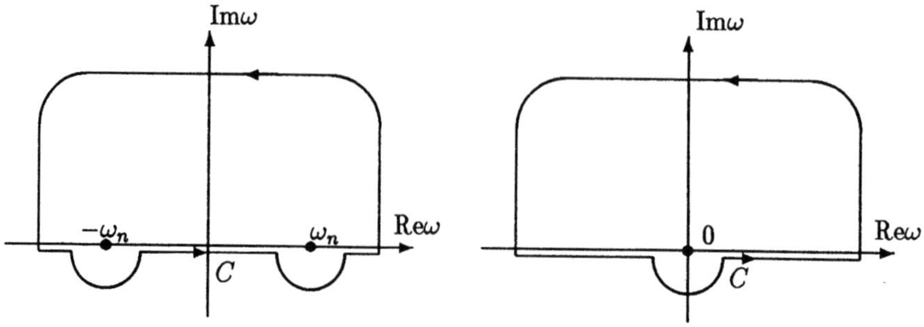

Figure 29: The contour C of the Fourier integral

lower half plane. The second term of the impedance, which is due to the scalar potential Φ, does not contribute to the wake potential. Hence this term can be dropped a posteriori for the impedance. This term has a pole for $\omega = 0$. Using Chauchy's theorem we have to calculate the voltage $v_n(x, y, \omega)$ for $\omega = 0$. But due to the boundary condition the voltage is zero:

$$v_n(x,y,0) = \int_{-\infty}^{\infty} dz \left(\frac{\partial}{\partial z}\phi_n\right)(x,y,z) = [\phi_n(x,y,z)]_{\text{boundary}} = 0.$$

Hence we have for $s > 0$

$$\begin{aligned} W_{||}(x,y,s) &= \sum_n \frac{1}{4U_n} \left(V_n(x,y,\omega_n) V_n^*(x_0,y_0,\omega_n) \exp(i\omega_n\, s/c) \right. \\ &\quad \left. + V_n(x,y,-\omega_n) V_n^*(x_0,y_0,-\omega_n) \exp(-i\omega_n\, s/c) \right). \end{aligned}$$

Since all eigen vectors $\boldsymbol{a}_n$ can be chosen as completely real vectors the following relation holds:

$$V_n^*(x,y,-\omega) = V_n(x,y,\omega).$$

For that choice of eigen vectors one obtains ($s > 0$)

$$W_{||}(x,y,s) = \sum_n \frac{1}{4U_n} 2\, \mathrm{Re}\left(V_n(x,y,\omega_n) V_n^*(x_0,y_0,\omega_n) \exp(i\omega_n\, s/c) \right).$$

For the special case $x = x_0, y = y_0$ the wake potential is given by

$$W_{||}(x,y,s) = \sum_n 2 \frac{|V_n(x,y,\omega_n)|^2}{4U_n} \cos(\omega_n\, s/c),$$

where $|V_n(x,y,\omega_n)|^2/(4U_n)$ is the loss parameter k_n of mode number n.
For $s = 0$ we have

$$W_0(x,y,0) = \frac{1}{2\pi} \int_C d\omega\, Z_0(x,y,\omega).$$

Since $Z_0(x,y,\omega) = -Z_0(x,y,-\omega)$ the integral has to be calculated for the two contours C_1 and C_2 with radius ε (see figure 30).

In the limit $\varepsilon \to 0$ we obtain for C_1

$$
\begin{aligned}
\frac{1}{2\pi}\int_{C_1} d\omega \frac{-i\omega}{\omega^2-\omega_n{}^2}\frac{|V_n(x,y,\omega)|^2}{2U_n} &= \frac{1}{2}\frac{1}{i\pi}\int_{C_1} d\omega \frac{1}{\omega+\omega_n}\frac{|V_n(x,y,\omega)|^2}{4U_n} \\
&= \frac{1}{2}\frac{1}{i\pi}\int_{\pi}^{2\pi} d\varphi\, i\, \frac{|V_n(x,y,\omega_n-\varepsilon\, e^{i\varphi})|^2}{4U_n} \\
&\to \frac{1}{2} k_n .
\end{aligned}
$$

The same result is obtained for the contour C_2.
The complete result can be written as

$$
W_{\|}(s) = \theta(s) \sum_n 2\, k_n \cos(\omega_n\, s/c),
$$

$$
\theta(s) = \begin{cases} 0 & s<0 \\ 1/2 & s=0 \\ 1 & s>0 \end{cases} .
$$

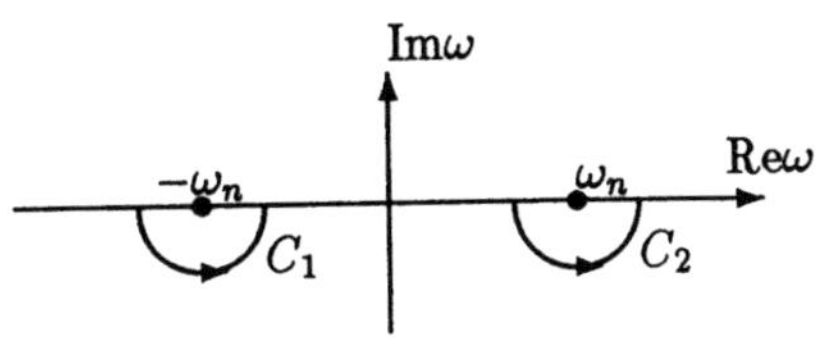

Figure 30: Contours C_1 and C_2

BENCH METHODS FOR BEAM-COUPLING IMPEDANCE MEASUREMENT

F. Caspers
CERN, Geneva, Switzerland

1. INTRODUCTION

The coaxial-wire method is a wide-spread tool for bench measurements of beam-coupling impedance. By introducing a wire into the vacuum chamber under test, a surface current distribution on the inner surface of the beam pipe will be obtained which corresponds approximately to the current distribution produced by a passing particle bunch. When this image-current distribution has been perturbed by a discontinuity (e.g. change in pipe diameter) a reaction on the wire takes place similar to the reaction of a wake field on the particle bunch. Since the wire, which should be located on the axis of the particle beam, transforms the empty beam pipe into a coaxial line, a new boundary value problem is created and one has to be aware of the limits of validity of this simulation method. The maximum characteristic impedance of such a coaxial line can be, at best, several hundred ohms. This has to be compared with the impedance of the beam, which is infinite, and thus represents a current source. Furthermore, the phase velocity of the line must approximate to the beam velocity. Therefore, the simulation is rather easy for a highly relativistic beam ($\beta = v/c \approx 1$) but may lead to some difficulties for small β (at least in the case of longitudinal impedance measurements).

For longitudinal impedance and loss-factor measurements, the set-up consists of a single wire (thin, in order to approximate $Z_L = \infty$) in the center of the vacuum chamber (at the position of the beam). For transverse measurements one may use either two parallel wires (dipole excitation, high common-mode rejection required) or an image plane (metal plate), with a single wire *on one* side. Both methods have been applied for pick-up and kicker measurements. Both have advantages and disadvantages and the decision for one or the other depends on the particular problem. For the field simulation of slow beams where the ratio of E/H is not equal to 377Ω, a technique using individually excited probes and loops is described. Since it is tedious to install the coaxial wire in certain cases, measurements on waveguide modes may be carried out. For reflection this is analogous to the well-known time-domain reflectometry, except that the lowest frequency is given by the cut-off wavelength of the beam pipe. In transmission using TM_{0n} modes valuable information on the beam-coupling impedance can be revealed.

In this paper, mainly a discussion of practical aspects of several measurement methods will be given, including limits of validity, pitfalls, and a number of examples. Since the coaxial-wire method is in general not very well suited to high impedance and/or high-Q cavity-like objects, the bead-pull perturbation measurement technique to determine R/Q is also briefly discussed.

2. WAKE POTENTIALS AND IMPEDANCE

The electromagnetic field of a point charge moving in free space with constant velocity $\vec{v}$ is given as [1,2]

$$\vec{E} = \frac{q\,\vec{r}_0}{4\pi\,\varepsilon_0\,r_0^3}\,\frac{1 - v^2/c^2}{(1 - (v^2/c^2)\,\sin^2\psi)^{3/2}} \tag{2.1}$$

$$\vec{B} = \frac{1}{c^2}\,\vec{v} \times \vec{E} \tag{2.2}$$

where $\vec{r}_0$ = vector from the charge to the test particle position,
ψ = angle between $\vec{r}_0$ and $\vec{v}$.

For the case $v \to c$ this field appears like an infinitely thin disk, to an observer in the laboratory. The ratio E/H in this field disk approaches 377 Ω. If the charge is travelling inside a smooth and perfectly conducting beam pipe the shape of the field is not changed. Only its diameter is limited by the dimensions of the beam pipe. It can be shown that the same field structure is present in a lossless coaxial line with an infinitely thin inner conductor and a δ-pulse propagating in the same direction [1] (Green's function).

The longitudinal wake function $w_\parallel$ for a point charge is defined as the energy loss ΔU of a test particle (q_2) following a point-like bunch (q_1) at a distance s [3-5].

$$\Delta U = q_1 q_2 \cdot w_\parallel(s) \tag{2.3}$$

$$w_\parallel(s,r) = -\frac{1}{q_1}\int_{-\infty}^{+\infty} v \cdot E_z(z,r,t)dt\Big|_{z=vt-s} \tag{2.4}$$

where E(z,r,t) is the electric field of the point-like bunch q_1. The longitudinal impedance $Z_\parallel(\omega,r)$ is defined as the Fourier transform of the point wake function (time dependence $e^{+j\omega t}$),

$$Z_\parallel(\omega,r) = \frac{1}{v}\int_{-\infty}^{+\infty} w_\parallel(s,r)ds\, e^{-j\omega s/v}. \tag{2.5}$$

Very often, one is just considering the case $\vec{v} = \vec{j}_z v_z$ at $r = 0$ (small transverse dimensions of the bunch, v_z only). Then the longitudinal loss factor $k_\parallel$ may be written approximately as

$$k_\parallel = \frac{1}{2\pi}\int_{-\infty}^{+\infty} Z_\parallel(\omega,0) \cdot \left|\frac{\hat{\varrho}(\omega)}{\varrho(0)}\right|^2 d\omega \tag{2.6}$$

with $\hat{\varrho}(\omega)$ being the Fourier transform of a real bunch charge distribution $\varrho(z)$. In what follows the indices of $Z_\parallel(\omega,r)$ are omitted for simplicity.

3. MEASUREMENT SYSTEMS

The shape of the wake field of a charged-particle bunch can be seen from the computation results of Weiland [6] in Fig. 1. Considering the electric field lines as being linked to free charges (beam) and surface currents (beam pipe), one gets some qualitative insight into the interactions of the wake field. Due to a discontinuity, a certain amount of electromagnetic energy is extracted from the bunch and deposited into the near vicinity of, say, the step, or into a cavity. The associated electromagnetic field reacts back on the tail of the particle bunch (causality) and, for cavity-like objects, also on later bunches which are sufficiently close. Since for frequencies below the cut-off frequency (for waveguide modes) of the beam pipe the deposited field energy cannot travel away, it will finally be absorbed due to metallic losses.

a) b)

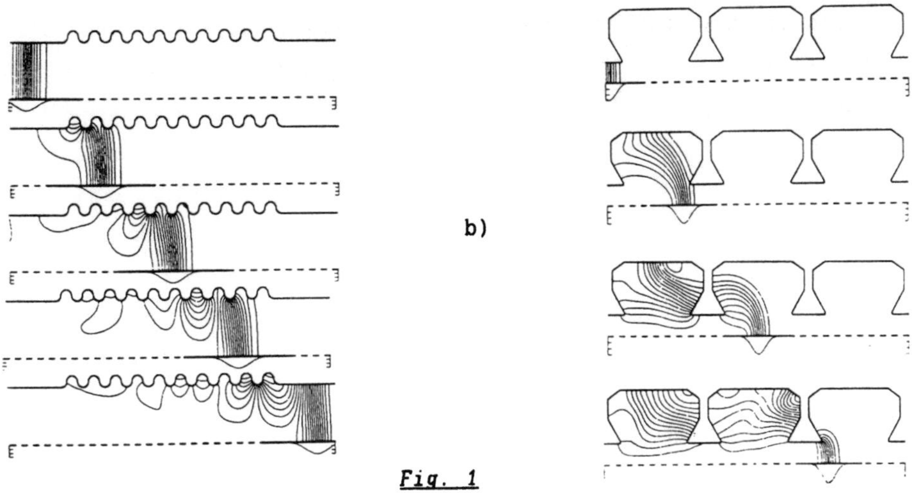

Fig. 1

Electromagnetic field of Gaussian (σ = 1 cm) bunch of charged particles traversing a bellows at the speed of light (side tube radius 6 cm, maximum radius of the bellows section 7 cm). The charge density on-axis is plotted below the symmetry axis [6].

Electric field of a Gaussian bunch (σ = 2cm) traversing 3 cells of a PETRA cavity at the speed of light. Displayed lines (rH_φ = const) show the direction of the electric field and their density is proportional to $\vec{r}\cdot|E|$. The bunch charge density on-axis is plotted below the symmetry axis [6].

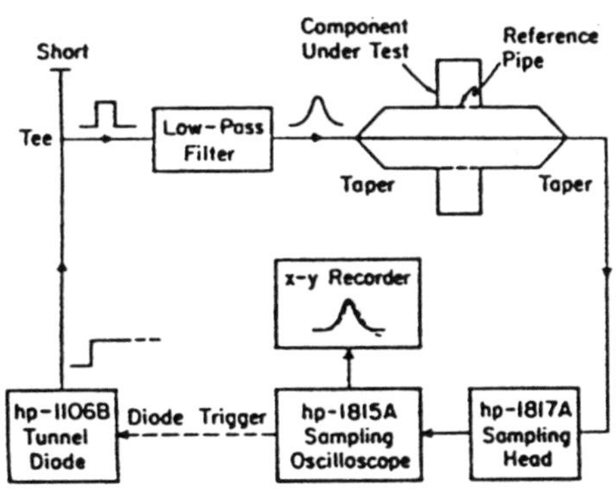

Fig. 2 - Instrumentation for measuring parasitic mode loss [7].

A measurement system which is capable of producing a field configuration very closely related to the situation shown in Fig. 1 a) and b) is presented in Fig. 2 [8].

The basic measurement method has been described by Sands and Rees [7]. A fast step with about 20 ps rise time (tunnel-diode generator) passes a pulse-forming network (shorted tee and low-pass filter). The purpose of the low-pass filter is to adjust the pulse length to the given bunch length. Thus a nearly Gaussian pulse is generated. It travels either through a reference pipe consisting of a straight section of vacuum chamber and a thin wire on the axis of the beam, or the component under test which preferably has the same mechanical length. Both signals, $I_0(t)$ = reference signal = unperturbed case = solid line in Fig. 3, and $I_1(t)$ = object signal = dashed line in Fig.3, are recorded sequentially and stored. Note that a pulse generated this way may have a rather low amplitude (50 to 100 mV). If higher amplitudes are required, pulse generators with an avalanche transistor and sometimes an additional step recovery diode have been used.

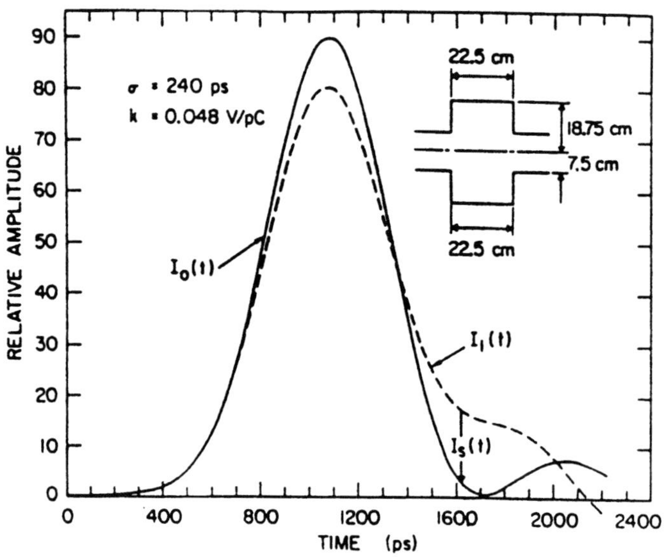

Fig. 3 - Recorded output pulses ($I_0(t)$ for the reference pipe and $I_1(t)$ for a cylindrical cavity having the dimensions shown [8].

For evaluation of the loss factor k (having dimensions volt/coulomb) the following relations hold:

$$k(\sigma) = \frac{2Z_0}{Q^2} \int I_0(t) I_s(t) dt \tag{3.1}$$

with $Q = \int I_0(t)dt$ = charge in the pulse,

$I_s(t) = I_0(t) - I_1(t)$ = subtracted signal,

Z_0 = characteristic impedance of the coaxial line formed by the vacuum chamber with the wire,

σ = rms length of the Gaussian pulse (bunch length ≈ 4σ).

In practice the limits of the integral in Eq. (3.1) extend from $\tau-2\sigma$ to $\tau+2\sigma$ with τ = position of the maximum of $I_0(t)$. This truncation already implies a time-filtering process omitting wake-field effects coming later than $\tau+2\sigma$. Since the loss factor is related only to the real part of the coupling impedance [8] by

$$k = \sum_n (\omega_n/4)(R/Q_n) \exp(-\omega_n^2\sigma^2), \tag{3.2}$$

the influence of the imaginary part, which would contribute to a delay between $I_0(t)$ and $I_1(t)$ has to be removed by superimposing both pulses in the knee region (i.e. 400 ps to 600 ps, Fig. 3) using some adequate correlation technique. It turns out that for properly positioned $I_1(t)$ the loss factor k is not sensitive to small shifts between $I_0(t)$ and $I_1(t)$. $I_0(t)$ and $I_1(t)$ may be recorded simultaneously with a modified measurement system (two sampling heads). In this case drifts of the timing system in the sampling scope do not need to be removed by the correlation procedure.

It should be pointed out that by omitting the influence of the imaginary part of the beam-coupling impedance, valuable information may be lost. In certain cases (e.g. bellows) the imaginary part at lower frequencies is of interest (for determination of Z/n). Using this method it can be deduced from the real part measured around some resonances at higher frequencies and applying Hilbert transforms.

A rather similar measurement system to that depicted in Fig. 2 (operated in the time domain) is shown in Fig. 4 for frequency domain measurements [9] which are straightforward. The output signal of an rf generator of variable frequency is split symmetrically. One part passes the reference line, the other part the measuring line or device under test. The configuration depicted in Fig. 4 may be considered as a primitive network analyzer. Here as well (compared to loss-factor measurements) the difference between reference and object is used to evaluate (approximately) the unknown impedance $Z(\omega)$ (only valid if $Z \ll R_0$). In fact, for simultaneous recording of $I_0(t)$ and $I_1(t)$ the type of measurement bridge shown in Fig. 2 was used at Cornell University in the early eighties [10]. In this case, the correla-

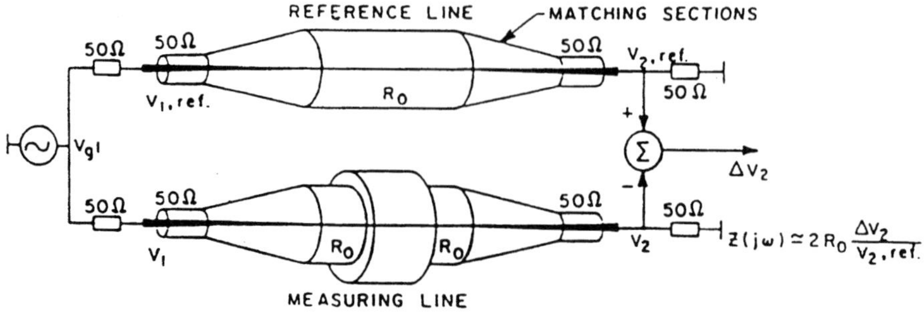

Fig. 4 - Beam-coupling impedance measurement system (frequency domain) [9].

tion method mentioned above has not always been applied and the signal delay was used to evaluate the imaginary part of the beam coupling impedance. High mechanical stability (to 0.1 mm) is required to obtain meaningful results. For a discussion of limitations and possible sources of errors, a brief description of the measurement procedure will now be given.

4. S-PARAMETERS AND IMPEDANCE

Consider a single localized impedance Z (Fig. 5) in a coaxial line (characteristic impedance Z_L)[9]. The S-matrix (S = scattering) of this 2-port or 4-pole system has in general four complex and frequency-dependent elements, namely S_{11} = input reflection coefficient, S_{12} = reverse transmission, S_{21} = forward transmission, and S_{22} = output reflection coefficient.

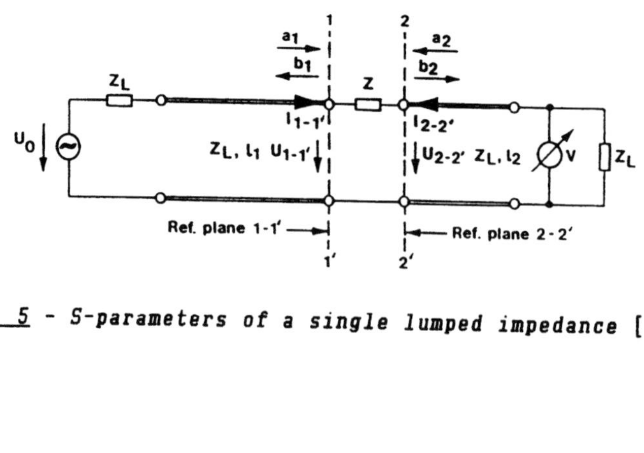

Fig. 5 - S-parameters of a single lumped impedance [11].

The relation between waves travelling towards the 2-port (a_1, a_2) with $|a_1| = (U_0/2)(1/\sqrt{Z_L})$ and those travelling away (b_1, b_2) is given by:

$$\begin{pmatrix} b_1 \\ b_2 \end{pmatrix} = \begin{pmatrix} S_{11} & S_{12} \\ S_{21} & S_{22} \end{pmatrix} \begin{pmatrix} a_1 \\ a_2 \end{pmatrix} \quad \text{or (b)} = S \quad \text{(a)}.$$

For determination of $S_{11} = S_{22}$ and $S_{21} = S_{12}$ the well-known formula for the reflection coefficient ϱ (here $\varrho = S_{11}$) applies and yields

$$S_{11} = \frac{(Z + Z_L) - Z_L}{(Z + Z_L) + Z_L} = \frac{Z}{2Z_L + Z} \quad ; \quad Z = \frac{S_{11} \cdot 2Z_L}{1 - S_{11}} \tag{4.1}$$

$$S_{21} = U_{2-2'}/(U_0/2) = 2Z\ /(2Z_L + Z\ _L).$$

Thus the unknown Z can be determined from the measured S_{21} as

$$Z = \frac{2Z_L(1 - S_{21})}{S_{21}} \tag{4.2}$$

or approximately $Z \approx 2Z_L(1 - S_{21})$ for $|Z| \ll Z_L$(cf. Fig. 4).

It should be noted that for a convenient mathematical treatment of cascaded elements the T-matrix (Transfer matrix [12]) instead of the S-matrix may be used:

$$\begin{pmatrix} b_1 \\ a_1 \end{pmatrix} = \begin{pmatrix} T_{11} & T_{12} \\ T_{21} & T_{22} \end{pmatrix} \begin{pmatrix} a_2 \\ b_2 \end{pmatrix}. \tag{4.3}$$

It appears from Eq. (4.2) that it might be rather convenient to determine Z from S_{11} rather than from S_{21} since S_{11} is more sensitive to variations of small Z than S_{21}. But for a proper measurement of S_{11} the precise knowledge of the reference-plane position 1-1' is required. This may be mechanically tedious and sometimes difficult to define. In contrast, for a transmission measurement the location of Z on the line is of no relevance [9]. Furthermore, for a real beam there will be no reflected energy for spectral components below the cut-off frequency of waveguide modes in the beam pipe.

As a somewhat more complicated case a double step in cross section will now be considered [9].

Normally, the electric field energy stored in an abrupt change of diameter of a coaxial line is represented by a parallel capacitor as shown

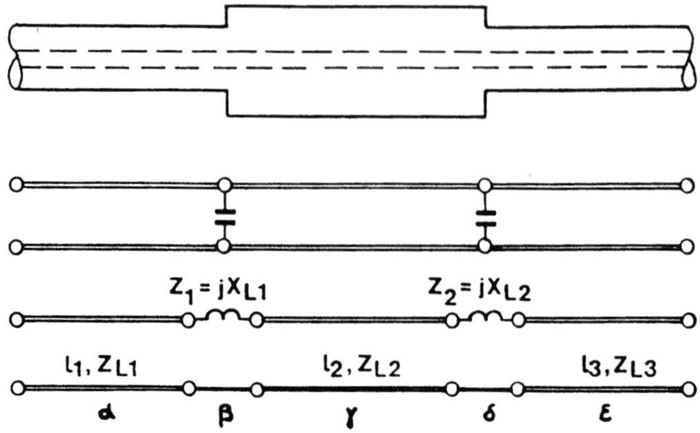

Fig. 6 - *Double-step and equivalent circuits [9,11].*

in Fig. 6. This equivalent circuit is often used to describe the S-parameters for this type of discontinuity on a coaxial line both in transmission and reflection. The parallel element has to be converted into a serial one to represent the interaction with a particle beam of infinite impedance [9,13]. Now the serial inductance is only approximately valid for small perturbations (with respect to Z_L) and only in transmission. In a mathematical model for sections α, β, j, δ and ε either the S- or T-matrices may be evaluated. As shown in Ref. [9] the sum of X_{L_1} and X_{L_2} appears rather clearly from a transmission measurement as an additional phase shift for reasonably small changes in diameter, and the measured results agree well with theoretical data. For reflection measurements, errors of a factor 30 have been found. It turns out that, as a general rule [9], the S-parameter for beam coupling impedance is better measured by transmission than by reflection. However, one has to ensure that transmission data are not perturbed by parasitic effects. For the example in Fig. 6 a considerable amount of reflected energy is due to the change in characteris impedance $(Z_{L_1};Z_{L_2})$ in addition to jX_{L_1} and jX_{L_2}, respectively. Even if generator and load are perfectly matched some scattered energy will travel back and forth between X_{L_1} and X_{L_2} which would not be the case for the empty beam pipe.

For small perturbations one may write

$$|S_{21}^{\beta}| = \sqrt{(1 - |S_{11}^{\beta}|^2)} \approx 1 - \frac{|S_{11}^{\beta}|^2}{2} \,. \tag{4.4}$$

For example, $S^{\beta}_{11} = 0.1 \Rightarrow S^{\beta}_{21} = 0.995 \Rightarrow \Delta U = 0.5\%$. Thus there is little change in the voltage of the transmitted wave, but a rather large reflection (voltage of the reflected wave). In the case of two discontinuities (β and δ) and $Z_{L_1} = Z_{L_2}$ one obtains in transmission $\Delta U \approx 1\%$. But the reflected wave ($b_{1\delta}$) from jX_{L_2} travels back to jX_{L_1}, is reflected again ($b_{2\beta}$) (now going forward), and appears superimposed on the transmitted wave after $jX_{L_2}(b_{2\delta})$. Taking the numbers from above with $b_{1\delta} = 0.1$ and $a_{2\beta} = 0.01 = 1\%$ the measurement error of $\Delta U = 1\%$ is just equal to twice the change in amplitude relevant for the beam-coupling impedance. This implies that measurement errors can be easily induced by multiple reflections.

In the measurement system of Fig. 4 with distributed transformers on both ends of the beam pipe, one must keep in mind that at lower frequencies the cones look like a step. Often the inner conductor is not tapered in order to maintain a constant impedance over all the length (like an adaptor from an N connector to a 10-cm-diameter coaxial power line). Serious errors due to multiple reflections from the generator or load will then occur for a single discontinuity. Thus the double-step problem in Fig. 6 really becomes a quadruple-step problem. Since normally 50 Ω rf instrumentation is used, some sort of (resistive) broadband matching network to adapt to, say, 150 Ω is required in addition to a tapered center conductor [14]. This type of construction is limited by practical difficulties to about 150 Ω characteristic impedance Z_L. In most cases the highest possible values for Z_L should be chosen. A simple very thin wire may be used, but then one has to work either in the time domain (loss factor) with implicit time filtering (Fig. 2) or, if frequency domain results (impedance) are required, the frequency domain data can be used only after time filtering [14].

It should be mentioned that the rule of taking the highest characteristic impedance wherever possible has one exception. That is when the object being measured has such a small impedance that the change in amplitude of the transmitted wave disappears in the measurement noise and uncertainty. In such a case (e.g. short bellows) a large-diameter inner conductor (50 Ω) will give better results than a very small one. Of course, measurements may be carried out using different diameters of the center conductor as a cross check for the validity of results obtained.

With the availability of computer-controlled network analyzers the measurement system of Fig. 4 and Fig. 2 may be improved. The technique shown in Figs. 7 and 8 consists of generating a synthetic pulse in the time domain via FFT from measurements taken in the frequency domain. This leads to higher spectral density than real time or sampling oscilloscope pulse measurements, thus giving a higher dynamic range and better reproducibility. The method (synthetic pulse) has been known since about 1960 [15] but became more readily accessible in 1984 with the introduction of computer-controlled network analyzers. An early version (1983) consisting of separate units from different manufacturers (signal generator, vector analyzer, etc.) was replaced in 1984 by a single network analyzer with a calibration facility and was capable of storing and comparing sequential object and reference measurements (Fig. 7). The corresponding flow diagram (Fig. 8) explains the measurement procedure. In contrast to Fig. 4, the complex quotient of

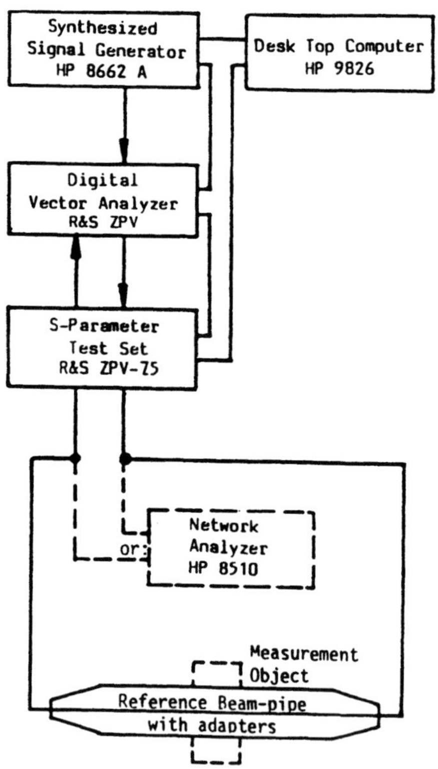

Fig. 7 - Computer-controlled network analyzer system for frequency and time domain measurements [14].

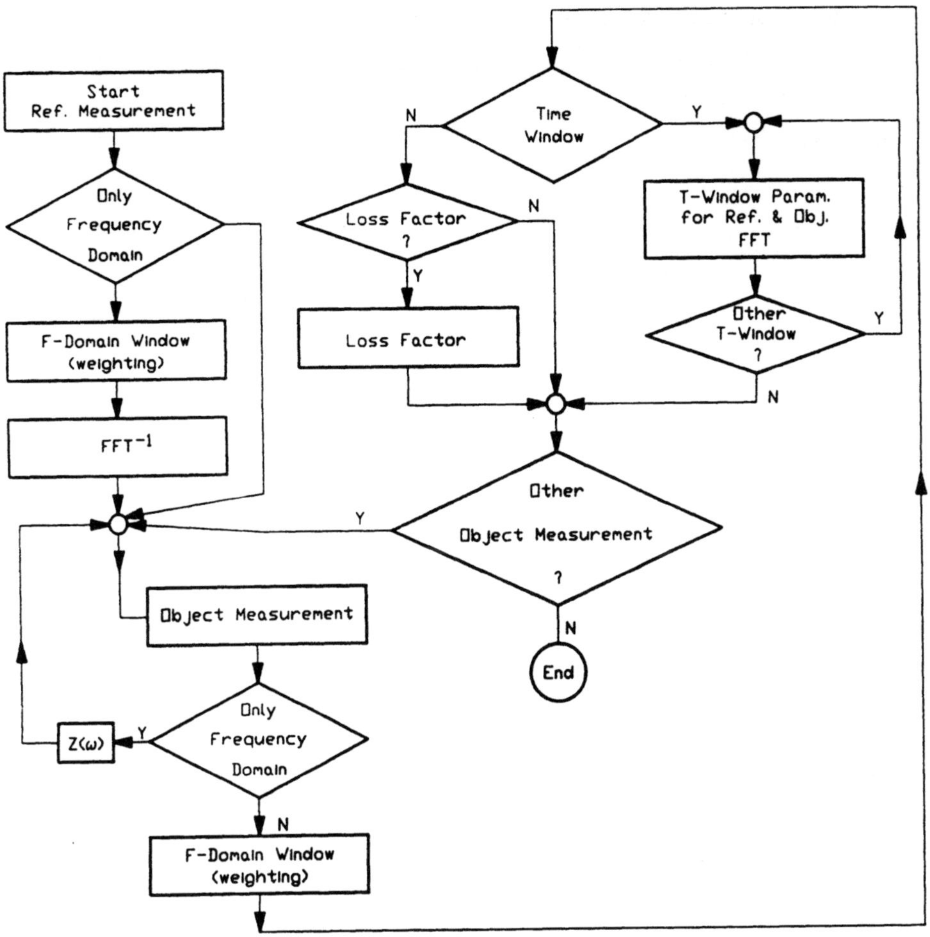

Fig. 8 - Simplified flow diagram for measurement procedure [14].

object and reference measurement can now be evaluated, instead of the difference which gives acceptable results for small impedances only. The possible use of the synthetic pulse method for loss-factor measurements was already mentioned by Sands and Rees in 1974 [7].

As modern network analyzers offer the application of "de-embedding" procedures using different error models, one may remove the influence of badly matched tapers or other transitions with a properly chosen calibration routine [16-26]. De-embedding essentially means that the electromagnetic properties of non-perfect cables, connectors, tapers, etc., are

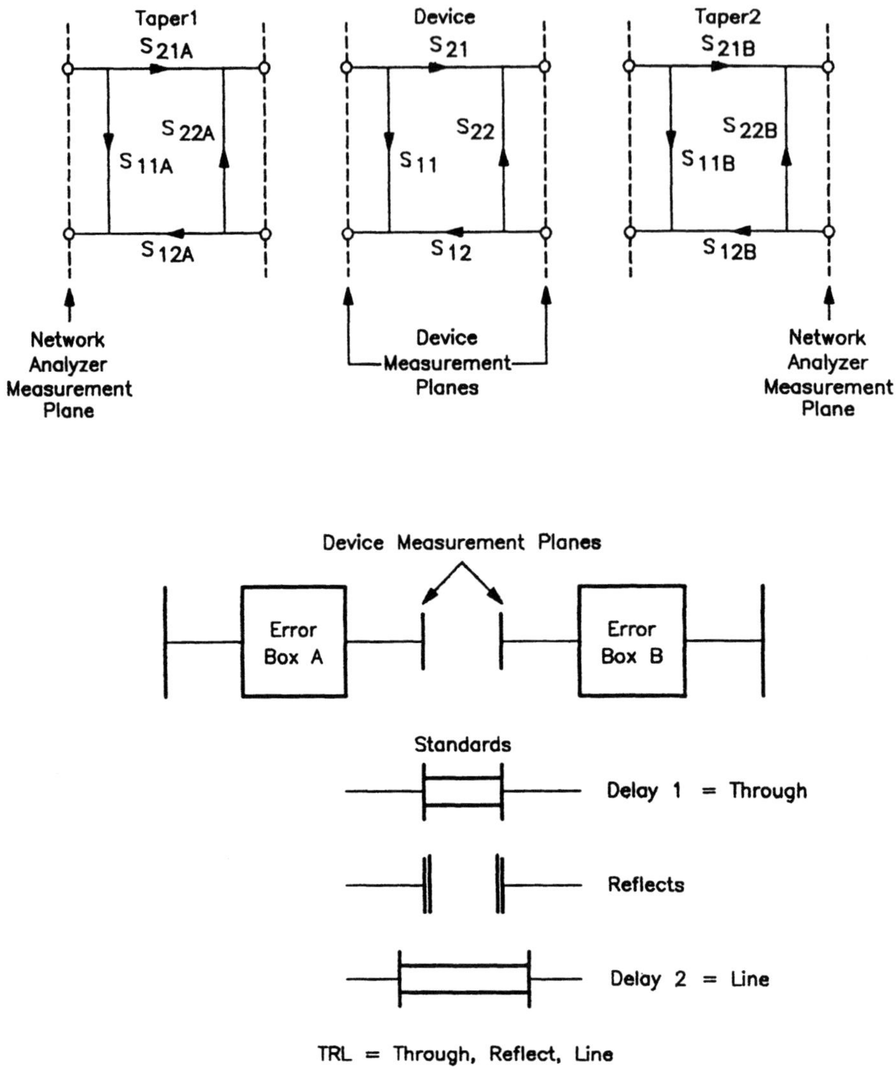

Fig. 9 - A functional block diagram of two-port error-corrected measurement system. TRL-calibration procedure.

determined by measuring several calibration standards (often: open, short, load). With all elements of the T-matrix known, that T-matrix is multiplied with its inverse, resulting in a perfect transmission line (Fig. 9). However, since the de-embedding procedures are not always easy to carry out, gating (= time filtering) of the first pulse arriving in transmission is the preferred method. Only if considerable overlapping occurs between the first pulse, which may have been lengthened by dispersive elements, and following pulses caused by multiple reflections, can gating not be used any more.

This calibration method is particularly useful for transverse impedance measurements (two parallel wires in opposite phase [17]). Here it turns out that a good broadband match from a 50-Ω, 180° hybrid to some parallel-wire line is not easy to realize. But, as mentioned above for longitudinal impedance, the matching problem always exists for a single wire. The influence of this mismatch may either be removed by time-filtering (implicitly done for loss factors and kick-parameters) or by applying the de-embedding method. Independent of all software (time filter, de-embedding) procedures, the mismatch in the hardware should be made as small as possible. This may be accomplished for frequencies up to 1 GHz by placing small (to avoid inductance) carbon resistors in series with the wire in order to match the high impedance of the beam-pipe wire configuration.

In addition, a few lossy ferrite beads on the wire increase the attenuation easily by 10 to 20 dB. This is to prevent the exchange of electromagnetic energy between discontinuities in the beam pipe, which would not "talk" to each other if the wire were not there. Of course, the additional losses of the ferrite beads have also to be taken into account in the reference measurement. However, the use of such ferrite beads as reasonably matched attenuators is limited to frequencies below cut-off (waveguide modes) of the beam pipe.

For transverse impedance or kick-parameter measurements, basically the same systems as already shown in Fig. 2 or Fig. 7 are applied, except that for the excitation one should use two parallel wires (off-phase) (Fig. 10) or a single wire over a metallic image plane [14,18].

Transverse beam-coupling impedance measurements have been carried out by simply displacing the center conductor in one plane and recording the change of S_{21} as a function of the lateral position. But, due to the presence of a strong signal caused by the longitudinal impedance, the evaluation of the much smaller effect from the transverse impedance on S_{21} is difficult to carry out.

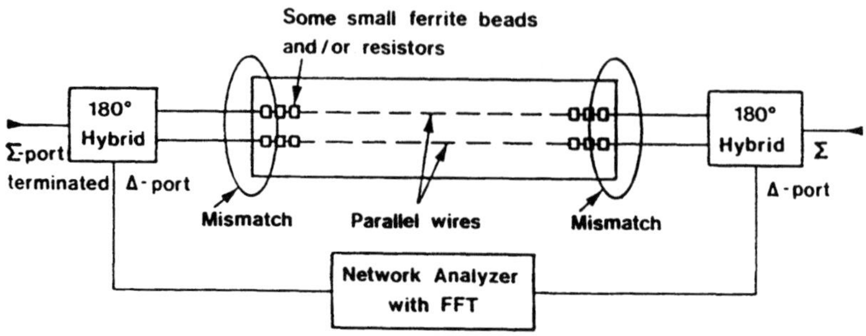

Fig. 10 - *Transverse impedance measurement system (2 wires)* [11].

A different matter is the transverse response [27] measurement of a pick-up or kicker, where the transmission from the coaxial wire to the pick-up electrodes is considered. Here, it is common practice to displace the wire (provided the coupling impedance is not too high) and to look at the Σ and Δ signal from the pick-up outputs.

Thus, it is important to ensure that for transverse IMPEDANCE measurements a high common mode rejection can be maintained. Broadband (2 MHz to 5 GHz) 180° hybrids show a common-mode suppression of about 30 dB each. With the set-up of Fig. 10 roughly 60 dB common-mode rejection is achievable. The 180° hybrids can easily be included in FFT and also in calibration (de-embedding) routines, in particular those shown in Fig. 9. For the evaluation of Z_T the following approximative relation [18,28] can be used:

$$Z_T = cZ/\omega\Delta^2 \tag{4.5}$$

where Δ = separation of wires, Z = measured change in transmission line impedance. Note that this relation to evaluate Z_T from a two-wire measurement given in Refs. [28,29] has been derived for the case of a S_{11} measurement with a short-circuit at the far end.

The image plane approach for transverse measurements has the advantage that just one half of the structure to be tested is required. Furthermore, less matching problems occur as a 50-Ω line on a ground plane is easy to realize. Also it is easier to adjust the propagation velocity of the quasi-TEM wave to the β of the beam (usable for β > 0.5) by means of dielectric material between ground plane and wire. On the other hand, introducing an image plane into a given structure may cause serious difficulties. Therefore it depends on the user which method will be finally chosen. The image plane method has been extensively used to measure (amongst others) the sensitivity characteristic of CERN-AA (Antiproton Accumulator) pick-ups and kickers (loop PUs). Due to the weak coupling of the fringeing field to the pick-up structure, reliable measurements could be carried out for the transverse response. Here, direct longitudinal measurements with a single wire gave meaningless results, as the longitudinal coupling impedance to be measured was much higher than Z_L. In this case, conclusions on $Z_{\parallel}$ and the longitudinal transfer impedance, derived from the transverse response, were confirmed later from beam observation. The relation between longitudinal and transverse response in pick-ups and kickers have been described in detail by G. Lambertson [30].

5. REAL-PULSE AND SYNTHETIC-PULSE METHODS

Comparison of real-pulse and synthetic-pulse measurement techniques shows that the key element, a Schottky diode (or four of them), is just the same. When using a sampling oscilloscope this is the gate of the sampling head, sometimes also referred to as a harmonic mixer, and for the synthetic pulse it is one or more balanced mixers in the rf front end of the network analyzer. The maximum applicable signal for both structures is roughly 1 volt, and due to the duty cycle, an average signal power of the order of microwatts or even less is obtained. From this already low power more than 99% is lost due to the sampling process and the small amount left is spread over a wide frequency range (e.g. 20 GHz). As a consequence, one has to face a rather low spectral density, 1/f noise of the sampling gate, non-linearities, problems with time stability, drifts and jitter. In practice, measurements are carried out over an averaging period of the order of 1 minute, with a usable dynamic range of about 70 dB and a time stability of 1-2 ps. But the measurement process requires careful observation (sudden drifts) and up to 30% of the measured traces have to be rejected because of instabilities during the averageing period [31].

On the other hand, a synthetic-pulse test set-up to 20 GHz or more is much more expensive since a vector network-analyzer, a synthesizer and an S-parameter test set are needed. One has also to ensure that low-frequency spectral components can be measured properly (low f_C of S-parameter test set = directional couplers), otherwise the base line of the time domain trace might be very unstable. Trying to save money by using a sweep-generator instead of a synthesizer may not pay off as, due to frequency drifts of the sweeper (via calibration), the amplitude and time (∝ phase) stability of the system deteriorates. The dynamic range of a properly assembled synthetic-pulse test set can be 100 dB, with an amplitude stability of 0.01 dB (30 minutes) and a time stability of 1 ps (30 minutes).

6. EXAMPLES OF MEASUREMENTS

As a first case, results for a "bumper" magnet [18] are given in Fig. 11. This example is typical for certain kicker magnet structures whose resonances (obviously) depend on the external loading condition (terminations). One can recognize very clearly the mutual dependence of the real and imaginary part around resonance. Measurements were carried out with a 1.6-mm-diameter wire.

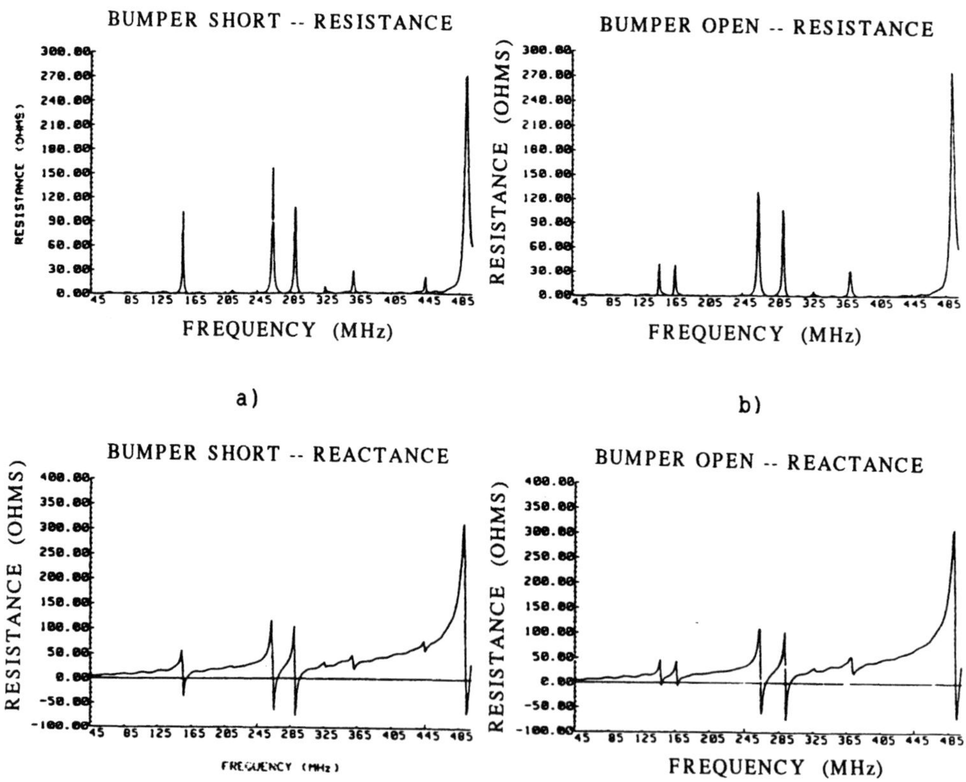

Fig. 11 - *Longitudinal impedance of bumper magnet with plates disconnected and magnet electrodes shorted (a) and left open (b) [18].*

The results shown in Fig. 12 [18] demonstrate a good agreement between measurement and theory for a device with a single low-Q resonance. A typical bellows (ID 100 mm, OD 120 mm, length 38 mm with six undulations) was measured in transmission and reflection mode using time filtering. The beam coupling impedance (imag. part) resulting from the transmission measurement was checked to be independent (within ±10%) of Z_L (50 Ω < Z_L < 150 Ω) [14]. Good agreement was also found with results obtained by analytical and numerical approaches [32]. The apparent energy loss indicated by the decrease of $|S_{21}|$ (0.02 dB at 1 GHz) is in fact explained by the measured reflected power. In this case, the changes in S_{21} due to the device under test were 1° in phase and 0.02 dB in amplitude. Very good connectors and a stable mechanical and electrical system (semi-rigid cables) are required here.

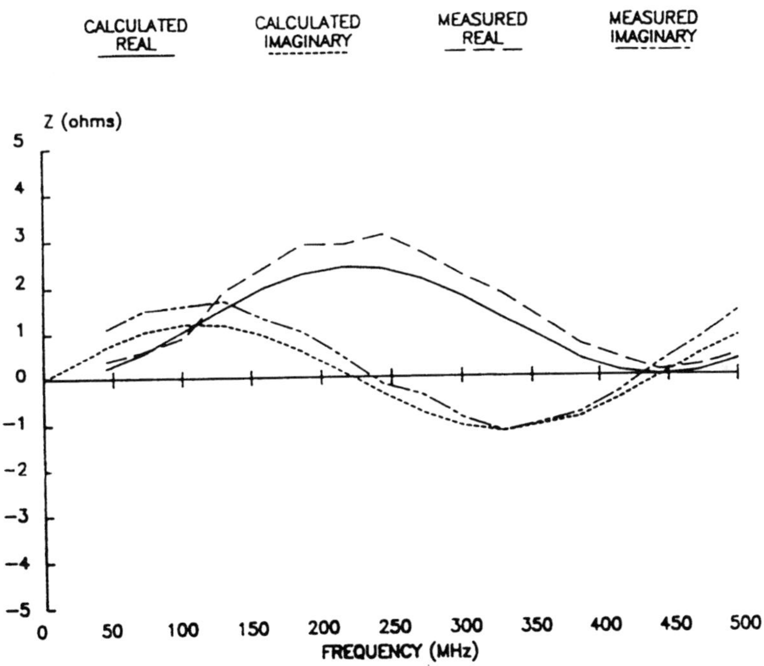

Figure 12 - Longitudinal impedances measured by stripline BPM [17].

A more detailed discussion and other examples are given in Refs. [11,18, 33-35].

7. ELECTROMAGNETIC FIELDS OF SLOW BEAMS

The well-known coaxial-wire method permits the simulation of the electromagnetic field of beams with β close to unity. One may slow down the phase velocity of the simulator line by dielectric loading, but in this case the ratio of E and H does not correspond in general to that of the slow beam. In this section, small loop and probe antennas developed for transverse pick-up response measurements will be discussed.

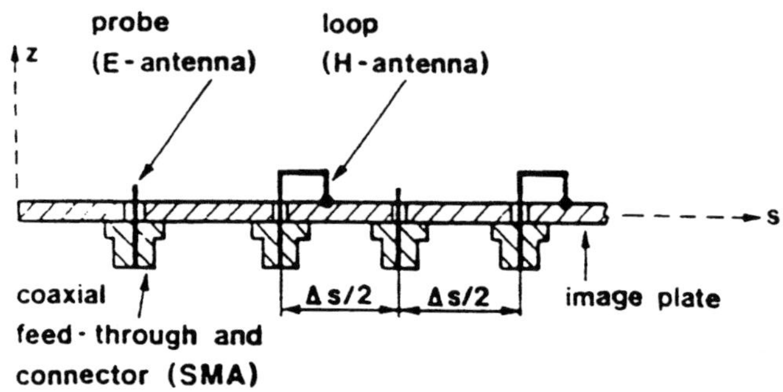

Fig. 13 - Cross section of the beam simulator for slow beams [36].

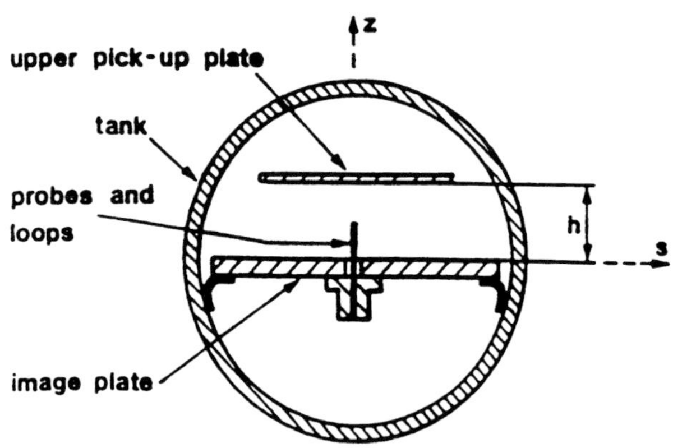

Fig. 14 - Beam simulator installed in pick-up tank [36].

The loops and probes are lined up on a metal plate with a close spacing between adjacent elements (Fig. 13). The installation in the PU tank is sketched in Fig. 14. The elements are fed via delay lines of properly adjusted length $\ell \propto s/\beta$ to create the $e^{i\omega(t - s/\beta c)}$ dependence. The loop and probe power levels are separately adjusted while attenuators at the entrance to the elements avoid reflections. The signal is coupled out at the normal exit port of the pick-up and displayed using the network analyzer. A simple loop-coupler type of pick-up with known response is installed downstream in the same tank for calibration.

A detailed mathematical discussion of the electromagnetic fields in this system as compared to the fields of a beam is given in Ref. [36]. It can be shown that for very slow beams ($\beta < 10\%$) most of the field energy is due to the electric field and thus only probes (E-field) are required. The other condition to be met is a sufficient line density (spatial density) of the probes and loops, which is also a function of the distance h. Their spacing Δs should not exceed $\lambda_0/2\beta$ of the highest spectral component (λ_0 = free-space wavelength).

A beam simulator with probe antennas spaced by Δs = 17 mm was used to measure the sensitivity of a pick-up for stochastic cooling at $\beta = 0.065$ (2 MeV kinetic energy) in LEAR. The device [37] tested is a slow-wave structure formed by a meander-shaped stripline coated on dielectric plates placed 30 mm above and below the beam. The phase velocity in the beam direction is matched to 0.065. In the measurement, the required electrical delay length from one probe to the next, $\Delta\ell = \Delta s/\beta$, is therefore about 260 mm.

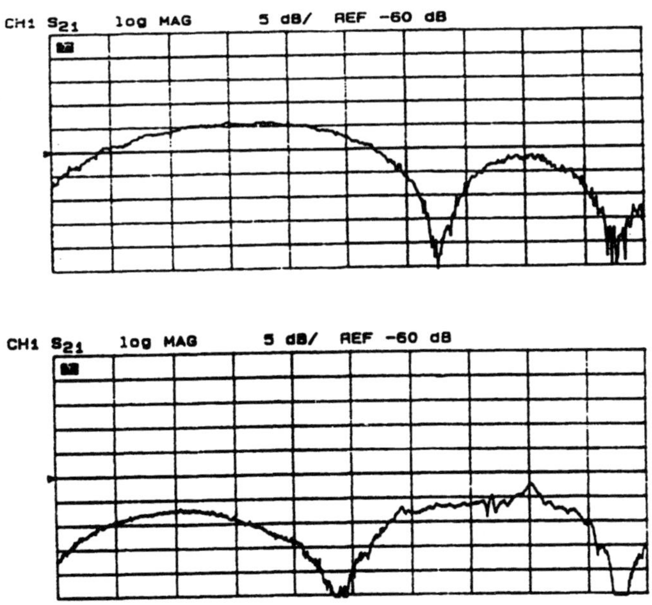

Fig. 15 - *Sensitivity of meander coupler (upper curve) and loop coupler vs. frequency. Vertical scale: 5dB/div [36].*

In Fig. 15 the response of the 600-mm-long meander coupler is compared to that of a 200-mm-long, solid electrode installed as a loop coupler at the same distance above the beam but downstream in the PU tank. One concludes that the longer meander structure has a 13-dB (factor 20) higher average sensitivity. At equal length, i.e. assuming several short loop couplers in series with a proper delay, this reduces to 4 dB (factor 2.5). One also notes the expected $\sin^2[\omega(\ell/2c)(\beta^{-1}-1)]$ response of the loop with the first maximum at f ≈ 50 MHz compared to f ≈ 80 MHz for the longer meander structure.

It should be stressed that with the set-up depicted in Fig. 14 nearly any ratio of E/H and any β value (including β > 1) may be adjusted independently. The probes and loops do not form a transmission line that could transport electromagnetic energy below waveguide cut-off. Thus no signal degradation, as with the conventional stripline/image plane approach, appears.

8. MEASUREMENT ABOVE THE CUT-OFF FREQUENCY

In a beam pipe with a coaxial wire the geometric dimensions define the cut-off frequency of TE and TM modes. For the empty beam pipe (without wire) these cut-off frequencies are slightly different.

The cut-off wavelength of the first waveguide mode (TE_{11} type) on a circular coaxial line is roughly equal to the mean circumference i.e. $\lambda_{c(TE_{11})} \approx \pi(r_1+r_2)$. When increasing the frequency of the network analyzer (Fig. 16) beyond this cut-off, one will already note a number of deep notches in the transmission response (S_{21}) of the reference measurement of the beam pipe. These notches are due to multipath interference effects from one or more waveguide modes with the TEM field. It is obvious that under such circumstances a "response" calibration (as the easiest calibration routine) may become meaningless. More sophisticated calibration procedures, as used for de-embedding, are by definition restricted to a single-mode case (TEM on a coaxial line and normally the fundamental mode for waveguides). The reason why these modes show such bothersome effects is that they are inevitably excited at any discontinuity, such as kinks of a taper or support disks of the center conductor. One possible way to fight the problem is by very careful mechanical construction, e.g. smooth tapers with all edges rounded off. A serious difficulty is to position the center conductor without causing mode conversion. A very thin wire center conductor may be kept under strong mechanical tension in a special fixture at the end of the tapers [38]. Another possibility for supporting the wire with a minimum of dielectric material is by using two pairs of very thin nylon strings. But this method requires holes to be drilled into the vacuum chamber. In a recent paper [39] Lambertson et al. have proposed to introduce lossy material (absorber pads) at the ends of the tapers in order to suppress in particular the TM modes (Fig. 16). This rather easy manipulation can suppress the notches very efficiently. However, the TEM field also experiences a certain amount of additional attenuation which is, however, within an

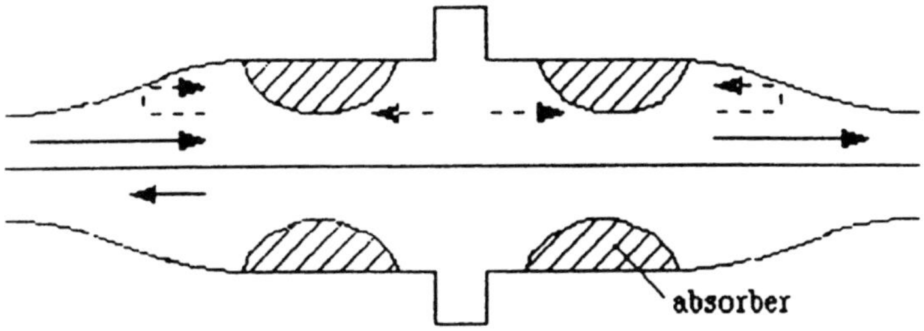

Fig. 16a - *Wire with pads to absorb TM modes, shown dashed. TEM signals are shown in solid [39].*

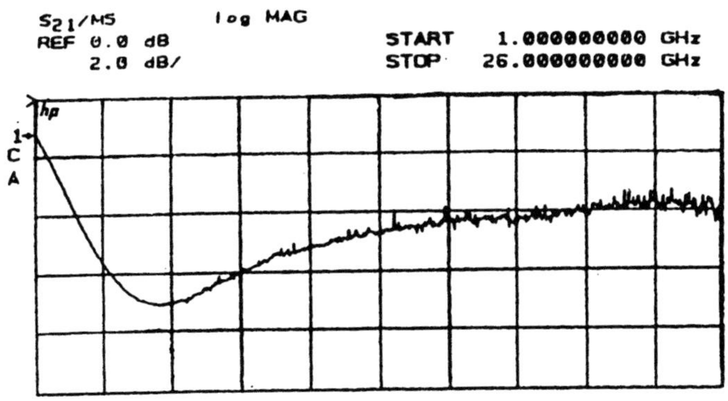

Fig. 16b - *Attenuation of wire current by absorbers [39].*

acceptable range (Fig. 16b) and can easily be removed by "response" calibration which is now possible. The absorber pads may be made from anechoic-chamber cladding material. Their size and position have to be determined experimentally for a given beam pipe.

For some applications it is very desirable to have a remote probe in the beam pipe without the need to pull a wire through [40,41]. In this case, the choice of selectively exciting (by means of a specially designed mode launcher) TM or TE modes exists. Since the TM_{01} waves exhibit a surface current distribution with longitudinal current only, they are better related to the TEM field than TE modes. By relating the surface current and the power flux in such TM waves, a waveguide mode impedance (frequency dependent) can be defined. It is used in the same way as Z_0 (= characteristic coaxial line impedance) to evaluate the unknown longitudinal beam-coupling impedance (transmission measurement).

If one wants just to localize a small discontinuity in a beam pipe this method may be applied similarly to the well known TDR (Time-Domain Reflectometry) on coaxial lines or optical fibers. Such measurements may be useful for acceptance tests of beam pipes of several meters length, or pipes already installed in a machine. The beam pipe is measured in analogy to a waveguide including S_{11} calibration to remove the effect of imperfections of the coaxial waveguide (= beam pipe) adapter (valid only for one mode). Depending on the cross section of the beam pipe, one often has to use TE modes for practical convenience in this band-limited waveguide TDR measurement with synthetic pulses (bandpass mode). Thus there is no direct relation to the beam-coupling impedance, but the method shows with great sensitivity the location and the approximate size (in terms of reflection coefficient) of discontinuities [42,43]. As a closing remark it should be added that during the construction of the CERN-ISR similar acceptance tests on sections of the beam pipe were carried out. However, at that time sophisticated instrumentation for synthetic pulses and waveguide S_{11} calibration was not available [44]. The beam pipe was converted into a wave-guide resonator by means of shorts at either end. Then with suitable couplers the frequencies and Q-values of the resonances were measured and compared to a reference. Significant deviations from this reference indicated that there was something different in the device tested, but normally this did not indicate the location of the fault.

9. PERTURBATION MEASUREMENTS ON CAVITIES

For experimental determination of the field distribution in a resonant cavity one usually uses a perturbation method: a small perturbing object is moved along some path through the cavity, and the resonance frequency is measured as a function of the position of this object. Perturbing objects may be dielectric or metallic bodies of revolution depending on the type of field to be measured [45]. The resonance frequency can be obtained by transforming the cavity into an oscillator by means of an external amplifier or by measuring the phase of a transmitted signal at the unperturbed frequency f_0. Q measurements are performed by taking two or more points of the resonance curve and applying an appropriate formula [46].

The frequency shift of the cavity resonance is given by

$$\frac{\omega - \omega_0}{\omega_0} = \frac{1}{W}\left(-K_1\varepsilon_0\hat{E}_\parallel^2 - K_2\varepsilon_0\hat{E}_\perp^2 + K_3\mu_0\hat{H}_\parallel^2 + K_4\mu_0\hat{H}_\perp^2\right) \tag{9.1}$$

where W is the energy stored in the cavity; $E_\parallel$, $E_\perp$, $H_\parallel$, $H_\perp$ are the electric and magnetic fields parallel and transverse to the perturbing object. The coefficients K_1 to K_4 are proportional to the electric and magnetic polarizability of the perturbing object. If the latter is elongated, K_1 is much larger than K_2, K_3, K_4: the frequency shift is then due essentially to the component of E which is parallel to the long axis of the perturbing object. If $P_\parallel$ is the electric dipole moment induced by an external electric field $E_\parallel$ in a perturbing object of length 2ℓ, we may write

$$P_\parallel = \varepsilon_0 E_\parallel \frac{4\pi}{3}\ell^3 G$$

where G is a dimensionless geometrical factor. The ratio $P_\parallel/(\varepsilon_0 E_\parallel)$ is the electric polarizability of the object. For a prolate ellipsoid [47] with semi-axes ℓ and a

$$G_{ell} = [1 - (a^2/\ell^2)]^{3/2}[\mathrm{arc\ th}\sqrt{(1-a^2/\ell^2)} - \sqrt{(1-a^2/\ell^2)}]^{-1}. \tag{9.3}$$

When $\ell/a \to \infty$,

$$G_{ell} \approx [\log(2\ell/a) - 1]^{-1}. \tag{9.4}$$

The coefficient K_1 in Eq. (9.1) is directly related to the electric polarizability of the perturbing object by [45]

$$K_1 = \frac{1}{4}\frac{P_\parallel}{\varepsilon_0 E_\parallel} = \frac{\pi}{3}\ell^3 G \; . \tag{9.5}$$

In practice the elongated object is often a cylindrical needle. Hahn and Halama [45] made a proposal to calculate K_1 of a cylindrical needle by applying a correction factor 3/2 to the electric ease of polarizability of the inscribed ellipsoid. But this correction factor depends on the shape of the needle; its correct value can be determined only by computing exactly the ease of polarizability of a needle considered as a hollow metallic cylinder [48].

The frequency shift can be measured either directly, by measuring ω at each position of the perturbing object by a fast tuning technique, or indirectly, by measuring the phase Φ of a transmitted signal at ω_0. If Φ is set to zero at the unperturbed frequency, the perturbed frequency is derived from

$$\frac{\omega - \omega_0}{\omega_0} = \frac{1}{2Q_\ell}\,\mathrm{tg}\Phi \; . \tag{9.6}$$

When Φ is not small, because of the non-linearity of $\mathrm{tg}\Phi$, considerable errors may result on R/Q if the frequency of the transmitted signal is not exactly ω_0; this means that the unperturbed resonance frequency ω_0 must be accurately determined when the phase method is used. The same reasoning applies for Q measurements, since they are usually based on Eq. (9.6).

In practical measurements the loaded Q of the cavity should be close to the unloaded Q (= Q_0), which implies weak coupling of the input and output coupler. Single-cell and multiple-cell cavities are measured essentially in the same way, i.e. in transmission on the resonance to be considered.

For excitation of TM modes with cylindrical symmetry, on-axis probes are used, adjusted at resonance to a dip of 0.1 to 0.3 dB in reflection coefficient. In order to distinguish dipole and quadrupole modes, probe pairs and quartets may be taken for excitation and pick-up, together with broadband 3-dB, 180° hybrids. In the case of degeneracy (two modes having the same frequency), the suppression of unwanted mode may exceed 60 dB.

To measure zeropole (= cylinder symmetric) modes the perturbing object is moved along the axis of the cavity, and off-axis for dipole and higher-order modes. The sign of the electric field may be determined by comparing the phase signals measured with a probe or loop at a set of holes drilled in the different cells (multi-cell cavity).

If it is not allowed to drill such holes into the cavity wall, usually the correct sign has to be guessed by the operator after examination of the trace that was recorded. Obviously, this is a weak point in the whole procedure and sometimes ambiguities occur. For dipole and higher-order modes, the plane of polarization must also be determined since, in general, it is not horizontal or vertical due to asymmetries on the cavity (coupling ports, etc.). An extensive description of measurements carried out on single- and multiple-cell cavities (including a discussion of many practical aspects) has been given by J. Jacob [49].

Nowadays, these measurements are usually carried out with a vectornetwork analyzer and a synthetisizer generator using the phase method. When determining the R/Q knowledge of Q is not required. The R/Q value is obtained as the result of a Fourier transform on the measured phase values as a function of the position [48-50].

With the availability of high-performance computer codes to work out electromagnetic fields in accelerator structures, often only the measurement of Q is required and R/Q is obtained from numerical calculations. But these types of perturbation methods are still needed to carry out final tuning of multiple-cell structures or RFQ's for example. The purpose of the tuning is to obtain a prescribed field pattern or a certain degree of field homogeneity in a given structure. This includes also the adjustment of the resonance frequency and coupling of each individual cell. Such measurements on multiple cells are often done in reflection and travelling-wave mode, i.e. with a terminating load at the end of the structure [51]. For adjustments of an RFQ a very special perturbing object has been used, namely a ping-pong ball [52]. This ping-pong ball could be slid along the vanes, its exact position being very critical for this type of measurement.

10. CONCLUSION

The coaxial wire method is a useful and efficient tool for (simulated) beam-coupling impedance and loss-factor measurements. Usually one obtains reliable results for a single, localized impedance with $Z \ll Z_L$. This does not mean that in other cases (several discontinuities $Z > Z_L$) the results are incorrect, but caution and cross checking are required. It should be kept in mind that the wire permits coupling (= energy exchange) between two or more discontinuities, which would not be coupled in the case of an empty beam pipe. In general, transmission rather than reflection measurements are better. The effect of multiple reflections due to mismatch in adapter pieces can be eliminated by time filtering (= gating).

As a second choice de-embedding (calibration) procedures may be used, as they are more difficult to handle than gating, but sometimes also more effective in rejecting undesired signals. In all cases a high electrical and mechanical repeatability is required.

The beam-coupling impedance and the loss factor can be measured with the same instrumentation using the synthetic-pulse technique. Synthetic pulses are more powerful in terms of stability and signal-to-noise ratio than real pulses but the set-up may be more expensive. Caution should be exercised for frequencies beyond waveguide cut-off, and the effect of waveguide modes must be carefully considered. Impedance measurements without a coaxial wire relying on TM modes appear to be possible within certain limitations. Band-limited Time Domain Reflectometry (TDR) using TM or TE modes can return useful information on beam pipe discontinuities. In order to simulate transverse fields of very slow beams, an array of probe and loop antennas gave promising results.

Bead-pull measurements on cavity-like objects are nowadays used rather for tuning purposes (multiple cell, RFQ) than for R/Q determination. With the availability of powerful computer codes for field calculation, often only the Q measurement is still required.

ACKNOWLEDGEMENTS

The author would like to thank G. Dôme, G. Jackson, D. Möhl, C. Taylor and B. Zotter for many helpful and enlighting discussions and contributions, as well as A. Molat-Berbiers for typing the manuscript.

REFERENCES

[1] J.D. Jackson, Classical Electrodynamics, John Wiley and Sons, New York 1975, Sect. 9.1.

[2] E. Karantzoulis, An Overview on Impedance Measuring Methods for Accelerators, Sincrotrone Trieste, ST/M-91/1, January 1991.

[3] S.A. Heifets and S.A. Kheifets, Coupling Impedance in Modern Accelerators, SLAC-PUB-5297, Sept. 1990.

[4] L. Palumbo and V.G. Vaccaro, Wake Fields Measurements, Joint US-CERN School on Particle Accelerators, Lecture Notes in Physics, No. 343, Springer-Verlag, 1988.

[5] R.L. Gluckstern and R. Li, Analysis of Coaxial Wire Measurement of Longitudinal Coupling Impedance, 4th Int. Conf. On High-Energy Accelerators, KEK, Japan, 1990.

[6] T. Weiland, On The Numerical Solution of Maxwell's Equations and Applications in the Field of Accelerator Physics, DESY M-84-06, Jan. 1984.

[7] M. Sands, J. Rees, SLAC-report PEP-95, August 1974.

[8] P.B. Wilson, J.B. Styles and K.L.F. Bane, Comparison of Measured and Computed Loss to Parasitic Modes in Cylindrical Cavities with Beam Ports, SLAC-PUB-1908, PEP-240, March 1977.

[9] H. Hahn and F. Pedersen, On Coaxial Wire Measurements of the Longitudinal Coupling Impedance, BNL 50870, Particle Accelerators and High-Voltage Machines - TID-4500, April 1978.

[10] M. Billing (Cornell), Private Communication.

[11] F. Caspers, Beam Impedance Measurement Using the Coaxial Wire Method, Proc of the Workshop on Impedance and Current Limitations, ESRF Grenoble 1988, Sect. 2.1, and CERN PS/88-59 (AR).

[12] K. Lange, K.H. Löcherer, Taschenbuch der Hochfrequenztechnik (4 Auflage, 1986), Springer Verlag, p. C20, 1986.

[13] D. Boussard, H. Henke, J. Gareyte, Academic Training Programme, CERN, 1986, Unpublished.

[14] F. Caspers, Beam Impedance Measurement by the Wire Method Using a Synthetic Pulse Technique, IEEE, NS-32, pp. 1914-1916, 1985.

[15] M.E. Hines, H.E. Stinehelfer, Time-Domain Oscillographic Microwave Network Analysis Using Frequency-Domain Data, IEEE Trans. on Microwave Theory and Techniques, Vol. MTT-22, No. 3, March 1974.

[16] J. Curran, Applying TRL Calibration for Non-Coaxial Measurements, Microwave System News, pp. 91-98, March 1988.

[17] L. Walling, LANL, Private Communication.

[18] L. Walling, D. McMurray, D. Neuffer, H.A. Thiessen, Transmission Line Impedance Measurements for an Advanced Hadron Facility, LANL 1988.

[19] H.J. Eul and B. Schiek, Experimental Results of New Self-Calibration Procedures for Network Analyzers, Proc. of the 20th European Microwave Conf., Budapest, 1990, 1461-1466.

[20] H.J. Eul and B. Schiek, Broadband Self-Calibration Techniques for Network Analyzers, Frequenz 44 (1990), 5.

[21] H.J. Eul, Comments on "A Unified Mathematical Approach to Two-Port Calibration Techniques and Some Applications", IEEE Trans. on Microwave Theory and Techniques, Vol. 38, No. 8, August 1990.

[22] G.F. Engen and C.A. Hoer, "Thru-Reflect-Line": An Improved Technique for Calibrating the Dual Six-Port Automatic Network Analyzer, IEEE Trans. on Microwave Theory and Techniques, Vol. MTT-27, No. 12, Dec. 1979.

[23] R.R. Pantoja, M.J. Howes, J.R. Richardson and R.D. Pollard, Improved Calibration and Measurement of the Scattering Parameters of Microwave Integrated Circuits, IEEE Trans. on Microwave Theory and Techniques, Vol. 37, No.11, Nov. 1989.
[24] R.A. Soares, P. Gouzien, P. Legaud and G. Follot, A Unified Mathematical Approach to Two-Port Calibration Techniques and Some Applications, IEEE Trans. on Microwave Theory and Techniques, Vol. 37, No. 11, Nov. 1989.
[25] R.A. Speciale, A Generalization of the TSD Network-Analyzer Calibration Procedure, Covering n-Port Scattering-Parameter Measurements, Affected by Leakage Errors, IEEE Trans. on Microwave Theory and Techniques, Vol. MTT-25, No. 12, Dec. 1977.
[26] H.J. Eul and B. Schiek, Thru-Match-Reflect: One Result of a Rigorous Theory for De-embedding and Network Analyzer Calibration, Proc. of the 18th European Microwave Conf., Stockholm, 1988.
[27] King-Yuen Ng, Impedances of Stripline Beam-Position Monitors, Particle Accelerators, 1988, Vol. 23, 93-102.
[28] G. Nassibian, The Real Part of the Low Frequency Beam Impedance of Travelling Wave Kickers with Arbitrary Terminations, CERN/PS/85-68 (BR), 1985.
[29] G. Nassibian and F. Sacherer, A Method for Measuring Transverse Coupling Impedance, CERN/ISR-TH/77-61, 1977.
[30] G. Lambertson, Dynamic Devices-Pickups and Kickers, FNAL Acc. School 1984, AIP Conf. Proc. 153, 1413-1442.
[31] H. Henke, Private Communication.
[32] S. Kheifets and B. Zotter, Longitudinal and Transverse Impedance of Bellows in the Low Frequency Range, Nucl. Instr. and Methods in Physics Res., A243, 13-27, 1986.
[33] F. Caspers, J.P. Delahaye, J.C. Godot, K. Hübner and A. Poncet, EPAC, Rome, 1988, CERN PS/88-37 (LP).
[34] A. Faltens, E.C. Hartwig, D. Möhl, and A.M. Sessler, An Analog Method for Measuring the Longitudinal Coupling Impedance of a Relativistic Particle Beam with its Environment, Proc. 8th International Conf. on High Energy Acc. CERN 1971.
[35] L.J. Laslett, K. Neil, A. Sessler, Transverse Resistive Instabilities Rev. Sci. Instr. 36, pp. 436-448, 1965.
[36] F. Caspers, D. Möhl, and A. Schwinn, A New Bench Method to Simulate Electromagnetic Fields of Slow Beams, EPAC 90 and CERN/PS 90-35 (AR).
[37] N. Tokuda, Travelling Wave Pick-ups and Kickers for Stochastic Cooling at Low Beam Velocity, in: Proc. IVth LEAR Workshop, Villars (CH), 1987, (Edited by C. Amsler et al.) Harwood Academic Publishers, Chur 1988, pp. 135-138.
[38] G. Jackson, FNAL, Review of Impedance Measurements at FNAL, 1990, unpublished.
[39] G.R. Lambertson, A.F. Jacob, R.A. Rimmer, F. Völker, Techniques for Beam Impedance Measurements Above Cut-off, LBL-28190, EPAC Rome 1990.
[40] R.A. Rimmer, D.A. Goldberg, A.F. Jacob, G.R. Lambertson, F. Völker, Beam Impedance Measurements on the ALS Curved Sector Tank, LBL-28192, EPAC Rome 1990.
[41] A. Jacob and G.R. Lambertson, Impedance Measurements on Button Electrodes, IEEE Part. Accel. Conf., Chicago 1989, LBL-25955.
[42] P.I. Somlo, The Locating Reflectometer, IEEE Trans. on Microwave and Techniques, Vol. MTT-20, No. 2, Feb. 1972.

[43] D. Stone, K.L. Felch and S.T. Spang, Mode-Specific Reflectometry in a Multimode Waveguide, IEEE Trans. on Microwave Theory and Techniques, Vol. MTT-31, No. 9, Sept. 1983.
[44] C. Taylor, Private Communication.
[45] H. Hahn and H.J. Halama, Perturbation Measurement of Transverse R/Q in Iris-Loaded Waveguides, IEEE Trans. on Microwave Theory and Techniques, MTT-16, Jan. 1988, 20-29.
[46] L.G. Bernier, T. Sphicopoulos and F.H. Gardiol, An Automatic System for the Measurement of the Field Distribution in Resonant Cavities, IEEE Instrum. and Methods, IM-32, Dec. 1983, 462-466.
[47] L.C. Maier and J.C. Slater, Field Strength Measurements in Resonant Cavities, J. Appl. Physics, Vol. 23, No. 1, Jan. 1962.
[48] F. Caspers and G. Dôme, Precise Perturbation Measurements of Resonant Cavities and Higher Order Mode Identification, CERN SPS/85-46 (ARF), 1985.
[49] J. Jacob, Measurement of the Higher-Order Mode Impedances of the LEP Cavities, ESRF-MAC-06/10, 1988.
[50] P. Marchand and D. Proch, Higher Order Mode Measurements in a 5-Cell Copper Cavity at 2 GHz and Application to a Superconducting Cavity for Petra, CERN/EF/RF 82-7, Oct. 1982.
[51] Lapostolle and Septiev, Linear Accelerators, 1970 North-Holland Publishing, 91-100.
[52] W. Pirkl, Tuning of the RFQ, CERN-PS/RF Note 83-4, 1983.

COHERENT BEAM INSTABILITIES

A. Hofmann
CERN, Geneva, Switzerland

1 Introduction

The motion of a single particle in a storage ring is well determined by the external guide fields (dipole and quadrupole magnets, RF system, etc.), by the initial conditions and the synchrotron radiation. The many particles in a high intensity beam represent a sizeable charge and current which act as a source of electromagnetic fields (self-fields). These fields are modified through the boundary conditions imposed by the beam surroundings (vacuum chambers, cavities, etc.) and act back on the beam. This can lead to a *frequency shift* (change of the betatron or synchrotron frequency), to an increase of a small disturbance of the beam, i.e. an *instability* or a *change of the particle distribution*, e.g. bunch lengthening. These phenomena are usually called *collective effects* since a collective or a coherent action of the many particles in the beam is involved.

As an example we consider bunches in a storage ring going through a cavity. Each bunch induces electromagnetic fields in this cavity which oscillate and slowly decay away. The next bunch, or the same bunch on the next turn, might find some field left and will be influenced by it. Such a case is illustrated in Fig. 1. The phase of the field seen in the next turn can be such that a small initial synchrotron oscillation of the bunch is increased. In each turn the oscillation is amplified, resulting in an exponentially growing instability.

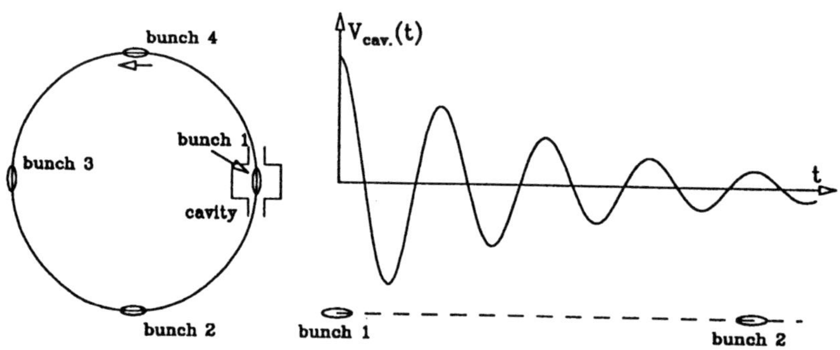

Figure 1: Field acting from one bunch to the next one

In many cases the fields created by the beam are small compared to the guide fields and their effects can be treated as a *perturbation*. This is done in three steps:

a) We calculate the stationary particle distribution which is given by the guide field, initial condition and synchrotron radiation.

b) We consider small disturbances of the bunch from the stationary distribution (different modes of betatron and synchrotron oscillations). We calculate the fields due to such a disturbance taking into account the boundary condition imposed by the beam surroundings (impedance).

c) We calculate the effect of these fields and investigate if the initial disturbance is increased *(instability)* or decreased *(damping)* or if the frequency of the mode of oscillation is changed *(frequency shift)*.

For the case of small self-fields considered here the particle distribution in the bunch is given by external conditions (machine parameter, initial condition, synchrotron radiation) and is usually Gaussian for electron machines. As disturbances of the stationary distribution we consider some modes of oscillation which are orthogonal to each other such that the stability of each mode can be treated independently.

For strong self-fields, however, the particle distribution is modified and the modes of oscillation are influenced such that they are no longer independent. A self-consistent solution has to be found in this case which is much more difficult. It is usually attempted only for the case of bunch lengthening where the longitudinal particle distribution in the presence of an impedance is to be obtained. The beam position monitors and the transverse impedance can usually not "resolve" the transverse particle distribution. It is therefore of importance for beam instabilities only in exceptional cases.

We distinguish between *single* and *multi-traversal* collective effects. For the first kind no memory of the induced field over one revolution or over the time distance between the passage of adjacent bunches is assumed. An example of a single-traversal effect is bunch lengthening. For multi-traversal effects the impedance has to have a memory such that one bunch can influence the next one or itself after one revolution. This is provided by cavity-like objects with a relatively large quality factor Q.

Finally, we distinguish between *longitudinal* and *transverse* effects. In the first case a longitudinal impedance influences the synchrotron oscillation such that its amplitudes grows or decays or changes its frequency. The transverse impedance has a corresponding effect on the horizontal or vertical betatron oscillations.

2 Impedance and wake potential of a resonator

2.1 Cavity resonance

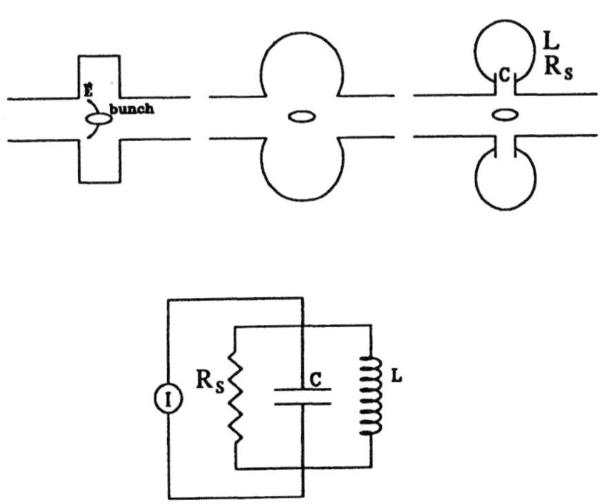

Figure 2: RLC circuit equivalent to a cavity resonance

Impedances and wake potentials have been treated extensively in an earlier lecture. As an introduction to the interaction between the beam and a cavity resonance we recapitulate here briefly some of the essential properties.

Cavities are the most important objects which can cause coupled-bunch mode instabilities since the induced fields oscillate for a relatively long time and provide a memory over one revolution. Such a cavity can be of a form which resembles an RCL circuit, Fig. 2, and can, in good approximation, be treated as such. The RCL circuit has a shunt impedance R_s, an inductance L and a capacity C. In a real cavity these three parameters cannot easily be separated. For this reason one uses some other parameters which can easily be measured directly: The *resonance frequency* ω_r, the *quality factor* Q and the *damping rate* α:

$$\omega_r = \frac{1}{\sqrt{LC}}\ , \quad Q = R_s\sqrt{\frac{C}{L}} = \frac{R_s}{L\omega_r} = R_s C\omega_r\ , \quad \alpha = \frac{\omega_r}{2Q}. \tag{1}$$

If this circuit is driven by a current I we have the voltages across each element

$$V_R = I_R R_s,\ V_C = \frac{1}{C}\int I_C dt\,,\ V_L = L\frac{dI_L}{dt} \tag{2}$$

and the relations between the voltages and currents

$$V_R = V_C = V_L = V\ , \quad I_R + I_C + I_L = I. \tag{3}$$

Differentiating with respect to t gives

$$\dot{I} = \dot{I}_R + \dot{I}_C + \dot{I}_L = \frac{\dot{V}}{R_s} + C\ddot{V} + \frac{V}{L}. \tag{4}$$

Using $L = R_s/(\omega_r Q)$ and $C = Q/(\omega_r R_s)$ we get the differential equation

$$\ddot{V} + \frac{\omega_r}{Q}\dot{V} + \omega_r^2 V = \frac{\omega_r R_s}{Q}\dot{I}. \tag{5}$$

The solution of the homogeneous equation represents a damped oscillation

$$V(t) = \hat{V}e^{-\alpha t}\cos\left(\omega_r\sqrt{1-\frac{1}{4Q^2}} + \phi\right) \tag{6}$$

or

$$V(t) = e^{-\alpha t}\left(A\cos\left(\omega_r\sqrt{1-\frac{1}{4Q^2}}\,t\right) + B\sin\left(\omega_r\sqrt{1-\frac{1}{4Q^2}}\,t\right)\right). \tag{7}$$

2.2 Wake potential

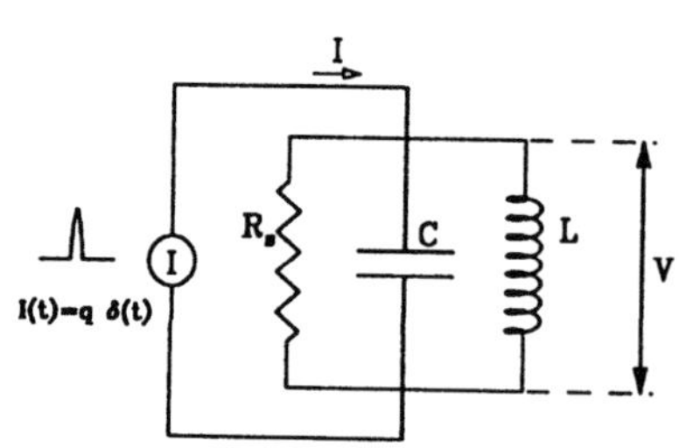

$$\omega_r = \frac{1}{\sqrt{LC}}$$

$$Q = R_s\sqrt{\frac{C}{L}}$$

$$\alpha = \frac{\omega_r}{2Q}$$

$$\ddot{V} + \frac{\omega_r}{Q}\dot{V} + \omega_r^2 V = \frac{\omega_r R_s}{Q}\dot{I}$$

Figure 3: Equivalent RLC circuit driven by a delta pulse

We now calculate the response of the RCL-circuit (representing a cavity) to a delta function pulse (very short bunch), Fig. 3,

$$I(t) = q\delta(t) \tag{8}$$

The charge q will charge up the capacity to a voltage

$$V(0^+) = \frac{q}{C} = \frac{\omega_r R_s}{Q} q. \tag{9}$$

The energy stored in the capacitor is equal to the energy lost by the charge

$$U = \frac{q^2}{C} = \frac{\omega_r R_s}{2Q} q^2 = \frac{V(0^+)}{2} q = k_{pm} q^2, \tag{10}$$

where we introduced the *parasitic mode loss factor*

$$k_{pm} = \frac{\omega_r R_s}{2Q}. \tag{11}$$

The charged capacitor C will now discharge first through the resistor R_s and then also through the inductance L

$$\dot{V}(0^+) = -\frac{\dot{q}}{C} = -\frac{I_R}{C} = \frac{1}{C}\frac{V(0^+)}{R_s} = -\frac{\omega_r^2 R_s}{Q^2} q = \frac{2\omega_r k_{pm}}{Q} q. \tag{12}$$

The resonance circuit has now the initial conditions

$$V(0^+) = 2k_{pm} q \text{ and } \dot{V}(0^+) = \frac{2\omega_r k_{pm}}{Q} q\,. \tag{13}$$

Taking the solution of the homogeneous differential equation and its derivative

$$V(t) = \mathrm{e}^{-\alpha t}\left(\mathrm{A}\cos\left(\omega_r\sqrt{1-\frac{1}{4\mathrm{Q}^2}}\,\mathrm{t}\right) + \mathrm{B}\sin\left(\omega_r\sqrt{1-\frac{1}{4\mathrm{Q}^2}}\,\mathrm{t}\right)\right) \tag{14}$$

$$\begin{aligned}\dot{V}(t) = \mathrm{e}^{-\alpha t} \quad &\left(\left(-A\alpha + B\omega_r\sqrt{1-\frac{1}{4Q^2}}\right)\cos\left(\omega_r\sqrt{1-\frac{1}{4Q^2}}\,t\right)\right.\\ &\left.-\left(B\alpha + A\omega_r\sqrt{1-\frac{1}{4Q^2}}\right)\sin\left(\omega_r\sqrt{1-\frac{1}{4Q^2}}\,t\right)\right)\end{aligned}$$

and satifying the above initial conditions with

$$A = 2k_{pm} q \quad \text{and} \quad -A\alpha + B\omega_r\sqrt{1-\frac{1}{4Q^2}} = -\frac{2\omega_r k_{pm}}{Q} q \tag{15}$$

gives for the voltage in a resonator circuit excited at the time $t = 0$ by a δ-pulse $I(t) = q\delta t$

$$V(t) = 2qk_{pm}\mathrm{e}^{-\alpha t}\left(\cos\left(\omega_r\sqrt{1-\frac{1}{4\mathrm{Q}^2}}\,\mathrm{t}\right) - \frac{\sin\left(\omega_r\sqrt{1-\frac{1}{4\mathrm{Q}^2}}\,\mathrm{t}\right)}{2\mathrm{Q}\sqrt{1-\frac{1}{4\mathrm{Q}^2}}}\right). \tag{16}$$

This voltage is induced by a charge q going through the cavity at the time $t = 0$. A second point charge q' going through the cavity will gain or lose the energy $U = q'V(t)$. This energy gain/loss per unit source and unit probe charge is called the *wake potential* of a point charge or also the *Green function* $G(t)$. For our resonator (cavity resonance) we have

$$G(t) = 2k_{pm}\mathrm{e}^{-\alpha t}\left(\cos\left(\omega_r\sqrt{1-\frac{1}{4\mathrm{Q}^2}}\,\mathrm{t}\right) - \frac{\sin\left(\omega_r\sqrt{1-\frac{1}{4\mathrm{Q}^2}}\,\mathrm{t}\right)}{2\mathrm{Q}\sqrt{1-\frac{1}{4\mathrm{Q}^2}}}\right) \tag{17}$$

which for a large quality factor $Q \gg 1$ simplifies to

$$G(t) \approx 2k_{pm}\mathrm{e}^{-\alpha t}\cos\left(\omega_r \mathrm{t}\right). \tag{18}$$

2.3 Impedance

We assume now a *harmonic* excitation of the circuit with a current $I = \hat{I}\cos(\omega t)$, Fig. 4.

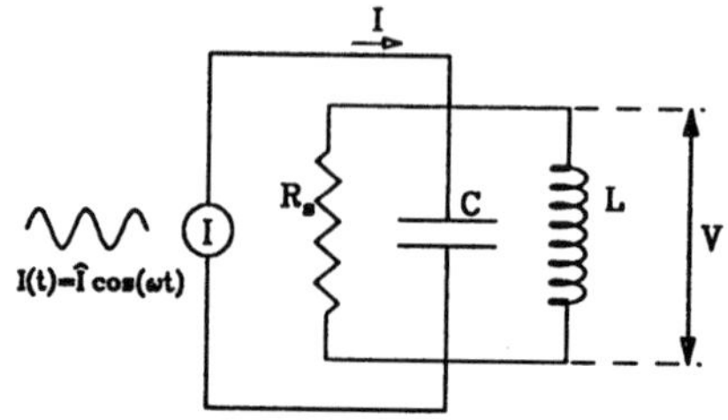

Figure 4: Equivalent RLC driven by a harmonic excitation

The differential equation of the harmonic excitation is

$$\ddot{V} + \frac{\omega_r}{Q}\dot{V} + \omega_r^2 V = -\frac{\omega_r R_s}{Q}\hat{I}\omega\sin(\omega t). \tag{19}$$

The solution of the homogeneous equation is a damped oscillation which disappears after some time. We are left with the particular solution of the form $V(t) = A\cos(\omega t) + B\sin(\omega t)$. Inserting this into the differential equation and separating cosine and sine terms gives

$$(\omega_r^2 - \omega^2)A + \frac{\omega_r\omega}{Q}B = 0 \text{ and } (\omega_r^2 - \omega^2)B - \frac{\omega_r\omega}{Q}A = -\frac{\omega_r\omega R_s}{Q}\hat{I}. \tag{20}$$

The voltage induced by the harmonic excitation of the resonator becomes

$$V(t) = \hat{I}R_s\frac{\cos(\omega t) - Q\frac{\omega_r - \omega}{\omega_r\omega}\sin(\omega t)}{1 + Q^2\left(\frac{\omega_r^2-\omega^2}{\omega_r\omega}\right)^2}. \tag{21}$$

The voltage has a cosine term which is *in phase* with the exciting current. It can absorb energy and is called *resistive* term. The sine term of the voltage is *out of phase* with the exciting current and does not absorb energy, it is called *reactive*. The ratio between the voltage and current is the *impedance*. It is a *function of frequency* ω and has a resistive part $Z_r(\omega)$ and a reactive part $Z_i(\omega)$

$$Z_r(\omega) = R_s\frac{1}{1 + Q^2\left(\frac{\omega_r^2-\omega^2}{\omega_r\omega}\right)^2}\ ,\quad Z_i(\omega) = (\pm)R_s\frac{Q\frac{\omega_r^2-\omega^2}{\omega_r\omega}}{1 + Q^2\left(\frac{\omega_r^2-\omega^2}{\omega_r\omega}\right)^2}. \tag{22}$$

2.4 Complex notation

We have used a harmonic excitation of the form

$$I(t) = \hat{I}\cos(\omega t) = \hat{I}\frac{e^{j\omega t} + e^{-j\omega t}}{2} \text{ with } 0 \leq \omega \leq \infty. \tag{23}$$

It is often more convenient to use a complex notation

$$It) = \hat{I}e^{j\omega t} \text{ with } -\infty \leq \omega \leq \infty \tag{24}$$

which leads to more compact expressions. Taking the differential equation

$$\ddot{V} + \frac{\omega_r}{Q}\dot{V} + \omega_r^2 = \frac{\omega_r R_s}{Q}\dot{I} \tag{25}$$

of the excited resonance with $I(t) = \hat{I}\exp(j\omega t)$ and seeking a solution of the form $V(t) = V_0\exp(j\omega t)$, where V_0 is in general complex, one gets

$$-\omega^2 V_0 e^{j\omega t} + j\frac{\omega_r - \omega}{Q} V_0 e^{j\omega t} + \omega_r^2 V_0 e^{j\omega t} = j\frac{\omega_r \omega R_s}{Q}\hat{I}e^{j\omega t} \tag{26}$$

and for the impedance which is defined as the ratio V/I,

$$Z(\omega) = \frac{V_0}{\hat{I}} = R_s \frac{j\frac{\omega_r\omega}{Q}}{\omega_r^2 - \omega^2 + jQ\frac{\omega_r\omega}{Q}} = R_s \frac{1 - j\frac{\omega^2-\omega_r^2}{\omega\omega_r}}{1 + Q^2\left(\frac{\omega^2-\omega_r^2}{\omega\omega_r}\right)^2}. \tag{27}$$

For a large quality factor the impedance is only large for $\omega \approx \omega_r$ or $|\omega - \omega_r|/\omega_r = |\Delta\omega|\omega_r \ll 1$ and can be simplified

$$Z(\omega) \approx R_s \frac{1 - j2Q\frac{\Delta\omega}{\omega_r}}{1 + 4Q^2\left(\frac{\Delta\omega}{\omega_r}\right)^2}. \tag{28}$$

The resonator impedance has some specific properties:

$$\begin{aligned} \omega &= \omega_r \rightarrow Z_r(\omega_r) \text{ has a maximum, } Z_i(\omega_r) = 0 \\ |\omega| &< \omega_r \rightarrow Z_i(\omega) > \omega_r \text{ (inductive)} \\ |\omega| &> 0 \rightarrow Z_i(\omega) < 0 \text{ (capacitive)} \end{aligned} \tag{29}$$

and some properties which apply to any impedance or wake potential

$$\begin{aligned} Z_r(\omega) &= Z_r(-\omega)\ ,\ Z_i(\omega) = -Z_i(-\omega), \\ Z(\omega) &= \int_{-\infty}^{\infty} G(t)e^{-j\omega t}dt,\ Z(\omega) = \text{Fourier transform of } G(t), \text{Fig.5}, \end{aligned} \tag{30}$$

$$t < 0 \rightarrow G(t) = 0 \text{ no fields before particle arrives.} \tag{31}$$

Caution: sometimes one uses $I(t) = \hat{I}e^{-i\omega t}$ instead of $I(t) = \hat{I}e^{j\omega t}$, this reverses the sign $Z_i(\omega)$.

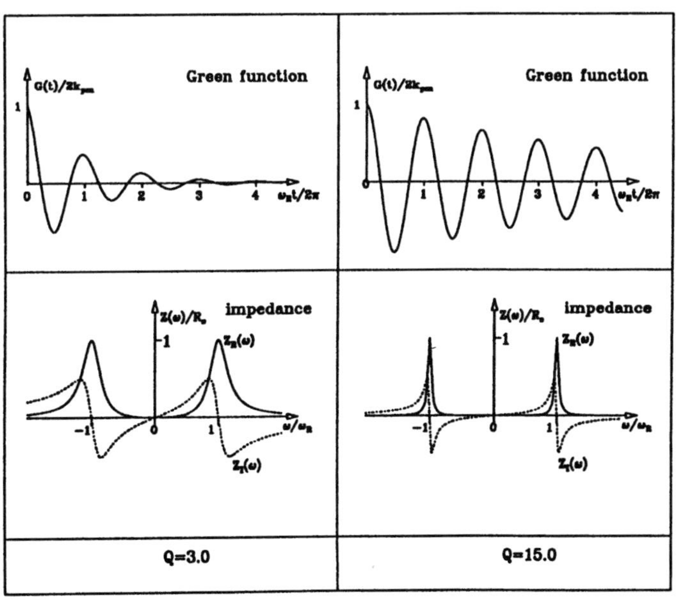

Figure 5: Green function and impedance of a resonance

3 Interaction of a stationary bunch with an impedance

3.1 Spectrum of a stationary bunch

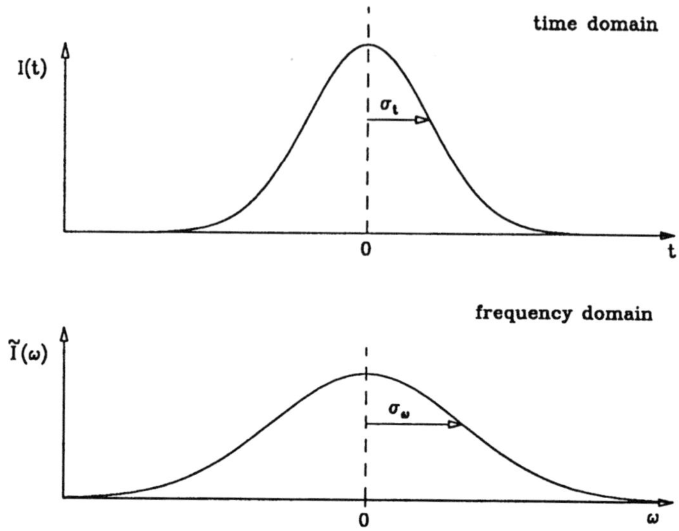

Figure 6: Single passage of a bunch in time and frequency domain

We consider a single traversal at the time $t = 0$ of a bunch having the form $I(t)$ which we assume for convenience to be symmetric in t. The spectrum of this single bunch traversal is given by the *Fourier transform*, Fig. 6,

$$\tilde{I}(\omega) = \frac{1}{\sqrt{2\pi}} \int_{-\infty}^{\infty} I(t) e^{-j\omega t} dt. \tag{32}$$

Had this bunch passed by the observer at an earlier time T its current and spectrum would be of the form $I_T(t) = I(t-T)$ and

$$\begin{aligned}\tilde{I}_T(\omega) &= \frac{1}{\sqrt{2\pi}} \int_{-\infty}^{\infty} I(t-T) e^{-j\omega t} dt \\ &= \frac{e^{-j\omega T}}{\sqrt{2\pi}} \int_{-\infty}^{\infty} I(t-T) e^{-j\omega(t-T)} d(t-T) = e^{-j\omega T} \tilde{I}(\omega).\end{aligned}$$

A shift of the time domain function produces a phase factor in the Fourier transform (*Fourier shift theorem*).

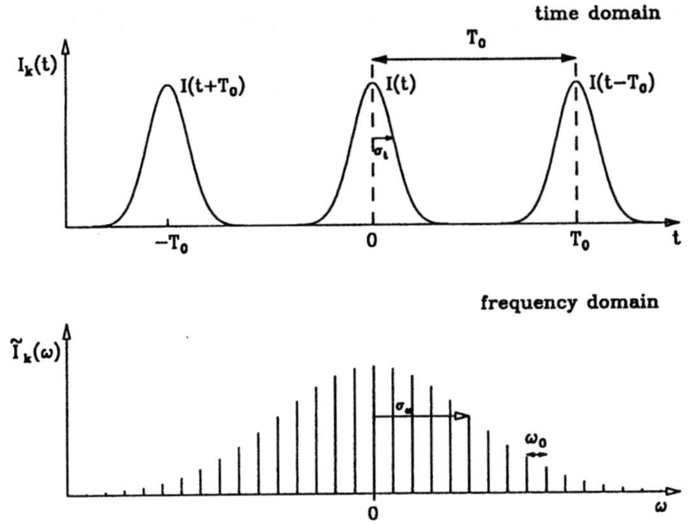

Figure 7: Multiple passage of a bunch in time and frequency domain

Next we take the case of a bunch circulating in a storage ring with revolution time T_0. It represents a current

$$I_k(t) = \sum_{k=-\infty}^{k=\infty} I(t - kT_0). \tag{33}$$

The spectrum is obtained using the shift theorem

$$\tilde{I}_k(\omega) = \sum_{k=-\infty}^{\infty} \mathrm{e}^{-j\omega kT_0}\, \tilde{I}(\omega)\,. \tag{34}$$

Using the relation

$$\sum_{k=-\infty}^{\infty} \mathrm{e}^{-jkx} = 2\pi \sum_{p=-\infty}^{\infty} \delta(x - 2\pi p) \text{ and } \delta(ax) = \frac{1}{a}\delta x \tag{35}$$

we get

$$\sum_{k=-\infty}^{\infty} \mathrm{e}^{-jk\omega T_0} = 2\pi \sum_{p=-\infty}^{\infty} \delta(\omega T_0 - 2\pi p) = \frac{2\pi}{T_0} \sum_{p=-\infty}^{\infty} \delta(\omega - p\omega_0), \tag{36}$$

where $\omega_0 = 2\pi/T_0$ is the revolution frequency. The Fourier transformed current is

$$\tilde{I}_k(\omega) = \omega_0 \sum_{p=-\infty}^{\infty} \tilde{I}(\omega)\delta(\omega - p\omega_0). \tag{37}$$

An inverse Fourier transform gives the current in time domain, Fig. 7,

$$I_k(t) = \frac{1}{\sqrt{2\pi}} \int_{-\infty}^{\infty} \tilde{I}_k(\omega)\mathrm{e}^{j\omega t} d\omega = \frac{\omega_0}{\sqrt{2\pi}} \sum_{p=-\infty}^{\infty} \int_{-\infty}^{\infty} \delta(\omega - p\omega_0)\tilde{I}(\omega)\mathrm{e}^{j\omega t} d\omega \tag{38}$$

$$I_k(t) = \frac{\omega_0}{\sqrt{2\pi}} \sum_{p=-\infty}^{\infty} \tilde{I}(p\omega_0)\mathrm{e}^{jp\omega_0 t}. \tag{39}$$

Since we assumed $I(t)$ to be symmetric we have $\tilde{I}(p\omega_0)$ is real and $\tilde{I}(-p\omega_0) = \tilde{I}(p\omega_0)$ so that combining positive and negative frequency terms gives

$$I_k(t) = I_0 + \frac{2\omega_0}{\sqrt{2\pi}} \sum_{p=1}^{\infty} \tilde{I}(p\omega_0)\cos(p\omega_0 t). \tag{40}$$

We used here a Fourier transform in view of later applications. Since the bunch current $I_k(t)$ is a periodic function it would have been easier to develop it into a Fourier series

$$I_k(t) = I_0 + \sum_1^{\infty} I_p \cos(p\omega_0 t) \text{ with } I_p = \frac{2\omega_0}{\sqrt{2\pi}}\tilde{I}(p\omega_0). \tag{41}$$

For low frequencies $I_p \approx 2I_0$.

3.2 Voltage induced by the stationary bunch

In the presence of a cavity resonance or any *general impedance* $Z(\omega)$ the circulating stationary bunch induces a voltage in frequency domain

$$\tilde{V}_k(\omega) = \tilde{I}_k(\omega)Z(\omega) = \omega_0 \sum_{p=-\infty}^{\infty} \tilde{I}(\omega)\delta(\omega - p\omega_0)Z(\omega) \tag{42}$$

and in time domain

$$\begin{aligned} V_k(t) &= \frac{\omega_0}{\sqrt{2\pi}} \sum_{p=-\infty}^{\infty} \int_{-\infty}^{\infty} \tilde{I}(\omega)\delta(\omega - p\omega_0)Z(\omega)e^{j\omega t}d\omega \\ &= \frac{\omega_0}{\sqrt{2\pi}} \sum_{p=-\infty}^{\infty} \tilde{I}(p\omega_0)Z(p\omega_0)e^{jp\omega_0 t} \end{aligned}$$

or, by combining positive and negative frequencies and observing the symmetry conditions $Z_r(-\omega) = Z_r(\omega)$, $Z_i(-\omega) = -Z_i(\omega)$ and the fact that $Z(0) = 0$

$$V_k(t) = \frac{2\omega_0}{\sqrt{2\pi}} \sum_{p=1}^{\infty} \tilde{I}(p\omega_0)\left(Z_r(p\omega_0)\cos(p\omega_0 t) - Z_i(p\omega_0)\sin(p\omega_0 t)\right). \tag{43}$$

3.3 Energy loss per turn of a stationary circulating bunch

The energy lost by the circulating stationary bunch in one turn due to the impedance $Z(\omega)$ is

$$W_b = \int_0^{T_0} I_k(t)V_k(t)dt. \tag{44}$$

From Sections 3.1 and 3.2

$$I_k(t) = \frac{\omega_0}{\sqrt{2\pi}} \sum_{p=-\infty}^{\infty} \tilde{I}(p\omega_0)e^{jp\omega_0 t} \tag{45}$$

$$V_k(t) = \frac{\omega_0}{\sqrt{2\pi}} \sum_{p'=-\infty}^{\infty} \tilde{I}(p'\omega_0)Z(p'\omega_0)e^{jp'\omega_0 t} \tag{46}$$

so that

$$W_b = \frac{\omega_0^2}{2\pi} \sum_{p=-\infty}^{\infty} \sum_{p'=-\infty}^{\infty} \tilde{I}(p\omega_0)\tilde{I}(p'\omega_0)Z(p'\omega_0) \int_0^{T_0} e^{j\omega_0(p+p')t}dt. \tag{47}$$

The integral

$$\int_0^{T_0} e^{j\omega_0(p+p')t}dt = \int_0^{T_0} \left(\cos(\omega_0(p+p')t) + j\sin(\omega_0(p+p')t)\right)dt \tag{48}$$

vanishes except for $p+p' = 0$ in which case it takes the value $T_0 = 2p/\omega_0$. We have therefore $p = -p'$

$$W_b = \omega_0 \sum_{p=-\infty}^{\infty} \left|\tilde{I}(p\omega_0)\right|^2 Z(p\omega_0) = 2\omega_0 \sum_{1}^{\infty} \left|\tilde{I}(p\omega_0)\right|^2 Z_r(p\omega_0) \tag{49}$$

where we used $\tilde{I}(\omega)\tilde{I}(-\omega) = \tilde{I}(\omega)\tilde{I}^*(\omega)) = |\tilde{I}(\omega)|^2$, which is true for the Fourier transform of a real function $I(t)$, and the symmetry relations $Z_r(\omega) = Z_r(-\omega)$, $Z_i(\omega) = -Z_i(-\omega)$.

This is the energy loss of the whole bunch. The average loss per particle U is

$$U = \frac{W_b}{N} = \frac{\omega_0 e U_b}{2\pi I_0} \tag{50}$$

where N is the number of particles and I_0 the current of the bunch.

4 Interaction of an oscillating bunch with a cavity

4.1 Review of the longitudinal dynamics

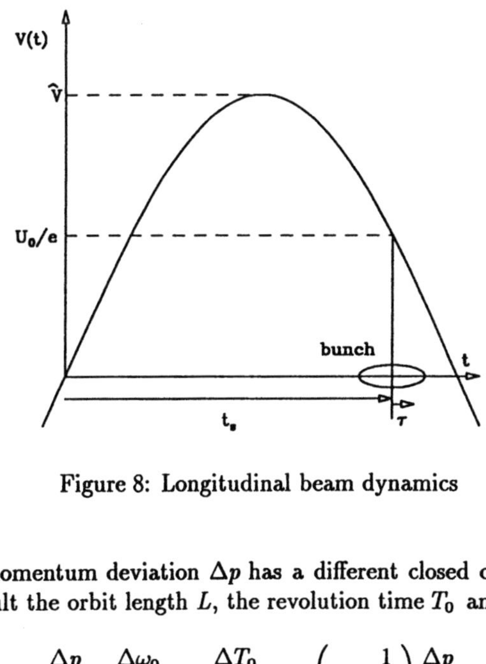

Figure 8: Longitudinal beam dynamics

A particle with a momentum deviation Δp has a different closed orbit which is displaced by $\Delta x = \alpha \Delta p/p$. As a result the orbit length L, the revolution time T_0 and the revolution frequency ω_0 are changed:

$$\frac{\Delta L}{L} = \alpha\frac{\Delta p}{p}\ , \ \frac{\Delta\omega_0}{\omega_0} = -\frac{\Delta T_0}{T_0} = -\left(\alpha - \frac{1}{\gamma^2}\right)\frac{\Delta p}{p} = -\eta\frac{\Delta p}{p} \tag{51}$$

with $\eta = \alpha 1/\gamma^2$. There is a transition energy $E_T = m_0c^2\gamma_T$ with $\gamma_T = 1/\alpha^2$ for which the dependence of the revolution frequency on momentum (or energy) changes sign:

$$E > E_T \rightarrow \frac{1}{\gamma^2} < \alpha \rightarrow \eta > 1 \rightarrow \omega_0 \text{ decreases with } \Delta E \tag{52}$$

$$E < E_T \rightarrow \frac{1}{\gamma^2} > \alpha \rightarrow \eta < 1 \rightarrow \omega_0 \text{ increases with } \Delta E\,. \tag{53}$$

Many machines work far above transition energy and $\eta \approx \alpha$. We will assume that the particles are ultra relativistic in which case $\Delta p/p \approx \Delta E/E = \epsilon$. In the presence of an RF system, of synchrotron radiation loss U_s and of the loss due to impedance U, a circulating particle has an energy gain δE per turn

$$\delta E = e\hat{V}\sin(h\omega_0(t_s+\tau) - U_s(E) - U(t) \tag{54}$$

with t_s being the synchronous arrival time of the particle in the cavity and τ the deviation from it, and U_s the energy loss per turn due to synchrotron radiation. Introducing the synchronous phase angle $\phi_s = h\omega_0 t_s$ and assuming the $\tau \ll T_0$, Fig. 8, we get

$$\delta E \approx e\hat{V}\left[\sin(\phi_s) + \cos(\phi_s)h\omega_0\tau\right] - U_0 - \frac{dU_s}{dE}\Delta E - \frac{dU}{dt}\tau. \tag{55}$$

The energy gain per turn is very small $\delta E \ll E$ and we can make a smooth approximation

$$\frac{\Delta E}{E} = \dot{\epsilon}T_0 = \dot{\epsilon}\frac{2\pi}{\omega_0} \tag{56}$$

$$\dot{\epsilon} = \frac{e\hat{V}\sin\phi_s\omega_0}{2\pi E} + \frac{\omega_0^2 he\hat{V}\cos\phi_s}{2\pi E}\tau - \frac{\omega_0}{2\pi}\frac{U_0}{E} - \frac{\omega_0}{2\pi}\frac{dU_s}{dE}\epsilon - \frac{1}{E}\frac{\omega_0}{2\pi}\frac{dU}{dt}\tau. \tag{57}$$

To have equilibrium for the synchronous particle $\epsilon = 0$, $\tau = 0$ we have $U_0 = e\hat{V}\sin\phi_s$. Using $\dot{\tau} = \frac{\omega_0}{2\pi}\Delta T_0 = \eta\epsilon$ we get

$$\begin{aligned}\dot{\epsilon} &= \omega_0^2\frac{he\hat{V}\cos\phi_s}{2\pi E}\tau - \frac{\omega_0}{2\pi}\frac{dU_s}{dE}\epsilon - \frac{1}{E}\frac{\omega_0}{2\pi}\frac{dU}{dt}\tau \\ \dot{\tau} &= \eta\epsilon.\end{aligned}$$

These two first-order differential equations can be combined into a second-order one

$$\ddot{\epsilon} + \frac{dU_s}{dE}\frac{\omega_0}{2\pi}\dot{\epsilon} - \frac{\omega_0^2 h\eta e\hat{V}\cos\phi_s}{2\pi E}\epsilon - \frac{\eta}{E}\frac{\omega_0}{2\pi}\frac{dU}{dt}\epsilon = 0 \tag{58}$$

which is the equation of a damped oscillation. Using

$$\omega_{s0}^2 = -\omega_0^2\frac{h\eta e\hat{V}\cos\phi_s}{2\pi E}, \quad \alpha_s = \frac{1}{2}\frac{\omega_0}{2\pi}\frac{dU_s}{dE}, \tag{59}$$

seeking a solution of the form $\exp(j\omega t)$ and assuming $\alpha_s \ll \omega_{s0}$ we get

$$-\omega^2 + j\omega\alpha_s + (\omega_{s0}^2 + \frac{\omega_0}{2\pi}\frac{\eta}{E}\frac{dU}{dt}) = 0 \tag{60}$$

$$\omega = j\alpha_s \pm \sqrt{(\omega_{s0}^2 + \frac{\omega_0}{2\pi}\frac{\eta}{E}\frac{dU}{dt}) - \alpha_s^2} \approx j\alpha_s \pm (\omega_{s0} + \frac{1}{2}\frac{\omega_0}{2\pi}\frac{\eta}{\omega_{s0}E}\frac{dU}{dt}). \tag{61}$$

Calling

$$\Delta\omega_i = \frac{1}{2}\frac{\omega_0}{2\pi}\frac{\eta}{\omega_{s0}E}\frac{dU}{dt} \tag{62}$$

gives

$$\epsilon = A\left(e^{(-\alpha_s + j(\omega_{s0}+\Delta\omega_i)t} + Be^{(-\alpha_s - j(\omega_{s0}+\Delta_i)t}\right). \tag{63}$$

For the initial conditions $\epsilon(t) = \hat{\epsilon}$, $\dot{\epsilon}(0) = -\alpha_s\hat{\epsilon}$ we get $A = B = \hat{\epsilon}/2$ and

$$\epsilon(t) = \hat{\epsilon}e^{-\alpha_s t}\cos((\omega_{s0} + \Delta\omega_i)t). \tag{64}$$

In order to get a stable oscillation we need

$$E > E_T \rightarrow \cos\phi_s < 0 \;,\; E < E_T \rightarrow \cos\phi_s > 0. \tag{65}$$

4.2 Qualitative treatment of the Robinson instability

The most important effect of the interaction between a longitudinally oscillating bunch and a cavity is the so-called Robinson instability [1] which is treated here in some detail since it can be generalized to describe all multi-turn instabilities in storage rings. We start with a qualitative treatment of the Robinson instability by considering a single bunch circulating in a storage ring and exciting a cavity resonance with resonance frequency ω_r and impedance $Z(\omega)$ of which we consider only the resistive part Z_r.

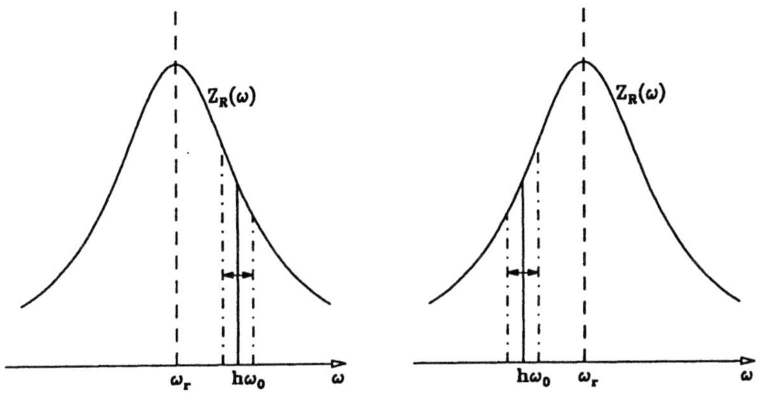

Figure 9: Qualitative treatment of the Robinson instability

The revolution frequency ω_s of the circulating bunch depends on its energy deviation ΔE

$$\frac{\Delta\omega_0}{\omega_0} = -\alpha\frac{\Delta E}{E} \text{ or } \omega_0 = p\omega_{0c}\left(1 - \alpha\frac{\Delta E}{E}\right). \tag{66}$$

While the bunch is executing a coherent dipole mode oscillation $\epsilon(t) = \hat{\epsilon}\cos(\omega_s t)$ its revolution frequency is modulated. *Above transition* the revolution frequency ω_0 is *small* when the *energy is high* and ω_0 is *large* when the *energy is small.* If the cavity is tuned to a resonant frequency being slightly smaller than the RF frequency $\omega_r < p\omega_0$, Fig. 9 left, the bunch sees a higher impedance and *loses more energy* when it has an *energy excess* and it *loses less energy* when it has a *lack of energy.* This leads to a *damping* of the oscillation. If $\omega_r > p\omega_0$ this is reversed, Fig. 9 right, and leads to an *instability.* Below transition energy the dependence of the revolution frequency is reversed, which changes the stability criterion.

4.3 Semi-quantitative treatment of the Robinson instability

We consider a narrow band cavity with a circulating bunch as before. The bunch executes a synchrotron oscillation which is approximately described as $\tau = \hat{\tau}\cos(\omega_s t)$. This will produce sidebands to the revolution frequency harmonics of the bunch. Taking now only the revolution harmonic h, i.e. the RF frequency, we have

$$\begin{aligned} I_k(t) &= I_h\cos(h\omega_0 t - h\omega_0\hat{\tau}\cos(\omega_s t)) \\ &\approx I_h\left(\cos(h\omega_0 t) + \sin(h\omega_0 t)h\omega_0\sin(h\omega_s t)\right) \end{aligned}$$

for $\hat{\tau} \ll T_0$, or

$$I_k(t) = I_h\left[\cos(h\omega_0 t) + \frac{h\omega_0\hat{\tau}}{2}\left(\sin((h\omega_0+\omega_s)t) + \sin((h\omega_0-\omega_s)t)\right)\right]. \tag{67}$$

We calculate the voltage induced by the bunch in the cavity and take *only the resistive part* of the impedance. Using

$$Z^+ = Z(h\omega_0 + \omega_s)\ ,\ Z^- = Z(h\omega_0 - \omega_s)\ ,\ Z^0 = Z(h\omega_0) \tag{68}$$

we get

$$\begin{aligned}
V_k(t) &= I_h\,[\,Z_r^0 \cos(h\omega_0 t) \\
&+ \frac{h\omega_0}{2}\left(Z_r^+ \sin((h\omega_0+\omega_s)t) + Z_r^- \sin((h\omega_0-\omega_s)t)\right)\,] \\
&= I_h\,[\,Z_r^0 \cos(h\omega_0 t) \\
&+ \frac{h\omega_0\hat{\tau}}{2}\,(\,Z_r^+(\sin(h\omega_0 t)\cos(\omega_s t) + \cos(h\omega_0 t)\sin(\omega_s t)) \\
&+ Z_r^-(\sin(h\omega_0 t)\cos(\omega_s t) - \cos(h\omega_0 t)\sin(\omega_s t))\,)\,]\,.
\end{aligned}$$

We use

$$\tau = \hat{\tau}\cos(\omega_s t)\ ,\ \dot{\tau} = -\omega_s\hat{\tau}\sin(\omega_s t). \tag{69}$$

so that

$$\begin{aligned}
V_k(t) &= I_h\,[\,Z_r^0 \cos(h\omega_0 t) \\
&+ \frac{h\omega_0}{2}\left(Z_r^+(\sin(h\omega_0 t)\tau - \cos(h\omega_0 t)\frac{\dot{\tau}}{\omega_s}\right) \\
&+ \frac{h\omega_0}{2}\left(Z_r^-(\sin(h\omega_0 t)\tau + \cos(h\omega_0 t)\frac{\dot{\tau}}{\omega_s}\right)\,]\,.
\end{aligned}$$

We calculate now the energy loss per turn

$$W_b = \int_0^{T_0} V_k(t) I_k(t)\,dt = \frac{I_h^2 T_0}{2}\left[Z_r^0 - \frac{h\omega_0}{2}\frac{\dot{\tau}}{\omega_s}(Z^+ - Z^-)\right]. \tag{70}$$

The average loss per particle is

$$U = \frac{W_b}{N} = W_b\frac{e\omega_0}{2\pi I_0} = \frac{eI_h^2 Z_r^0}{2} - \frac{eI_h^2 h\omega_0}{4I_0\omega_s}(Z_r^+ - Z_r^-)\dot{\tau} \tag{71}$$

and with the relation $\dot{\tau} = \eta\epsilon$ we get

$$U_0 = \frac{eI_h^2 Z_r^0}{2} \text{ and } \frac{dU}{dE} = \frac{eh\eta\omega_0 I_h^2(Z_r^+ - Z_r^-)}{4I_0 E\omega_s}. \tag{72}$$

With

$$\epsilon = \hat{\epsilon}e^{-\alpha_s t}\cos(\omega_s t + \phi)\ ,\alpha_s = \frac{1}{2}\frac{\omega_0}{2\pi}\frac{dU}{d\epsilon} \tag{73}$$

we get for the damping or growth rate of the Robinson instability

$$\alpha_s = \frac{\omega_0^2 h\eta e I_h^2(Z_r^+ - Z_r^-)}{16\pi\omega_s I_0 E} = \frac{\omega_s I_h^2(Z_r^+ - Z_r^-)}{8I_0\hat{V}\cos(\phi_s)} \tag{74}$$

where we used

$$\omega_{s0}^2 = -\omega_0^2\frac{h\eta e\hat{V}\cos\phi_s}{2\pi E}. \tag{75}$$

Since the bunch length is usually much shorter than the RF wavelength we have $I_h \approx 2I_0$

$$\alpha_s \approx \frac{\omega_s I_0(Z_r^+ - Z_r^-)}{2\hat{V}\cos(\phi_s)}. \tag{76}$$

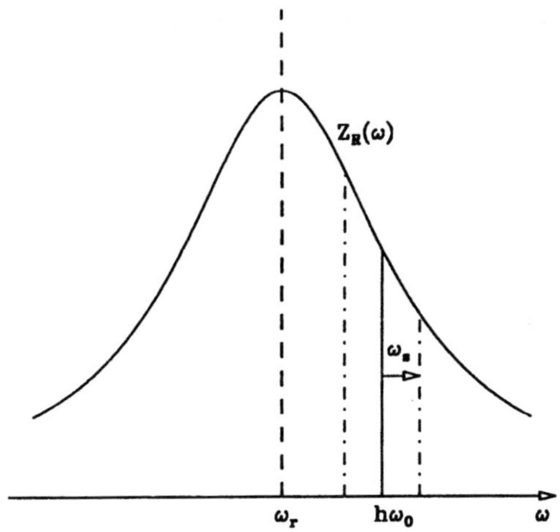

Figure 10: Semi-quantitative treatment of the Robinson instability

The growth rate of the Robinson instability is given by the difference of the resistive impedance at the upper and lower synchrotron side band. Above transition energy we have $\cos\phi_s < 0$ and $\alpha_s > 0$, i.e. stability if $Z_r^- > Z_r^+$ as we found already from qualitative arguments, Fig. 10. We neglected here the effect of the reactive part of the impedance. It leads to a shift of the synchrotron frequency [6].

4.4 Discussion and generalization of the Robinson instability

We have derived the Robinson instability for the case of a single bunch and a single, narrow band and relatively weak resonance. We will here briefcly discuss the way this instability can be extended to cover more general cases.

A more general impedance will cover not just a single revolution harmonic with the two synchrotron oscillation sidebands but many such frequency lines. In this case the voltage induced in the impedance by each such line contained in the spactrum of the bunch current has to be considered. The growth rate will no longer be given by the difference between the impedance at the upper and lower synchrotron side band but between the sums of the impedance times the spectral power taken at each upper and each lower synchrotron side band contained in the spectrum of the oscillating bunch.

This can be extended to the case of many bunches [2, 3]. With M equidistant bunches in the machine we have M different modes of oscillation each having a different phase between the oscillations executed by adjacient bunches. The spectrum of each such coupled bunch mode has synchrotron side bands at distinct revolution harmonics. In calculating the stability of a certain coupled bunch mode we have to sum over these side bands.

So far we considered only dipole oscillations where the bunch makes a rigid oscillation around the nominal phase without changing the form. There are higher modes of oscillation, called bunch shape oscillations, which can be classified as quadrupole ($m = 2$), sextupole ($m = 3$), octupole ($m = 4$), etc. modes. Each mode has a spectrum with side bands at $m\omega_s$ from the revolution harmonics. Again, to calculate the stability of these modes we have to sum over these side bands.

We have so far assumed that the effect of the impedance is relatively weak such that the changes of the synchrotron frequency and the growth rate of the instability are small compared to the synchrotron frequency itself. For very narrow band cavities with high shunt impedance, e.g. superconducting cavities, this might no longer be true. In this case we have to evaluate the impedance not at the unperturbed side band ω_{s0} but at the shiftet synchrotron frequency ω_s. Furthermore, if we are interested in the growth rate we have to consider the cavity impedance for a growing oscillation which is different as soon as the growth time of the oscillation becomes comparable to the filling time of the cavity. Taking all this into account one arrives at a 4th order equation for the shifted synchrotron frequency and the growth rate for which a more general stability criterion can be derived, often called the second Robinson instability [1].

We have so far considered stability for the case of infinitesimally small oscillations and calculated their growth or damping time. If the oscillation amplitude becomes large some non-linear effects should be considered. The modulation index of the phase oscillation will become large leading to side bands at twice the synchrotron frequency. They have to be included when forming the sum over the impedance contribution. This can lead to a situation where the beam is unstable for small oscillation amplitudes but becomes stable again at large amplitudes. Such cases represent themselves in practice as bunches oscillating with finite but more or less constant amplitudes [4, 5].

5 Transverse Stability of a Coasting Beam and Landau Damping

5.1 Introduction

We investigate now the transverse stability of a coasting, i.e. an unbunched, beam which has no longitudinal time structure. Such beams can only exist over longer time for particles and conditions where the energy loss due to synchrotron radiation is negligible. The application of the following treatment is therefore somewhat limited but, on the other hand, it is relatively simple and tranparent. We use the case of the transverse stability of such a coasting beam mainly to explain the concept of Landau damping which is much more complicated for bunched beams. We follow here basically the method used in earlier Accelerator Schools [6, 7].

5.2 Transverse oscillation modes of a coasting beam

We consider now a coasting (unbunched) beam circulating without energy loss or gain in a storage ring. We assume that this beam has a central momentum p_0 with a corresponding revolution frequency ω_{0c}. The particles themselves have a distribution in momentum with deviations Δp and

$$\Delta\omega_0 = -\omega_{0c}\eta\frac{\Delta p}{p} \tag{77}$$

from the central values with distribution function

$$F_p(\Delta p) = \frac{1}{N}\frac{dN}{dp}, \tag{78}$$

where $\eta = \alpha - 1/\gamma^2$, $\alpha =$ momentum compaction and N is the number of particles in the beam.

The beam executes at the same time a vertical betatron oscillation such that the motion of each particle consists of a rotation $\theta = \omega_0 t$, and a vertical oscillation $y = \hat{y}\cos(Q\omega_0 t)$, see Fig. 6. This beam as a whole can execute such a motion in a very large number of modes as long as we do not fix a phase relation between the betatron oscillation of the individual particles. We are, of course, interested in modes which are very simple, e.g. where all the particles go up and down together or with a simple phase relation. Such simple modes contain only relatively low frequencies and are more likely to interact with realistic impedances of the beam surroundings. We classify the modes as closed waves with a different number n of undulations around the circumference. They can be described as

$$y_n = \hat{y}\cos(n\theta - \omega t)\,,\ \ or\ \ y_n = \hat{y}e^{i(n\theta-\omega t)}. \tag{79}$$

For $n = 0$ all particles move together. If the motion is frozen by fixing the time t, the mode number n gives the number of waves around the ring ($n = 4$ for Fig. 11).

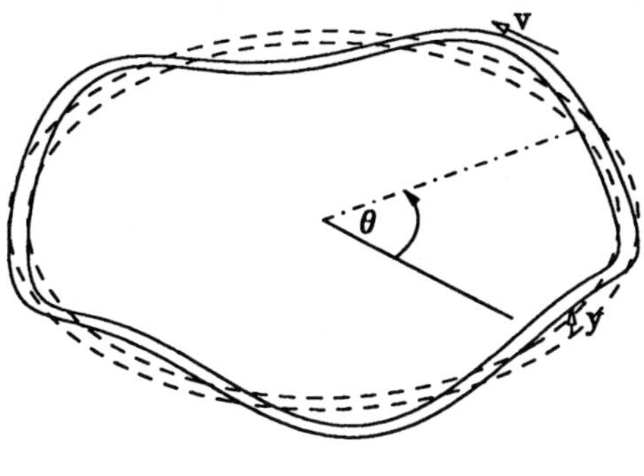

Figure 11: Oscillation modes of a coasting beam

The mode number n can take positive or negative values depending on a phase advance or a phase lag of the oscillation executed by a particle in front of a reference particle. These two classes of modes are often called fast or forward, and slow or backward, waves. We now wish to determine the frequency ω which is seen by a stationary observer at a fixed location θ. We know that the vertical oscillation frequency is $Q\omega_0$ for an individual particle. Such a particle executes the azimuthal motion $\theta = \theta_0 + \omega_0 t$. This gives

$$e^{i(n\theta_0 + n\omega_0 t - \omega t)} = e^{n\theta_0 - iQ\omega_0 t} \tag{80}$$

or

$$\omega = (n + Q)\omega_0 = \omega_\beta, \tag{81}$$

where n goes through positive and negative integers. It is convenient to divide these modes into fast waves

$$\omega_{\beta f} = (n_f + Q)\omega_0\,, \quad n_f > -Q \tag{82}$$

and slow waves

$$\omega_{\beta s} = (n_s - Q)\omega_0\,, \quad n_s > Q. \tag{83}$$

The spread in momentum of the particles in the beam results in a spread in the betatron frequency ω_β through two effects. The revolution frequency ω_0 depends on Δp

$$\Delta\omega_0 = -\omega_0\eta\frac{\Delta p}{p}\,, \quad \eta = \alpha - \frac{1}{\gamma^2}. \tag{84}$$

Furthermore the betatron tune Q depends on the momentum through the chromaticity ξ

$$\Delta Q = \xi Q\frac{\Delta p}{p} = Q'\frac{\Delta p}{p}\,, \quad \xi = \frac{\Delta Q/Q}{\Delta p/p}. \tag{85}$$

This leads to a spread in betatron frequency observed at a stationary location

$$\Delta\omega_{\beta f} = (Q' - \eta(n_f + Q))\omega_0\frac{\Delta p}{p}\,, \quad \Delta\omega_{\beta s} = (Q' - \eta(n_s - Q))\omega_0\frac{\Delta p}{p}. \tag{86}$$

5.3 Response of the beam to a transverse excitation

We consider now an experiment where a periodic transverse force is applied to the beam by an electromagnetic device of length L_G. This force results in an acceleration G averaged over one revolution

$$G = \frac{F}{m} = \hat{G}e^{-i\omega t} = \frac{e[\vec{E} + [\vec{\beta}c \times \vec{B}]]_T}{m_0\gamma}\frac{L_G}{2\pi R}. \tag{87}$$

The equation of motion for a particle with betatron wave number Q and revolution frequency ω is

$$\frac{d^2y}{dt^2} + \omega_0^2Q^2y = \hat{G}e^{-i\omega t}. \tag{88}$$

We are seeking a solution where a string of particles appear to a stationary observer as

$$y = \hat{y}e^{i(n\theta - \omega t)}. \tag{89}$$

Using this and the relation between absolute and partial differentiation we get

$$\frac{dy}{dt} = \frac{\partial y}{\partial t} + \frac{\partial y}{\partial \theta}\dot{\theta} = \frac{\partial y}{\partial t} + \frac{\partial y}{\partial \theta}\omega_0 = i(n\omega_0 - \omega)y \tag{90}$$

and the equation of motion becomes

$$[-(n\omega_0 - \omega)^2 + \omega_o^2Q|s^2]\hat{y}e^{i(n\theta-\omega t)} = \hat{G}e^{-i\omega t}. \tag{91}$$

We solve this for $\hat{y}$ and assume that $\theta = 0$ which means that the oscillation is observed close to the exciter

$$\hat{y} = \frac{\hat{G}}{\omega_0^2Q^2 - (n\omega_0 - \omega)^2} = \frac{-G_0}{(\omega - \omega_0(n+Q))(\omega - \omega_0(n-Q))} \tag{92}$$

$$\frac{\hat{y}}{\hat{G}} = \frac{-1}{(\omega_{\beta f} - \omega)(\omega_{\beta s} - \omega)}. \tag{93}$$

This is the response of a single particle. It is only large if the exciting frequency ω is either close to the frequency $\omega_{\beta s}$ of the fast, or to the frequency $\omega_{\beta s}$ of the slow, wave. We will approximate for these two cases and get for the single particle responses

$$\left(\frac{\hat{y}}{\hat{G}}\right)_f \sim \frac{-1}{2Q\omega_0}\frac{1}{(\omega_{\beta f} - \omega)}, \quad \left(\frac{\hat{y}}{\hat{G}}\right)_s \sim \frac{1}{2Q\omega_0}\frac{1}{(\omega_{\beta s} - \omega)}. \tag{94}$$

This approximation takes only the response to the mode n (being closest in frequency) while the contributions due to the other modes are ignored [8]. In this form the single particle response is the same as that for the case of a set of oscillators used earlier. Therefore, we can easily find the response for the centre of charge for the fast wave case by an integration over the single particle response weighted with the distribution $f(\omega_{\beta f})$ which is normalized such that $\int f(\omega_{\beta f})d\omega_{\beta f} - 2\pi$

$$<\hat{y}>_f = \int f(\omega_{\beta f})\hat{y}(\omega_{\beta f})d\omega_{\beta f} = -\frac{\hat{G}}{2Q\omega_0}\int\frac{f(\omega_{\beta f})}{\omega_{\beta f} - \omega}d\omega_{\beta f} = \frac{\hat{G}}{2Q\omega_0}\left[PV\int\frac{f(\omega_{\beta f})}{\omega_{\beta f} - \omega}d\omega_{\beta f} \pm i\pi f(\omega)\right]. \tag{95}$$

This integral goes over a pole at $\omega = \omega_{\beta f}$ where the denominator vanishes. Such an integral has two parts, a principle value "PV" which is obtained by excluding the pole with a small gap and an imaginary residue the sign of which is ambiguous since we did not speczify the initial conditions. The physical meaning of the two parts can be made more clear by calculating the beam response in velocity $<\dot{y}>$ of the centre-of-mass motion

$$<\dot{y}>_s = -i\omega <y> = -\frac{\hat{G}\omega e^{-i\omega t}}{2Q\omega_0}[\pi f(\omega) - iPV\int\frac{f(\omega_{\beta f})}{\omega_{\beta f} - \omega}d\omega_{\beta f}]. \tag{96}$$

The corresponding expression for the slow wave is obtained by changing the sign and replacing $\omega_{\beta f}$ by $\omega_{\beta s}$. In both cases the limits of the integration are such that only the side band in question is covered. This is the response of a coasting beam to a transverse harmonic excitation with a frequency ω being close to the one of a betatron side band ω_β. In our approximation the integration has to cover just such a single side band and we neglect the influence of other side bands. This response has a real part which is — apart from a common factor — equal to $\pi f(\omega)$. The fact that this part is real means that for this term the velocity is *in phase* with the exciting acceleration and can absorb energy. It can therefore lead to damping of the centre-of-mass oscillation. The second term with the principle value integral is imaginary which means that the velocity is out of phase compared to the acceleration and no energy is exchanged. For this reason the two terms are also called resistive and reactive response. The resistive term can lead to a damping called Landau damping. This effect rests essentially on the existance of the residue. Since this is not very transparent we will derive the beam response again using real notation only but specifying the initial conditions clearly.

5.4 Time evolution of the response using real notation

We saw that the resistive part of the beam response to a harmonic excitation is due to the residue of the integral. Since it is this term which leads to Landau damping we would like to better understand the underlying physics. For this purpose we now discuss the time evolution of the response using real notation [6]. This leads to derivations which are more lengthy but more transparent. The differential equation describing the excitation of the beam is, in analogy with Eq. (96),

$$\frac{d^2y}{dt^2} + \omega_\beta^2 y = \hat{G}\sin(\omega t). \tag{97}$$

This equation has a homogeneous solution

$$y_h = A\sin(\omega_\beta t) + B\cos(\omega_\beta t) \tag{98}$$

and a particular solution

$$y_p = \frac{\hat{G}}{\omega_\beta^2 - \omega^2}\sin(\omega t). \tag{99}$$

The general solution is a combination of the two,

$$y = A\sin(\omega_\beta t) + B\cos(\omega_\beta t) + \frac{\hat{G}}{\omega_\beta^2 - \omega 2}\sin(\omega t). \tag{100}$$

To determine the integration constants A and B we have to specify the initial conditions. We take the case where the oscillators are at rest until the time $t = 0$ when the excitation starts,

$$y(0) = \dot{y}(0) = 0 \rightarrow A = -\frac{\omega}{\omega_\beta}\frac{\hat{G}}{(\omega_\beta^2 - \omega 2)}\ ; \quad B = 0. \tag{101}$$

This gives for the general solution

$$y = \frac{\hat{G}}{\omega_\beta 2 - \omega 2}\left(\sin(\omega t) - \frac{\omega}{\omega_\beta}\sin(\omega_\beta t)\right) = \frac{\hat{G}}{2\omega}\left(\frac{1}{(\omega_\beta - \omega)} - \frac{1}{\omega_\beta + \omega}\right)\left(\sin(\omega t) - \frac{\omega}{\omega_\beta}\sin(\omega_\beta t)\right). \tag{102}$$

We differentiate with respect to t to get the velocity of the single oscillator response

$$\dot{y} = \frac{\hat{G}}{2}\left(\frac{\cos(\omega t) - \cos(\omega_\beta t)}{(\omega_\beta - \omega)} - \frac{\cos(\omega t) - \cos(\omega_\beta t)}{\omega_\beta + \omega}\right) \tag{103}$$

and rewrite this equation by substituting ω_β with $\omega_\beta = \omega + (\omega_\beta - \omega)$ in the first term and $\omega_\beta = (\omega_\beta + \omega) - \omega$ in the second term of the parenthesis,

$$\begin{aligned}\dot{y} &= \frac{\hat{G}}{2}\left[\cos(\omega t)\left(\frac{1-\cos((\omega_\beta-\omega)t)}{(\omega_\beta-\omega)} - \frac{1-\cos((\omega_\beta+\omega)t)}{\omega_\beta+\omega}\right)\right. \\ &+ \left.\sin(\omega t)\left(\frac{\sin((\omega_\beta-\omega)t)}{(\omega_\beta-\omega)} + \frac{\sin((\omega_\beta+\omega)t)}{\omega_\beta+\omega}\right)\right]. \end{aligned} \tag{104}$$

The above equation gives the velocity response for a single oscillator with resonant frequency ω_β. For the first term in the square bracket velocity and acceleration are basically out of phase while for the second term they are in phase. This statement is not exact since both terms also contain an oscillatory term with frequency $(\omega_\beta - \omega)$ which has to be discussed further. Before we integrate to obtain the centre-of-mass response we discuss these two components of the single oscillator response Eq. (104). We concentrate on a region in the vicinity of the exciting positive frequency ω, i.e. $\omega_\beta \approx \omega$ where the first term inside each of the two round brackets is dominant. A corresponding discussion for $\omega_\beta \approx -\omega$ could easily be carried out in addition. In Fig. 12 the envelope of the oscillation executed is shown as a function of the difference $(\omega_\beta - \omega)$ between the resonant and the exciting frequency for different times t after the start of the excitation. As this time increases particles oscillating with opposite phase are close together in frequency leading to some cancellation in the integration to follow. For the resistive (in phase) term the oscillators with resonant frequency close to the exciting frequency gain large amplitudes.

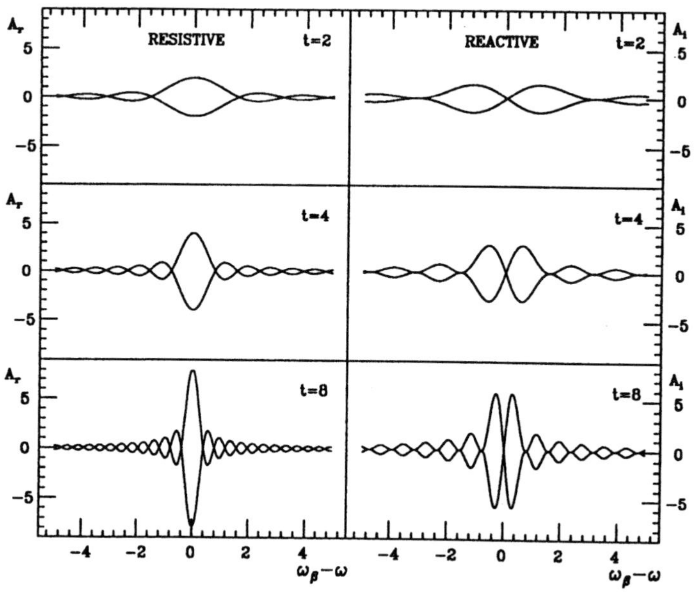

Figure 12: Time evolution of the beam response to a harmonic excitation

We calculate the centre-of-mass response by integrating the single-particle response Eq. (104) over the resonant frequency ω_β weighted with the distribution $f(\omega_\beta)$. Using again the symmetry property of the distribution $f(\omega_\beta) = f(-\omega_\beta)$ we can show that

$$\int_{-\infty}^{\infty} \frac{1-\cos((\omega_\beta-\omega)t)}{(\omega_\beta-\omega)} d\omega_\beta = -\int_{-\infty}^{\infty} \frac{1-\cos(\omega_\beta+\omega t)}{\omega_\beta+\omega} d\omega_\beta \tag{105}$$

and

$$\int_{-\infty}^{\infty} \frac{\sin((\omega_\beta - \omega)t)}{(\omega_\beta - \omega)} d\omega_\beta = \int_{-\infty}^{\infty} \frac{\sin(\omega_\beta + \omega t)}{\omega_\beta + \omega} d\omega_\beta \tag{106}$$

which gives for the velocity response

$$< \dot{y} >= \frac{\hat{G}\cos(\omega t)}{2\pi} \int_{-\infty}^{\infty} f(\omega_\beta) \frac{1 - \cos((\omega_\beta - \omega)t)}{\omega_\beta - \omega} d\omega_\beta + \frac{\hat{G}\sin(\omega t)}{2\pi} \int_{-\infty}^{\infty} f(\omega_\beta) \frac{\sin((\omega_\beta - \omega)t)}{\omega_\beta - \omega} d\omega_\beta. \tag{107}$$

We start with the out of phase (reactive) part which has an oscillatory term of the form $(\cos((\omega_\beta - \omega)t))$. As the time t increases this term will have opposite phase for smaller and smaller frequency differences in ω_β. The integration over ω_β will therefore vanish and we can replace the denominator of the reactive term by unity $(1 - \cos((\omega_\beta - \omega)t) \rightarrow 1$ except for the oscillators with $\omega_\beta \approx \omega$. This central part $\omega_\beta \approx \omega$ becomes more and more narrow as the time t increases and we can replace the integral over the reactive term by the principle value integral Eq. (95)

$$\lim_{t\to\infty} \int f(\omega_\beta) \frac{1 - \cos((\omega_\beta - \omega)t)}{\omega_\beta - \omega} d\omega_\beta = \lim_{\epsilon\to 0} \left(\int_{-\infty}^{-\epsilon} + \int_{\epsilon}^{\infty} \right) = PV \int f(\omega_\beta) \frac{1 - \cos((\omega_\beta - \omega)t)}{\omega_\beta - \omega} d\omega_\beta.. \tag{108}$$

The resistive (in phase) term of Eq. (107) contains an oscillatory term under the integral of the form

$$\frac{\sin((\omega_\beta - \omega)t)}{\omega_\beta - \omega} = t \frac{\sin((\omega_\beta - \omega)t)}{(\omega_\beta - \omega)t} = t \operatorname{sinc}((\omega_\beta - \omega)t). \tag{109}$$

For $\omega_\beta \neq \omega$ and large t the above expression oscillates with opposite phase for small changes in ω_β. The integration over this frequency will vanish as long as the distribution $f(\omega_\beta)$ is sufficiently smooth. For $\omega_\beta \approx \omega$ and very large t the function $\operatorname{sinc}((\omega_\beta - \omega)t)$ will be about unity and the above expression grows with increasing time t without limits. Furthermore the integral [9]

$$\int_{-\infty}^{\infty} \frac{\sin((\omega_\beta - \omega)t)}{(\omega_\beta - \omega)} d\omega_\beta = \pi \tag{110}$$

is independent of t. We can therefore replace Eq. (109) by the δ-function

$$\lim_{t\to\infty} \frac{\sin((\omega_\beta - \omega)t)}{(\omega_\beta - \omega)} = \pi\delta(\omega_\beta - \omega). \tag{111}$$

Collecting the results obtained for the reactive and resistive part of the beam response we get

$$\lim_{t\to\infty} < \dot{y} >= \frac{\hat{G}}{2\pi} \left[\sin(\omega t)\pi f(\omega) + \cos(\omega t) PV \int \frac{f(\omega_\beta)}{(\omega_\beta - \omega)} d\omega_\beta \right]. \tag{112}$$

This is the same result as that already obtained more quickly using complex notation. However, in this subsection we learned basically three things:

a) The result Eq. (96) is only correct if the excitation has lasted for a long time. How long this time has to be depends on the resolution with which the distribution $f(\omega_\beta)$ has to be considered. If this distribution does not change significantly over a frequency range of $\Delta\omega_\beta$ it is sufficient to excite for a time $t \gg 1/\Delta\omega_\beta$.

b) The sign of the residue can be determined from the initial conditions. Usually one excites a set of oscillators being initially at rest. However, it is in principle possible to have a set of particles oscillating initially with a particular distribution in amplitude and phase such that the exciting acceleration takes energy out of the beam. Since in the complex notation the initial conditions were not specified, both possibilities are contained in the beam response equation.

c) From Fig. 1 it is clear that a few oscillators, having resonant frequencies close to the exciting frequency $\omega_\beta \approx \omega$, attain large amplitudes. The energy absorbed by the beam from the exciter goes, therefore, into large oscillation amplitudes obtained by a small fraction of the oscillators.

5.5 Transverse impedance

It is well known that a beam can excite longitudinal modes in a cavity which then react onto the beam. It is also possible that the beam excites so-called deflecting modes which give a transverse force to the beam. A simple case is illustrated in Fig. 13 where a bunch is going through a cavity with a displacement y from the axis. This can excite a mode having a longitudinal electric field which increases with distance from the cavity axis. A quarter of an oscillation later this mode has a transverse magnetic field which can deflect particles. The transverse impedance is defined as the integrated deflecting field per unit dipole moment of the exciting current

$$Z_T(\omega) = -i\frac{\int_0^{2\pi R}[\vec{E}(\omega) + [\vec{\beta}c \times \vec{B}(\omega)]]_T ds}{Iy(\omega)}. \tag{113}$$

The right hand side is multiplied with i which indicates that the driving dipole moment Iy is out of phase with the deflecting field. The above expression is matched to a complex description of an oscillation with $e^{-i\omega t}$. If one uses the convention $e^{j\omega t}$ instead, i has to be replaced by $-j$. This impedance might become more clear if we relate the fields to the vertical velocity $\dot{y}$ rather than to the position y. With $y = \hat{y}e^{j\omega t}$ we have $\dot{y} = j\omega y$ and

$$Z_T(\omega) = -\omega\frac{\int_0^{2\pi R}[\vec{E}(\omega) + [\vec{\beta}c \times \vec{B}(\omega)]]_T ds}{I\dot{y}(\omega)}. \tag{114}$$

A real transverse impedance means that the transverse deflecting fields are in phase with the transverse velocity and transfer energy to or from the transverse motion of the beam.

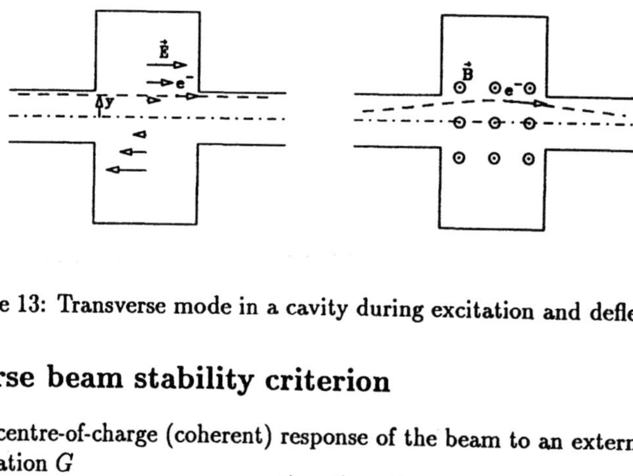

Figure 13: Transverse mode in a cavity during excitation and deflection

5.6 Transverse beam stability criterion

We calculated the centre-of-charge (coherent) response of the beam to an external transverse excitation with acceleration G

$$G = \frac{< e[\vec{E} + [\vec{\beta}c \times \vec{B}]] >}{m_0\gamma} \tag{115}$$

which gives for the fast wave

$$< y >= -\frac{\hat{G}e^{-i\omega t}}{2Q\omega_0}\int\frac{f(\omega_{\beta f})}{\omega_{\beta f} - \omega}d\omega_{\beta f}. \tag{116}$$

If the beam surroundings represent a transverse impedance Z_T,

$$Z_T(\omega) = -i\frac{\int_0^{2\pi R}[\vec{E}(\omega) + [\vec{\beta}c \times \vec{B}(\omega)]]_T ds}{Iy(\omega)}, \tag{117}$$

the oscillating beam will induce fields in it which will act back on the beam. We assume now that the acceleration G is not external but due to the self-fields induced

$$G = \frac{ieZ_T I < y >}{\gamma m_0 2\pi R}. \tag{118}$$

Substituting this in the expression for the beam response we get for the fast and the slow wave

$$1 = -\frac{icIZ_T}{4\pi Q m_0 c^2 \gamma}\int \frac{f(\omega_{\beta f})}{\omega_{\beta f} - \omega} d\omega_{\beta f} \text{ and } 1 = \frac{icIZ_T}{4\pi Q m_0 c^2 \gamma}\int \frac{f(\omega_{\beta s})}{\omega_{\beta s} - \omega} d\omega_{\beta s}. \tag{119}$$

This gives a condition for which the beam is just at the limit of stability [10, 11]. By replacing the external excitation with the self induced one we assume that an oscillation, once started, is just kept going by the self-forces. To make it more applicable we introduce some normalization to separate the term which depends on the different beam parameters from the one which is just given by the form of the distribution. We introduce the half width at half height S of the distribution $f(\omega_\beta)$ and normalize the two frequencies ω and ω_β with it

$$\xi_f = \frac{\omega_{\beta f}}{S}\ \xi_s = \frac{\omega_{\beta s}}{S},\ \xi = \frac{\omega}{S},\ f(\xi_f) = Sf(\omega_{\beta f}),\ f(\xi_s) = Sf(\omega_{\beta s}) \tag{120}$$

and get

$$1 = \frac{iecIZ_T}{4\pi EQS}\int \frac{f(\xi_f)}{\xi - \xi_f} d\xi_f \text{ and } 1 = -\frac{iecIZ_T}{4\pi EQS}\int \frac{f(\xi_s)}{\xi - \xi_s} d\xi_s. \tag{121}$$

This expression allows us to find, for given beam parameters, for each value of the driving frequency ω (or its normalized value ξ), the maximum impedance Z_T which still does not lead to an instability. In this expression the integral and the impedance are complex numbers. It is therefore convenient to visualize the stability criterion by mapping the impedance to the frequency ω or ξ_1. As long as this frequency has no imaginary part there is no growing instability since we expressed the oscillation as $e^{-i\omega t}$. We write the above condition slightly differently

$$\frac{ecIZ_T}{4\pi EQS} = V_T + iU_T = \frac{1}{i\int \frac{f(\xi_f)}{\xi - \xi_f} d\xi_f} \text{ and } \frac{ecIZ_T}{4\pi EQS} = V_T + iU_T = -\frac{1}{i\int \frac{f(\xi_s)}{\xi - \xi_s} d\xi_s}. \tag{122}$$

Plotting the above equality for a real frequency ξ_1 gives the stability diagram. We give in Fig. 14 the example of a Gaussian distribution. The diagram consists of two curves; at the right, one for the slow wave with positive values of the impedance; at the left, one for the fast wave with negative values for the resistive impedance. These curves represent a situation which is at the limit of stability. A slightly larger impedance than the one corresponding to this limit will lead to an instability. For the slow wave this means that values for the complex impedance Z_T giving a reduced impedance $U_T + iV_T$ lying on the right of the stability diagram lead to instability while for corresponding values lying on the left we still have stability. For the fast wave only negative impedances lead to instabilities if the resulting reduced impedance $U_T + iV_T$ lies on the left of the fast wave stability curve. To summarize the situation for both waves, we have stability as long as the reduced impedance $U_T + iV_T$ is inside the stability diagram bounded by the fast and the slow wave stability limit curves.

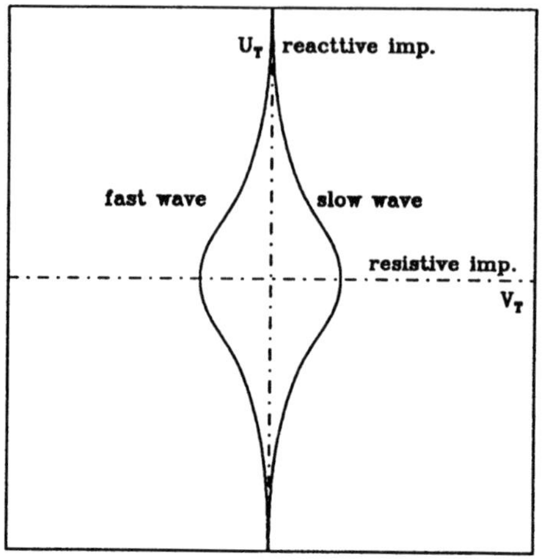

Figure 14: Stability diagram

5.7 Landau damping for bunched beams and for longitudinal instabilities

Without going into any details concerning the Landau damping for bunched beams and for longitudinal oscillations we should make a few remarks on the limitation of the treatment presented here. The Landau damping against transverse instabilities in an unbunched beam which has a betatron frequency spread due to the energy spread combined with sextupoles Eq. (86) is a special case. The betatron oscillation we excite will not influence the frequency spread in the beam. In other words, the frequency ω_β of a particle is given by parameters like Δp, η and Q' which are not influenced by the excitation of betatron oscillations. This situation is different and more complicated in cases where the betatron frequency spread is determined by octupole fields which give a dependence of the betatron frequency on amplitude. This is always the case for a bunched beam but it can also be the dominant effect in unbunched beams. Exciting a betatron oscillation will, at the same time, influence the betatron frequency distribution. Going through this calculation one finds that the integrals, Eq. (95), determining the beam response and the stability diagram do not contain the distribution $f(\omega_\beta)$ but rather its derivative [10-12] The same situation is present for longitudinal stability for unbunched as well as for bunched beams [12-16]

* * *

References

[1] K.W. Robinson, Stability of beam in radiofrequency system, Cambridge Electron Accel. CEAL-1010 (1964).

[2] F. Sacherer, A longitudinal stability criterion for bunched beams, Proc. of the 1977 Particle Accel. Conf., IEEE Trans. on Nucl. Sci. NS 20-3 (1973) 825.

[3] J.L. Laclare, Bunched beam coherent instabilities, CAS CERN Accelerator School, Advanced Accelerator Physics, 1985; ed. S. Turner. CERN 87-03, p. 264.

[4] F. Pedersen, CERN, Private communication.

[5] S. Krinsky, Saturation of a longitudinal instability due to nonlinearity of the wake field, Proc. of the 1985 Particle Accel. Conf., IEEE Trans. on Nucl. Sci. NS 32-5 (1985) 2320.

[6] A. Hofmann, Physics of beam instabilities, Proc of a Topical Course 'Frontiers of Particle Beams' held by the Joint US-CERN School on Particle Accel. at South Padre Island, Texas, Oct. 1986, ed. M. Month and S. Turner, Lecture Notes in Physics 296. Springer (1988) 99.

[7] A. Hofmann, Landau damping, CAS CERN Accelerator School, Advanced Accelerator Physics, 1987; ed. S. Turner. CERN 89-01, p. 40.

[8] S. van der Meer, A different formulation of the longitudinal and transverse beam response, CERN/PS/AA/80-4 (1980).

[9] I.S. Gradshteyn, I.M. Ryzhik, Table of Integrals Series and Products, Academic Press 1980, p. 406.

[10] L.J. Laslett, V.K. Neil, A.M. Sessler, Transverse resistive instabilities of intense coasting beams in particle accelerators, Rev. Sci. Instr. 36 No. 4 (1965) 436.

[11] K. Hübner and V. Vaccaro, Dispersion relations and stability of coasting particle beams, CERN/ISR-TH-RF/69-23 (1969).

[12] V.K. Neil, A.M. Sessler, Longitudinal resistive instabilities of intense coasting beams in particle accelerators, Rev. Sci. Instr. 36 No. 4 (1965) 429.

[13] H.G. Hereward, Landau damping by non-linearities, CERN/MPS/DL 69-11 (1969).

[14] D. Boussard, Schottky noise and beam transfer function diagnostics, CAS CERN Accelerator School, Advanced Accelerator Physics, 1985; ed. S. Turner. CERN 87-03, p. 416.

[15] J. Gareyte, Landau damping of the longitudinal quadrupole mode in SPEAR II, SLAC SPEAR-207 (1977).

[16] R. Boni, S. Guiducci, M. Serio, F. Tazzioli, F.H. Wang, Investigations on beam longitudinal transfer function and coupling impedance in ADONE, Frascati LNF-Div. Report, RM-23,1981.

Observation and Correction of Instabilities in Circular Accelerators

J. Gareyte
CERN, Geneva, Switzerland

Abstract : The correction of beam instabilities in circular accelerators requires a good knowledge of some basic concepts which form the basis of a widely accepted Standard Model. These concepts are introduced here in a simple semi-empirical way, and their relevance to the observation and understanding of collective phenomena is illustrated by practical examples.

1. Introduction

The knowledge we have at present of Particle Physics is well described by the Standard Model. Similarly, in the field of Accelerator Physics, a general model of beam instabilities has been progressively built up over the years, which allows us to describe almost every observation. Many individuals contributed to the construction of this model, from the first observations of an electron beam instability in 1962 in the Mk II accelerator of Stanford and the first theory of resistive wall instability in 1965 by Neil and Sessler, to the recent developments on mode-coupling instabilities. A decisive step in the unification of the known phenomena was made around 1974 by F. Sacherer. This illuminating work had a profound influence on the author of this paper, who had the privilege of collaborating closely with F. Sacherer at this time. The aim of this lecture is to show how the Standard Model of Beam Instabilities, as understood by the author, can be used to describe, classify, understand and cure the different collective instabilities that we encounter in circular accelerators.

The only justification of this exercise is of course the correction of these phenomena, which hopefully leads to better performance of the machines. This can be obtained by suppressing the cause of the instability (coupling impedance, phase shift), by active feedback damping or by Landau damping. This very important final step is however trivial once the phenomenon has been properly understood. The most important phase is the process of understanding, which requires good observation skills coupled to a thorough knowledge of the theoretical models. This will be reflected in the organization of this course : emphasis will be put primarily on the description of the coherent modes of oscillation of the beams. The case of the continuous beam will be examined first, because it is simple and allows the introduction of many basic concepts. Then the description will be extended to bunched beams, for which the dynamics is dominated by the RF focusing. As long as the forces due to the beam-environment interaction, which are responsible for the instabilities, are small compared to the RF focusing force, destructive instabilities can arise only from multibunch effects (very similar to continuous beam instabilities, although with some differences) and Head-Tail effects. For strong enough interactions, the RF focusing is no longer able to hold the bunch together, and single-bunch instabilities known as Mode-Coupling, Microwave instability, or Beam Break-Up can arise.

2. Continuous beams

2.1 Transverse modes

Let us examine the case of a continuous beam in a circular machine, as shown in Fig.1. Suppose that at time t = 0 a perturbation is imposed on the beam so that each azimuthal slice is displaced transversely by an amount $y_o(\theta)$. This pattern $y(\theta)$ is necessarily closed around the circumference and therefore it can be decomposed into a Fourier series:

$$y(t=0,\theta) = \sum_{n=-\infty}^{+\infty} y_n e^{-jn\theta} \tag{1}$$

so that it is only necessary to consider the evolution of these sinusoidal waves. The betatron oscillation of a particle with a tune Q and a revolution frequency ω_o, and which was at azimuth θ_o at time t = 0 is given by

$$y_{\theta_o}(t) = y_n e^{j(Q\omega_o t - n\theta_o)} \tag{2}$$

but since for any particle the azimuth θ after a time t is given by $\theta = \theta_o + \omega_o t$ we can describe the position of the whole beam in azimuth and time as follows:

$$y(\theta,t) = y_n e^{j[(Q+n)\omega_o t - n\theta]}. \tag{3}$$

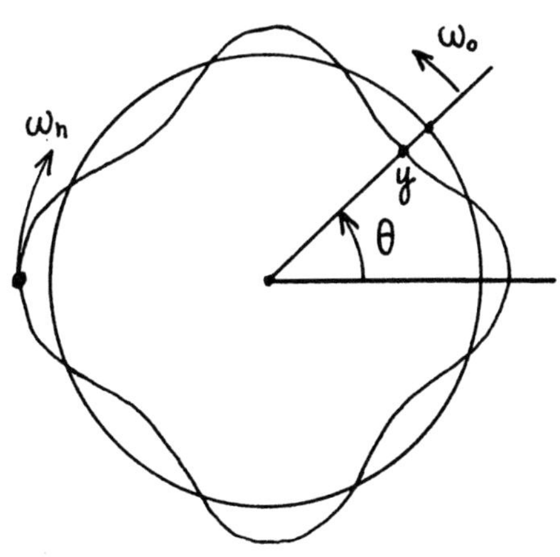

Fig. 1

Transverse mode pattern

With this formula it is easy to see that the initial wave pattern rotates with an angular velocity $\omega_n = (1 + \frac{Q}{n})\omega_0$. Thus different wave patterns rotate at different velocities; for positive values of n the angular velocity of the wave is larger than the velocity of the particle, whereas for negative values of n the wave travels slower than the particles. For this reason the waves associated with positive orbital harmonics are called <u>fast waves</u>, and those associated with negative orbital harmonics are called <u>slow waves</u>. A small subset of the slow waves travel backwards with respect to the particles: these are associated with orbital harmonics of order $n< 0$ such that $|n| < Q$.

Although a snapshot of the whole beam at a given time will reveal exactly the same pattern for waves of opposite harmonic number n and -n these waves are of fundamentally different nature. The slope of the wave at azimuth θ and time t is given by $\partial y/\partial\theta = -jny$, while the slope of the particle trajectory at the same time and azimuth is $dy/dt = jQ\omega_o y$. For $n > 0$ the amplitude of a particle increases although the associated wave has a negative slope at the same azimuth; this is possible because the wave has a phase velocity larger than the particle velocity. For negative values of n the wave has a positive slope and the particle amplitude can increase because the phase velocity of the wave is smaller than the particle velocity (see Fig. 2). This is very reminiscent of the evolutions of a surfer on ocean waves.

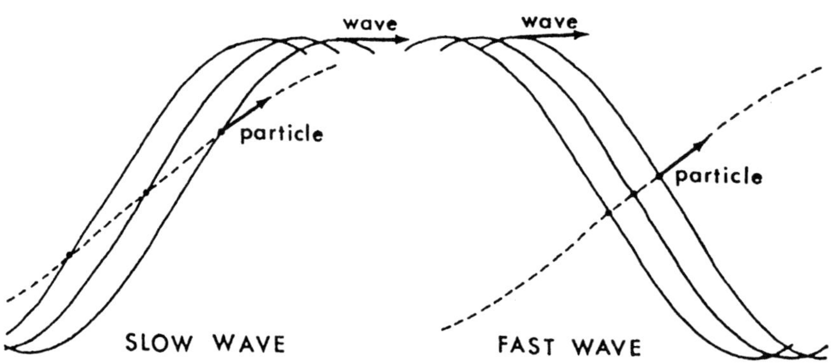

Fig. 2

Coherent wave velocity

The wake fields generated by such a beam displacement wave in the surrounding structure exert a force on the particles. For a fast wave this force has the right phase to produce a damping of the betatron wave. On the contrary slow waves generate wake fields which are capable of increasing the amplitude of the wave, leading to an instability. Backward waves cannot lead to instabilities.

Up to now we have described the propagation of betatron waves around the machine, and we have seen how it is related to the motion of the individual particles which constitute the beam. What are the observable properties of such a wave ? The signal generated in a transverse detector at a fixed azimuth is (see equation (3)).

$$y\,(\theta_o, t) = y_n e^{j[(Q+n)\omega_o t - n\theta_o]}. \qquad (4)$$

It consists of a single line at frequency (Q + n) ω_o for a fast wave (n > 0), (|n| - Q) ω_o for a slow wave (n < 0 and |n| > Q) and (Q - |n|)ω_o for a backward wave (n < 0 and |n| < Q). The amplitude of this single line is proportional to the displacement y_n of the beam and to the number N of particles.

The different lines corresponding to the case of an accelerator with Q = 2.25 are shown in Fig. 3. In reality, when observed on a spectrum analyzer the lines with a negative frequency are folded onto the positive frequency axis as indicated. Around each revolution harmonic n'ωo one observes two lines at (n'+q) ωo and (n'-q) ωo, where q is the fractional part of Q. However, the lines n'+q and n'-q are not associated with n' but with n = n'-k and n = -n'-k respectively (if k is the integer part of Q). Although the nature of these lines cannot be distinguished on a spectrum analyzer their associated phases contain the information on the sign of the wave harmonic n.

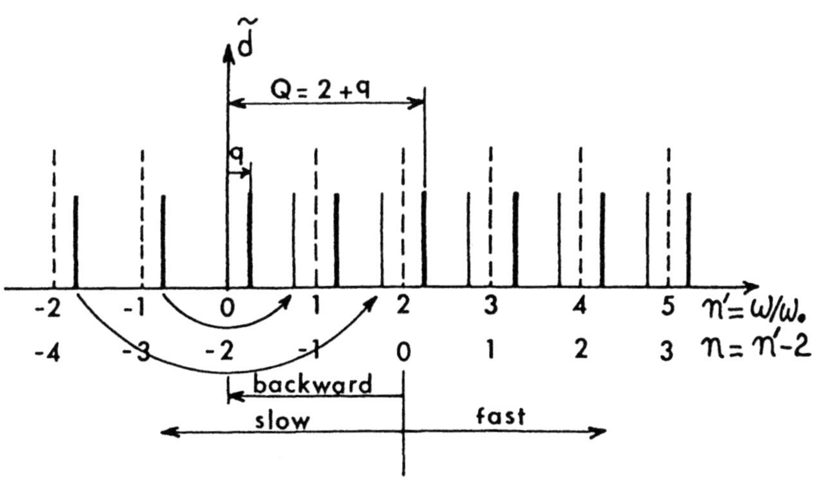

Fig. 3

Betatron lines for an accelerator with Q = 2.25

2.2 Instability mechanism and damping

Any interacting structure which presents a non-zero resistive impedance at the frequency corresponding to any slow wave can drive the beam unstable. Since all the wave patterns of a continuous beam are independent, there can be no cancellation effects, contrary to what happens for more complicated systems that will be analyzed later on.

As a consequence the frequency range of a feedback system capable of stabilizing such a beam has to be as large as that of the coupling impedance itself, a requirement very difficult to meet in practical cases. In fact, the stability of continuous beams is assured by Landau damping, and feedback is only used for the few most unstable modes.

Landau damping is a very powerful damping mechanism in continuous beams. It arises from the spread in the betatron frequencies of the single particles which constitute the beam. The largest contribution to the frequency spread comes from the momentum dispersion, which induces both a spread of the revolution frequencies given by

$$\frac{\Delta\omega}{\omega_o} = -\eta\,\frac{\Delta P}{P_o} \qquad (5)$$

and a spread in the betatron tunes given by

$$\frac{\Delta Q}{Q_o} = \xi\,\frac{\Delta P}{P_o} \qquad (6)$$

where $\eta = \frac{1}{\gamma_{tr}^2} - \frac{1}{\gamma^2}$, with γ_{tr} the relativistic factor at transition energy, and ξ the machine chromaticity.

With these formulae the frequency dispersion on a betatron sideband at $(n+Q)\,\omega o$ can be written:

$$\Delta\,(n + Q)\,\omega_o = (Q_o\,\xi - n\eta)\,\frac{\Delta P}{P_o}\,\omega_o. \qquad (7)$$

There is always a value of n for which the betatron frequency spread induced by momentum dispersion vanishes. In this case the stability can only be assured by an additional frequency spread induced by octupoles, but this term can be insufficient if the beam has small transverse dimensions. It is therefore imperative to choose the sign of ξ and η such that the cancellation in formula (7) only happens for a positive value of n (this corresponds to a fast wave, which is always stable). This means that above transition energy ($\eta > 0$) ξ must be positive, and that below transition ($\eta < 0$) ξ must be negative.

The frequency around which this cancellation occurs, $\omega_\xi = Q_o\omega_o\frac{\xi}{\eta}$, is called the chromatic frequency. As we will see later on, this quantity also plays a very important role in the stability of bunched beams.

2.3 The resistive wall instability

The transverse coupling impedance of a smooth beam pipe of radius b and resistivity ρ is given as a function of the frequency ω by:

$$Z_T = \frac{c^2}{\omega_o b^3}\sqrt{2\mu_o\rho/|\omega|}\,((\mathrm{sign}\,(\omega) + j)) \tag{8}$$

where c is the speed of light.

This function is sketched in Fig. 4 together with the frequency lines of the coherent modes. If the tune value lies just below an integer, there is one slow wave close to the origin, where the value of the resistive wall impedance is large, leading to a large growth rate of the instability. When this phenomenon is dominant, as was the case in the CERN PS Booster during its commissioning, it is preferable to choose a tune value just above an integer, since this pushes the frequency of the most dangerous mode to higher values and reduces its growth rate.

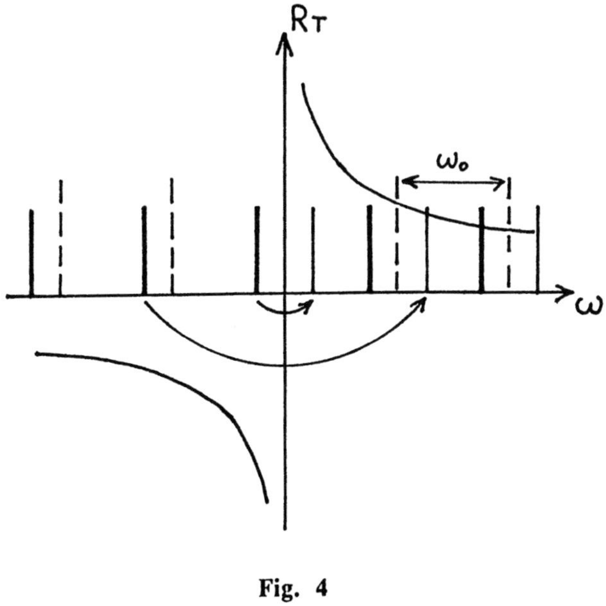

Fig. 4

Resistive wall impedance and betatron lines for Q = k + 0.75.

One of the slow waves sees a large impedance.

2.4 Longitudinal effects

In addition to the transverse displacement waves analyzed above, a debunched beam can sustain longitudinal charge density waves. These generate single coherent lines at frequencies $n\omega_0$ corresponding to the number n of wavelengths of the perturbation around the machine.

3. Bunch modes

In bunched beams the particles describe synchrotron orbits around the synchronous particle as sketched in Fig. 5. The time of arrival of a particle at a fixed azimuth varies like

$$\tau = \hat{\tau} \cos(\omega_s t + \psi_o) \tag{9}$$

where $\hat{\tau}$ is the amplitude and ψ_o the phase of the synchrotron oscillation. As a result the spectral lines of this particle are split into synchrotron sidebands.

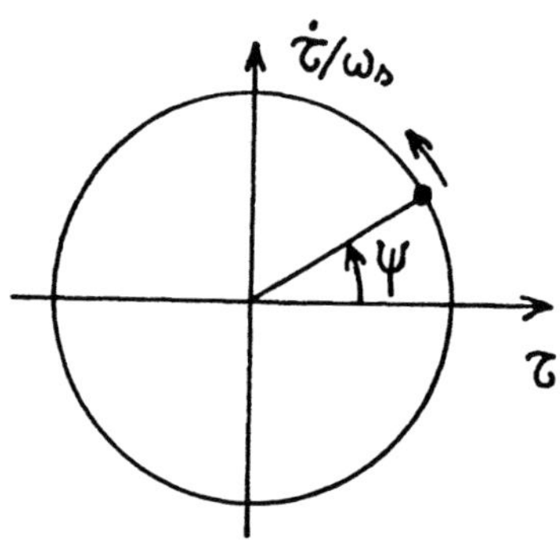

Fig. 5
The synchrotron orbit

3.1 Longitudinal modes

Let us treat first the case of the longitudinal motion, which is somewhat simpler than the transverse one.

In absence of RF focusing a particle generates spectral lines at each harmonic $n\omega_o$ of the revolution frequency ω_o [1]. In presence of synchrotron motion these lines are split into an infinite series of sidebands of amplitude

$$S = J_m(n\omega_o\hat{\tau}) e^{j[(n\omega_o + m\omega_s)t + m\psi_o]} \tag{10}$$

where m is the sideband number and J_m is the Bessel function of order m.

For a bunch containing a large number N of particles the signal is obtained by adding for each sideband the contribution of the individual particles. If the motion of the particles are uncorrelated the phases ψ_o are randomly distributed and the global signal is very small. This signal, which does not vanish completely because the number N of particles, although usually very large, is not infinite, constitutes the Schottky signal of the bunch. By introducing a correlation between the particles one can increase the signal on one of the sidebands. For instance it is easy to see that with a particle distribution of the form

$$\rho\,(\psi_o) = e^{-jm\psi_o} \tag{11}$$

the signal is maximum for the sideband of order m. We will define a normal mode m of the bunch as a special distribution of particle phases which enhances coherently the signal on sideband number m.

In a real bunch the particles are also distributed over a certain range of amplitudes $\hat{\tau}$. When the bunch is unperturbed the particles are evenly distributed in phase and the distribution can be written:

$$\rho_o(\hat{\tau}, \psi) = \rho_o\,(\hat{\tau}). \tag{12}$$

When applying a perturbation Liouville's theorem has to be satisfied, that is the bunch distribution in the two dimensional phase space with coordinates τ and $\dot{\tau}/\omega_s$ (Fig. 5) can be deformed provided this preserves the local density around individual particles. Such a deformation can be described by adding a term $\rho'(\hat{\tau}, \psi)$ to the unperturbed distribution so that

$$\rho_o\,(\hat{\tau}, \psi) = \rho_o\,(\hat{\tau}) + \rho'(\hat{\tau}, \psi) \tag{13}$$

and ρ' can be Fourier analyzed in azimuth ψ and written :

$$\rho' = \sum_{m=-\infty}^{+\infty} \rho_m\,(\hat{\tau}) e^{-jm\psi}. \tag{14}$$

Isodensity contours in phase space relative to terms m = 1 to 3 are drawn on Fig. 6a while the corresponding line charge density appears in Fig. 6b. By subtracting the contribution of the unperturbed distribution, one obtains the signal induced in a sum pick-up as a result of the perturbation. This signal is given by the projection on the time axis of the distribution ρ'. The exact form of this signal depends on the distribution $\rho'(\hat{\tau})$ but in any case for m = 1 it has one node in the middle, for m = 2 it has two nodes and so on. As a consequence, its Fourier Transform is peaked around a frequency Ω_m which is close to $\pi(m + 1)/\tau_L$ where τ_L is the bunch length.

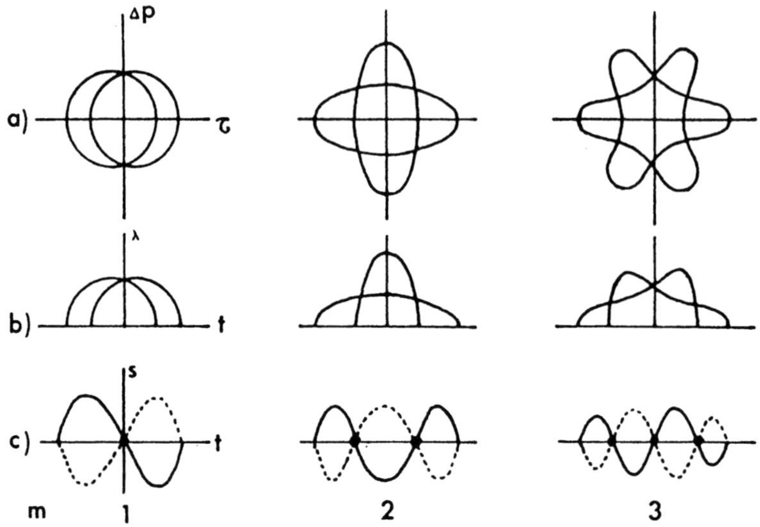

Fig. 6

Longitudinal dipole, quadrupole, and sextupole modes.

a) Isodensity contours in phase space, b) projection on the time axis, c) coherent signal.

This signal oscillates in time at the frequency $m\omega_s$ and this generates coherent synchrotron sidebands of order m around each orbital harmonic $n\omega_o$.

In summary any perturbation of the bunch in the longitudinal phase plane can be decomposed into an infinite number of normal modes m each characterized by an oscillation frequency $m\omega_s$. The envelope of the coherent synchrotron satellites generated by a mode m is peaked around $\pi(m + 1) / \tau_L$.

Up to now the density distribution in amplitude $\rho_m(\hat{\tau})$ has not been specified, and a smooth enough law has been assumed. For more complicated distributions the signal can have more nodes than specified by the azimuthal mode number m. For instance the dipole mode (m = 1) can be obtained by shifting the unperturbed distribution $\rho_o(\hat{\tau})$ in one direction and letting it rotate in phase space at frequency ω_s. But by displacing different parts of the bunch by different amounts while retaining the symmetry of order 1 in ψ as sketched in Fig. 7 it is possible to generate a mode which oscillates at ω_s (m = 1) but induces a signal which has three nodes. Its Fourier transform will then be peaked at $\Omega_3 = \pi \times 4/\tau_L$ as for the mode m = 3. Therefore one must in fact consider a two-fold infinity of modes ρ_{mr} where m is the azimuthal and r the radial mode number. We will usually refer in the following to the basic modes for which m = r and which are called dipole, quadrupole, sextupole etc. modes.

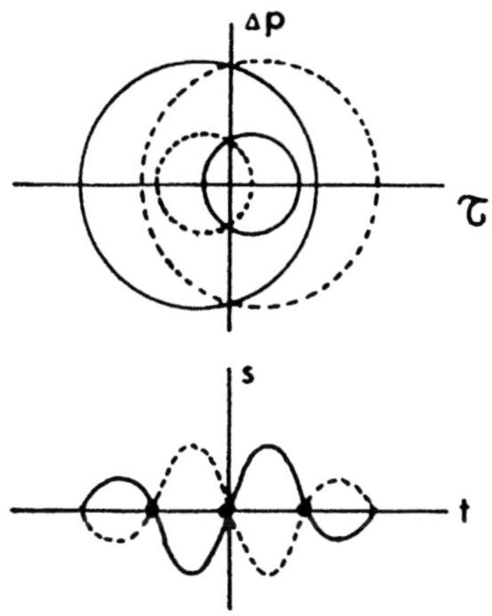

Fig. 7
Mode m = 1, r = 3

3.2 Transverse modes

As in the case of the longitudinal motion we are going to construct a coherent mode by arranging particle phases so that only one set of synchrotron satellites are coherently enhanced. The amplitudes of the synchrotron satellites of the betatron spectral lines of a single particle are given by (see for instance ref. [1])

$$S_T = J_m(\Omega_n \hat{\tau})\, e^{j(\omega_{nm} t + \varphi_o + m\psi_o)} \tag{15}$$

where $\Omega_n = (n + Q)\,\omega_o$, $\omega_{nm} = (n + Q)\,\omega_o + m\omega_s$ and φ_o and ψ_o are the initial phases of the betatron motion and of the synchrotron motion respectively.

Clearly the signal from all particles on the same synchrotron orbit will add up coherently on the synchrotron satellites of order m if $\varphi_o + m\psi_o = 0$. For instance the lines with m = 0 are coherent if the initial betatron phase φ_o is the same for all the particles around the synchrotron orbit. This is the case when all the particles are displaced at time t = 0 in the same direction, in other words the mode m = 0 corresponds to a rigid, parallel motion of the bunch. On the contrary for the m = 1 satellites to be coherent initial phases must be opposite in the head and the tail of the bunch. For m = 2 the head and the tail have equal phases but the center is displaced in the opposite direction, and so on. Modes of oscillation corresponding to these special arrangements of the betatron phase are called head-tail modes of order m.

Let us follow what happens to the particles in a bunch which executes a coherent head-tail mode. The case of the rigid bunch mode is straightforward : since all particles are in phase, the synchrotron motion, while shuffling them back and forth along the bunch, has no observable coherent effect. On the contrary it plays a very important role for all the other modes $m \neq 0$. Let us study in detail the case of the head-tail mode m = 1. This mode corresponds to a special arrangement of the betatron phases along the bunch, such that the head and the tail oscillate 180^0 out of phase. Figure 8 will help us to follow four particles which have the same maximum betatron amplitude and are evenly distributed around the synchrotron orbit.

At time t = 0 (Fig. 8) particles 1 and 3 have maximum opposite displacement y, whereas particles 2 and 4 are centered. The small circles close to each particle show the betatron phase space while the large circle represents the synchrotron phase space.

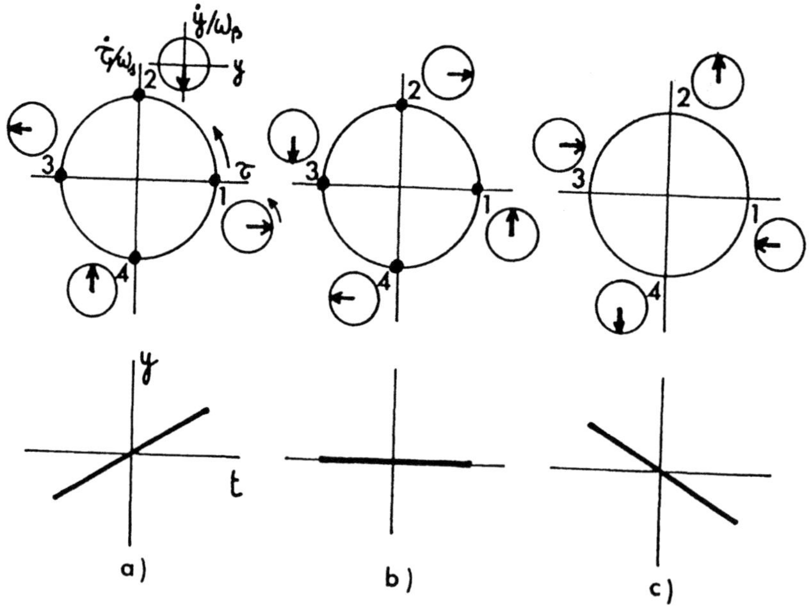

Fig. 8

Head-tail mode m =1

The resulting transverse dipole moment of the bunch is shown below. The particles have all the same frequency (chromaticity is zero) and therefore the situation after a quarter, respectively half a betatron period is described by Fig. 8b and 8c : the extremities of the bunch oscillate but the dipole moment is always zero at the center. On a position pick-up the signal will always have a node in the middle : this is characteristic of head-tail mode 1. In general the signal corresponding to head-tail mode m has m nodes, and for this reason the envelope of its spectral lines peak at approximately $\omega_m = \pi\,(m + 1)/\tau_L$ for $m \neq 0$.

While executing their betatron oscillations our reference particles migrate slowly (ω_s is usually much smaller than ω_β) anticlockwise. After a quarter of a synchrotron period particle 1 has replaced particle 2 and so on. The betatron phases of the particles which occupy the reference positions have been advanced by a quarter of a betatron period as a consequence of the synchrotron motion. After a complete synchrotron period the coherent signal has made one betatron oscillation more than it would have in absence of synchrotron motion : we see that for an external observer the coherent signal of mode m = 1 oscillates at $\omega_\beta + \omega_s$. Generally, the coherent signal of mode m oscillates at $\omega_\beta + m\omega_s$.

Now let us come back to Fig. 8a : we have chosen the phase of particle 2 to be retarded by $\pi/2$ with respect to particle 1, and the phase of particle 4 to be advanced by $\pi/2$ with respect to particle 1. We could equally well have chosen the opposite : particle 2 advanced by $\pi/2$ and particle 4 retarded by $\pi/2$, and the aspect of the coherent mode would have been preserved. The very important difference is that in this last case the coherent frequency of oscillation turns out to be $\omega_\beta - \omega_s$: the mode m = -1 has been generated instead of mode m = 1. This can be generalized to all modes $m \neq 0$.

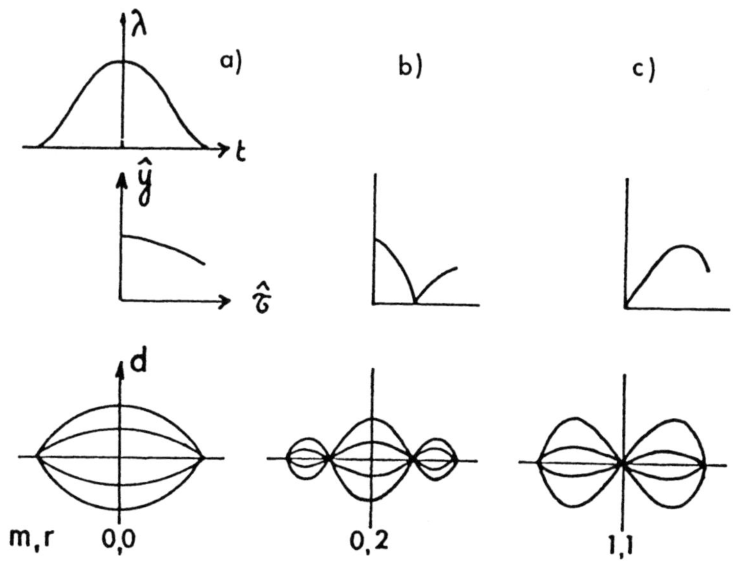

Fig. 9

A few head-tail modes. For the signal d several consecutive traces are drawn to show the nodes.

Up to now we have considered particles evenly distributed in phase around a single synchrotron orbit of amplitude $\hat{\tau}$. In this case a transverse coherent motion was created with all the particles having the same betatron amplitude $\hat{y}$, and their betatron phases arranged in different patterns corresponding to head-tail modes m. In fact particles are distributed over a range of synchrotron amplitudes from zero to $\tau_L/2$, and we must specify the betatron amplitude on each synchrotron orbit. If the distribution of betatron amplitudes as a function of synchrotron amplitudes is smooth, almost uniform, as in Fig. 9a, the behavior of the modes is not changed very much. However, as for the longitudinal modes one can find distributions which create additional nodes in the coherent signal. For instance with a distribution function $\hat{y}(\hat{\tau})$ which goes to zero at some intermediate amplitude as in Fig. 9b, the coherent signal has necessarily two nodes in the case of the mode m = 0. In fact each head-tail mode m is split into an infinity of radial modes m,r which induce signals in a transverse pick-up with r = m, m + 2, m + 4, etc... nodes. The signal of mode m = 0, r = 2 which is displayed as an example in Fig. 9b looks very much like the signal of the basic head-tail mode with azimuthal mode number m = 2, r =2. The difference is that m = 2 oscillates at $\omega_\beta + 2\omega_s$ whereas mode m = 0, r = 2 is a head-tail mode zero and oscillates at ω_β. On a spectrum analyzer the overall image of their signal will look the same, but a closer look at any one of the spectral lines will show the difference of $2\omega_s$.

In general modes with high radial mode numbers are more difficult to drive, and it is often sufficient to consider only the basic modes with r = m.

4. Stability of bunch modes

The normal modes defined in the preceding chapter are standing-wave patterns. Two different points arbitrarily chosen along the bunch coherent signal are either in phase or 180° out of phase. The resistive wake fields generated by such modes in the surrounding structures are proportional to the signal and therefore cannot drive these modes unstable : for this to happen a 90° phase shift between the perturbing force and the displacement is necessary.

However, such a phase shift can arise from different mechanisms. When the value of the chromaticity is different from zero a phase shift is introduced between the head and the tail of the bunch. This phenomenon is responsible for the so-called Head-Tail instability, a single-bunch effect. In the case of a multibunch beam, the coupling between successive bunches provides the necessary 90° component of the force.

4.1 The Head-Tail instability

When chromaticity is not zero the instantaneous betatron frequency of a particle varies along its synchrotron orbit. Let us consider the case of $\xi > 0$, $\eta > 0$ and follow a particle as it performs a phase oscillation (Fig. 10). It starts at point 1 with a betatron phase $\varnothing = 0$. Relative to the phase it would have maintained if held at point 1, $\varnothing$ is constantly increasing as the particle migrates along the top of the synchrotron orbit, where ΔP is positive (see equation (6)).

At point 2 the particle has returned to the nominal momentum and thereafter, along the bottom part of the orbit, its betatron phase will start to decrease because now the betatron frequency is smaller than nominal, and at point 1 it will have recovered its initial phase $\varnothing = 0$. This phase excursion can easily be calculated (see for instance ref. [1]). Its linear variation as a function of the position along the bunch is represented on Fig. 10 : we see that at a certain position $\Delta\tau$ the phase is the same for the particles which are on the upper part of the orbit and for those which are on the lower part. The maximum phase excursion from the head to tail of the bunch is $\chi = \tau_L\, \omega_\xi$, where τ_L is the bunch length in time units, and $\omega_\xi = Q_o \omega_o \frac{\xi}{\eta}$ is the chromatic frequency already defined in section 2.

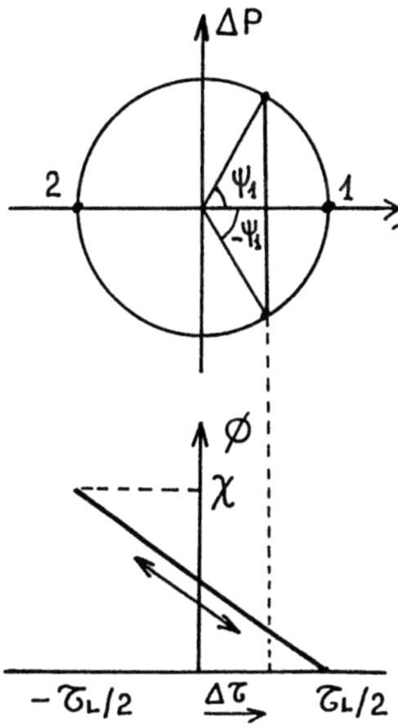

Fig. 10

Head-tail phase shift due to chromaticity

If we now distribute particles along the synchrotron orbit with betatron phases just matching the above curve, they will trade places due to the synchrotron motion but always arrive at a given position with the phase appropriate to that point. Therefore a coherent mode can still be created with a finite chromaticity but at the expense of introducing a phase slewing $\varnothing = \omega_\xi \Delta\tau$ along the bunch. The appearance of the head-tail modes is modified by this phase shift. As an example Fig. 11 shows the deformation of modes m = 0 and m = 1 due to a phase shift $\chi = \pi$ between the head and the tail. The mode m = 0 no longer corresponds to a rigid displacement, whereas mode m = 1 shows a non-negligible center-of-mass motion. In the frequency domain the spectra of all modes are shifted by the chromatic frequency ω_ξ. For this reason high order modes can have spectral lines of large amplitude at low frequency (this is the case of mode 1 in Fig. 11b).

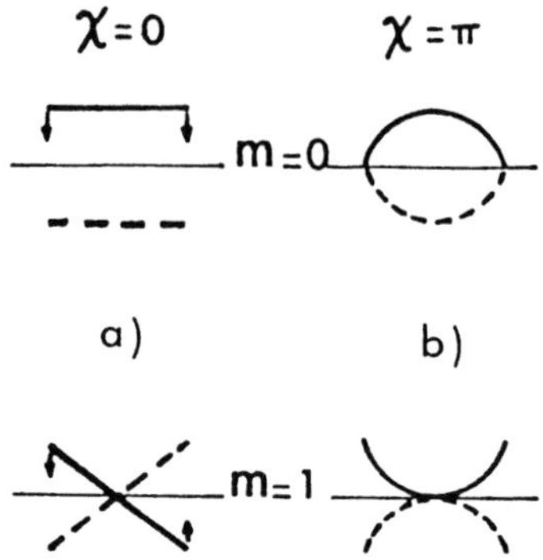

Fig. 11

Head-tail modes for a phase shift $\chi = 0$ and $\chi = \pi$

Signals from high order head-tail modes were first recorded in the CERN PS Booster around 1974 [2]. Typical patterns are shown in Fig. 12 for different values of χ. Several successive traces have been superimposed to clearly show the existence of the nodes for $m \neq 0$ and the phase slewing for $\chi \neq 0$.

As far as stability is concerned the phase shift from the head to the tail of a bunch due to the finite chromaticity has important consequences : it allows an interaction between these two parts (mediated by wake fields) with the right quadrature component of the phase so that energy transfer and therefore instability is possible.

In the frequency domain the situation is sketched in Fig. 13. For a zero value of the chromaticity the power spectra of the head-tail modes are symmetric around the origin and their interaction with the resistive part of the coupling impedance, which is antisymmetric, vanishes (the broad-band, $Q = 1$ resonator model is assumed here). For a non-zero chromaticity the mode spectra are shifted in frequency by an amount ω_ξ, so that the interaction no longer vanishes. It is easy to see that the most dangerous mode $m = 0$ is maximally unstable if $\omega_\xi = -\omega_r$ (ω_r is the resonant frequency of the broad band resonator). This is reminiscent of a result already found in continuous beams : although the theoretical treatment is very different for a single bunch and a continuous beam, it is natural to find similar properties in both cases, since by reducing adiabatically the RF voltage to zero it is possible to go from one state to the other.

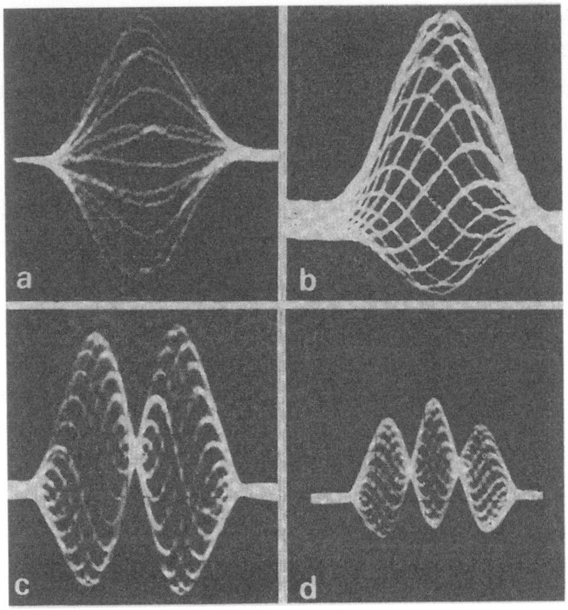

Fig. 12

Head-tail modes 0, 1 and 2 observed in the CERN PS Booster on a transverse position pick-up electrode

a) $m = 0, \chi = 0$; b) $m = 0, \chi = \pi$; c) $m = 1$; d) $m = 2$

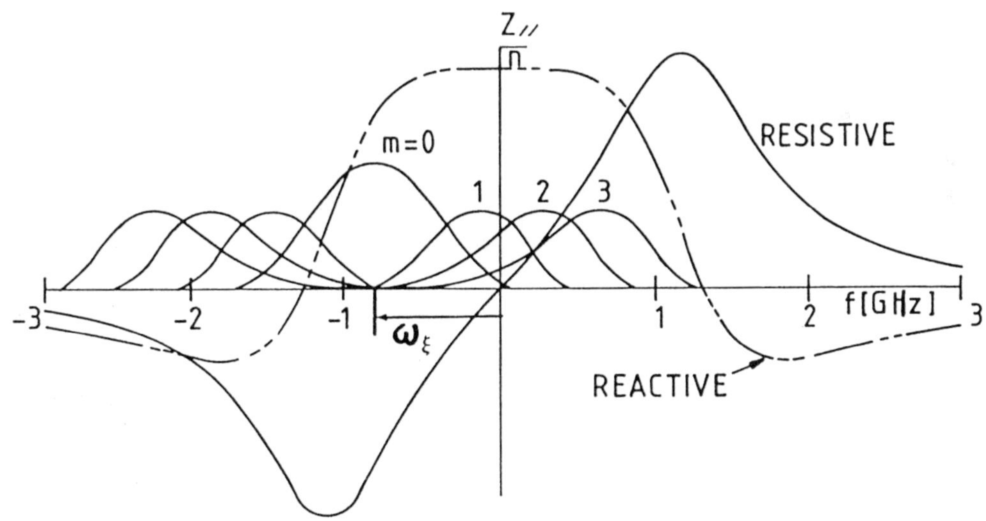

Fig. 13
Chromatic frequency shift

In accelerators in which the transition energy is crossed the head-tail instability is a serious problem. Close to transition the value of η becomes very small and as a consequence the chromatic frequency ω_ξ assumes large values even for a well corrected chromaticity. This allows a strong coupling of the m = 0 bunch mode to the maximum of the resistive part of the broad-band coupling impedance which occurs near to the beam pipe cut-off frequency around a few GHz. Figure 14 illustrates the difficulty of accelerating dense proton bunches through transition in the CERN SPS. In Fig. 14a the chromaticity is slightly positive and a fast head-tail instability occurs before transition. The signature of this phenomenon is a fast increase of the signal detected on a high frequency vertical monitor, followed by partial beam loss. When the value of ξ is reduced the instability occurs closer to transition (Fig. 14b) and when the sign of ξ changes to negative the instability occurs after transition. In such a situation the head-tail instability can only be cured by swiftly changing the sign of the chromaticity at transition energy. Such a ξ jump has been successfully implemented in the CERN PS, where the same phenomena occured. Together with a fast γ_{tr} jump it allowed dense bunches to be accelerated through transition. In the SPS, such schemes were too delicate and costly and it was decided instead to increase the injection energy well above transition when the machine was upgraded into a proton antiproton collider.

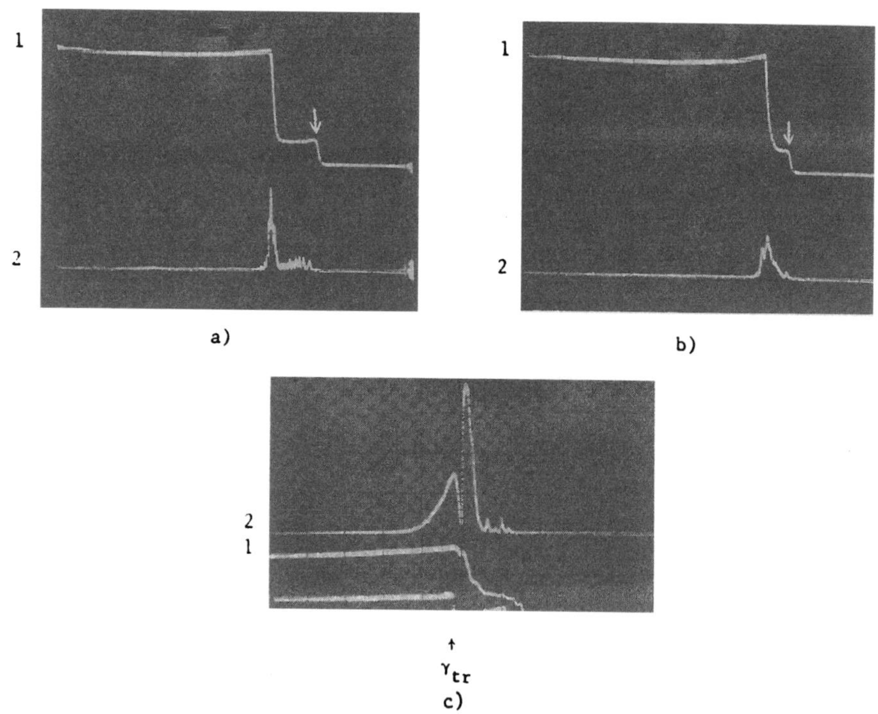

Fig. 14

Head-tail instability around transition in the SPS for $\xi > 0$ (Fig. 13a and b) and $\xi < 0$ (Fig. 13c). Trace 1 shows the evolution of the bunch intensity and trace 2 the signal from a wide-band vertical monitor. The arrows indicate transition crossing.

4.2 Longitudinal Head-Tail

An effect similar to the transverse Head-Tail instability but concerning the longitudinal motion was conjectured long ago by H. Hereward [3]. This is a very weak instability because there is no chromatic effect in the longitudinal plane to first order ; only second order effects can provide a variation of the synchrotron frequency with momentum. Johnsen [4] described this effect in the following formula

$$\frac{\Delta L}{L} = \alpha_1 \frac{\Delta P}{P} (1 + \alpha_2 \frac{\Delta P}{P}). \tag{16}$$

where $\Delta L/L$ is the relative change of the orbit length of a particle with a relative momentum deviation $\Delta P/P$. Here α_1 is the usual term from which the transition energy is obtained ($\alpha_1 = 1/\gamma^2_{tr}$) while α_2 gives the variation of the transition energy with momentum :

$$\frac{\gamma_{tr} - \gamma_{tr_o}}{\gamma_{tr_o}} = - (0.5 + \alpha_2) \frac{\Delta P}{P}. \tag{17}$$

The term α_2 which depends on the sextupoles used to correct chromaticity can be measured by recording the change in transition energy for different radial positions of the beam. In the CERN SPS $\alpha_2 = - 0.7$ in normal operating conditions. The synchrotron frequency depends on the transition energy and therefore the instantaneous synchrotron frequency varies along the synchrotron orbit due to the second order effect described above. This effect is relatively stronger close to transition energy.

In the CERN SPS collider transition occurs at 21 GeV/c. Dense proton bunches are sequentially injected at 26 GeV/c and circulate at this energy for up to 28 seconds. A slow longitudinal dipole instability with an e - folding time of the order of a few seconds is observed, as shown in Fig. 15. This is a single bunch effect which has been attributed to a longitudinal Head-Tail instability [5].

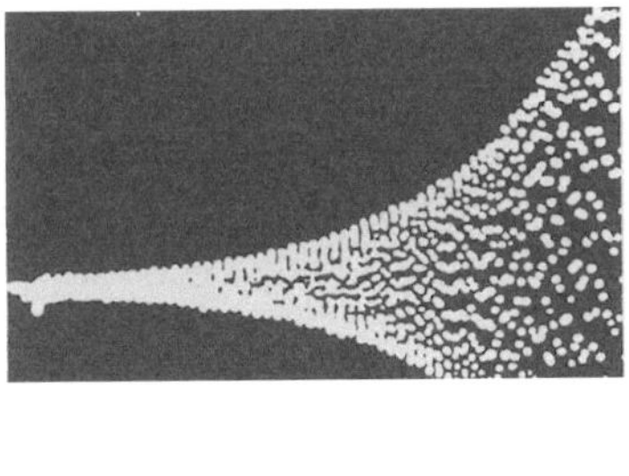

Fig. 15

Slow single-bunch dipole instability at 26 GeV/c in the SPS (output of phase detector, 20 s full horizontal scale)

This very weak instability can be observed in the SPS because Landau damping is completely suppressed as a result of the interaction of the dense bunches with the inductive part of the broad-band coupling impedance : the frequency shift of the coherent mode is larger than the spread of the incoherent particle frequencies within the bunch. This instability is cured by feedback on individual bunches.

4.3 The coupled-bunch instability

In the case of a continuous unbunched beam, a mode corresponds to a closed sinusoidal pattern with n wavelengths around the machine circumference. The signal is a sinusoid of frequency $\omega_n = n\omega_0$ (for longitudinal density waves) or $\omega_0 = (n + Q)\omega_0$ (for transverse displacement). It seems legitimate to assume that gathering the beam particles into a large number N of closely spaced bunches should not dramatically change the coherent behavior of the beam. This is indeed true, although some important differences appear which distinguish a bunched from an unbunched beam, as we shall see. For short, well separated bunches it is relatively easy to visualize a transverse coupled-bunch mode : we start from the closed sinusoidal pattern of a continuous beam mode with its n wavelengths around the circumference, and we distribute the bunches like beads on top of this wave. For N equispaced bunches the betatron phase shift $\Delta\varphi$ from bunch to bunch must therefore be $2\pi n/N$. If $\Delta\varphi$ is different from zero or π, which always happen if the number of bunches N is larger than 2, the resistive wake fields have components in quadrature with displacement, and an instability can occur.

When bunches are long compared to the wavelength of the continuous beam pattern, we can still play this game but now the bunches must be bent to closely follow the sinusoidal curve : this is possible if either we make use of high order Head-Tail modes, or if we introduce a non-zero chromaticity which, as we know, allows different parts of the bunch to oscillate with different phases even for mode m = 0. In the frequency domain this means that the frequency of the coupled-bunch mode pattern we want to create corresponds to some part of the Head-Tail mode spectrum of a single bunch shifted by ω_ξ (Fig. 16).

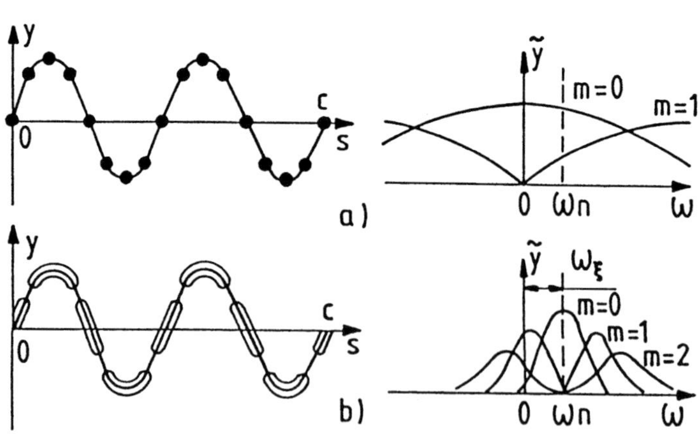

Fig. 16 **Multibunch modes**

a) short bunches

b) long bunches

This can be illustrated by the case of the transverse coupled bunch instability induced by the resistive wall in the CERN PS Booster. Figure 17 shows the low frequency wave pattern generated around the ring by the displacement of the five bunches and the exponential growth of the instability. The single-bunch modes which contribute to this signal for different values of the chromatic frequency shift χ have already been shown in Fig. 12.

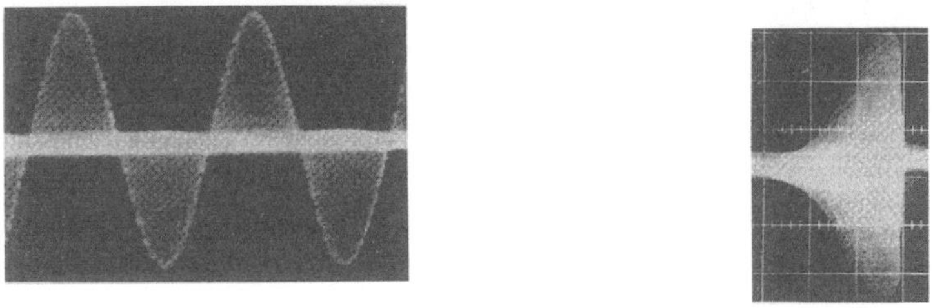

a) 10 μs/div.

b) 5 ms/div.

Fig. 17

Coupled-bunch mode pattern a) and growth rate b) of the resistive wall instability in the CERN PS Booster

In the longitudinal case it is the phase of the bunches relative to the synchronous particles which is distributed according to a sinusoidal law. Again for short bunches such a mode pattern can be created by a rigid displacement of the bunches (dipole mode m = 1). But for long bunches the phase displacement has to vary within the bunch to fit the sinusoidal density variation of a continuous beam mode : in this case high-order single-bunch modes (quadrupole, sextupole ...) come into play, in very much the same way as Head-Tail modes in the transverse case (with the difference that here chromaticity effects are negligible).

In the case of a continuous unbunched beam, a transverse mode n corresponds to one single frequency $\omega_n = (Q + n)\omega_o$. In this situation what matters for the determination of the instability growth rate or frequency shift is the value of the coupling impedance at that frequency.

In bunched beams the spectra of coupled-bunch modes contain aliased frequencies $\omega = k\,N\omega_o + \omega_n$ in addition to the basic mode frequency ω_n. This is because, as the bunch occupies a very short fraction of the ring circumference, its spectrum extends to frequencies much higher than the revolution frequency, whereas its position is sampled only once per revolution. Therefore, many wave patterns of different frequencies can be invoked to account for the measured signal as can be seen easily from Fig. 18.

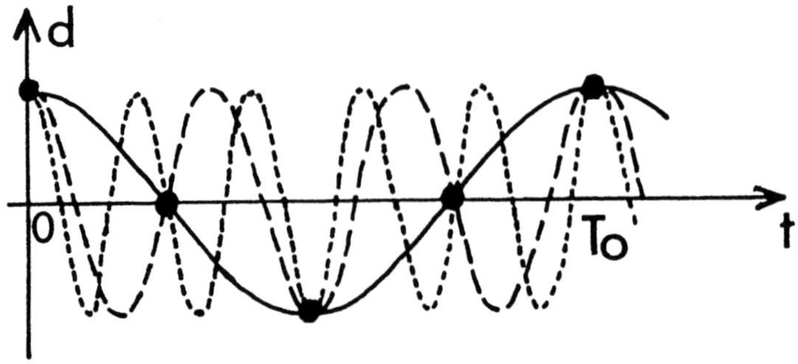

Fig. 18
Aliasing effect

The stability of this mode is therefore influenced by the value of the impedance at all aliased frequencies on which consequently a summation has to be performed. These spectral lines do not extend to infinite frequencies : they fall off at a frequency roughly equal to the inverse bunch length expressed in seconds.

Future large hadron colliders will use a large number N of bunches. In this case the aliased frequencies are far apart and it is often true that a given structure only influences the beam on one of its spectral lines : this is the case for the resistive wall and for high Q resonators. In this limit the bunched beam behaves very much like a continuous beam, in particular the growth rates are the same in both cases for the same total beam intensity. However, in a realistic environment particular features emerge for a bunched beam, which we are now going to study.

As we have already mentioned, the resistive wall instability is very important in large hadron colliders in particular because of the existence of transverse modes at very low frequency. However, it is easy to build powerful active feedback systems covering all modes up to the MHz range to damp this instability. At higher frequencies the growth rate of the resistive wall instability decreases like $\omega^{-1/2}$ and one could hope that a moderate octupole would be sufficient to Landau damp all the modes outside the bandwidth of the low frequency damper. Unfortunately, this is not true because the short bunches also interact strongly with the broad-band coupling impedance. This interaction produces a real frequency shift which tends to suppress Landau damping : as a consequence, even a very small growth rate can lead to the eventual destruction of the beam. In such a situation the damping system must cover all the possible collective modes, which is equivalent to saying that each bunch must be individually damped.

As an example, in the CERN Large Hadron Collider now under design the bunches are typically 30 cm long and are regularly spaced at about 5 m intervals. This means that the bunch spectrum extends over 1.25 GHz, and in this range a collective mode n produces of the order of 20 aliased frequency lines. All these lines sample the inductive part of the broad-band coupling impedance, which is constant in this frequency range, whereas the resistive-wall impedance mostly affects the line with the lowest frequency. Although it is not responsible for the growth rate, the broad-band coupling impedance thus dominates the stability properties of a large number of the coupled bunch modes.

5. Mode-Coupling and Beam Break-Up

We are now going to study the stability of single, dense bunches interacting with short range wake fields (broad-band coupling impedance). We have already seen that due to the symmetry of the power spectra of the bunch modes, one expects single bunches to be stable when chromaticity is suppressed. However, even in this case one observes destructive phenomena in both the longitudinal and transverse phase planes at high bunch densities. In the longitudinal case the microwave instability manifests itself by bursts of microwave signals followed by an increase of the energy spread. In the transverse case a fast increase of the transverse emittance followed by beam losses is observed.

Two complementary approaches are used to explain these phenomena. Historically the first idea was applied to the longitudinal motion : in the case where the wake fields are so strong that they dominate the RF focusing force it is legitimate to ignore the synchrotron motion and treat the bunch as a piece of coasting beam. This is the Hereward-Boussard approach [6] which leads to the prediction of the longitudinal microwave threshold by applying the Keil-Schnell criterion for coasting beams [7] to single dense bunches. In the same spirit the theory of the transverse Beam Break-Up developed long ago for Linear Accelerators can be applied to circular machines if the synchrotron motion is so slow compared to the instability that it can be ignored.

Another approach to the same phenomena analyses the evolution of the bunch modes as the wake fields are increased, until the very concept of modes breaks down : this is the Mode-Coupling theory [8].

We will successively describe the Mode Coupling and the Beam Break Up approaches, and illustrate with examples their very good predictive power. The Mode-Coupling theory is to be used below the instability threshold, while the Beam Break-Up theory describes best phenomena occuring above threshold.

5.1 Mode-Coupling

5.1.1 Theoretical considerations

We have defined a normal mode in absence of interaction as a distribution of particle phases and amplitudes which ensures global coherence on one synchrotron satellite. Any bunch oscillation can then be described by using an infinite set of orthogonal functions and their Fourier transforms $g_m(\omega)$, each corresponding to a mode m.

However in presence of wake fields these modes are no longer independent since the wake field generated by each mode acts on all the others : in this case the real, observable modes must be calculated self-consistently. The eigenvalues of the interaction matrix

$$M_n^m = C \; i \sum_{p=-\infty}^{+\infty} i^{m-n} \, g_m(p) \, Z(p) \, g_n(p) \tag{18}$$

give the new, perturbed frequencies, from which the beam stability can be evaluated. In this expression C is a real quantity, Z(p) is the coupling impedance and p is the frequency normalized to the revolution frequency. The coefficient M_n^m represents the action of the wake field induced by mode m (which is expressed by $g_m Z$) on mode n. Figure 19 shows the different mode spectra g_m superimposed on the broad -band coupling impedance represented by the usual Q = 1 resonator model. It is clear that the interaction term between two modes of the same parity contains only a reactive component, which can be responsible for a frequency shift but cannot lead to instability. On the contrary two modes of different parity couple through the resistive part of the impedance, and this can produce an instability.

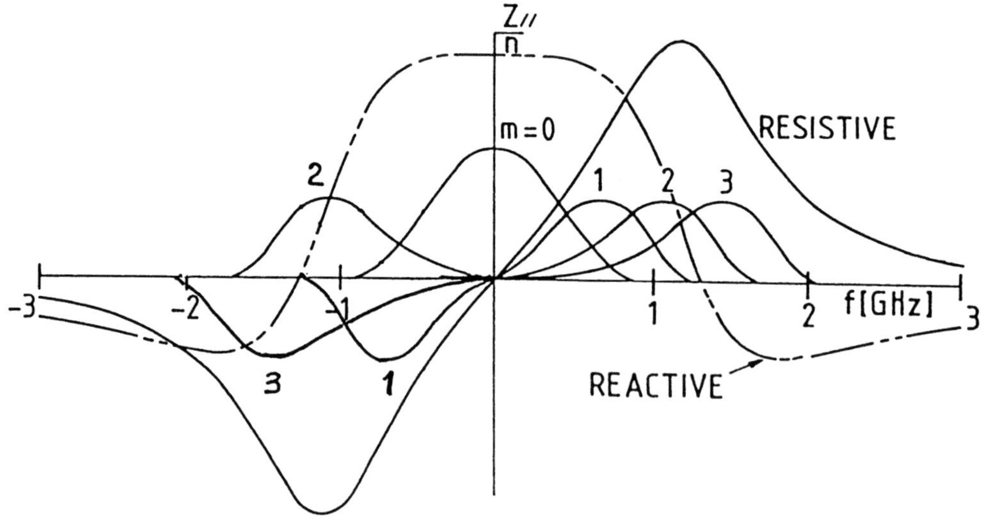

Fig. 19

Head-tail mode spectra superimposed on the broad-band coupling impedance (Q = 1 resonator model)

An important feature whose demonstration will not be given here but can be found elsewhere [9] is that the interaction matrix is real and has the property :

$$M_n^m = (-1)^{m-n} M_m^n . \tag{19}$$

Two adjacent modes m and m + 1 oscillate at different frequencies separated by ω_s. Therefore for small interactions one can usually neglect the off-diagonal terms which couple different mode numbers. In this case no single bunch instability can happen if the value of the chromaticity is zero. However this is no longer true for strong interactions. As an illustration, let us consider the case of two coupled modes, described by the following matrix and its eigenvalues λ :

$$\begin{vmatrix} \lambda - m - M_m^m & - M_n^m \\ - M_m^n & \lambda - n - M_n^n \end{vmatrix} \tag{20}$$

$$\lambda = \frac{1}{2}\left(m + M_m^m + n + M_n^n\right) \pm \sqrt{\left(m + M_m^m - n - M_n^n\right)^2 - 4\, M_n^m M_m^n} \; . \tag{21}$$

Here m and n are the mode frequencies in absence of interaction divided by ω_s.

Two different behaviors can be observed, depending on the relative parity of m and n. Using formula (19) one sees that for modes of different parities $M_n^m = M_m^n$ and the eigenvalues can acquire an imaginary part if the coupling terms become comparable to the frequency difference between the two modes. The corresponding evolution of the eigenvalues as the beam intensity is increased is sketched in Fig. 20a. On the contrary, for two modes of the same parity $M_n^m = -M_m^n$ and the eigenvalues remain real : one sees on Fig. 20b how the two modes repel each other, whereas on Fig. 20a they are attracted towards each other.

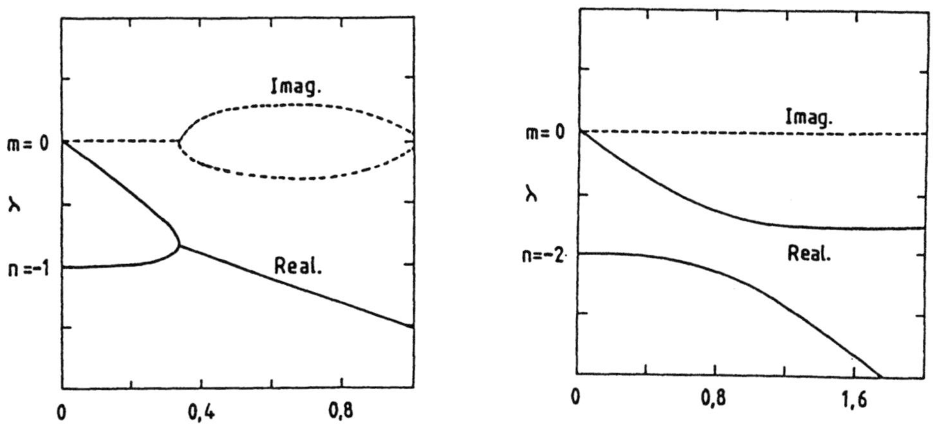

Fig. 20

Eigenvalues as a function of bunch intensity for the cases (m - n) odd a) and (m - n) even b).

5.1.2 Observation of Mode-Coupling

The manifestation of the Mode-Coupling mechanism depends very much on the overlap of the bunch spectrum (related to the bunch length) with the broad-band coupling impedance.

The most unfavorable situation occurs in the case shown in Fig. 21 : the bunch length is such that two modes (here the m = 0 and m = 1 modes) sample mainly, one the inductive part of the broad-band impedance, the other the capacitive part. As a consequence their frequencies move in opposite directions and become closer with increasing intensity. In addition both modes sample an important resistive term. At a certain point the two frequencies are close enough for the resistive coupling term to dominate : the two modes lose their identity, each becoming a mixture of m = 0 and m = 1 mode. At a slightly higher intensity there is only one coherent frequency : maximum stable coupling has been reached. Above this threshold, one of the mode frequencies acquires a positive imaginary part while the other acquires a negative imaginary part and the mode becomes unstable.

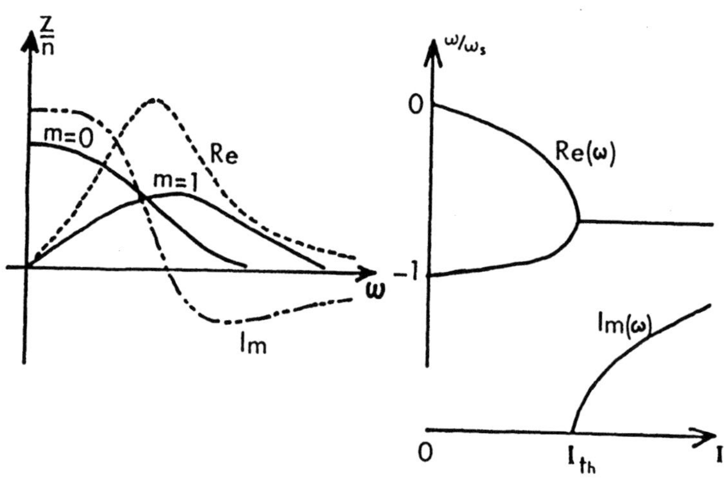

Fig. 21

Mode-coupling for short bunches

The situation described above occurs when the wake field decays in about a bunch length, and this leads to the lowest threshold. It is characteristic of short bunches in electron colliders.

A rule of thumb which is often used to predict this instability threshold is the following : a mode coupling instability can arise when the frequency shift of the lowest order mode, extrapolated from measurements made at low intensity, becomes comparable to the synchrotron frequency. In the worst case described above this criterion is too optimistic : the true threshold intensity is about half that predicted by this criterion. On the contrary, as we shall now see, for long proton bunches it is too pessimistic.

The case of long proton bunches is described in Fig. 22. The spectra of many of the low-order modes lie in the region where the broad-band impedance is mainly inductive : their frequencies move in the same direction when the intensity is increased, although by different amounts since higher-order modes are fundamentally more difficult to drive. Eventually, different mode frequencies merge, but the resistive coupling terms are always small compared to the reactive terms. For this reason, at an intensity slightly above threshold, the modes are again decoupled, and no important growth rates can be reached.

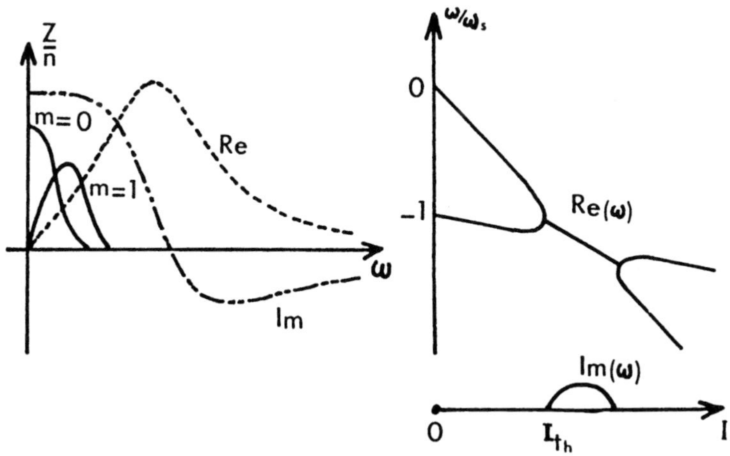

Fig. 22

Mode coupling for long bunches

The CERN SPS is well suited to illustrate the difference between long and short bunches, since both hadrons and leptons are accelerated in this machine. Proton bunches containing up to 1.5 10^{11} particles and injected at 26 GeV/c with an r.m.s. length $\sigma_s = 30$ cm are stable, while positron bunches containing 10^{10} particles and injected at 3.5 GeV with $\sigma_s = 16$ cm are unstable.

The reason for this different behavior is explained in Figs. 23 and 24. For protons (Fig. 23) many modes cross as the intensity is increased but the growth rates acquired remain small. For positrons (Fig. 24) modes 0 and -1 cross at about 0.4 10^{10} particles, but again the growth rates are limited in this region. It is only when modes -2 and -3 cross at about 0.8 10^{10} that the instability is observed. The reason is that the spectra of these two modes overlap the resistive peak of the coupling impedance around 1.3 GHz.

5.2 Beam Break-Up

5.2.1 Theoretical considerations

In linear accelerators the head of the bunch of particles induces wake fields in the accelerating structure which in turn deflect the tail of the bunch. This effect accumulates along the machine length and is more severe for long structures. When the amplitude of the bunch tail reaches the iris aperture part of the bunch is lost, hence the name of Beam Break-Up instability given to this phenomenon which was observed long ago [10].

Since the particles go only once through the structure a simpler theoretical approach than for circular machines can be used. It is interesting to extend this approach to circular accelerators and this is possible whenever the wake fields decay completely in one turn, or between successive bunches : in this case a circular accelerator of radius R is equivalent to a linear machine of length $L = 2\pi R n$, where n is the number of turns over which the instability develops. However a fundamental property of circular accelerators limits the number of turns over which this comparison with Linacs can be made : this is the existence of a bunch stretching mechanism described by formula (5) which can be rewritten

$$\frac{dT}{T} = \eta \frac{dP}{P} \tag{22}$$

where T is the revolution period.

This provides a natural damping mechanism: a Beam Break-Up instability can only be observed in a circular accelerator if it is faster than the bunch stretching effect.

Whereas the beam environment interaction is treated in the frequency domain in circular machines, the use of the time domain is more natural in linear structures. Considering our usual broad-band impedance model described by a low-Q resonator with a resonant frequency ω_r and a transverse maximum coupling resistance R_T the wake field generated by a point charge passing in the structure at time zero is

$$W_T = \omega_r \frac{R_T}{Q} e^{-\varepsilon t} \sin S\, \omega_r t \tag{23}$$

with $\varepsilon = \omega_r/2Q$ and $S = (1 - 1/4Q^2)^{1/2}$.

For bunches which are short compared to the oscillation period of the wake field it is instructive to use a two-particle model, as sketched in Fig. 25 : half of the bunch charge is gathered in particle 1 (the bunch head) and half of the charge in particle 2 (the tail). If the head particle enters the structure with a transverse displacement δ (injection error, misalignment) the

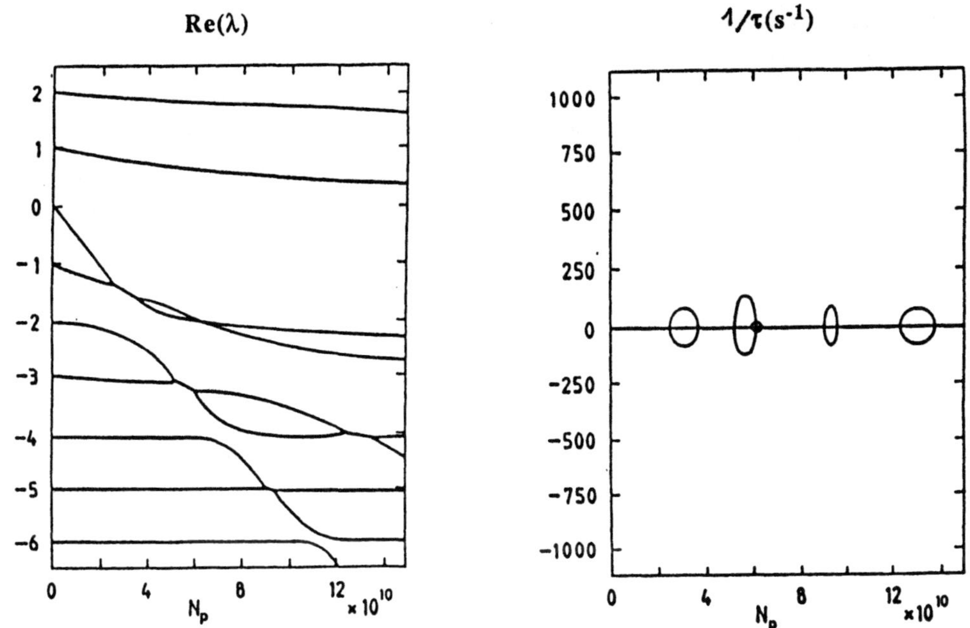

Fig. 23

Mode frequencies versus number of particles for proton bunches in the SPS at 26 GeV/c (σ_s = 30 cm)

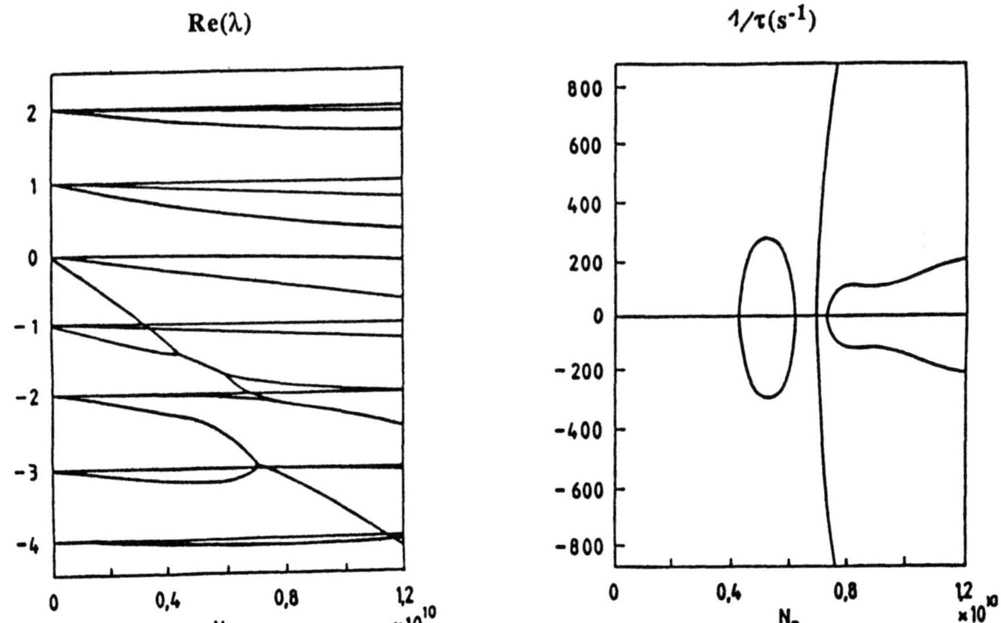

Fig. 24

Mode frequencies versus number of particles for positron bunches in the SPS at 3.5 GeV/c (σ_s = 16 cm)

maximum force exerted on the tail through the transverse wake field is

$$f = e\frac{Ne}{2}\,\delta W_1 \tag{24}$$

provided particle 2 arrives just at the maximum W_1 of the wake field. If the betatron oscillation of particle 1 is given by $y_1 = \delta \cos \omega_\beta t$ the oscillation of particle 2 is given by the equation :

$$\ddot{y}_2 + \omega_\beta^2\, y_2 = \frac{Ne^2 W_1}{2m_o\gamma}\, y_1 \tag{25}$$

with the result :

$$y_2(t) = \delta\left[\cos \omega_\beta t + \frac{Ne^2 W_1}{4\omega_\beta m_o\gamma}\, t \sin \omega_\beta t\right]. \tag{26}$$

At the extremity of a structure of length L the amplitude of the tail reaches :

$$y_2 = \delta\, \frac{Ne^2 W_1 L}{4\omega_\beta m_o\gamma c}\,. \tag{27}$$

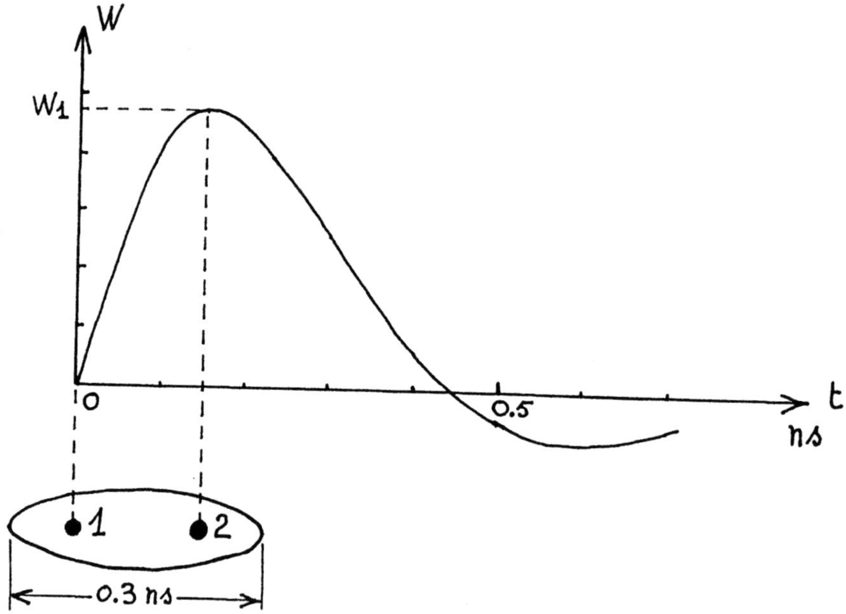

Fig. 25

Transverse wake field and "two-particle" model

This simple treatment cannot be applied in the case of the CERN SPS because the bunches are longer than the oscillation period of the wake fields. A more sophisticated theory due to Yokoya [11] is interesting in this case : the bunch is split into a large number m of bunchlets, each one generating its own wake field. The effect of all these individual wake fields is summed up along the bunch and this results in an amplitude growth of the last bunchlet proportional to L^{m-1}. If m is very large the amplitude growth is given by :

$$y(n) = \frac{\delta c}{2\sqrt{2\pi}\omega_r \ell}\left(\frac{\Omega\ell}{c}\right)^{1/4} \exp\left[-\varepsilon\frac{\ell}{c}+\sqrt{\frac{\Omega\ell}{c}}\right] \tag{28}$$

where ℓ is the bunch length ($4\sigma_S$) and $\frac{\Omega\ell}{c} = \frac{Nec}{\omega_\beta E/e}\,\frac{\omega_r R_T}{Q}n$.

In the SPS a series of measurements performed over many years in different beam conditions have been used to adjust the parameters of the broad-band impedance model : the best fit is obtained with a Q = 1 resonator with $R_T = 23\ M\Omega m^{-1}$ and $\omega_r/2\pi = 1.3$ GHz. Using this model and applying formula (28) with a bunch length $\ell = 64$ cm and a number $N = 4\ 10^{10}$ particles per bunch gives $y(n)/\delta = 20$ for n = 15 turns. From this we conclude that a bunch of leptons with the above characteristics injected 1 mm off center would be lost to the vertical aperture of about 20 mm after 15 turns. As we shall see later on a loss is observed after 11 turns, in reasonably good agreement with the above prediction.

Another approximate prediction which can be made using very simple arguments concerns the intensity threshold below which the Beam Break-up instability should not be observed. Using formula (22) with $\eta = 1.85\ 10^{-3}$ in the SPS one sees that the length of a bunch injected with $\ell = 64$ cm and a total energy spread $\Delta E/E = 2.4\ 10^{-3}$ doubles in about 40 turns. This defines an approximate threshold for the instability : if the initial growth rate is not large enough the competing effect of bunch stretching eventually wins and the amplitude does not grow significantly. Experimentally as the bunch intensity is decreased the losses become erratic between 20 and 40 turns after injection and are never observed after 40 turns. The threshold corresponds to a bunch population of about 10^{10} particles.

5.2.2 Observations of Beam Break-Up in the SPS [12]

Figure 26 shows that a bunch of 4 10^{10} positrons is partially lost soon after injection. Figure 27 displays the signal of a wide-band vertical monitor filtered between 1.4 and 1.7 GHz : the signal grows exponentially after injection for about 10 turns and then decays (particles are lost, see Fig. 26). Subsequently the signal grows again until turn number 30, where additional losses are visible in Fig. 26.

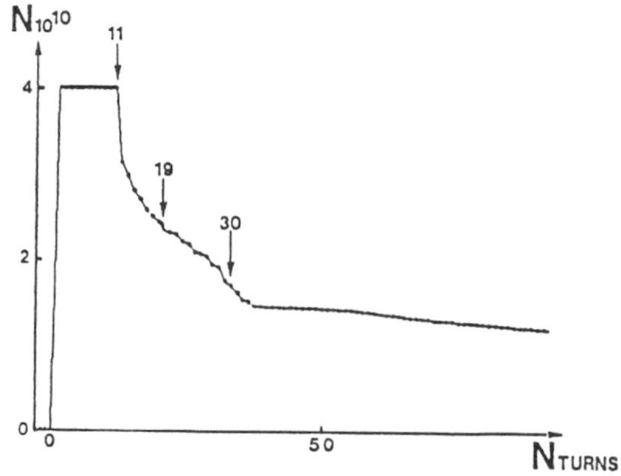

Fig. 26

Beam loss after injection as a function of the number of turns

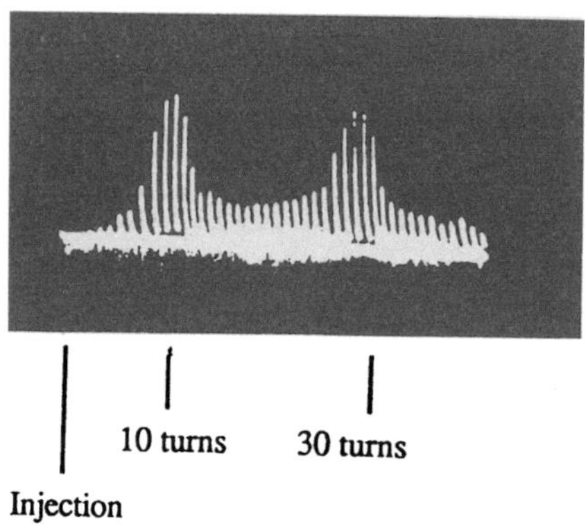

Fig. 27

Signal from a vertical wide-band beam position monitor filtered between 1.4 and 1.7 GHz

If the SPS coupling impedance can be represented by a broad-band resonator peaked at 1.3 GHz as represented in Fig. 28, one expects the corresponding wake fields to distort the 64-cm-long bunch as indicated. The signal induced by such a wriggling bunch has a wide spectrum centered around the peak of the resonator impedance. On the contrary the frequency spectrum of the bunch just after injection extends only to about 500 MHz and cannot be seen by the 1.4 to 1.7-GHz receiver : this is why the signal grows from zero after injection.

a)

b)

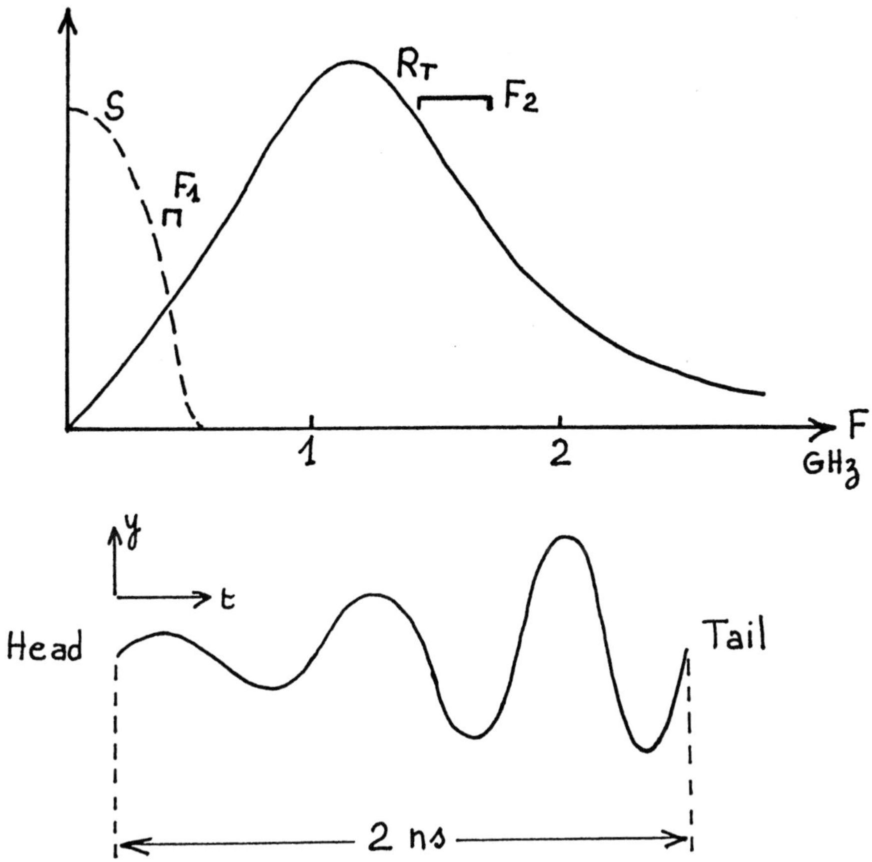

Fig. 28

a) The resistive component R_T of the SPS broad-band coupling impedance model and the bunch spectrum S at injection. The frequencies of the filters used, F_1 (460 MHz) and F_2 (1.4 – 1.7 GHz) are indicated.

b) The wriggling bunch after a few turns in the SPS

By tuning the receiver around 460 MHz, one observes a large signal at injection, because this frequency lies within the bunch spectrum. This signal, which is caused by the injected bunch passing off-center in the monitor, does not grow on subsequent turns.

The synchrotron period at injection in the SPS is 70 turns, and therefore one expects no influence of the RF system on this fast instability above threshold : indeed switching the RF on or off does not change the loss pattern.

5.3 Feedback damping of the Mode-Coupling instability

The Mode-Coupling instability develops when the frequency shift of the fundamental mode becomes comparable to the synchrotron frequency. As a consequence the growth rate just above the threshold can reach values as high as the synchrotron frequency ω_s; in other words, since Q_s is often of the order of 0.1 in electron machines, the e-folding time can be as small as a few turns, and a very fast feedback system has to be used. In the case of the SPS described previously, a somewhat slower feedback would suffice since Q_s is smaller. However, with long bunches there is no sizeable component of the oscillating bunch signal at low frequency, so that a high frequency wide-band feedback is needed.

In any case it seems difficult to fight the Mode-Coupling Instability with a classical resistive feedback. For this reason another approach has been proposed : it consists in holding the frequency of the mode m = 0 constant with a low frequency, reactive feedback. However, this works only for short bunches (for which the mode m = 0 is involved in the instability) and one cannot hope to increase the threshold in this way by a factor larger than about two, because in the end the resistive interaction, which is of the same order as the reactive one, dominates. Such a feedback has been tried in the storage ring PEP at Stanford [13].

References

[1] J. Gareyte, Beam observation and the nature of instabilities, Physics of Particle Accelerators, Fermilab summer school 1987, AIP Conference proceedings 184, p. 343.

[2] J. Gareyte, F. Sacherer, Head-Tail type instabilities in the CERN PS and Booster, Proc. 9th Int. Conf. on High Energy Accelerators, Stanford 1974, P. 341.

[3] H. G. Hereward, Longitudinal Head-Tail effect by the bunch length variation method, RL-74-062, EPIC/MC/48 Rutherford Lab. April 1974.

[4] K. Johnsen, Effects of non-linearities on phase transition, CERN Symposium on High Energy Accelerators, Geneva 1956, Vol. 1, p. 106.

[5] D. Boussard, T. Linnecar, Longitudinal Head-Tail instability in the CERN-SPS Collider, 2nd EPAC, Nice, June 12-16, 1990.

[6] D. Boussard, Observation of microwave longitudinal instabilities in the CPS, CERN Lab. II/RF/Int. 75-2 (1975),
H.G. Hereward, ISR performance report ISR/DI/HGH/amb (1975).

[7] E. Keil, W. Schnell, Concerning longitudinal stability in the ISR,
ISR-TH-RF/69-48 (1969).

[8] F. Sacherer, Bunch lengthening and microwave instability, IEEE Trans. Nucl. Sci. Vol. NS-24 N°3 (1977), p. 1393,
R. D. Kohaupt, Transverse instabilities in Petra, Proc. 11th Int. Conf. on High Energy Accelerators, CERN (1980) EXS Vol. 40, Birkhäuser Verlag, p. 562.

[9] Y. H. Chin, Transverse mode coupling instabilities in the SPS, CERN/SPS/85-2 (DI-MST).

[10] R. Helm, G. Loew, Beam Break-Up in Linear Accelerators (eds. P.M. Lapostolle and A.L. Septier) North Holland Book Co, Amsterdam, 1970, chapter B.

[11] K. Yokoya, Collective beam break-up in large scale linacs, DESY 86-084.

[12] D. Brandt, J. Gareyte, Fast instability of positron bunches in the SPS, 1st EPAC, Rome 7-11 June 1998, World Scientific Publishing Co. Pte. Ltd, Singapore, p. 690.

[13] S. Myers, Stabilization of the fast Head-Tail instability by feedback, 1987 IEEE Particle Accelerator Conference, March 16-19, Washington, DC, 87CH2387-9, p. 503.

Some Longitudinal Dynamics of Bunched Beams

Shane R. Koscielniak
TRIUMF, 4004 Wesbrook Mall, Vancouver B.C., Canada V6T 2A3

Abstract

We show the relation of the Rees G-function to the conventional complex impedance Z. Then we consider the Vlasov equation and beam instability, and present an example calculation of growth rate and coherent frequency shift. The working is presented in detail to show how the G-function enters the problem. A novel feature of the calculation is the use of purely real (i.e. non-complex) quantities and solely positive (i.e. physically measurable) frequencies throughout. As an aid to deeper understanding of the problem, we demonstrate that a single density wave, circulating clockwise in phase space, gives rise to two modulation sidebands; and further, by considering the driven response of the beam, we show that only clockwise waves are unstable because driven anti-clockwise waves cannot grow.

1 The 'G' function and algebra

1.1 Introduction

The response of an impedance to a carrier excitation $\cos(pt)$ is

$$\mathcal{R}[Z(p)e^{jpt}] = Z_r^p \cos(pt) - Z_i^p \sin(pt) \tag{1.1}$$

where Z_r^p and Z_i^p are real and imaginary parts of the impedance evaluated at frequency p. The operator $\mathcal{R}$ is an instruction to exclude terms containing j.

It would be nice if we could write the response to a modulated carrier $\cos(pt)\cos(wt)$ as

$$\mathcal{R}[G(p,w)e^{jpt}e^{iwt}] \tag{1.2}$$

and this facility is just the intention of the Rees[1] 'G' function.

Let us think how we might generalise from (1.1) to equation (1.2). The complex exponential $\exp(jpt)$ is a two-dimensional vector $[\cos(pt), j\sin(pt)]$ in the complex plane (i.e. Argand diagram). The action of the impedance $Z(p)$ is to mix the fundamental components, so that a single component at input (say $\cos pt$) yields two components at the output, as in equation (1.1). For the modulated carrier, there are *four* fundamental components: $\cos(pt)\cos(wt)$, $\sin(pt)\cos(wt)$, $\cos(pt)\sin(wt)$, $\sin(pt)\sin(wt)$, which we shall write for convenience sake as $c^p c^w$, $s^p c^w$, $c^p s^w$, $s^p s^w$, respectively. The action of $G(p,w)$ is to mix these components so that if one is input then four are output. This property requires a dual-complex algebra, akin to W.R. Hamilton's "Quaternion" Calculus. For an exposition of the generalization of complex numbers to higher dimensions, the reader is referred to Kantor[2].

1.2 Dual-complex Description of Single-Modulated Carrier

First, let us find the response to $\cos(pt)\cos(wt)$ by conventional means. The product is expressed $\frac{1}{2}[\cos(p+w)t + \cos(p-w)t]$, and the response is

$$(1/2)[Z_r^{p+w}\cos(p+w)t - Z_i^{p+w}\sin(p+w)t \quad + \quad Z_r^{p-w}\cos(p-w)t - Z_i^{p-w}\sin(p-w)t]\,.$$

This may be written

$$\frac{1}{2}[c^pc^w(Z_r^{p+w}+Z_r^{p-w}) - s^pc^w(Z_i^{p+w}+Z_i^{p-w}) \;+\; c^ps^w(Z_i^{p-w}-Z_i^{p+w}) + s^ps^w(Z_r^{p-w}-Z_r^{p+w})]\,. \quad (1.3)$$

The dual-complex solution to the problem is a straight-forward generalization from two to four dimensions. In the 2-D case we formed the product $Z(p)e^{jpt} = (Z_r + jZ_i)(c^p + js^p)$. The response to $\cos(pt)$ was found by taking the terms with no j occurring, while the response to $\sin(pt)$ was found by retaining only those terms containing j. In the 4-D case we form the product of $G(p,w) = (G_1 + jG_2 + iG_3 + ijG_4)$ and $e^{jpt}e^{iwt} = (c^pc^w + js^pc^w + ic^ps^w + ijs^ps^w)$. The products $i\times i = -1$ and $j \times j = -1$. The response to $\cos(pt)\cos(wt)$ is found by taking terms with no i or j occurring. Inspecting

$$(G_1 + jG_2 + iG_3 + ijG_4) \times (c^pc^w + js^pc^w + ic^ps^w + ijs^ps^w) \quad (1.4)$$

we see the result is

$$G_1c^pc^w - G_2s^pc^w - G_3c^ps^w + G_4s^ps^w\,. \quad (1.5)$$

Comparing expression (1.5) with (1.3) we make the following identifications:

$$\begin{aligned} G_1 &= (Z_r^{p+w} + Z_r^{p-w})(1/2) \\ G_2 &= (Z_i^{p+w} + Z_i^{p-w})(1/2) \\ G_3 &= (Z_i^{p+w} - Z_i^{p-w})(1/2) \\ G_4 &= (Z_r^{p-w} - Z_r^{p+w})(1/2)\,. \end{aligned} \quad (1.6)$$

Evidently, $G(p,w)$ can *always* be found from $Z(p \pm w)$, though knowledge of Z merely at the carrier is insufficient. It is worth pointing out that though we have spoken of carrier (p) and modulation (w), the mathematics treats both frequencies on an equal footing. Now that the components of G are known, it is an easy matter to find the response to any modulation. For instance, the response to $\sin(pt)\cos(wt)$ is found by taking from (1.4) only the terms in j. The response is

$$(G_2c^p + G_1s^p)c^w - (G_4c^p + G_3s^p)s^w\,. \quad (1.7)$$

Similarly, the response to $\cos(pt)\sin(wt)$ is given by the terms in i; and response to $\sin(pt)\sin(wt)$ by terms in ij.

It is worth noting that dual-complex quantities are sometimes written in a form that emphasizes their nature, for instance

$$e^{jpt\times iwt} = [(c^p + js^p)c^w + i(c^p + js^p)s^w]\,.$$

Finally, we give the generalization for complex conjugation, which follows from the basis vectors. For the 2-D case $(1+j)^* = (1-j)$. For the 4-D case $(1+j+i+ij)^* = (1-j-i+ij)$. Visualizations of the 2-D and 4-D spaces are given below, showing the mutually orthogonal fundamental components for each space.

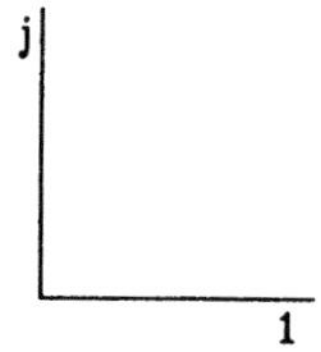

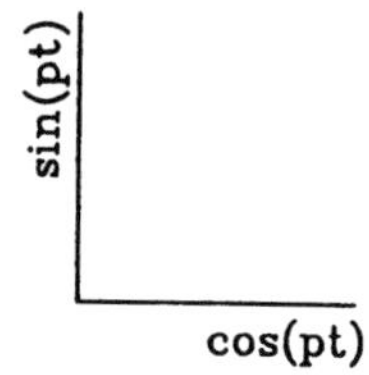

Single–complex space

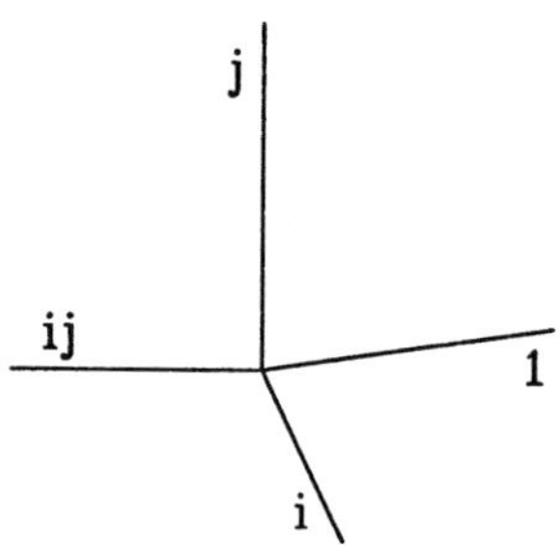

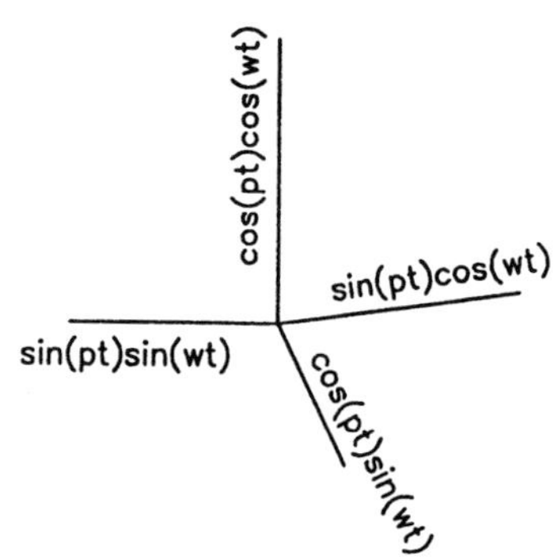

Dual–complex space

1.3 Triple-complex Description of Double-Modulated Carrier

This section is not for the novice, and may be skipped over on a first reading. A complete theory of coherent synchro-betatron instability[1] would include doubly modulated carriers. Such modulations require triple-complex vectors with 8 components. Before embarking, let us take a 'refresher' on complex numbers.

Single complex numbers

The basis vector is $I = (1+i)$ with conjugate $I^* = (1-i)$. Their product is $I \times I^* = 2$. The complex excitation is $e^{iq} = (c^q + is^q)$. The impedance is $Z(q) = (Z_r^q + iZ_i^q)$.

[1]Though modulation of the betatron tunes by the synchrotron motion is included as a chromaticity effect in the "head-tail" transverse bunched-beam instability, the effect of the betatron motion does not feed back into the longitudinal plane; and so this is not a complete synchro-betatron theory in the sense used by Rees.

Dual complex numbers

The basis vector is $I = (1+i)(1+j) = (1+i+j+ij)$ with conjugate $I^* = (1-i)(1-j) = (1-i-j+ij)$. Their product is $I \times I^* = 4$ which is the dimension of the space. The singly-modulated excitation is $e^{iq}e^{jp} = (c^q + is^q)(c^p + js^p)$. The impedance *is not* $Z(q,p) = (Z_r^q + iZ_i^q)(Z_r^p + jZ_i^p)$. The 'impedance' is $G(q,p) = (G_1 + iG_2) + j(G_3 + iG_4)$.

Triple complex numbers

The basis vector is $I = (1+i)(1+j)(1+k) = (1+i+j+k+ij+jk+ki+ijk)$ and defines an 8-D space. The conjugate $I^* = (1-i)(1-j)(1-k)$ and the product $I \times I^* = 8$. We should like to be able to write the response to $\cos(qt)\cos(pt)\cos(rt)$ as $\mathcal{R}[G(q,p,r)e^{iqt}e^{jpt}e^{krt}]$. We note that

$$e^{iqt}e^{jpt}e^{krt} = [c^qc^pc^r + is^qc^pc^r + jc^qs^pc^r + kc^qc^ps^r + ijs^qs^pc^r + jkc^qs^ps^r + kis^qc^ps^r + ijks^qs^ps^r]$$

defines 8 fundamental components which are to be mixed by G, and so define

$$G(q,p,r) = (G_1 + iG_2 + jG_3 + kG_4 + ijG_5 + jkG_6 + kiG_7 + ijkG_8) .$$

We form their product and take the real part, that is ignore terms containing i or j or k. Hence the response to $c^qc^pc^r$ is $\mathcal{R}[Ge^{(iq+jp+kr)t}] =$

$$G_1c^qc^pc^r - G_2s^qc^pc^r - G_3c^qs^pc^r - G_4c^qc^ps^r + G_5s^qs^pc^r + G_6c^qs^ps^r + G_7s^qc^ps^r - G_8s^qs^ps^r. \quad (1.8)$$

The components of G are found by comparison with a conventional derivation. We write

$$\cos(qt)\cos(pt)\cos(rt) = \frac{1}{4}[\cos(q+p+r)t + \cos(q+p-r)t + \cos(q-p+r)t + \cos(q-p-r)t] .$$

The frequency spectrum of this signal is shown below. There are two side-bands above and below the carrier q.

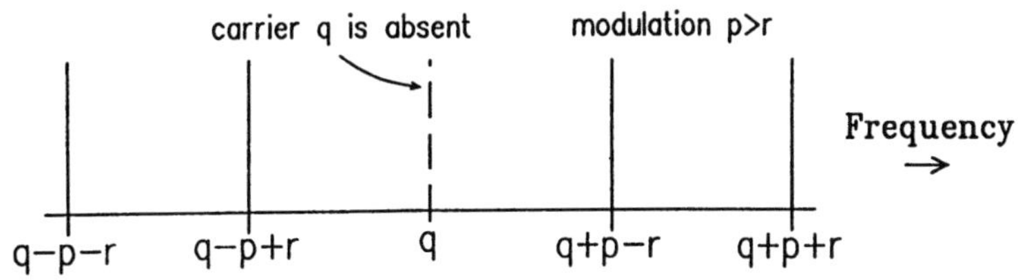

The response is found by summing over the frequency spectrum:

$$\begin{aligned}
&Z_r^{q+p+r}\cos(p+q+r)t &-& \quad Z_i^{q+p+r}\sin(q+p+r)t \ + \\
&Z_r^{q+p-r}\cos(q+p-r)t &-& \quad Z_i^{q+p-r}\sin(q+p-r)t \ + \\
&Z_r^{q-p+r}\cos(q-p+r)t &-& \quad Z_i^{q-p+r}\sin(q-p+r)t \ + \\
&Z_r^{q-p-r}\cos(q-p-r)t &-& \quad Z_i^{q-p-r}\sin(q-p-r)t \ .
\end{aligned} \quad (1.9)$$

The trigonometric terms can be expanded into products $c^qc^pc^r$ etc.. There are 8 terms. Comparison with expression (1.8) gives the components of $G(q,p,r)$:

$$
\begin{aligned}
G_1 &= [+Z_r^{q+p+r} + Z_r^{q+p-r} + Z_r^{q-p+r} + Z_r^{q-p-r}](1/4) \\
G_2 &= [-Z_i^{q+p+r} - Z_i^{q+p-r} - Z_i^{q-p+r} - Z_i^{q-p-r}](1/4) \\
G_3 &= [-Z_i^{q+p+r} - Z_i^{q+p-r} + Z_i^{q-p+r} + Z_i^{q-p-r}](1/4) \\
G_4 &= [-Z_i^{q+p+r} + Z_i^{q+p-r} - Z_i^{q-p+r} + Z_i^{q-p-r}](1/4) \\
G_5 &= [-Z_r^{q+p+r} - Z_r^{q+p-r} + Z_r^{q-p+r} + Z_r^{q-p-r}](1/4) \\
G_6 &= [-Z_r^{q+p+r} + Z_r^{q+p-r} + Z_r^{q-p+r} - Z_r^{q-p-r}](1/4) \\
G_7 &= [-Z_r^{q+p+r} + Z_r^{q+p-r} - Z_r^{q-p+r} + Z_r^{q-p-r}](1/4) \\
G_8 &= [+Z_i^{q+p+r} - Z_i^{q+p-r} - Z_i^{q-p+r} + Z_i^{q-p-r}](1/4) \,. \qquad (1.10)
\end{aligned}
$$

2 Vlasov equation

Consider the one-dimensional motion of N particles. If the particles are mutually interacting, a $2N$-dimensional phase-space is required for the complete description of motion. However, if the individual inter-particle forces can be replaced by a single collective force-field derivable from a scalar potential, then the complete description collapses to a 2-D phase-space. In this case, the motion is derivable from a Hamiltonian $H = H(q,p)$. Suppose that conjugate canonical coordinates for the 2-D space are q and p, and that the particle dispositions through time t are given by a function $\psi(q,p,t)$. If the number of particles is conserved, then there must be a continuity equation:

$$\frac{D}{Dt}\psi = \left[\frac{\partial}{\partial t} + \dot{q}\frac{\partial}{\partial q} + \dot{p}\frac{\partial}{\partial p}\right]\psi \equiv 0 \,. \qquad (2.1)$$

Whenever canonical variables are used, the continuity equation takes on this simple form. Equation (2.1) is Liouville's theorem, and D/Dt is sometimes called the convective derivative. The equation tells us that if we imagine ourself in a reference frame that moves along with the flow of particles in phase-space, then the number of particles in a small test volume of phase-space does not change.

The *Vlasov* equation derives from simply replacing $\dot{q}$ and $\dot{p}$ with explicit expressions for $\partial H/\partial p$ and $-\partial H/\partial q$, respectively. This means replacing $\dot{q}$ with a velocity, and replacing the acceleration $\dot{p}$ with a force.

2.1 Stationary distribution

In a synchrotron or storage ring, the longitudinal motion of each particle (in a bunched beam) with respect to the bunch centre approximates to that of a harmonic oscillator. The angular frequency is called the synchrotron frequency ω_s. The individual particle Hamiltonian is

$$H = x^2/2 + \omega_s^2 y^2/2 \,. \qquad (2.2)$$

A suitable collection of particles can give rise to a bunch whose (statistical) properties are time invariant. This is called the ***stationary distribution*** ψ_0. Any collective electromagnetic fields that it produces can be included into the equations of motion by renormalizing the synchrotron frequency and synchronous phase angle. We shall assume this has already been done; for details refer to Koscielniak[3].

The phase-space paths are ellipses, or circles. Hence it is appropriate to use a coordinate system which better shows the symmetry of the problem. We take Action-Angle variables J, θ defined as follows:

$$x = \sqrt{2J/\omega_s}\cos\theta \qquad -y = \sqrt{2J\omega_s}\sin\theta\,. \tag{2.3}$$

The relation between the coordinate systems is sketched in figure 1. Since both sets of coordinates are canonical, the element of phase-space area takes the simple form $dxdy = dJd\theta$. Liouville's equation for the stationary distribution ψ_0 is

$$\left[\dot{\theta}\frac{\partial}{\partial\theta} + \dot{J}\frac{\partial}{\partial J}\right]\psi_0 = 0\,. \tag{2.4}$$

The Hamiltonian becomes $H = \omega_s J$, so that $\dot{\theta} = \omega_s$ and $\dot{J} = 0$. Hence the Vlasov equation is $\omega_s(\partial\psi_0/\partial\theta) = 0$ with solution $\psi_0 = \psi_0(J)$. There are infinitely many of these distributions, in which the ensemble spins as a rigid disc or bell or annulus etc. at angular frequency ω_s. For instance, figure 2 shows the normalized distribution $\psi_0 = (\hat{J} - J)^2 \times (3/2\pi\hat{J}^3)$.

Fig.1

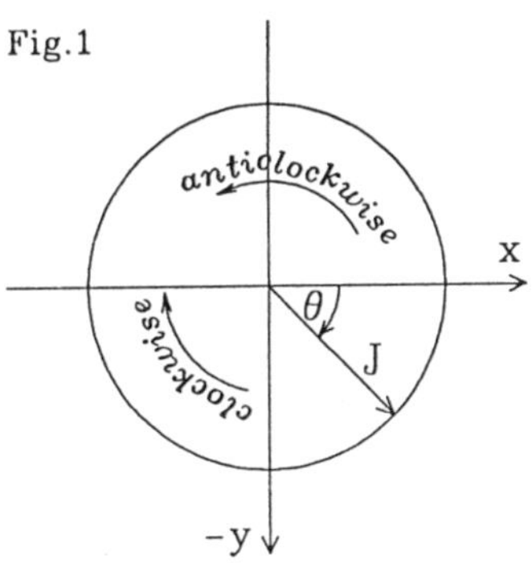

Fig.2

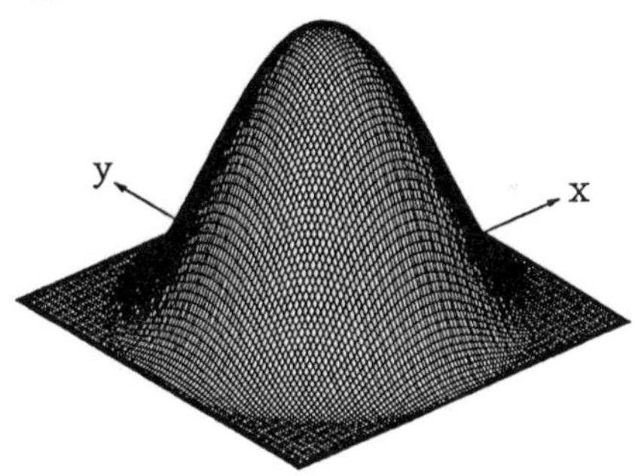

Figure 1: Relation between Cartesian and Action-Angle coordinates.
Figure 2: Example stationary distribution, $\psi_0(r) \propto (a^2 - r^2)^2$.

2.2 Wave-like distributions

There are also stable wave-like solutions to the Vlasov equation which are periodic in θ and t; they obey the *homogeneous* equation

$$\frac{D}{Dt}\psi_1 = \left[\frac{\partial}{\partial t} + \omega_s\frac{\partial}{\partial\theta}\right]\psi_1 = 0\,. \tag{2.5}$$

Let us take trial solutions of the form

$$\psi_1(J,\theta,t) = R(J)\cos(m\theta \pm wt) \qquad \text{with} \quad m > 0 \quad \text{and} \quad w > 0\,. \tag{2.6}$$

Periodicity in θ forces m to be an integer, that is $m = 1, 2, 3, \ldots$, and m is often called the mode number. Substituting into the Vlasov equation generates the ***dispersion*** relation:

$$(\pm w + m\omega_s) = 0\,. \tag{2.7}$$

Taking the positive sign yields the contradiction that w is negative, in which case $w = m = 0$. So free anti-clockwise waves are not permitted.

Taking the negative sign, we find $w = m\omega_s$, in which case the solutions spin rigidly at $\dot{\theta} = \omega_s$ clockwise; the stationary distribution and density wave spin at the same rate. There is a double infinity of such solutions since we can pick (almost) any $R(J)$ to go with any m. The restriction on R comes from the fact that the sum of ψ_0 and ψ_1 must yield a positive number of particles at all points in phase-space. Example modes are shown in figures 3 and 4. The left-hand column displays 'snap-shots' of an $m = 1$ mode, and the right-hand column gives an $m = 2$ mode. The flow of times is taken to run down the page.

We have seen that the 'free' solutions of Vlasov's equation consist of density waves with a single sense of rotation in phase-space. Under the influence of a small perturbation the rotation frequency w/m may change a little, but we should not expect to see any ***self-sustaining*** wave change direction when the perturbation is added. This conjecture is confirmed by the exposition in sections 4 and 5.

3 Signal due to perturbation

In this section we shall find a few properties of a non-stationary distribution: the time variation of the bunch shape and, related, bunch centroid and rms length variation.

3.1 Perturbation distribution

We shall superpose on the stationary phase-space distribution (ψ_0) a density wave ψ_1. The form of the perturbation must preserve the total number of particles, and so a suitable form comes from the solutions of the homogeneous Vlasov equation:

$$\psi_1 = R(J)\cos(m\theta - wt) \qquad \text{with} \quad m > 0 \quad \text{and} \quad w > 0\,. \tag{3.1}$$

This is a clockwise circulating wave in phase-space. The angular frequency w does not necessarily equal $m\omega_s$, since waves can have velocities greatly different from the media that support them. The projection over momenta gives the bunch shape perturbation. In the beam frame of reference this is a stably oscillating standing wave, as will be shown below.

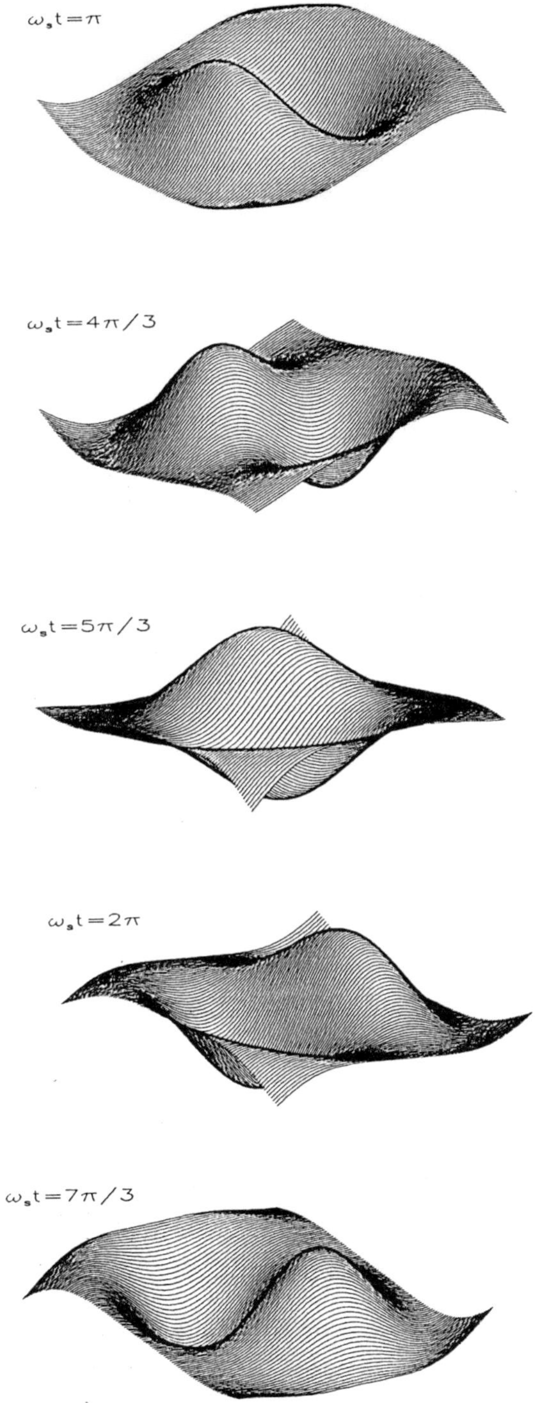

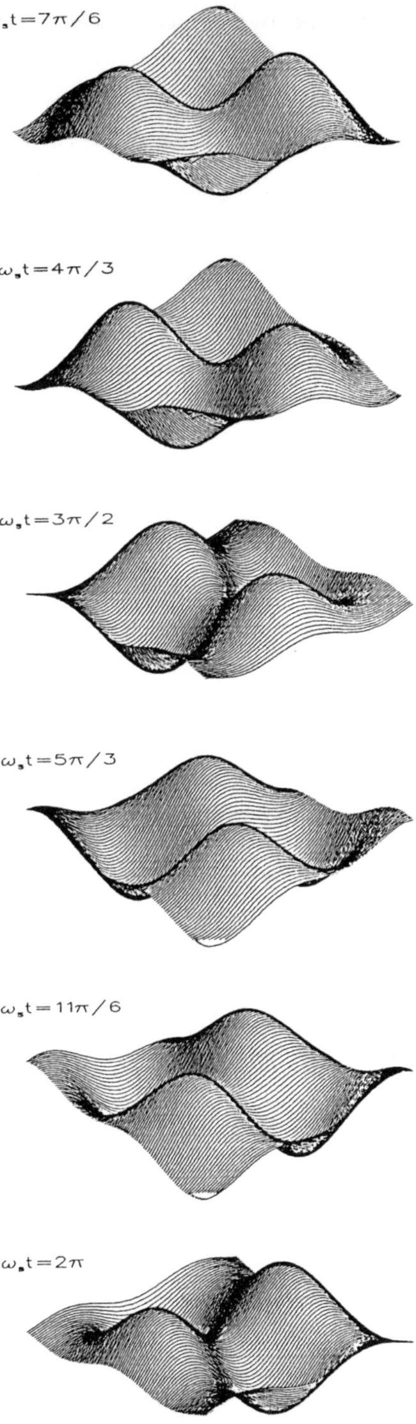

Fig.3: Phase-space views of an $m = 1$ density wave. Mode shown is $J_1(r)\cos(\theta - \omega_s t)$.

Fig.4: Phase-space views of an $m = 2$ density wave. Mode shown is $J_2(r)\cos 2(\theta - \omega_s t)$.

3.2 Current perturbation

As before, let x and y be Cartesian coordinates corresponding to position and momentum; and J, θ action-angle variables. Let the average circulating current be $\langle I \rangle$. Then the current perturbation is

$$I(x,t) = \langle I \rangle \int R[J(x,y)] \cos[m\theta(x,y) - wt] dy \; . \tag{3.2}$$

This can be analysed into a set of standing waves:

$$I(x,t) = \sum_{p=1}^{\infty} [I_p^c(t) \cos(px) \; + \; I_p^s(t) \sin(px)] \; . \tag{3.3}$$

The coefficients I_p^c and I_p^s are found by the normal methods for Fourier series. For instance,

$$\pi I_p^c(t) = \int_{-\pi}^{+\pi} \cos(px) I(x,t) dx \; .$$

This double integral is readily performed when the integration is over $dJd\theta$ rather than $dxdy$:

$$\pi I_p^c(t) = \langle I \rangle \int\int \cos[p\sqrt{2J/\omega_s} \cos\theta] R(J) \cos[m\theta - wt] dJ d\theta .$$

Let $z = p\sqrt{2J/\omega_s}$ and substitute the Fourier-Bessel series for $\cos[z\cos\theta]$, from (4.15):

$$\pi I_p^c(t) = \langle I \rangle \int\int [J_0(z) + 2\sum_{k=1}^{\infty}(-1)^k J_{2k}(z)\cos(2k\theta)] R(J) [\cos m\theta \cos wt + \sin m\theta \sin wt] dJ d\theta .$$

We perform the integration over θ and find that $I_p^c(t) = 0$ unless m is even, in which case only the $2k = m$ term remains.

Similarly, for the sine coefficient,

$$\pi I_p^s(t) = \langle I \rangle \int\int \sin[p\sqrt{2J/\omega_s} \cos\theta] R(J) \cos[m\theta - wt] dJ d\theta \; .$$

Again put $z = p\sqrt{2J/\omega_s}$ and substitute the Fourier series for $\sin[z\cos\theta]$ from (4.16):

$$\pi I_p^s(t) = \langle I \rangle \int\int 2\sum_{k=0}^{\infty}(-1)^k J_{2k+1}(z) \cos[(2k+1)\theta] R(J) [\cos m\theta \cos wt + \sin m\theta \sin wt] dJ d\theta .$$

We perform the θ integration and find that $I_p^s(t) = 0$ unless m is odd, in which case only the $2k+1 = m$ term remains.

Hence to summarize,

$$I(x,t) = \cos(wt) \sum_p I_p^c(0) \cos(px) \quad \text{for } m \text{ even} \tag{3.4}$$

$$= \cos(wt) \sum_p I_p^s(0) \sin(px) \quad \text{for } m \text{ odd} \tag{3.5}$$

with the coefficients given by

$$I_p^c(0) = 2\langle I \rangle (-1)^{m/2} \int R(J) J_m(p\sqrt{2J/\omega_s}) dJ \quad \text{with } m \text{ even} \tag{3.6}$$

$$I_p^s(0) = 2\langle I \rangle (-1)^{(m-1)/2} \int R(J) J_m(p\sqrt{2J/\omega_s}) dJ \quad \text{with } m \text{ odd.} \tag{3.7}$$

We can think of equations (3.4) and (3.5) as giving us the bunch-shape time variation as seen by strobing at the revolution frequency ($\omega_0/2\pi$), provided $w << \omega_0$. By way of example, figures 5 and 6 present 'snap-shots' of the signals due to $m = 1$ and $m = 2$ modes as a time series; the respective radial functions chosen are $R = J_1(r)$ and $R = J_2(r)$. The phase-space flows which generate the signals are indicated by the motion of equi-density contours.

The $\cos(wt)$ time dependence in (3.4) and (3.5) comes from aligning one of the perturbation maxima with the x-axis at time $t = 0$; other choices of initial phase would give a mix of $\sin(wt)$ and $\cos(wt)$. There are some other points worth making about the form of signals $I(x,t)$.

Firstly, they cannot drive the perturbation that generates them; and so there is no cause for instability. For the case m is odd, the signal induced in a pure resistance gives a periodic variation of the restoring force. This translates into a variation of bucket height, and could drive an even-order mode if w were a multiple of $2\omega_s$. For the case m is even, the voltage across a pure resistance gives a periodic variation of the stable phase-angle; and this could drive an odd mode if the angular frequency w is an odd multiple of ω_s.

Secondly, the *single* clockwise wave given in (3.1) gives rise to *both* upper and lower sidebands about the carrier frequency; an anti-clockwise wave is not required. Consider, for example, the case m is odd. We use sum and difference relations to write each of the standing waves as a superposition of forward and backward waves travelling along the bunch:

$$I(x,t) = \sum_p I_p^s(0)[\sin(px + wt) + \sin(px - wt)](1/2)\,. \tag{3.8}$$

The frequency spectrum comes from transforming to laboratory frame coordinates. Let the revolution frequency about the ring be be ω_0 and the synchronous radio-frequency $\omega_{\rm rf} = h\omega_0$. The laboratory azimuthal position about the ring ($|X| < 2\pi$) is given by $X = \omega_0 t + (x/h)$. Hence the signal is

$$I(X,t) = \sum_p I_p^s(0)\{\sin[phX - (ph\omega_0 - w)t] + \sin[phX - (ph\omega_0 + w)t]\} \tag{3.9}$$

and consists of a fast wave of frequency $(p\omega_{\rm rf} + w)$ and slow wave at frequency $(p\omega_{\rm rf} - w)$. We recognise the carrier to be $p\omega_{\rm rf}$. At a fixed point of observation, say $X = 0$,the signal becomes

$$I(t) = \sum_p I_p^s \sin(p\omega_{\rm rf}\, t)\cos(wt) \tag{3.10}$$

and the response of the beam environment is found from $G(p\omega_{\rm rf}, w)$ using equation (1.7).

Let us make one final point. Originally we considered a density wave circulating clockwise in phase-space. Had we chosen an anti-clockwise wave, $R(J)\cos(m\theta + wt)$, the bunch shape variation and current perturbation would have been identical, since it is the *projection* of the distribution onto the x-axis which counts.

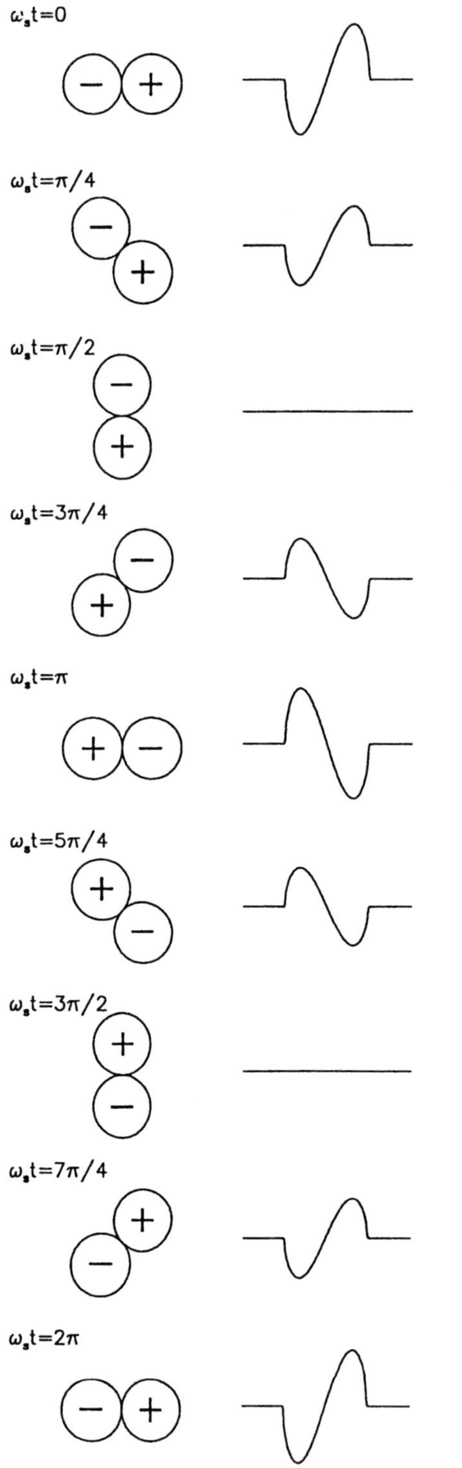

Fig.5: Time series of a phase-space contour and the signal due to an $m = 1$ density wave.

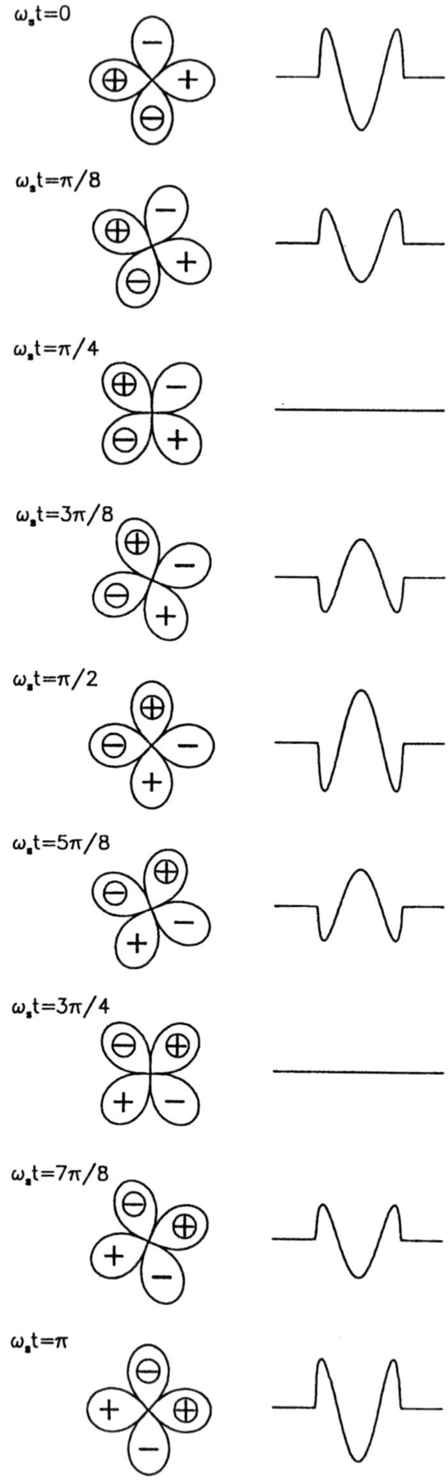

Fig.6: Time series of a phase-space contour and the signal due to an $m = 2$ density wave.

3.3 Bunch moments

Density waves $\psi_1(J,\theta)$ are fine from an abstract stand-point, but what counts practically is whether or not the wave makes the bunch walk out of the rf-bucket. For an $m = 1$ mode, the bunch centroid oscillates from side to side about the synchronous phase. So, instead of $R_1(J)\cos(\theta - \omega_s t)$ we should prefer the amplitude of the phase-oscillation M_1. Similarly, for an $m = 2$ mode what we care about is the increase of the bunch length. These quantities relate directly to the bunch moments. Commonly employed means of observing the first and second moments are phase and peak detection, respectively. The nth moment is defined:

$$\overline{x^n} = \int \psi(J,\theta) x^n(J,\theta) dJ d\theta \Big/ \int \psi(J,\theta) dJ d\theta\,. \tag{3.11}$$

Let $\psi = \psi_0(J) + \psi_1(J,\theta)$ with $\psi_1 = \sum_m R_m(J)\cos[m\theta - (w_m t - \phi_m)]$, and substitute $x = \sqrt{2J/\omega_s}\cos\theta$. Then the first moment becomes $\bar{x}(t) = M_1\cos(w_1 t - \phi_1)$ with

$$M_1 = \int R_1(J)\sqrt{2J/\omega_s} dJ \Big/ 2\int \psi_0(J) dJ\,. \tag{3.12}$$

If the wave ψ_1 is a solution of the homogeneous Vlasov equation, then $w_1 = \omega_s$ and so the bunch centroid oscillates from side to side at the synchrotron frequency. This is commonly called a dipole oscillation.

The second moment of the stationary distribution is

$$\overline{x_0^2} = \int \psi_0(J)(J/\omega_s) dJ \Big/ \int \psi_0(J) dJ. \tag{3.13}$$

After adding the perturbation, the second moment becomes

$$\overline{x^2}(t) = \overline{x_0^2} + M_2\cos(w_2 t - \phi_2) \quad \text{with}$$

$$M_2 = (1/2)\int R_2(J)(J/\omega_s) dJ \Big/ \int \psi_0(J) dJ\,. \tag{3.14}$$

The rms bunch length is

$$L_{\text{rms}}(t) = 2[\overline{x_0^2} + M_2\cos(w_2 t - \phi_2)]^{1/2} \approx 2(\overline{x_0^2})^{1/2} + \frac{M_2}{(\overline{x_0^2})^{1/2}}\cos(w_2 t - \phi_2)\,.$$

If the wave is a solution of the homogeneous equation, then $w_2 = 2\omega_s$ and the bunch length contracts and dilates at twice the synchrotron frequency. This is called a quadrupole oscillation.

4 Instability theory

Thanks to the efforts of Sacherer, Besnier, Zotter and Laclare etc., the calculation of bunched beam stability is in many cases a mechanical task, and derivations of the stability conditions abound. In this section I hope to give a simplistic[2] but careful treatment which (for the first time) avoids the use of complex exponentials, sums over positive and negative frequencies, etc.; so that it is completely transparent as to which density waves (clockwise or anti-clockwise) are being assumed and which modulation sidebands are present. We shall also show the utility of the 'G' function.

4.1 Linearized Vlasov equation

We superpose a particle density perturbation $\psi_1(J,\theta)$ on the stationary distribution, calculate the additional electromagnetic field due to the perturbation, and find if this field causes the perturbation to grow or damp down. Formally, we write

$$\begin{aligned} \psi_0 &\rightarrow \psi_0 + \psi_1 \\ H &\rightarrow \omega_s J + \Delta H(\psi_1)\,. \end{aligned} \tag{4.1}$$

The Vlasov equation for the joint distribution is

$$\left\{\frac{\partial}{\partial t} + \left[\omega_s + \frac{\partial \Delta H}{\partial J}\right]\frac{\partial}{\partial \theta} - \frac{\partial \Delta H}{\partial \theta}\frac{\partial}{\partial J}\right\}(\psi_0 + \psi_1) = 0\,. \tag{4.2}$$

We subtract equation (2.4), which leaves

$$\left[\frac{\partial}{\partial t} + \omega_s \frac{\partial}{\partial \theta}\right]\psi_1 = \frac{\partial \Delta H}{\partial \theta}\frac{\partial \psi_0}{\partial J} + \left\{\frac{\partial \Delta H}{\partial \theta}\frac{\partial \psi_1}{\partial J} - \frac{\partial \Delta H}{\partial J}\frac{\partial \psi_1}{\partial \theta}\right\}\,. \tag{4.3}$$

Suppose now that the perturbation ψ_1 is very very small. Since ΔH is of order ψ_1, it follows that the term in braces $\{\cdots\}$ is the product of infinitesimals and so negligible. What results is the *linearized* Vlasov equation:

$$\left[\frac{\partial}{\partial t} + \omega_s \frac{\partial}{\partial \theta}\right]\psi_1 = \frac{\partial \Delta H}{\partial \theta}\frac{\partial \psi_0}{\partial J}\,. \tag{4.4}$$

Because of the approximation made, any growth rate we may obtain from solving this equation does not necessarily apply to large finite perturbations nor is the rate necessarily constant in time. Further, for growing modes, our solutions tell us nothing about the approach to equilibrium; for instance whether over-shoot occurs.

---

[2] I shall not consider large coherent frequency shifts, nor the coupling of bunches one to another.

4.2 Sacherer's equation

To close the equation we have to find an expression for $\partial\Delta H/\partial\theta$ which is consistent with ψ_1. The assumption which led to equation (2.1) implies that the coherent electromagnetic field due to the perturbation is derivable from a potential, in which case $\Delta H = \Delta H(x)$. Provided there are no frictional forces, equation (2.3) allows us to write

$$\frac{\partial\Delta H}{\partial\theta} = \frac{\partial x}{\partial\theta}\frac{\partial\Delta H}{\partial x} = \sqrt{\frac{2J}{\omega_s}}\sin(\theta)\times\xi\,F(x) \tag{4.5}$$

where $\xi\,F(x)$ is the 'force' on a particle at position x. The constant $\xi = \omega_s/(\hat{V}_{\rm rf}\cos\phi_s)$ where $\hat{V}_{\rm rf}$ is the total rf-voltage per turn and ϕ_s is the synchronous phase angle. $F(x)$ is given below. The substitution of (4.5) into (4.4) gives Sacherer's equation:

$$\left[\frac{\partial}{\partial t} + \omega_s\frac{\partial}{\partial\theta}\right]\psi_1 = \sqrt{\frac{2J}{\omega_s}}\frac{\partial\psi_0}{\partial J}\sin(\theta)\times\xi\,F(x)\,. \tag{4.6}$$

It is an equation which mixes coordinates J, θ and x; and from this derives the principle difficulty of solution. The appearance of $\sin\theta$ on the right-hand side is tied to the original definition of the action-angle variables, equation (2.3).

4.3 Solution of Sacherer's Equation

In section 3 we found the beam current perturbation due to a density wave circulating in phase-space. With a small modification, we shall search for solutions of Sacherer's equation of this form. The trial perturbation is a clockwise wave:

$$\psi_1 = R(J)\exp(\alpha t)\cos(m\theta - wt) \qquad \text{with} \quad m>0 \quad \text{and} \quad w>0. \tag{4.7}$$

The growth rate is α and the *coherent* angular rotation frequency is w/m radian per second. The left-hand side of the Sacherer equation is

$$\left[\frac{\partial}{\partial t} + \omega_s\frac{\partial}{\partial\theta}\right]\psi_1 = R(J)e^{\alpha t}[\alpha\cos(m\theta - wt) \;+\; (w - m\omega_s)\sin(m\theta - wt)]\,. \tag{4.8}$$

For the moment we shall take m to be *odd*. From equation (3.10), the beam current perturbation is

$$I(x,t) = \exp(\alpha t)\sum_{p=1}^{\infty} I_p^s(0)\sin(px)\cos(wt)\,. \tag{4.9}$$

From section 1 on the 'G' function, we can use equation (1.7) to immediately write down the response of the beam environment: $F(x,t) =$

$$\begin{aligned} &+\exp(\alpha t)\sum_{p=1}^{\infty} I_p^s(0)[G_2(p,w)\cos(px) \;+\; G_1(p,w)\sin(px)]\cos(wt) \\ &-\exp(\alpha t)\sum_{p=1}^{\infty} I_p^s(0)[G_4(p,w)\cos(px) \;+\; G_3(p,w)\sin(px)]\sin(wt)\,. \end{aligned} \tag{4.10}$$

This signal contains lower ($p\omega_{\rm rf} - w$) and upper ($p\omega_{\rm rf} + w$) sidebands. Note the introduction of terms in $\cos(px)$ through G_2 and G_4; they have the correct spatial symmetry to produce a periodic variation of the stable phase and hence can drive an odd-order mode. As will be seen below, the G_1 and G_3 terms are superfluous. There are two more things we should say at this point. Firstly, strictly speaking 'G' does not give the response to an exponentially growing (or decaying) oscillation since the impedance 'Z' from which it is derived is the response to a stable sinusoid of infinite duration. However, if the incremental growth (or decrement) per cycle of the carrier frequency $p\omega_{\rm rf}$ is small then the approximation is good. Secondly, we should have calculated the response to the forward-travelling waves equation (3.9) and then transformed back into a frame co-moving with the bunch. However, the symmetries of the fields with respect to the bunch centre ($x = 0$) must be the same in both frames; so we were able to miss a step.

For brevity, we shall proceed with the notation introduced in section 1: c and s replace cosine and sine respectively, and superfices will indicate the argument of the trigonometric function. Substituting (4.8) and (4.10) into the Sacherer equation we find

$$R(J)\alpha[c^{m\theta}c^{wt} + s^{m\theta}s^{wt}] + R(J)(w - m\omega_s)[s^{m\theta}c^{wt} - c^{m\theta}s^{wt}] =$$
$$\sqrt{2J/\omega_s}\,\psi_0'\sin\theta \times \xi\sum_p I_p^s[G_2^p c^{px}c^{wt} + G_1^p s^{px}c^{wt} - G_4^p c^{px}s^{wt} - G_3^p s^{px}s^{wt}]\,. \tag{4.11}$$

The solutions have to be valid for all θ and all t. We begin by comparing coefficients of $\cos(wt)$ and $\sin(wt)$, which leads to 2 conditions:

$$R(J)\alpha[c^{m\theta} + (w - m\omega_s)s^{m\theta}] = \sqrt{2J/\omega_s}\,\psi_0'\sin\theta \times \xi\sum_p I_p^s[G_2^p c^{px} + G_1^p s^{px}] \quad \text{and} \tag{4.12}$$

$$R(J)\alpha[s^{m\theta} - (w - m\omega_s)c^{m\theta}] = \sqrt{2J/\omega_s}\,\psi_0'\sin\theta \times \xi\sum_p I_p^s[G_4^p c^{px} + G_3^p s^{px}](-1)\,. \tag{4.13}$$

Since $x = \sqrt{2J/\omega_s}\cos\theta$, the terms $\cos(px)$ and $\sin(px)$ are both even (or symmetric) in θ. Consequently, both have Fourier series with only cosine terms present. Suppose we integrate the equations over θ. The only way to avoid generating zero on the right-hand side is to multiply by $\sin(m\theta)$ before the integration. For brevity let $z = p\sqrt{2J/\omega_s}$.

4.3.1 Growth rate

The integral $\int$[equation (4.13)] $\times \sin(m\theta)\,d\theta$ gives $\alpha\pi R(J) =$

$$\sqrt{\frac{2J}{\omega_s}}\psi_0'(-\xi/2)\sum_p I_p^s\int_{-\pi}^{+\pi}[\cos(m-1)\theta - \cos(m+1)\theta][G_4^p\cos(z\cos\theta) + G_3^p\sin(z\cos\theta)]d\theta\,. \tag{4.14}$$

$$\text{Now}\quad \cos(z\cos\theta) = J_0(z) + 2\sum_{k=1}^{\infty}(-1)^k J_{2k}(z)\cos(2k\theta) \tag{4.15}$$

$$\text{and}\quad \sin(z\cos\theta) = 2\sum_{k=0}^{\infty}(-1)^k J_{2k+1}(z)\cos[(2k+1)\theta]\,. \tag{4.16}$$

Since we consider the case m odd, it follows that $m \pm 1$ is even, and so only terms from the series for $\cos(z \cos\theta)$ remain after the integration. Thus

$$\alpha R(J) = \sqrt{2J/\omega_s}\,\psi_0'(-\xi)\sum_p I_p^s G_4(p,w)(-1)^{(m-1)/2}\left\{J_{m-1}(z) + J_{m+1}(z)\right\}\,. \tag{4.17}$$

We use the recurrence relations for Bessel functions to give

$$\alpha R(J) = (z/p)\,\psi_0'(-\xi)\sum_p I_p^s G_4(p,w)(-1)^{(m-1)/2}(2m/z)J_m(z)\,. \tag{4.18}$$

The radial function $R(J)$ has still to be found self-consistently, but this can be avoided if in place of R we are content to find the spatial Fourier spectrum due to R. Following Laclare[4], we multiply both sides by $2\langle I\rangle(-1)^{(m-1)/2}J_m(q\sqrt{2J/\omega_s})$ and integrate over the action J. Recalling the definition of I_p^s given in equation (3.7), we find

$$\alpha I_q^s = m\,\xi\,\langle I\rangle\sum_{p=1}^{\infty}\mathcal{G}_{qp}\frac{2}{p}G_4(p,w)I_p^s \tag{4.19}$$

with

$$\mathcal{G}_{qp}(m) = \int \frac{\partial\psi_0}{\partial J}J_m(q\eta)J_m(p\eta)dJ \qquad \text{and} \quad \eta = \sqrt{\frac{2J}{\omega_s}}\,. \tag{4.20}$$

This is an eigen-equation. Using the results of section 1, $G_4(p,w)$ can be expressed in terms of $Z(p \pm w)$. For a narrow band impedance centred at frequency $p\omega_{\text{rf}}$ we find the *growth rate*:

$$\alpha = \xi\,\langle I\rangle\,\mathcal{G}_{pp}\frac{m}{p}\left(Z_r^{p+w} - Z_r^{p-w}\right)\,. \tag{4.21}$$

Note, $\xi = \omega_s/(\hat{V}_{\text{rf}}\cos\phi_s)$ so this is dimensionally correct provided that ψ_0 is normalized according to $\int \psi_0 dJ d\theta = 1$.

4.3.2 Coherent frequency

Now let us determine the coherent frequency. We multiply equation (4.12) by $\sin(m\theta)$ and integrate over θ. The working is identical to that for equation (4.13): the only terms surviving come from $G_2(p,w)\cos(px)$. Then we multiply throughout by $2\langle I\rangle(-1)^{(m-1)/2}J_m(q\eta)$ and integrate over J. Hence we find

$$(w - m\omega_s)I_q^s = m\,\xi\,\langle I\rangle\sum_{p=1}^{\infty}\mathcal{G}_{qp}\frac{2}{p}G_2(p,w)I_p^s\,. \tag{4.22}$$

Using equation (1.6) G_2 can be replaced by $Z(p \pm w)$. For a narrow band impedance that spans only one value of the carrier frequency, we find that the density wave ψ_1 rotates in phase-space at the *coherent* angular frequency:

$$\omega_c \equiv (w/m) = \omega_s + \xi\,\langle I\rangle\,\mathcal{G}_{pp}\frac{1}{p}\left(Z_i^{p+w} + Z_i^{p-w}\right)\,. \tag{4.23}$$

Note how it is the terms $G_2 c^{px} c^{wt}$ and $G_4 c^{px} s^{wt}$ which are *even* in x that drive the *odd*-order mode. Had we considered the case of m *even*, we should have found the mode driven by the terms *odd* in x; namely $G_2 s^{px} c^{wt}$ and $G_4 s^{px} s^{wt}$, from equation (1.5). In nearly every other respect the details of solution, for α and ω_c, are identical.

4.3.3 Radial function

For a mode driven by a single harmonic line $(ph)\omega_0$ it is a simple matter to find the radial function $R(J)$. Consider equation (4.19). If we divide the equation for component I_q by the equation for I_p, we find

$$I_q = \int \frac{\partial \psi_0}{\partial J} J_m(p\eta) J_m(q\eta) dJ \times [I_p/\mathcal{G}_{pp}] \,. \tag{4.24}$$

The value of I_p depends on the initial conditions, and determines the starting amplitude or 'seed' of the instability. Recall equations (3.6) and (3.7), which are of the form

$$I_q \propto \int R(J) J_m(q\eta) dJ \,. \tag{4.25}$$

By comparison, we see the radial function of the mth polar mode is

$$R_m(J) \propto \frac{\partial \psi_0}{\partial J} J_m[p\eta(J)] \,. \tag{4.26}$$

This function should have the property that for 'large' J its modulus tends toward zero more quickly than does $\psi_0(J)$ else there is a net negative particle density. The radial functions are, perhaps, more readily visualized in the more familiar polar coordinates (r, θ). Let $r = \eta(J)$ and hence $H = \omega_s^2 r^2/2$.

$$\text{In polar coordinates} \qquad R_m(r) \propto \frac{1}{r}\frac{\partial \psi_0}{\partial r} J_m(pr) \,. \tag{4.27}$$

Different stationary distributions support different forms of radial modes, and this is analogous to the fact that differing confining potentials in the Schrödinger wave equation give rise to different forms of eigen-modes. For instance, the stationary distribution $\psi_0 = (a^2 - r^2) \times (4/a^4)$ gives radial modes of the form $\psi_1 \propto J_m(pr)$. Solutions of this type were sketched previously. Figure 3 shows $J_1(r)\cos(\theta - \omega_s t)$ and figure 4 shows $J_2(r)\cos[2(\theta - \omega_s t)]$.

4.3.4 Anti-clockwise wave

Suppose we substitute an anti-clockwise wave of the form

$$\psi_1 = R_m(J)\exp(\alpha_m t)\cos(m\theta + wt) \qquad \text{with} \;\; m > 0 \;\; \text{and} \;\; w > 0 \tag{4.28}$$

into the self-consistent Sacherer equation (4.6). We follow through the working in an identical manner to the previous case, and find the following equations for growth rate and coherent frequency:

$$\begin{aligned} -\alpha_m &= \xi \langle I \rangle \mathcal{G}_{pp} \frac{m}{p} \left(Z_r^{p+w} - Z_r^{p-w} \right) \\ -(w/m) &= \omega_s + \xi \langle I \rangle \mathcal{G}_{pp} \frac{1}{p} \left(Z_i^{p+w} + Z_i^{p-w} \right) . \end{aligned}$$

The second equation implies a contradiction, since we initially assumed $w > 0$. Clearly, the Sacherer equation will simply not admit anti-clockwise waves with $\exp(\alpha t)$ time dependence.

This is a little mysterious. Given that waves circulating in either direction ($\pm\theta$) produce the same double-sideband frequency spectrum, why should not the anti-clockwise wave be able to drive itself? In the next section, we explain why this cannot be so; and by way of introduction we think first about the simple 1-D harmonic oscillator.

5 Driven Systems

5.1 Driven Harmonic Oscillator

It is well known that the response of a driven harmonic oscillator is very different when the excitation frequency is either on or off resonance. Away from resonance, the response is in phase with the drive, and the amplitude of the response is constant in time. At resonance, the response is in quadrature with the drive and the amplitude grows linearly with time. The details of the problem are less widely appreciated, but knowledge of them will prepare us for a following section. Consider the driven oscillator

$$y'' + \omega^2 y = f(x) \qquad \text{with initial conditions} \quad y(0) = \dot{y}(0) = 0 \,. \tag{5.1}$$

The Green's function solution for these particular boundary conditions is

$$y(x) = \frac{1}{\omega}\int_0^x f(t)\sin[\omega(x-t)]dt \,. \tag{5.2}$$

Suppose $f(x) = \sin(px)$ for $t \geq 0$ and $f(x) = 0$ for $t < 0$. The off-resonance ($p \neq \omega$) solution is

$$y(x) = \frac{\sin(px)}{(\omega^2 - p^2)} + \frac{p}{\omega}\frac{\sin(\omega x)}{(p^2 - \omega^2)} \,. \tag{5.3}$$

The second term represents 'ringing'. The on-resonance ($p = \omega$) solution is

$$y(x) = \frac{1}{\omega}\left[-x\cos(\omega x) + \frac{1}{\omega}\sin(\omega x)\right] \tag{5.4}$$

which exhibits the growing quadrature term, as anticipated. Examples of these solutions are shown below.

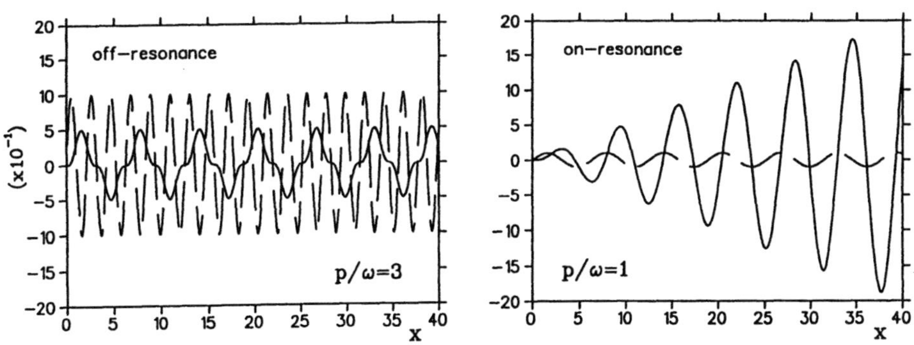

Solutions to $y'' + y = \sin(px/\omega)$ for two values of p.

The response is indicated by the solid line, and the drive function by the dashed line.

5.2 Vlasov Equation with harmonic drive

Let us suppose we drive the harmonic oscillators in the stationary distribution with some *known* small perturbing force $\varepsilon F(x,t)$ so that the particle trajectories are not greatly distorted, that is $\varepsilon << \omega_s^2$ and $|F| \leq 1$. Practically, this means exciting the beam with a high harmonic cavity. We assume the new distribution can be found by expanding about the stationary distribution. Accordingly we let $\psi_0 \to \psi_0 + \psi_1(J,\theta)$ and $H \to H + \Delta H$ with $\Delta H = \varepsilon \int^x F(r,t)dr$. The Vlasov equation for ψ_1 is found by the methods of section 4.1. Provided ψ_1 is less than or of order ε, the product terms can be neglected leading to the approximate evolution equation:

$$\left[\frac{\partial}{\partial t} + \omega_s \frac{\partial}{\partial \theta}\right] \psi_1 = \sqrt{2J/\omega_s}\, \psi_0' \sin(\theta) \times \varepsilon F(x,t)\,. \tag{5.5}$$

Consider the case $F(x,t) = \sin(px)\cos(wt)$, that is the excitation is odd in x. We substitute $x = \sqrt{2J/\omega_s}\cos\theta$ and expand $\sin(px)$ as a Fourier-Bessel series in θ. Let $\eta = \sqrt{2J/\omega_s}$, then the coefficients are a function of $p\eta$. After multiplying by $\sin\theta$, we use the trig' sum and difference rules to write the right-hand side of (5.5) as

$$\eta(J)\psi_0'(J) \times \varepsilon \cos(wt) \sum_{k=0}^{\infty}(-1)^k J_{2k+1}(p\eta)\left\{\sin[(k+1)2\theta] - \sin(2k\theta)\right\}\,. \tag{5.6}$$

5.2.1 Off resonance

Firstly, consider the case that w is *not* equal to ω_s. Hence we are to find the off-resonance response. Inspecting the RHS of expression (5.6), we see that our solution ψ_1 must have a symmetry so that the difference of partial derivatives $(\partial/\partial t - \omega_s \partial/\partial\theta)\psi_1$ yields a product of the form $\cos(wt)\sin(2n\theta)$ with $n = 1, 2, 3, \ldots$. Accordingly we try a solution of the form

$$\psi_1 = \sum_{m=1}^{\infty} R_m(J) \times \left[\frac{\cos[m(\theta - wt)]}{m(w-\omega_s)} - \frac{\cos[m(\theta + wt)]}{m(w+\omega_s)}\right]\,. \tag{5.7}$$

Substituting (5.7) into (5.5), we find

$$\sum_m R_m(J) 2\sin(m\theta) = \eta\, \psi_0' \times \varepsilon \sum_{k=0}^{\infty}(-1)^k J_{2k+1}(p\eta)\left\{\sin[2(k+1)\theta] - \sin[2k\theta]\right\}\,. \tag{5.8}$$

Immediately we see that all m must be even. We multiply by $\sin(m\theta)$ and integrate over θ to find the individual radial functions. The surviving terms come from $2(k+1) = m$ and $2k = m$.

$$R_m(J) = (1/2)\eta\, \psi_0'\, \varepsilon\, (-1)^{m/2}\left\{J_{m-1}(p\eta) + J_{m+1}(p\eta)\right\} = \eta\, \psi_0'\, \varepsilon\, (-1)^{m/2}(m/p\eta) J_m(p\eta)\,. \tag{5.9}$$

Hence, when the drive $\sin(px)\cos(wt)$ is off-resonance ($w \neq \omega_s$) the particular integral is

$$\psi_1 = \varepsilon\, \psi_0'(J) \sum_m^{\infty} \frac{m}{p} J_m(p\eta)(-1)^{m/2}\left[\frac{\cos[m(\theta - wt)]}{m(w-\omega_s)} - \frac{\cos[m(\theta + wt)]}{m(w+\omega_s)}\right] \tag{5.10}$$

with $m = 2, 4, 6, 8, \ldots$ and $\eta = \sqrt{2J/\omega_s}$. The solution is periodic in time, with constant amplitudes. Notice also that the solution consists of clockwise *and* anti-clockwise circulating density waves in phase-space. The medium on which both waves travel, namely $\psi_0(J)$, does, however, continue to spin clockwise at $\dot{\theta} = \omega_s$. Far from resonance the two waves are of almost equal amplitude. Close to resonance ($w \approx \omega_s$), the clockwise wave dominates and the rate of θ phase-slippage between ψ_0 and this wave is small. Very close to resonance, that is $(w - \omega_s) < |\varepsilon|$, our original approximations break down and equation (5.5) becomes inappropriate.

Note that to find the complete solution, for some particular set of initial conditions, we might have to add in multiples of the complementary functions, which are solutions of the homogeneous or free Vlasov equation. Had the excitation been an even function of x, such as $\cos(px)\cos(wt)$, then the response would consist of only odd polar waves so that the index $m = 1, 3, 5, 7, \ldots$

There is no resonance at $w = -\omega_s$, since in this text we never use the fiction[3] of negative frequency, but only what is physically measurable with, say, a spectrum analyser; that is positive frequencies.

5.2.2 On-resonance

When the ensemble is driven at the resonant frequency $w = \omega_s$, we should expect linear growth of the excited modes. From our experience with the harmonic oscillator, we should also expect the response to be in quadrature with respect to the off-resonance case. Hence we try solutions of the form $\psi = \alpha t \sin[m(\theta - \omega_s t)]$.

$$\text{Now} \quad \left[\frac{\partial}{\partial t} + \omega_s \frac{\partial}{\partial \theta}\right] \psi = \alpha[\sin(m\theta)\cos(\omega_s t) \; - \; \cos(m\theta)\sin(\omega_s t)] \, .$$

Comparing with the drive, $\sin(px)\cos(\omega_s t)$, we see that term in $\sin(\omega_s t)$ is unwanted; but this can be cancelled by adding to ψ an anti-clockwise wave of constant amplitude. Accordingly, we seek a solution of the form

$$\psi_1 = \sum_{m=1}^{\infty} R_m(J) \left[\alpha t \sin[m(\theta - \omega_s t)] \; - \; \frac{\alpha}{2m\omega_s} \cos[m(\theta + \omega_s t)]\right] \, . \tag{5.11}$$

Substituting this form in (5.5) and expanding $\sin(px)$ as a series in $\cos[(2k+1)\theta]$, we find

$$\sum_m \alpha R_m(J) \sin(m\theta) = \sqrt{2J/\omega_s}\, \psi_0' \, \varepsilon \sin\theta \sum_k (-1)^k J_{2k+1} \cos[(2k+1)\theta] \, . \tag{5.12}$$

This equation is solved for R_m as in the off-resonance case. Hence, when driven at resonance, the ensemble response (particular integral) is

$$\psi_1 = \varepsilon\, \psi_0'(J) \sum_m^{\infty} \frac{m}{p} J_m(p\eta)(-1)^{m/2} \left[t \sin[m(\theta - wt)] \; - \; \frac{\cos[m(\theta + wt)]}{2m\omega_s}\right] \tag{5.13}$$

with $m = 2, 4, 6, 8, \ldots$ and $\eta = \sqrt{2J/\omega_s}$. Importantly, we see that only the clockwise wave grows; the anti-clockwise wave has constant amplitude. Had the excitation been even [e.g. $\cos(px)$], then the response would be odd with only $m = 1, 3, 5, 7, \ldots$ allowed. Finally, note the solution (5.13) is not valid for long time scales, since for large t the perturbation ψ_1 may become large enough to invalidate the original approximations.

[3]Negative frequency only has meaning in the Argand diagram, which is itself an abstraction.

5.3 Conclusion

In the 1-D case of a harmonic oscillator driven at resonance, only the quadrature oscillation grows; the in-phase response is constant. In the 2-D case of the Vlasov equation driven on-resonance, only the clockwise wave grows; the anti-clockwise response is constant. In the case of a beam instability, the resonance frequency is no longer ω_s but, nevertheless, only one wave responds to the drive by growing: the clockwise wave. In fact, just for finding which of the density waves grow, it is immaterial that the drive is produced by the waves themselves. The resonance frequency is the coherent frequency (ω_c) defined in equation (4.23).

References

[1] G. Rees:"Intensity limitations in Circular Particle Accelerators", these proceedings.

[2] I.L. Kantor and A.S. Solodovnikov: *Hypercomplex Numbers, an Elementary Introduction to Algebras*, Springer-Verlag, 1989.

[3] S. Koscielniak: *Phase and Incoherent Frequency Shifts*, TRI-DN-90-K136 (Triumf Design Note).

[4] J.L. Laclare: *Bunched Beam Instabilities, Memorial Talk for F.J. Sacherer*, Proceedings of 11th International Conference on High-Energy Accelerators, Geneva, Switzerland, 1980.

LOW ENERGY ASPECTS OF CIRCULAR ACCELERATORS

Stephen D. Holmes
*Fermi National Accelerator Laboratory**
P.O. Box 500
Batavia, IL 60510

1 Introduction

Performance in circular accelerators can be limited by some of the same sorts of phenomona described by Miller[1] and Wangler[2] in their lectures on low energy behavior in linear accelerators. In general the strength of the perturbation required to degrade performance is reduced in circular accelerators because of the repetetive nature of the orbits. For example, we shall see that space-charge can severely limit performance in circular accelerators even when operating far from the "space-charge dominated regime" as defined in linear accelerators.

We will be discussing two particular aspects of low energy operation in circular accelerators--*space-charge* and *transition*. "Low energy" is defined within the context of these phenomona. We shall see that the phenomona are really relevant only in hadron accelerators. First, for space-charge the low energy regime is given approximately by

$$\gamma^2 \lesssim \frac{N_p r_p}{\varepsilon_N}\left(\frac{Z^2}{A}\right) .$$

Here γ is the relativistic factor of the beam, N_p is the total number of particles in the accelerator, r_p is the classical radius of the proton, ε_N is the transverse normalized beam emittance, and Z and A are the particle charge and atomic weight. A circular accelerator which operates in this regime is typically the initial circular ring in a hadron accelerator complex. Examples include the Fermilab, CERN, and Brookhaven Boosters, the Brookhaven AGS, the SSC Low Energy Booster, various Kaon Factories, and, perhaps surprisingly, the CERN Sp̄pS.[3] The primary impact of space-charge is to limit the transverse phase-space densities that can be delivered from these machines. This is obviously of vital importance in hadron colliders where luminosity is directly proportional to transverse phase space density.

Transition crossing takes place in accelerators in which the beam attains the condition

---

* Operated by Universities Research Association under contract with the U.S. Department of Energy

$$\gamma^2 = \frac{R}{<\eta>}$$

sometime during the acceleration cycle. Here R is the average radius of the machine and $<\eta>$ is the average dispersion function in the bending magnets. Such a condition has no analog in linear accelerators and is again typically encountered in the injector stages of hadron accelerator complexes. Examples here include the Fermilab Booster and Main Ring, the Brookhaven AGS and Relativistic Heavy Ion Collider (RHIC), the CERN PS, and, potentially, the SSC Low and Medium Energy Boosters. In contrast to space-charge, transition is primarily a longitudinal effect which can limit one's ability to accelerate high intensity beams. It is important in any accelerator complex which either contains a longitudinal bottleneck, or one in which short bunch length is important, e.g. in a collider with non-zero crossing angle.

2 Space-Charge

We will start by examining data obtained in the Fermilab Booster.[4] Figure 1 shows the beam emittance delivered from the Booster as a function of beam intensity. The emittance displayed here is the *95% normalized transverse emittance*. The relationship between the beam emittance (ε_N), beam sigma (σ), lattice function (β_L), and relativistic factors (γ and β) is given by

$$\sigma = \sqrt{\frac{\varepsilon_N \beta_L}{6\pi\beta\gamma}} \quad . \tag{1}$$

The data show a dependence of the delivered emittance on the delivered intensity of the form

$$\varepsilon_N/\pi \text{ (mm-mr)} = \text{Max}(7, 6N_p \text{x} 10^{-12})$$

where N_p is the total number of particles delivered. The maximum number of particles delivered is $3.3\text{x}10^{12}$ and corresponds to an emittance of 20π mm-mr-- the aperture of the machine at the injection energy of 200 MeV. Since the Booster utilizes an H^- multi-turn injection, the emittance should, in principle, be independent of the intensity. The observed correlation is believed to be due to space-charge forces at injection into the Booster.

The data shown in Figure 1 can be interpreted in terms of the Laslett tune-shift.[5] The Laslett formula relates the detuning in a circulating charged particle beam to basic beam intensity and kinematic parameters. Consider a bunch of charged particles (we assume q=e here) of length L, moving with a common velocity v, as shown in Figure 2. In order to simplify the discussion we will, for

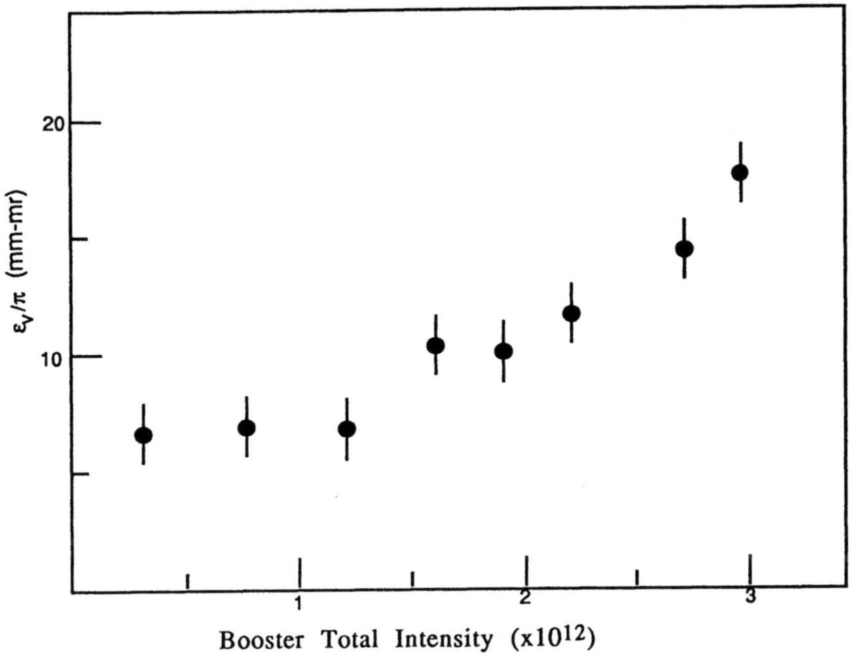

Figure 1: Transverse emittance delivered from the Fermilab Booster as a function of beam intensity.

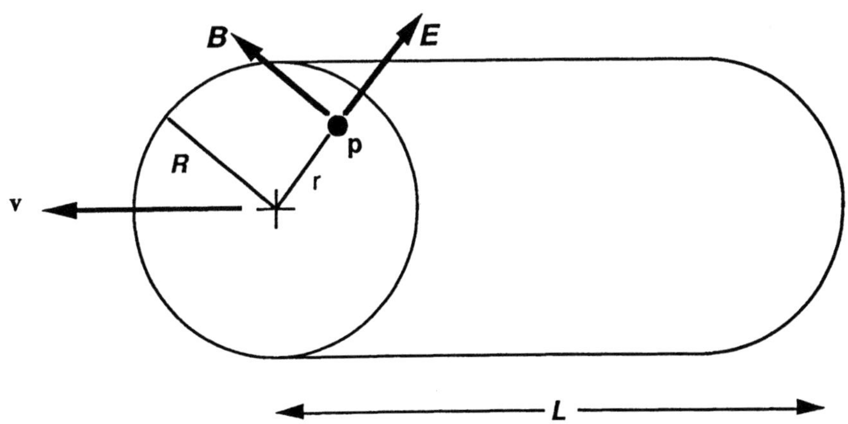

Figure 2: The electric and magnetic fields seen by a proton comoving at a velocity v with a bunch of length L and radius R, containing N protons.

the moment, assume a uniform charge distribution, and will let R be the radius and N the total number of protons in the bunch. The electric field seen by a comoving proton a distance r off axis is given by Gauss' Law:

$$\vec{E} = \frac{eN}{2\pi\varepsilon_o LR^2}\ \vec{r}\ .$$

Likewise, the magnetic field is

$$\vec{B} = \frac{v}{c^2}\,|\vec{E}|\ \hat{\phi}\ .$$

The force experienced by the proton is then given by

$$\vec{F} = e\vec{E} + \vec{v}\,x\vec{B} = \left(\frac{e^2N}{2\pi\varepsilon_o LR^2}\right)\left(1-\beta^2\right)\ \vec{r}\ .$$

Note that the force is linear in the displacement r and defocussing in nature. The effect on the proton is the same as that of a quadrupole with a strength parameter, k (=B'/[Bρ]), of

$$k = -\left(\frac{e^2}{2\pi\varepsilon_o mc^2}\right)\frac{N}{LR^2}\frac{1}{\gamma^3\beta^2} = \frac{2r_pN}{LR^2}\frac{1}{\gamma^3\beta^2}\ .$$

The tune shift due to a perturbation, Δk, over a distance l is given by

$$\Delta\nu = \frac{\beta_L \Delta k l}{4\pi}$$

where β_L is the lattice beta at the position of the perturbation. So for a complete traversal of the accelerator, the tune shift seen by the proton is

$$\Delta\nu = -\frac{\beta_L r_p}{2\pi R^2\gamma^3\beta^2}\left(\frac{NC}{L}\right)\ . \qquad (2)$$

Here C is the total circumference of the accelerator.

Note that the tune shift is independent of r under the assumption of a uniform charge distribution. In reality the charge distribution is more like a Gaussian in the transverse coordinates,

$$\frac{dN}{dxdy} = \frac{N}{2\pi\sigma^2}\exp\left(\frac{-(x^2+y^2)}{2\sigma^2}\right)\ .$$

So we can easily write the tune shift experienced by protons near the center of the beam by making the replacement $\pi R^2 \rightarrow 2\pi\sigma^2$ in eq. (2) and use eq. (1) to express it in terms of the emittance.We will also change from particles per bunch, N, to total particles, N_p, by rewriting the quantity (NC/L) as (N_p/B) where B is called the "bunching factor,"

$$\boxed{\Delta\nu = -\frac{3r_pN_p}{2\varepsilon_N\beta\gamma^2 B}\,.} \qquad (3)$$

Equation (3) is the Laslett tune shift formula. It gives the detuning of particles near the (transverse) center of the bunch. The tune shift is incoherent in the sense that different particles in the beam experience different tune shifts with large betatron amplitude particles experiencing no shift at all, i.e. $\Delta\nu$ is really a *tune spread.* The tune shift as a function of betatron amplitude has been given by Evans[6] in a previous accelerator school, and the interested reader is referred to those proceedings. In addition to the transverse dependence, if the longitudinal charge distribution is non-uniform a particle will experience a tune shift that is modulated at twice the synchrotron frequency as it travels through regions of varying charge density. There can also be coherent contributions to the tune shift arising from image currents/charges in the vacuum chamber. In general these are small compared to the incoherent tune shift for bunched beams and will be ignored here.

2.1 Observations

It has been observed in existing proton accelerators that there appears to be a limit on how large a $\Delta\nu$ can be obtained. For example the linear increase of transverse emittance with intensity in the Fermilab Booster shown in Figure 1 corresponds to a tune shift $\Delta\nu$ = 0.4 at the injection energy of 200 MeV. Other accelerators such as the Brookhaven AGS and the CERN PS Booster have attained $\Delta\nu$'s in the range 0.5-0.8. Note that in addition to limiting the phase-space density achievable, for an accelerator with a given aperture, any limit imposed on N_p/ε_N also translates into a limit on total intensity. It is believed today that the limit on acheivable $\Delta\nu$'s represents the most fundamental intensity limit in low energy proton synchrotrons.

The following characteristics of eq. (3) should be noted: First, the space-charge tune shift is proportional to the total charge in the accelerator. In particular this means that smaller circumference rings can support higher line charge densities for the same value of $\Delta\nu$. Such reasoning has provided the impetus for the construction of the AGS Booster, and for the inclusion of the SSC Low Energy Booster into the SSC injector design. Second, the kinematic factor,

$\beta\gamma^2$, in the denominator of (3) insures that space-charge is usually significant only at injection energy into the lowest energy circular accelerator in a hadron accelerator complex. Finally, (3) provides a prescription for curing an accelerator suffering from a space-charge limit--raising the injection energy. This reasoning has provided the impetus for raising the Fermilab Booster injection energy from 200 MeV to 400 Mev through the Fermilab Linac Upgrade, for the choice of 600 MeV for the SSC Linac energy, and again for construction of the AGS Booster.

In the spring of 1990 a series of measurements were completed in the Fermilab Booster to understand in greater detail how the beam emittance becomes diluted to creáte the correlation observed in Figure 1. Beam profiles were measured for a wide range of beam intensities at times varying between 40 and 3000 μs after injection. (The revolution period at injection is about 2.8 μs.) Representative results are shown in Figures 3 and 4. Figure 3 shows beam profiles measured 1000 μs after injection for six different beam intensities varying between $4\text{x}10^{11}$ and $3\text{x}10^{12}$. The increase in beam width with intensity is observed as is a change in the distribution shape. The data have been analyzed in terms of 38%, 68%, and 95% emittances. Figure 4 shows the time evolution of the emittance for an intensity of $2\text{x}10^{12}$. The following conclusions were drawn:

1. Any beam blowup appears to occur within the first twenty revolutions of the accelerator.
2. The beam profile changes, with the core being affected more than the tails.
3. The degree of blowup depends weakly on the bare tune of the machine over the range 6.6 to 6.9.

2.2 Phenomonological Descriptions

The question naturally arises, "Whence the limit?" For some time people have felt that the limit was connected with overlap of the beam with low-order machine resonances. Qualitatively this reasoning goes as follows: Since $\Delta\nu$ is really a tune spread, and clearly the half and integer resonances need to be avoided, then the largest $\Delta\nu$ tolerable must lie in the range 0.5 to 1.0, depending upon the selection of the bare tune. Obviously such a statement does not represent a very deep understanding. However, it has provided a criterion for designing new accelerators which are potentially space-charge limited, as long as one designs to a $\Delta\nu$ on the conservative side.

More recently the problem has been attacked using computer tracking simulations.[7,8,9,10,11] The codes have become increasingly sophisticated over the last five years and now typically track thousands of macroparticles over a few

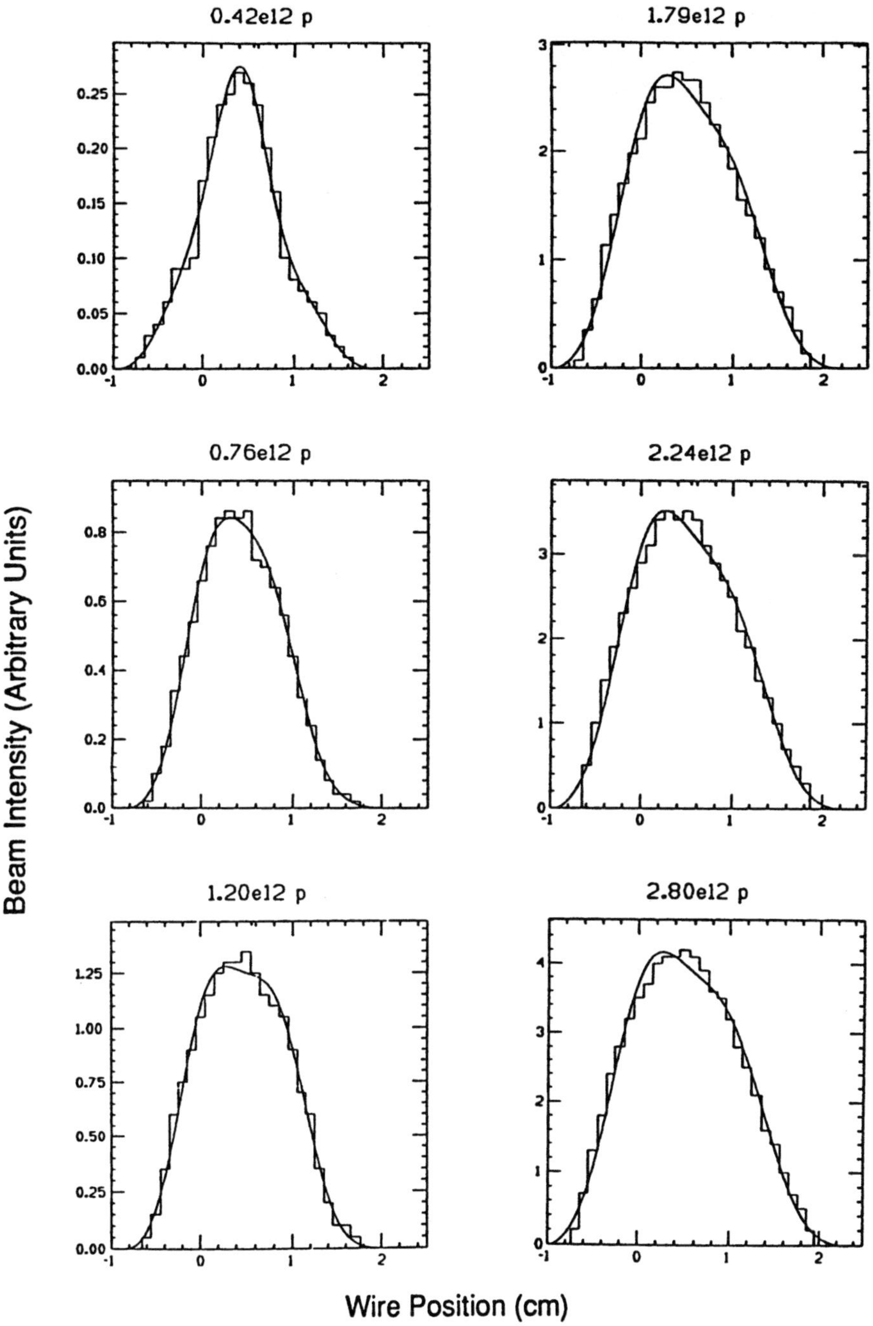

Figure 3: Beam profiles measured in the Fermilab Booster 1 msec after injection for a range of beam intensities.

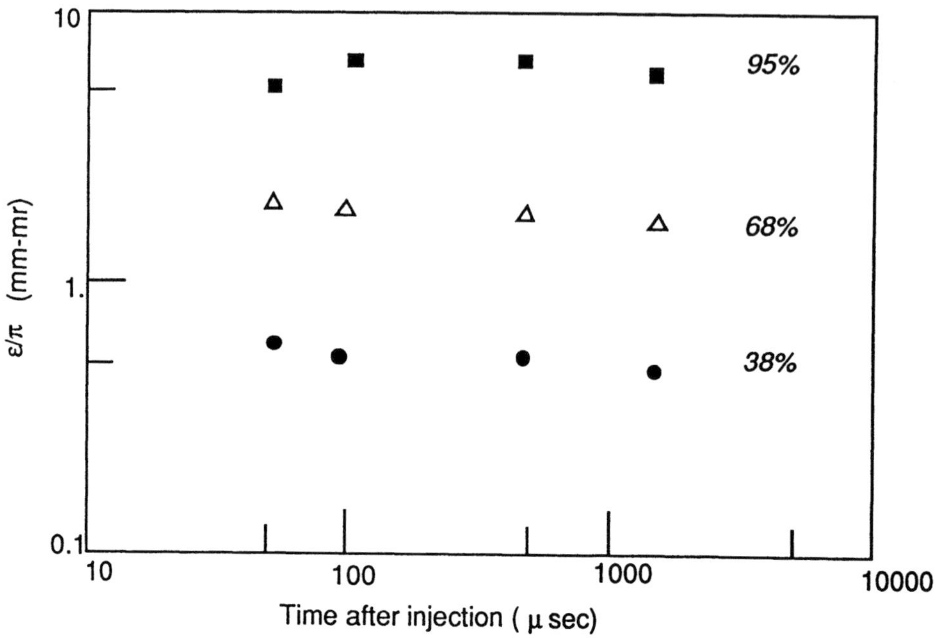

Figure 4: Time evolution of the 38%, 68%, and 95% (unnormalized) transverse emittance for $2x10^{12}$ protons ($2.5x10^{10}$/bunch) injected into the Fermilab Booster.

thousand turns, incorporating machine imperfections but not sychrotron oscillations.

Among the more advanced simulations are those of Machida,[10] who simulated the SSC Low Energy Booster in a self-consistent manner using 2000 macroparticles tracked with space-charge kicks applied 100 times per revolution. Machida is able to reproduce the characteristic emittance growth with intensity as shown in Figure 5. The $\Delta\nu$ represented here is in the range 0.3 to 0.4. He also investigated the time evolution of the emittance, the dependence on the bare tune, and the effect of machine errors driving second- and third-order resonances.

Among the conclusions drawn from the simulations are the following.

1. Emittance dilution is very fast, occurring typically in tens of turns. (Note that this conclusion justifies the exclusion of synchrotron oscillations from the simulations.)
2. Machine imperfection resonances other than the half and integer are irrelevant.
3. Location of the bare tune with respect to the half-integer can change achievable $\Delta\nu$'s by 50%.

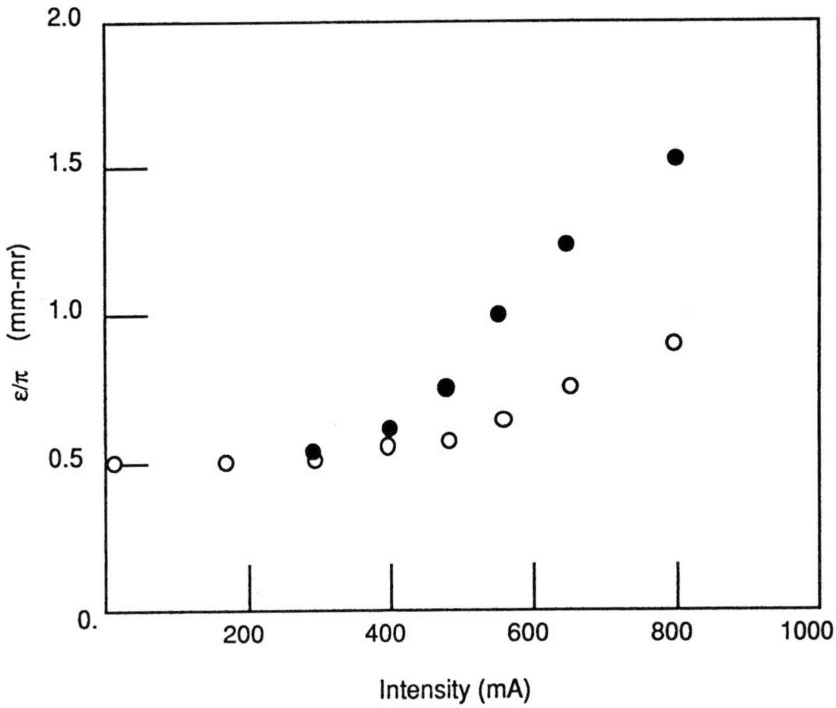

Figure 5: Transverse emittance (rms) as a function of beam intensity from the SSC Low Energy Booster simulation of Machida. The two data sets show the effect of corrected (open circles) and uncorrected (closed circles) half integer resonances.

4. Dilution is caused by the nonlinear fields of the beam itself. In particular the fourth-order resonance appears to be important. As a corollary intrinsic resonances are important and machine tunes approximating $4\nu = nS$ need to be avoided. (S is the superperiodicity of the accelerator. This point is particularly emphasized by both Parzan and Machida.)

Conclusion 1 is in obvious agreement with the observations described above.

It is worth mentioning that alternative descriptions, not given here, based on envelope equations potentially provide an alternative avenue toward understanding the behavior of circulating beams under the influence of space-charge.[12]

3 Transition

The primary concern in hadron accelerators in which the beam is required to pass through transition during the acceleration cycle is longitudinal emittance growth, possibly accompanied by beam loss. Beam loss can occur either if the longitudinal emittance becomes larger than the rf bucket area or if the product of the dispersion times the momentum spread becomes larger than the physical aperture of the accelerator.

One could reasonably ask at this point what difference, in the absence of beam loss, dilution of the longitudinal phase space could make, since there is not an obvious connection to collider luminosity as in the case of transverse emittances. At Fermilab the proton beam passes through transition both in the 8-GeV Booster and in the Main Ring, and will likely cross transition in the proposed Main Injector accelerator. The luminosity in the Tevatron proton-antiproton collider is sensitive to the preservation of longitudinal phase space in at least two respects: First, the antiproton production uses a bunch rotation scheme in the Main Ring, followed by debunching of the secondary antiproton beam in the Debuncher Ring. With this $\bar{p}$ production technique the production rate is nearly proportional to N_p/ε_L, where N_p is the number and ε_L the longitudinal emittance of the targeted, 120-GeV, protons. Second, both proton and antiproton bunches in the collider are created by a longitudinal bunch coalescing method. The efficiency of this operation, and hence the intensity of the bunches formed in the collider, is sensitive to the longitudinal emittance of the pre-coalesced bunches. As a result we at Fermilab are very sensitive to the transmission of beams through transition and become worried when we see behavior such as that shown in Figure 6.

Preservation of longitudinal emittance through transition is also important in other machines. For example, all heavy ion beams will pass through transition in RHIC. Because of the finite crossing angle in that machine the luminosity depends critically on the bunch length, i.e. the longitudinal emittance. In Kaon Factories preservation of longitudinal phase-space density in itself is not of overwhelming importance; however, acceleration of very large beam intensities with minimal beam loss is. Uncertainties in the ability to accelerate large quantities of beam through transition efficiently have led the TRIUMF people to propose a scheme in which beams never pass through transition within the complex.

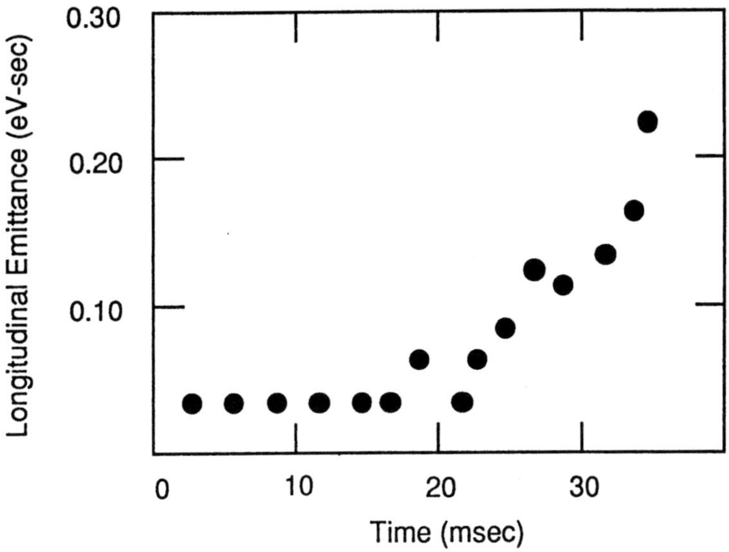

Figure 6: Longitudinal beam emittance as a function of time during the Fermilab Booster acceleration cycle. The beam intensity is about 2×10^{10}/bunch. Transition occurs at about 19 msec.

3.1 Consequences of Transition

Let us start by defining transition. Transition is related to the variation revolution periods acquired by particles having different momentum offsets relative to the nominal beam momentum. Two effects come into play here--the change in path length, and the change in velocity with momentum:

$$\frac{\Delta T}{T} = \frac{\Delta C}{C} - \frac{\Delta \beta}{\beta} .$$

We will write the path length for a particle to complete one revolution of the accelerator as

$$\frac{\Delta C}{C_0} = \alpha_0 \delta + \alpha_1 \delta^2 + \ldots \qquad (4)$$

where $\delta = \Delta p / p_0$ is the momentum offset relative to the central momentum, p_0, and C_0 is the path length corresponding to p_0. The parameters α_0 and α_1 are related to the optics of the accelerator lattice. Note that

$$\alpha_0 = \left. \frac{dC/C}{dp/p} \right|_{\delta=0}$$

is also known as the *momentum compaction.* The transition gamma is defined in terms of the momentum compaction and is related to the dispersion and mean radius of the accelerator by

$$\gamma_t \equiv \frac{1}{\sqrt{\alpha_0}} = \sqrt{\frac{R}{<\eta_H>}}.$$

Here R is the mean radius and $<\eta_H>$ is the average dispersion in the dipole magnets. The change in speed can be easily worked out and is given by

$$\frac{\Delta\beta}{\beta} = \frac{1}{\gamma^2}\left(\frac{\Delta p}{p}\right).$$

So, our final result is

$$\boxed{\frac{\Delta T}{T} = \left(\frac{1}{\gamma_t^2} - \frac{1}{\gamma^2}\right)\left(\frac{\Delta p}{p}\right)}.$$

The quantity multiplying $\Delta p/p$ is called the *phase slip factor,* η. As the beam is accelerated through transition η changes sign, attaining the value $\eta=0$ when $\gamma=\gamma_t$.

Crossing through transition has several well-known consequences:

1. $\phi_s \rightarrow \pi - \phi_s$. The stable synchronous phase moves from one side of the rf wave to the other.
2. $\Delta p/p \rightarrow \infty$. The momentum spread in the beam gets large.
3. $\sigma_L \rightarrow 0$. The bunch length gets short.
4. $f_s \rightarrow 0$. The synchrotron frequency goes to zero.
5. $A_{Bucket} \rightarrow \infty$. The rf bucket area gets large.

None of the above listed consequences directly interferes with the acceleration of beam through transition in circular accelerators. It is common practice to flip the phase of the rf at transition to deal with 1. and to accelerate quickly so that 2. is never attained. Consequences 1 to 5 are not entirely benign, however, when one accelerates substantial quantities of beam through transition.

3.1.1 Microwave Instability

The longitudinal microwave threshold for a Gaussian bunch is given by[13]

$$\frac{Z_{\parallel}}{n} \leq \frac{2\pi|\eta|(E/e)}{I_p}\left(\frac{\sigma_E}{E}\right)^2. \qquad (5)$$

Here I_p is the peak current in the beam ($eN/\sqrt{2\pi}\,\sigma_t$ for a Gaussian bunch). As can be seen from the formula, the threshold impedence goes to zero as the Landau damping disappears at transition. This is shown graphically in Figure 7, where the threshold limit is given for the Fermilab Main Injector parameters as a function of time during the acceleration cycle.

Fortunately the expression (5) is derived only for a non-accelerating beam, i.e. $\dot{\gamma}=0$. A more general expression has been derived by J. Wei[14] for the case of a parabolic charge distribution and $\dot{\gamma} \neq 0$:

$$\frac{Z_{\|}}{n} \leq \frac{3V_{rf}\cos\phi_s}{8hI_p} L_\phi^2 .$$

Here h is the harmonic number and L_ϕ is the bunch length in radians. For the Fermilab Main Injector the limit here is 10Ω, not the 0Ω indicated by eq. (5). The microwave instability at transition could become a problem if bunch intensities of order 5×10^{11} were ever realized in the Main Injector.

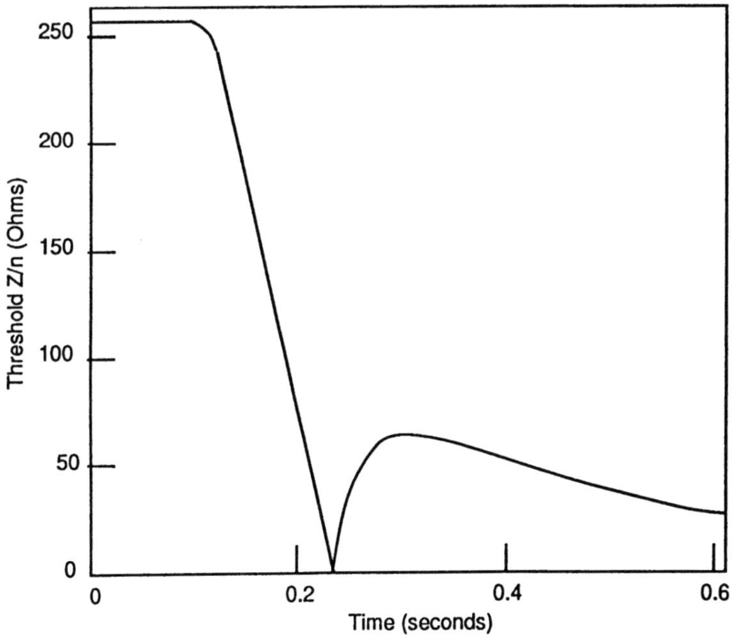

Figure 7: Longitudinal microwave instability threshold as a function of time in the Fermilab Main Injector acceleration cycle. Transition is at about 0.24 seconds. Parameters used are 5×10^{10} protons/bunch, 0.25 eV-sec bunch area before transition, 0.5 eV-sec after.

3.1.2 Non-adiabatic Time

As the beam approaches transition η approaches 0, which means that the particles become locked into a particular phase relative to the rf wave. As a consequence, close to transition the particles constituting the bunch cannot follow as the rf bucket shape changes rapidly. This results in a so-called *non-adiabatic time* during which the concept of an rf bucket has no meaning. This was recognized in the very early days by Courant and Snyder.[15] The non-adiabatic time is given by

$$\boxed{T_{na} = \left(\frac{\pi \beta_t^2 \gamma_t^3}{h \omega_0^2 \dot{\gamma}} \frac{E_t/e}{V_{rf} \cos\phi_s} \right)^{1/3}} . \qquad (6)$$

In the Main Injector the non-adiabatic time will be about 2.0 msec.

3.1.3 Johnsen Effect

The Johnsen effect arises from the chromatic nonlinearity represented by a nonzero α_1 in eq. (4). We will write the expression for the transition gamma of different particles in the beam to second order in δ. In doing so we must make sure we are always defining γ_t in terms of local derivatives, rather than differences from the nominal circumference, C_0:

$$\frac{1}{\gamma_t^2(\delta)} = \frac{p}{C} \frac{dC}{dp} .$$

After taking appropriate derivatives and some algebra, we obtain the expression for the variation of γ_t with δ:

$$\gamma_t(\delta) = \gamma_t(0) \left[1 - \frac{1}{2} \left(1 + \frac{2\alpha_1}{\alpha_0} - \alpha_0 \right) \delta \right] .$$

The variation of γ_t over the particles making up the beam, coupled with their different energies, means that all particles within the beam do not cross transition at the same time. We will define the *Johnsen*[16] or *nonlinear time* as the time difference between the passage through transition of a particle at $\delta = 0$ and that of a particle at $\delta = \sigma_p/p$. The reader can work it out and will find

$$\boxed{T_J = \frac{\gamma_t(0) \left(\frac{3}{2} + \frac{\alpha_1}{\alpha_0} - \frac{\alpha_0}{2} \right)}{\dot{\gamma}} \left(\frac{\sigma_p}{p} \right)} . \qquad (7)$$

In the Main Injector the nonlinear time will be about 2.7 msec.

A problem in the transmission of beams through transition can arise if the nonlinear time is larger than the nonadiabatic time. In this case some particles in the beam will be above, while others will be below, transition for a period of time during which the motion of particles in phase space is describable by the standard set of trajectories, i.e. some particles will be found on bounded and others on unbounded trajectories. This situation will persist until the non-adiabatic time is entered and the particles become locked onto a certain rf phase. The ultimate evolution of the distribution is shown in the simulation[17] of Figure 8.

Wei[14] has given a semi-analytic description of the emittance dilution arising from the Johnsen effect. He finds that the emittance dilution can be parameterized approximately as

$$\frac{\Delta\varepsilon}{\varepsilon} = \begin{cases} 0.76 T_J/T_{na} & T_J << T_{na} \\ e^{4/3(T_J/T_{na})} - 1 & T_J > T_{na}. \end{cases}$$

For the Main Injector the formula predicts emittance growth of about a factor of five, in reasonable agreement with the simulation. The Johnsen effect is expected to be the dominant mechanism for longitudinal emittance dilution through transition in the Main Injector.

3.1.4 Space-Charge (Bunch shape mismatch)

For intense beams longitudinal space-charge forces can distort the rf waveform over the length of the bunch. The effect is a loss of longitudinal focussing below transition and a gain above. (The effect is the same as that described by Hoffman[18] except that the space-charge associated impedence is capacitive rather than inductive.) The change in longitudinal focussing as the beam passes through transition is accompanied by a change in the rf bucket area. If the change in bucket area is significant the mismatch between the bunch and bucket shapes will cause dilution of the longitudinal phase space. The mismatch is proportional to the ratio of the self-field of the beam to the externally applied rf voltage. An approximate formula is given by Wei for a parabolic distribution,

$$\frac{\Delta\varepsilon}{\varepsilon} = \frac{2hI_p|Z_{\parallel}/n|}{V_{rf}\cos\phi_s\, L_\phi^2}$$

where the space-charge impedence is given by

$$\frac{Z_{\parallel}}{n} = \frac{ig_0 Z_0}{2\beta\gamma^2} \; .$$

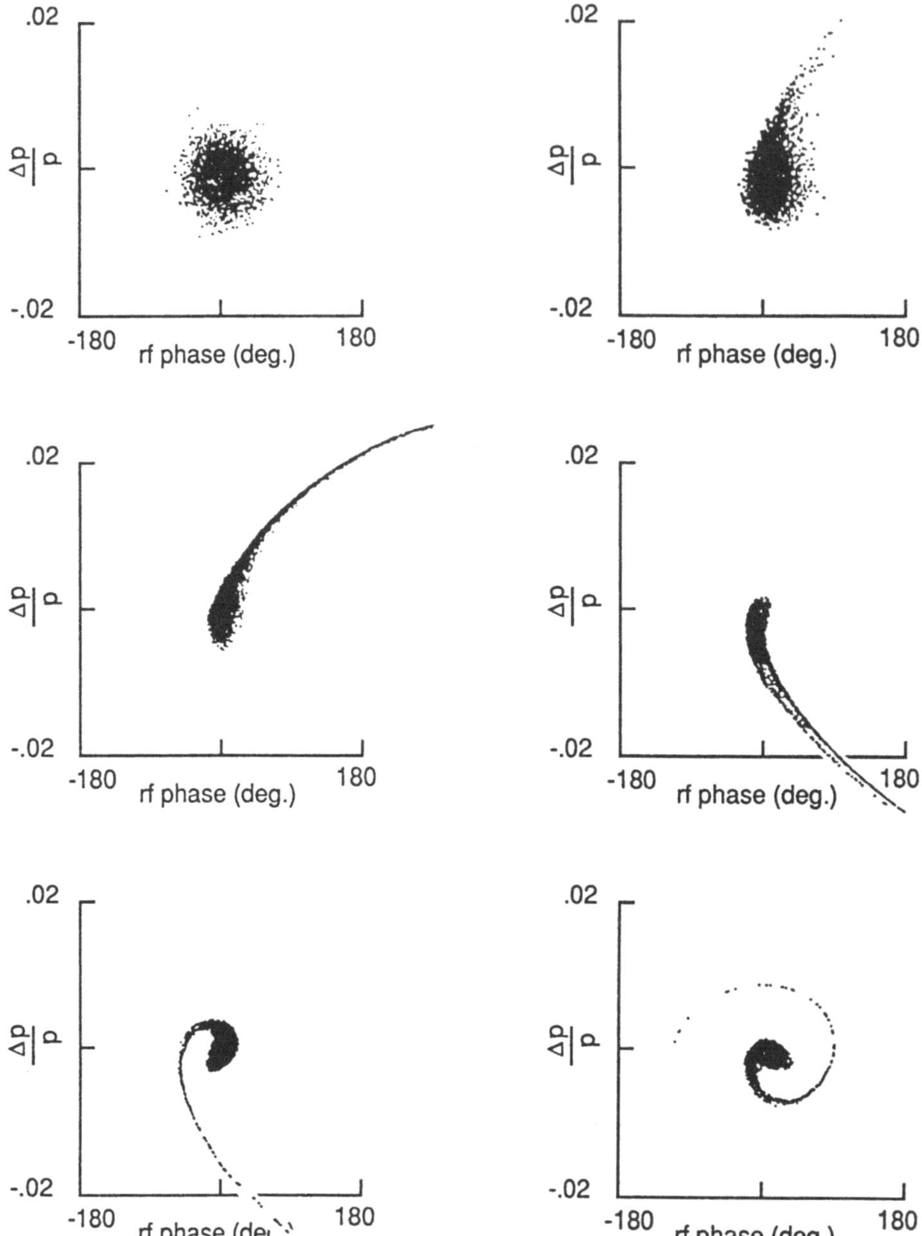

Figure 8: Simulation of the longitudinal phase space for transition crossing in the Main Injector with a 0.4-eV-sec bunch. Time progresses from left to right and top to bottom. Transition is crossed between the second and third frames. For this example $\alpha_0 = 2.5\text{x}10^{-3}$ and $\alpha_1 = 5\text{x}10^{-3}$.

Here g_o is a geometric factor (≈4) and Z_o is the impedence of free space ($1/\varepsilon_o c$ = 377Ω). Note the presence of the same $\beta\gamma^2$ in the denominator as in the transverse space-charge expression (3). For intensities characteristic of the Main Injector this effect is small. However it could become significant if the intensity were to be increased by a factor of 5 or so.

3.1.5 Umstätter Effect

The final effect we will consider for high intensity beams is connected with the transverse space-charge described earlier. The change in betatron tune given in eq. (3) is accompanied by a change in γ_t. Particles at the head of the bunch will see a higher γ_t than those at the center. As in the case of the Johnsen effect all particles do not cross transition at the same time, and a "nonlinear" time can be defined.[19] This time is

$$T_{\mathrm{Umstatter}} \propto \frac{N_p}{\dot{\gamma}\,\beta^2\gamma_t^3}$$

and is small compared to the Johnsen time in the Main Injector. As in the case of the longitudinal space-charge the Umstätter effect would become appreciable only if the intensity in the Main Injector were to increase to ~$3\text{x}10^{11}$/bunch.

3.2 Strategies for Coping with Transition

A number of strategies have been conceived of to ameliorate the effects of transition crossing in circular accelerators. The most straightforward approach usually involves accelerating through transition as quickly as possible so that the effects described above don't have time to come into play. The basic limitation here is the availability of rf accelerating voltage. A second simple approach involves tailoring the rf voltage program through transition in a manner that trades off the various effects listed above. This is generally done empirically on existing machines. However, a number of more sophisticated measures can be taken when an accelerator is still in the design stage.

3.2.1 Transition Jump

One way to increase the effective acceleration rate through transition is to implement a system of pulsed quadrupole magnets in the accelerator that will instantaneously change the optics of the ring in such a way that the tunes stay fixed but γ_t drops as transition is approached. This is shown schematically in Figure 9. Such schemes are implemented in the Fermilab[20] and CERN PS Boosters. Typically γ_t is dropped by one unit in 100 μsec, resulting in an effective $\dot{\gamma}$ of

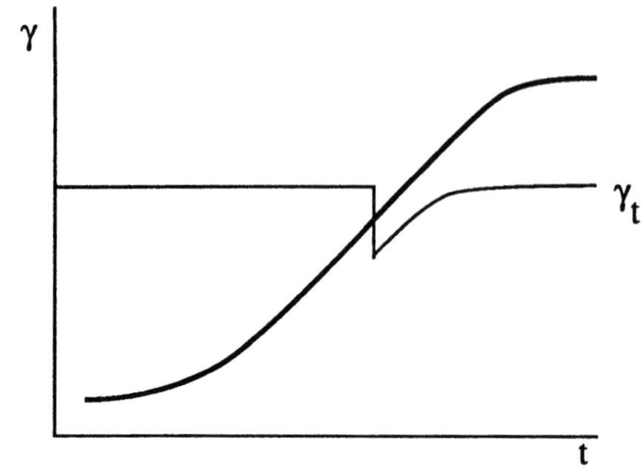

Figure 9: Schematic description of a γ_t jump scheme.

10000/sec. This represents an increase of a factor 10 to 100 in $\dot{\gamma}$ and reduces the associated nonlinear times by a corresponding amount. In addition, the longer bunches maintained by not allowing the beam to get too close to transition during the adiabatic period minimize effects that are proportional to peak currents such as the microwave instability and space-charge. The downside of these schemes is that they typically introduce a large dispersion wave into the lattice precisely at the time when the momentum spread is getting large.

3.2.2 Transition Avoidance

Avoiding transition by removing γ_t from the acceleration range obviously solves the problems described above. The most straightforward way to do this is to design the accelerator so that the transition energy lies either below the injection energy or above the extraction energy. This condition is met in most high energy hadron accelerators and in all existing electron machines. However, since FODO cell based lattices exhibit approximate equality between γ_t and the horizontal tune, hadron accelerators in the 5- to 20-GeV range often cannot be designed to avoid transition in a natural manner.

One approach which has been investigated is the use of different types of lattice cells to produce a γ_t that is imaginary. (An imaginary γ_t simply means that the average value of the dispersion is less than zero.) Such lattices have been examined as potential designs for the Fermilab Main Injector,[21] the TRIUMF Kaon Factory,[22] and the SSC Low Energy Booster.[23] An example of a cell for a possible Main Injector design is shown in Figure 10. The idea is to create negative dispersion by providing a dipole-free straight section with a phase advance of around 150°.

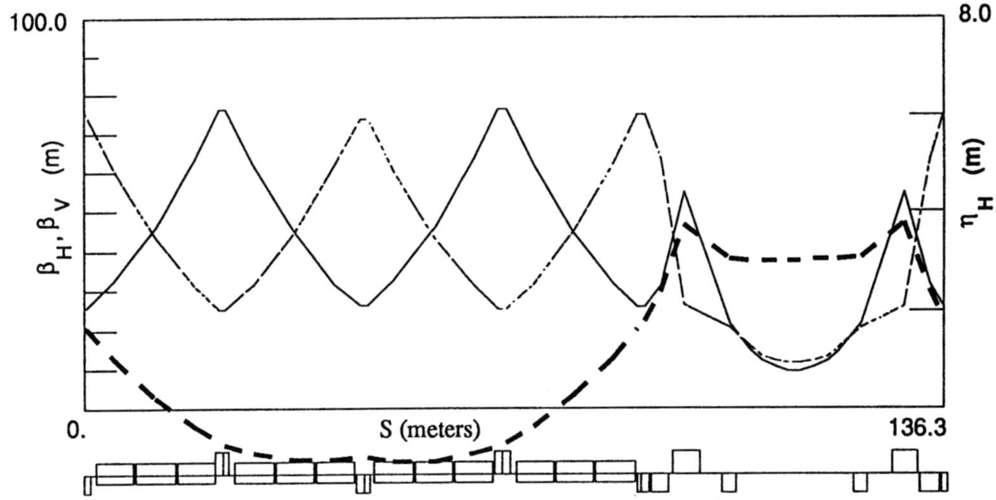

Figure 10:A cell that will produce an imaginary γ_t (39i). The phase advance across the cell is 257°. Twenty-four such cells could be used to construct the 150-GeV Main Injector. Note the negative average dispersion through the dipoles.

The scheme shown does a good job of keeping the maximum dispersion and beta functions low while not creating abnormal chromaticities. It has the disadvantage of a relatively low dipole packing factor, 53% in this example, and a variety of quadrupole strengths. Further work at Fermilab has indicated that packing factors approaching 67% might be attained with this scheme.

3.2.3 α_1 Control

Note that in eq. (7), for $\alpha_1 = -\frac{3}{2}\alpha_0$, the nonlinear time is approximately zero. So, one can contemplate eliminating the Johnsen effect by designing a lattice with the appropriate α_1. As might be expected, the value of α_1 depends on the distribution of sextupole fields within the ring. Peggs and Bogacz[24] have derived an analytic expression for α_1 as a function of sextupole strength in a FODO cell. A comparison between the analytic prediction and a simulation using the program MAD is given in the table below.

In the table "f" represents the strength of two families of chromaticity-correcting sextupoles located at the F and D quadrupoles, and "g" represents the strength of a third sextrupole at the half-cell midpoint. The point (0,0) has all sextupoles turned off. The point (1,0) has the horizontal and vertical chromaticities adjusted to zero with the third sextupole family off. Finally, (1,5) represents the correction of horizontal and vertical chromaticities with the third sextupole running at five times the strength of the chromaticity sextupoles.

	$\alpha_0(\times 10^{-3})$		$\alpha_1(\times 10^{-3})$	
(f,g)	predicted	simulated	predicted	simulated
(0,0)	2.956	2.956	3.213	3.332
(1,0)	2.956	2.956	0.129	0.244
(1,5)	2.956	2.956	-0.638	-0.512

As can be extrapolated from the table, the third sextupole family would be required to run about a factor of 12 higher than the chromaticity-correcting sextupoles in order to attain $\alpha_1=-\frac{3}{2}\alpha_0$. However, Peggs and Bogacz have found that the introduction of a small modulation of η_H around the ring can reduce the required strength of the third sextupole family by about a factor of ten. This technique is being pursued at Fermilab and may be incorporated into the Main Injector design.

3.2.4 Higher Harmonic Cavity

Griffin and MacLachlan at Fermilab have suggested the use of a higher harmonic cavity[25] both in the existing Main Ring and in the proposed Main Injector as a means of flattening the rf voltage wave form around transition. In the proposed implementation a second harmonic cavity is turned on at the beginning of the nonlinear time and the sychronous phase is simultaneously switched to 90°. This scheme offers two distinct advantages: First, it keeps the bunches long, and I_p low, reducing susceptibility to the micorowave instability and space-charge effects. Second, it combats the Johnsen effect by providing all particles with the same voltage during the nonlinear and nonadiabatic times.

A simulation using the program ESME is shown in Figure 11. The simulation is done for the nominal Main Injector parameters and includes longitudinal space-charge as well as an external impedence with $Z/n=5\Omega$ centered at 1.7 GHz. For the case shown the emittance dilution is 2% with $N_p=6\times10^{10}$/bunch and $\alpha_1=3\times10^{-3}$! This technique obviously shows great promise for transmission of high intensity beams through transition.

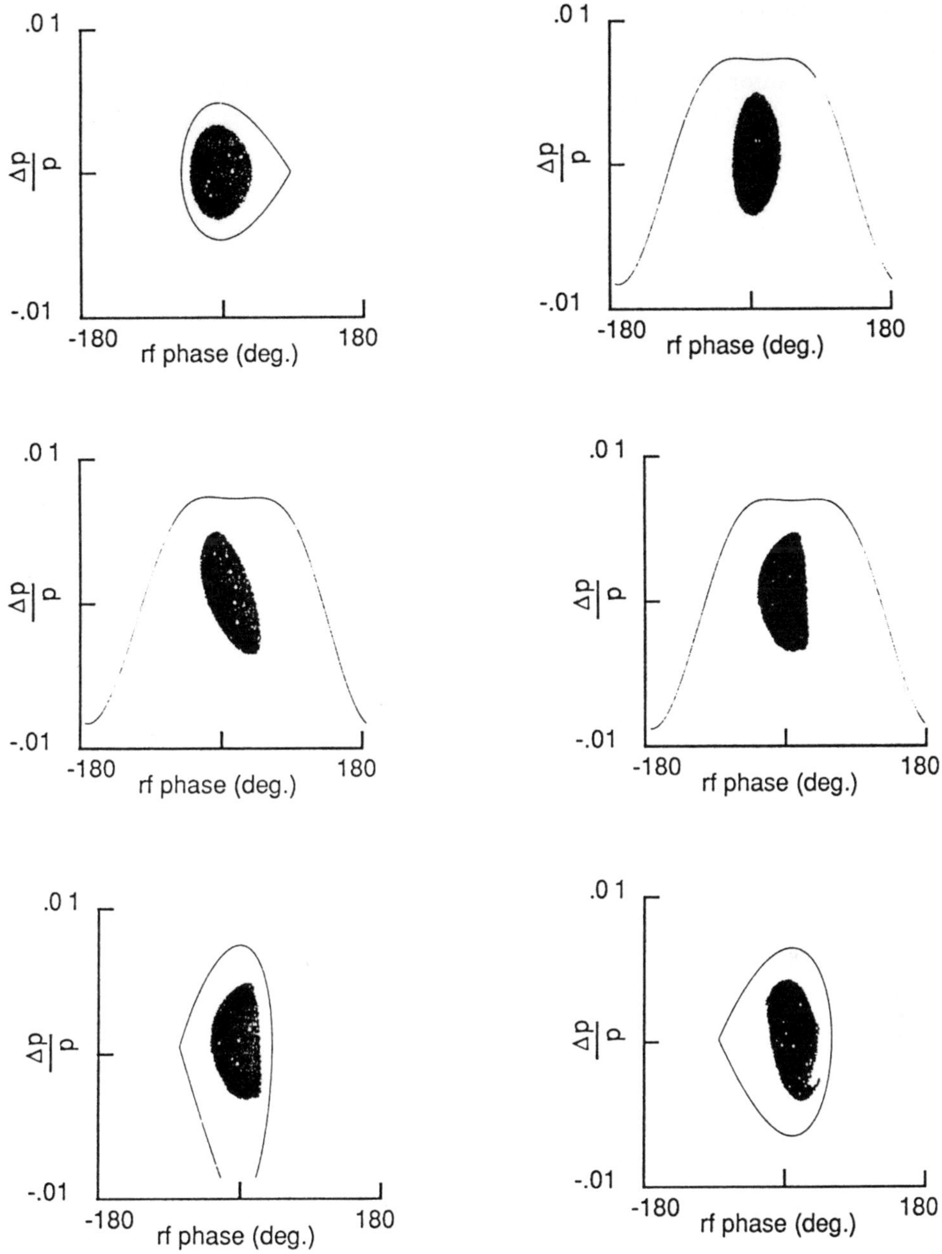

Figure 11: ESME simulation of transmission of beam through transition in the Main Injector using a second harmonic (106-MHz) cavity. The number of protons/bunch is 6×10^{10} with a longitudinal emittance of 0.5 eV-sec. The rf voltage wave form is shown during the period that the second harmonic cavity is on.

4 Summary

Particle intensities, N, and densities, N/ε, emanating from circular accelerators can be limited for $\gamma<<\infty$. Limits occur in both transverse and longitudinal phase space. These effects are generally present only in proton and heavy ion accelerators/storage rings.

The primary limit to the transverse densities achievable in a low energy hadron accelerator is due to space-charge forces at injection. Values of the Laslett tune shift parameter achieved in existing accelerators range from 0.4 to 0.9. Our understanding of the phenomona is still rudimentary but developing rapidly with the help of simulations. This limit affects essentially all hadron accelerator complexes. The cures include raising injection energies, increasing the degree of cascading in the complex, and lengthening the bunches.

Transition crossing represents an important limitation in the achievement of high longitudinal phase-space densities in circular accelerators. A variety of effects need to be contended with while passing through transition, including the microwave instability, space-charge, and the Johnsen effects. These are exacerbated by the tendency of the bunch length to approach zero, accompanied by a loss of Landau damping as transition is approached. Transition crossing impacts essentially all proton accelerator complexes, and even the high energy storage ring RHIC. Potential cures include avoidance, γ_t jumps, α_1 control, and higher harmonic cavities.

5 References

1. R. Bruce Miller, these proceedings.

2. T. Wangler, these proceedings.

3. L. Evans, these proceedings.

4. C. Ankenbrandt and S.D. Holmes, "Limits on the Transverse Phase Space Density in the Fermilab Booster," in Proc. 1987 IEEE Particle Accelerator Conf., Washington, DC, p. 1066.

5. L.J. Laslett, "On Intensity Limitations Imposed by Transverse Space-Charge Effects in Circular Accelerators," in Proc. 1963 Summer Study on Storage Rings, Brookhaven National Laboratory, BNL7534, p. 324.

6. L. Evans, "Beam Effects in Hadron Colliders," in Physics of High Energy Particle Accelerators, AIP Conf. Proc. 127, p. 243 (1985).

7. I. Hofmann and K. Beckert, "Resonance Crossing in the Presence of Space Charge," IEEE Trans. on Nucl. Sci., **NS-32**, 2264 (1985).

8. M. Furman, "Effect of the Space-Charge Force on Tracking at Low Energy," in Proc. 1987 IEEE Particle Accelerator Conf., Washington, DC, p. 1034.

9. G. Parzan, Nucl. Instrum. and Methods **A281**, 413 (1989); "The Effects of the Intrinsic Space Charge Resonances on the Space Charge Limit," BNL44562, Brookhaven National Laboratory (1990).

10. S. Machida, "Space Charge Effects in Low Energy Synchrotrons," in Proc. 2nd European Particle Accelerator Conf., Nice, France (1990), p. 1655; "Space Charge Effects in Low Energy Proton Synchrotrons," U. of Tokyo/Texas Accelerator Center PhD. Thesis (1990).

11. S.R. Mane, "Space-Charge Effects in the Fermilab Main Ring at 8 GeV," in Proc. 1989 IEEE Particle Accelerator Conf., Chicago, p. 1804.

12. J. Rosensweig and P. Zhou, "Envelope Instability in Low Energy Proton Synchrotrons," Fermilab FN-539 (1990).

13. S. Krinsky and J.M. Wang, Particle Accelerators **17**, 109 (1985).

14. J. Wei, "Longitudinal Dynamics of the Non-Adiabatic Regime on Alternating Gradient Synchrotrons," State U. of New York at Stony Brook PhD Thesis (1990); "Transition Crossing in the Main Injector," in Proc. Fermilab III Instabilities Workshop, Fermilab TM-1696 (1990).

15. E.D. Courant and H.S. Snyder, Ann. Phys. **3**, 1 (1958).

16. K. Johnsen, in Proc. CERN Symp. on High-Energy Accelerators and Pion Physics, Geneva (1956), **Vol. 1**, p. 106.

17. A. Bogacz, "Transition Crossing in the Main Injector - ESME Simulation," in Proc. Fermilab III Instabilities Workshop, Fermilab TM-1696 (1990).

18. A. Hoffman, these proceedings.

19. A. Sørenssen, Particle Accelerators **6**, 141 (1975).

20. W. Merz, C. Ankenbrandt, and K. Koepke, "Transition Jump system for the Fermilab Booster," in Proc. 1987 IEEE Particle Accelerator Conf., Washington, DC, p. 1343.

21. D. Trbojevic, D. Finley, R. Gerig, and S. Holmes, "Design Method for High Energy Accelerator Without Transition Energy," in Proc. 2nd European Particle Accelerator Conf., Nice, France (1990), p. 1536.

22. TRIUMF-KAON Project, "KAON Factory Study-Accelerator Design Report," TRIUMF, Vancouver, B.C., Canada (1990).

23. R. Gerig, SSC Laboratory, Private Communication (1990).

24. A. Bogacz and S. Peggs, "Comments on the Behavior of α_1 in the Main Injector γ_t Jump Schemes," in Proc. Fermilab III Instabilities Workshop, Fermilab TM-1696 (1990).

25. J. Griffin and J. MacLachlan, abstract submitted to 1991 IEEE Particle Accelerator Conference, San Francisco.

HIGH CURRENT BEAM TRANSPORT IN LINEAR ACCELERATOR STRUCTURES

R. Bruce Miller

TITAN/SPECTRON DIVISION
2340 Alamo SE, Suite 300, Albuquerque, NM 87106

ABSTRACT

In recent years, a number of advanced linear induction accelerators (LIAs) have been developed which have the capability for accelerating very high beam currents, provided that satisfactory answers are obtained to a number of beam transport and stability issues. Examples of this new technology include the advanced technology accelerator (ATA), the RADLAC accelerators, and the LIU accelerators of the Soviet Union. In this review we analyze several important beam transport issues pertaining to the propagation of very high current electron beams in and through the accelerating structures. In particular, we develop criteria for various equilibrium configurations, injector designs for producing the equilibria, general beam stability criteria, zero-order analyses of accelerating gap designs and considerations of various multiple-gap instabilities, including transverse beam break-up and image displacement. We also briefly describe the new transport technique of ion channel guiding. Finally, we analyze the special problems associated with transporting high current ion beams (protons) through such linacs.

1 PARAXIAL EQUATION

This paper pertains to the transport of relatively high current beams in linear accelerator structures. Since much of this discussion may not be very familiar, we begin with a discussion of beam transport, as described by the paraxial equation, Eq. (1).[1] It is generally familiar, and it will serve as a good point of departure for the more detailed treatments to follow.

Equation (1) describes the envelope of a beam with mean radius a. The transverse coordinates are x and y, and primes denote differentiation with respect to the axial coordinate z. Paraxial (non-crossing) trajectories are assumed, enabling separation of the transverse and longitudinal motion:

$$a'' + ka - (K/a) = \varepsilon^2/a^3 \qquad (1)$$

where k is the force constant of an external focusing system (which is a function of z, in general). For alternating gradient (AG) or thin lens solenoidal focusing, k can be a rapidly varying function. We will later use a "smooth approximation" in which k represents the average inward focusing force.

For K=0, ε=0, the solutions of Eq. (1) are sinusoidal with a wavelength λ_o defined by $(2\pi/\lambda_o)^2 = k$. The high frequency ripple associated with AG focusing is superposed on the basic envelope solutions; we usually ignore it when the smooth approximation is invoked.

K is the dimensionless perveance defined by

$$K = 2(I/I_o)(m/M)(Q^2/\beta^3\gamma^3) \tag{2}$$

with I denoting the particle current, I_o = 17 kA, m being the electron rest mass, M the mass of the particle of interest, and Q being the particle's charge state. β and γ are the usual relativistic factors given by $\beta = v/c$ and $\gamma = (1-\beta^2)^{-1/2}$. As defined here, perveance is useful for ions of any mass and charge state, and for electrons as well.

For ions in the nonrelativistic limit, K reduces to

$$K = (I/15.5\ \mathrm{MA})\ Q^2\ A^{1/2}\ (466/T)^{3/2}$$

where T is the particle kinetic energy in MeV and A is the atomic weight in amu. Note that this form is readily identified with the perveance used in conventional (microwave) electron guns, $K = I/V^{3/2}$, with I in amps and V in volts.

The perveances of some electron and ion sources are given below:

Gun	V(kV)	I(amp)	K(x10^{-3})	$\beta\gamma$
HERA(p^+)	18	0.02	5	0.006
LBL(Cs^{+1})	160	0.02	2.4	0.0016
Klystron	300	300	19	1.2
ATA inj.	2500	10000	6	5.8

The parameter ε is the emittance, which is a measure of the internal beam temperature. The rms emittance ε_x (or ε_y) is given by

$$\varepsilon_x = 4[\langle x^2\rangle\langle x'^2\rangle - \langle xx'\rangle^2]^{1/2} \tag{3}$$

where the averages include each particle in the beam. We can also define a normalized emittance given by $\varepsilon_n = \beta\gamma\varepsilon$. For linear systems, the normalized emittance remains constant as the particles are accelerated. Thus, the absolute emittance decreases as $(\beta\gamma)^{-1}$.

Equation (1) has several limiting cases of practical importance.

1. No focusing, small beam space charge (k=K=0)
   $a'' = \varepsilon/a^2$. The beam expands along a hyperbolic trajectory.
2. No focusing, small emittance (k=ε=0)
   $a'' = K/a$. The beam expands according to Dawson's integral.
3. Equilibrium ($a'' = 0$). There are two special cases:
   a. Emittance dominated: $(Ka^2/\varepsilon^2) \ll 1$; $a^2 = \varepsilon/k^{1/2} = \varepsilon\lambda_o/2\pi = \varepsilon_n\lambda_o/2\pi\beta\gamma$.
   b. Space-charge dominated: $(Ka^2/\varepsilon^2) \gg 1$; $a^2 = K/k$.

$Ka^2 = \varepsilon^2$, for example, for a typical ion source (K = 0.01) when ε = 50 cm-mrad for a 1-cm-diameter beam. A good proton source with ε_n = 0.1 cm-mrad at 100 kV has ε = 7 cm-mrad. Such a beam can be either space-charge dominated or emittance dominated, depending on the beam radius.

The beam space charge causes an increase in the wavelength of the particle motion. To demonstrate this, we first define the "phase advance" in the absence of space charge as

$$\sigma_o = 2\pi L/\lambda_o = Lk^{1/2} \tag{4}$$

where L is the period of an external AG focusing lattice. Thus, σ_o depends only on the lattice period and the focusing strength. To see the wavelength increase, first write the envelope equation as $a'' + a(k-K/a^2) = \varepsilon^2/a^3$, from which it follows that $\lambda = 2\pi/(k-K/a^2)^{1/2}$. We can therefore define a "depressed" phase advance σ as

$$(\sigma/L)^2 = k - K/a^2 = (\sigma_o/L)^2 - K/a^2. \tag{5}$$

Thus, in the space-charge-dominated case $\sigma_o^2 >> \sigma^2$ and $a^2 = K(L/\sigma_o)^2$. What clearly matters is the generalized current density, or "perveance density" K/a^2. The maximum value possible in this type of linear transport is $K/a^2 = (\sigma_o/L)^2$, independent of the emittance. Finally, note that from the envelope equation $a^2 = \varepsilon L/\sigma$. Thus, $\sigma/\sigma_o = (\sigma_o\varepsilon)/(KL)$.

We now consider a numerical example of a space-charge-dominated beam. We assume that a beam with normalized perveance of 0.006 and an absolute emittance of 7 cm-mrad is captured in an AG channel. We also assume the zero-current phase advance is 90°, corresponding to four quadrupole pairs per wavelength. We further assume that the phase advance is depressed by a factor ten due to the beam space charge. Then, the required lattice period L is about 18 cm, and the beam radius is about 0.9 cm. The ratio of the space-charge term to the emittance term is about 100. If this beam is accelerated at constant current, it is easily verified that the perveance drops rapidly, and the beam quickly becomes emittance dominated.

In much of the following material we will use the term beam plasma frequency. This parameter is simply related to the depressed phase advance. Note that $\omega/\beta c = 2\pi/\lambda = \sigma/L$. Thus, we can write the depressed phase advance as $\omega^2 = \omega_o^2 - K(\beta c/a)^2$, and we can identify $K(\beta c/a)^2 = 1/2\ \omega_p^2$, where $\omega_p^2 = 4\pi e^2 n/m$ is the beam plasma frequency.

The emittance growth of space-charge-dominated beams is a very important topic with relevance to applications ranging from heavy ion fusion accelerators to free electron lasers. A recent theory of Wangler (discussed elsewhere in these proceedings) indicates that such emittance growth can be minimized provided that (1) $\sigma_o \le \pi/2$, and (2) the particle density is uniform across the beam cross section. Of course, these conditions assume that lens misalignments, image forces, etc., are small.

2 BEAM EQUILIBRIA CRITERIA[2]

Having established an initial familiarity with beam transport in accelerators via the paraxial equation, we next discuss several other preliminary topics, including how individual particles move in a strong magnetic field and the limitation to beam transport posed by the potential depression resulting from the beam space charge. We also derive the beam envelope equations for cases involving cylindrical symmetry, and discuss additional important special cases.

2.1 INDIVIDUAL PARTICLE ORBITS

Consider a cylindrically symmetric electron beam with velocity $v = \beta c$ and density n_e propagating along a uniform, axial magnetic field $\underline{B} = B_o\hat{e}_z$. From Gauss' law and Ampere's law, E_r and B_θ due to the beam space charge and current can be found as (in Gaussian units)

$$E_r = -\frac{4\pi e}{r}\int_0^r n_e r\, dr\ , \tag{6}$$

$$B_\theta = -\frac{4\pi e}{r}\int_0^r \beta\, n_e\, r\, dr\ . \tag{7}$$

Assuming that the beam density is uniform out to a radius r_b, in Cartesian coordinates the beam self-fields are given by

$$E_x = \frac{2Ix}{\beta c r_b^2}, \quad E_y = \frac{2Iy}{\beta c r_b^2}, \tag{8}$$

$$B_x = -\frac{2Iy}{c r_b^2}, \quad B_y = \frac{2Ix}{c r_b^2}, \tag{9}$$

where the beam current $I = -\pi r_b^2 (e n_o)\,\beta c$. The motion of a single particle of (relativistic) mass γm and charge $-e$ is described by the Lorentz force law

$$\frac{d}{dt}\underline{p} = -e\left[\underline{E} + \frac{1}{c}(\underline{V} \times \underline{B})\right] \tag{10}$$

where $\underline{p} = \gamma m \underline{v}$ is the relativistic momentum. Writing the transverse components explicitly gives

$$\ddot{x} = -(\Omega^2/\gamma^3)\,x - (\Omega_o/\gamma)\,\dot{y}, \qquad \ddot{y} = -(\Omega^2/\gamma^3)\,y + (\Omega_o/\gamma)\,\dot{x}$$

where $\Omega_o = eB_o/mc$ and $\Omega^2 = 2eI/m\,\beta c r_b^2$. Defining the parameter $s = x + iy = re^{i\theta}$, the transverse equations can be combined to yield

$$\ddot{s} = -(\Omega^2/\gamma^3)\,s + i\,(\Omega_o/\gamma)\,\dot{s}. \tag{11}$$

Equation (11) has the formal solution $s(t) = Ae^{i\omega_+ t} + Be^{i\omega_- t}$, where

$$\omega_\pm = \frac{\Omega_o}{2\gamma}\left[1 \pm \left(1 + \frac{4\Omega^2}{\gamma\Omega_o^2}\right)^{1/2}\right].$$

With the initial conditions $s(0) = s_o$, $\dot{s}(0) = 0$, in the limit of large Ω_o, the solution becomes

$$s(t) \approx s_o\left(e^{i\omega_- t} + \frac{\Omega^2}{\gamma\Omega_o^2}e^{i\omega_+ t}\right). \tag{12}$$

Hence, the single-particle motion can be considered as the sum of two rotations: a large amplitude rotation at the slow E x B rotation frequency, and a small amplitude rotation at the fast cyclotron frequency. If $\Omega^2/\gamma\Omega_o^2 << 1$, then the detailed fine structure of the particle orbits can be ignored. (As will be shown later, there are also fast and slow rotation modes of the electron fluid equilibrium.)

2.2 BEAM ENVELOPE EQUATION DERIVATION

The perpendicular components of the equation of motion, Eq. (10), are given by

$$\dot{\gamma}\, m\underline{v}_\perp + \gamma m\dot{\underline{v}}_\perp = -e\left[\underline{E}_\perp + \frac{1}{c}(\underline{v} \times \underline{B})_\perp\right] \tag{13}$$

where the dot notation implies total time differentiation. With the assumed cylindrical symmetry, the important field components are E_r, B_θ, and B_z. We further assume that the longitudinal magnetic field is generated by external magnetic field coils only and is constant and uniform across the beam radial profile, i.e., $B_z = B_o$. In this case, the radial component of Eq. (13) yields

$$\ddot{r} + \frac{\dot{\gamma}\dot{r}}{\gamma} - \frac{v_\theta^2}{r} = -\frac{e}{\gamma m}\left[E_r + \frac{1}{c}(v_\theta B_o - v_z B_\theta)\right] \tag{14}$$

while the azimuthal component becomes a statement of conservation of canonical angular momentum

$$\gamma r v_\theta - \frac{\Omega_o r^2}{2} = P_\theta/m = \text{const.} \tag{15}$$

Restricting attention to a single particle at the beam edge, and assuming constant beam kinetic energy, substitution of Eq. (15) into Eq. (14) gives an equation that describes the motion of the radial envelope of the beam:

$$\ddot{r}_b + \frac{\dot{r}_b \dot{\gamma}}{\gamma} + \left(\frac{\Omega_o}{2\gamma}\right)^2 r_b = -\frac{e}{\gamma m}\left(E_r - \frac{v_z B_\theta}{c}\right) + \frac{(P_\theta/m)^2}{\gamma^2 r_b^3} . \tag{16}$$

As an example of the use of Eq. (16), again consider the case of a beam of uniform density and energy with radius r_b propagating along a uniform longitudinal magnetic field. From Eqs. (6) and (7), the beam self-fields are given by

$$E_r = -2\pi e n_o r_b , \qquad B_\theta = -2\pi e n_o \beta r_b = \beta E_r ,$$

and Eq. (16) can be written as

$$\ddot{r}_b + \left(\frac{\Omega_o}{2\gamma_o}\right)^2 r_b - \frac{\omega_e^2}{2\gamma_o^3} r_b = \frac{(P_\theta/m)^2}{\gamma_o^2 r_b^3} \tag{17}$$

where $\omega_e = (4\pi n_o e^2/m)^{1/2}$ is the beam plasma frequency. Hence, the un-neutralized beam self-fields cause the beam to expand, while the external magnetic field supplies a restoring force. The equilibrium radius r_{bo} ($\ddot{r}_b = \dot{r}_b = 0$) is given by

$$r_{bo}^2 = \frac{2(P_\theta/m)}{(\Omega_o^2 - 2\omega_e^2/\gamma_o)^{1/2}} \tag{18}$$

and there can be no equilibrium unless

$$\Omega_o^2 > 2\omega_e^2/\gamma_o . \tag{19}$$

For a 1-MeV electron beam of density $10^{12}/\text{cm}^3$, the magnetic field strength required for equilibrium is only 2.6 kG; however, for a 1-MeV proton beam of the same density B_o must exceed 188 kG. Hence, while solenoidal magnetic fields can easily satisfy the equilibrium condition for transport of intense electron beams in vacuum, some charge neutralization is usually required for intense ion beam transport.

Before leaving this topic, note that the general solution of Eq. (17) is oscillatory. Eliminating ω_e^2 in favor of the total beam current magnitude, which does not vary as the beam expands or contracts, Eq. (17) becomes

$$\ddot{r}_b + \left(\frac{\Omega_o}{2\gamma_o}\right)^2 r_b - \frac{2eI}{\gamma_o^3 m \beta c r_b} = \frac{(P_\theta/m)^2}{\gamma_o^2 r_b^3} \tag{20}$$

[Note the similarity of this equation with the paraxial equation, Eq. (1).] Assuming small perturbations about the equilibrium radius, $r_b = r_{bo} + \delta r$, the linearization of Eq. (20) yields

$$\ddot{\delta r} + \left[\left(\frac{\Omega_o}{\gamma_o}\right)^2 - \frac{4eI}{\gamma_o^3 m \beta c r_{bo}^2}\right] \delta_r = 0$$

which has sinusoidal solutions with frequency

$$\omega = \left[\left(\frac{\Omega_o}{\gamma_o}\right)^2 - \frac{4eI}{\gamma_o^3 m\beta c r_{bo}^2}\right]^{1/2}$$

and wavelength $\lambda = 2\pi\beta c/\omega$.

2.3 SPACE-CHARGE-LIMITING CURRENT

Even if the confining longitudinal magnetic field is infinitely large, there is a limit to the amount of current (space charge) that can propagate in an evacuated drift space. Suppose that a beam is injected into the space between two grounded, conducting plates. As the charge enters the cavity, a space-charge potential barrier will form. However, if the current is low, the maximum potential will occur at the mid-plane between the conducting plates and all of the current will cross the gap. Now suppose that the injected current is raised until the magnitude of the potential barrier becomes equal to the injected beam kinetic energy. Any further increase in the beam current will cause an increase in the potential barrier height, resulting in a reflection of injected beam. In this case, a virtual cathode is said to form while the largest current that can flow without reflection is termed the space-charge-limiting current.

There are a few important special cases for which analytical solutions for the space-charge-limiting current can be derived. To calculate the limiting current, we must relate the beam density and the beam velocity (through the continuity equation), the beam density and the electrostatic potential (via Poisson's equation), and the beam velocity and the electrostatic potential (through the conservation of energy). Assuming one-dimensional motion (implying a very large longitudinal magnetic field), in the steady state these relations become

$$\frac{\partial}{\partial z}(nv_z) = 0, \quad \nabla \cdot \underline{E} = -4\pi e n\,, \quad (\gamma - 1)\, mc^2 - e\phi = (\gamma_o - 1)\, mc^2.$$

When the beam density and velocity are eliminated in favor of the electrostatic potential, Poisson's equation becomes

$$\frac{1}{r}\frac{\partial}{\partial r}\left(r\frac{\partial \phi}{\partial r}\right) + \frac{\partial^2 \phi}{\partial z^2} = -\frac{4\pi j}{c}\left[1 - \left(\gamma_o + \frac{e\phi}{mc^2}\right)^{-2}\right]^{-1/2} \tag{21}$$

where $E = -\nabla\phi$, $(\gamma_o - 1)\, mc^2$ is the injected beam kinetic energy, and $j = -env_z$. The solution of Eq. (21) subject to the appropriate boundary conditions describes the space-charge potential (and, hence, the electron kinetic energy) as a function of position in the cavity.

As an example, consider an annular beam whose thickness is much smaller than both the beam radius r_b and the distance between the beam and the chamber wall. In this case, the electrostatic potential is essentially constant across the thin beam and the problem reduces to the solution of the homogeneous equation

$$\frac{1}{r}\frac{\partial}{\partial r}\left(r\frac{\partial \phi}{\partial r}\right) = 0\,,$$

subject to the boundary conditions

$$\phi(R) = 0\,, \quad \frac{\partial \phi}{\partial r}\bigg|_{r = r_b} = \frac{2I}{r_b v_b},$$

where I is the total beam current and v_b is the longitudinal electron velocity (uniform across the thin beam). The solution is

$$\phi(r) = -2\frac{I}{v_b}\begin{cases} \ln(R/r), & r_b < r < R \\ \ln(R/r_b), & r < r_b \end{cases} \tag{22}$$

where v_b is given by

$$v_b = c\left[1 - \left(\gamma_o + \frac{e\phi_b}{mc^2}\right)^{-2}\right]^{1/2}. \tag{23}$$

Combining Eqs. (22) and (23) yields the following result for $\phi_b = \phi(r = r_b)$:

$$F(\phi_b) = \frac{-e\phi_b}{mc^2}\left[1 - \left(\gamma_o + \frac{e\phi_b}{mc^2}\right)^{-2}\right]^{1/2} = \frac{2eI}{mc^3}\ln(R/r_b). \tag{24}$$

$F(\phi_b)$ has a maximum at $|e\phi_b| = \gamma_o^{1/3}\, mc^2$ given by $F_{max}(\phi_b) = (\gamma_o^{2/3} - 1)^{3/2}$. Since maximizing F also maximizes the current allowed to propagate, the space-charge-limiting current for this system is given by

$$I_\ell = \frac{mc^3}{2e}\frac{(\gamma_o^{2/3} - 1)^{3/2}}{\ln(R/r_b)}. \tag{25}$$

For currents $I < I_\ell$, there are two possible solutions for ϕ_b according to Eq. (24): (1) $0 < |e\phi_b|/mc^2 < \gamma_o - \gamma_o^{1/3}$, corresponding to particles of high velocity and low density; and (2) $\gamma_o - \gamma_o^{1/3} < |e\phi_b|/mc^2 < \gamma_o - 1$ corresponding to particles of lower velocity and higher density. For currents $I > I_\ell$, there is no solution to Eq. (24), and the existence of a steady-state flow of current in excess of the space-charge limit is not possible.

Calculation of the limiting current in the general case is very difficult; however, a relatively simple procedure allows a quick estimate that is generally accurate to within 10%. To illustrate this method, consider a cylindrically symmetric, uniform density, annular electron beam of inner and outer radii r_1 and r_2 propagating in an evacuated drift tube of radius R along a very strong longitudinal magnetic field. We first use Gauss' law to show that the radial electric field is given by

$$E_r = -\frac{2\pi e n_o}{r} \times \begin{cases} 0, & r < r_1 \\ (r^2 - r_1^2), & r_1 < r , r_2 \\ (r_2^2 - r_1^2), & r_2 < r < R . \end{cases}$$

A radial integration of E_r then gives the electrostatic potential in the drift tube as

$$\phi(r) = -2\pi e n \times \begin{cases} (r_2^2 - r_1^2)\ln(R/r_2) - r_1^2\ln(r_2/r_1) + (r_2^2 - r_1^2)/2, & 0 < r < r_1 \\ (r_2^2 - r_1^2)\ln(R/r_2) - r_1^2\ln(r_2/r) + (r_2^2 - r^2)/2, & r_1 < r < r_2 \\ (r_2^2 - r_1^2)\ln(R/r), & r_2 < r < R \end{cases}$$

An approximate expression for the space-charge limit is then obtained by setting the magnitude of the electrostatic potential energy equal to the injected beam kinetic energy. This yields the correct expression for the extreme ultrarelativistic limit

$$I_\ell = \beta(\gamma_o - 1)\left(\frac{mc^3}{e}\right)\left[1 - \frac{2r_1^2}{r_2^2 - r_1^2}\ln(r_2/r_1) + 2\ln(R/r_2)\right]^{-1} . \tag{26}$$

To extend the range of validity to non-relativistic energies, replace the factor $\beta(\gamma_o - 1)$ by $(\gamma_o^{2/3} - 1)^{3/2}$ obtained. in Eq. (26). The final result of this ad hoc procedure is

$$I \approx (\gamma_o^{2/3} - 1)^{3/2}\,(mc^3/e)\left[1 - \frac{2r_1^2}{r_2^2 - r_1^2}\ln(r_2/r_1) + 2\ln(R/r_2)\right]^{-1} . \tag{27}$$

By taking appropriate limits of Eq. (27), it is easy to derive the following expressions:

Solid Beam: $$I_\ell \approx \frac{(\gamma_o^{2/3} - 1)^{3/2}\,(mc^3/e)}{1 + 2\ln(R/r_b)} ;$$

Annular Beam: $$I_\ell \approx \frac{(\gamma_o^{2/3} - 1)^{3/2}\,(mc^3/e)}{[a/r_b + 2\ln(R/r_b)]}, \quad a = r_2 - r_1, \quad r_b = \frac{r_2 + r_1}{2} ;$$

Annular Beam (close to wall): $$I_\ell \approx (\gamma_o^{2/3} - 1)^{3/2}\,(mc^3/e)\left[\frac{R}{a + 2\delta}\right], \quad \delta = R - r_2 .$$

Inspection of these expressions indicates that the highest beam currents can be transported if the beam is annular and located near the drift tube wall.

2.4 LAMINAR FLOW EQUILIBRIA

Having first considered several preliminary topics, we will now develop the general equations that describe the self-consistent equilibria of un-neutralized intense beams propagating in a metallic drift tube along a uniform magnetic field. The analysis is performed in cylindrical geometry (r, θ, z) with the z axis aligned with the beam and the external magnetic field. The beam is assumed to have axisymmetric equilibrium radial density and velocity profiles (no θ dependence) and is assumed to be uniform in the z direction. The beam front physics are not included, and the beam pulse is assumed to be much shorter than the time required for diffusion of the beam self-fields through the drift tube walls. The assumption of laminar flow implies that all particles (fluid elements) move on nonintersecting helical paths. Hence, the amplitude of the fast gyrational electron motion about the external field lines must be small compared with the beam radius.

The basic equations for examining the equilibrium are the fluid and Maxwell equations. For the case of an electron beam these are

$$\frac{\partial}{\partial t}n + \nabla\cdot(n\underline{v}) = 0, \qquad \frac{\partial}{\partial t}\underline{p} + v\cdot\nabla p = -e\left(\underline{E} + \frac{1}{c}(\underline{v}\times\underline{B})\right);$$

$$\nabla\times\underline{E} = -\frac{1}{c}\frac{\partial \underline{B}}{\partial t}, \qquad \nabla\times\underline{B} = \frac{-4\pi e n \underline{v}}{c} + \frac{1}{c}\frac{\partial \underline{E}}{\partial t};$$

$$\nabla\cdot\underline{E} = -4\pi e n, \qquad \nabla\cdot\underline{B} = 0 .$$

In these equations n, $\underline{v}$, and $\underline{p}$ represent the macroscopic equilibrium electron fluid density, velocity, and momentum, and $\underline{E}$ and $\underline{B}$ represent the self-consistent equilibrium electric and magnetic fields. Following the assumptions presented in the preceding paragraph and assuming the steady state, the equation of motion yields the radial force balance given by

$$\gamma(r)\omega_\theta^2(r) + \frac{1}{r^2}\int_o^r dr'\, r'\, \omega_p^2(r') - \omega_\theta\Omega_o - \frac{\beta_z(r)}{r^2}\int_o^r dr'\,(r')\,\beta_z(r')\,\omega_p^2(r') = 0 \tag{28}$$

where $\Omega_o = eB_o/mc$, $\omega_p{}^2 = 4\pi ne^2/m$, and $\omega_\theta(r) = \beta_\theta(r)c/r$. If it is further assumed that the axial velocity profile of the electron beam is independent of radius r, i.e., $\beta_z(r)c \equiv \beta_o c$ = const, then

$$\gamma(r) \approx [1 - \beta_z^2(r)]^{-1/2} \equiv \gamma_o = \text{const.}$$

With these simplifying assumptions Eq. (28) reduces to

$$\gamma_o\omega\theta^2(r) + \frac{1}{r^2}(1-\beta_o^2)\int_o^r dr'\, r'\, \omega_p^2(r') - \omega_\theta(r)\Omega_o = 0\,. \tag{29}$$

Solving Eq. (29) for the angular velocity $\omega_\theta(r)$ yields

$$\omega_\theta(r) = \omega_\theta^\pm(r) = \frac{\Omega_o}{2\gamma_o}\left\{1 + \left[1 - \frac{4}{\gamma_o\Omega_o^2 r^2} \times \int_o^r dr'\,(r')\,\omega_p^2(r')\right]^{1/2}\right\}. \tag{30}$$

For a constant density profile

$$\omega_p^2(r) = \begin{cases} \dfrac{4\pi n_o e^2}{m} = \omega_{po}^2\,, & 0 < r < r_b \\ 0, & r > r_b\,. \end{cases}$$

Equation (30) reduces to

$$\omega_\theta(r) = \omega_\theta^\pm = \frac{\Omega_o}{2\gamma_o}\left[1 \pm \left(1 - \frac{2\,\omega_{po}^2}{\gamma_o\,\Omega_o^2}\right)^{1/2}\right]. \tag{31}$$

The condition for the radial to be real (the existence of an equilibrium) is

$$\Omega_o{}^2 > 2\,\omega_{po}{}^2/\gamma_o \tag{32}$$

which is identical to Eq. (19). The beam rotates as a rigid body with two allowable rotation rates given approximately by

$$\omega_\theta^+ \approx \frac{2\gamma_o\Omega_o^2 - \omega_{po}^2}{2\gamma_o^2\Omega_o}\,, \tag{33}$$

$$\omega_\theta^- \approx \frac{\omega_{po}^2}{2\gamma_o^2\Omega_o}\,. \tag{34}$$

For the slow mode, the rotation frequency is $\omega_\theta{}^- \approx cE(r)/r\gamma^2 B_o$, which corresponds to a beam precession in θ at the E x B drift rate with relativistic corrections, and the fast mode rotation corresponds to the fast cyclotron motion about the axial magnetic field lines with $\omega_\theta{}^+ \approx \Omega_o/\gamma_o$.

As a second application of Eq. (29), consider the case of a hollow beam with density profile given by

$$\omega_p^2(r) = \begin{cases} 0, & 0 < r < r_o \\ \dfrac{4\pi n_o e^2}{m} = \omega_{po}^2, & r_o < r < r_b \\ 0, & r > r_b\,. \end{cases}$$

Substituting into Eq. (29) yields the solution for the angular velocity $\omega_\theta(r)$ given by

$$\omega_\theta(r) = \omega_\theta^\pm(r) = \frac{\Omega_o}{2\gamma_o}\left\{1 \pm \left[1 - \frac{2\omega_{po}^2}{\gamma_o\Omega_o^2}(1 - r_o^2/r^2)\right]^{1/2}\right\}, \quad r_o < r < r_b\,. \tag{35}$$

Since $\partial\omega_\theta/\partial r \neq 0$, the hollow beam no longer rotates as a rigid body but has angular velocity shear. This difference in rotation rate across the beam provides a source of free energy that can drive a filamentation instability.

For both the rigid rotor and hollow beam equilibria there is considerable arbitrariness; any two of the three variables $n(r)$, $v_\theta(r)$, and $v_z(r)$ could be specified. However, since the beam density is related to the beam kinetic energy through Poisson's equation, conservation of energy provides one independent relation, i.e.,

$$(\gamma - 1)mc^2 - e\phi(r) = (\gamma_o - 1)mc^2. \tag{36}$$

In addition, if the source geometry is known, then the variation of canonical angular momentum, $P_\theta(r)$, is specified as

$$P_\theta(r) = \gamma(r)mr\, v_\theta(r) - \frac{er_c}{c}A_\theta(r) = -\frac{er_c}{c}A_{\theta s}(r_c) \tag{37}$$

where $A_{\theta s}$ is given by the axial magnetic field B_z at the cathode according to

$$A_{\theta s} = \frac{1}{r}\int_o^r B_z^c(r')r'dr'\,.$$

Under the assumption of laminar flow, Eq. (37) implies that all particles at a given radius were emitted from the source at the same radius r_c and, hence, have the same canonical angular momentum.

There are three source geometries of practical importance. The first is the non-immersed source for which the cathode is shielded from the axial magnetic field in the downstream region. In this case, $P_\theta = B_z{}^c = 0$.

A second important case is that of an immersed source whose emitting area coincides with a flux surface; i.e., an annular cathode that produces a thin hollow beam. In this instance, all particles have the same canonical angular momentum.

The final important case is that of an immersed source for which the axial magnetic field is uniform. In this instance, P_θ varies across the beam. If it is also assumed that the beam expands or contracts in self-similar fashion, then the radii of particles in the downstream region must be related to their emission radii according to $r = \alpha r_c$, where α is a constant that depends on the actual system geometry.

The resulting equation set is complete and it uniquely specifies the equilibrium in a self-consistent way. However, analytical solutions for this general problem are very difficult and have been obtained in only a few isolated special cases. In practice, the usual procedure is to assume one of the simple radial force balance equilibria, such as the rigid rotor, as a first guess, and then iterate until the necessary accuracy is achieved.

2.5 GENERAL BEAM INSTABILITIES

In the preceding sections, the equilibria of intense relativistic beams were considered; now, attention will be focused on the stability of small amplitude oscillations about various equilibrium configurations. The mean azimuthal motion of an electron beam is assumed to be non-relativistic. In this case the axial diamagnetic field contribution can be neglected and the radial force balance is approximated by

$$\gamma_o\omega_\theta^2(r) + \frac{1}{r^2}(1-\beta_o^2)\int_o^r dr'r'\omega_p^2(r') - \omega_\theta(r)\Omega = 0\,. \tag{38}$$

To determine the stability of the various equilibrium configurations, the macroscopic fluid and field quantities are expressed as the sum of their equilibrium values plus a perturbation. In the electrostatic approximation, the perturbed magnetic field $\delta\underline{B}(\underline{r},t)$ is assumed to remain negligibly small so that the perturbed electric field $\delta\underline{E}(\underline{r},t)$ can be expressed as the gradient of a scalar potential

$$\delta\underline{E}(\underline{r},t) = -\nabla\phi(\underline{r},t)\,. \tag{39}$$

Since the axial motion of the beam may be relativistic, the two-mass approximation is assumed for the perturbed momentum according to

$$\delta p_r = \gamma_o m\delta v_r, \qquad \delta p_\theta = \gamma_o m\delta v_\theta, \qquad \delta p_z = \gamma_o^3 m\delta v_z\,.$$

As a further simplification, all perturbed quantities are assumed to have harmonic dependence in time, the azimuthal coordinate, and the axial coordinate; i.e., $\delta\psi = \delta\psi(r)e^{-i\omega t}e^{i\ell\theta}e^{ik_z z}$, where ω is the characteristic frequency of oscillations, ℓ is the azimuthal harmonic number, and k_z is the axial wave number. With this assumption, the linearized Poisson's equation can be written as

$$\frac{1}{r}\frac{\partial}{\partial r}\left[r\left(1-\frac{\omega_{po}^2}{\gamma_o v^2}\right)\frac{\partial}{\partial r}\delta\phi\right] - \frac{\ell^2}{r^2}\left(1-\frac{\omega_{po}^2}{\gamma_o v^2}\right)\delta\phi - k_z^2\left[1-\frac{\omega_{po}^2/\gamma_o^3}{(\omega-k_z v_{zo}-\ell\omega_{\theta_o})^2}\right]\delta\phi$$

$$= -\frac{\ell\delta\phi}{r}\,\frac{(\partial/\partial r)[(\omega_{po}^2/v^2)(-\Omega/\gamma_o+2\omega_{\theta_o})]}{(\omega-k_z v_{zo}-\omega_{\theta_o})} \tag{40}$$

where

$$v^2(r) = (\omega-k_z v_{z_o}-\ell\omega_{\theta_o})^2 - \left(-\frac{\Omega}{\gamma_o}+2\omega_{\theta_o}\right)\left[-\frac{\Omega}{\gamma_o}+\frac{1}{r}\frac{\partial}{\partial r}(r^2\omega_{\theta_o})\right]$$

and

$$\omega_{po}^2(r) = 4\pi n_o(r)e^2/m\,.$$

2.6 RIGID ROTOR STABILITY

Analytical solutions for Eq. (40) are tractable for only a few simple cases. As a first example, consider the case of constant beam density, corresponding to the rigid rotor relativistic electron beam equilibrium. It is assumed that the beam radius is equal to the radius of a perfectly conducting wall, i.e., $[\delta\phi]_{r=r_b} = 0$.

Since the equilibrium rotation velocity $\omega_\theta(r)$ is independent of radius, it follows that $v^2(r)$ is given by

$$v^2 = (\omega - k_z v_{z_o} - \ell\omega_{\theta_o})^2 - (-\Omega/\gamma_o + 2\omega_{\theta_o})^2$$

which is also independent of radius. Under these conditions, Poisson's equation reduces to

$$\frac{1}{r}\frac{\partial}{\partial r} r \frac{\partial}{\partial r}\delta\phi - \frac{\ell^2}{r^2}\delta\phi + k_\perp^2 \delta\phi = 0, \qquad 0 < r < r_b \tag{41}$$

where

$$k_\perp^2 = -k_z^2 \frac{\left[1 - \dfrac{\omega_{po}^2/\gamma_o^3}{(\omega - k_z v_{zo} - \ell\omega_{\theta_o})^2}\right]}{1 - \omega_{po}^2/\gamma_o v^2} \, . \tag{42}$$

Equation (41) is recognized as a form of Bessel's equation. The solution that remains finite at the origin is

$$\delta\phi = A J_\ell(k_\perp r)$$

where A is a constant and J_ℓ is the Bessel function of the first kind of order ℓ. Enforcing the vanishing of the perturbed potential at the conducting wall leads to the condition

$$J_\ell(k_\perp r_b) = 0 \tag{43}$$

from which it follows that

$$k_\ell^2 r_b^2 = \rho_{\ell m}^2, \qquad m = 1, 2, \ldots \tag{44}$$

where $\rho_{\ell m}$ is the mth zero of $J_\ell(x) = 0$, and $k_\perp$ is recognized as the effective perpendicular wave number quantitized by the finite radial geometry.

Equation (42) is the dispersion relation for electrostatic waves in the electron-beam-filled guide tube. It relates the characteristic oscillation frequency ω to the azimuthal harmonic numbers ℓ, the axial wave vector k_z, and the properties of the rigid rotor equilibrium configuration.

Equation (42) may be expressed in the equivalent form

$$1 - \frac{k_z^2}{k^2}\frac{\omega_{po}^2/\gamma_o^3}{(\omega - k_z v_{zo} - \ell\omega_{\theta_o})^2} - \frac{k_\perp^2}{k^2}\frac{\omega_{po}^2/\gamma_o}{(\omega - k_z v_{zo} - \ell\omega_{\theta_o})^2 - \omega_v^2} = 0 \tag{45}$$

where $k^2 = k_z^2 + k_\perp^2$ and ω_v is the vortex frequency defined by

$$\omega_v = -(-\Omega_o/\gamma_o + 2\omega_{\theta_o}) \, . \tag{46}$$

The solution to Eq. (45) can be written as

$$(\omega - k_z v_{zo} - \ell\omega_{\theta_o})^2 = \left[\omega_v^2 + (\omega_{po}^2/\gamma_o)\left(\frac{k_z^2/\gamma_o^2 + k_\perp^2}{k^2}\right)\right]\left\{1 \pm \left[1 - \frac{4(k_z^2/k^2)\omega_v^2\omega_{po}^2/\gamma_o^3}{\omega_v^2 + (\omega_{po}^2/\gamma_o)\left(\dfrac{k_z^2/\gamma_o^2 + k_\perp^2}{k^2}\right)}\right]^{1/2}\right\} . \tag{47}$$

For the analysis to be valid, the value of the beam density must be less than the maximum density for which equilibrium solutions exist. An upper bound on the density can be determined by requiring that the radical in Eq. (47) be real; the condition that obtains (assuming the slow rotation mode) is also

$2\omega_{po}{}^2/\gamma_o\Omega^2 \leq 1.$

Provided this criterion is satisfied, $\text{Im}\omega_{\theta_o} = 0$ (i.e., ω_{θ_o} contains no imaginary component). Hence, this condition implies the existence of undamped stable oscillations for the four characteristic eigenmodes of the dispersion relation. Examination of the character of the eigenmodes is simplified by restricting the analysis to the axisymmetric ($\ell = 0$) modes and assuming the slow rotation frequency. In this case, the dispersion relation can be approximately factored and the eigenmodes can be identified as follows:

(1) Two Doppler-shifted plasma modes explicitly displaying the longitudinal electron mass effects:

$$\omega = k_z v_{zo} \pm \left(\frac{k_z^2 \omega_{po}^2}{k^2 \gamma_o^3} \right)^{1/2} . \tag{48}$$

(2) Two Doppler-shifted cyclotron modes:

$$\omega = k_z v_{zo} \pm \left(\frac{\Omega^2}{\gamma_o^2} \right)^{1/2} . \tag{49}$$

The plasma modes correspond to longitudinal bunching of the beam space charge, and the cyclotron waves are transverse modes corresponding to a traveling constriction in the beam.

Before leaving this section, it should be emphasized that the relatively simple form of the dispersion relation, Eq. (46), could be obtained only by considering the radially homogeneous rigid rotor equilibrium. In general, radial inhomogeneities, particularly the variation of $\gamma(r)$ across the beam, strongly influence the structure of the wave spectra. Specific numerical solutions of the linear theory for more realistic solid beam equilibria indicate several significant differences.

3 LINEAR INDUCTION ACCELERATORS[3]

Linear induction accelerators are inherently capable of efficiently accelerating very high current beams without the power limitation commonly associated with RF linacs. For example, typical accelerating gradients achievable with S-band linacs (~3 GHz) are 10 MV/meter. Standard SLAC klystrons are capable of producing about 30 MW of power. Assuming no copper losses, the maximum beam current (averaged over the klystron macropulse) that can be accelerated is about 3 amps/L, when L is now the length of accelerator serviced by a single klystron. Thus, the average macropulse currents are typically a few amperes at most.

In contrast, beam currents of several tens of kiloamperes have been accelerated with induction linacs. The disadvantages of such devices include smaller accelerating gradients (0.5 to 3 MV/m), relatively short beam pulses (≤ 100 ns) and (usually) larger sizes and weights for the same final kinetic energy per accelerated particle.

All linear induction accelerators operate by producing voltage around a circuit enclosing a time-varying magnetic flux. Mathematically, this is described by Eq. (50), where $\underline{B}$ and $\underline{E}$ represent the magnetic induction and the induced electric field, and V is the voltage induced around the circuit C which encloses the area A:

$$V = \int_C \underline{E} \cdot dl = -\frac{1}{c}\frac{d}{dt}\int_S \underline{B} \cdot dA . \tag{50}$$

Time is represented by t, and c is the speed of light. Note that Eq. (50) allows the generation of voltage by changing either the enclosed magnetic field or the area normal to the field lines. For example, an external circuit can be used to establish a magnetic field through a second circuit. If the current through the external circuit is time-varying, a voltage will be established at the output connections of the second circuit, i.e., this is a single-turn transformer. The output voltage can be increased by inserting a core of ferromagnetic material since the B-field enclosed is increased by the magnetic permeability of the material according to $B = \mu H$, where H represents the macroscopic magnetic field, and μ is the magnetic permeability.

As a practical embodiment of this concept, consider the acceleration gap geometry of Figure 1. A high voltage pulse is injected into an accelerating cavity containing an annular core of magnetic material. It is assumed that the magnetic induction pre-existing in the core is saturated at the value $+B_0$. Note that without the core the input pulse would be short-circuited; however, as the current begins to flow around the core, the magnetic induction in the material begins to change until it reaches the value $-B_0$. During this time T, the accelerating voltage is effectively applied directly to the accelerating gap. From Eq. (50), these parameters are related according to $VT = \Delta BA = 2\,B_0A$, where A is the cross-sectional area of the annular core. In other words, the product of the accelerating voltage and the pulse time is equal to the flux swing times the area of the annular core. Note that the voltage does not appear external to the accelerating gap; i.e., the exterior of the accelerator remains at ground potential and the high voltage is summed by only the beam as it propagates through the LIA structure.

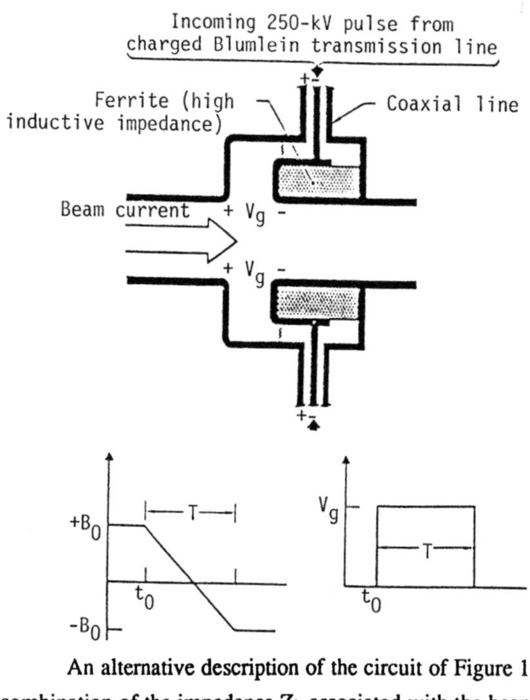

Figure 1. A linear induction accelerator cavity that uses ferromagnetic material.

An alternative description of the circuit of Figure 1 is that the voltage pulse appears across the parallel combination of the impedance Z_b associated with the beam and the impedance Z_c associated with the flow of current around the magnetic core. When the core is saturated at the level $+B_0$, the inductance associated with the current path around the magnetic core is very high, and Z_c is very large; hence, the pulse energy is delivered directly to the beam. During the time T, however, the flow of leakage current around the core drives the core into saturation at the level $-B_0$. Z_c then drops to essentially zero, and the accelerating pulse is shorted.

Equation (50) can also be used to describe the behavior of waves in transmission lines. Consider the strip transmission line of Figure 2. The voltage between the plates can be thought of as arising from a time-changing magnetic flux as the electromagnetic wave propagates down the line. With the current flow on the surfaces of the strip lines as shown, the magnetic induction is described by

$$B = \frac{4\pi\mu I}{cW} \tag{51}$$

where μ is the magnetic permeability of the dielectric between the electrodes, and W is the width of the lines. The area transverse to the magnetic field is simply $A = a\ell$, where a is the separation of the lines and ℓ is the linear dimension. As the wave propagates down the line, the length ℓ increases linearly with the constant of proportionality being equal to $c/(\mu\varepsilon)^{1/2}$, where ε is the material dielectric constant. Hence, in this case, the voltage across the gap arises from the time-changing area transverse to the magnetic field and is given by

$$V = \frac{4\pi}{c}\sqrt{\frac{\mu}{\varepsilon}}\,\frac{a}{W}I = Z_o I \tag{52}$$

where we have noted that the impedance of a strip transmission line is $Z_o = (4\pi/c)\,(\mu/\varepsilon)^{1/2}\,a/W$.

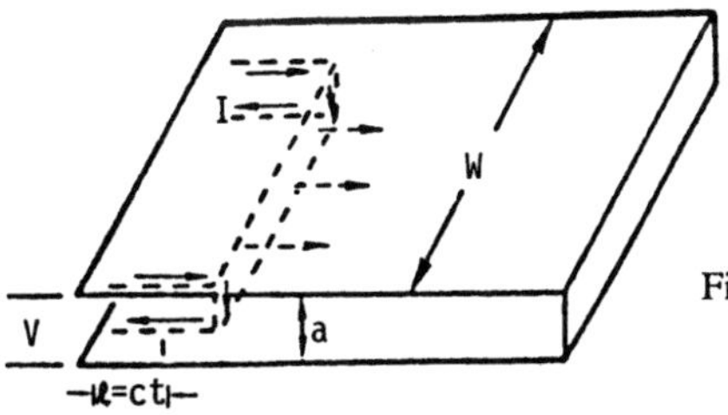

Figure 2. A strip transmission line in which the voltage across the plates arises from a time-changing magnetic flux (pulse-forming transmission line).

As first noted by Pavlovskii, closed transmission lines can also be used to form the accelerating cavities of linear induction accelerators. The impedance behavior of these structures obeys Eq. (52). Typical high voltage dielectric materials are transformer oil and de-mineralized water with dielectric constants of approximately 2.1 and 81, respectively. The cavity dimensions necessary to avoid high voltage insulator breakdowns for charging pulses of typically a megavolt generally imply a characteristic cavity impedance of a few ohms for water-insulated cavities, and a few tens of ohms for oil-insulated cavities. As a result, efficient transfer of energy from the cavity to the beam requires that the beam current be relatively high (around 100 kA). At these current levels, the self-fields of the beam are very large and must always be considered in describing the beam transport through the accelerator. A summary of several successful induction linacs is given in Table 1.

Table 1. High Current Linear Induction Accelerators

DEVICE	I (kA)	V (MV)	P (GW)
ETA	10	5	50
ATA	10	50	500
RADLAC I	20	10	200
RADLAC II	40	20	800
LIU-10	50	14	700

3.1 RADLAC BEAM TRANSPORT ANALYSES

As an introduction to some of the important problems associated with designing beam transport lines in high current linacs, we will consider some of the issues encountered in the development of the RADLAC series of accelerators. These include the injector, the accelerating gap design, and several high current beam instabilities.

3.1.1 The Foilless Diode Injector

An immersed source configuration that has proven to be extremely useful for injecting high current beams into vacuum regions is the foilless diode. In this configuration, beam expansion and transport in the diode is controlled by extending the solenoidal guide field of the accelerator into the injector diode region (Fig. 3). Theoretical analyses of Ott et al. and Chen and Lovelace have adopted the somewhat idealized geometry of Figure 4. A semi-infinite cathode with surface potential $-\phi_c(r)$ is joined at a right angle to a grounded anode. Electron emission is assumed to occur over a limited region of the cathode surface, and the electron motion is constrained to be one-dimensional by application of a very strong axial magnetic field. In addition, the applied voltage is assumed to be ultrarelativistic so that the narrow cathode sheath region can be ignored. Under these conditions, a self-consistent solution for the electron current density can be found which satisfies the criterion of space-charge-limited emission.

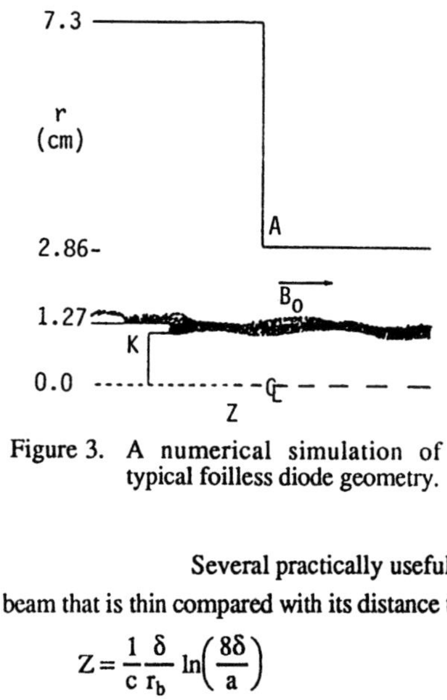

Figure 3. A numerical simulation of a typical foilless diode geometry.

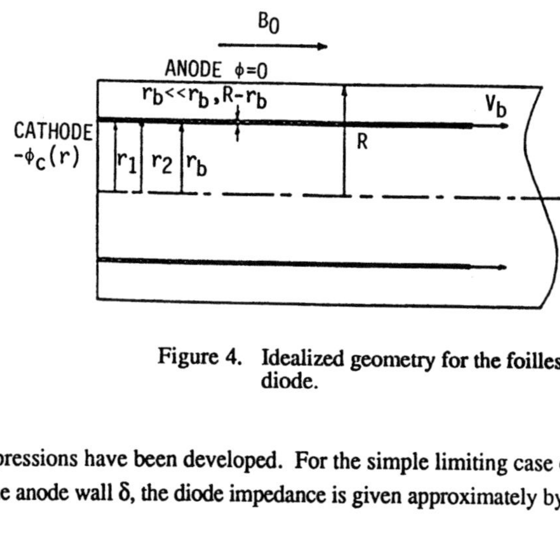

Figure 4. Idealized geometry for the foilless diode.

Several practically useful expressions have been developed. For the simple limiting case of a beam that is thin compared with its distance to the anode wall δ, the diode impedance is given approximately by

$$Z = \frac{1}{c}\frac{\delta}{r_b}\ln\left(\frac{8\delta}{a}\right) \tag{53}$$

and the total diode currently is approximately

$$I_t = (\gamma_o - 1)\,(mc^3/e)\,\frac{r_b}{\delta}\left[\ln\frac{8\delta}{a}\right]^{-1} . \tag{54}$$

An ad hoc approximation that extends the validity of Eq. (54) to non-relativistic voltages is to replace $(\gamma_o - 1)$ by the factor $(\gamma_o^{2/3} - 1)^{3/2}$. Relatively simple corrections to account for finite magnetic field effects (non-zero

Larmor orbits and radial oscillations of the beam envelope) and the inclusion of a longitudinal anode-cathode gap to provide additional control of the diode impedance are also available in the literature.

A nominal foilless diode design used in the RADLAC program is shown in Figure 5.

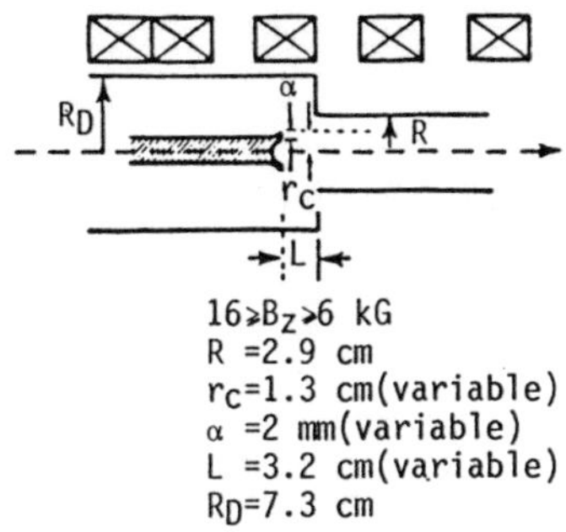

Figure 5. Schematic diagram of IBEX foilless diode injector.

To achieve beam currents in excess of ~50 kA with these relatively small dimensions requires relatively high diode voltages (> 4 MeV). A systematic study of the diode performance and beam vacuum propagation was undertaken by Mazarakis using the 4-MeV IBEX electron beam generator. The free variables of the experiments were anode-cathode voltage, axial magnetic field, cathode shank radius, and anode-cathode spacing. Extreme care was taken to precisely align the diode axis with the magnetic field axis as well as the vacuum pipe axis. In all cases, the foilless diode produced a low emittance, thin annular beam with practically no radial oscillations, as indicated by beam damage patterns on a set of brass witness plates that scanned an entire cyclotron wavelength.

3.1.2 Accelerating Gap Effects

During the course of multi-gap experiments with RADLAC I, it became apparent that the accelerating gap dimensions, if not carefully chosen, could result in serious disruption of the beam equilibrium. Consider the schematic accelerating gap geometry illustrated in Figure 6. The zero-order gap dimensions are constrained by (1) virtual cathode formation, (2) electron emission, and (3) radial oscillations of the beam envelope. We consider these constraints in turn.

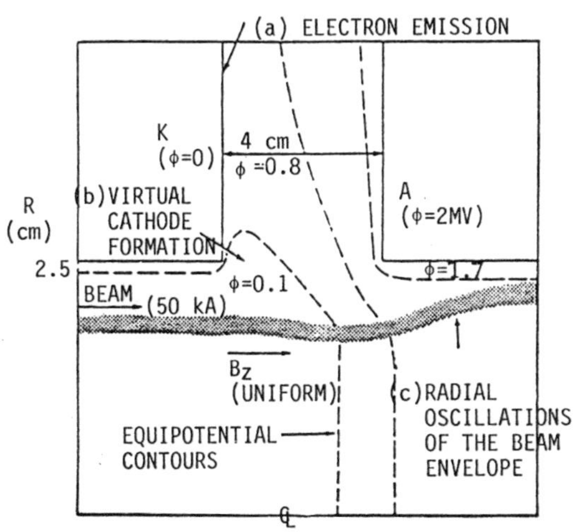

Figure 6. Schematic accelerating gap geometry.

3.1.2.1 Electron Emission

As is apparent from Figure 6, the beam space charge distorts the equipotential contours away from the cathode side of the gap; however, since this distortion typically extends radially into the gap a distance of only a gap width L, a general rule is that the gap spacing must be chosen such that

$$L \geq \phi_o/E_c \tag{55}$$

where ϕ_o represents the accelerating potential applied to the gap and E_c is the critical electric field for explosive electron emission. Equation (55) thus establishes the minimum acceptable gap spacing for entirely avoiding electron emission.

3.1.2.2 Space-Charge Limits

Even though accelerating voltage is applied to the gap, the equipotential contours of Figure 6 indicate the possibility of virtual cathode formation at the gap entrance if the beam current is too high. In fact, the limiting current can easily decrease by 50% at the entrance to the gap region. At the low voltage end of the accelerator it will often be the case that $4I_\ell \ln(R/r_b) << \phi_o c$. In this limit

$$\frac{I_\ell}{I_{\ell o}} \approx 1 - 2\ln\frac{R}{r_b}\left(\frac{I_{\ell o}}{\phi_o c}\right)[\ln(1 + L/R)]^2 . \tag{56}$$

As the applied gap voltage ϕ_o is raised, the limiting current increases.

3.1.2.3 Radial Oscillations of the Beam Envelope

The motion of the beam envelope in the smooth drift tube sections is well described by Eq. (20). However, the presence of an accelerating gap introduces electric field variation δE_r and δE_z. In this case, the linearization of Eq. (20) yields

$$\delta\ddot{r} + \frac{4}{\gamma_b^2}\left[\left(\frac{\Omega_o}{2}\right)^2 - \frac{eI}{\gamma_b m v_z r_o^2}\right]\delta r = -\frac{2eI}{\gamma_b^4 m v_z r_o}\,\delta\gamma - \frac{e\delta E_r}{\gamma_b m} , \tag{57}$$

$$\delta\gamma = -\frac{(\gamma_b^2 - 1)^{1/2}}{\gamma_b}\left(\frac{e}{mc}\right)\int^t \delta E_z dt . \tag{58}$$

For even moderately relativistic electron beams, the δ_γ term in Eq. (57) is negligible. Defining the quantity

$$k_c^2 = \frac{\Omega_o^2 - 4eI/\gamma_b m v_z r_o^2}{\gamma_b^2 c^2} , \tag{59}$$

Eq. (57) can be rewritten as

$$\delta r'' + k_c^2\,\delta r = -\frac{e\delta E_r}{\gamma_b m c^2} \tag{60}$$

where the primes denote differentiation with respect to the axial coordinate. Hence, radial electric field variations in the accelerating gap region will excite radial oscillations of the beam envelope with a wavelength given by $\lambda_o = 2\pi/k_c$. The physical origin of the radial oscillations is the strong excitation of zero-frequency cyclotron waves due to radial electric field variations when the gap width is approximately equal to half the beam cyclotron wavelength.

To avoid this resonant behavior, accelerating gaps must always be designed such that $L \neq \pi/k_c \approx 5.36\ \gamma_b/B_z$ (kG). for linear induction accelerators with constantly increasing γ_b, this relation indicates that simply increasing B_z may not be sufficient to control the oscillations. However, if the injector kinetic energy is sufficiently high, the resonance can always be avoided. To be sufficiently removed from the resonance, we require

$$L \leq \frac{\pi}{2k_c} . \tag{61}$$

Defining an average accelerating field stress by $E_a = \phi_o/L$, and assuming that $k_c \approx \Omega/\gamma_b c$, then if

$$\frac{e\phi_o}{\gamma_b mc^2} < \frac{\pi E_a}{2B_o} \tag{62}$$

the resonance condition can be entirely avoided, and the largest amplitude excitation will occur at the first accelerating gap. As an example, for a magnetic field strength of 10 kG, we must have $(\Delta\gamma/\gamma_b)_1 < 0.05\ \pi$, for $E_a = 300$ kV/cm.

3.1.2.4 Contoured Magnetic Fields

In principle, the radial oscillations can be completely eliminated by eliminating radial force imbalances, e.g., by adding an additional term (or terms) to the right-hand side of Eq. (58) which exactly counterbalances the variations in δE_r. For example, if the axial magnetic field strength B_z were varied in the gap region, an approximate radial force balance might be achieved. The envelope analysis gives the simple solution

$$B_z(z) = \frac{2mc}{e}\left\{\frac{C^2}{r_b^4} + \frac{2eI}{\gamma_b mcr_b^2} - \frac{\gamma_b e\delta E_r}{mr_b}\right\}^{1/2} , \tag{63}$$

where C is a constant that depends on the source configuration. (If the source is immersed, $C = P_\theta/m$, where $P_\theta \approx -(er_c^2B_c/2mc)$, with P_θ representing the canonical angular momentum, and r_c and B_c denoting the cathode radius and the applied magnetic field strength at the cathode tip.)

Equation (63) is only an approximate solution to the required magnetic field variation because it was derived assuming that the magnetic field was axially uniform; a rigorous solution of this problem is presented by Genoni et al. Nevertheless, Eq. (63) does indicate that this force balance method is restricted in applicability to beams of relatively small kinetic energy. Note that δE_r variations are magnified by γ_b. Hence, this method will generally work provided that the beam cyclotron wavelength is typically shorter than or comparable to the axial scale length for changes in the radial force.

The correctness of this "magnetic tailoring" approach has been verified by numerically simulating the beam transport through a single accelerating gap and through multiple accelerating gaps (Fig. 7). However, for beam kinetic energies in excess of nominally 10 MeV, the required rapid variation in axial magnetic field strength cannot be obtained with practical solenoidal field coil designs. Accordingly, we have investigated alternative gap geometries that substantially reduce the electric field variations.

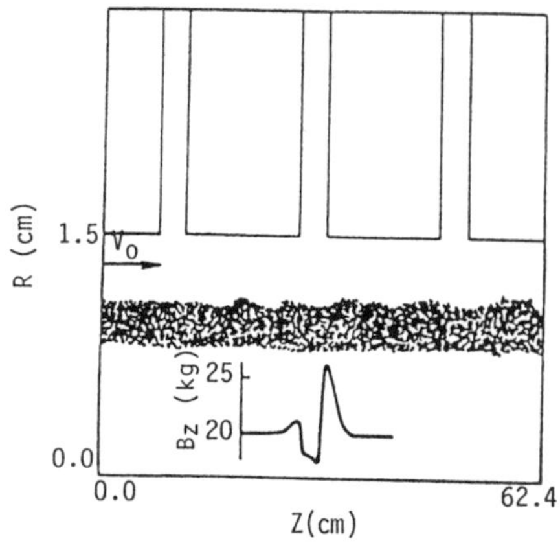

Figure 7. Particle simulation of radial oscillation suppression by contoured magnetic field profiles. A 50-kA beam from a 4-MV diode traverses three gaps (each with L = 4 cm, ϕ_0 = 3 MV). In each gap, the $B_z(z)$ is as shown; the field is uniform (20 kG) between gaps.

For example, one alternative gap geometry that totally eliminates radial electric field variation is shown in Figure 8. This structure has no axial gap, and the voltage is applied across concentric cylinders. The condition that eliminates radial electric field variations is that the beam must not cross any equipotential contours, i.e., the beam is not accelerated in the gap region. Far from the gap region, the radius of the outer wall is slowly decreased such that the beam is gradually accelerated over an axial length that is much greater than a beam cyclotron wavelength.

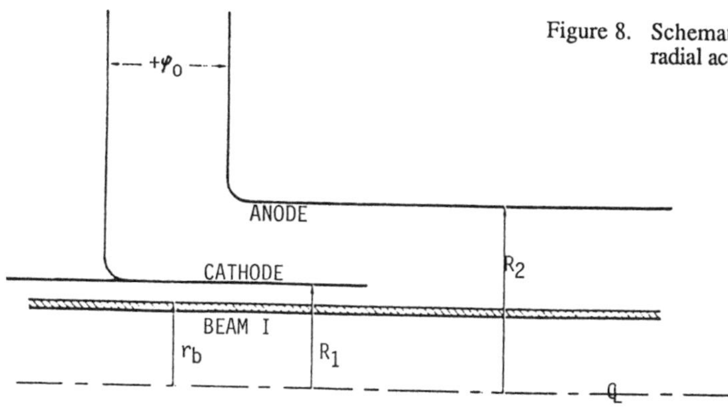

Figure 8. Schematic geometry of a radial accelerating gap.

To illustrate the RADLAC gap design procedure, we begin with a square gap geometry (Fig. 6, for example) with the gap spacing set by the explosive emission criterion, Eq. (55). The behavior of the beam envelope is also indicated. Note that for the 50-kA beam current, a virtual cathode could form near the entrance of the gap, depending on the beam kinetic energy. The virtual cathode can be eliminated by contouring the cathode and the anode of the gap as shown in Figure 9. A Poisson solver code is then rerun for the new gap design and the results are compared with the emission and virtual cathode criteria. If these comparisons are satisfactory, another beam envelope calculation is performed to determine the oscillation amplitudes. The procedure is then repeated until the electron emission, limiting current, and oscillation amplitude criteria are satisfied. The first two accelerating gaps of RADLAC II, designed by this procedure, are shown in Figure 10, together with the beam envelope solution. Note that this design also employs the radial gap concept. The predicted oscillation amplitudes are <1 mm.

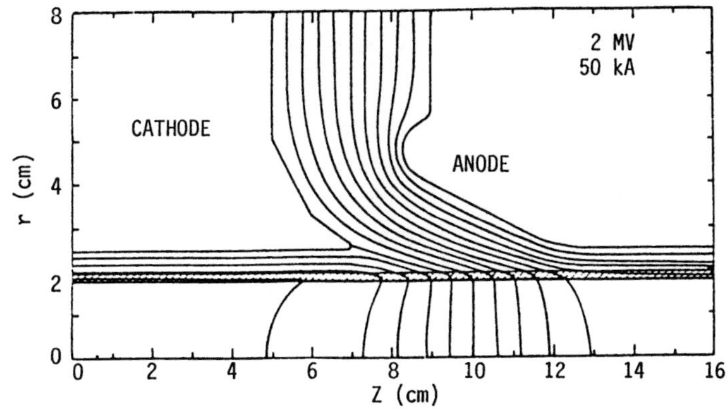

Figure 9. Intermediate stage of contouring the anode and cathode surface of the accelerating gap.

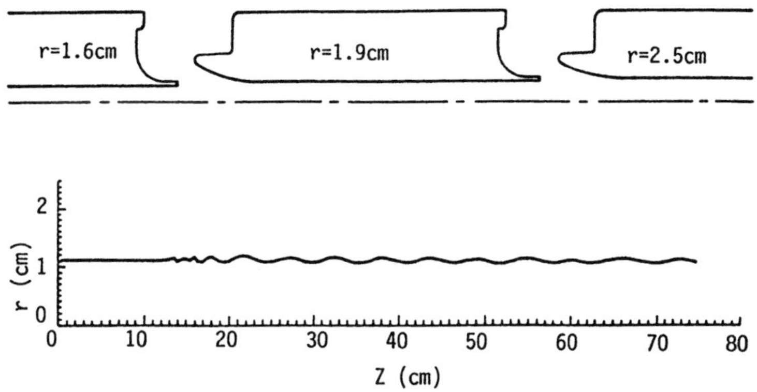

Figure 10. Final design of the first two accelerating gaps of RADLAC II. The final oscillation amplitude is <1 mm.

The validity of this gap design procedure has been checked and verified in a detailed series of post-acceleration experiments by Mazarakis et al.

3.1.3 Resistive Wall Instabilities

We have previously indicated that an unneutralized, solid electron beam propagating in an infinitely conducting drift tube is stable to small perturbations. Physically, the beam induces image charges and image currents of opposite sign on the surface of the conducting wall. The attractive force due to the image charge is canceled to order γ^{-2} by the magnetostatic repulsive force due to the image current, with the residual countered by the solenoidal guide field. If the drift tube is not perfectly conducting, however, the slow decay of the image currents can lead to the growth of resistive wall instabilities. To illustrate the instability mechanism consider a step displacement of the beam from the drift tube axis. The restoring force due to the resulting asymmetric image current rises very rapidly [in a time $\sim(4\pi\sigma)^{-1}$] to the same value that it would have for a perfect conductor, and then decays, on a slower, magnetic diffusion time, $4\pi\sigma a^2/c^2$. At the end of the step displacement, a decaying wake is produced which exerts a force on subsequent particles. Depending on the wave phase velocity, sinusoidal beam displacements can grow as a result of this resistive drag. The phase velocities ($v_{ph} = \omega/k$) associated with the space-charge and cyclotron waves are given by

$$v_{ph} = v_{zo} \pm \left(\frac{\omega_{po}^2}{k_z^2 \gamma_o^2} \right) \quad \text{(space–charge)}, \quad v_{ph} = v_{zo} \pm \left(\frac{\Omega^2}{k_z^2 \gamma_o^2} \right)^{1/2} \quad \text{(cyclotron)}. \tag{66}$$

Hence, there are both "slow" space-charge and cyclotron waves in which the wave phase velocity is less than the velocity of the particles, and "fast" waves for which the phase velocity exceeds the particle velocity. As shown in Figure 11, for a fast wave an observer moving with the beam velocity sees both wave and particles moving inward together; for a slow wave, however, the wave and particles move in opposite directions. Therefore, for a slow wave and a resistive wall, the wakefields can accelerate the particles outward while the particles are moving outward, and the wave grows. Correspondingly, dissipation produces damping of fast waves. Because slow waves can grow even though the system energy is decreasing because of dissipation in the resistive walls, they are often termed negative energy waves.

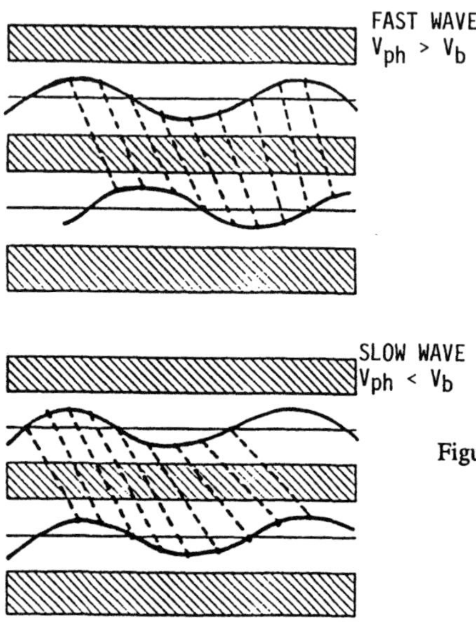

Figure 11. Diagram illustrating the motion of particles and waves in fast and slow transverse waves. For a fast wave, an observer moving with the beam velocity sees both wave and particles moving inward together. For a slow wave, the wave and particles appear to move in opposite directions.

The first studies of the resistive wall instability were performed for low current beams accelerated to high energies in circular accelerators; although the currents were low and the characteristic growth lengths were long, the effective distance traversed by the beam was also very long. In high current linear induction accelerators, the case is somewhat different; the effective beam transport length in the accelerator is typically short (tens of meters), but the high currents (tens of kiloamperes) can drastically decrease the characteristic growth length of the instability. For this case, Godfrey has obtained the dispersion relation of the slow cyclotron instability for an annular beam of mean radius r_b and thickness δ propagating in a drift tube of radius R and conductivity σ as

$$\omega - \omega_o = \pm (4\pi\sigma i\, \omega)^{-1/2} \left(\frac{2\omega_p^2\, \delta c}{\Omega_c R r_b} \right) f_\pm \tag{67}$$

with $k_z v_b = \Omega_c/\gamma$. In Eq. (67), ω_p is the beam plasma frequency, $\Omega_c = eB_z/mc$, and $f_\pm$ is a geometrical factor given by

$$f_\pm = (\ell^2 + k_z^2 R^2) \left[\frac{I_{\ell\pm1}(k_z r_b)}{I_{\ell+1}(k_z R) + I_{\ell-1}(k_z R)} \right]^2 \tag{68}$$

where ℓ is the azimuthal wave number and I denotes the modified Bessel function. Waves of helicity matching that of the particles grow most rapidly. For such waves in the long wavelength limit ($kR << 1$) and $\omega_0 \approx 0$, an approximate expression for the temporal growth rate is

$$\Gamma_R \approx e^{i\pi/6} (r\pi\sigma)^{-1/3} \left\{ \frac{2\omega_p^2 \delta c}{\Omega_c R r_b} f_\pm \right\}^{2/3} . \tag{69}$$

The predictions of Eq. (69) have been essentially verified with the computational results of Godfrey's linear theory code GRADR for several sets of parameter values relevant to high current linacs: beam kinetic energies and currents in the range of E = 4 to 20 MeV and I = 10 to 100 kA, guide fields in the range of 10 to 20 kG, and dimensions of around R = 2.5 cm, r_b = 2 cm, and δ = 0.2 cm. The agreement between the analytic theory and GRADR is typically better than 5%. One important result, the variation of $|\ell| = 1$ wave growth versus beam energy, is shown in Figure 12. Also depicted in Figure 12 is the low frequency, slow space-charge wave resistive wall growth rate. For this mode, instability occurs at very long wavelengths, $k_z r_b \approx 0$, so helicity is not important. At low energies, this instability is strongly affected by azimuthal velocity shear and is often masked by the diocotron instability for high azimuthal wave numbers.

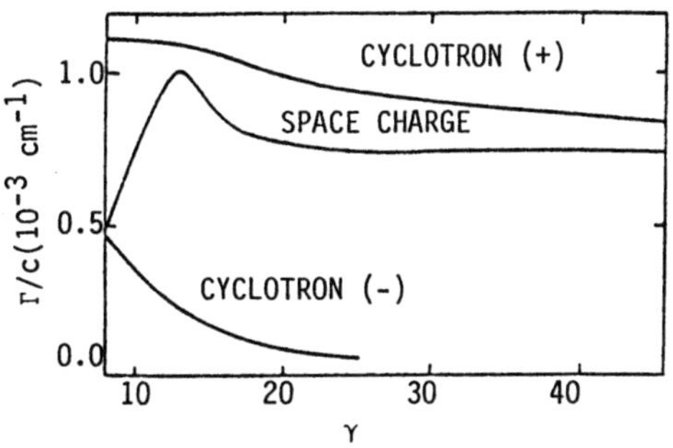

Figure 12. Peak growth rate versus beam energy for low frequency $\ell = 1$ resistive wall instabilities.

3.1.4 Image Displacement Instability

In the previous section, we indicated that an uncentered beam in a uniform drift tube is attracted to the nearest point in the wall by its electrostatic image charge. Similarly, it is repelled by its image current, with the two forces canceling to order γ^2. At an accelerating gap, however, the image current is "displaced" to large radius, while the image charge becomes concentrated on the gap and drift tube walls nearest the beam. The resulting loss of image force cancellation can give rise to an abrupt transverse force on the beam in the direction of its initial offset. Just as symmetric radial force imbalances can produce radial oscillations of the beam envelope, this transverse force can excite zero-frequency transverse oscillations as the beam travels to the next gap. Depending on the oscillation phase at the next gap, further growth can occur.

Assuming that each axial beam segment displaces rigidly in the transverse direction and the displacement is small compared to the drift tube radius R, the equation of transverse motion for the beam centroid in a highly conducting drift tube is

$$\frac{d}{dt}\gamma\frac{d}{dt}\xi = i\Omega_c\frac{d}{dt}\xi + \frac{2eI}{\gamma^2 mcR^2}\xi \tag{70}$$

where $\xi = x + iy$ is the complex sum of transverse displacements.

Under equilibrium conditions, the magnetic focusing force dominates the residual image forces, and the spatial variation of ξ between gaps is approximately

$$\xi = \xi_1 + \xi_2\, e^{ik_o z}$$

with $k_o = \Omega_c/\gamma c$. In words, as the beam travels from one gap to the next, the (complex) amplitude of the low frequency space-charge mode is essentially constant, while the amplitude of the high frequency cyclotron mode is multiplied by exp $(i\Omega_c L/\gamma c)$, where L is the separation between gaps.

In crossing a thin accelerating gap of width $\omega << \gamma c/\Omega_c$, the beam will experience an impulsive force $F\xi$ proportional to the product of the displacement from the axis and the space-charge image force, which changes the beam transverse velocity by $iF\xi/\gamma k_o c^2$ but leaves the displacement constant.

The transverse motion of the beam in the drift tube after passing through the gap can then be related to the motion prior to the gap by using the above equation and the matching condition in the gap. Assuming for the moment that γ is constant, the equations for the mode amplitudes can be recast into matrix format as

$$\begin{pmatrix}\xi_2\\ \\ \xi_1\end{pmatrix}_+ = e^{ik_oL/2}\begin{pmatrix}\left(1-\dfrac{iF}{\gamma k_o c^2}\right)e^{ik_oL/2} & -\dfrac{iF}{\gamma k_o c^2}\\ \\ \dfrac{iF}{\gamma k_o c^2} & \left(1+\dfrac{iF}{\gamma k_o c^2}\right)e^{-ik_oL/2}\end{pmatrix}\begin{pmatrix}\xi_2\\ \\ \xi_1\end{pmatrix}_- . \tag{71}$$

The eigenvalues of Eq. (71) describe the growth of the gap-induced image displacement instability: these are

$$\lambda = \psi \pm i(1-\psi^2)^{1/2}\,, \tag{72}$$

$$\psi = \cos\frac{k_oL}{2} + \frac{F}{k_o c^2 \gamma}\sin\frac{k_oL}{2}\,, \tag{73}$$

where

$$\frac{F}{\gamma k_o c^2} = \frac{2eI\,fw}{\gamma m^2 R^2 \Omega_c}\,.$$

The image displacement force constant f is the ratio of the transverse force in the gap to the transverse electric force in the drift tube; it is typically around 1/2.

Returning to Eq. (72), growth corresponds to $|\lambda| > 1$. Since $(F/k_o c^2\gamma)$ is purely real, the condition for stability becomes $\psi \le 1$. The dependence of ψ on the phase advance per gap $(k_oL/2)$ is presented in Figure 13 for several values of the parameter F. It is apparent that $F/k_o c^2\gamma < 1/2$ is required to assure minimal growth over several periodically spaced gaps with γ constant.

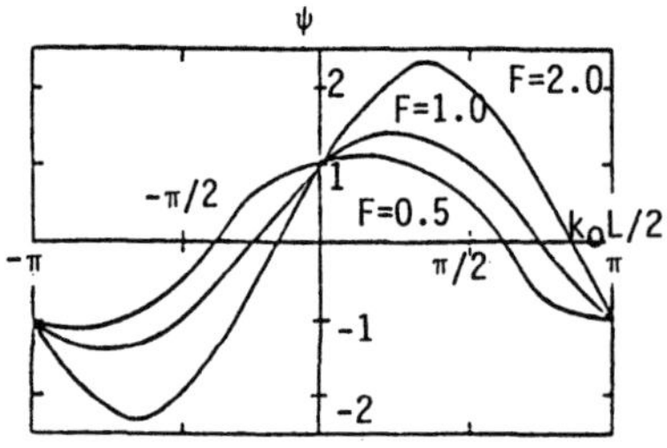

Figure 13. Image displacement instability growth parameter ψ as a function of the phase advance per gap.

To take into account the variation in γ, the amplification after several gaps is determined by the eigenvalue of the matrix formed by multiplying the matrices for each gap. In general, the eigenvalue of the product is much smaller than the product of the eigenvalues; i.e., beam acceleration decreases image-displacement instability growth.

For high current linacs, the thin-gap approximation is not always well satisfied. To the extent that $F/k_o c^2 \gamma$ is less than $k_o \ell/2$, increasing gap width weakens the instability. Accurate treatments of this important case are now available in the literature.

In order to estimate the growth of the image displacement instability, Eq. (70) and the following equation were numerically integrated for a few sets of idealized conditions in the projected RADLAC II parameter regime by Godfrey. The results for a 100-kA beam are summarized in Table 2. Cases 1 to 5 are without acceleration, and cases 6 to 10 are for acceleration through gaps of nominally 2 MV each. Γ is the total growth for ten such gaps.

Table 2. Calculated Image Displacement Instability Growth

CASE	B_z (kg)	ω/R^2 (cm^{-1})	Γ
1	20.0	0.8	7.5
2	19.88	0.8	1600.0
3	19.74	0.8	340.0
4	19.01	0.8	2.4
5	19.47	0.8	2.2
6	20.0	0.8	470.0
7	20.46	0.8	48.0
8	20.0	0.29	4.5
9	19.83	0.29	7.3
10	20.46	0.29	1.6

The phase sensitivity of the instability is clearly evidenced by the resonance behavior with respect to variations in magnetic field strength. Varying the intergap spacing as a function of beam kinetic energy might also help control the phase angle of the oscillation, but a more straightforward approach is to reduce the separation of the image charges and currents by carefully designing the accelerating gap structures. For example, the radial accelerating gap design of Figure 8 should substantially decrease image displacement growth. This fact is reflected in the contoured gap geometries of RADLAC II; modeling these gaps gives the reduced growths reflected in cases 8 to 10.

3.1.5 Beam Break-Up (BBU) Instability

In addition to the zero-frequency image displacement instability, high current induction linacs with periodic accelerating cavities can be susceptible to a microwave instability termed transverse beam break-up. Physically, the instability arises from the coupling between a beam oscillating off-axis and accelerating cavity modes with a transverse magnetic field on axis (T_{lmo} modes, for example). If the frequency of the beam oscillation occurs at a natural resonance of the cavity, then cavity mode energy will increase and be in temporal phase with the oscillations, thereby causing the oscillations to increase in amplitude during the pulse duration. Moreover, if every cavity is identical, the peak amplitude of the oscillations will also grow from cavity to cavity.

The normal modes of the accelerating cavities can be conveniently described by damped harmonic oscillator equations, driven by the off-axis beam current. For example,

$$\frac{d^2}{dt^2}A + \frac{\omega_n}{Q}\frac{d}{dt}A + \omega_n^2 = \alpha_1 \xi \tag{74}$$

where A is the axial vector potential of the mode, ω_n and Q are the mode frequency and quality factor, ξ is the transverse displacement of the beam centroid, and a is a coupling coefficient.

In turn, the beam transverse oscillation is driven by the transverse magnetic field of the accelerating cavity normal mode according to

$$\frac{d}{dt}\gamma\frac{d}{dt}\xi = \alpha_2 A \tag{75}$$

where α_2 is a second coupling coefficient. Hence, the physically significant quantity is the product of the coupling coefficients α_1, α_2. This product is conveniently expressed in terms of the beam current I and the scaled transverse impedance $Z_\perp/Q$ of the particular cavity mode as

$$\alpha_1\,\alpha_2 = \omega_n^3 \frac{Z_\perp}{Q}\frac{Ie}{mc^2} \tag{76}$$

Hence, the strength of the coupling between beam oscillations, cavity fields, and later beam oscillations depends on the beam current, cavity constants, and the oscillation frequency.

For a single cavity, coherent transverse oscillations can be selectively excited from transverse beam noise. What makes the BBU instability so dangerous is that subsequent identical cavities are then resonantly driven by coherent transverse oscillations. Unlike the case of image displacement, BBU always grows, apart from phase mixing.

In fact, the image displacement instability can be viewed as the low frequency limit of the beam break-up instability. Setting the time derivatives to zero in Eq. (74) and inserting the resulting expression for the vector potential into Eq. (75) yields

$$\frac{d}{dt}\gamma\frac{d}{dt}\xi = \frac{\omega_n Z_\perp}{Q}\frac{eI}{mc^2}\xi\,. \tag{77}$$

This equation simply states that all gap modes tend to increase an initial transverse displacement. The combined effect is found by summing all the $(\omega_n Z_\perp/Q)$ for the various modes, which is of order ω/R^2.

Analogous to the treatment used in the case of image displacement, the thin gap approximation leads to a matrix equation for the evolution of the mode amplitudes with the matrix eigenvalues describing the growth of the gap-induced instabilities. The impulsive force F due to beam break-up can be formally obtained from the Laplace transform of Eq. (74). However, if it is assumed that the beam oscillates uniformly at $\omega_n \tau >> Q$, then the Laplace transform variable can be replaced by i ω_n. This procedure yields

$$\frac{F}{c^2 k\gamma} \approx \frac{iQ}{c}\frac{Z_\perp}{Q}\frac{\omega_n}{\Omega_c}\frac{eI}{mc^3}\,. \tag{78}$$

Substituting Eq. (78) into Eq. (71) gives the amplification of beam break-up oscillations at the gap, in a manner similar to the case of image displacement. However, F is now purely imaginary, and $|\gamma| > 1$ for all $|F| > 0$ and $kL/2 \neq n\pi$, i.e., the oscillations always grow. Hence, for high current linacs, controlling beam break-up requires large focusing fields (Ω_c), and lossy cavities (low Q) with low transverse shunt impedance ($Z_\perp/Q$). Also, the beam rise time τ_r should be long compared to the cavity field ringup time, Q/ω_n, to avoid shock excitation of the cavities.

From the previous analysis, beam break-up growth rates are most strongly affected by the transverse impedance and the quality factor of the accelerating cavity normal modes. High voltage standoff is provided by a vacuum interface consisting of alternating rings of metal and plastic dielectric. Inside the vacuum interface, metallic fieldshapers serve to localize and define the accelerating fields; a high dielectric constant fluid (water or ethylene glycol) is usually located outside the vacuum interface. The results of test bench measurements of the transverse impedance and cavity quality factor give very favorable values ($Z_\perp/Q < 10$ ohms and $Q < 10$) for the lowest frequency, most dangerous $\ell = 1$ mode. Essentially, the insulator stack geometry is a very leaky cavity; since the fields extend outside the rings, they are greatly damped by the liquid dielectric medium because of the loss tangent at the frequencies of most concern (>1 GHz).

With the results of these measurements, the rigid disk code BALTIC has been used to evaluate growth of the beam break-up instability in RADLAC II. The results of these studies are summarized in Table 3 for a 100-kA beam and 5-cm gaps separated by 100 cm. Case 1 indicates acceptably small BBU growth (through ten gaps) even without acceleration; however, the sensitivity to increases in $Z_\perp/Q$ or Q are illustrated in cases 2 and 3. The damping effect of beam acceleration is evident in cases 4 and 5. Also (as expected) no strong magnetic field resonant effects are seen (cases 5, 6, and 7). Finally, the effect of non-zero risetime clearly reduces instability growth, since the cavities are not shock excited (cases 8 and 9). In the latter cases, the growth is sufficiently small that additional suppression mechanisms are not required.

Table 3. Calculated Beam Break-up Instability Growth

CASE	B_z (kg)	$Z_\perp/Q$ (ohm)	Q	$\Delta\gamma$	τ_R(cm)	Γ
1	20.	10	10	0	0	7.0
2	20.	10	20	0	0	350.
3	20.	20	10	0	0	700.
4	20.	10	10	4	0	3.3
5	20.	20	10	4	0	35.
6	19.83	10	10	4	0	4.2
7	19.66	10	10	4	0	3.1
8	20.	10	10	4	10	1.7
9	20.	20	10	4	40	3.6

3.2 NEW CONCEPTS IN HIGH CURRENT BEAM TRANSPORT: ELECTROSTATIC CHANNEL GUIDING

Although great care was taken in the design of the accelerating cavities of the Advanced Test Accelerator, growth of the beam break-up instability resulted in severe pulse disruption for beam currents exceeding about 5 kA. Although higher solenoidal fields would help to decrease this growth, an alternative is to create a continuous electrostatic focusing system consisting of a positive line of charge through the entire accelerator length. The symmetric radial electric field of the positive charge would focus the electron beam onto the accelerator axis; moreover, since the focusing electric field varies inversely with radius outside the ion channel, the beam electrons are confined with an anharmonic potential well, which can lead to orbital phase mix damping of any transverse beam oscillation (at the expense of an emittance increase).

The first test of this concept used the beam from the experimental test accelerator (ETA) to induce positive charge on a resistive graphite thread that was supported along the length of the accelerator. Although successful, this technique was not feasible for creating a positive line charge along the full 80-m length of the ATA. Instead, a small amount of benzene was introduced into the accelerator and a low energy KrF laser was used to partially photoionize a small-diameter channel in the gas. The subsequent injection of the ATA electron beam rapidly expelled the plasma electrons, and the residual ion core then focused and guided the electron beam pulse. With use of this technique, virtually no growth of the BBU instability was apparent and successful operation at the full ATA design (10 kA) was achieved.[4] Following these initial experiments, ion channel guiding has been successfully used in beam steering and bending experiments[5,6] and has been suggested for use in a recirculating linac concept.[7] In the following, the basic physics of electron beam guiding by ion focusing is briefly summarized.

If the electron beam is injected into an ionized plasma channel, the resulting motion of the plasma charges will tend to neutralize the beam self-fields, as indicated in Figure 14. The beam space charge will become neutralized by the expulsion of plasma electrons in approximately one plasma oscillation period $\tau_p \sim (4\pi e^2 n_b/m_e)^{-1/2} = (\omega_p)^{-1}$, assuming that the beam density and the plasma electron density are approximately equal.

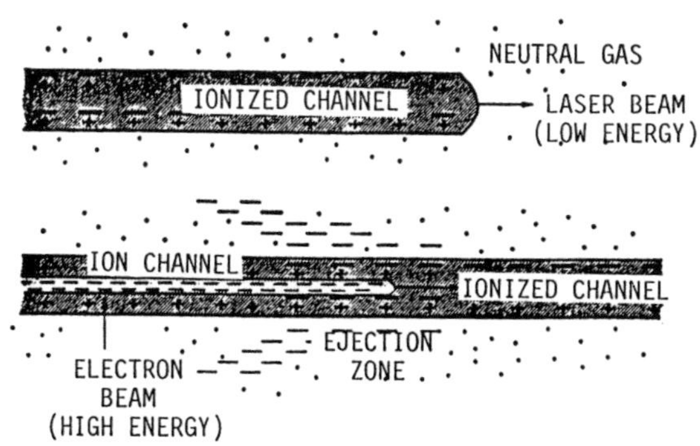

Figure 14. A laser preionizes a plasma column; the space-charge fields of the beam expel the plasma electrons leaving a strongly focusing positive ion channel. (For accelerator applications, the ion channel is usually smaller in radius than the electron beam.)

To avoid possible electron-electron streaming instabilities, the plasma electron line density should not exceed the beam electron line density; however, there is a minimum plasma line density necessary for a radial force balance equilibrium. To calculate this density, assume that all plasma electrons have been expelled so that the beam is propagating along a stationary ion background of linear density, $n_i r_c^2$; the fractional beam electron space-charge neutralization is denoted by $f_n = n_i r_c^2/n_c r_b^2$. Solving for the radial space-charge electric field and the azimuthal magnetic field and substituting these quantities into the equation of motion yields an expression for the equilibrium radius

$$r_b^2 = \frac{\varepsilon^2}{c^4(\gamma^2 - 1)} \frac{I_A}{2I_b(f_n - 1/\gamma^2)} \tag{79}$$

where $I_A = \beta\gamma\, mc^3/e$ is the Alfven current and $I_b = \pi r_b^2\, en_b\, \beta c$ is the beam current. The beam emittance is denoted by ε. It is apparent that there is no equilibrium unless $f_n \geq 1/\gamma^2$. In the limit of large beam kinetic energy, the beam self-fields essentially cancel, and the beam is radially confined by the electrostatic force of the ion channel. In this case,

$$r_b \cong \frac{\varepsilon}{\gamma c^2}\left(\frac{I_A}{2f_n I_b}\right)^{1/2} . \tag{80}$$

Having calculated the zero-order beam equilibrium requirements, we now return to the details of how the positive ion channel becomes established and we examine the behavior of the electrons at the beam front during this process. At least two significant effects are expected:*

(1) emittance-driven beam expansion until $f_n \sim 1/\gamma^2$,

(2) loss of energy from beam electrons to the plasma electrons that are expelled.

Each of these processes will cause a loss of beam particles/energy at the beam front (erosion), decreasing the effective pulse length as the beam propagates. We now estimate the magnitude of each of these effects and derive appropriate "erosion" formulae.

In the case of emittance-driven erosion, the natural scale lengths are the length of beam pulse necessary to establish the force balance criterion and the axial propagation distance necessary for the beam to expand outside the channel capture radius. Roughly, the rate at which the beam is lost will be given by dividing the first of these scale lengths by the second. Simple estimates give

$$\left.\frac{dx}{dz}\right|_{\text{emittance}} \cong \frac{2}{\gamma R_c}\,\frac{f_n^{1/4}\,\varepsilon}{r_b^{3/2}} . \tag{81}$$

The erosion resulting from beam energy loss caused by expelling electrons from the channel can also be simply estimated by calculating the difference in electrostatic potential for an unneutralized beam compared with a beam that is neutralized by the channel ion space charge. Since this difference comes only from the ions, the average electric field acting to slow the beam electrons is just $\Delta V/\ell = E \cong (I_b\, f_n/\beta_o c)\,(1 + 2 \ln R/r_c)$, where R is the drift-tube radius. Consequently, the rate at which beam electrons will lose energy is just $eE/(\gamma-1)mc^2$, or

$$\left.\frac{dx}{dz}\right|_{\text{energy loss}} \cong \frac{\nu f_N}{(\gamma-1)}\,(1 + 2 \ln R/r_c) \tag{82}$$

where $\nu = I_b/(\beta mc^3/e)$ is the dimensionless beam current.

The slowness of the beam erosion phenomena suggests that efficient propagation can be achieved by simply using long pulses. However, possible growth of various streaming instabilities between the beam electrons and the plasma species may limit the practical pulse length. If the beam itself does not generate significant ionization, then the instability most likely to occur is a transverse streaming instability between the beam electrons and the channel ions. Conventional instability analyses indicate that the characteristic growth time must scale as the ion bounce period, i.e., the time required for an ion to oscillate radially through the beam. In this case, the number of e-foldings of (linear) instability growth during the pulse length t_p should scale approximately as

$$n_{cf} \sim t_p / r_c\,(I_b / m_i)^{1/2} \tag{83}$$

where m_i is the ion mass. It is also expected, however, that this instability should saturate nonlinearly if the amplitude of the oscillation should become of the order of the beam radius. This instability has now been clearly observed,[8] and the experimental data have essentially verified these analytical estimates.

---

*Additional erosion processes can occur if there is ion channel curvature or if transverse magnetic fields are present.

Because of the newness of this concept, it is reasonable to expect that its potential, as well as possible limitations, are not yet fully known. For example, the betatron oscillations and emittance growth resulting from phase mix damping may not allow this transport technique to be used for high current free-electron laser applications. Yet, it was once thought that transporting currents > 17 kA would be difficult because of plasma electron trapping in the magnetic self-field of the beam. By arranging the beam risetime such that $f_n \sim 1$ early in the pulse, but $f_n \sim 1/2$ at the current peak, 23-kA beams have been successfully transported via ion channels in recent autoacceleration experiments by Coleman et al. Further work on these and several other issues is clearly necessary.

4 BEAM TRANSPORT ANALYSES FOR A HIGH CURRENT ION ACCELERATOR[9]

P3T was an ion accelerator concept for the production of tritium based on a pulse power/induction linac approach. Although this device will probably never be built, the beam transport considerations included many of the limiting features already discussed, as well as several new issues. In this section, we discuss several of these new issues. In particular, we develop criteria for various equilibrium configurations, injector specifications, general beam stability criteria, zero-order analyses of accelerating gap designs, transverse beam break-up analyses, pulse-compression techniques, and longitudinal instability analyses.

4.1 BEAM TRANSPORT SYSTEM APPROACHES

The general current parameter range of interest is 1 to 10 kA. To provide some perspective, a simple measure of the difficulty of transporting a bare ion beam of current I and radius r_b is the radial electric field given by

$$E_r = 2I/(vr) \quad \text{for} \quad r > b$$

where v is the average axial velocity. For example, this field is responsible for radial space-charge blow-up of the beam and can give rise to electron emission from the drift tube wall, assuming that the beam and the wall radii are of comparable magnitude. A useful practical limit for the latter effect is typically 100 kV/cm. For a 10-MeV proton beam, v = 0.14c, and

$$E_r\,(\text{kV/cm}) = 414\, I(\text{kA})/r(\text{cm}) \ . \tag{84}$$

Thus for currents exceeding 10 kA, the beam (wall) radius must become quite large (>40 cm), a much higher voltage injector must be used, or the ion space charge must be neutralized by a background electron density.

4.1.1 Non-Neutral Beam Transport

Consideration of Eq. (84) indicates that for reasonable radii (10 cm) the general magnitude of the unneutralized beam current could be a few kiloamperes or less. Whether or not such beam parameters can be achieved depends on several criteria, including space-charge limits and limitations imposed by the focusing system.

The space-charge potential depression on the beam axis of symmetry for a solid, cylindrically symmetric, un-neutralized beam is given by

$$\phi(0) - \phi(R) = (I/v)[1 + 2\ln(R/r_b)]$$

where R is the drift tube radius. For a 1-kA, 10-MeV proton beam with $R/r_b = 1.6$, the resulting potential depression is about 0.4 MeV.

The space-charge limit for such a beam is described by

$$I_L = (mc^3/e)(\gamma^{2/3} - 1)^{3/2}[1 + 2\ln(R/r_b)]^{-1} . \tag{85}$$

With the same beam parameters as above, the space-charge limiting current is about 9.6 kA. Thus, on the basis of simple space-charge consideration, non-neutral beam transport does appear feasible, provided the beam current does not exceed a few kiloamperes.

The other primary limitations on the current arise from constraints associated with the focusing system. The two principal non-neutral transport options use either solenoidal magnetic field focusing or magnetic quadrupole focusing. Solenoidal transport has been used on all high current electron induction linacs constructed to date, and magnetic quadrupole focusing is being considered in the induction linac approach to heavy ion fusion. For small-radius, high-β ion beams it is well known that "strong focusing" with quadrupoles is more effective than "weak focusing" with solenoids. However, for the present case of large-radius, low-β ion beams, solenoids are competitive and can have advantages over quadrupoles. We briefly summarize the important features of each approach and develop a useful qualitative comparison of quadrupole and solenoidal focusing.

4.1.2. Solenoidal Transport

The details of ion-beam transport in vacuum along a solenoidal magnetic field depend importantly on the extent to which the anode is immersed in the axial magnetic field. Since this depends in turn on the details of the ion injector design, the following discussion will be somewhat general.

The motion of a single charged particle in a solenoidal magnetic field can be considered as the sum of a (usually) large amplitude rotation at the slow E x B rotation frequency and a (usually) small amplitude rotation at the ion cyclotron frequency. In the case of Brillouin flow, which might be obtained from a totally shielded source, these modes are identical. More generally, the ratio of the amplitude of the fast rotation to that of the slow rotation is approximately given by $2Imc^2/(\gamma v r_b^2 eB^2)$. For an immersed source configuration, if this ratio is small, then the detailed fine structure of the particle orbits can be ignored, and a simple envelope equation analysis can be used to obtain the equilibrium condition given by

$$B \leq B_{eq} = [8Imc^2/(\gamma e v r_b^2)]^{1/2} . \tag{86}$$

Writing $B = \alpha B_{eq}$ and substituting for B in the amplitude ratio expression gives $(2\alpha)^{-2}$. Hence, if the strength of the applied magnetic field exceeds about 1.5 times the strict equilibrium field value as given by Eq. (86), then the cyclotron orbit radius will be about one-tenth the large amplitude E x B rotation, or less, and use of the envelope equation analysis is justified.

Evaluation of B_{eq} for a 1-kA, 10-MeV proton beam using an equilibrium radius of 10 cm gives 13.1 kG. Thus, an applied field strength of 20 kG should be sufficient to transport such a beam. Although this value is not insignificant, superconducting magnets capable of providing larger fields have been built many times. (In fact, the primary magnet for the proposed Compact Ignition Tokamak is being designed to operate at maximum field strengths in the range of 80 to 100 kG.) In addition, the field strength required for a shielded source equilibrium of equal radius at the same beam current is usually somewhat smaller.

4.1.3 Quadrupole Transport

For quadrupoles, focusing is achieved with a magnetic field of the form $B_\theta(r/r_b)$, and an alternating-gradient system is used (focus, drift, defocus, drift, etc.) that focuses on the average in both transverse coordinates. A simple way to model a quadrupole system is to assume a constant azimuthal magnetic field (as in a Z-pinch transport channel) with a peak value B_θ / k at $r = r_b$, where $k > 1$ and typically $k >> 1$. (The factor k depends on several parameters including the phase advance per lattice period, the spacing between magnets, etc.) Examination of the radial force balance equation for quadrupole focusing indicates that the equilibrium criterion has the form $B_\theta \geq (2kI/vcr_b)$.

In order to compare the focusing effectiveness of quadrupoles and solenoids, we examine the magnitude of the magnetic restoring force arising from the two systems, setting $B_q = B_z = B$. The resulting ratio is

$$F_r(\text{quad})/F_r(\text{sol}) = 4\beta mc^2/(eBr_bk)\ . \qquad (87)$$

Quadrupoles are thus more effective for small-radius beams and they become increasingly effective as β increases. However, for relatively large-radius (10-cm) proton beams of a few kiloamperes and field strengths of a few tens of kilogauss, $F_r(\text{quad}) \approx F_r(\text{sol})$ for $k \approx 4$. (Under these conditions, it is more probable that $k = 10$.) Thus, at the low energy end of the accelerator it is likely that solenoidal and quadrupole focusing will be comparable, with quadrupoles becoming more desirable as the kinetic energy increases.

These results are summarized in Figure 15, where the radial forces due to solenoids, quadrupoles (assuming $k = 4$), and space charge are plotted. Also plotted is the line where the radial electric field due to space charge is 100 kV/cm. Note that for the injection parameters considered, solenoids and quadrupoles are about equally effective. However, as b increases, F_r due to the solenoid stays fixed, F_r due to the quadrupole increases, and F_r due to the space charge decreases.

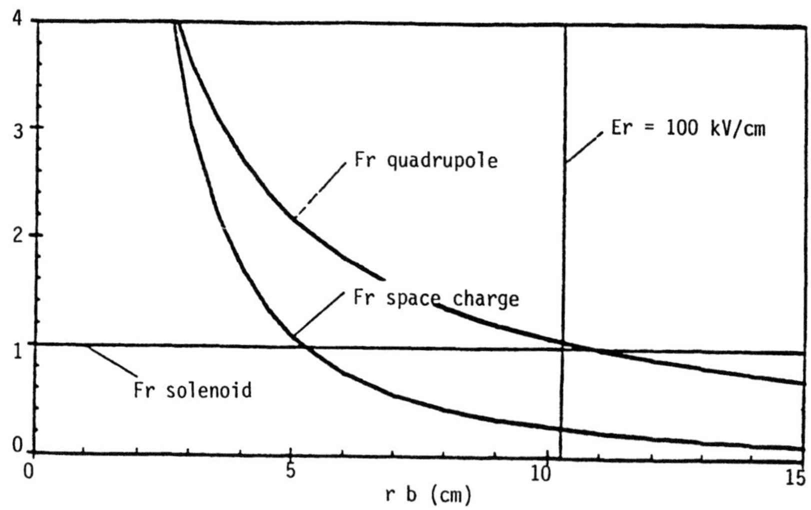

Figure 15. Variation of the radial focusing forces due to solenoids and quadrupoles (k=4) and space-charge force with ion-beam velocity.

For parameter ranges in which solenoids and quadrupoles are roughly comparable, solenoids are to be preferred because (1) they are simpler, (2) they are θ-symmetric, and (3) they will permit the use of bunching schemes that do not require voltage ramping (to be discussed later). Thus, at the low energy end of the accelerator it is preferable to use solenoidal focusing, while at the high energy end, quadrupole focusing may offer some advantage. Since many of the beam transport problems are most severe at the low energy end of the accelerator where space-charge effects are most severe, we choose solenoidal focusing for further analysis.

4.2 PARAMETER SELECTION

The magnetic field and space-charge limit constraints can be conveniently summarized in terms of a current vs magnetic field diagram, as schematically shown in Figure 16. We have used $\alpha = 1.5$ to ensure that the ion cyclotron orbits are about one-tenth of the beam radius. Thus, for a nominal 10-cm-radius beam, the small orbit radius will be about 1 cm. To be conservative, we have allowed a 5-cm spacing between the outermost radial excursion of the orbit and the drift tube wall, implying a drift tube radius of 16 cm. One should not exceed about half the space-charge-limiting current I_L, and it is very desirable (especially for non-relativistic ions) to stay below $I_L/4$. Finally, we have assumed a practical upper limit on the magnetic field strength of 40 kG. The allowable operating points lie in the shaded region of Figure 16. The highest allowable beam current at the lowest magnetic field strength is thus about 2.4 kA at 30 kG. The pulse duration at the low energy end of the accelerator required to give 1.9 mC is 0.81 μsec.

At 2.4 kA, the electric field at the drift tube wall is 62 kV/cm, from Eq. (84). The space-charge depression is 0.96 MeV, and the potential difference between the center and the edge of the beam is I/v = 0.5 MV. This produces a velocity shear of 2.5%, which is tolerable.

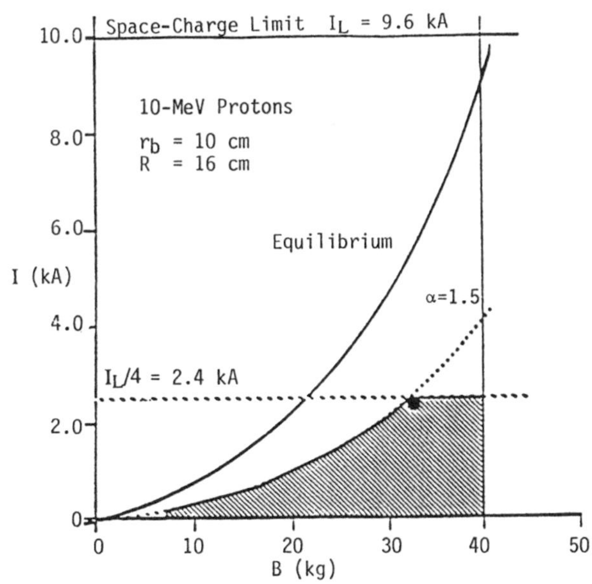

Figure 16. Comparison of various equilibrium constraints imposed by space-charge limits, solenoidal focusing, and magnetic field strength.

4.3 ACCELERATING GAP CONSIDERATIONS

In experimentation with high current electron linacs, it was discovered that the presence of an accelerating gap could destroy the zero-order equilibrium of the beam if care was not exercised in the design of the gap. The key criteria include avoiding field emission, avoiding virtual anode formation, and ensuring that the beam cyclotron wavelength is significantly different from the axial gap length. To practically eliminate any problem of electron emission, we require that the electric field stress in the gap region be <100 kV/cm. Also, from rather detailed analyses of space-charge limits in accelerating gap regions we estimate that virtual anode formation should not occur unless the beam current exceeds 10 kA at 10 MeV.

Linearization of the radial force balance equation indicates that radial oscillations of the beam envelope will occur at the beam cyclotron wavelength ($\lambda = 2\pi v/\Omega$), increased somewhat by beam space-charge effects. For 10-MeV protons and a 40-kG magnetic field strength, this wavelength is 72 cm. If the axial gap dimension is much shorter than half this length, then gap-excited envelope oscillations should be minimal. Assuming an average field stress of 100 kV/cm, the gaps at the low-voltage end of the accelerator should be about 5 cm (and 500 kV). As the beam kinetic energy increases (and the beam velocity increases), the oscillation wavelength will also increase and the gap width (and applied voltage per gap) can be increased (e.g., 10-cm gap with 1 MV at 40 MeV).

These accelerating gap analyses were tested by performing particle-in-cell calculations of a proton beam propagating through an accelerating gap using a modified version of Sandia's quasi-static diode code. Virtual anodes do not form provided the beam current is only a few kiloamperes at the low energy end of the accelerator, and radial oscillations of the beam envelope are not excited provided the axial gap width is much smaller than half the cyclotron wavelength. These conditions are easily satisfied, assuming a field stress level of 100 kV/cm.

4.4 TRANSVERSE BEAM DYNAMICS

Using the baseline parameters determined from the previous sections, we have examined a number of potentially critical transverse beam dynamics issues, including transverse beam confinement, the effects of misalignments, and the transverse beam break-up instability. We will discuss each of these in some detail and summarize our results in this section.

4.4.1 Beam Misalignments and Emittance Growth

The ion-beam emittance growth resulting from cumulative effects of small misalignments must be considered, especially since the number of accelerating gaps and transport cells is very large (many hundreds). While many types of misalignments are possible for estimating gross effects, the worst case scenario seems to be the one in which a transverse displacement Δx results in a configuration space increase of the beam radius by $\Delta r \approx \Delta x$. Most of the cases examined require considerable mixing in phase space to produce this effect. If phase mixing occurs and the beam radius grows by Δr at each stage, then the total radius increase will be $\Delta R = \Sigma \Delta r = N\Delta r$, and the allowed average increase at each state is $\Delta r = \Delta R/N$. For an allowed ΔR of 6 cm and $N = 800$, this gives $\Delta r \approx 0.08$ mm (0.003 inches) for each stage. Alignment to within 0.125 mm (0.005 inches) is typical. Thus, even in this worst case scenario, only small improvements in current practices would be needed.

4.4.2 Transverse Beam Stability

The displacement of the centroid of an ion beam from the axis of an evacuated conducting wall generates asymmetric image charges and currents that can affect the transverse motion of the beam. In particular, asymmetric image charges attract the beam toward the wall, thus tending to increase the displacement. In contrast, asymmetric image currents repel the displaced beam toward the pipe axis. For a displacement $\zeta(z,t)$ in the x-direction, the fields produced by these image charges and currents have been calculated to be (Gaussian units)

$$\underline{\delta B} = (2I/cR^2)\,\zeta(z,t)\,\underline{e}_y\,, \qquad \underline{\delta E} = (2I/vR^2)\,\zeta(z,t)\,\underline{e}_x\,, \tag{88}$$

where propagation in the z-direction has been assumed.

Note that the magnitude of the magnetic restoring force is always less than the force due to the electric field, especially at the low energy end of the accelerator. Moreover, since the drift tube wall will not have infinite conductivity, the induced image current must diffuse into the pipe wall, and the magnetic restoring force resulting from the image current must decay to zero with distance backward from the head of the beam. Neglecting this restoring force entirely, the resulting equation for the beam displacement is

$$(d^2/dt^2)\zeta + (\Omega/\gamma)^2\zeta = (2eI/vR^2\gamma m)\zeta \tag{89}$$

in which $\Omega = eB/mc$ is the ion cyclotron frequency, and it has been assumed that the beam is propagating in a uniform solenoidal field of strength B. It is clear that we must require

$$(\Omega/\gamma)^2 > (2eI/vR^2\gamma m) \tag{90}$$

for stability to transverse displacements. Equation (90) may be rewritten as

$$B_{st} = \frac{\gamma r_b}{2R}\,B_{eq} \tag{91}$$

where B_{st} is the minimum field for stability and B_{eq} is the minimum equilibrium field from Eq. (86). If $\gamma < 2R/r_b$, then $B \geq B_{eq}$ assures that $B > B_{st}$ and the transverse instability will not occur. Thus, for $R/r_b = 1.6$, the transverse instability should not develop for proton energies of 2 GeV or lower ($\gamma < 3.2$).

4.4.3 Transverse Beam Break-up Instability

The beam break-up (BBU) instability is always a concern for high current, multiple gap, induction accelerators. Physically, the instability arises from the coupling between a beam undergoing transverse oscillations and accelerating cavity modes with a transverse component of magnetic field on-axis. If the frequency of the beam oscillation occurs at a natural resonance of the cavity, then cavity mode energy will increase and be in temporal phase with the oscillations, thereby causing the oscillations to increase in amplitude during the pulse duration. Moreover, if every cavity is identical, the peak amplitude of the oscillations will also grow from cavity to cavity.

The asymptotic BBU growth per gap is $\exp(\nu Z\omega_0/\Omega)$, although the growth is typically slower in the first several gaps. In this expression, ν is the ion beam current normalized to 31 MA, Z is the transverse coupling impedance normalized to 30 ohms, ω_0 is the instability radian frequency, and Ω is the ion cyclotron frequency.

For comparison, the asymptotic growth in the 10-kA ATA electron accelerator is about 12% per gap, assuming a 2-kG magnetic guide field. In ATA simulations, the asymptotic state is reached after about 80 gaps, and a total growth of 3×10^4 is predicted after 180 gaps. Experimentally, the growth of BBU limits the achievable beam current to the 4- to 6-kA level. To reach the full 10 kA requires the use of ion-focused transport.

In the present case, there are several significant differences: (1) The larger ion mass will delay the asymptotic state. (2) Because of temporal pulse compression, the beam current varies from about 2.4 kA at the injector to almost 20 kA at the end of the accelerator. (3) The axial magnetic field strength in P3T is substantially higher (≥ 35 kG). (4) The factor $Z\omega_0$ scales approximately as d/R^2, for $d/R < 1$, with d being the nominal width of the accelerating gap. In the case of ATA, $d/R = 0.34$, $d/R^2 = 0.044\ cm^{-1}$, and $Z\omega_0 = 5 \times 10^9$. In our case, $d/R = 0.625$, $d/R^2 = 0.039\ cm^{-1}$, and we estimate $Z\omega_0 = 4.4 \times 10^9$.

Substituting for all of these parameter values (using an average current value of 12 kA) gives an asymptotic growth rate of about 0.5% per gap. Assuming asymptotic growth in 690 one-MV gaps gives a worst case total growth of about 30, which is probably acceptable.

An additional, powerful BBU detuning effect results from temporal pulse compression. Suppose that early in the accelerator the beam pulse duration is τ and the transverse oscillation frequency is f. If the beam is uniformly compressed to a new pulse duration τ', then the new oscillation frequency is given by $f' = f(\tau/\tau')$. If f' is not in the Q envelope of the cavity resonance ($\Delta f = f/Q$), then BBU cannot grow.

For example, assume a 10-MeV proton beam with $v = 0.14\ c$ and $\tau = 1\ \mu sec$. Also, assume $Q = 10$ and that the cavity resonant frequency is 1 GHz. A 10% compression of the pulse will produce a 10% upward shift in the oscillation frequency which will be out of resonance. The increase in average ion kinetic energy which permits this pulse compression (to $v/c = 0.16$) is about 12 MeV. Thus, instead of 690 cavities, the BBU instability will grow coherently in only 12 cavities, and the gain will be only 0.06. This level of growth would not be observable. Although this situation will be somewhat worse at the high energy end of the accelerator, our tentative conclusion is that the beam break-up instability will not be a problem.

4.5 LONGITUDINAL BEAM DYNAMICS

Several important longitudinal effects must be considered for a nonrelativistic ion pulse. First, the ions must be confined longitudinally so that they cross the acceleration gaps during the applied voltage pulses. Second, a good (i.e., relatively constant) beam current profile must be maintained, both to satisfy transverse equilibrium constraints and to ensure good acceleration efficiency from a beam loading standpoint. Third, it is probably desirable to use a significant amount of temporal pulse compression to save volt-seconds of core in the high energy stages of the accelerator. Finally, the beam can be unstable to axial bunching modes that result from the resistive interaction of the beam with the acceleration gaps. In this section we analyze several of these issues.

4.5.1 Longitudinal Space-Charge Forces

The on-axis space-charge potential in the middle of the ion beam is repeated here for convenience:

$$\phi(0) = (I/v)[1 + 2\ln(R/r_b)]\ .$$

In the body of the beam pulse, the axial electric field is very small because of the effect of wall image charges. If the beam has blunt ends, then the peak space-charge field is approximately $E_z \approx \phi(0)/2R$. Otherwise, the space-charge electric field is approximately determined by the beam rise (or fall) time t_r as

$$E_z = \phi(0)/vt_r. \tag{92}$$

For our proposed injection parameters, assuming a 50-ns pulse risetime, evaluation of Eq. (92) gives E_z = 4.75 kV/cm, which is more than an order of magnitude smaller than E_r. However, it is also about half the average gradient of the P3T accelerator (~1 MV/m). Therefore, it is clear that a constant accelerating voltage pulse of duration equal to the current pulse cannot provide longitudinal confinement of the beam. Assuming minimal velocity spreads, the accelerating voltage waveform required to accelerate and longitudinally confine and maintain a flat beam current pulse waveform is shown in Figure 17.[10]

Figure 17. Schematic accelerating voltage waveform required to longitudinally confine and maintain a flat beam current pulse.

As the beam becomes more energetic, the longitudinal space-charge force is offset by the inductive electric field of the rising beam current. The reduction factor is ~$(1/\gamma)$, which is about a factor of 2 at the high energy end of the accelerator.

4.5.2 Beam Loading Effects

To minimize operating costs, it is essential that the beam be efficiently accelerated. In the simplest accelerating gap model, the efficiency is 100% if the impedance Z of the driving pulse line is matched to the beam impedance $Z_g = V_g/I$; i.e., $Z = Z_g$.

More generally, the efficiency (using the simple gap model) is

$$\eta = 4(Z/Z_g)(1 + Z/Z_g)^{-2} . \tag{93}$$

Consideration of this efficiency relationship is especially important in the context of pulse compression and preservation of the pulse waveform. For example, if the voltage per gap is held constant, then the impedance of the pulse line driver must decrease by a factor of 10 to maintain high efficiency for the assumed factor of 10 pulse compression. In addition, it will also be important to consider the dynamic beam loading problem during the rise and fall of the current pulse.

4.5.3 Pulse Compression

Both the equilibrium condition and the space-charge limit criterion are more easily satisfied as the beam kinetic energy increases. It should, therefore, be possible to decrease the beam pulse duration, or radially compress the beam, or perform some suitable combination of the two approaches. Let $\delta = \gamma - 1$. In the small δ limit (low kinetic energy), it can be shown that the allowable beam current as determined by the equilibrium condition scales as $(2\delta)^{1/2}$. In contrast, the allowable current as determined from the space-charge limit criterion scales as $(2\delta/3)^{3/2}$. Hence, as the ion kinetic energy increases, the equilibrium requirement becomes the controlling condition. That is, pulse compression is allowed such that

$$I \le (B/\alpha)^2 (e r_b^2/8mc^2) (\gamma v) . \tag{94}$$

At the low and high energy extremes, $\gamma v = 0.146$ c for 10-MeV protons, and $\gamma v = 1.43$ c for 700-MeV protons. As a result, the beam can be temporally compressed to increase the beam current by a factor of about 10 (while holding the magnetic field strength and the beam radius constant). Alternatively, the beam current could be held constant, and the equilibrium beam radius could be decreased by about a factor of 3 to about 3 cm. Since the cost of the accelerator cores is a strong function of the pulse duration, we decided to decrease the pulse duration to a minimum of 100 ns. It can be verified that this maximum amount of pulse compression gives a radial electric field at the drift tube wall of 62 kV/cm at the low energy end and 89 kV/cm at the high energy end of the accelerator, if all other parameters are held constant. Thus, the wall breakdown criterion (≤ 100 kV/cm) is satisfied along the entire length of the accelerator.

Consider a low current beam pulse of duration t_p moving through an accelerating gap with constant applied voltage V_o. If the longitudinal velocity of the ion is v_s, then the axial length of the ion pulse is $L = v_s t_p$. If the gap is narrow, then every ion entering the gap leaves it immediately and the temporal pulse length is not changed. If the new ion velocity is v_s', then the axial beam length is increased to

$$L' = L(v_s'/v_s) \tag{95}$$

where v_s' is determined from conservation of kinetic energy according to (nonrelativistically)

$$v_s' = [v_s^2 + 2eV_o / m]^{1/2} . \tag{96}$$

In order to achieve temporal pulse compression, it is necessary to provide more acceleration for ions in the pulse tail and less acceleration for ions in the pulse head relative to ions in the middle of the beam pulse. The allowable beam current scales like v_s, which implies that the allowed temporal pulse compression scales like $(1/v_s)$. Thus, at the low energy end of the accelerator it is desirable that the axial pulse length remain approximately constant.

Consider the ramped accelerating voltage waveform of Figure 18. The average accelerating voltage is V_o, and the maximum voltage appearing at the pulse tail is $V_o + \Delta V$. We assume that $v_s^2 \gg (2e/m)(V_o + \Delta V)$. Let D be the separation between gaps and assume the thin gap limit. Then, the velocity of the first ion to leave gap 1 is

$$v_{s-}' \approx v_s[1 + e(V_o - \Delta v)/mv_s^2]$$

and it will arrive at gap 2 after a time t_- given by $t_- = D/v_{s-}'$. The last ion leaving gap 1 will have a velocity

$$v_{s+}' \approx v_s[1 + e(V_o + \Delta V)/mv_s^2] ,$$

and it will arrive at gap 2 after a time t_+ given by $t_+ = t_p + D/v_{s+}'$. The new temporal pulse duration as seen at gap 2 is thus

$$t_+ - t_- = t_p' = t_p + D[(1/v_{s+}') - (1/v_{s-}')] \approx t_p - 2De\Delta V/mv_s^3 .$$

If the axial pulse length is to be held constant, then we require $v_s' = v_s(1 + eV_o/mv_s^2) = L/t_p'$. Solving this relationship for $\Delta V/V_o$ gives

$$\Delta V/V_o = L/2D . \tag{97}$$

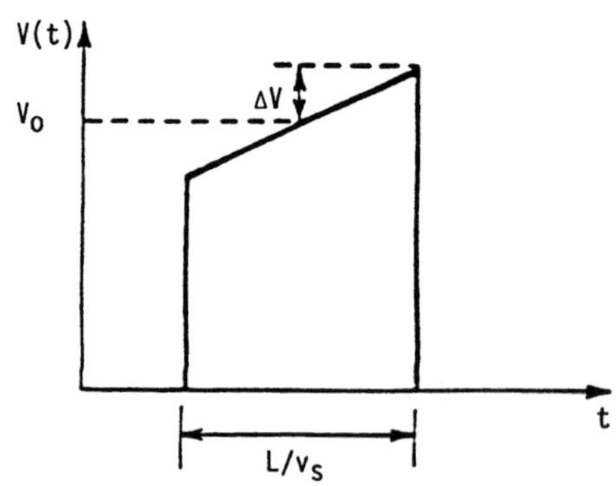

Figure 18. Accelerating voltage waveform with a voltage ramp.[10]

The physical length L of a 1-μs pulse of 10-MeV protons is 43 m, and the anticipated gap separation is about 4 m. Thus, the required voltage ramp would be unacceptably large.

Rather than trying to compress the beam to match the $(1/v_s)$ criterion at each individual gap, it is thus more reasonable to have modestly ramped acceleration pulses applied to several gaps in an acceleration section. The beam temporal pulse duration would remain approximately constant throughout this region. This section would then be followed by a drift-compression section, in which the beam would axially compress, prior to entering another section of accelerating gaps with shorter pulse durations.

Perhaps a more attractive alternative strategy for producing longitudinal pulse compression without voltage ramping has been suggested by E. Lee. It can be described as follows. Consider a low current beam for which space-charge and beam loading effects can be neglected. also, consider only square voltage pulses so that each acceleration gap gives a fixed energy increase to each ion as it passes. Then, if all gaps are turned on before the injected beam arrives, the space-time history of the beam will be as plotted in Figure 19. This case is the passive case of a spatial start for which the beam temporal length t_p stays fixed and the beam axial length L increases.

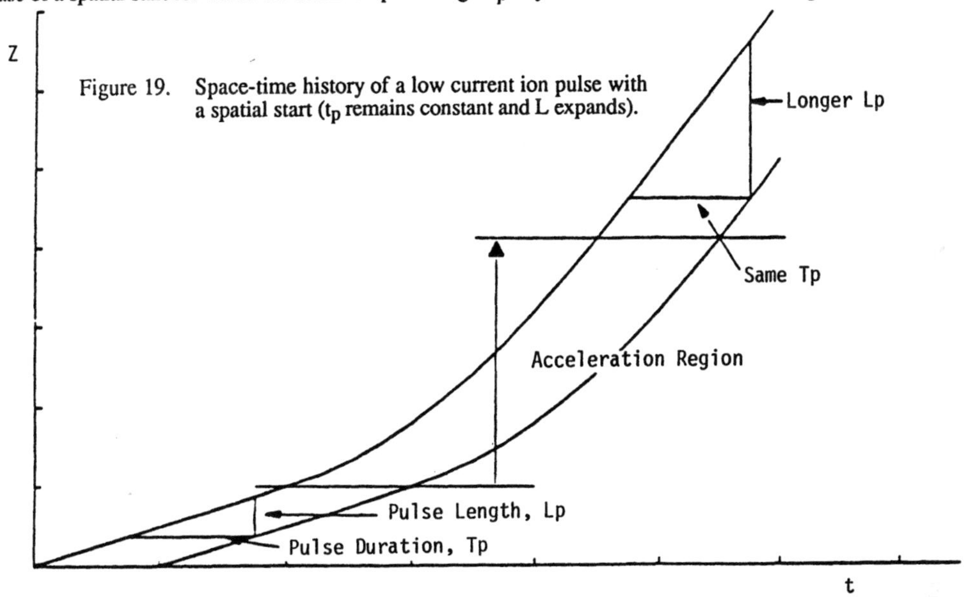

Figure 19. Space-time history of a low current ion pulse with a spatial start (t_p remains constant and L expands).

Now imagine a temporal start, as shown in Figure 20, in which the injected beam is allowed to drift into the acceleration gap region first, and then the gap voltages are all turned on simultaneously. For this case, the beam spatial length L will stay fixed, but as the beam accelerates, its temporal length t_p will decrease. If the gap voltages are all turned off simultaneously, then the final output beam will be monoenergetic, but compressed in time. A simple extension of this idea is to have a programmed start as shown in Figure 21. This is similar to the case of Figure 20, but now the gaps are turned on in a timed sequence as shown. For this case, the beam will compress both spatially and temporally and still be monoenergetic at the output. In fact, by programming the turn-on sequence, any desired compression or spreading may be achieved.

In these schemes, it is apparent that each focusing element in the acceleration region will see particles with many different velocities. Since the magnetic field will likely be fixed, the focusing properties might vary with energy if a quadrupole focusing system is used. However, for solenoidal focusing, the restoring force does not depend on the particle velocity. This means that for different velocities the radial amplitude of the trajectories will stay the same and only the "betatron" axial wavelength will change. This means solenoidal transport may be

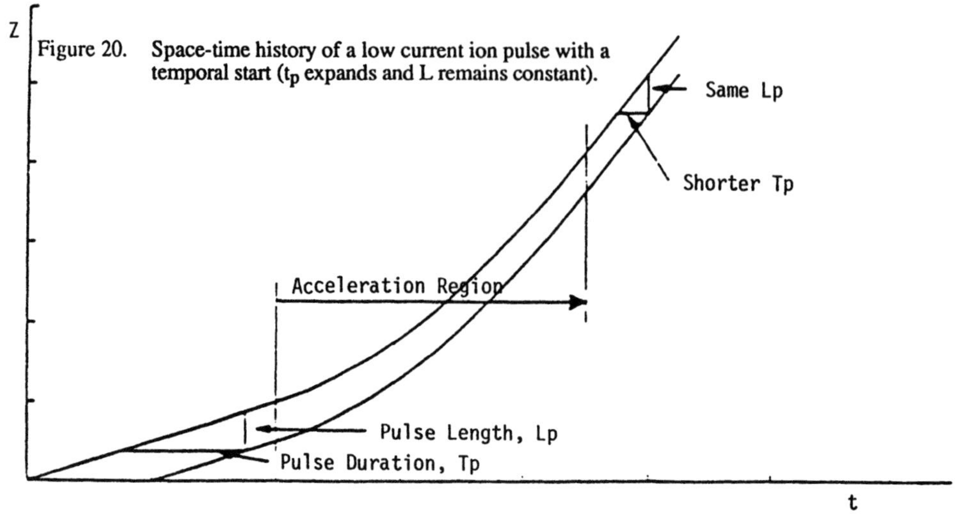

Figure 20. Space-time history of a low current ion pulse with a temporal start (t_p expands and L remains constant).

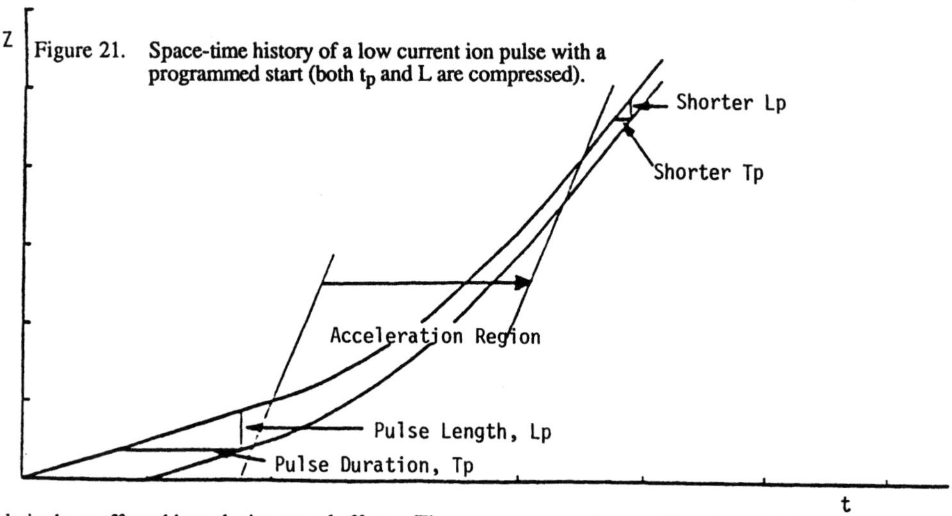

Figure 21. Space-time history of a low current ion pulse with a programmed start (both t_p and L are compressed).

relatively unaffected by velocity spread effects. Thus, a programmed start with solenoidal focusing appears to be an attractive method for achieving the desired pulse compression.

4.5.4 Longitudinal Stability

Longitudinal instabilities have never been observed in electron induction linacs, presumably because the electrons are relativistic during the acceleration process with an axial velocity almost independent of kinetic energy. On the other hand, the velocity of the ions depends on the ion kinetic energy, and the beam can therefore be susceptible to growth of a longitudinal bunching instability with the following mechanism. An ion beam with an excess localized current perturbation loads the driving pulse line circuit of an accelerating gap, thereby lowering the gap voltage. The ion bunch emerges from the gap with reduced velocity and can be overtaken by trailing particles that received a higher accelerating voltage. As a result, the current perturbation appearing at the next gap is further enhanced, and the perturbation grows as the beam moves through the accelerator. The instability thus results from the resistive character of the gap impedance, and exhibits spatial amplification from pulse head to pulse tail. Growth of this instability is naturally inhibited by the repulsive forces of the excess space charge, and by dynamic effects such as particle rotation, axial velocity shear, and kinetic energy spreads that create axial momentum spreads.

The growth of this instability has been estimated by using a simplified accelerator model[11] consisting only of an applied field E_a, distributed gap impedance per meter R, and beam-pipe capacity per meter C. The beam is treated as a cold, one-dimensional, nonrelativistic fluid. Thus, the effects of beam rotation and axial velocity shear and thermal effects are neglected, although the effects of excess space charge due to axial bunching are included. It is further assumed for simplicity that the unperturbed velocity is held constant and that the unperturbed field of the beam is exactly cancelled by the programmed field of the gaps, i.e., the net unperturbed field vanishes.

The results of this calculation indicate that an initially small disturbance, when expanded in terms of the system normal modes, is found to oscillate with maximum amplitude proportional to an amplification factor e^{λ}, with λ given by

$$\lambda = (RCLv_o/2) \tag{98}$$

where L is the axial pulse length and v_o is the drift velocity.

As previously mentioned, R and C are the continuous representation of the effects of gaps and the drift tube. The capacity is taken to be that of the coaxial beam given by (MKS units)

$$C = 4\pi\varepsilon_o/[1 + 2\ln(R/r_b)]\ . \tag{99}$$

The resistance per unit length can be estimated from the average accelerating gradient divided by the beam current. Thus, at the low energy end of the accelerator we have $C = 5.7 \times 10^{-11}$ F/m, $R = (2.5 \times 10^5$ V/m)/2.4 kA = 100 ohms/m, assuming a matched resistance for maximum efficiency, L = 35 m for a 0.8-μs pulse, and v_o = 4.3 x 107 m/s. Substituting for these values gives $\lambda = 4.4$, for a total growth factor of about 80. This amount of growth is uncomfortably large, because the anticipated perturbation amplitudes are expected to be as high as a few percent.

This calculated growth rate has been independently obtained by Humphries using a one-dimensional fluid code. Initially, at each z location the beam is treated as a series of discs that are constrained to move rigidly along the z-axis. In the physical picture of the instability, a single disc with larger current than normal will receive a smaller energy gain than normal as it loads an acceleration gap. After exiting the gap, this disc will fall behind so that, when it reaches the next gap, it will overlap a normal disc and cause even worse loading. The effect continues and the instability grows. However, an additional parameter that can be obtained from Lee's theory is a damping length, which depends strongly on the amount of space charge in the beam. Using our injection parameters, the computed damping length is only 225 m, which is roughly comparable to the growth length (~200 m). If this damping factor is really independent of the growth, it is difficult to see how this instability can significantly disrupt the beam for the envisioned operating parameters, even assuming rather large current perturbations (few percent) originating from the injector. These results are in marked contrast to the results obtained using heavy ion fusion parameters. In these calculated HIF cases, the growth rates are larger and damping is insignificant, but a rather modest velocity spread (~1%) is sufficient to damp the disturbance.

Clearly, a more comprehensive longitudinal instability model is necessary. At present, the calculational models are somewhat simplistic and do not include several important effects that could also provide damping of the instability. For example, radial dynamic effects will tend to wash out the instability. Consider the spread in actual particle trajectory path lengths as the particles undergo betatron oscillations. A "disc" of particles will spread axially even though all particles have the same energy. This spreading effect means that an initial bump in the beam current can wash out substantially before the next gap is reached, thus tending to quench the instability effect. For our injection parameters (10-MeV protons, r_b = 10 cm, B = 35 kG), this effect alone produces a velocity spread on-axis of $\Delta v/v \approx 0.08$. An additional, but smaller, effect is that of radial velocity shear due to the beam space-charge potential depression. This will also cause a small radial variation in velocity that will again tend to wash out the instability. For the baseline injection parameters, this effect produces an axial velocity shear of $\Delta v/v \approx 0.03$.

5 SUMMARY

In this paper we have analyzed several important issues pertaining to the transport of high current electron and ion beams through linear induction accelerator structures. In particular, several laminar flow equilibrium configurations were considered, including the effects of space-charge limits and source geometries. Various aspects of the design of high voltage, high current injectors and accelerating gaps were discussed in some detail, including the problems associated with space-charge limits, field emission criteria, and excitation of radial oscillations (zero-frequency cyclotron waves). Several potentially disruptive beam instabilities were discussed, including the diocotron instability for annular beams, resistive wall instabilities, the image displacement instability, and the transverse beam break-up instability. To illustrate the use of these analyses, several aspects of the RADLAC II and the P3T beam transport lines were analyzed in detail.

REFERENCES

1. Our treatment of the paraxial equation follows that of T.F. Godlove, "Summary of Linear Beam Transport," Proceedings of the NATO ASI on High Brightness Accelerators, A.K. Hyder, M.F. Rose, and A.H. Guenther, eds., Plenum (1988).

2. For more detailed discussions of several of these topics see R.B. Miller, An Introduction to the Physics of Intense Charge Particle Beams, Plenum (1982), and the references cited therein.

3. For more discussions of the topics of this section see R.B. Miller, "High-Current Electron-Beam Transport in Linear Induction Accelerators," Proceedings of the NATO ASI on High Brightness Accelerators, A.K. Hyder, M.F. Rose, and A.H. Guenther, eds., Plenum (1988), and the references cited therein.

4. D.S. Prono, IEEE Trans. Nuclear Science NS-32, 3144 (1985).

5. C.A. Frost, S.L. Shope, R.B. Miller, G.T. Leiteste, and C.E. Crist, IEEE Trans. Nuclear Science NS-32, 2754 (1985).

6. S.L. Shope, C.A. Frost, G.T. Leiteste, C.E. Crist, P.D. Kickel, and J.W. Poukey, IEEE Trans. Nuclear Science NS-32, 3092 (1985).

7. R.B. Miller, IEEE Trans. Nuclear Science NS-32, 3149 (1985).

8. J.R. Smith, I.R. Shokair, K.W. Struve, E. Schamiloglu, P.W. Werner, and R.J. Lipinski, "Transverse Oscillations of a Long-Pulse Electron Beam on a Laser-Formed Channel," 14th Int'l Symp. Discharge and Electrical Insulation in Vacuum, 1990.

9. Much of the material of this section is contained in the document "Pulsed Power Produced Tritium," SAND89-1451, which was a Scoping Study/Feasibility Assessment. The beam transport analysis was led by this author, with substantial inputs from M.J. Clauser, J.R. Freeman, C.L. Olson, J.W. Poukey, J.P. Quintenz, S.A. Slutz, R.W. Stinnett, and S. Humphries, Jr.

10. S. Humphries, Jr., Principles of Charged Particle Acceleration, Wiley (1986).

11. E.P. Lee, "Resistance-Driven Bunching Mode of an Accelerated Ion Pulse," UCRL-86452 (1981).

OBSERVATION OF HIGH CURRENT EFFECTS IN HIGH ENERGY LINEAR ACCELERATORS

John T. Seeman
Stanford Linear Accelerator Center
Stanford University
Stanford, California 94309

1. Introduction

The acceleration and transport of electron beams with high currents and low emittances are subjects studied very carefully for many operating accelerators and most future accelerators. For example, several Free Electron Lasers (FEL) [1] and the Stanford Linear Collider (SLC) [2] presently operate with high energy intense beams. The next generation of synchrotron light sources [3] and future linear colliders [4,5,6,7] require a much higher degree of component precision and beam control. In this note the basic concepts governing the effects of high currents in accelerators are explored, including observations of high currents in present accelerators.

The effects of longitudinal wakefields on acceleration and energy spreads are studied first. Then, the transverse equations of motion are developed including the technique of BNS damping to control wakefield growth. These wakefield effects are investigated to characterize their influence on the control and observation of beam oscillations. Finally, the reduction of transverse wakefield effects is important for maintaining the emittances of intense beams.

2. Longitudinal Effects

2.1 Acceleration

The energy E of particles accelerated in a linac after n klystrons is given by [8]

$$E(z) = E_{inj} + \sum_{i=1}^{n} [\ \Delta E_i \cos(\phi_i+\phi(z)) + \Delta s_i \int_{z}^{\infty} W_l(z'-z)\ \rho(z')\ dz'] \tag{1}$$

where E_{inj} is the injection energy, ϕ_i are the klystron phases-which are free parameters-, $\phi(z) = 2\pi z/\lambda_{RF}$, where λ_{RF} is the RF wavelength, z is the longitudinal position along the bunch, ΔE_i is the maximum energy gain in the distance Δs_i (one klystron), $\rho(z)$ is the longitudinal density distribution, and the last term on the right-hand side is the longitudinal wakefield contribution to the particle energy. The longitudinal wakefield contribution W_l arises from all particles in the bunch preceding the one of interest. The longitudinal wakefield depends on the construction of the RF cavities and surrounding vacuum hardware. The longitudinal wakefield for the SLAC structure is shown in Figure 1 [9]. Typical accelerator parameters are $\Delta E_i = 20$ MeV/m, $\lambda_{RF} = 10$ cm, and $\Delta s_i = 12$ m. $\rho(z)$ is often a gaussian with sigma σ_z.

2.2 Energy spread

The mean energy of the bunch <E> is an integral over the energy distribution.

$$\langle E\rangle = \int_{-\infty}^{\infty} E(z)\,\rho(z)\,dz \tag{2}$$

The root-mean-square (rms) energy spread can also be calculated.

$$\sigma_E / E = \frac{1}{\langle E\rangle}\left[\int_{-\infty}^{\infty} [E(z) - \langle E\rangle]^2\,\rho(z)\,dz\right]^{1/2} \tag{3}$$

The energy spread for a low current beam is minimized when the bunch is centered on the crest of the RF wave form. An example of the resulting spectrum is shown in Figure 2 for the LIL linac at CERN [10]. The core of the beam has the highest energy and both the longitudinal head and tail particles have lower energies making a long low energy tail.

At high currents the longitudinal wakefield decelerates the longitudinal tail of the bunch (generally). Thus, in order to minimize the energy spread the mean phase of the bunch is moved forward of the RF crest. This process can be seen schematically in Figure 3. The energy distribution in the bunch develops a complicated shape due to the nonlinear wakefield, particle density distribution, and the RF sine wave. These distributions have been calculated for the SLC linac for low and high currents and are shown in Figure 4. The fact that the minimum spectrum is obtained at a phase forward of the RF crest is easy to verify in a linac. An example for the SLC is shown in Figure 5 [11]. The "double horned" energy distributions indicated in Figure 3 for high currents have also been seen in the SLC (Figure 6) at about 3 x 10^{10} e^- with $\sigma_z = 1$ mm. The changes in the optimum linac phase for the minimum energy spread as a function of current are shown in Figure 7. The required phase shift can be on the order of 10 to 15 degrees. This shift allows phase jitter in the klystrons to produce energy jitter in a linac.

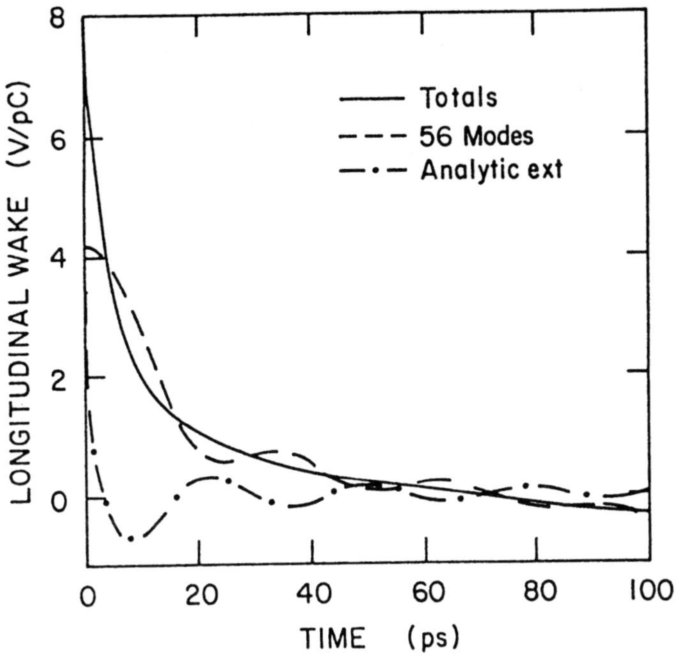

Figure 1 Longitudinal wake potential of the SLAC RF structure for a single cell as experienced by a test particle following a single electron.

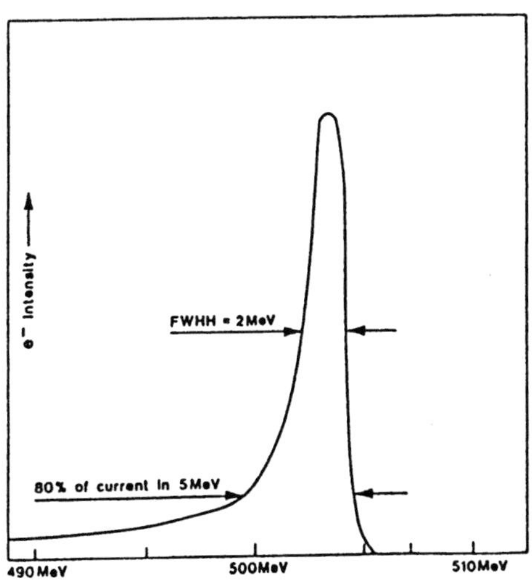

Figure 2 Measured energy spectrum for a low charge bunch in the LIL linac at CERN.

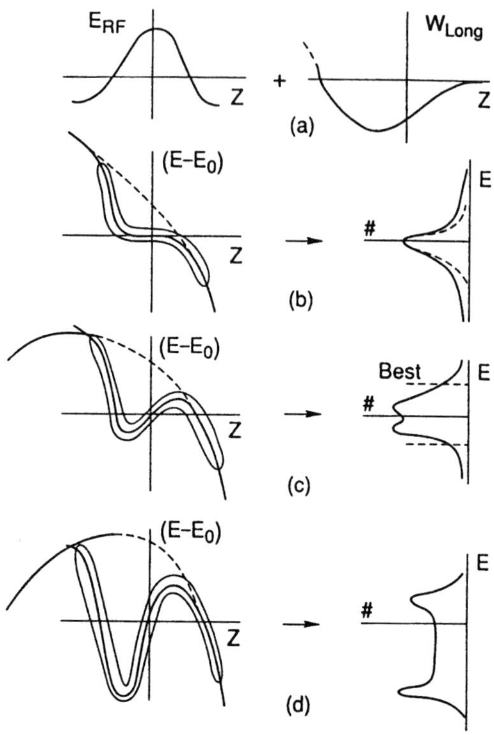

Figure 3 Schematic view of the effect of the overall linac RF phase on the energy spectrum of bunches with high charges including longitudinal wakefields (a). The energy versus longitudinal position displays on the left (b,c,d) lead to the energy density distributions on the right. The optimum phase places as many particles as possible near the mean energy. The optimum produces a double peaked distribution (c). With longitudinal wakefields the optimum phase of the bunch is noticeably ahead of the RF crest.

Occasionally, the longitudinal bunch distributions are not true gaussians and careful integrals are required to extracted the energy spreads [12]. In practice near gaussians and gaussians are difficult to distinguish experimentally.

2.3 Multibunch longitudinal effects

When a linac accelerates multiple bunches, the longitudinal loading affects the energy differences between the bunches. Trailing bunches see a lower accelerating field. The energy extracted from the RF cavities by the beam must be replenished by an RF power source but there is a finite filling time. The methods to equalize the bunch energies are to stagger time the RF pulses to add extra energy between bunches or to change the klystron pulse amplitude with time. The former is being considered for the next

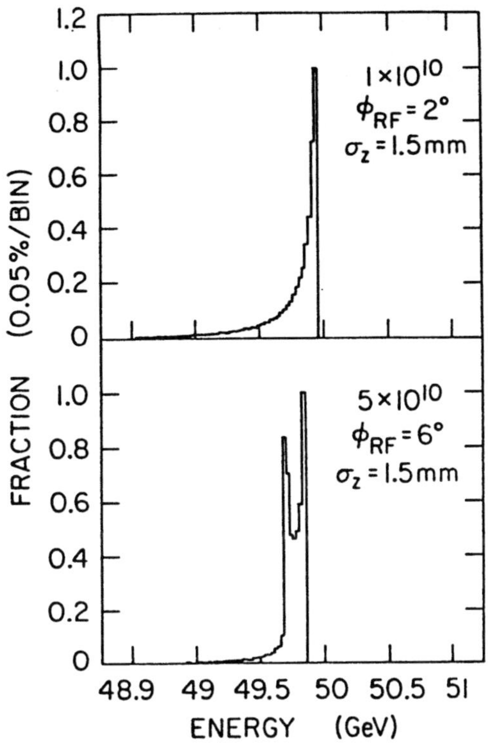

Figure 4 Simulated energy spectra for the SLC linac showing optimized cases for low charge (upper) and high charge (lower). The high charge spectrum has a complicated shape affecting the transport of low emittance beams.

linear collider [13] and the latter is used for making the two colliding bunches in the SLC have equal energies [11]. An example of multibunch longitudinal loading without compensation is shown in Figure 8 where each trailing bunch has lower and lower energy. The center bunches in the beam have higher charge than either end and the corresponding energy spacings are larger. The method to compensate the two colliding bunches in the SLC is shown in Figure 9, where the pulsed SLED system [14] provides a time varying field.

3. Transverse Oscillations

3.1 Equation of motion with transverse wakefields

The transverse equation of motion for particles under the influence of transverse wakefields is

$$\frac{d^2}{ds^2}\, x(z,s) + k^2\,(z,s)\; x(z,s) = \frac{r_e}{\gamma(z,s)} \int_z^{\infty} dz'\; \rho(z') W_T(z'-z)\; x(z',s), \tag{4}$$

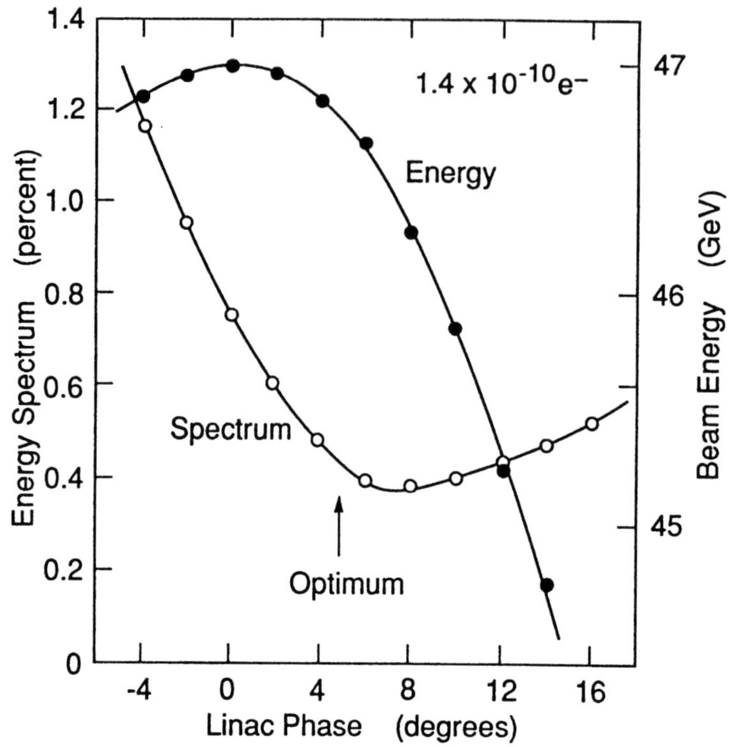

Figure 5 Measured energy and energy spectrum of a bunch in the SLC linac with medium charge as a function of RF phase. The peak energy and smallest rms energy spread are separated by several degrees as expected from longitudinal wakes.

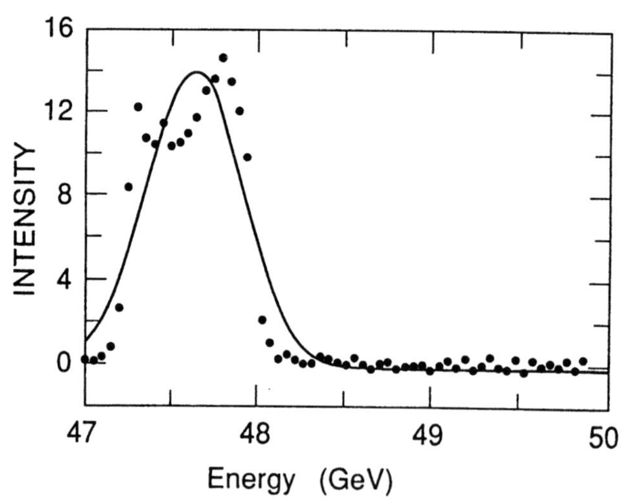

Figure 6 At high charge the measured energy spectrum in the SLC has the characteristic double peaked structure as schematically illustrated in Figure 3.

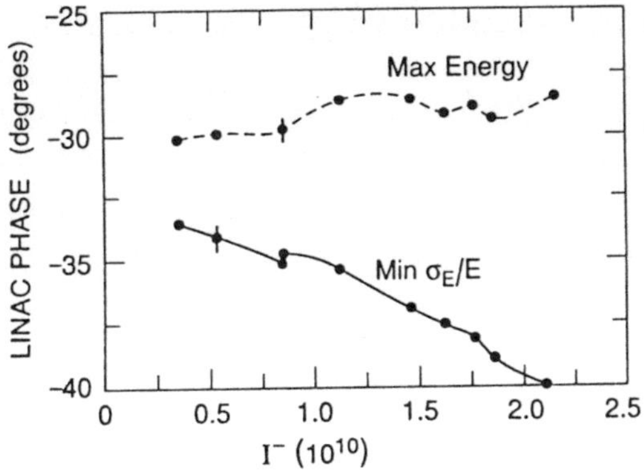

Figure 7 Measured RF phases for the maximum energy of the core of the beam as well as the minimum energy spread of the whole bunch as a function of charge. The phase of the peak energy remains nearly the same but the phase for the minimum energy spread shifts considerably forward at high intensity.

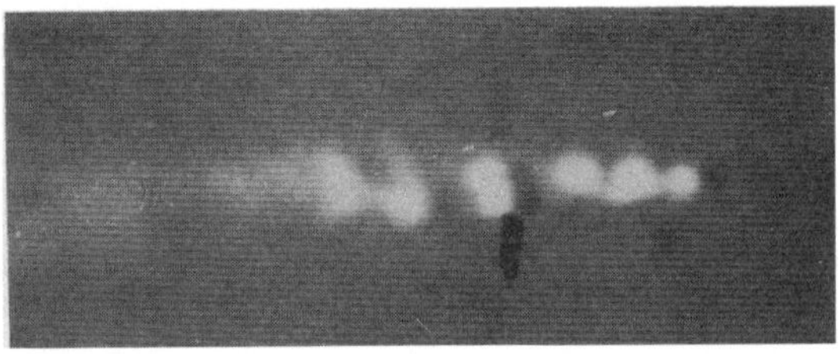

Figure 8 Photographed position of a train of six electron bunches in the SLAC linac at 8 GeV on a profile monitor in a dispersive region (η = 1m) after 800 m of RF accelerator. The bunches fill six adjacent RF buckets. The position offsets indicate the long range longitudinal loading of each bunch on all trailing bunches. Each bunch charge is about 0.5 - 1.0 x 10^9 e^-. The horizontal full scale is 2.5 cm. The dark mark is an artifact of the measurement.

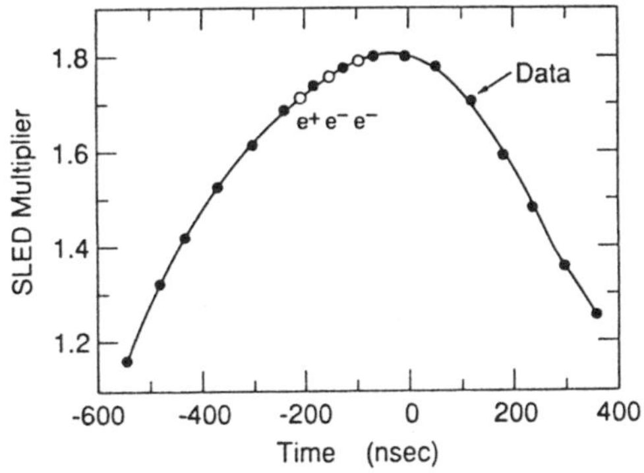

Figure 9 Measured SLED RF wave form indicating the timing of the three SLC bunches. The global timing of the three bunches is adjusted to alleviate any energy changes from charge differences and longitudinal loading between the bunches.

where x is the transverse particle displacement which depends on where in the bunch the particle is located and where along the accelerator the bunch is located, k is the lattice strength, ρ is the longitudinal particle density, $\gamma = E/mc^2$, r_e is the classical electron radius, s is the distance long the accelerator, and W_r is the transverse wakefield which depends on the separation of the leading and trailing particles. W_r grows nearly linearly for most bunches with millimeter bunch lengths as can be seen in Figure 10 calculated for the SLAC structure [9]. The acceleration is generally linear, $\gamma(s)=\gamma(0)+Gs$, where G is the accelerating gradient. In Equation (4) the distance to double the energy is assumed to be long compared to the betatron wavelength, or k >> G.

The driving force on the right hand side is proportional to the transverse offset of all the leading particles in the RF structure of the accelerator. Thus, there is the potential for resonant excitation of the back of the bunch by the head. The wakefield excitation of the bunch through an oscillation can be seen schematically in Figure 11. There are many mechanisms to start an oscillation and the resulting resonant excitation. The energy spread in the bunch affects the equation of motion. Also, the motion will develop differently if the initial deflection is made at different locations in the accelerator.

If the transverse wakefields are set to zero, then the solution to Equation (4) is the standard betatron oscillation. The expected beam oscillation given a dipole deflection $\Delta\theta$ at location '0' with acceleration is

$$x(s) = \Delta\theta \; \beta(s)\sqrt{E(0)\;\beta(0)\;/E(s)\;\beta(s)}\;\sin(\psi(s)-\psi(0)) \tag{5}$$

where β is the beta function, and $\psi(s)$ is the betatron phase to position s. Note that the betatron oscillation damps with the square root of the energy.

The equation of motion involves only one geometrical plane and assumes no horizontal-vertical coupling. This is often a reasonable assumption for round beam accelerators (see Figure 12) and is a strong requirement for a future collider [7].

An instructive exercise is to solve Equation (4) for a beam with three longitudinal slices without acceleration, i.e. a three particle model. If the bunch has a total charge of N, then the center 'c' slice is assigned N/2 charges and the head 'h' and tail 't' slices have N/4 charges each. They are all spaced apart by σ_z. The three equations of motion assuming a linear wakefield become

$$\begin{aligned} \ddot{x}_h + k_\beta^2\, x_h &= 0 \\ \ddot{x}_c + k_\beta^2\, x_c &= B\, x_h \\ \ddot{x}_t + k_\beta^2\, x_t &= 2\,B\,x_h + 2\,B\,x_c \end{aligned} \tag{6}$$

where B is proportional to the bunch charge and the transverse wakefield.

$$B = -\,e\,\frac{N\;W_r\;\sigma_z}{4\;E_0} \tag{7}$$

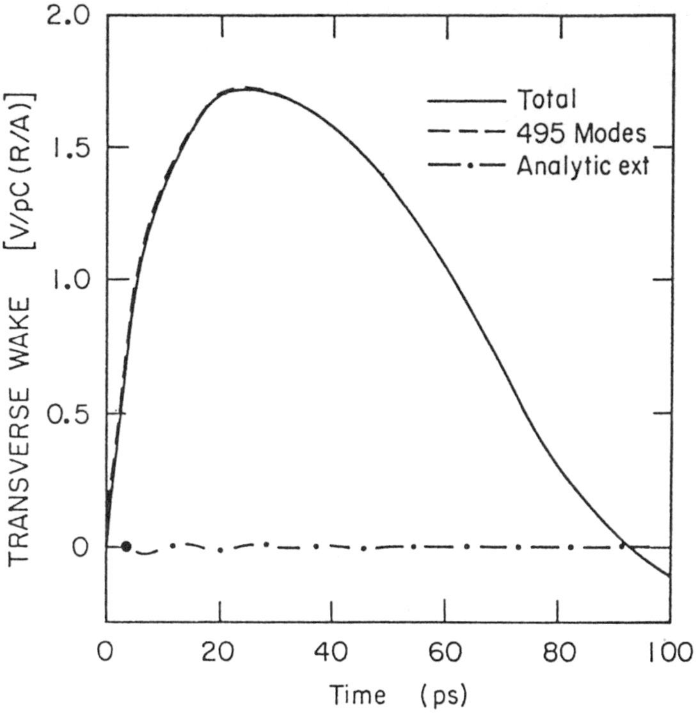

Figure 10 Transverse wake potential produced by a single electron in one RF cell of the SLAC linac.

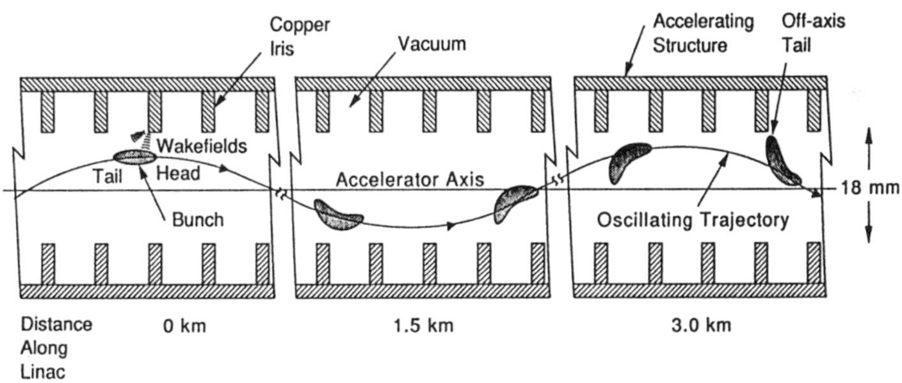

Figure 11 Schematic illustration showing the growth in oscillation amplitude of the longitudinal tail of a beam driven by transverse wakefields from the leading portions of the bunch. Trajectory correction downstream has difficulty converging as some part of the beam is always off axis driving trailing particles to larger amplitudes.

The solutions to the above equations assuming an initial deflection at $s = 0$ of $\theta_0 = k_\beta x_0$ are

$$\frac{x_h}{x_0} = \sin(k_\beta s)$$

$$\frac{x_c}{x_0} = \left(1+\frac{B}{4k_\beta^2}\right)\sin(k_\beta s) - \frac{B}{2k_\beta} s\cos(k_\beta s) \qquad (8)$$

$$\frac{x_t}{x_0} = \left[1+\frac{2B}{k_\beta^2}\right]\sin(k_\beta s) - \frac{2B}{k_\beta} s\cos(k_\beta s) - \frac{B^2}{4k_\beta^2} s^2 \sin(k_\beta s)$$

There are three important conclusions: 1) The variable x_0 factors from all the solutions, which means that the solutions are linear in the amplitude. 2) Each additional slice adds one higher power of B (or W_r N) and s, indicating the nonlinearity in bunch charge and distance along the accelerator. 3) The core lags 90 degrees behind the head. The tail, ultimately, lags 180 degrees behind the head. Conclusion 2 suggests an exponential growth of the tail of the beam in the limit of many slices.

This resonant blowup at high currents was observed in the SLC as is shown in Figure 13. Simulations and observations show good agreement indicating an approximate exponential trajectory increase with distance [15]. In this condition the SLC was very susceptible to injection jitter.

3.2 Transverse wakefield (BNS) damping

A technique to control this resonant growth is called BNS damping (named after the inventors: V. Balakin, A. Novokhatsky, and V. Smirnov)[16]. BNS damping lowers the energy of the back of the bunch by accelerating the bunch behind the crest of the RF early in the linac and then ahead of the RF crest downstream to restore the energy spread at the end. With the back of the bunch low in energy, the wakefields forces in Equation (4), which act to defocus particles, are counterbalanced by the increased quadrupole focusing of the back of the bunch because of its low energy. The resonant excitation is then reduced. It is not possible to perfectly cancel the forces over the whole bunch so some enlargement is expected.

Again, it is instructive to find the solutions to the three particle model but with BNS damping added. BNS damping enters the equations in the lattice constants k_β. The energies of the three particles are different and so too are their lattice constants. k_β for the head is set to $k_\beta + \alpha$, for the core k_β, and for the tail $k_\beta - \delta$. The solutions to the three equations with $x_0 = \theta_0 / (k_\beta+\alpha)$ are

$$\frac{x_h}{x_0} = \sin((k_\beta + \alpha)s)$$

$$\frac{x_c}{x_0} = \sin(k_\beta s) + D\left[(k_\beta + \alpha)\sin(k_\beta s) - k_\beta \sin((k_\beta + \alpha)s)\right] \qquad (9a)$$

where

$$D = \frac{B}{k_\beta((k_\beta + \alpha)^2 - k_\beta^2)}$$

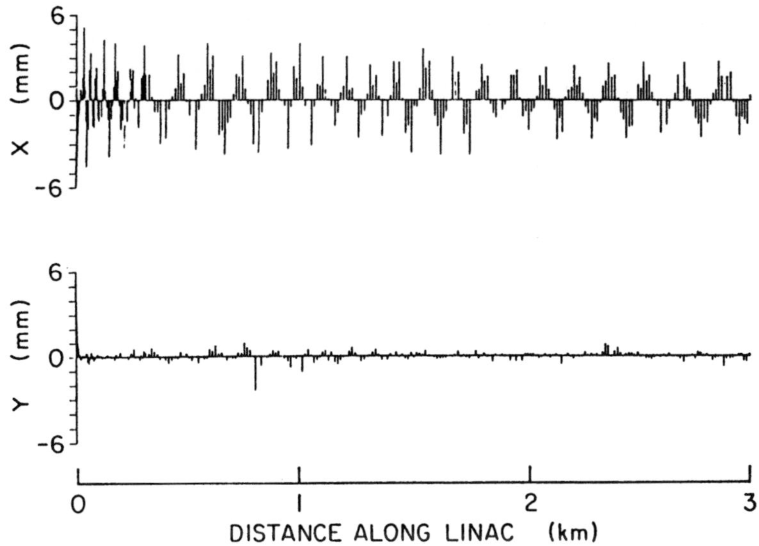

Figure 12 A measured horizontal betatron oscillation in the linac induced by a dipole magnet early in the linac. Note that the oscillation remains coherent and does not couple into the vertical plane showing the independence of the transport equations (6 x 10^9 electrons at 42 GeV).

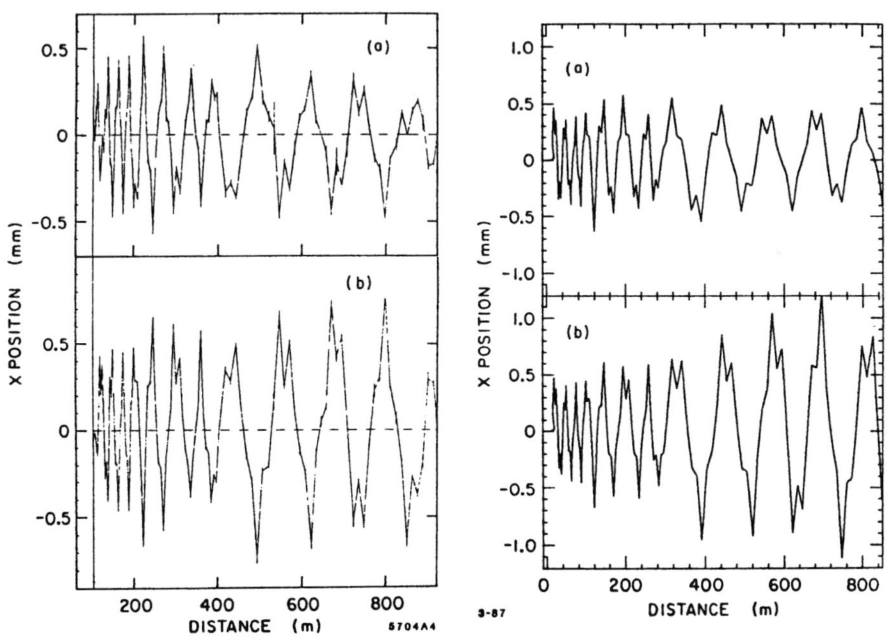

Figure 13 Left: Measured trajectories of an accelerated electron beam due to a 3 gauss-m dipole deflection. Right: Simulated trajectories of the electron beams on the left. The bunch lengths are 1 mm. The upper curves are for 8 x 10^9 e^- ; the lower curves are for 2.2 x 10^{10} e^-. No BNS damping phases are used. Note that the simulation and measurements agree and that at 2.2 x 10^{10} e^- the trajectory is starting to grow exponentially with distance.

$$\frac{x_t}{x_0} = \sin((k_\beta - \delta) s) + G [(k_\beta + \alpha) \sin((k_\beta - \delta) s) - (k_\beta - \delta) \sin((k_\beta + \alpha) s)] + F [k_\beta \sin((k_\beta - \delta) s) - (k_\beta - \delta) \sin(k_\beta s)]$$

where

$$G = \frac{2 B[1 - B / [(k_\beta + \alpha)^2 - k_\beta^2]}{(k_\beta - \delta) [(k_\beta + \alpha)^2 - (k_\beta - \delta)^2]}$$

$$F = \frac{2 B (1 + H)}{(k_\beta - \delta) [k_\beta^2 - (k_\beta - \delta)^2]} \tag{9b}$$

and

$$H = \frac{B (k_\beta + \alpha)}{k_\beta [(k_\beta + \alpha)^2 - k_\beta^2]}.$$

After this tedious work, the conditions for BNS damping can be determined. To make the head and core particles oscillate together, α is adjusted to make identical trajectories.

$$\alpha = e N W_r \sigma_z / (4 E_0 k_\beta) \qquad \text{(BNS condition)} \tag{10}$$

This is the standard BNS condition for a two particle model. The core wants to have an energy that makes the lattice appear stronger, i.e. a lower energy. To make the third particle oscillate like the first two then

$$\delta = 4 \alpha. \qquad \text{(optimum)} \tag{11}$$

This solution indicates that the energy differences over the bunch need careful tailoring. This is the essence of the notion of autophasing described in Section 3.4. Finally, RF focusing can be used to provide different lattice strengths to the various parts of the bunch instead of a head-tail energy difference. RF focusing may be used in a future collider [17].

The strong effect of this transverse damping is shown in Figure 14 where beams with and without BNS damping are made to oscillate in the SLC [18]. BNS damping works very well and is used continuously. With full BNS damping conditions, oscillations of a bunch have been studied in detail. Four measured oscillations, with identical initial conditions, are shown in Figure 15 for four different bunch currents. The BNS settings are for 2.5 X 10^{10} particles, where the first 48 klystrons have a phase of -20 degrees and the remaining klystrons have a phase of +15 degrees. In all cases the overall linac phase was adjusted to make the energy spectrum correct at the end of the accelerator. At low current the oscillation decoheres because of the large energy spread. At 2.5 X 10^{10} particles BNS damping is nearly correct because the oscillation damps approximately was the square root of the energy, as expected. At this setting the bunch resists wakefield emittance enlargement. At the highest current the BNS conditions are no longer effective and the oscillation starts to grow strongly at the end of the linac, similar to the examples in Figure 14. Thus, the BNS conditions must be matched to the beam current and lattice to make the technique most successful. However, if the klystron phases are significantly misphased from the desired values, bizarre oscillations can result. An example is shown in Figure 16 where strong beating is observed.

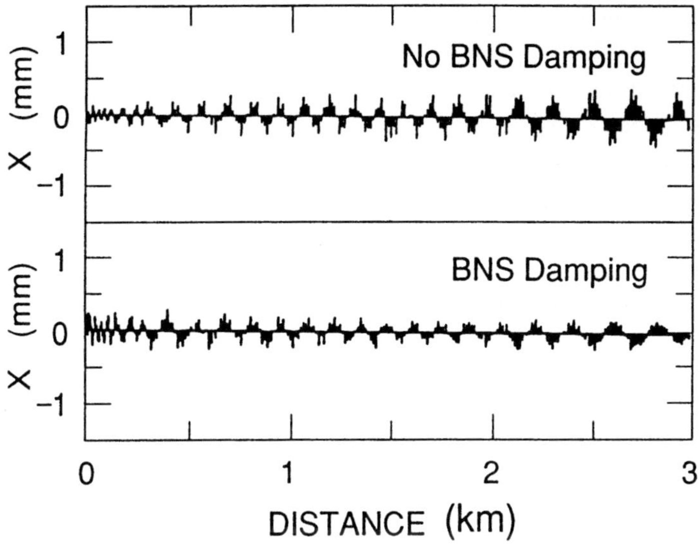

Figure 14 Measured oscillations in the SLC linac with and without transverse wakefield damping (BNS). The bunch charges are 2×10^{10} e^-. The upper plot shows the exponential growth of the oscillation without BNS damping and the lower plot shows that with BNS damping. In this example, the improvement in the ratio from front to end of the linac is a factor of ten.

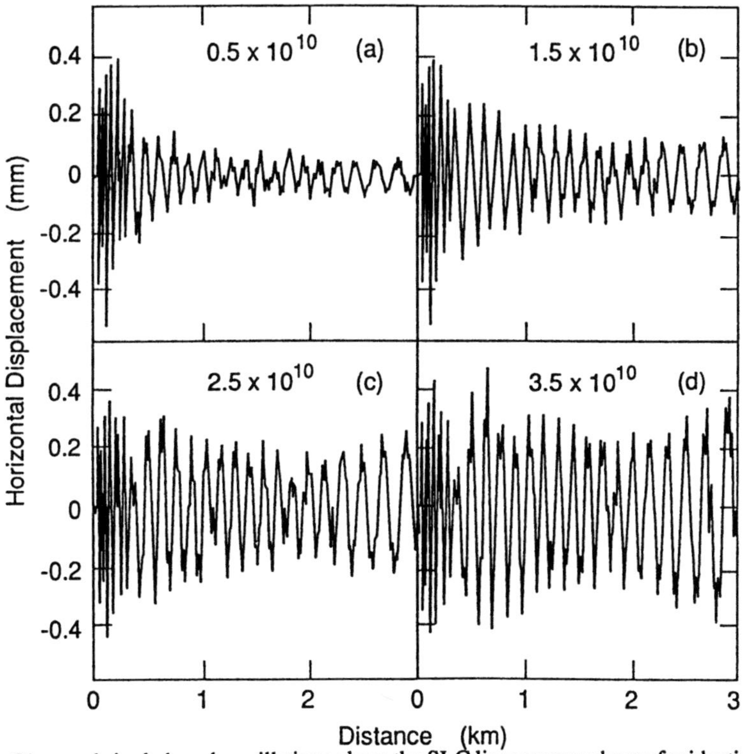

Figure 15 Observed single bunch oscillations along the SLC linac versus charge for identical dipole changes. Standard BNS conditions for 3×10^{10} e^- are used in all cases with the overall linac phase adjusted to make a small energy spectrum at 47 GeV. Note that at high charge, damping from acceleration is not as strong as wakefield growth in the latter half of the linac.

3.3 Analysis of oscillations

A beam oscillation can be used to determine lattice and beam parameters. An example of an oscillation with is fit using a global energy factor as a parameter is shown in Figure 17 [19]. This type of analysis can aid in determining if klystrons need phasing, if quadrupoles are misadjusted, or if transverse wakefield effects are proper.

By recording a beam trajectory and fitting the oscillation using Equation (4), an effective lattice strength can be determined. A plot of the change in the effective lattice strength with current for the SLC is displayed in Figure 18. In an accelerator with 30 betatron wavelengths, a strength change of a few percent is quite noticeable. Conversely, Equation (4) can be used to determine an effective beam current, assuming a longitudinal density profile. This study was done and the currents from the fits versus the actual beam currents for a particular SLC accelerator condition are plotted in Figure 19. Although the data and the predictions do not fully agree, there is no doubt about the strong effect on the apparent betatron frequency [19].

3.4 Autophasing

The cancellation of wakefield forces by BNS damping may be exploited further, if a concept called autophasing can be made practical [20]. By careful arrangement of the bunch charge density based on knowledge of the local beam energy, lattice, bunch length, and RF structure, nearly all particles in the bunch can be made to follow exactly the same trajectory. The conditions for this behavior can be derived by substituting an identical oscillation ($x=x_0\sin(ks)$) into Equation (4) for all particles, and cancelling position terms on both sides [8]. The resulting autophasing condition is

$$k^2(z,s) = k_0^2 + \frac{r_0}{\gamma(z,s)}\int_z^{\infty} dz'\ \rho(z')\ W_r(z'-z) \tag{12}$$

where k_0 is the nominal lattice strength and $k(z,s)$ would be the lattice strength for all particles if transverse wakefield forces were removed. With wakefields and the above condition, all particles have the same effective betatron frequency. The difficulty is that this condition must hold for all particles in the bunch, even the ones far forward off the RF crest, and at all locations along the accelerator. The adjustable parameters available are the bunch current, the lattice strength, the klystron energy profile, the klystron phases, and the longitudinal density profile. However, of all these variables, the longitudinal density profile is the most useful.

Using the two particle model in Equation 10, one can calculate the reduction in energy needed by the second particle to achieve autophasing. This calculation was done for the SLC and is shown in Figure 20 [19]. The early part of the SLC has several lattice discontinuities which require beam energy spread changes which are more rapid than can be practically made.

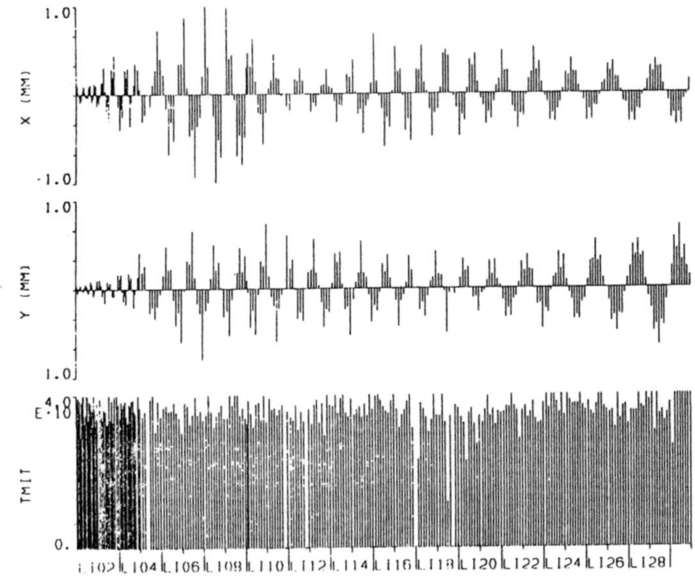

Figure 16 Observed dipole driven oscillations along the SLC linac of a beam with an abnormally large energy spectrum. The beating of the observed positions arises from the head and tail positions being on the same or opposite side of the axis. Large chromatic emittance enlargement also results.

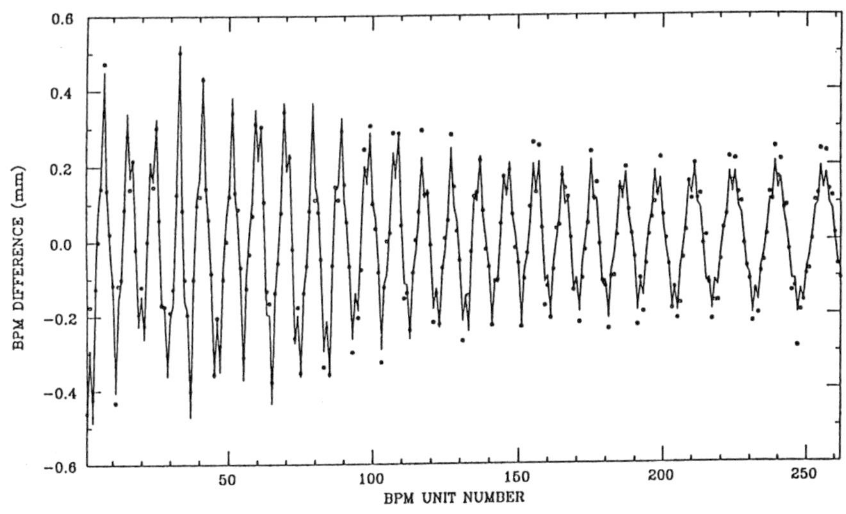

Figure 17 An observed oscillation (dots) is fit to a betatron oscillation (solid line) using the equation of motion with a single adjustable energy parameter over the whole linac. Transverse wakefields in the SLC produce apparent beam energy errors of a few percent.

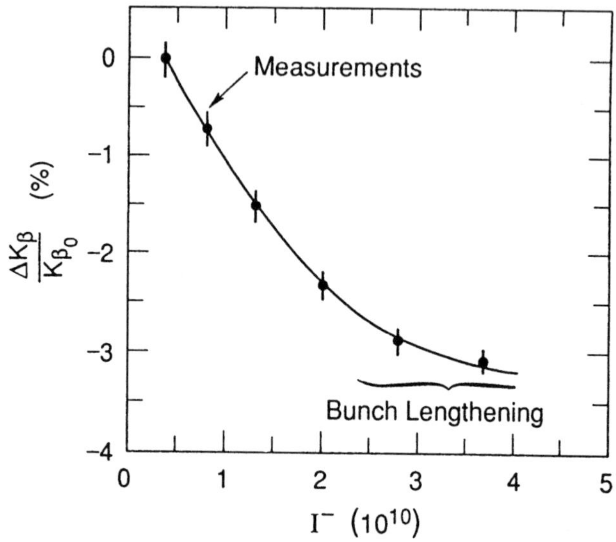

Figure 18 Change of the measured betatron spatial frequency due to transverse wakefields as a function of beam current in the SLC. These wakefields retard apparent betatron oscillations.

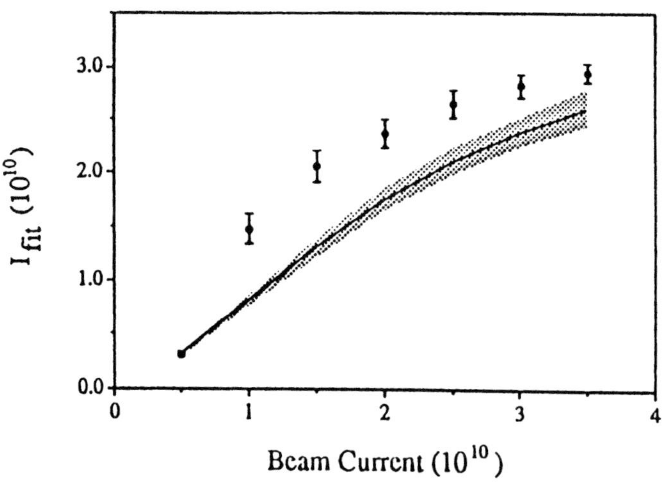

Figure 19 The fit value for the beam current from measured (circles) and simulated (solid lines) BPM difference trajectories as a function of the actual beam current in the SLC. The measured and simulated trajectories were fit with the same algorithm. The curves show that even at relatively low bunch intensities the transverse wakefield effects influence the trajectories. The differences between the measurements and simulations arise from the uncertainty of the longitudinal bunch profile, the incoming beam transverse profile, and an uncertainty in the transverse wake potential.

A forty-one slice simulation of autophasing for the SLC has been done [8]. Some of the results are shown in Figures 21 and 22. The difficulties in achieving the conditions of Equation 12 are apparent as Plots 21c and 21e must be matched over their full width. The irregularities of the lattice, the strong curvature of the RF voltage, and the need for a small energy spread at the end of the linac have made practical autophasing not fruitful to date, although studies continue. Autophasing without acceleration has also been studied [21]. Nevertheless, if this condition can be satisfied in a future collider, the beam intensity threshold for wakefield emittance growth will increase significantly .

3.5 Injection Tolerance

The tolerance on injection parameters can be calculated with several goals in mind. One goal is to have the emittance of the beam not increase unacceptably during acceleration.

$$\Delta\varepsilon / \varepsilon < 0.1 \qquad (13)$$

Another goal might be to keep the transverse position of the beam at the end of the accelerator within some fraction of the beam size from the accelerator axis to maintain luminosity.

$$\Delta x < \sigma_x / 3 \qquad (14)$$

Conversely, the goal may be to limit the number of particles a certain distance from the axis, as if to protect a device from destruction (i.e. collimator).

$$\int_d^{\infty} \rho(x)\, dx \leq n \qquad (15)$$

where ρ is the particle density and d is the off axis distance of interest.

A tolerance on injection position (and angle) jitter can be calculated for the SLC from the data in Figure 15, with proper BNS conditions at each current. To maintain luminosity using Equation 14, the oscillation must be kept below about $\sigma_{x,y} / 3$ at the end of the linac. The results of these calculations are shown in Figure 23 [22]. At 7.5 x 10^{10} particles, the displacement at the beginning of the linac must be less that 40 μm, within a factor of two of present operational values. Similar tolerances for a future collider will be significantly smaller, as the emittances are much lower.

3.6 Transverse jitter

Feedback systems that work pulse-by-pulse are essential for keeping the beam parameters within acceptable limits given the many possible sources of transverse jitter. Several sources are varying power supplies of dipoles and off axis quadrupoles, vibrating quadrupoles, klystron phase and amplitude jitter,

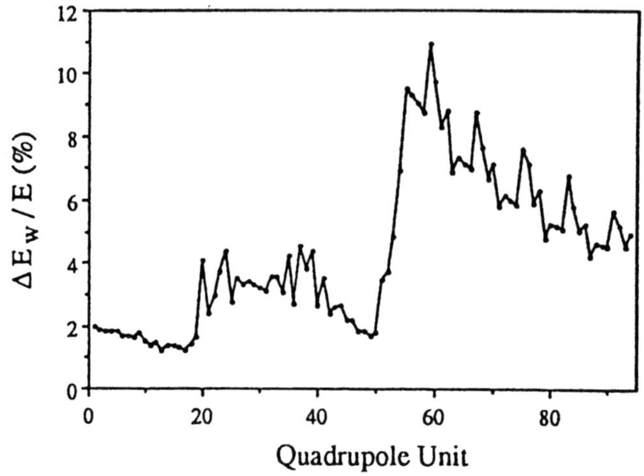

Figure 20 An equivalent energy change at the back of the bunch which cancels the phase advance per cell from transverse wakefields at 3 X 10^{10} electrons plotted versus quadrupole unit number resulting from a two particle model. The variations result from the uneven spacing of the quadrupoles. Adjustments in the energy spectrum using klystron phases can not be induced in the beam as rapidly as the discontinuities require. However, the general trends can be followed.

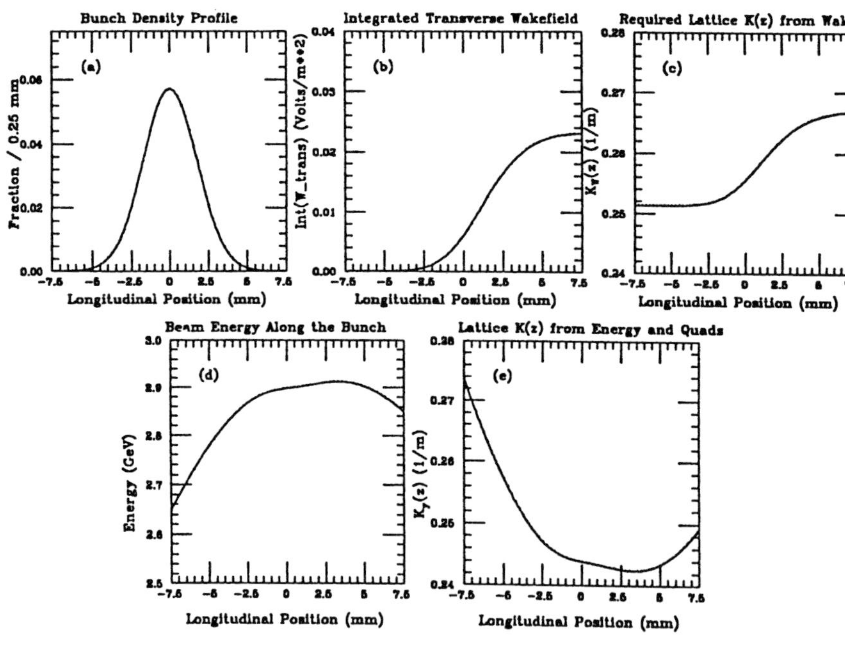

Figure 21 Schematic representation of the mutual compensation of wakefield and chromatic effects over the entire bunch in the SLC linac. The bunch density is shown in (a). The integrated transverse wakefield over the bunch is shown in (b). The required betatron frequency determined by transverse wakefields over the bunch is shown in (c). The energy is shown in (d). Finally, the betatron frequency along the bunch as determined from lattice and bunch energy considerations is shown in (e). The task for autophasing is to match (c) with (e) by adjusting the energy spectrum, bunch length, and lattice parameters.

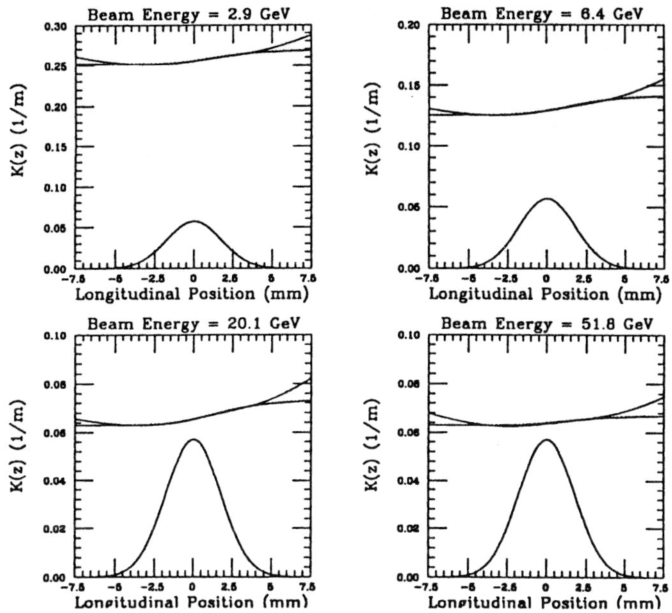

Figure 22 An optimized autophasing condition is shown at four locations along the SLC where the wakefield required and lattice generated betatron frequencies are superimposed over the length of the bunch. In this solution the klystron phases are specially set to -7^0, -10^0, -20^0, -1^0, and 0^0 for five linac sections of length 100m, 200m, 400m, 800m, and 1400m, respectively. The resulting matches are reasonably good over the length of the bunch. The bunch charge is 3.5×10^{10} electrons with a length of 1.75 mm in a 90 degrees per cell lattice. The final energy spread is 1.4 %, which is too large for the SLC final focus to chromatically correct. By adjusting the phase of the last section of the linac, the energy spread can be reduced but the auto phasing condition is then violated leading to emittance growth.

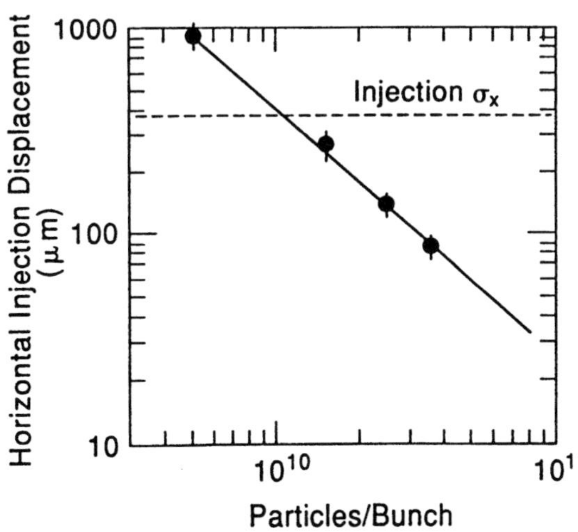

Figure 23 Injection tolerances versus bunch population as determined for the SLC using the data in Figure 14 and the requirement that the oscillation be smaller than the final bunch size.

dark current in the RF structure, unstable kicker magnets, temperature changes in the tunnel [23], and changes in the beam intensity. In the SLC, over 80 beam parameters (beam positions, angles, and energies) are controlled by feedback routinely with many corrections each second. Not all parameters need rapid feedback. For example, the energy spectrum feedback has proven not to need pulse-by-pulse control. Modern control theory is used to provide cascaded control of position and angle loops from the beginning to the end of the accelerator that minimally interfere with each other and provide maximum control [24]. Reduction of oscillations with frequencies up to one sixth of the accelerator pulse rate can be expected.

Injection jitter too rapid for feedback control must be isolated and fixed. The sources of jitter are generally not stable, as can be seen in Figure 24. Thus, sophisticated analysis techniques must be developed.

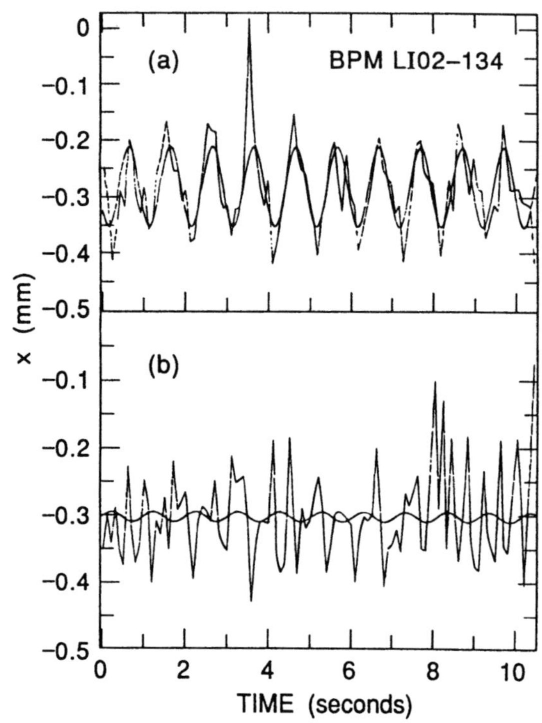

Figure 24 Measured bunch position in the early SLC as a function of time (straight lines) for two periods separated by 3 minutes. The sine curve is a fit to the data at 1.2 Hz. Note that the frequency spectrum of the beam jitter can change rapidly, even though the root-mean-square jitter is about the same.

The magnitude of the jitter can depend strongly on current. An observation in the SLC is shown in Figure 25 where the horizontal jitter becomes rapidly stronger above 3.5 x 10^{10} e^- per pulse. The nature of this jitter can be seen in Figure 26 and 27. The probable cause for this rapid change is the large dispersion

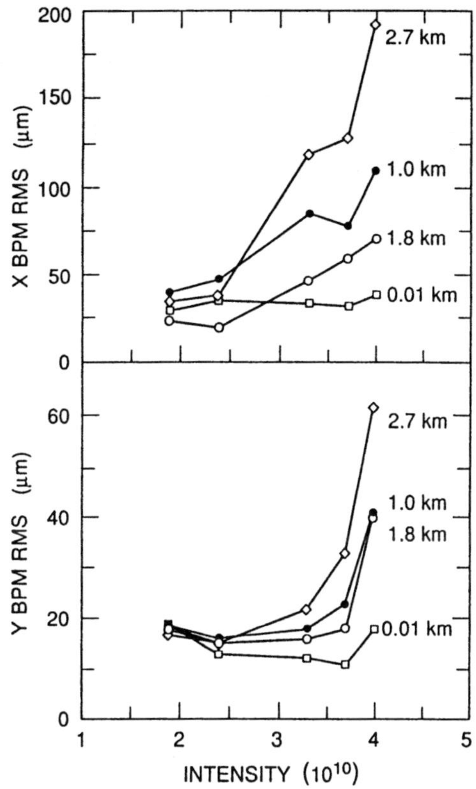

Figure 25 Measured transverse position jitter in the SLC beam as a function of current and position along the linac. At low currents, all positions have similar root-mean-square jitter values. However, at high currents the downstream locations indicate much stronger oscillations. Studies suggest that the injected beam develops a quadratic offset profile at high currents which causes an increased jitter in the linac.

in the transport line between the damping ring and the linac combined with the lengthened bunch in the damping ring at high currents. At the resulting large amplitudes the particles at the front and back are focused incorrectly. These particles then enter the linac far off axis and far ahead or behind the main core of the beam. The head particles excite wakefields downstream [25,26]. Correction of the second-order optics in the transport line is warranted.

If a bunch has a skew profile as it enters the accelerator, it is not possible to correct the bunch with steering to remove all transverse wakefield effects. Some of the possible injection conditions (errors) of a bunch into an accelerator are shown in Figure 28, indicating the importance of a well corrected and controlled launch. Plot (a) is for a perfect bunch, (b) shows a transverse offset, (c) shows a head-tail energy spread and dispersion η, (d) indicates upstream wakefields, (e) characterizes second order terms (T_{166} for example), and (f) is similar to (c) but with η' added.

Dispersion effects will be discussed in more detail. Dispersion at the entrance to the accelerator can cause the beam to have a transverse-longitudinal tilt, similar in shape to Figure 28c. $\Delta x = \eta\, \Delta E/E$. The

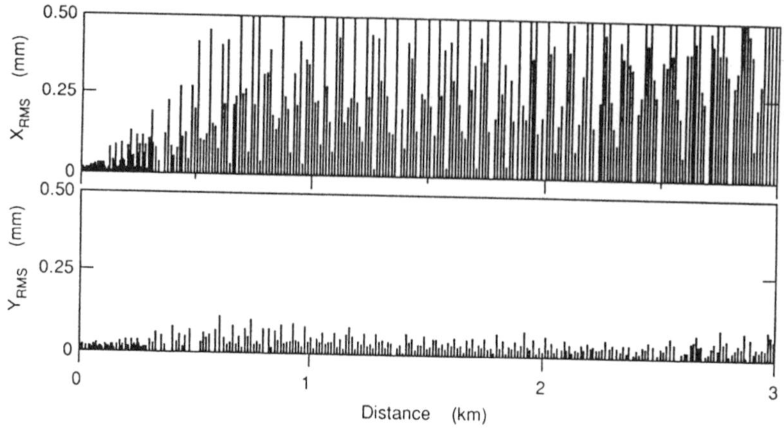

Figure 26 The rms trajectory averaged over 10 pulses is shown along the SLC linac for 4×10^{10} e^-. The jitter causes the x amplitude to grow rapidly. The y enlargement is acceptably small.

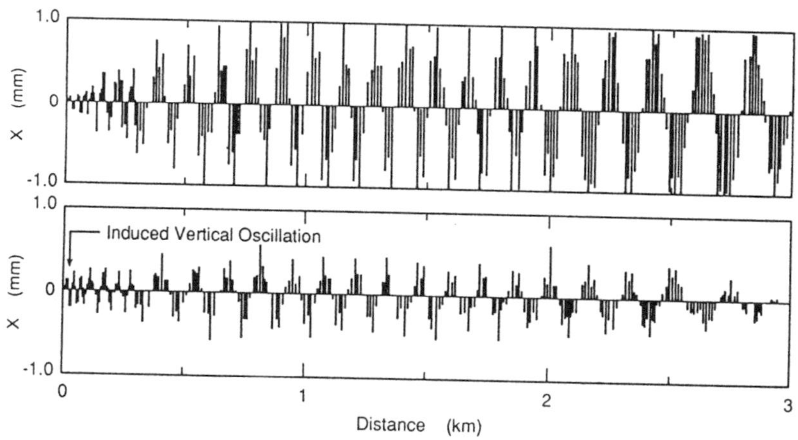

Figure 27 A horizontal oscillation resulting from the unknown jitter shown in Figure 25 and 26 at 4×10^{10} e^- in the SLC. A vertical oscillation was also produced but with a change in a dipole magnet early in the linac. The vertical growth is not the same as the horizontal, indicating that the transverse offsets over the length of the bunch are not like those from a dipole deflection but more likely a higher moment. A possible configuration has the core on axis but the head and tail off axis quadratically.

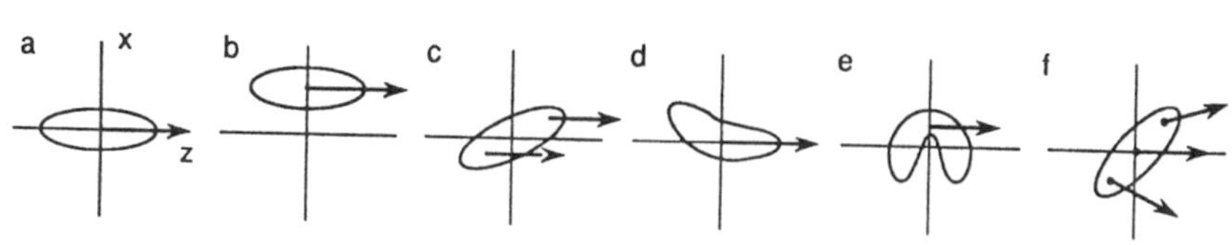

Figure 28 Possible injection orientations of a bunch entering the linac. All orientations have wakefield and dispersion implications except for plot (a), the ideal condition.

head of the bunch is not on axis even though the average position of the bunch is. The oscillation of the head downstream will most likely drive the trailing portions of the bunch to large amplitudes, increasing the emittance. Calculations of this effect for the SLC linac give a tolerance on the allowed dispersion as a function of current [27]. The results are shown in Figure 29. Besides the removal of the dispersion errors, careful placement of the bunch and the use of BNS damping are quite helpful. However, jitter in the energy of the beam entering the linac can make this problem especially troublesome.

The RF structures in the linac have several small asymmetries that generate transverse fields [28]. These transverse fields can be in- or out-of-phase with the accelerating fields and deflect (slightly) all or portions of each bunch, as can be seen in Figure 30. The accumulation of these kicks along the linac can enlarge the emittance. There are three effects. (a) The average deflection component can be removed by trajectory correction. However, changes in the RF phase or amplitude of a klystron cause changes in the deflecting fields and thus generate position jitter downstream [29]. (b) The steady state portion of the transverse RF kick causes local trajectory changes, even with steering, and produces off center beams leading to dispersion and wakefield growth [30]. (c) Finally, the magnitude of an RF deflection can vary over the length of the bunch, which causes direct emittance growth and further filamentation downstream [31]. Care must be taken during construction of the RF structure to minimize these asymmetries. They cannot be fixed after construction except possibly by cancellation using appropriate pairing.

4. Transverse Emittance Effects At High Currents

4.1 General emittance reduction techniques

In a linear accelerator many emittance enlargement effects may occur: some result from injection errors, some arise from accumulation of errors along the linac, and some occur at the end. The injection errors include betatron mismatches, dispersion mismatches, x-y coupling, static injection offsets, and launch jitter. Accumulating errors include transverse wakefields, misaligned quadrupoles and position monitors, RF deflections, and component vibration. Effects at the end of the linac include collimator wakefields and x-y coupling.

Betatron mismatches occur when the injection beam has a phase-space orientation (β,α) that does not match the linac lattice. Because of geometry the linac lattice cannot be chromatically corrected. Therefore, particles with different energies have different oscillation frequencies. Since the injected beam has an internal energy spread that changes during acceleration, the trajectories of the different energy portions of the beam rotate in phase space at different speeds and soon undergo filamentation. Given a beam β_b and α_b that are mismatched from the lattice design values β_l and α_l, the emittance enlargement after filamentation is given by a parameter Bmag [32].

$$(\gamma\varepsilon)_{final} = \text{Bmag}\,(\gamma\varepsilon)_{initial} \qquad (16)$$

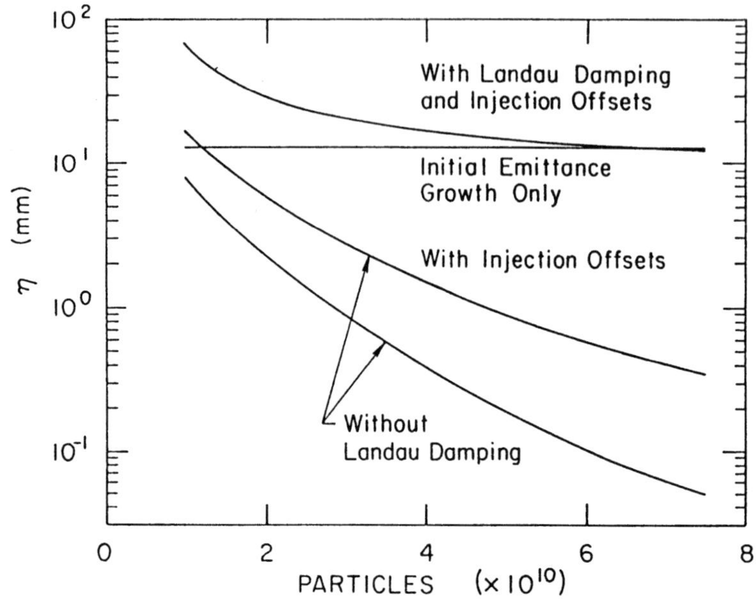

Figure 29 Upper limits of the allowed residual dispersion η entering the SLC linac as a function of bunch intensity. The limits are calculated for a partially compressed bunch (1.5 mm) where dispersion tilts the bunch transversely and excites transverse wakefields in the linac. Steering the head of the beam near the axis increases the tolerances. Also, BNS damping (Landau damping) increases the tolerances significantly. Higher order dispersion will have equivalent effects.

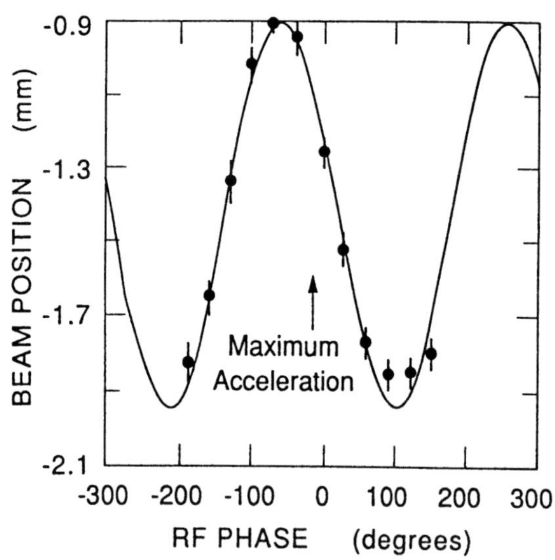

Figure 30 Measured RF deflections of a beam in the SLAC linac are seen as position changes downstream with phase changes. The beam energy was 1 GeV and the accelerating section with the changing phase was 12 m long with an accelerating gradient of 12 MeV/m. There are several consequences. Jitter in the phase and amplitude of the klystrons produces dipole deflections in the beam. The finite longitudinal length of the beam results in the beam shearing transversely from the RF deflection and having an increased emittance.

with

$$\mathrm{Bmag} = \frac{1}{2}\left[\frac{\beta_l}{\beta_b} + \frac{\beta_b}{\beta_l} + \beta_b\,\beta_l\left(\frac{\alpha_b}{\beta_b} - \frac{\alpha_l}{\beta_l}\right)^2\right] \tag{17}$$

To avoid emittance increases of order 10%, β must be matched to 30% or so. The α match has a similar constraint.

Dispersion mismatches are similar [33,34]. A dispersion η is a transverse position-energy correlation which adds to the apparent size and emittance of the beam: $\sigma^2 = \varepsilon\,\beta + \eta^2\,(\Delta E / E)^2$. Given the chromatic lattice, the beam particles displaced by dispersion undergo filamentation if allowed and the real emittance grows. The measured effective emittance $\varepsilon_{\mathrm{eff}}$ in the presence of dispersion [32] is given by

$$\varepsilon_{\mathrm{eff}}^2 = \varepsilon_\beta^2 + \frac{\varepsilon_\beta}{\beta_0}\left[\eta^2 + \left(\beta_0\,\eta' + \alpha_0\,\eta\right)^2\right]\langle\delta^2\rangle \tag{18}$$

and with filamentation the emittance increases further leading to a final value of

$$\varepsilon_{\mathrm{eff}} = \varepsilon_\beta + \frac{1}{2\,\beta_0}\left[\eta^2 + \left(\beta_0\,\eta' + \alpha_0\,\eta\right)^2\right]\langle\delta^2\rangle. \tag{19}$$

The errors in the dispersion on injection must usually be controlled to a few millimeters. Minimization of the measured emittance early in the linac is made by adjusting quadrupoles in a dispersive region upstream to remove the dispersion errors [35]. Future colliders require an order of magnitude smaller adjustments.

Emittance growth from x-y coupling occurs when particle trajectories in one plane, say x, are rotated into the other plane (y) by skew transport elements such as rotated quadrupoles or off-axis sextupoles. This problem is of more concern for a future collider, where flat beams are needed. Coupling can increase both effective emittances (x and y) or move emittance from one plane to the other [36]. Filamentation of the mixed beam can cause further growth.

4.2 Transverse profile changes with transverse wakefields

A betatron oscillation in the linac combined with transverse wakefields leads to the potential for strong wakefield enlargement. Studies of beam transverse enlargement at the end of the SLC linac from betatron oscillations generated by a dipole magnet are shown in Figure 31. These measurements were made with profile monitors using fluorescent screens and TV cameras producing images as shown in Figure 32. The observations in Figure 31 clearly show that leading particles which are off-axis in the accelerator excite trailing particles to larger amplitudes. It is evident that the transverse beam tails oscillate along the linac at the betatron frequency.

The effects discussed in Section 4.1 which affect the transverse beam profile are symmetrical, leaving the beam with left-right equality. Transverse wakefields, however, are asymmetrical and often produce skewed distributions. The particle distributions resulting from wakefields are not gaussian and

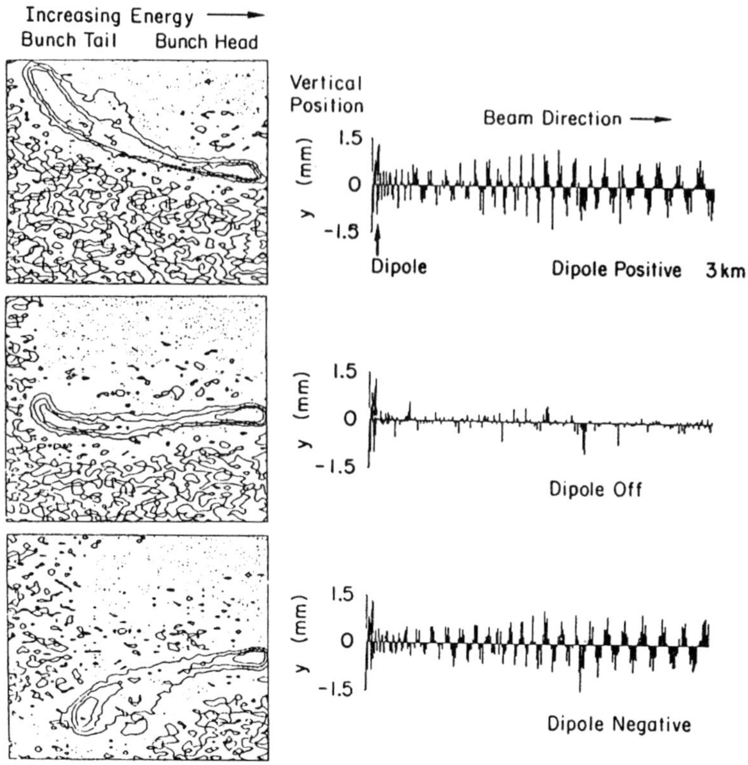

Figure 31 Experimental results demonstrating that the bunch head drives the tail to larger amplitudes during a betatron oscillation due to transverse wakefields. On the right side three trajectories (transverse positions along the linac) are shown. The upper plot has a 'positive' oscillation produced by a small change in a dipole magnet early in the linac, the center shows 'no' oscillation, and the bottom a 'negative' oscillation. On the left side the three resulting beam shapes are shown measured on a profile monitor. The accelerator conditions are such that the longitudinal head of the bunch is located on the right side of the measured profile and the longitudinal tail is on the left side. Note that the position of the bunch head is independent of the oscillation but that the tail of the bunch moves vertically with the sign of the oscillation. This example shows that leading particles in a bunch drive trailing particles to large amplitudes via transverse wakefields. It also shows that the phase space dimensions of each longitudinal slice of the beam remains the same even though the effective emittance of the beam as a whole has increased from transverse wakefield effects.

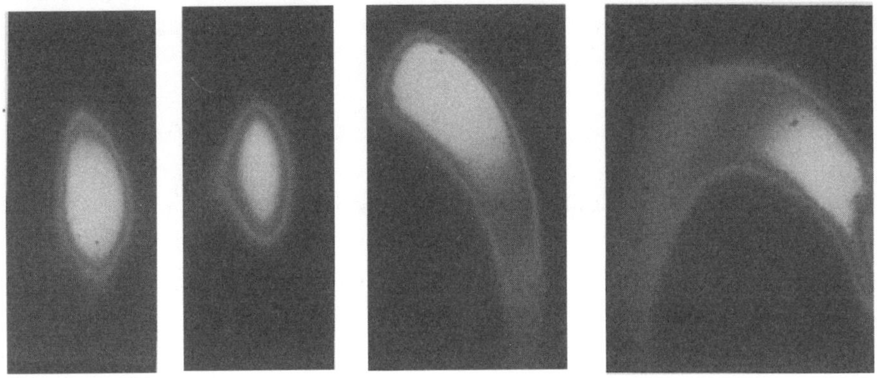

Figure 32 Images of an electron bunch on a profile monitor at 47 GeV showing wakefield growth with increasing oscillation amplitudes. The images from left to right are for a well-steered beam, a 0.2 mm oscillation, a 0.5 mm oscillation, and a 1.0 mm oscillation, respectively. Both horizontal and vertical oscillations are present. The beam intensity is 2 x 10^{10} electrons. The core sizes σ_x and σ_y are about 120 mm.

must be described in a new way. Special methods are available to model and control these changes [38]. Multiple gaussian fits, skewness, kurtosis, and fitting left and right halves separately have been tried with various degrees of success. An intuitive accelerator physics model for tail propagation is described here.

A bunch executing a betatron oscillation in the quadrupole lattice of the linac experiences transverse wakefields in the accelerating structure. The head of the bunch drives the core and back of the bunch to ever increasing amplitudes producing a non-gaussian tail. Simulations of this growth have been made where a bunch is divided into longitudinal slices and traced through the linac. A nearly exponential growth from head to back is apparent. The transverse particle distribution $\rho(x)$ of each slice is equal to that of the initial phase space distribution. The initial particle distribution is a gaussian with width σ.

$$\rho(x) = \exp(-[x - x_0]^2 / 2\sigma^2) / \sigma (2\pi)^{1/2} , \tag{20}$$

where x_0 is the mean of the distribution. The overall distribution must be integrated over the slices with different transverse positions x_0. The position x_0 of each slice is represented by an exponential which initiates at position z_0 along the length of the bunch and has a growth distance of τ. This tail rotates in phase space with the betatron phase ϕ_i but has an initial phase ϕ_0. The emittance of each slice of the beam is ε and the betatron function at each profile measurement 'i' is β_i. $\sigma_i^2 = \varepsilon \beta_i$. The tail extension is scaled locally by σ_i.

$$x_0(\phi_i, z) = \sigma_i U(z_0 - z) [\exp((z_0 - z)\tau / \sigma_z) - 1] \cos(\phi_i + \phi_0) \tag{21}$$

where U is the unit step function and the bunch length is σ_z. The transverse distribution of each slice is given by :

$$\rho\ (x, \phi_i, z) = \exp(-\ (x - x_0(\phi_i, z))^2 / 2\sigma_i^2) \,/\, (2\ \pi)^{1/2} \sigma_i \tag{22}$$

Finally, the overall transverse distribution called $f(x,\phi_i)$ is given by

$$f\ (x, \phi_i) = \int_{-\infty}^{+\infty} \rho\ (x, \phi_i, z)\ h\ (\ z\ )\ dz \tag{23}$$

where h(z) is the longitudinal profile, usually assumed to be a gaussian with length σ_z. By choosing ϕ_0, τ, and z_0, the beam shape can be calculated at any location over a reasonably short region of the linac (less than several betatron wavelengths so that wakefields do not change the distributions). Measured beam shapes can be analyzed to determine the tail structure of the beam and measure the effective ϕ_0, z_0, and τ. For a test, an oscillation was induced in the SLC electron beam with a charge of about 2.5 x 10^{10} e^- by a dipole magnet and the resulting trajectory is shown in Figure 33. The associated beam profiles on four wire scanners are shown in Figure 34. The observed beam shapes have a definite tail with a particular phase, readily fit with the above algorithm.

This model assumes that the transverse tails are not large. As an experiment, an SLC beam was lengthened to about 2.5 mm and accelerated. The expected transverse wakefield effects were large. The internal energy spread was also exaggerated. The resulting beam profile is shown in Figure 35. Clearly, a full model of all parameters must be used to diagnose beams with distributions like the one in Figure 35.

4.3 Emittance observations

The emittance of the beam at injection into the linac is important as it gives the scale of the tolerances to be studied. For example, the measured emittances at the entrance to the SLC linac are shown in Figure 36. The horizontal emittance increases with intensity because of bunch lengthening with current in the damping ring [12,39].

The beam launch parameters into the linac must be correct in order to avoid filamentation and wakefield effects downstream. To repeat, the dispersion functions must be properly adjusted, the betatron mismatches removed, and the second-order optics checked. Operation and studies of the SLC have shown that these emittance enlargement effects including chromatic and wakefields effects can be controlled during colliding beam operations to provide significant luminosity [40]. The best measured single bunch emittance conditions of the SLC to date are shown in Figure 37. These results were also obtained during collisions with three bunch operation (e+,e-, e- scavenger) up to 3 x 10^{10} particles per bunch. The extension of these techniques to higher currents is now under active study as intensities near 4.5 x 10^{10} are needed and the excessive horizontal beam jitter at high currents mentioned in Section 3.6 must be reduced.

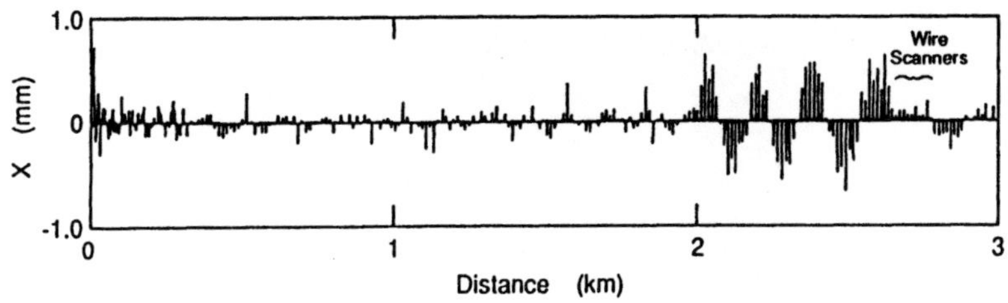

Figure 33 A particular beam oscillation used to generate the transverse beam tails displayed in Figure 34.

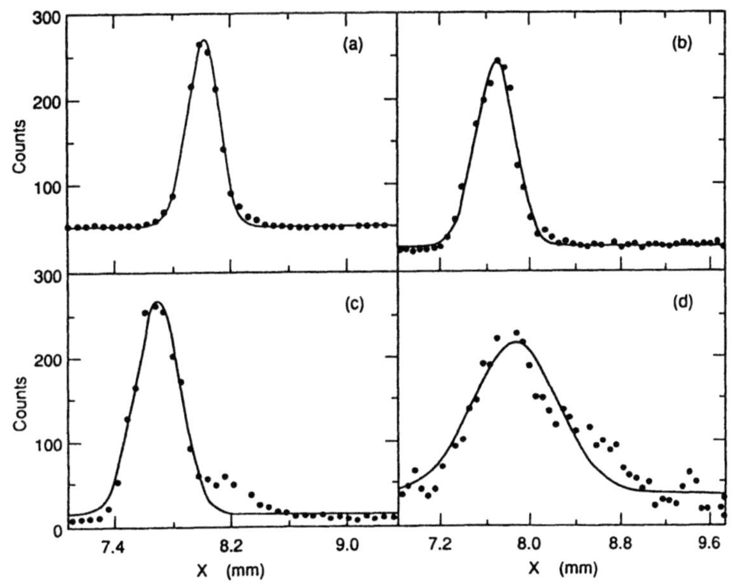

Figure 34 Measured beam profiles (47 GeV) in the SLC for the betatron oscillation in Figure 33. The projections were taken with four wire scanners at (a) 0, (b) 22.5, (c) 90., and (d) 112.5 degrees in betatron phase. Note that the induced transverse tail has a particular phase such that the tail does not appear at position (a). The produced profiles have different shapes depending on the betatron phase. Consequently, the transverse tail rotates in phase space at the betatron frequency and makes the beam appear to have a betatron mismatch.

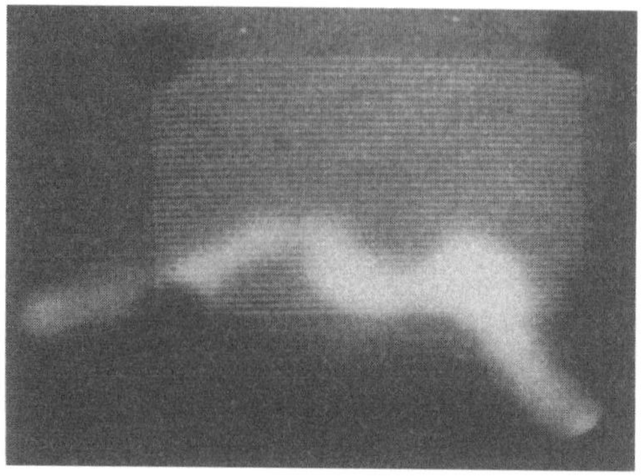

Figure 35 An SLC electron bunch was generated with an unusually long length (of order 2.5 mm) and was made to oscillate both horizontally and vertically. The resulting beam was photographed on a profile monitor at the end of the linac. The image displays a "wormlike" correlation of particle positions along the bunch with the bunch head at the lower right. This particle distribution is characteristic of very strong wakefields. The beam length in this picture is 4 mm.

4.4 Emittance minimization

After the input parameters are optimized and the trajectory nominally corrected, the beam experiences emittance growth during acceleration because of alignment errors of the accelerator components. This results from the trajectory being steered through misaligned quadrupoles and accelerator structures onto beam position monitors with finite residual offset errors. Consequently, transverse wakefields excite the beam.

Methods to reduce these effects have been theoretically studied [41,42]. It was shown that the addition of appropriate injection errors (Δx, $\Delta x'$, Δy, and $\Delta y'$) could cancel most of the emittance enlargement. Since the advent of BNS damping, a more global scheme of distributing short range oscillations along the accelerator has been shown to be satisfactory [43]. Oscillations have been added to the SLC linac to test this procedure. Examples of these oscillations are shown in Figure 38. The emittance at full energy was measured as a function of the oscillation amplitude. The results are shown in Figures 39 and 40. A good choice of the amplitude of the short range oscillation in the appropriate linac location significantly reduced the emittance enlargement. Furthermore, the betatron match of the beam was left properly maintained. Combinations of short range (200 m) oscillations in the SLC were used with two beams during particle physics collisions, the emittances were reduced to the design values at 3×10^{10} e^-. The required electron trajectories for reducing the emittances are shown in Figure 41. Note that significant trajectory offsets were needed for this optimized case.

At high currents the alignment of the RF structure relative to the quadrupoles and position monitors dominates all other alignment issues. The effects of structure misalignments are schematically shown in Figures 42 and 43. It is a daunting task to align the RF structure at the one meter scale to 100 μm or less over many kilometers of linac. Great efforts have been made to develop techniques which will align the structures to these tolerances and to keep them aligned over a long period with minimal effort.

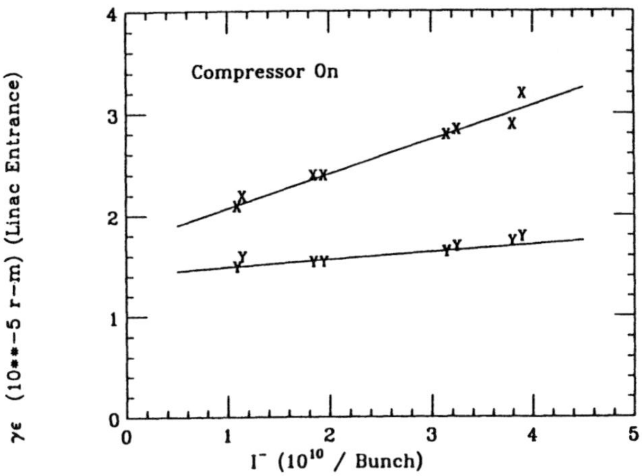

Figure 36 Measured invariant emittances at the entrance to the SLC linac (1.15 GeV) as a function of the beam intensity with the bunch length compressor excited in the standard configuration. The vertical emittance remains constant but the horizontal emittance grows. This growth is associated with first- and second-order chromatic errors in the transport line ($\sigma_E/E = 1\%$) and the onset of bunch lengthening in the damping ring at about 1 x 10^{10} e^-.

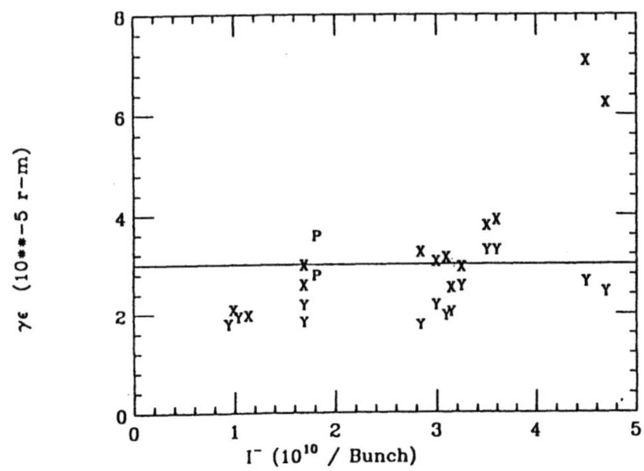

Figure 37 Best measured invariant emittances at the exit of the SLC linac (47 GeV) as a function of beam current. (The symbol P refers to positrons.) Emittances (x and y) at 1 X 10^{10} electrons have been measured at 2 x 10^{-5} r-m, which is 67% of the desired value of 3 x 10^{-5} r-m. At 3 x 10^{10} electrons the allowed emittance values were obtained. Above about 4.0 x 10^{10} a strong launch jitter is produced from the damping ring system and amplified in the linac which causes larger emittances. Bunch lengthening and wakefields effects in the linac may also contribute. Comparing with Figure 36, most of the emittance growth to 3 x 10^{10} e^- is caused by increases in the injected emittance.

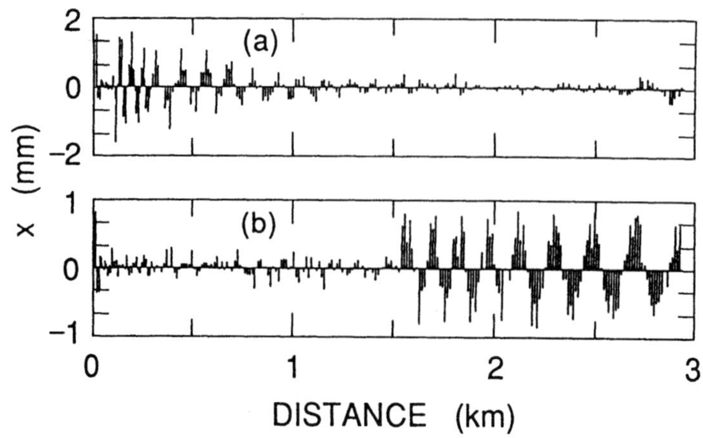

Figure 38 Two induced oscillations in the SLC accelerator used to cancel accumulated wakefield and dispersion errors in the linac, thus reducing the final emittance. See Figures 39 and 40.

Experiments have taken place to study wakefields in misaligned structures [44]. Experimentally, it has been shown that moving structure irises with no other changes can deflect the beam as is shown in Figure 44. These observations have naturally directed efforts towards a possible cure. The component of the structure alignment errors at the betatron spatial frequency is primarily that which drives the beam. Therefore, if a control mechanism can be made to move the structure at that spatial frequency (see Figure 45), then sine and cosine adjustments can be used to cancel the accumulation of alignment errors over reasonably short regions of the accelerator [45]. The alignment tolerances can then be markedly relaxed. This control can also be used to correct for any long term position drift of the accelerator or quadrupole supports due to temperature changes or floor creep in the tunnel.

4.5 Emittance collimation

Intense, low emittance beams at the end of a linac are often passed through small collimators to remove unwanted halos or to reduce the effective emittance. If the core of an intense bunch is slightly off-axis in these collimators, the bunch experiences deflecting forces that vary over the bunch depending on the local charge density [46]. A schematic illustration of this effect is shown in Figure 46.

The configuration of the collimation system is important. The desire is to have the sum of the deflections small compared to the beam size. The offset of the beam in the collimator is proportional to the square root of the local beta function β. The collimator aperture is proportional to the beam size which is proportional to the square root of β. Thus the kick which is proportional to the ratio of the beam offset in the collimator to the collimator bore is independent of β. The kick is, however, inversely proportional to the beam energy E. The beam angular divergence is inversely proportional to the square root of the product of the energy and β.

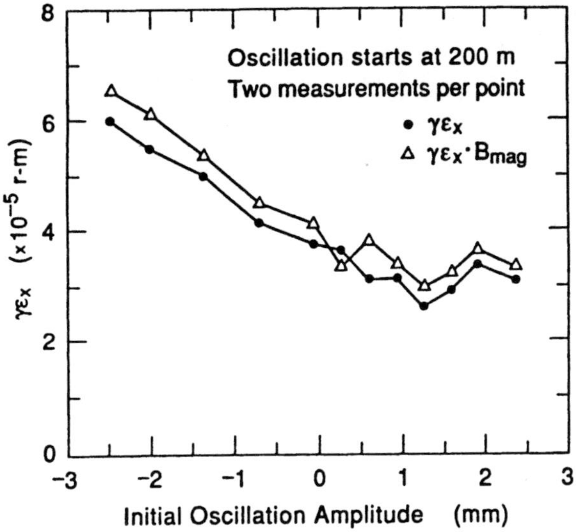

Figure 39 Invariant emittance changes at the end of the linac (47 GeV) as a function of the amplitude of an oscillation starting early in the linac (upper plot in Figure 38). The emittance measurements and the emittance times Bmag measurements track each other very well. Since Bmag is a measure of the expected filamentation from betatron mismatches, the beam at the end of the linac has already filamented. A decrease in the transverse emittance (25%) is observed with a finite oscillation added to the beam. The error which caused the original emittance enlargement is thus near the beginning of the linac.

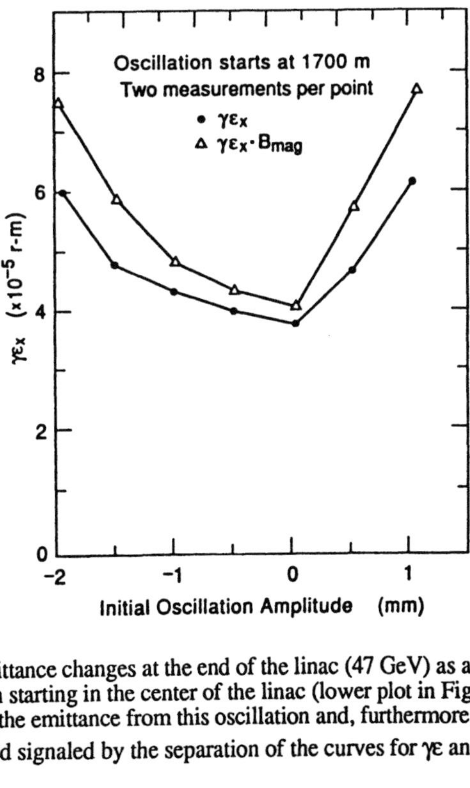

Figure 40 Invariant emittance changes at the end of the linac (47 GeV) as a function of the amplitude of an oscillation starting in the center of the linac (lower plot in Figure 38). There is no reduction of the emittance from this oscillation and, furthermore, a large betatron mismatch has developed signaled by the separation of the curves for $\gamma\varepsilon$ and $\gamma\varepsilon$ x Bmag.

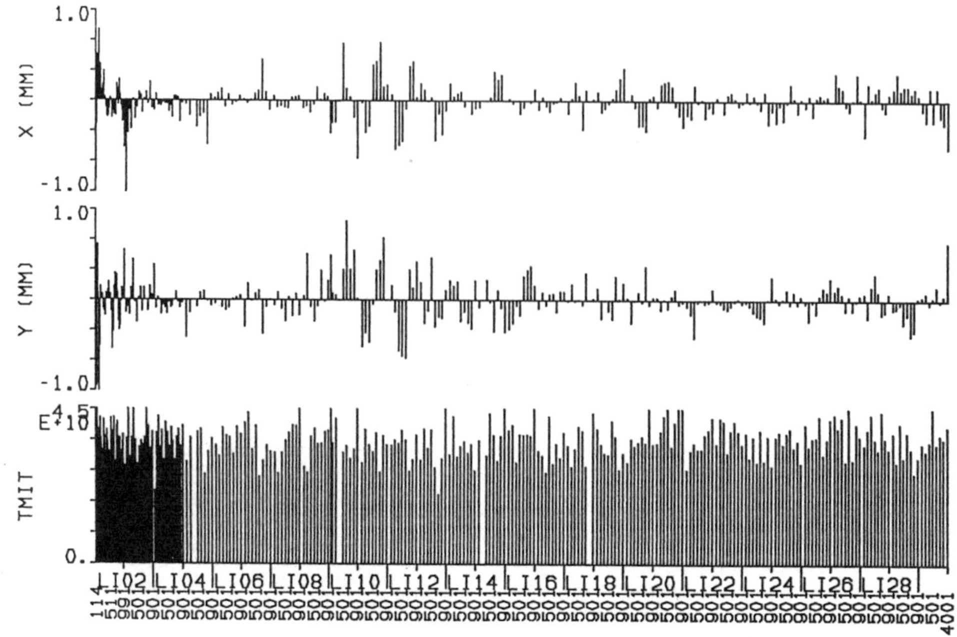

Figure 41 Empirically determined linac trajectory along the 3 km SLC linac which cancels the errors from the accumulation of dispersion and transverse wakefield errors for a bunch with 3 x 10^{10} electrons. These oscillations lowered the invariant horizontal emittance from 4.5 to 3.0 (x 10^{-5} r-m). This trajectory is not unique as other trajectories with similar oscillations can produce the same effect. This non-zero trajectory remained the optimum trajectory during several weeks of collisions.

Therefore,

$$\frac{\theta_{kick}}{\theta_{beam}} \alpha \left[\frac{\beta}{E}\right]^{1/2} \tag{24}$$

where a low β and a high E minimize the emittance growth. Unfortunately, a low β value leads to small beam sizes and the potential for component radiation damage. For the next linear collider special collimation sections must be designed [47].

4.6 Multibunch transverse effects

The acceleration of multiple bunches requires the RF structures to have sufficiently small long range transverse wakefields so that trailing bunches do not experience significant transverse forces. The forces could generate or increase oscillation amplitudes of trailing bunches and ultimately increase their emittances. The solutions to this problem are to space the bunches far enough apart to allow the higher order transverse modes to decohere or to damp the modes with absorbers. The new structures for a future linear collider will likely include a combination of transverse waveguides to damp higher order modes and cell-to-cell detuning to dephase the higher order fields. Furthermore, the captured modes in the structure, cell-to-cell fields, that are driven by long trains of bunches must be detuned to avoid unwanted multiple beam breakup.

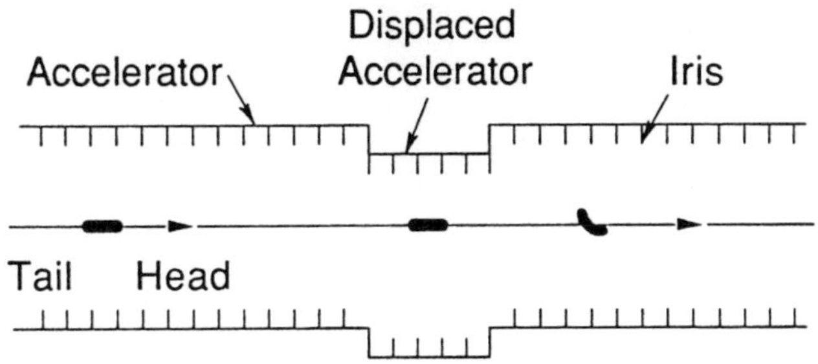

Figure 42 A displaced accelerating structure will induce a transverse offset in the beam from transverse wakefields. The head of the bunch remains on axis but the core continues to drive the tail to ever increasing amplitudes.

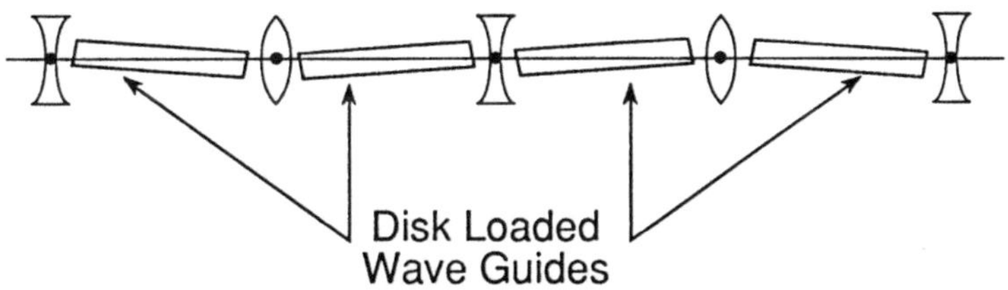

Figure 43 Due to alignment errors, the RF structure displacements in an accelerator, even though small, collectively contribute to the transverse wakefield effects of the beam shape.

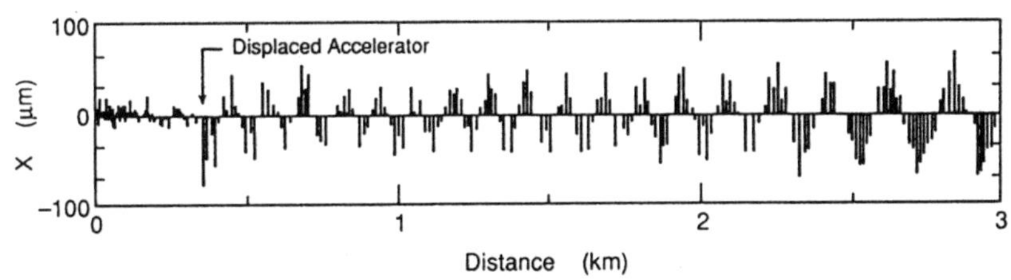

Figure 44 The center of an RF accelerating girder (12 m) of the SLAC linac was displaced by 1 mm while leaving all other parameters fixed. The observed transverse oscillation of a beam downstream resulted from the displacement of copper irises only.

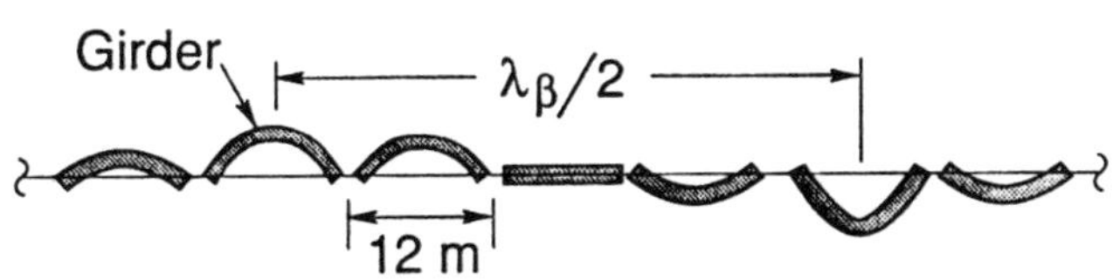

Figure 45 Deliberate distortions of the accelerating structure at the betatron spatial frequency can be used to control random natural misalignments. Quadrupoles are located at the nodes.

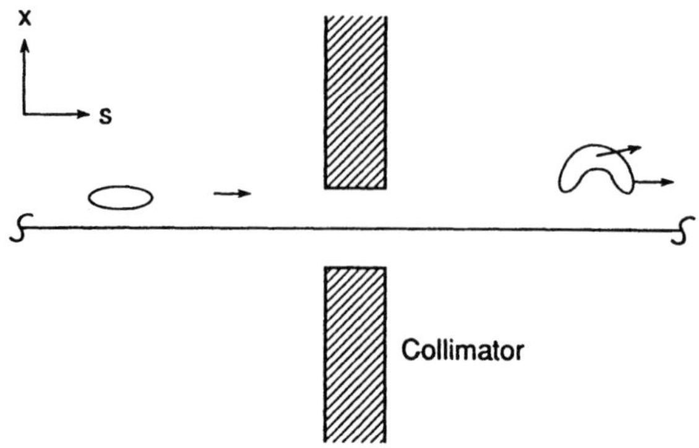

Figure 46 Schematic illustration of the increase in transverse emittance from a close but asymmetric encounter of an intense electron bunch with a beam halo collimator.

Oscillations at high currents have been used to study long range wakefields. Two bunches were made to oscillate along the SLC linac with variable spacings. The results are shown in Figure 47. The trajectory of the second bunch was measured with and without the leading bunch. No noticeable change can be seen for the nominal 60 ns spacing. Effects were seen, however, with a 10 ns spacing, as expected [48].

Studies of very closely spaced bunches were also made at the KEK linac as is shown in Figure 48. Their findings were that the deflections are linear in the initial offset, they depend on the charge per bunch, and they depend strongly on the length of the accelerator [49].

5. Acknowledgments

The author wishes to thank many people at the Stanford Linear Accelerator Center and from other laboratories around the word for stimulating discussions about high current effects in accelerators. Special thanks are giv^n to the SLC Linac and Emittance Study Groups.

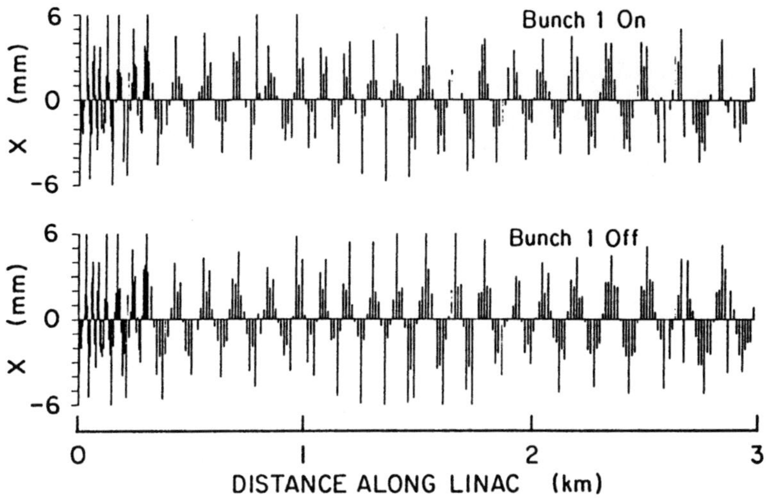

Figure 47 Oscillations of a trailing bunch when the leading bunch in the SLAC linac is present or not. With a spacing of 60 ns as shown, there is negligible effect. Effects begin to occur with spacings of about 10 ns.

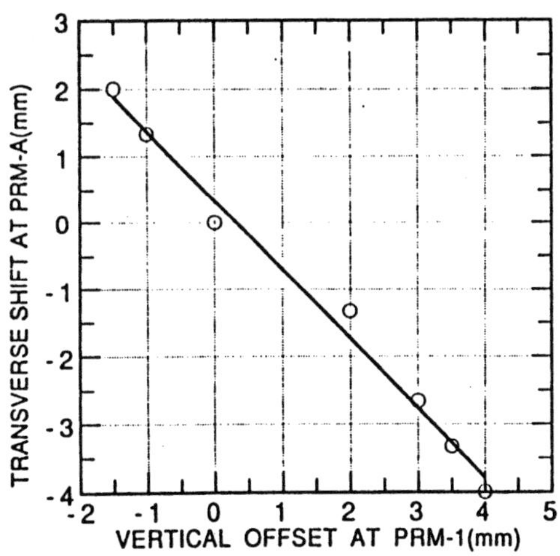

Figure 48 Transverse shift of the beam at the end of the KEK accelerator as a function of initial offset.

6. References

1. C. Pellegrini, 1990 Linear Accelerator Conference, Albuquerque, p.273 (1990).
2. J. Seeman, *Annu. Rev. Nucl. Part. Sci.* 1991. 41: 389-428.
3. A. Hofmann and H. Winick, private communication.
4. B. Richter, *Part. Accel.* 1990. 26: 33-50.
5. R. Palmer, *Annu. Rev. Nucl. Part. Sci.* 1990. 40:529-592.
6. K. Takata and Y. Kimura, *Part. Accel.* 1990. 26: 87-96.
7. R. Ruth, Proc. of Joint CERN/US Particle Accelerator School, Hilton Head (1990).
8. J. Seeman and L. Merminga, 1990 Linear Accelerator Conference, Albuquerque, p. 387 (1990).
9. K. Bane and P. Wilson, 11th Int. Conf. on High Energy Accelerators, Birkhauser, p. 592 (1980).
10. D. Warner, 1988 Linear Accelerator Conference, CEBAF, M01-4 (1988).
11. J. Seeman, et al., SLAC-PUB-5438, Stanford (1991).
12. K. Bane, SLAC/AP-80, Stanford (1990).
13. K. Thompson and R. Ruth, *IEEE 89CH2669-0*, Chicago, p. 792 (1989).
14. Z. D. Farkas, et al, 9th Int. Conf, on High Energy Accel., p. 576 (1974).
15. J. Seeman, et al., *IEEE CH2387-9/87/0000*, Washington PAC, p. 73 (1987).
16. V. Balakin, et al., 12th Int. Conf. on High Energy Accel., FNAL, p. 119 (1983).
17. H. Henke, 1988 Linear Accelerator Conference, CEBAF, p. 481 (1988).
18. J. Seeman, et al., SLAC-PUB-4968 (to be published).
19. C. Adolphsen, et al., SLAC-PUB-5581, Stanford (1991).
20. V. Balakin, 1988 Workshop on Linear Colliders, SLAC, p. 119 (1988).
21. R. Gluckstern, et al., 1990 Linear Accelerator Conference, Albuquerque, p. 309 (1990).
22. T. Limberg, et al., EPAC, Nice, p. 1506 (1990).
23. J. Seeman, et al., SLAC-PUB-5439, Stanford (1991).
24. T. Himel, et al., SLAC-PUB-5470, Stanford (1991).
25. M. Ross, private communication.
26. J. Seeman, et al., SLC-Reports-249 and 252, Stanford (1991).
27. J. Seeman, SLAC-CN-330, Stanford (1986).
28. J. Seeman, et al., *IEEE Trans. Nucl. Sci.* 32:2629 (1985).
29. R. Stiening, SLAC-CN-181, Stanford (1982).
30. J. Seeman, *IEEE Trans. Nucl. Sci.* 87-CH2387-9: p. 1267 (1987).
31. J. Seeman, *Part. Accel.* 30:7, p. 73 (1990).
32. W. Spence, et al., SLAC-PUB-5276, Stanford (1991).
33. J. Sheppard, SLAC-CN-298, Stanford (1985).
34. J. Seeman, SLAC-CN-330, Stanford (19886)
35. L. Merminga, et al., SLAC-PUB-5514, Stanford (1991).
36. K. Brown and R. Servranckx, SLAC-CN-350, Stanford (1986).
37. J. Seeman, SLAC-PUB-5294, Stanford (1991).
38. J. Seeman, SLAC-PUB-5440, Stanford (1991).
39. K. Bane, SLAC-PUB-5177, Stanford (1990).
40. J. Seeman, et al., SLAC-PUB-5437, Stanford (1991).
41. A. Chao, B. Richter, and C. Yao, *Nucl. Inst. and Meth.* 178, p. 1 (1980).
42. K. Bane, *IEEE Trans. Nucl. Sci.* NS-32, No. 5, p. 2389 (1985).
43. J. Seeman, et al., SLAC-PUB-5705, Stanford (1991)
44. J. Seeman, SLAC-PUB-5706, Stanford (1991)
45. J. Seeman, 1990 Linear Accelerator Conference, Albuquerque, p. 387 (1990).
46. K. Bane and P. Morton, 1986 Linear Accelerator Conference, SLAC, p. 490 (1986).
47. L. Merminga, et al., SLAC-PUB-5436, Stanford, (1991).
48. J. Seeman, et al., *IEEE CH2387-9/87/000*, Washington PAC, p. 73 (1987).
49. Y. Ogawa, T. Shidara, and A. Asami, *Phys. Rev. D* 43: 1, p. 258 (1991).

Observations of the Beam-Beam Interaction in Hadron Colliders

S. Saritepe, G.Goderre, S.Peggs

Fermi National Accelerator Laboratory*

Batavia, Illinois, USA

1 Introduction

Experimental data on hadron beam-beam interactions come from CERN ISR, CERN SPS and FERMILAB TEVATRON colliders. They all successfully collided hadron beams and produced luminosities that resulted in physics discoveries. The ISR pioneered the stochastic cooling techniques paving the way for proton-antiproton colliders, achieving a maximum luminosity of 1.3×10^{32} $cm^{-2}sec^{-1}$ colliding stochastically cooled coasting proton beams at energies up to 31 GeV. The SPS and the TEVATRON in their 1988-89 runs both reached 2×10^{30} $cm^{-2}sec^{-1}$ and delivered integrated luminosities of 8100 nb^{-1} and 9600 nb^{-1}, respectively, while colliding 6 proton bunches with 6 antiproton bunches.

The SPS was built with conventional magnets, limiting its beam energy to 315 GeV for stored beams. The TEVATRON, with approximately the same circumference as the SPS, was the first large accelerator built with superconducting magnets. This allowed beams to be accelerated to 900 GeV. In the very near future the TEVATRON beam energy will be raised to 1 TeV by reducing the temperature of the cooling system from 4.6°K to 4.2°K.

In bunched beam hadron colliders, the beam brightness (intensity/emittance) and therefore the luminosity is limited by head-on beam-beam interactions. However, the effect of the beam-beam interaction can be reduced by separating the closed orbits of protons and antiprotons. Unnecessary head-on interactions are eliminated, making it possible to increase the luminosity. In the SPS, beams were horizontally separated around 3/4 of the ring. Three interaction points were located

*Operated by Universities Research Association,Inc., under contract with the United States Department of Energy.

in the remainder of the ring. In the TEVATRON beams will be separated in both planes (helical separation) so that the machine can be operated with more than 6 bunches per beam [1]. Beams will only be brought into collision at 2 locations, B0 and D0. The helical separation scheme will be operational for the first time during the 1992 TEVATRON collider run. Major components of the system have been tested (new low-beta lattice, electrostatic separators, feed-down circuits) during two study periods. These studies concentrated particularly on long-range beam-beam interactions.

This paper has three parts. In the first part the basic beam-beam theory will be reviewed. Theoretical issues relevant to e^+e^- colliders will not be mentioned. In the second part we summarize the operational experiences at FERMILAB and CERN. In the last part of the paper, experiments on long-range beam-beam interactions in the TEVATRON are reviewed.

2 Hadron Beam-Beam Theory

Resonances play an important role in periodic Hamiltonian systems such as hadron storage rings[2]. If the fractional tune of the betatron oscillation is close to a rational number the particle will experience correlated kicks from various nonlinear fields such as lattice imperfections, beam-beam interactions et cetera. Even if the perturbation is very weak the effect will be magnified, and the particle amplitude changes. In practice, in the absence of damping, the emittance grows.

The resonance condition cannot be sustained if the tune changes with amplitude. Therefore the curve showing the tune as a function of amplitude is very important in the understanding of resonances. The "detuning function" is uniquely determined by the source of the nonlinearity, and its strength.

A resonance is characterized by three attributes: the "resonance-width function", the strengths of the perturbations, and the phase advances between them. In practice, excited resonances cause emittance growth and beam lifetime effects. In theory, the amplitude growth mechanism (diffusion) is not well understood. The condition of "chaos" seems to be necessary but not sufficient for the diffusion of particles. There is no theory linking the beam lifetime to beam-beam resonances in hadron storage rings. The hadron beam-beam theory reviewed below is mostly concerned with the onset of chaos.

2.1 Beam-Beam Resonances

Consider a collider with a single beam-beam collision per turn. The betatron tune of a test particle depends on its amplitude, according to

$$Q(\alpha) = Q_0 + \xi D(\alpha) . \tag{1}$$

Here Q_0 is the unperturbed tune, $\alpha \equiv a/\sigma$ is the normalized amplitude where a is the betatron amplitude and σ is the (transverse) rms size of the source bunch. The source bunch generates the beam-beam force and perturbs the test particle.

$D(\alpha)$ is the "detuning function" and ξ is the "beam-beam parameter", defined as

$$\xi = \frac{N_p r_p \beta^*_{x,y}}{2\pi\gamma\sigma_{x,y}(\sigma_x + \sigma_y)} \tag{2}$$

Here $\beta^*_{x,y}$ are the betatron functions at the collision point, $\sigma_{x,y}$ are the rms beams sizes for the source bunch, r_p is the classical proton radius, N_p is the number of particles in the source bunch and γ is the relativistic factor. The beam-beam parameter ξ is equal to the tune shift experienced by a small-amplitude particle. For round beams ($\sigma_x = \sigma_y$) the beam-beam parameter ξ can be written as

$$\xi = \frac{N_p r_p}{\pi(\varepsilon\beta\gamma)} \tag{3}$$

where ε is the source bunch emittance. Beams in hadron colliders are almost round, both due to design and due to residual coupling between the x- and the y-planes. Therefore the round-beam expressions for detuning and resonance-width functions will be sufficient to describe the beam-beam dynamics. If colliding bunches have Gaussian transverse charge distributions, the detuning function for the beam-beam interaction of round beams has the exact analytic form [3]

$$D(\alpha) = 4\alpha^{-2}\left[1 - I_0(\alpha^2/4)\exp(-\alpha^2/4)\right] . \tag{4}$$

Here I_0 is a modified Bessel function. A beam-beam resonance of order N is present if the tune is equal to a rational fraction n/N at some amplitude α_N. A schematic representation of the phase-space flow of a resonance is shown in Fig.1.

The beam-beam resonance islands, seen in the phase-space flow diagrams, have a half width given by [3]

$$\Delta\alpha_w = 2\left[\frac{V_N(\alpha)}{\alpha D'(\alpha)}\right]^{1/2} . \tag{5}$$

$D'(\alpha)$ is the derivative of $D(\alpha)$ with respect to α. For round beams the "resonance width function" $V_N(\alpha)$ is (even-order N only)[4]

$$V_N(\alpha) = \int_0^\alpha \frac{8}{\alpha} I_{N/2}(\alpha^2/4)\exp(-\alpha^2/4)\, d\alpha . \tag{6}$$

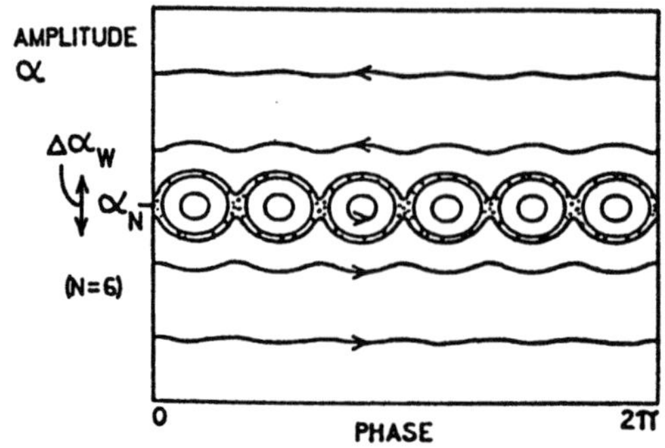

Figure 1: A sketch of phase-space flow near a beam-beam resonance. Note that the number of islands is equal to N, the order of the resonance.

The detuning function is shown in Fig.2. Beam-beam resonance half-widths for resonances up to the 12th order are shown in Fig.3.

One important observation from Fig.3 is that only large-amplitude particles can excite the high-order resonances. As will be discussed in the following sections, this theoretical result was used to explain the high background rates in CERN SPS collider operation when proton and antiproton bunches had unequal emittances.

2.2 Tune Modulation

Tune modulation caused by the synchrotron oscillations is practically unavoidable. If the chromaticity is not exactly zero, the oscillation in the particle energy is translated into an oscillation in tune. Another source of tune modulation is noise in the current supplied to the magnets. From experiment[5] and theory[6][7][8] it is known that tune modulation enhances the effect of resonances. The enhancement can be understood in terms of Chrikov's resonance overlap criterion [9].

It is supposed that, owing to an external modulating source, the perturbed betatron tune is given by

$$Q = Q_0 + q \sin(2\pi Q_s t) \tag{7}$$

where Q_0 is the unperturbed betatron tune, q is the amplitude of the modulation (modulation depth), Q_s is the modulation tune, and t is the turn number. The resonance analysis is done at a particular point in the ring and "time" for the purposes of this analysis is quantized.

Tune modulation causes a family of synchrobetatron sideband resonances to appear, at time-

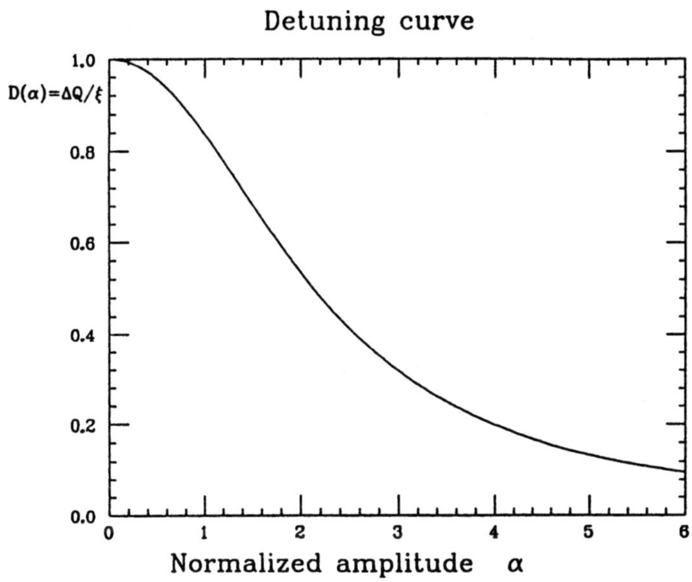

Figure 2: Beam-beam detuning function.

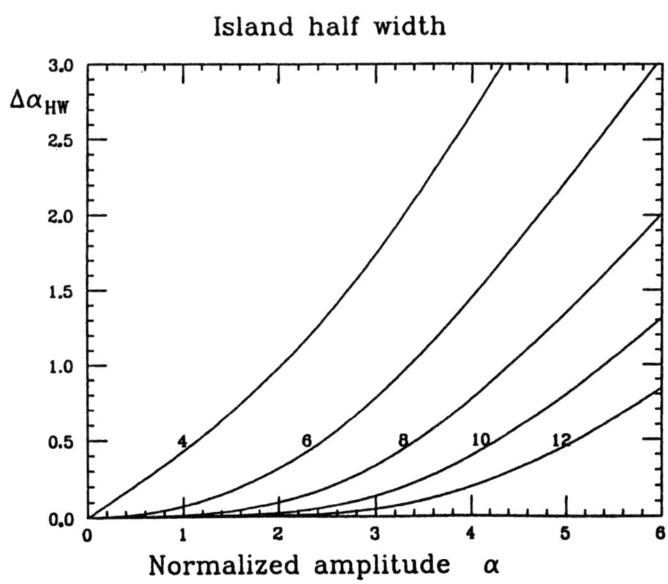

Figure 3: Beam-beam resonance island half-widths for resonances up to order 12.

averaged tunes of

$$Q(\alpha) = n/N + p\, Q_s/N \tag{8}$$

where p is an integer. This situation is depicted in Fig.4(a,b) where the sideband islands surround the fundamental islands.

The full width of the pth sideband is given (if the sidebands do not overlap) by

$$\Delta\alpha_{wp} = 4\left[\frac{V_N(\alpha_p)J_p(Nq/Q_s)}{\alpha_p D'(\alpha_p)}\right]^{1/2} . \tag{9}$$

Here J_p is the pth integer-order Bessel function, and α_p is the betatron amplitude corresponding to this sideband. The magnitude of J_p is of the order of

$$J_p(Nq/Q_s) \approx (Q_s/\pi Nq)^{1/2} \tag{10}$$

if

$$\frac{n}{N} - q < Q(\alpha_p) \ < \ \frac{n}{N} + q \tag{11}$$

and very small if condition 11 is violated. The physical interpretation of this condition is as follows. Because of the tune modulation, the "instantaneous" tune varies between $Q(\alpha) - q$ and $Q(\alpha) + q$. For the resonance to have effect, this tune must cross n/N. So, if $Q(\alpha) < (n/N - q)$ or $Q(\alpha) > (n/N + q)$, the tune never reaches the resonance condition and the sidebands are suppressed. Sidebands are separated in amplitude from each other by

$$\Delta\alpha_s \equiv \frac{(Q_s/N)}{Q'(\alpha)} = \frac{Q_s}{N\xi D'(\alpha)} . \tag{12}$$

As the beam-beam tune shift parameter ξ is increased, the sidebands remain constant in size while their separations decrease. When $\Delta\alpha_s \ < \ \Delta\alpha_{wp}$, the sidebands overlap and a chaotic layer is formed in phase-space flow as shown in Fig.4 d. In other words, there is overlap if

$$\xi \ > \xi_{\max} \equiv \frac{1}{4}(\pi q)^{1/4}(Q_s)^{3/4}\left(\frac{\alpha}{N^{3/2}V_N(\alpha)D'(\alpha)}\right)^{1/2} . \tag{13}$$

This needs to be generalized to multiple collisions. The generalized $\Delta\alpha_s$ and $\Delta\alpha_{wp}$ are [8]

$$\Delta\alpha_s = Q_s \Big/ \left[N \cdot \sum_{i=1}^{m} |\ \vec{\xi}_{Ni}\ | \cdot D'(\alpha)\right] , \tag{14}$$

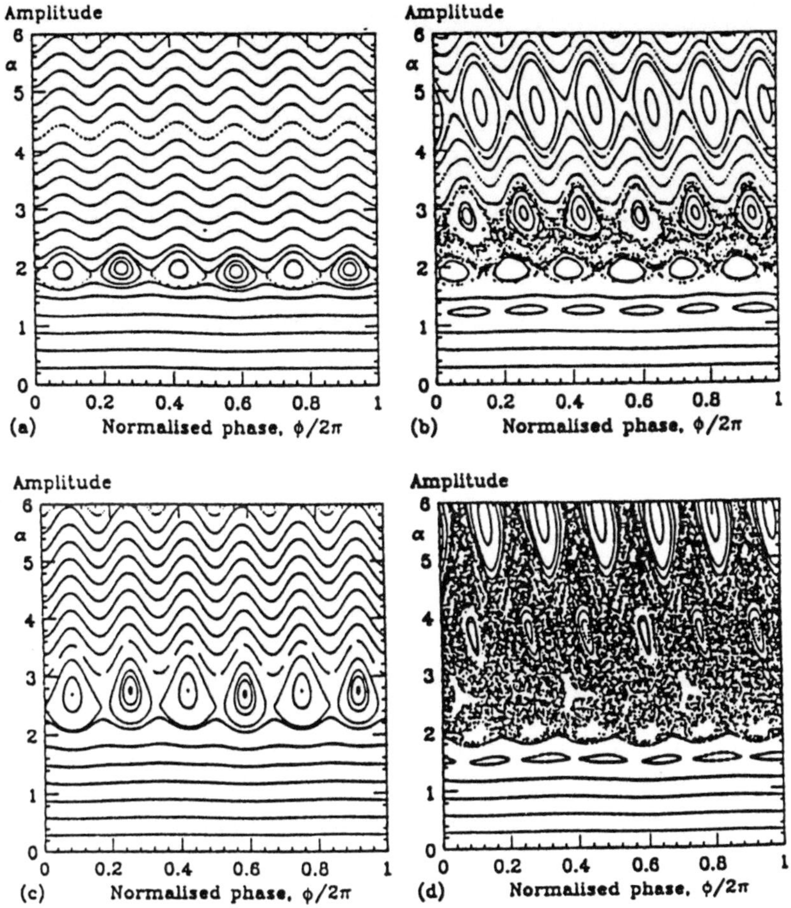

Figure 4: Simulated trajectories tracked for 2000 modulation periods, with $Q_s = 0.005$ and an unshifted tune of 0.331, near a sixth-order beam-beam resonance. The two left figures have no tune modulation, while the two right figures have modulation amplitude $q = 0.001$. The two top figures have a tune shift parameter of $\xi = 0.0042$, while the two bottom figures have a value $\xi = 0.0060$. Side bands $p = +1, 0, -1$, and -2, visible in (b) at increasing amplitudes, overlap and are submerged in a chaotic sea in (d).

$$\Delta\alpha_{wp} = 4\left(\left[|\sum_{i=1}^{m} \vec{\xi}_{Ni}| \cdot V_N(\alpha)J_p(Nq/Q_s)\right] / \left[\sum_{i=1}^{m}|\vec{\xi}_{Ni}| \alpha D'(\alpha)\right]\right)^{1/2} . \tag{15}$$

Using the overlap condition $\Delta\alpha_s < \Delta\alpha_{wp}$ and Eq.(14), Eq.(15) we obtain

$$\left(|\sum_{i=1}^{m} \vec{\xi}_{Ni}| \cdot \sum_{i=1}^{m}|\vec{\xi}_{Ni}|\right)^{1/2} > \xi_{\max} = \frac{1}{4}(\pi q)^{1/4}(Q_s)^{3/4}\left[\frac{\alpha}{N^{3/2}V_N(\alpha)D'(\alpha)}\right]^{1/2} \tag{16}$$

where m is the number of head-on beam-beam interactions. The magnitude of the beam-beam resonance vector $\vec{\xi}_{Ni}$ is equal to the beam-beam parameter given in Eq.(2).

The calculation of $|\sum \vec{\xi}_{Ni}|$ requires the knowledge of phases at crossing points. Typically there is a several percent error in the lattice functions, and it is difficult to know the phase at the crossing points to sufficient accuracy. It is usual to simply take the root-mean-square average of the resonance vectors $\vec{\xi}_{Ni}$, namely, we approximate

$$|\sum_{i=1}^{m} \vec{\xi}_{Ni}| \approx (m)^{1/2}\xi . \tag{17}$$

The other summation is easier since the phase information is not needed:

$$\sum_{i=1}^{m}|\vec{\xi}_{Ni}| = m\xi . \tag{18}$$

Eq.(16) gives the "threshold equation". Given the order of the betatron resonance N, the particle amplitude α, the tune modulation frequency Q_s and depth q, the threshold equation tells whether the beam-beam parameter ξ is large enough to cause an overlap of sideband resonances. The threshold condition, Eq.(16), also defines the highest-order betatron resonance that allows sideband overlap in the presence of tune modulation for a particular amplitude. From here on these will be called "critical resonances".

2.3 Beam-Beam Tune Shift and Spread

The tune shifts and spreads arising from beam-beam interactions can be calculated numerically or analytically. Tune shift from head-on interactions is well understood and given by [10], [11], [12]:

$$\Delta\nu_x = \frac{1}{4\pi}\frac{N_p r_p}{\gamma}\frac{\beta_x}{\sigma_x\sigma_y}\int_0^1 \frac{dw}{\sqrt{\lambda_x^3\lambda_y}}[Z_0(\zeta_x) - Z_1(\zeta_x)]Z_0(\zeta_y), \tag{19}$$

$$\begin{aligned}
\lambda_x &\equiv 1+(\frac{\sigma_x}{\sigma_y}-1)w \\
\lambda_y &\equiv 1+(\frac{\sigma_y}{\sigma_x}-1)w \\
\zeta_x &\equiv \frac{\beta_x J_x}{2\sigma_x\sigma_y}\frac{w}{\lambda_x} \\
\zeta_y &\equiv \frac{\beta_y J_y}{2\sigma_x\sigma_y}\frac{w}{\lambda_y} \\
Z_n(\zeta) &\equiv e^{-\zeta} I_n(\zeta)
\end{aligned}$$

where γ is the relativistic factor, r_p is the classical proton radius, I_n are modified Bessel functions. J_x and J_y are the so-called action variables ($J_x = \alpha_x^2/2$, $J_y = \alpha_y^2/2$, here α_x and α_y are the normalized amplitudes for the x- and the y-planes, respectively). A similar expression can be written for $\Delta\nu_y$ by interchanging x and y subscripts.

Analytical expressions for tune shifts arising from a long-range beam-beam interaction are more complicated. Expressions calculated from the multipole expansion of the long-range beam-beam kick are given in Ref.[11]. Other theoretical issues related to hadron beam-beam interactions can be found in references[13].

3 Experience at CERN

3.1 ISR Operation

The ISR was a high current, high luminosity collider consisting of two interleaved rings [14] [15]. It collided unbunched beams. Protons were brought into collision in 1971. Later operations stored alpha particles, deuterons and antiprotons [16] [17]. Beam energy was variable between 26 Gev and 31 GeV. High luminosity was possible when the beams were stochastically cooled. The maximum beam current was 60 Amps producing the maximum luminosity of 1.3×10^{32} cm^{-2}s^{-1} [18].

ISR beams crossed horizontally with an angle of 14.77° at 8 interaction regions around the ring. There was no tune shift in the horizontal plane. The vertical tune shift was of order 0.001 per crossing. The fractional tune (working point) was normally chosen to be between the 7th and the 9th-order resonances where the beam-beam interaction did not affect the beam lifetimes [21].

Tune modulation, which plays an important role in bunched beam colliders, did not influence the beam behaviour in the ISR. In a debunched beam there is no mechanism for tune modulation other than external sources. Review articles on ISR can be found in Ref.[19] and Ref.[20].

3.2 Beam-Beam Interactions in the SPS

The CERN SPS was the first hadron collider that operated with bunched beams (first operation in 1981). At the end of the second run (1982) a peak luminosity of 5.3×10^{28} cm^{-2}s^{-1} was achieved. In subsequent runs the peak luminosity was increased, culminating in peak luminosities consistently above 2×10^{30} cm^{-2}s^{-1} during 1988-89 operation. Prior to the 1987 run 3 proton bunches collided head-on with 3 antiproton bunches. Horizontal separators were installed for the 1987-1988 runs, allowing 6 bunches per beam and 3 head-on interaction regions.

Early beam-beam experiments in the SPS (a single antiproton bunch colliding head-on with 3 proton bunches, no separators) showed that the antiproton intensity lifetime is very sensitive to the tune. In particular, the experiment demonstrated that the 7th-order resonance was excited by beam-beam interactions [18]. Under ideal head-on conditions only even-order resonances are excited. The excitation of an odd resonance can be explained as follows. The average tunes and therefore the closed orbits of the proton and antiproton beams were different during this experiment, causing a small displacement at the interaction regions. Beam-beam interactions of transversely displaced beams excite odd-order resonances. Small residual dispersions at the interaction regions also excite odd resonances.

The 7th-order resonance did not affect the proton intensity lifetime significantly in this experiment (Fig.5). Lattice nonlinearities affect proton and antiproton beams equally. If the proton intensity lifetime is longer than that of the antiproton, one suspects another source of nonlinearity —the beam-beam interaction. From this it was concluded that the SPS lattice nonlinearities did not significantly excite the 7th-order resonance. The early beam-beam experiments at the SPS also demonstrated that the 10th-order resonance was excited by the head-on beam-beam interactions. These results established the operating point (unshifted, fractional tunes) for the SPS to be near 0.68, a safe distance from 0.70 (10th-order resonance).

An important aspect of head-on beam-beam interactions is that the particles in the transverse tails sample a nonlinear force (see Fig.6) and thus become sensitive to resonances; while small-amplitude particles sample linear beam-beam forces, therefore suffer only a tune shift. The significance of the beam-beam nonlinearity was demonstrated in SPS experiments studying the weak-strong case. Normally, the antiproton (weak beam) emittance is lower than the proton (strong beam) emittance. It was shown that when the antiproton emittance was larger than the proton emittance (due to malfunction or deliberate intervention) the large-amplitude antiprotons diffused out faster. In one study [18], 3 antiproton bunches with successively larger emittances were injected into the SPS, and collided with 3 proton bunches of lower emittance. The antiproton bunch with the largest

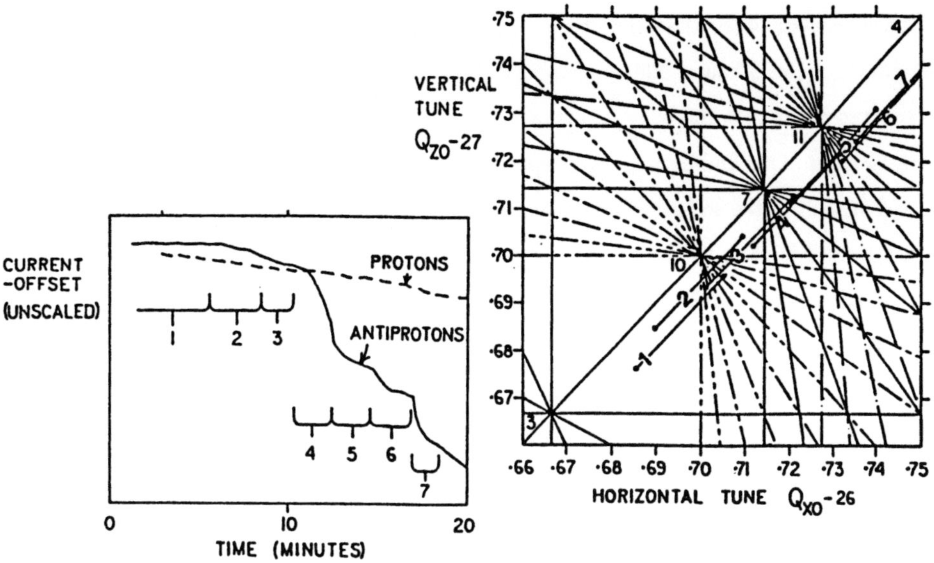

Figure 5: Measured proton and antiproton currents as a function of time in the early SPS beam-beam experiment.

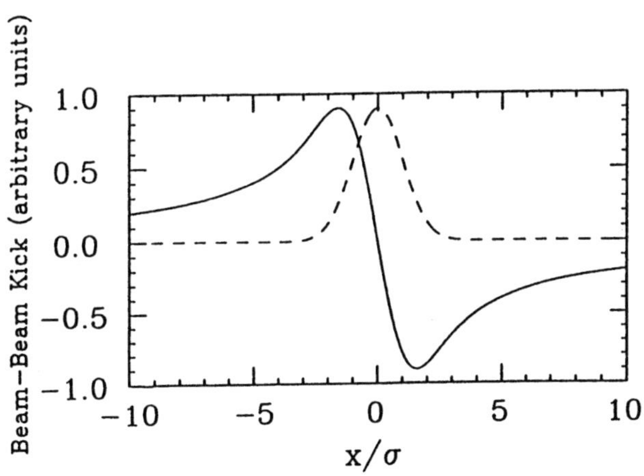

Figure 6: The kick experienced by an oppositely charged probe particle as it passes by a round Gaussian source bunch. The source bunch distribution is indicated with the dashed curve. The symbol σ is the rms size of the source bunch while x is the distance between the center of the source bunch and the probe particle.

emittance decayed faster than the other bunches initially, and its lifetime approached that of the other bunches after 5 hours. This phenomenon is sometimes refered to as "self-scraping", reflecting the observed effect that the beam emittance can actually decrease due to loss of particles from the transverse tails. This finding was later reproduced in the TEVATRON.

A variant of the self-scraping phenomenon is observed in proton beams when the antiproton bunch intensity is close to the proton bunch intensity (strong-strong case). In early SPS collider runs, the antiproton bunch intensity was typically 1.9×10^{10} particles per bunch while the proton bunch intensity was 15×10^{10} particles per bunch, clearly a weak-strong situation where the proton (strong) beam is not perturbed by the antiproton (weak) beam. In the 1988-89 SPS collider run the intensities were 5×10^{10} and 11×10^{10} for antiproton and proton bunches, respectively. This meant that the protons were significantly perturbed by the antiproton bunches, resembling the strong-strong case in e^+e^- colliders. In addition to a measurable beam-beam proton tune shift (0.0049 per crossing in the horizontal plane) there was also a decreased intensity lifetime for protons and very large background rates in the physics detectors[22].

This effect was caused by the difference in proton and antiproton beam emittances. What works in the weak-strong case does not work in the strong-strong case. In the weak-strong case the antiproton beam size had to be kept smaller than the proton beam size in order prevent antiproton losses from the tails. The tail antiprotons (large-amplitude particles) experience nonlinear forces from head-on beam-beam interactions. In the strong-strong case, if the proton emittance is larger than the antiproton emittance, the moderate-amplitude protons as well as the tail protons feel a nonlinear beam-beam force. This causes the diffusion of protons. Background rates decreased greatly when the proton emittance was comparable to the antiproton emittance.

Since the operating point of the SPS was chosen to avoid resonances lower than the 10th order, the proton diffusion in the case of unequal emittances must have been caused by higher-order resonances. Theory states that the resonance width of high-order resonances increases with amplitude (see Fig.3). The particles in the larger emittance beam have larger amplitudes, and therefore become more sensitive to higher-order resonances. It was determined that resonances of order 13 and 16 were affecting the large-amplitude protons[23]. This phenomenon was further studied in a series of experiments [24] which concluded that the linear beam-beam tune shift parameter is not sufficient to assess the strength of beam-beam effects in hadron colliders. Their conclusion was that the ratio of proton and antiproton emittances should be used in the parameterization.

More insight can be gained about relevant parameters from the threshold equation (Eq.16) which stresses the importance of vector addition. The magnitude of a resonance vector is equal to the beam-beam tune shift parameter for that interaction point. Simply adding the magnitudes of the vectors, i.e. using the total tune shift in the parametrization is not sufficient to describe the onset of chaos. The vector addition requires the knowledge of phase advances between the beam-beam kicks. Therefore the phase advances between the kicks should also be used in the parameterization: The threshold equation was used to study the combined effect of tune modulation and beam-beam interactions in the SPS in Ref.[8]. The results of this study regarding critical resonances are summarized in Fig.8

When the closed orbits of protons and antiprotons are separated the situation is completely different. Small amplitude particles as well as large amplitude particles can sample the nonlinear part of the beam-beam force. Fig.7 illustrates this situation. SPS data on long-range beam-beam interactions come from the 1988 and 1989 runs [22] [23] and experiments that investigated the effects of separation [25]. In one experiment a single proton bunch collided with two antiproton bunches. There were two head-on and two long-range interactions. The antiproton beam was scraped to achieve a small emittance in order to cause maximum effect on the proton beam. High background rates from large amplitude protons were the biggest concern in the SPS. This is why the proton beam was used as the probe.

In the SPS experiment, the proton emittance was larger than the antiproton emittance by a factor of 3, and intensities were 1.4×10^{10} and 7.5×10^{10} particles per bunch for antiproton and proton beams, respectively. At the long-range interaction points the separations were 5.9σ and 6.7σ when the separators were powered to full strength (100% separation). Tune scans were performed at 100% and 50% separations. During the tune scan with 50% separation, resonances of order 13 and 16 affected the background rates. Presence of the 13th proved that long-range beam-beam interactions can indeed excite the odd-order resonances. During the tune scan with 100% separation, proton intensity lifetime decreased from an initial value of 100 hours to 80 hours on the 16th and to 60 hours on the 13th. This demonstrated that the 13th order resonance can influence the tail particles even when beams are separated by 6σ.

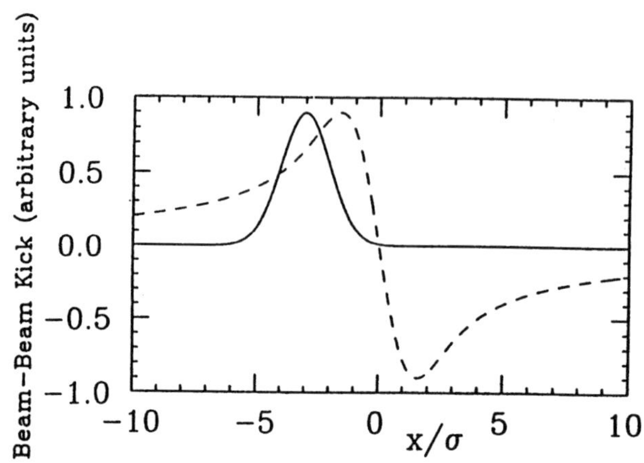

Figure 7: Long-range beam-beam interaction. The solid curve is the probe bunch positioned to ilustrate a 3σ separation between the centers of source and probe bunches. σ is still the rms size of the source bunch. Note that 1σ-amplitude particles in the probe bunch are sampling a nonlinear force.

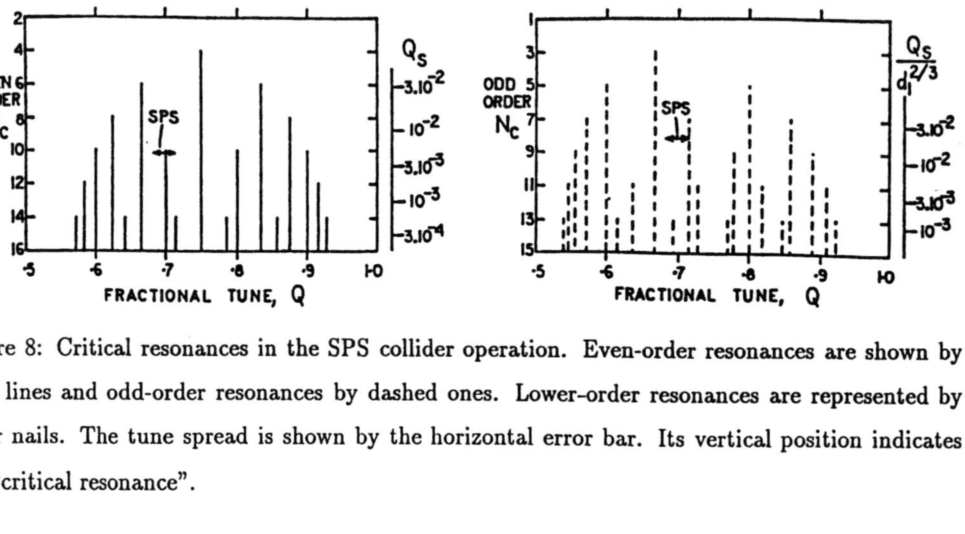

Figure 8: Critical resonances in the SPS collider operation. Even-order resonances are shown by solid lines and odd-order resonances by dashed ones. Lower-order resonances are represented by taller nails. The tune spread is shown by the horizontal error bar. Its vertical position indicates the "critical resonance".

4 Experience at FERMILAB

The first physics run of the TEVATRON collider was in 1987. During this run the luminosity lifetime was unexpectedly low (8 hours), due to transverse emittance growth rates of 8 πmm-mr/hr (95% definition). The causes of this emittance growth were found and fixed. In the 1988-89 TEVATRON collider run the luminosity lifetime was 15 hours in the beginning of a typical 30 hours store, increasing to 40 hours at the end of the store. During the 1988-89 run a peak luminosity of

2×10^{30} cm^{-2} sec^{-1} was reached and a total integrated luminosity of 9600 nb^{-1} was delivered to the CDF detector [26] [27] [28].

4.1 TEVATRON 1988-89 Collider Run

The 1988-89 TEVATRON Collider Run involved only head-on beam-beam interactions. Six antiproton bunches collided with six proton bunches, at 12 crossing points symetrically distributed around the ring. Typical intensities were 7×10^{10} and 2.5×10^{10} particles per bunch, for protons and antiprotons, respectively. The normalized transverse proton emittance was typically 25 π mm-mr in both planes and the antiproton transverse emittance was typically 18 π mm-mr. The proton emittance was increased by artificial means to place the antiproton beam in the linear region of the beam-beam force. When the emittances were approximately the same the antiproton lifetime was shorter than the proton lifetime—antiprotons sampling the nonlinear part of the beam-beam force were influenced by resonances. This confirmed the self-scraping phenomenon observed in the SPS. By blowing up the proton emittance in a controlled manner the antiproton lifetime was improved, and a higher integrated luminosity was achieved. The unshifted horizontal and vertical tunes were near 19.41.

It is worth pointing out the difference between the operating points of the SPS and the TEVATRON colliders. The SPS operated between the 3rd and the 10th while the TEVATRON operated between 5th and 7th-order resonances. The SPS working space covers the 13th and the 16th while TEVATRON's covers the 12th and the 17th-order resonances. Fig.9 shows the tune spreads at various phases of the collider operations in the SPS and the TEVATRON.

The procedure of artificial enlargement of proton emittance caused proton loss from tails in the SPS while it was beneficial in the TEVATRON. The difference can be explained by the following facts.

In the SPS protons could not avoid the 13th and the 16th which were observed to be important for the large amplitude particles sampling the nonlinear beam-beam forces. In the TEVATRON, the proton tune spread was such that the large-amplitude protons did not touch the 12th. They touched the 17th-order resonance—the higher the resonance order the lower the impact on beam lifetime.

The TEVATRON under storage conditions exhibits less tune modulation than SPS. In the TEVATRON a single low voltage power supply provides the current to all superconducting magnets, except the low-beta quadrupoles. Also, the effect of tune modulation due to synchrotron oscillations is smaller since the synchrotron tune of TEVATRON at 900 GeV is 0.0008, much smaller than the

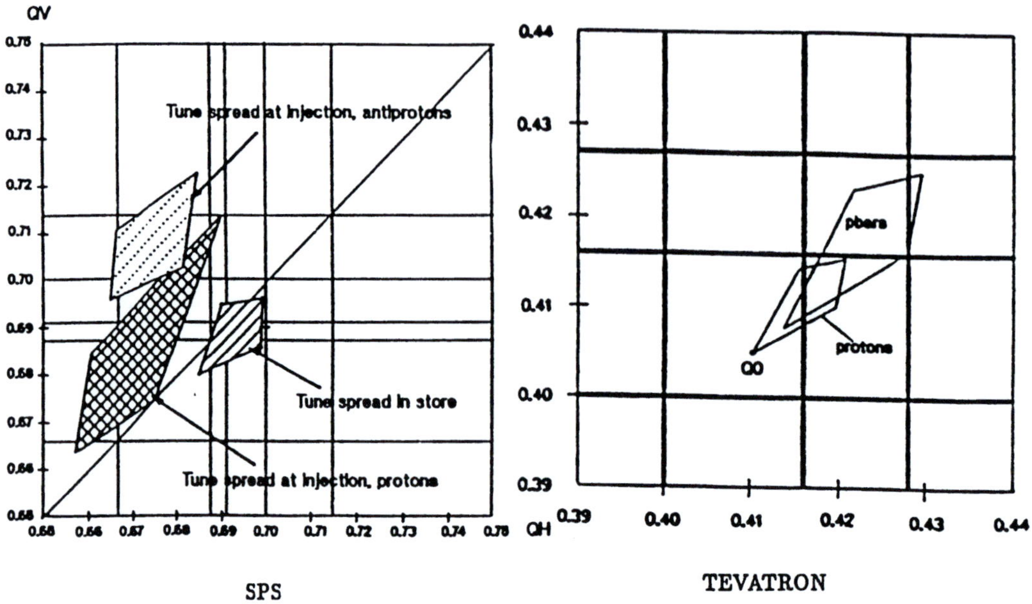

Figure 9: Tune spreads [23].

SPS value of 0.004. The combined effect of tune modulation and beam-beam interaction in the TEVATRON was studied in Ref.[29].

In the TEVATRON the tunes of protons and antiprotons were varied between the 5th and the 7th-order resonances to find the optimum working point. During this run there was no mechanism to control proton and antiproton tunes independently. From Fig.9 it can be seen that when the tunes were moved up antiprotons touched the 7th-order resonance, and when the tunes were moved down protons touched the 5th-order resonance. These resonances affected the intensity lifetimes. The proton lifetime decreased while the antiproton lifetime was unaffected when the tunes were lowered towards 0.4. The opposite effect was observed when the tunes were moved up towards 0.4286.

The excitation of the 7th-order resonance is dominated by the beam-beam interaction. The contribution from the TEVATRON lattice is negligible[30]. On the other hand the 5th-order resonance is significantly excited by the TEVATRON lattice.

In both SPS and TEVATRON a beam-beam tune spread of 0.025 proved to be the limiting value. Operational experience dictates that the tune spread must be kept smaller than the tune space between the limiting resonances. For stored beams the limiting resonances for the TEVATRON were 2/5 and 3/7, a tune space of 0.0286. In the SPS the limiting resonances were 2/3 and 7/10 , a tune space of 0.0400. However, the width of the 3rd-order resonance is large in the SPS, since it is a

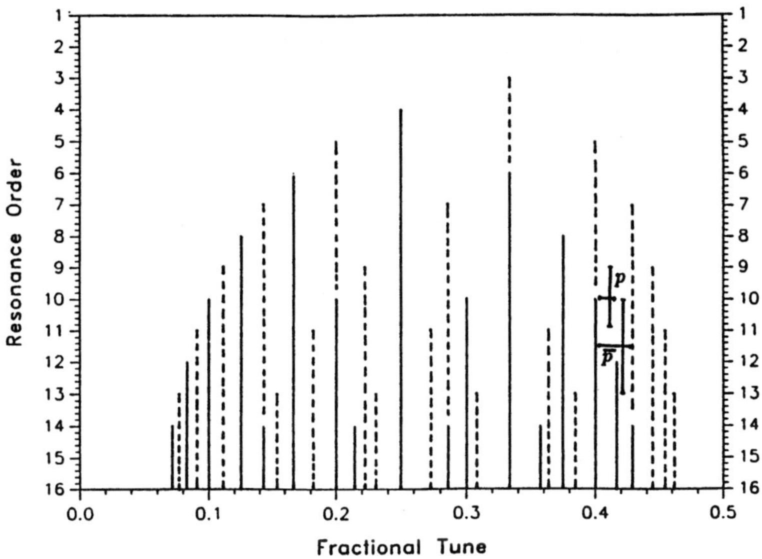

Figure 10: Critical resonances during the TEVATRON 1988-89 collider operation. The vertical error bar shows the range of critical resonances for particles in the range $\alpha = 2$ to $\alpha = 3$.

low-order resonance, driven to first order in sextupole strength. Therefore the SPS could not use the whole 0.0400 tune space. Similarly the widths of 2/5 and 3/7 resonances in the TEVATRON force the total tune spread to be less than 0.025. (See Fig. 10.)

4.2 Long-Range Beam-Beam Interactions in the TEVATRON

In the following sections we will review the TEVATRON experiments that concentrated on the long-range beam-beam interactions. Details of these experiments can be found in Ref.[31]. All experiments were conducted at 150 GeV using the fixed-target lattice configuration. Low-beta quadrupoles were not turned on. Two modules of electrostatic separators were available, at B17 and C48, providing 85μrad horizontal and vertical kicks, respectively. Since the phase advance between the B17 and C48 locations is approximately an integer multiple of 90° the orbits were helical, providing separation everywhere. The relative strength of these separators was not changed during the experiments. The helix amplitude and therefore the beam separation was changed by varying both separator voltages together.

When beams are separated in the TEVATRON they go through the chromaticity sextupoles off-axis and experience quadrupole fields (feed-down effect). Also, the superconducting TEVATRON

% Helix	Separation	Proton Lifetime [hrs]	Antiproton Lifetime [hrs]
0	0.0 σ	32.46 ± 0.50	-
20	0.6 σ	39.78 ± 1.09	7.69 ± 1.64
40	1.2 σ	48.43 ± 2.05	3.19 ± 0.22
60	1.8 σ	49.61 ± 5.50	1.79 ± 0.05
80	2.4 σ	-	6.92 ± 0.71
100	3.0 σ	48.96 ± 1.87	15.32 ± 3.08

Table 1: Lifetimes during the June 1989, 3x1 experiment.

magnets have a large sextupole moment (b_2) which dominates the feed-down effect [32]. To control the proton and antiproton tunes independently, "feed-down" sextupole circuits were instrumented in the TEVATRON. These "feed-down" sextupoles have no effect when beams circulate on the central orbit.

4.2.1 June 1989, TEVATRON 3x1 Experiment

A single antiproton bunch collided with 3 proton bunches at 150 GeV. Initial proton intensity was 7×10^{10} particles per bunch. Initial emittances were 20 and 10 πmm-mr for protons and antiprotons, respectively. The average value of β at 6 crossing points was approximately 31 meters and 91 meters in the horizontal and vertical planes, respectively. After injection separators were powered to 100% Helix, which corresponded to 3σ average separation. Beam lifetimes, emittance growth rates and tunes were measured at different separations. Table 1 reproduces the intensity lifetime data [31].

At 60% Helix (1.8σ average separation) the antiproton lifetime was significantly decreased. The beam-beam interaction is most nonlinear when the separation is near 1.6σ, therefore one expects a lifetime effect at 1.8σ separation if the tune is on one of the beam-beam resonances. The antiproton tune at 60% Helix was measured to be 0.412, close to the 5/12 resonance. However, the tunes were not adjusted by the feed-down sextupoles during this experiment. This meant that the beam-beam tune shift, and hence the instantaneous tunes, were changing according to the changes in beam intensity and emittance. For instance, the proton tune was 0.42, very different from the antiproton tune. Therefore, 3x1 data were inconclusive about the excitation of the 12th by long-range beam-beam interactions.

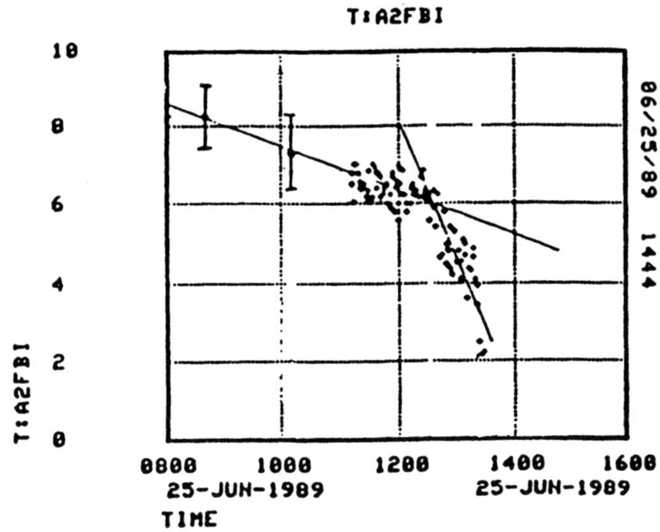

Figure 11: Antiproton intensity (particles per bunch, 10^9 scale) during the June 1989, 6x1 experiment.

4.2.2 June 1989, TEVATRON 6x1 Experiment

The 6x1 configuration was studied to answer questions regarding operational conditions that will exist in the 1992 Collider Run. Six proton bunches collided with a single antiproton bunch. Instead of a separation scan a tune scan was conducted. In other words the beam separation was held fixed at 100% (3σ average separation) and the tunes were adjusted using the feed-down sextupoles. Antiproton bunch intensity data from this experiment are presented in Fig.11.

The unperturbed bunch intensity lifetime is 13 hours in the TEVATRON at 150 GeV. This lifetime is caused by particles falling out of the rf buckets. With horizontal and vertical tunes sitting on the 12th order resonance and with 3σ average separation the antiproton lifetime was measured to be 13 hours. No lifetime effect due to the excitation of the 12th-order resonance by long-range beam-beam interactions was observed.

This experiment suggests that the long-range beam-beam interactions will be benign at 3σ separation with the nominal 1992 beam parameters.

4.3 TEVATRON April 1990 Studies

The goal of this study period was to study long-range beam-beam interactions under the conditions that will exist during the 36x36 operation. Ideally, 36 proton bunches would collide with 36 antiproton bunches. In reality there will be 34 bunches per beam, as dictated by the abort kicker timing requirements. Proton and antiproton bunch intensities will be 50 $\times 10^{10}$ and 3×10^{10} particles per bunch, emittances will be 30 and 22 πmm-mr for protons and antiprotons, respectively. During April 1990 studies, 34 proton bunches collided with a single antiproton bunch. At 100% Helix, the average separation was 4.5σ. The helix was collapsed from 100% to 0% in 20% steps. At each step, total beam current, proton and antiproton bunch intensities, emittances, beam sizes and tunes were measured.

4.3.1 34x1 Tune Shift Measurements

The tune shift data are presented in Table 2. The relevant beam parameters are summarized in Table 3. The measurement of antiproton tunes was difficult, due to low bunch intensities. In order to increase the signal to noise ratio the antiproton bunch was excited coherently in the horizontal plane, using the TEVATRON Superdamper System, and tunes were read from the Schottky plate signals using a spectrum analyzer. Tunes were cross-checked with calculation by using the known currents from the feed-down circuits. Since the feed-down circuits have no effect at 0% Helix, the antiproton tunes were outside the working space (i.e. greater than 19.4286). (See Fig. 12.)

Vertical excitation of the antiproton bunch was not possible due to problems associated with the vertical superdamper kicker. Proton and antiproton tunes were differentiated by turning on the feed-down circuit and watching the tune lines for protons and antiprotons move in opposite directions. The beam-beam tune shifts were obtained by subtracting bare tunes (tunes measured when there is no beam-beam interaction while the feed-down circuit is on) from actual tunes.

Tune shifts were also simulated using the beam-beam code HOBBI [33]. The results are compared to data in Fig.13.

% Helix	separation	Horz.Tune Shift (antiproton)
100	4.5σ	0.0025
80	3.6σ	-
60	2.7σ	0.0045
40	1.8σ	0.0131
20	0.9σ	0.0213
0	0.0σ	0.0425

Table 2: Antiproton tune shift data from April 1990, 34x1 experiment.

% Helix	ε_{p_x} [πmm-mr]	ε_{p_y} [πmm-mr]	$\varepsilon_{\bar{p}_x}$ [πmm-mr]	$\varepsilon_{\bar{p}_y}$ [πmm-mr]	protons/bunch [$\times 10^{10}$]
100	9.7	13.0	6.5	15.0	3.52
80	10.0	13.0	7.5	16.5	2.64
60	10.0	13.0	8.0	17.0	2.50
40	11.3	15.2	8.5	17.5	2.35
20	11.3	17.6	9.0	18.0	2.20
0	11.3	17.6	9.5	18.5	2.05

Table 3: Beam parameters used in the 34x1 simulation. Emittance values are given according to a 95% definition.

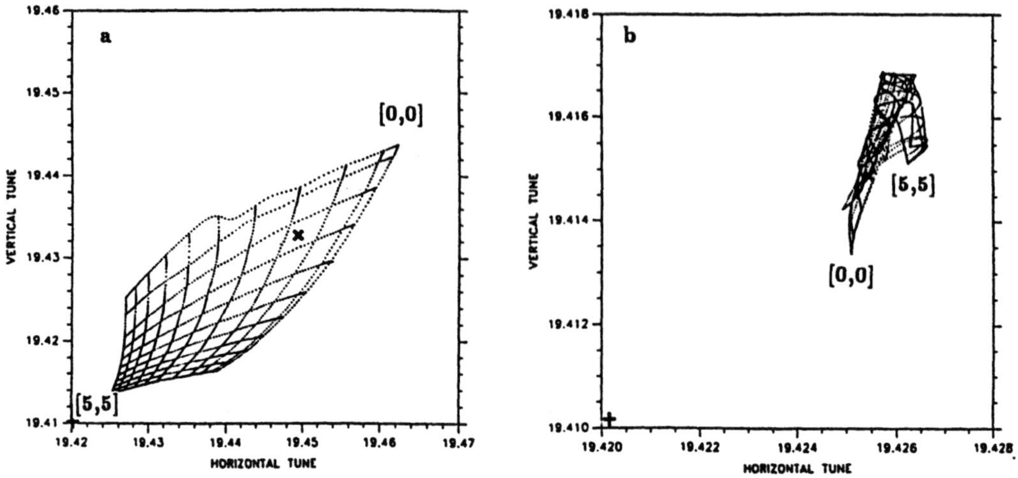

Figure 12: Tune shift footprint at a) 0% Helix and b) 60% Helix as calculated by HOBBI for the April 1990, 34x1 experiment. The footprint is generated for a mesh of amplitudes. The [0,0] and [5,5] refer to normalized amplitudes representing "small" and "large" amplitudes, respectively.

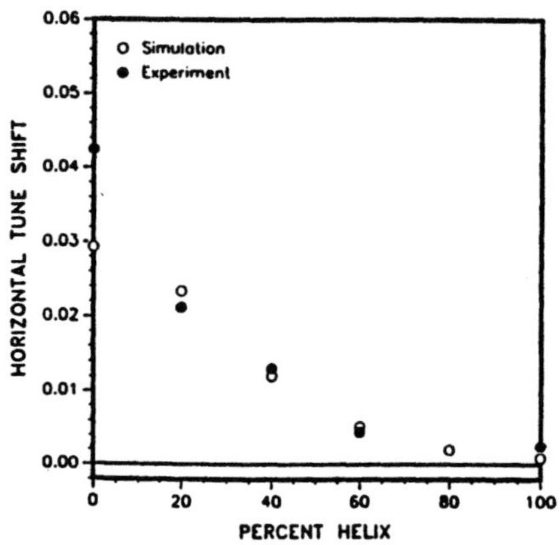

Figure 13: Comparison of tune shifts.

4.3.2 34x1 Lifetime and Emittance Measurements

The difference between proton and antiproton lifetimes was carefully observed during this experiment. Proton and antiproton tunes were kept on top of each other using the feed-down circuits. Therefore, any difference in lifetimes was due to long-range beam-beam interactions. Lifetime effects showed up at 60% and 40% Helix. Fig.14 and Fig.15 show the beam intensities as a function of time.

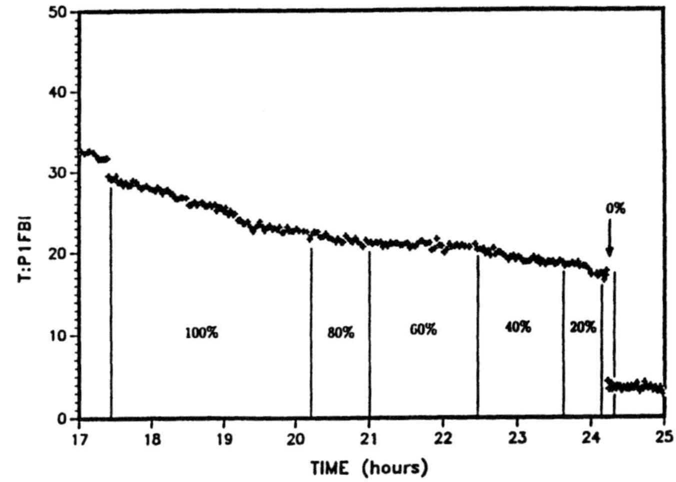

Figure 14: Proton intensity during the April 1990, 34x1 experiment. P1FBI is the number of protons in the P1 bunch (10^9 scale).

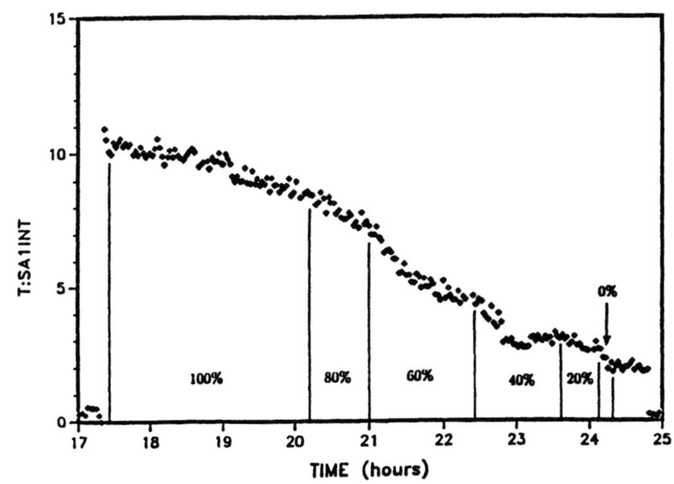

Figure 15: Antiproton intensity during the April 1990, 34x1 experiment. SA1INT is the number of antiprotons in the A1 bunch (10^9 scale).

The agreement between simulation of tuneshift and data is very good with separated orbits. The discrepancy in the case of 0% Helix (68 head-on collisions) is striking. It should be mentioned that the spectrum analyzer measurement of the antiproton tune at 0% Helix could not be confirmed using the feed-down currents. A plausible but unconfirmed explanation is that the tune corresponding to the π-mode of the coherent beam-beam oscillations was observed. Recall that the antiproton bunch was excited coherently during the experiment. Coherent beam-beam oscillations in the TEVATRON were observed[34] during the 1988-89 collider run.

The lifetime measurements showed clearly that at 60% Helix (2.7σ average separation) the antiproton lifetime decreased drastically. One does not expect a drastic lifetime effect at 2.7σ because in the simple beam-beam kick curve (Fig.6) this region is almost linear. One should also remember however that Fig.6 refers to a single interaction and that the separation is average. When the actual separation is taken into account for each interaction point the tuneshift-footprint shown in Fig.12.b results. This diagram illustrates a flip in the tuneshifts for small- and large-amplitude particles. When all beam crossings involve head-on interactions (0% Helix, Fig.12.a), small amplitude particles experience larger tune shifts. This relationship persists at 20% Helix and 40% Helix, but at 60% Helix the trend is reversed, and larger-amplitude particles experience larger tune shifts. How is this related to the beam lifetime? A plausible answer lies in connecting beam lifetimes and resonance widths. Beam-beam resonance widths are inversely proportional to the square root of the slope of the detuning curve Eq.(5). The detuning curve resulting from Fig.12.b would be very distorted, facilitating large resonance widths and, plausibly, short lifetimes.

Fig.15 shows that at 40% Helix (1.8σ average separation) there was also a lifetime effect. This effect can be attributed to the average separation being near 1.6σ, where the beam-beam kick is most nonlinear.

The 34x1 experiment has shown that the 36x36 scenario is viable provided the average separation is kept above 5σ separation. The proton bunch intensity will be much higher in the Collider Run with the Main Injector. This condition was studied in simulations which conclude that there is a beam-beam upper limit for proton bunch intensity, 40×10^{10} particles per bunch, coupled with a lower bound on proton emittance, 32 πmm-mr [35].

5 Conclusions

The beam-beam issues in three hadron colliders were reviewed. The beam-beam interaction was not a major problem in the ISR since it collided unbunched beams. The performance of bunched beam colliders, however, was seriously limited by the beam-beam interaction. Observations at the CERN SPS showed that the beam-beam interaction excites resonances up to order 16. The SPS experiments demonstrated the "self-scraping" phenomenon which was reproduced in the TEVATRON later. Other beam-beam experiments in the SPS showed the detrimental effects of having different emittances for protons and antiprotons, and the excitation of the 13th-order resonance in the case of separated beams.

Experiments on long-range beam-beam interactions in the TEVATRON have shown that there are two different processes that lead to beam loss. The first one occurs when the average separation is between 1.5-2σ. Particles experience very nonlinear fields and the resonances are excited. The second one occurs when the tune shift footprint exhibits a flip. Larger-amplitude particles experience larger tune shifts. The distorted shape of the detuning curve enlarges the resonance width.

Special attention was paid to the 12th-order resonance in the TEVATRON experiments. At 150 GeV, with proton and antiproton tunes sitting on the 12th and with 3σ average separation, no lifetime effect was observed in the 6x1 experiment.

The long-range beam-beam experiments in the TEVATRON also demonstrated that in the 6x6 mode, with nominal intensities and emittances given for the 1992 collider run, even 3σ average separation can be tolerated. In the 36x36 mode, with the nominal beam parameters that can be provided by the Fermilab Main Injector, the average separation must be kept above 5σ.

References

[1] FERMILAB Design Report: FERMILAB III Accelerator Upgrade, Phase 1 (1990/94).

[2] E.D.Courant and H.S.Snyder,"Theory of Alternating Gradient Synchrotron", Ann.Phys., **3**, 1 (1958).

[3] L.Evans, CERN SPS/DI(MST)/Note/81-2 (1981);
L.Evans, CERN SPS/83-38 (DI-MST) (1983);
L.Evans and J.Garayte, CERN SPS/82-8 (DI-MST) (1982);
L.Evans and J.Gareyte, IEEE Trans.Nucl.Sci, **NS-30**, 4 (1982);
L.Evans, "The Beam-Beam Interaction", CERN SPS/83-38 (DI-MST) (1983).

[4] S.Peggs and R. Talman, "Nonlinear Problems in Accelerator Physics", Ann.Rev.Nucl.Part.Sci., **36**, 287 (1986).

[5] D.Brandt et al,"Influence of Power Supply Ripple on the Dynamic Aperture of the SPS in the Presence of Strong Nonlinear Fields",
Proc. 1990 European Particle Accel. Conf. (Nice, France), p:1438.

[6] M.Month, BNL 19583 (1975).

[7] J.Tennyson, in Physics of High Energy Particle Accelerators, AIP Conf.Proc. No:87, p:345 (1981).

[8] S.Peggs,"Hadron Collider Behavior in the Nonlinear Numerical Model EVOL",
Particle Accelerators, **17**, 11 (1985).

[9] B. Chrikov, Phys.Repts., **52**, 263 (1979).

[10] Jejcic and J.Le Duff, VIII Int.Acc.Conf. CERN (1971).

[11] D.Neuffer and S.Peggs, "Beam-Beam Tune Shifts and Spreads in the SSC—Head On, Long Range, and PACMAN Conditions", SSC-63 (1986).

[12] L.Michelotti, "Beam-Beam Interaction Effects", FERMILAB TM-1738 (1989).

[13] L.Michelotti and S.Saritepe,"Exploratory Orbit Analysis of Tevatron Helical Upgrade:A First Look", FERMILAB TM-1603 (1989);
D.E.Johnson et al,"Lattice Function Perturbations Caused by the Beam-Beam Interaction", Proc. 1991 Particle Accel.Conf. (San Francisco);
D.E.Johnson,"Linear Beam-Beam Tune Shift Calculations for the Tevatron Collider", FERMILAB TM-1554 (1989) ;
S.R.Mane,"Beam-Beam Tuneshift During the Tevatron Squeeze", FERMILAB TM-1550 (1988);
M.Meddahi and R.Schmidt,"Calculation of the Tune Spread Induced by Beam-Beam Effects in the case of Partially Separated Beams",
CERN-SL/90-15 (AP) (1990).

[14] A.Hofmann, Proc. 1979 Workshop on Beam Current Limitations in Storage Rings, p:12, Ed:C.Pellegrini, BNL 51236 .

[15] J.Billan et al., IEEE Trans.Nucl.Sci., **NS-30**, 2036 (1983).

[16] P.Asboe-Hansen et al., IEEE Trans.Nucl.Sci., **NS-24**, 1557 (1977).

[17] P.J.Bryant, IEEE Trans.Nucl.Sci., **NS-30**, 2047 (1983).

[18] L.Evans,"Beam Effects in Hadron Colliders", in "Physics of High Energy Particle Accelerators", AIP Conf.Proc. No:127, p:244.

[19] E.Keil, CERN 72-14 (1972).

[20] K.Hubner, CERN 77-15 (1977).

[21] J.Borer et al, IEEE Trans.Nucl.Sci., **NS-28**, 2180 (1981).

[22] L.Evans et al, "Beam-Beam Effects in the Strong-Strong Regime at the CERN SPS", Proc. 1989 Particle Accel.Conf. (Chicago), p:1403

[23] M.Harrison and R.Schmidt, "The Performance of Proton Antiproton Colliders", Proc. 1990 European Particle Accel.Conf. (Nice), p:55.

[24] K.Cornelis, M.Meddahi and R.Schmidt, "The Beam-Beam Effect in the SPS Proton Antiproton Collider for Beams with Unequal Emittances", Proc. 1990 European Particle Accel.Conf. (Nice), p:1670.

[25] K.Cornelis, M.Meddahi and R.Schmidt, "Experiments on the Beam-Beam Effect in the CERN-SPS in the 1989 Collider Run", CERN-SPS/AMS/Note 89-13 (1989).

[26] D.Herrup, D.Finley and G.Jackson, "Luminosity Lifetime in the Tevatron Collider", Proc. 1989 Particle Accel. Conf. (Chicago), p:1824

[27] V.Bharadwaj et al,"Coupled-Bunch Instability and Longitudinal Emittance Growth in the Fermilab Booster", Proc. 14th Intl.Conf.High Energy Accel. (Tsukuba, Japan), p:537 (1989)

[28] G.Dugan, "Tevatron Collider:Status and Prospects", Proc. 14th Intl.Conf.High Energy Accel. (Tsukuba, Japan), p:121 (1989).

[29] S.Saritepe and S.Peggs,"Tune Modulated Beam-Beam Resonances in the Tevatron", Proc. 1991 Particle Accel.Conf. (San Francisco).

[30] G.P.Jackson, Private Communication.

[31] S.Saritepe et al, "Experimental Studies of Long-Range Beam-Beam Interactions in the Tevatron", FERMILAB EXP-175 (1991).

[32] G.Goderre and E.Malamud,"Helical Orbit Studies in the Tevatron", Proc. 1989 Particle Accel. Conf. (Chicago), p:1818.

[33] S.Saritepe, S.Peggs and L.Michelotti, "Long-Range Beam-Beam Interactions in the Tevatron: Comparing Simulation to Tune Shift Data", Proc. 1990 European Particle Accel. Conf. (Nice, France), p:1667.

[34] G.P.Jackson, "Tune Spectra in the Tevatron Collider", Proc. 1989 Particle Accel. Conf. (Chicago), p:860.

[35] S.Saritepe, "Future Collider Runs in the Tevatron: Beam-Beam Simulation Results", FERMILAB FN-563 (1991).

BEAM-BEAM EFFECTS IN ELECTRON-POSITRON STORAGE RINGS

A.A. Zholents
Institute of Nuclear Physics, 630090 Novosibirsk, USSR

INTRODUCTION

During the past 25 years, 15 electron-positron storage rings have been built and commissioned around the world, and substantial progress in luminosity has been achieved. Now a design considered for a new storage ring has gained approval both theoretically and experimentally in all aspects except one. The exception is the problem of beam-beam interaction, which is the subject of this paper. Our knowledge of this phenomenon is fairly extensive, but not at a level of good understanding. Why the beam-beam limit is so small for most existing storage rings is still unexplained.

The lack of good theory is hard to compensate, and this review of beam-beam effects is no exception. The choice of material for review and of the relative importance of different questions was made intuitively rather than on some systematic basis. For a more complete study of beam-beam effects, additional information is needed. Some excellent reviews are listed in the references.

1 EVALUATION OF THE BEAM-BEAM KICKS

Here we find the integrated transverse angular deflection received by a particle crossing a charged beam. The coordinate systems used below are shown in Fig. 1.

We assume that the distribution of the charge, Ne, in the oncoming bunch is Guassian in all three dimensions:

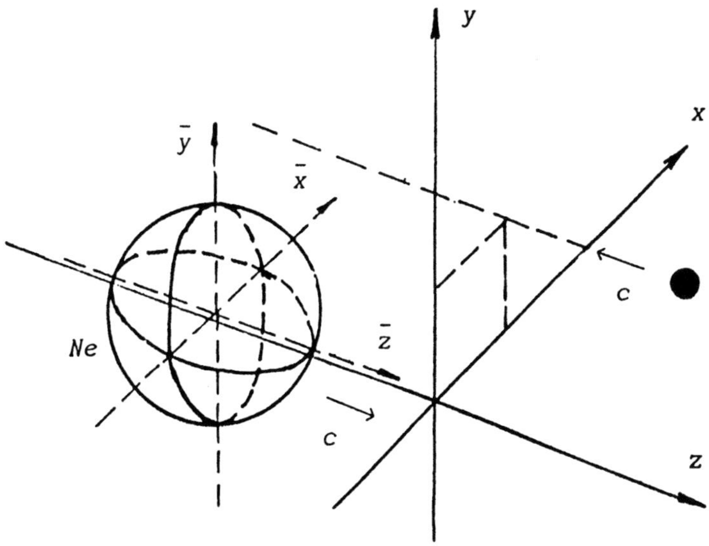

Fig. 1. The coordinate systems.

$$\rho(\bar{x},\bar{y},\bar{z}) = \frac{Ne}{(2\pi)^{3/2}abc} \exp\left\{-\frac{\bar{x}^2}{2a^2} - \frac{\bar{y}^2}{2b^2} - \frac{\bar{z}^2}{2c^2}\right\} \tag{1.1}$$

where a, b, and c are the standard deviations in $\bar{x}$, $\bar{y}$, and $\bar{z}$.

It has been shown[1] that the electric potential produced by this distribution is

$$U(\bar{x},\bar{y},\bar{z}) = \frac{Ne}{\sqrt{\pi}} \int_0^\infty \frac{\exp\left\{-\frac{\bar{x}^2}{2a^2+t} - \frac{\bar{y}^2}{2b^2+t} - \frac{\bar{z}^2}{2c^2+t}\right\} dt}{\left[(2a^2+t)(2b^2+t)(2c^2+t)\right]^{1/2}} . \tag{1.2}$$

The evaluation of the potential is given in Appendix A. The transverse electric and magnetic fields of this bunch moving with the speed of light c along the z-axis are given by

$$E_x = -\gamma\frac{\partial U}{\partial x} = \frac{2Ne}{\sqrt{\pi}}\gamma x \int_0^\infty dt \frac{\exp\left\{-\frac{x^2}{2\sigma_x^2+t} - \frac{y^2}{2\sigma_y^2+t} - \frac{\gamma^2(z+c\tau)^2}{\gamma^2 2\sigma_z^2+t}\right\}}{(2\sigma_x^2+t)\left[(2\sigma_x^2+t)(2\sigma_y^2+t)(\gamma^2 2\sigma_z^2+t)\right]^{1/2}}, \tag{1.3}$$

$$E_y = -\gamma\frac{\partial U}{\partial y} = \frac{2Ne}{\sqrt{\pi}}\gamma y \int_0^\infty dt \frac{\exp\left\{-\frac{x^2}{2\sigma_x^2+t} - \frac{y^2}{2\sigma_y^2+t} - \frac{\gamma^2(z+c\tau)}{\gamma^2 2\sigma_z^2+t}\right\} dt}{(2\sigma_y^2+t)\left[(2\sigma_x^2+t)(2\sigma_y^2+t)(\gamma^2 2\sigma_z^2+t)\right]^{1/2}}, \tag{1.4}$$

$$H_x = E_y , \tag{1.5}$$

$$H_y = E_x , \tag{1.6}$$

where γ is the Lorentz factor and σ_x, σ_y, σ_z are the standard deviations in the laboratory frame.

It is easy to calculate the transverse kicks by integrating the corresponding component of the Lorentz force over time τ:

$$\Delta x' \equiv \frac{\Delta P_x}{P} = -\int_{-\infty}^{\infty} \frac{d\tau e(E_x - H_y)}{P} \approx -\frac{2e}{\gamma mc}\int_{-\infty}^{\infty} d\tau E_x , \tag{1.7}$$

$$\Delta y' \equiv \frac{\Delta P_y}{P} = -\int_{-\infty}^{\infty} \frac{d\tau e(E_y - H_z)}{P} \approx -\frac{2e}{\gamma mc}\int_{-\infty}^{\infty} d\tau E_y . \tag{1.8}$$

We first assume that the variations of σ_x and σ_y with z in the interval $\pm\sigma_z$ are very weak, and we find

$$\begin{Bmatrix}\Delta x' \\ \Delta y'\end{Bmatrix} = -\frac{2Nr_e}{\gamma}\begin{Bmatrix}x \\ y\end{Bmatrix}\int_0^\infty dt \frac{\exp\left\{-\frac{x^2}{2\sigma_x^2+t} - \frac{y^2}{2\sigma_y^2+t}\right\}}{(2\sigma_{x,y}^2+t)\left[(2\sigma_x^2+t)(2\sigma_y^2+t)\right]^{1/2}} \tag{1.9}$$

where r_e is the classical electron radius.

The beam-beam effects associated with the finite bunch length are considered in Section 5. Equations (1.9) can also be written[2] as

$$\Delta x' = -\frac{Nr_e}{\gamma}\sqrt{\frac{2\pi}{\sigma_x^2-\sigma_y^2}}\,\mathrm{Im}\,\{F(x,y)\}\,, \tag{1.10}$$

$$\Delta y' = -\frac{Nr_e}{\gamma}\sqrt{\frac{2\pi}{\sigma_x^2-\sigma_y^2}}\,\mathrm{Re}\,\{F(x,y)\}\,, \tag{1.11}$$

where condition $\sigma_x > \sigma_y$ is assumed and

$$F(x,y) = W\left(\frac{x+iy}{\sqrt{2(\sigma_x^2-\sigma_y^2)}}\right) - \exp\left\{-\frac{x^2}{2\sigma_x^2}-\frac{y^2}{2\sigma_y^2}\right\} W\left(\frac{x\,(\sigma_y/\sigma_x) + iy\,(\sigma_x/\sigma_y)}{\sqrt{2(\sigma_x^2-\sigma_y^2)}}\right). \tag{1.12}$$

Here W( ) is the complex error function.

In the two extreme cases of a round beam cross-section $(\sigma_x = \sigma_y)$ and a very flat beam cross-section $(\sigma_x >> \sigma_y)$ the formulae (1.10) and (1.11) can be simplified significantly. Direct application of Gauss' law for evaluation of E_x, E_y gives, for round beams,

$$\Delta r' = -\frac{2Nr_e}{\gamma}\frac{1}{r}\left[1-e^{-r^2/2\sigma^2}\right], \tag{1.13}$$

and, for flat beams in the region $|y| << \sigma_x$,

$$\Delta x' = \frac{2r_eN}{\gamma}\frac{\sqrt{2}}{\sigma_x}F_D\left(\frac{x}{\sqrt{2}\,\sigma_x}\right),$$

$$\Delta y' = \frac{2r_eN}{\gamma}\frac{1}{\sigma_x}\sqrt{\frac{\pi}{2}}\,e^{-x^2/2\sigma_x^2}\,\Phi\left(\frac{y}{\sqrt{2}\,\sigma_x}\right), \tag{1.14}$$

where $F_D(z) = e^{-z^2}\int_0^z e^{t^2}dt$ is the Dawson function and Φ( ) is the error function.

2 LINEAR APPROXIMATION

We can obtain several useful results if we consider small oscillations with $|x| << \sigma_x$ and $|y| << \sigma_y$. Equation (1.9) becomes

$$\Delta x' = -\frac{2Nr_e}{\gamma\sigma_x(\sigma_x+\sigma_y)}x\,, \qquad \Delta y' = -\frac{2Nr_e}{\gamma\sigma_y\,(\sigma_x+\sigma_y)}y\,. \tag{2.1}$$

In this approximation x and y motions are decoupled and we can consider one dimension only. Let it be the y dimension.

The linearization makes the beam-beam kick equivalent to the effect of a thin quadrupole with focal length f:

$$\frac{1}{f} = \frac{2Nr_e}{\gamma\sigma_y(\sigma_x + \sigma_y)}. \tag{2.2}$$

The transformation of the (y,y') vector from one beam-beam collision to the next is described by the matrix

$$\begin{bmatrix} \cos(\mu + \Delta\mu) & \tilde{\beta} \sin(\mu + \Delta\mu) \\ -\frac{1}{\tilde{\beta}} \sin(\mu + \Delta\mu) & \cos(\mu + \Delta\mu) \end{bmatrix} = \begin{bmatrix} \cos\mu & \beta \sin\mu \\ -\frac{1}{\beta}\sin\mu & \cos\mu \end{bmatrix} \begin{bmatrix} 1 & 0 \\ -\frac{1}{f} & 1 \end{bmatrix} \tag{2.3}$$

where β, $\tilde{\beta}$ are unperturbed and perturbed beta-functions at the collision point and μ is the phase advance of betatron oscillations between two interactions.

If the perturbed motion is stable, then

$$\cos(\mu + \Delta\mu) = \cos\mu - 2\pi\xi \sin\mu, \qquad \tilde{\beta}\sin(\mu + \Delta\mu) = \beta \sin\mu, \tag{2.4}$$

where

$$\xi_{x,y} = \frac{Nr_e\beta_{x,y}}{2\pi\gamma\sigma_{x,y}(\sigma_x + \sigma_y)}. \tag{2.5}$$

For $\Delta\mu << 1$, we find for a tune shift of the betatron oscillation

$$\Delta\nu = \xi. \tag{2.6}$$

Therefore the parameter ξ is often called the linear beam-beam tune shift. The relationship (2.6) is valid only for tune values not too near the half-integer. The exact dependence of $\Delta\nu$ on ξ and μ is shown in Fig. 2.[3]

The conclusion evident from this plot is that, for a fixed value of ξ, one can achieve a smaller $\Delta\nu$ by choosing a betatron tune slightly above a half-integer.

From Eq. (2.4) we also find[4]

$$\tilde{\beta} = \frac{\beta}{\sqrt{1 + 4\pi\xi(\cot\mu) - 4\pi^2\xi^2}}. \tag{2.7}$$

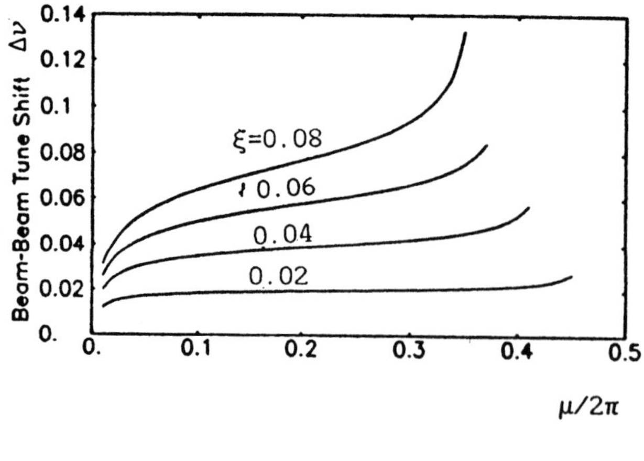

Fig. 2. Dependence of tune shift on ξ and μ.

The beam-beam force pinches the beta-function, and consequently the beam size at the collision point, to a smaller value if $\mu/2\pi < 0.1$. This is the so-called dynamic beta-function effect.[4] For large $\mu/2\pi$ the reverse is true. This means the luminosity would also benefit from having the betatron tune slightly above a half-integer. Several examples of dynamic beta behavior are shown in Fig. 3.

The above considerations lead to a recommendation for choosing betatron tunes slightly above a half-integer. SPEAR-I data[5] (Fig. 4) provide some experimental verification for this rule. Use of this strategy is not easy in practice, however, because of single-beam synchrobetatron resonances.

The beam-beam strength parameter ξ is much more relevant for calculating the luminosity than is the number of particles in the bunch N. The maximal ξ value corresponds to the maximal possible perturbation in the particle motion when it remains stable. Therefore a standard luminosity formula

$$L = f\,N^2/4\pi\sigma_x\sigma_y\,, \tag{2.8}$$

where f is the collision frequency, must be rewritten in the form, including ξ,

$$L = f\,\frac{\gamma N}{2r_e}\frac{\xi_y}{\beta_y}\left(1+\frac{\sigma_y}{\sigma_x}\right). \tag{2.9}$$

This form of the luminosity formula is consistent with experimental observations of luminosity and ξ behavior with beam intensity for many e^+e^- storage rings. (See for example CESR data in Fig. 5.[6])

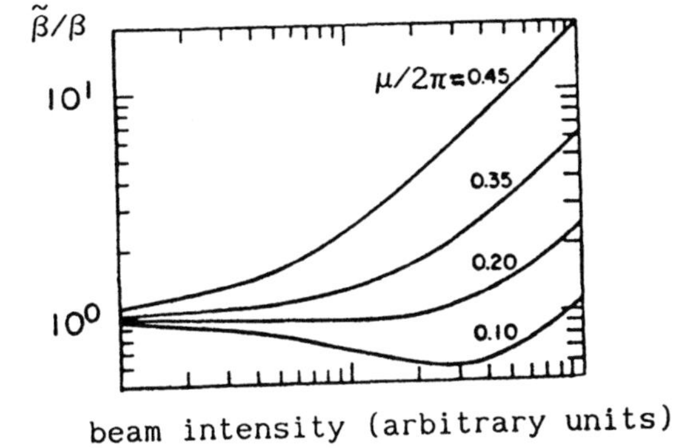

Fig. 3. The dynamic beta effect; μ is the unperturbed phase advance between collision points.

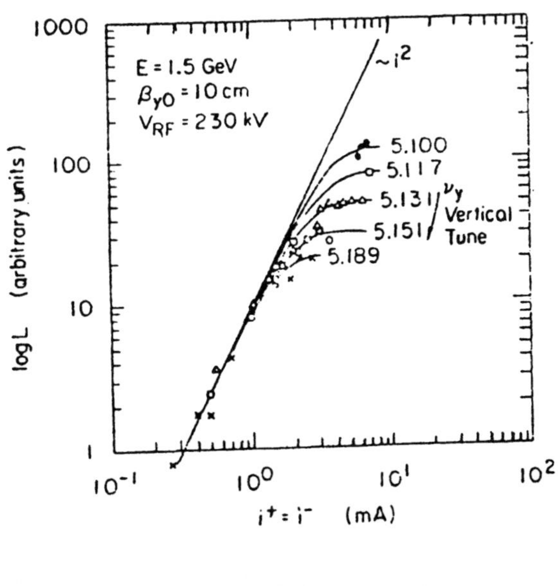

Fig. 4. SPEAR I data showing the dependence of tune and beta intensity on luminosity.

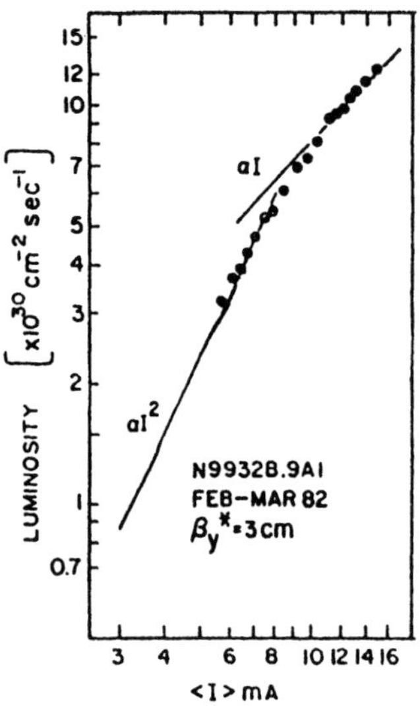

Fig. 5. Typical luminosity and ξ_y dependencies versus beam intensity for most e^+e^- storage rings.

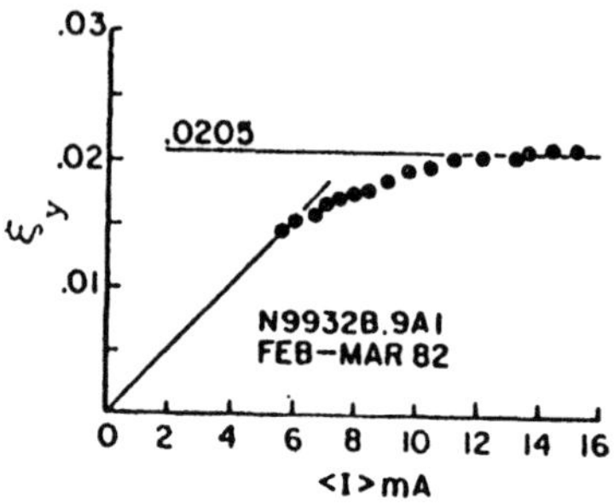

At low currents the luminosity is proportional to the current squared and the beam sizes are constant. At a certain current the vertical tune shift saturates, which forces the beam size product to grow in proportion to the current. Thus, the luminosity also grows linearly with current. Non-Gaussian transverse tails develop on the beams and subsequently increase linearly with current until the aperture limit is reached. At that value the beam lifetime is significantly reduced.

As well as the parameter ξ, three others determine the luminosity: crossing frequency, f; vertical beta-function at the interaction point (IP), β_y; and the ratio of beam dimensions, $r = \sigma_y/\sigma_x$.

Though the subject usually considered is the nonlinear dynamics of the beam-beam interaction, which involves the problem of the ξ limit, we cannot totally ignore these additional parameters. Attempts to increase the luminosity by using some of them are often limited by ξ reduction. Therefore at the appropriate points we shall return to consideration of these parameters. We will be most interested, however, in various possible mechanisms that cause the beam-beam limit.

3 AN ISOLATED RESONANCE IN TWO DEGREES OF FREEDOM

We consider first the two-dimensional motion of a single particle interacting with the charged bunch at the collision point and undergoing free betatron oscillations between the collisions: the so-called weak-strong case. The periodicity of the motion and the nonlinear character of the interaction cause two important effects: (1) excitation of nonlinear resonances and (2) dependence of the tune on the oscillation amplitude of the particle, which will result in tune spread if a weak beam contains particles of various amplitudes.

We begin with the Hamiltonian

$$H = \frac{K_x x^2}{2} + \frac{x'^2}{2} + \frac{K_y y^2}{2} + \frac{y'^2}{2} + V(x,y) \sum_{k=-\infty}^{\infty} \delta(z - kC)\,, \tag{3.1}$$

where K_x, K_y are the usual focusing functions; x, x' and y, y' are betatron coordinates and canonical momenta normalized on the standard deviations σ_x, σ_x', and σ_y, σ_y'; C is the period between interactions; and the potential

$$V(x,y) = \xi_y \frac{\sigma_y}{\beta} \int f_y(x,y)\, dy = \xi_x \frac{\sigma_x}{\beta} \int f_x(x,y)\, dx \tag{3.2}$$

is the perturbation, with f_x, f_y being dimensionless beam-beam forces. In this notation the beam-beam kicks are described in the form

$$\Delta x' = -\xi_x f_x, \qquad \Delta y' = -\xi_y f_y. \tag{3.3}$$

The transformation of the Hamiltonian (3.1) to action-angle variables $(x, x', y, y') \rightarrow (\Phi_x, I_x, \Phi_y, I_y)$ can be accomplished with the generation function

$$F(x,\Phi_x,y,\Phi_y) = \sum_{q=x,y} \frac{-q^2}{2\beta_q} \tan\left[\Phi_q - \frac{2\pi\nu_q s}{C} + \int_0^s \frac{ds'}{\beta_q(s')}\right] + \frac{q^2}{4\beta_q}\beta'_q, \tag{3.4}$$

which yields the transformation equations

$$q = \sqrt{2I_q\beta_q}\cos(\Phi_q), \qquad q' = \sqrt{2I_q/\beta_q}\left(\sin\Phi_q - \frac{\beta'_r}{2}\cos\Phi_q\right), \tag{3.5}$$

and the new Hamiltonian

$$H_1 = \nu_x I_x + V(I_x,\Phi_x,I_y,\Phi_y)\delta_{2\pi}(\theta) \tag{3.6}$$

where z has been replaced by a new dimensionless variable $\theta = 2\pi z/C$ and

$$\delta_{2\pi}(\theta) = \sum_{k=-\infty}^{\infty} \cos k\theta\,. \tag{3.7}$$

Then the problem can be reduced by expanding the potential V in the Fourier series

$$V = \sum\sum_{\ell,m=-\infty}^{\infty} V_{\ell,m}\cos(\ell\Phi_x)\cos(m\Phi_y)$$

and averaging the Hamiltonian over "time" θ. Only two terms are retained in the Hamiltonian, a phase-independent term

$$V_{00} = \frac{1}{(2\pi)^3}\iint_0^{2\pi} V\, d\Phi_x d\Phi_y \tag{3.8}$$

and a resonant term $V_{\ell m}\cos(\ell\Phi_x + m\Phi_y - k\theta)$, where

$$V_{\ell m} = \frac{1}{(2\pi)^2}\iint^{2\pi} V\cos\ell\Phi_x\cos m\Phi_y d\phi_x d\Phi_y\,. \tag{3.9}$$

All other Fourier terms are assumed to have phases that oscillate rapidly and can thus be averaged to zero.

Thus we come to the truncated Hamiltonian

$$\langle H_1\rangle = \nu_x I_x + \nu_y I_y + V_{00} + V_{\ell m}\cos(\ell\Phi_x + m\Phi_y - k\theta)\,. \tag{3.10}$$

The effect of V_{00} is to make the average frequencies of oscillation depend on the particle amplitudes defined as

$$A_x = \frac{\sqrt{2I_x\beta_x}}{\sigma_x}, \qquad A_y = \frac{\sqrt{2I_y\beta_y}}{\sigma_y}. \tag{3.11}$$

According to the standard definitions of the nonlinear tune shifts[7]

$$\Delta\nu_x = \frac{\partial V_{00}}{\partial I_x}, \qquad \Delta\nu_y = \frac{\partial V_{00}}{\partial I_y}, \tag{3.12}$$

substitution of Eqs. (3.2) and (3.8) into (3.12) gives

$$\Delta\nu_x(A_x,A_y) = \xi_x \frac{1}{(2\pi)^3 A_x} \iint_0^{2\pi} f_x \cos\Phi_x \, d\Phi_x d\Phi_y ,$$

$$\Delta\nu_y(A_x,A_y) = \xi_y \frac{1}{(2\pi)^3 A_y} \iint_0^{2\pi} f_y \cos\Phi_y \, d\Phi_x d\Phi_y . \tag{3.13}$$

The Hamiltonian (3.10) can be reduced to the one-dimensional case by using a canonical transformation to a rotating system in phase space. The generation function of this transformation is

$$F_2(\Phi_x,\Phi_y,K_1,K_2,\theta) = (\ell\Phi_x + m\Phi_y - k\theta)K_1 + (\ell\Phi_x - m\Phi_y)K_2 . \tag{3.14}$$

The relationship between old and new variables is

$$\psi_1 = \ell\Phi_x + m\Phi_z - k\theta , \quad \psi_2 = \ell_x - m\Phi_z ; \qquad I_x = \ell(K_1 + K_2) , \quad I_y = m(K_1 - K_2) . \tag{3.15}$$

The new Hamiltonian becomes

$$H_2 = (\ell\nu_x + m\nu_y - k)K_1 + (\ell\nu_x - m\nu_y)K_2 + V_{00}(K_1,K_2) + V_{\ell m}(K_1,K_2) \cos \psi_1 . \tag{3.16}$$

Since this Hamiltonian is independent of θ, it is a constant of motion. Since it is also independent of ψ_2, the new action

$$K_2 = \frac{I_x}{\ell} - \frac{I_y}{m} \tag{3.17}$$

is also an invariant.

As a result, the second term in Eq. (3.16) may be ignored since it just represents a constant energy term. Thus,

$$H_2 = (\ell\nu_x + m\nu_y - k) K_1 + V_{00}(K_1,K_2) + V_{\ell m}(K_1,K_2) \cos \psi_1 . \tag{3.18}$$

Let us define K_r as the action that yields the oscillation frequencies at resonance

$$\ell\nu_x + m\nu_y + \left.\frac{dV_{00}}{dK_1}\right|_{K_1=K_r} = k . \tag{3.19}$$

After expansion of V_{00} close to K_r we get from Eqs. (3.18) and (3.19)

$$H_2 \approx (\ell\nu_x+m\nu_y-k)\,K_r + \frac{1}{2}\frac{d^2V_{00}}{dK_1^2}(K_1-K_r)^2 + V_{\ell m}(K_r,K_2)\cos\psi_1 \,. \tag{3.20}$$

Here we can again omit the first term, which is now the constant energy term. Applying the third transformation with generation function

$$F_3 = (K_1 - K_r)\,\psi_1 \,, \tag{3.21}$$

we finally come to the pendulum-like Hamiltonian

$$H_3 = \alpha\frac{p^2}{2} + V_{\ell m}\cos\psi_1 \,, \tag{3.22}$$

where nonlinearity α is

$$\alpha = \frac{d^2V_{00}}{dK_1^2} = m^2\frac{\partial\Delta\nu_y}{\partial I_y} + 2m\ell\frac{\partial\Delta\nu_y}{\partial I_x} + \ell^2\frac{\partial\Delta\nu_x}{\partial I_x} \,. \tag{3.23}$$

The phase-space diagram for a pendulum is shown in Fig. 6.[8]

Because the energy of a pendulum is constant, the resonance width, which is equivalent to the size of the separatrix, can be easily found. It is

$$\Delta p = 4\sqrt{V_{\ell m}/\alpha} \,. \tag{3.24}$$

We consider below the flat beam case and will use the beam-beam forces in the same form as in Eqs. (1.14) (see also Ref. 9),

$$f_x = 4\pi\sqrt{2}\,F_D\,(x/\sqrt{2}\,)\,, \qquad f_y = (2\pi)^{3/2}e^{-x^2/2}\Phi\,(y/\sqrt{2}\,)\,. \tag{3.25}$$

Substitution of Eq. (3.25) in (3.13) gives

$$\Delta\nu_x(A_x,A_y) = \xi_x\frac{\sqrt{2}}{\pi}\int_0^{2\pi} F_D\left(\frac{A_x\cos\Phi_x}{\sqrt{2}}\right)\cos\Phi_x d\Phi_x \,, \tag{3.26}$$

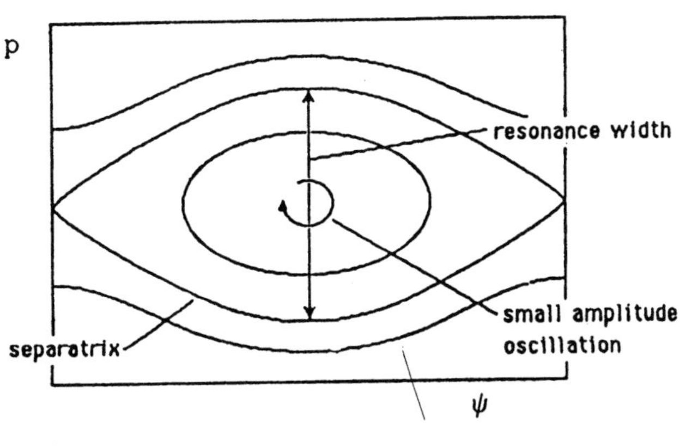

Fig. 6. Nonlinear resonance in phase space.

$$\Delta\nu_y(A_x,A_y) = \xi_y I_0\left(\frac{A_x^2}{4}\right)\left[I_0\left(\frac{A_y^2}{4}\right)+I_1\left(\frac{A_y^2}{4}\right)\right]\exp\left\{-\frac{A_x^2+A_y^2}{4}\right\}, \tag{3.27}$$

where I_0 and I_1 are modified Bessel functions of zero and first orders.

Figure 7A shows the tune spread (beam footprint) corresponding to Eqs. (3.26) and (3.27), and Fig. 7B shows it placed in the tune plane among the resonances.[10]

Figure 7B shows that the resonance lines inside the footprint are satisfied by different amplitude conditions. This means that the resonance can equally well be represented by certain lines in the amplitude plane. Their location within the amplitude space is determined by the resonant condition (3.19):

$$\ell\left[\nu_x+\Delta\nu_x(A_x,A_y)\right]+m\left[\nu_y+\Delta\nu_y(A_x,A_y)\right]=k\,, \tag{3.28}$$

where $\Delta\nu_x$ and $\Delta\nu_y$ are taken from Eqs. (3.26) and (3.27).

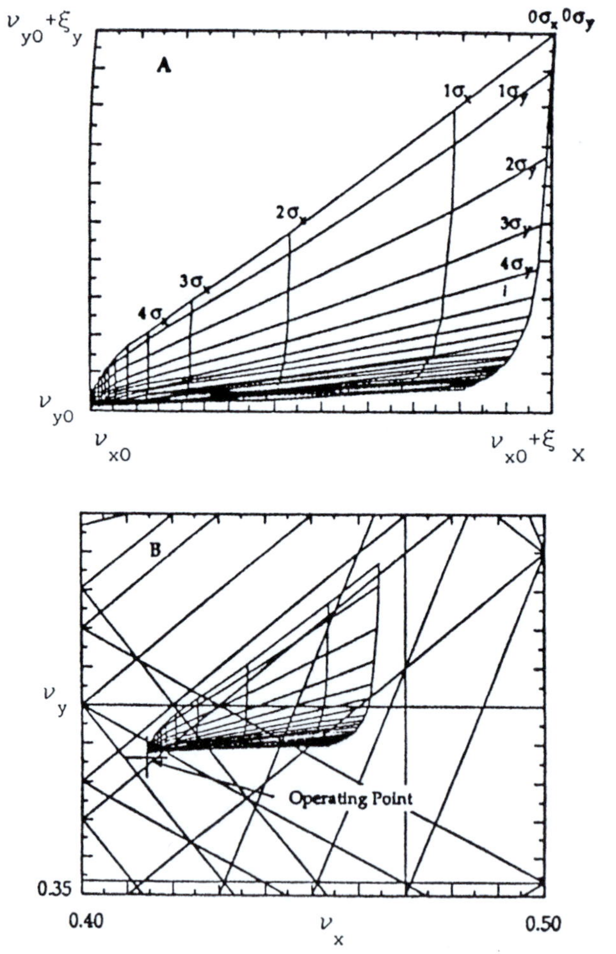

Fig. 7. (A) Tune spread of the beam, or beam footprint. The spacing between each grid line in the footprint is a 1σ step in either the vertical or the horizontal direction. The lower left corner of the footprint is at the machine tune. (B) The footprint placed in the tune plane among the resonances.

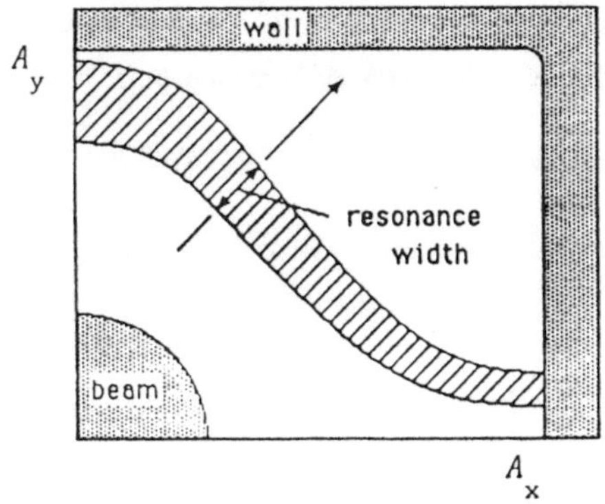

Fig. 8. Nonlinear resonance in the amplitude plane.

An example of the resonance in the amplitude plane is shown in Fig. 8. The resonance width here is measured by the value of the maximal beating of the amplitudes, which goes in the direction determined by the invariant (3.17).

In terms of amplitudes A_x, A_y the nonlinearity α from Eq. (3.23) has the form

$$\alpha = \frac{m^2\beta_y}{A_y\sigma_y^2}\left\{\frac{\partial\Delta\nu_y}{\partial A_y} + \frac{A_y}{A_x}\frac{\xi_x}{\xi_y}\frac{\sigma_y}{\sigma_x}\left[\frac{2\ell}{m}\frac{\partial\Delta\nu_x}{\partial A_x} + \frac{\ell^2}{m^2}\frac{\partial\Delta\nu_z}{\partial A_x}\right]\right\}. \tag{3.29}$$

In the case of a very flat beam, in the region of moderate-amplitude vertical oscillation, the first term in Eq. (3.20) is dominant. Therefore, in light of Eq. (3.27), one obtains

$$\alpha = \xi_y \frac{m^2\beta_y}{A_y^2\sigma_y^2}\exp\left(-\frac{A_x^2+A_y^2}{4}\right)I_0\left(\frac{A_x^2}{4}\right)I_1\left(\frac{A_y^2}{4}\right). \tag{3.30}$$

One can also rewrite Eq. (3.9) for $m \neq 0$ in the form

$$V_{\ell m} = \frac{\xi_y\sigma_y^2 A_y}{m\beta_y(2\pi)^2}\iint_0^{2\pi} f_y \cos\ell\Phi_x \sin m\Phi_y \sin\Phi_y \, d\Phi_x d\Phi_y\,. \tag{3.31}$$

The integrals in Eq. (3.31) can be evaluated for f_y in the form given in (3.25). Substitution of Eqs. (3.31) and (3.30) into (3.24) gives, for the resonance width,[11]

$$\Delta A_y = 2\sqrt{G_\ell(A_x)\,F_m(A_y)} \tag{3.32}$$

with the functions $G_\ell(A_x)$ and $F_m(A_y)$ determined as

$$G_\ell(A_x) = \frac{I_{\ell/2}\left(A_x^2/4\right)}{I_0\left(A_x^2/4\right)}, \tag{3.33}$$

$$F_m(A_y) = \frac{2A_y^2}{m^2-1}\,\frac{I_{m/2-1}\left(A_y^2/4\right) + \left(1-\dfrac{2(m-1)}{A_y^2}\right)I_{m/2}\left(A_y^2/4\right)}{I_1\left(A_y^2/4\right)}. \tag{3.34}$$

Note that ΔA_y does not depend on ξ_y.

The consideration above is valid for $\Delta p << K_r$ or for not too small nonlinearity α. Another important approximation is the assumption of a single isolated resonance.

4 CHIRIKOV OVERLAP CRITERION

In reality the set of resonances turns out to be very dense everywhere (see, for example, Fig. 9.[12])

A fairly general mechanism for the onset of instability of the motions is the so-called overlap of nonlinear resonances.[7] This instability has a rather peculiar nature resulting in an irregular, or stochastic, motion of the system. A plausible condition for the occurrence of stochasticity seems to be when the tunes approach the resonance within a distance of the order of a resonance size. To be precise, the overlap of resonances begins when their separatrices touch each other. How does one calcu-

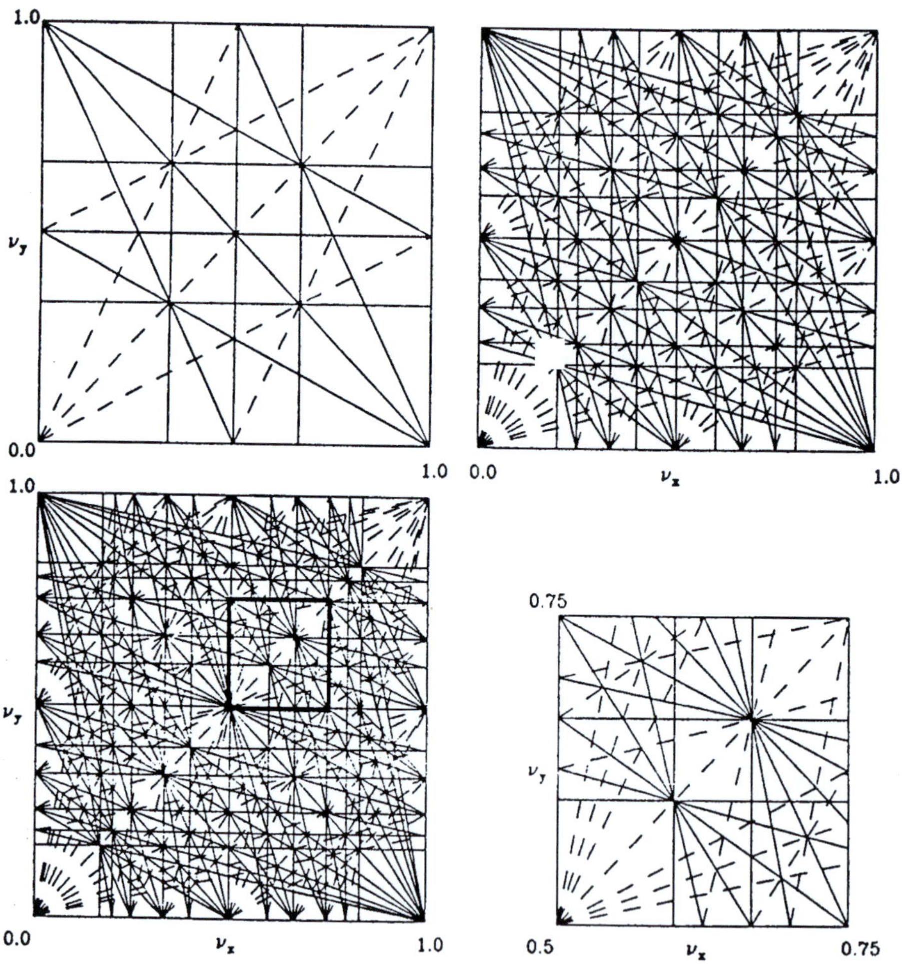

Fig. 9. Resonance lines in tune space. The figures include $|\ell| + |m| \leq 3$, 5, and 6. The outlined area in the sixth-order case is blown up to show detail.

late the condition of separatrix touching, taking into account deformation of the separatrix by a neighboring resonance? The simplest method, suggested by Chirikov, is to consider each resonance as if no other were present. This is what is meant by the overlap criterion. Although quite rough, this criterion is fairly efficient, especially in the case of rather complicated systems.

According to the oversap criterion, to avoid chaotic behavior in the simple one-dimensional case, we need to have

$$\delta\nu << \nu_1 - \nu_2 \tag{4.1}$$

where $\delta\nu$ is the resonance width in the tune plane, calculated as

$$\delta\nu = \alpha\Delta p = 4\sqrt{\alpha\, V_{\ell m}}\,, \tag{4.2}$$

and ν_1, ν_2 are the frequencies of two neighboring resonances

$$\nu_1 = n/m\,, \qquad \nu_2 = n'/m'\,, \tag{4.3}$$

with $nm' - n'm = \pm 1$.

Using Eqs. (4.2) and (4.3) we can rewrite (4.1) in the form

$$\sqrt{\alpha V_{\ell m}} << 1/4mm'\,. \tag{4.4}$$

Equation (4.4) sets a limit to the validity of the isolated resonance analysis. This condition requires that nonlinearity α not be too large since in this case the resonances do not separate.

Application of the Chirikov overlap criterion is convenient for a numerical study of beam-beam effects. With a computer it is easy to observe the modifications of the phase-space particle trajectory when the parameter ξ is increased, and to check when the isolated separatrices touch each other and regular motion is destroyed. Examples of phase space with regular and irregular motion for the one-dimensional case are shown in Fig. 10.[13]

Several authors have used the overlap criterion to determine the critical value ξ_{cr}.[13–15] Figure 11 shows the dependence of ξ_{cr} on the fractional part of the betatron tune for the one-dimensional beam-beam model. The calculated values of ξ_{cr} are much higher than experimental observations.

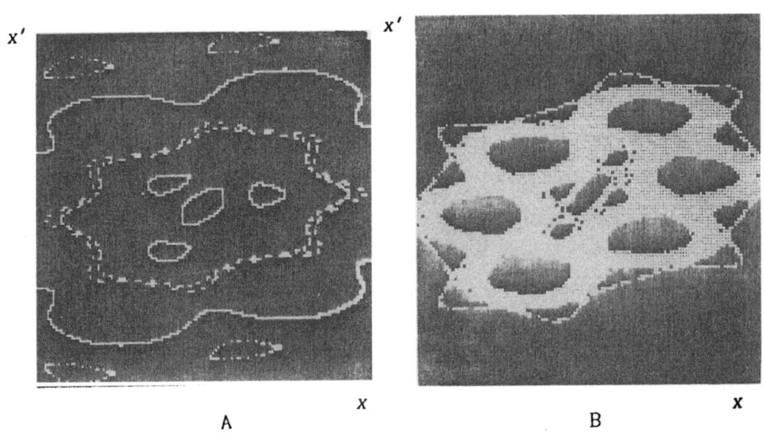

Fig. 10. Phase-space plot of a particle's motion. (A) an example of regular (stable) motion with parameter ξ below the threshold, determined by the overlap of the isolated resonances; (B) an example of irregular (unstable) motion, with ξ above the threshold.

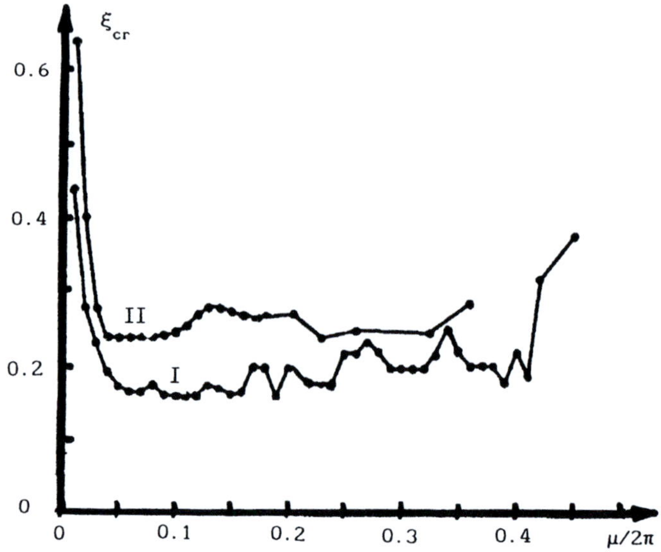

Fig. 11. The stochasticity threshold ξ_{cr} versus the fractional part of the betatron phase advance between interactions. (I) Round beam case; (II) flat beam case.

5 ADDING LONGITUDINAL MOTION

Taking the synchrotron oscillations into consideration will lead to new effects. We will show that interaction of the particle with the opposing bunch produces coupling between synchrotron and betatron oscillations. This coupling drives the multiplets of synchrobetatron sideband resonances near the primary beam-beam resonances. Now three regimes can characterize the resonance system[16] (see Fig. 12): (a) New resonances are isolated and do not overlap. (b) The distance between neighboring reso-

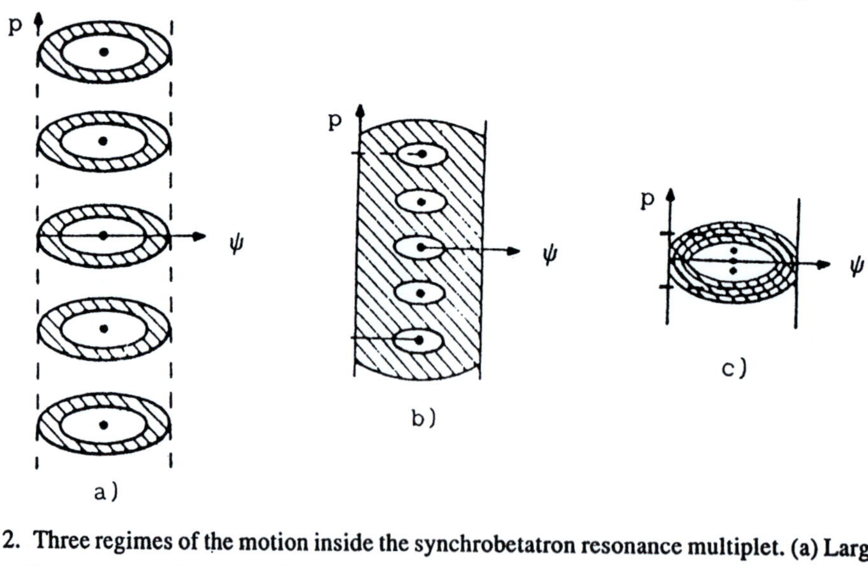

Fig. 12. Three regimes of the motion inside the synchrobetatron resonance multiplet. (a) Large modulation: the resonances do not overlap, and there is weak diffusion along the separate stochastic layer. (b) Intermediate modulation: there is strong diffusion along the wide stochasticity region. (c) Small modulation: all resonances merge into one, and there is weak diffusion along one stochastic layer.

nances is smaller and they overlap; strong instability arises here when the overlapping extends to fill the distances between primary betatron resonances. (c) All resonances of the multiplet merge together.

It is clear that the overlap of synchrobetatron resonances can decrease the value of the beam-beam limit.

Several beam-beam longitudinal-transverse coupling mechanisms are known. Our discussion begins by considering so-called force modulation. Such modulation arises from the two variations of the β-function at interaction, which depend on the synchrotron motion: β-function dependence (a) on the longitudinal deviation

$$\beta_y(z) = \beta_{y0} + \frac{z^2}{\beta_{y0}} \tag{5.1}$$

and (b) on the energy deviation $\Delta E/E$

$$\beta_y\left(\frac{\Delta E}{E}\right) = \beta_{y0} + \frac{d\beta_{y0}}{d(\Delta E/E)}\frac{\Delta E}{E} \tag{5.2}$$

where β_{y0} is the β-function at the interaction point (IP) and $d\beta_{y0}/d(\Delta E/E)$ is the lattice-dependent term.

We consider the flat beam case here and the round beam case in the next section. For a flat beam all the modulation effects are much more significant in the vertical direction. Because, in principle, the second term in Eq. (5.2) can be made negligible by proper choice of a chromatic correction scheme in the storage rings, we do not include it.

The longitudinal motion of a particle within the weak beam is assumed to have the following properties. It undergoes linear synchrotron oscillations with the synchrotron tune, ν_s:

$$s = s_0 \cos \nu_s\theta\ . \tag{5.3}$$

The longitudinal beam-beam impulse is assumed negligible; the direct effect of the rf cavity system on the motion of the particle is not considered. We also assume that the opposing bunch has no longitudinal extent. Therefore, if a particle within the weak bunch has a longitudinal offset s from the synchronous particle, it receives its transverse beam-beam kick at a distance $z = -s/2$ from the IP.

Equation (5.1) leads to ξ_y dependence of the particles on the longitudinal coordinate,

$$\xi_y = \xi_{y0}\left(1 + \frac{s^2}{4\beta_{y0}^2}\right)^{1/2}. \tag{5.4}$$

After substituting Eq. (5.3) into (5.4) we get

$$\xi_y = \xi_{y0}\left[1 + \left(\frac{\sigma_z}{2\beta_{y0}}\right)^2 A_z^2 \cos^2 \nu_s\theta\right]^{1/2} \tag{5.5}$$

where $A_z = S_0/\sigma_z$ is the amplitude of the longitudinal synchrotron oscillations normalized on the bunch length σ_z. For small A_z and $\sigma_z \sim \beta_{y0}$ we can write

$$\xi_y = \xi_{y0}\left[1 + \frac{A_z^2}{4}\left(\frac{\sigma_z}{2\beta_{y0}}\right)^2 (1 + \cos 2\nu_s\theta)\right]. \tag{5.6}$$

This shows the existence of two sideband resonances

$$m\nu_y \pm 2\nu_s = k\,. \tag{5.7}$$

In the case of large longitudinal oscillations the number of dangerous resonances is increased, and the resonance conditions become

$$m\nu_y + n\nu_s = k \tag{5.8}$$

where m and n are even.

The force modulation results also in tune shift dependence on the longitudinal oscillation. This directly follows after substitution of Eq. (5.5) into Eq. (3.13). This effect leads to dependence of the instantaneous betatron tune on the parameters of longitudinal motion:

$$\nu_y \cong \nu_{y0} + \Delta\nu_{y0}\left[1 + \left(\frac{\sigma_z}{2\beta_{y0}}\right)^2 A_z^2 \cos^2 \nu_s\theta\right]^{1/2}\,. \tag{5.9}$$

It also drives the synchrobetatron resonances, Eq. (5.8).[17]

The results of a numerical simulation of the effect of force modulation are shown on Fig. 13, which shows significant dependence a of ξ_{cr} on the amplitude of synchrotron oscillations.

Another source of longitudinal-transverse coupling in beam-beam interactions is the so-called **betatron tune modulation.**[18] This type of modulation arises from the longitudinal oscillations of the particles, but also important to it is the low β-function at the IP and the large accompanying betatron phase advance near the IP:

$$\Delta\Phi = \int_0^{s/2} dz/\beta_y(z)\,. \tag{5.10}$$

The instantaneous betatron frequency is

$$\nu_y = \nu_{y0} - \nu_s \frac{S_0}{2\beta_y} \sin \nu_x\theta\,, \tag{5.11}$$

and we can write for a canonical phase Φ in the Hamiltonian (3.10)

$$\Phi_y = \int_0^{\theta} \nu\, d\theta = \nu_{y0} + \frac{S_0}{2\beta_y} \cos \nu_s\theta\,. \tag{5.12}$$

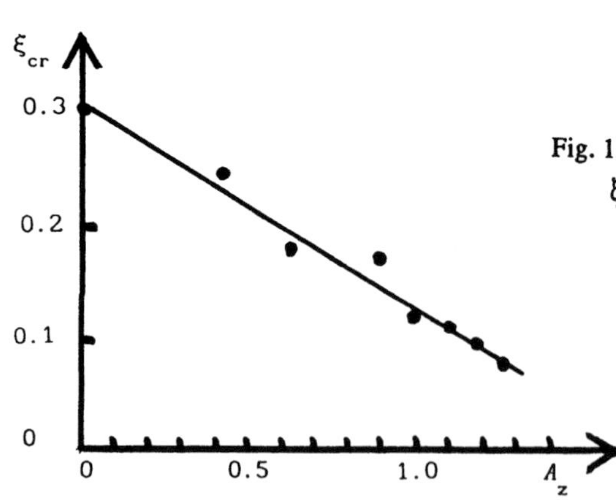

Fig. 13. Dependence of the critical value ξ_{cr} on the normalized amplitude of synchrotron oscillations, A_z.

Substitution of Eq. (5.12) into Eq. (3.10) yields new resonance conditions

$$\ell\nu_x + m\nu_y + n\nu_s = k \qquad (5.13)$$

with ℓ, m, n being even numbers, and new resonance harmonic amplitude

$$V_{\ell mn} = V_{\ell m} J_n \left[m(\sigma_z/2\beta_y)\, A_z\right] \qquad (5.14)$$

where J_n is the Bessel function of order n.

Recent studies[19] showed that, for relatively small amplitudes of the longitudinal oscillations, consideration of the finite longitudinal extent of the bunch results in an additional reduction factor

$$Y = \exp\left\{-\frac{1}{2}\left(\frac{m\sigma_z}{2\beta_y}\right)^2\right\}, \qquad (5.15)$$

in the resonance harmonic amplitude $V_{\ell mn}$. This is related to the effect of beam-beam kick averaging in the betatron phases and to the reduction of kick strength.

Two things contribute to the reduction of the beam-beam kick.[10] First, the deflection of the particle during interaction with the opposing bunch is proportional to the average charge density of the opposing beam over the length of interaction. Therefore, because of the β-function variation near the IP, the deflection caused by an integrated kick is less than that caused by an impulse-like kick at the symmetry point, where the β-function is at a minimum. Second, for a particle with small displacement, as the particle trajectory places it closer to the center of the opposing beam core it experiences less beam-beam force. Summed over the length of the interaction the net deflection is less. Figure 14 shows the results of calculations of particle trajectory for both the impulse-like and the integrated kick cases.

Numerical studies of the effect of tune modulation on the critical value ξ_{cr} usually show significant reduction of ξ_{cr} when the amplitude of modulation is increased. Figure 15[17] shows examples of such studies.

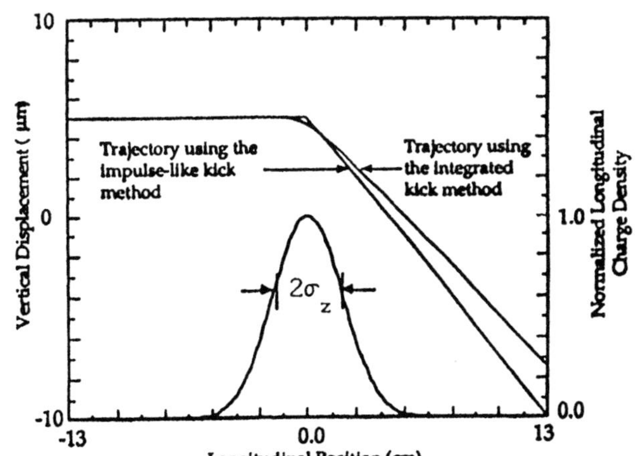

Fig. 14. The trajectory of a $1\sigma_y$ particle as it passes through the opposing beam. Two cases are shown: the impulse-like case and the smooth integrated kick case. The deflection of the trajectory produced by the integrated kick is less than that in the impulse-like case. The longitudinal distribution (Gaussian) of the opposing bunch is shown for reference.

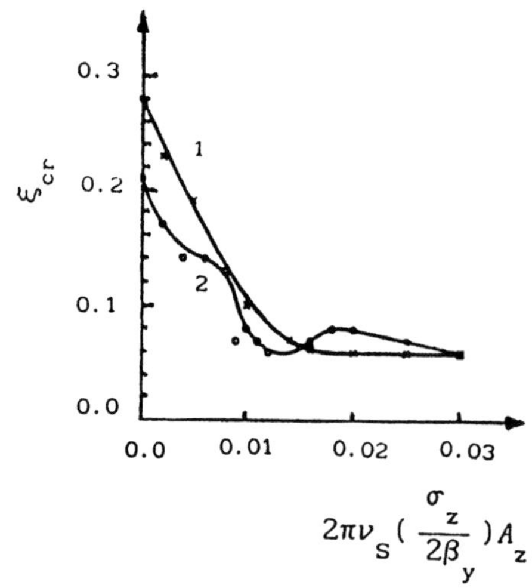

Fig. 15. Dependence of the critical value ξ_{cr} on the amplitude of the tune modulation. (1) Flat beam; (2) round beam.

Nonzero chromaticity is also a source of tune modulation, but it is usually compensated to a very small value. The next source of longitudinal-transverse coupling in beam-beam interaction–nonzero dispersion at the IP–is considered in Section 7. In principle, the dependence of the rotation period on particle energy also leads to excitation of the synchrobetatron resonances, but this source of the effect is very weak.

6 COLLISION OF ROUND BEAMS

The idea of colliding round e^+e^- beams has been considered since the early days of storage rings. Lately it has been studied intensively and considered promising as a means to advance towards high luminosity. What is so attractive about round beams?

First is the factor of two in luminosity, which directly follows from the luminosity formula (2.9) at $\sigma_y = \sigma_x$ with all other parameters being the same as for flat beams.

Second is the absolute elimination of the force modulation effect when $\beta_x \equiv \beta_y$. This follows from the lack of dependence of ξ on the beta-function in the round beam case,

$$\xi = Nr_e/4\pi\gamma\varepsilon \tag{6.1}$$

where ε is the beam transverse emittance.

Third is the reduction of the beam-beam problem from three to two degrees of freedom. Equal vertical and horizontal emittances and betatron tunes are implicit initial conditions here.

Several authors have simulated the beam-beam effects in the round beam configuration in a weak-strong case with the finite bunch length taken into account. Typical results show a large maximal value of ξ, determined by a threshold in beam core blowup. See, for example, Fig. 16.[20]

The maximal ξ for particles with large longitudinal excursions from a bunch center is substantially lower. This is related to the effect of tune modulation, discussed above. For large amplitude longitudinal oscillations, strong longitudinal-transverse coupling appears which results in rapid loss of these particles at the aperture. The dependence of the maximal attainable value of ξ on longitudinal amplitude is shown in Fig. 17.[20]

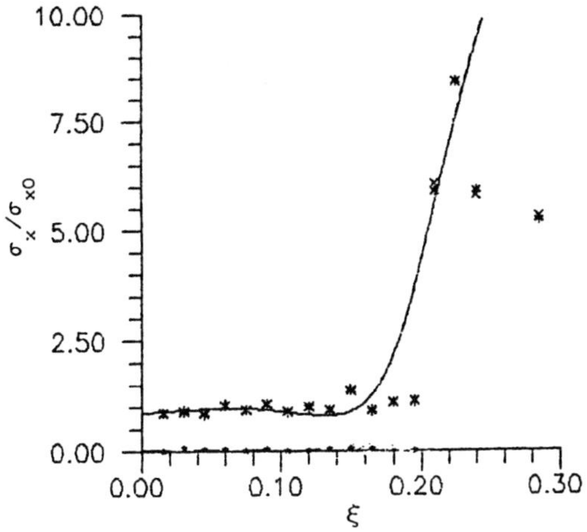

Fig. 16. Dependence of beam size on the parameter ξ. The simulation was carried out at $\sigma_z/\beta_y = 0.83$, $\nu_x = \nu_y = 0.08$, and $\nu_s = 0.02$.

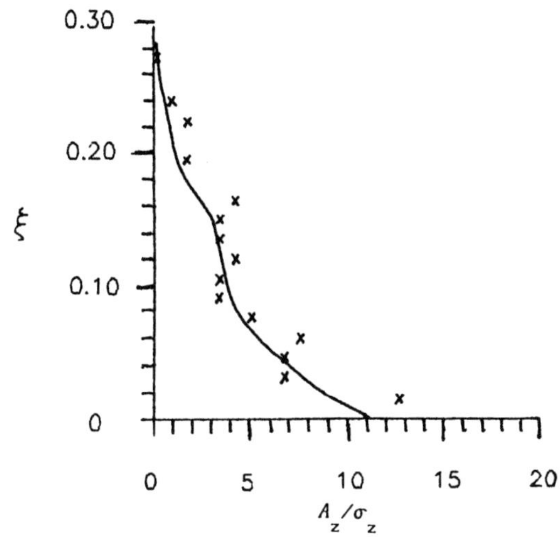

Fig. 17. Dependence of the maximal value of ξ on the amplitude of the longitudinal oscillation. The simulation was carried out at $\sigma_z/\beta_y = 0.83$, $\nu_x = \nu_y = 0.08$, and $\nu_s = 0.02$.

The realization of low β-function in both directions–as low as in the flat beam design–will demand sextupole lenses about twice as strong as for chromaticity correction. This reduces the attractiveness of round beams if one takes into account the possible interference of machine imperfections with beam-beam effects.

7 NONZERO DISPERSION AT THE INTERACTION POINT

For a long time accelerator physicists have been against using dispersion at the interaction point. This view has been confirmed both by experimental observations and by simulations.

All these studies were carried out with comparable contributions of the energy and of the betatron oscillations to the total transverse beam dimension in the dispersive direction, i.e. the magnitude of

$$\lambda = \sigma_{xs}/\sigma_{x\beta} \tag{7.1}$$

(where $\sigma_{x\beta}$ is the betatron beam size, $\sigma_{xs} = |\psi|\sigma_\varepsilon$ is the synchrotron beam size, ψ is the dispersion at the IP, and σ_ε is the beam energy spread) satisfied the condition

$$\lambda \lesssim 1\,. \tag{7.2}$$

The conclusion drawn is as follows. Dispersion at the IP causes excitation of the synchrobetatron resonances

$$\ell\nu_x + m\nu_y + n\nu_s = k\,, \tag{7.3}$$

with m and $(\ell + n)$ being even numbers, and results in deterioration of the maximal ξ.

If one makes

$$\lambda >> 1\,, \tag{7.4}$$

what happens? Will the deterioration process continue?

To answer this question, Gerasimov et al.[21] analyzed this case. They assumed a method of creating a large λ which used not only a high dispersion at the IP but also a low β-function and very small emittances. This approach results in a very flat beam configuration at the IP,

$$\sigma_x/\sigma_y \geq 250\,, \tag{7.5}$$

and a large difference between ξ_y and ξ_x,

$$\xi_y/\xi_x \geq 5\,. \tag{7.6}$$

For the case of $\lambda >> 1$ the particle excursion from the center in the horizontal direction is

$$X = \frac{A_x}{\lambda}\cos\phi_x + A_s\cos\phi_s\,, \tag{7.7}$$

where X is the coordinate normalized on the total beam size $\sigma_x = \sqrt{\sigma_{x\beta}^2 + \sigma_{sx}^2}$, and A_s is the synchrotron amplitude normalized by a condition

$$A_s = \frac{|\psi|\,\Delta E/E}{\sigma_{xs}} \tag{7.8}$$

where $\Delta E/E$ is the energy deviation.

Gerasimov et al.[21] estimated how the width of the resonances, Eq. (7.3), depends on the parameter λ. This can be done in a way similar to that used for the two-dimensional case in Section 3. Thus, according to Eq. (3.31), the resonance harmonic amplitude $V_{\ell mn}$ is

$$V_{\ell mn} = \frac{\xi_y A_y \sigma_y^2}{(2\pi)^3 m\,\beta_y}\int\!\!\int\!\!\int_0^{2\pi} d\phi_x\, d\phi_y\, d\phi_s\, f_y\left(\frac{A_x}{\lambda}\cos\phi_x + A_s\cos\phi_s,\, A_y\cos\phi_y\right)$$

$$\times \cos\ell\phi_x \cos n\phi_s \sin\phi_y \sin m\phi_y\,. \tag{7.9}$$

Here $m \neq 0$.

The nonlinearity α is expressed the same way as in Eq. (3.30) with A_s substituted for A_x.

It is natural to use, in Eq. (7.9), f_y from the flat beam presentation, Eq. (3.25). This allows factorization of the evaluation of the integrals in Eq. (7.9), which results in an expression analogous to (3.32) for the width of any particular resonance ℓ, m, n:

$$\Delta A_y = 2\sqrt{G_{\ell n}(A_x, A_s)\, F_m(A_y)}\,. \tag{7.10}$$

Here the functions F_m have exactly the same form as in Eq. (3.34), but the functions $G_{\ell n}$ differ from those in (3.33). They can be found for every value of ℓ by expanding f_y in a Taylor series with the small parameter A_x/λ. Several results of the evaluation of leading terms in $G_{\ell n}$ are presented below.

For $\ell = 0$:

$$G_{0n}(A_s) = \frac{I_{n/2}(A_s^2/4)}{I_0(A_s^2/4)} \tag{7.11}$$

where $I_{n/2}$ is the modified Bessel function of order n/2.

For $\ell = 1$ the leading term of $G_{\ell n}$ already contains a small factor $A_x/2\lambda$:

$$G_{1n} = \frac{A_x A_s}{4\lambda} \frac{I_{(n-1)/2}(A_s^2/4) - I_{(n+1)/2}(A_s^2/4)}{I_0(A_s^2/4)}. \tag{7.12}$$

For $\ell = 2$:

$$G_{2n} = \frac{1}{2}\left(\frac{A_s}{2\lambda}\right)^2 \frac{I_{n/2}(A_s^2/4)(A_s^2/2 - 1) + (A_s^2/2)\left[I_{n/2-1}(A_s^2/4) + I_{n/2+1}(A_s^2/4)\right]}{I_0(A_s^2/4)}. \tag{7.13}$$

For higher orders of ℓ:

$$G_{\ell n} \sim (A_x/2\lambda)^{\ell}. \tag{7.14}$$

Thus we come to the following conclusion. The width of the pure vertical resonances does not depend on the parameter λ. The width of all other resonances is reduced with increasing λ. The expected effect is estimated as

$$\Delta A_y \sim (2\lambda)^{-\ell/2}. \tag{7.15}$$

To clarify the importance of this effect for beam-beam interactions when the total number of resonances is increased by the new set of synchrobetatron satellites, we carried out an analytical and numerical study of the beam-beam effects. First we determined the dependence of the blowup effect in the weak beam on the working point on the tune plane. Then, for two good working points, we calculated the dependence of this effect on the intensity of the strong beam.

For fast analysis of a large number of betatron tunes a simple one-dimensional analytical model $(\xi_x = 0)$ was proposed for determining blown-up σ_y. The model is based on the crude approximation for the relaxed distribution function of the particle density in which the particle density within the nonlinear resonance is taken to be constant.

The validity of the model was checked by the conventional tracking technique. Figure 18 shows the results of the model prediction and the tracking results. The dependence on ν_y of the ratio σ_y/σ_{y0}, where σ_{y0} is the vertical size of the strong beam, is shown at a fixed horizontal betatron tune $\nu_x = 0.08$. Other parameters were chosen as $\nu_s = 0.025$, $\xi_y = 0.05$, $\xi_x = 0.01$, and $\lambda = 5$.

The analytical results were in satisfactory agreement with the simulation. We then extended the model calculations of vertical beam size to the whole tune place. For given values $\xi_y = 0.05$ and $\nu_s = 0.025$ and four values of λ (2.5, 5, 10, and 25) we determined σ_y/σ_{y0} at every point of the betatron tune on a mesh with steps $\Delta\nu_x = \Delta\nu_y = 0.005$. The result gave the areas of tunes where the blowup of σ_{y0} did

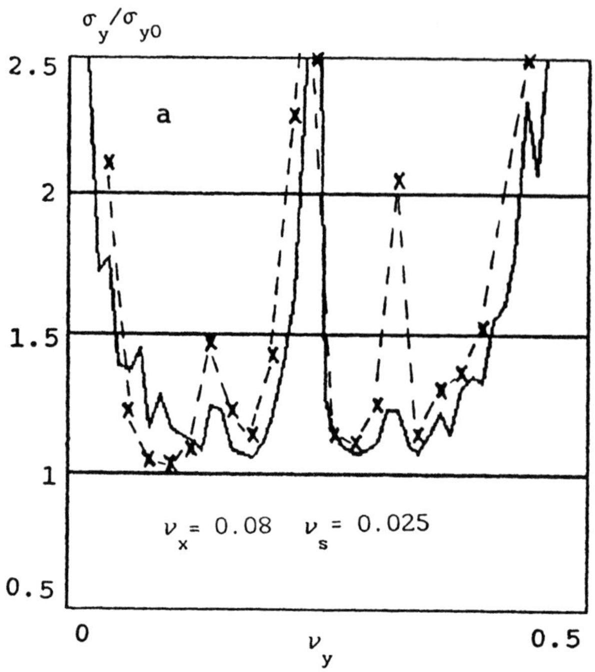

Fig. 18. Comparison of model calculations (solid line) of the weak beam vertical size with computer simulation using tracking (dashed line).

not exceed 15%. Figure 19 shows plots for each value of λ with these areas contoured and shaded. All resonances up to sixth order are also shown.

Our conclusion from the analysis of these four cases is as follows: Most destructive resonances are pure vertical betatron and synchrobetatron resonances with numbers $|m| = 0, 2, 4, 6$ and $|n| = 0, 2$. The effect of these resonances is observed independent of the parameter λ in all four plots. All pure horizontal and coupling resonances, however, are sensitive to λ. Their relative importance depends on the number ℓ. For example, at λ=5 the resonances $\pm 3\nu_x + 2\nu_y \pm \nu_s = k$ with $\ell=3$ are seen as the ones limiting the good areas, but they are less important at λ=10, where the limiting resonances are $\pm 2\nu_x + 2\nu_y = k$ and $\pm 2\nu_x + 2\nu_y \pm 2\nu_s = k$ with $\ell=2$. The good areas here are larger then in the previous case. The largest good areas are on the plot with λ=25. Here only the resonances $\pm 2\nu_x + 4\nu_y = k$ and $\pm \nu_x + 2\nu_y \pm \nu_s = k$ are considered as the destructive ones.

The dependence of the weak beam vertical size on the linear tune shifts ξ_y, ξ_x was studied numerically at a constant synchrotron tune $\nu_s = 0.025$ for two working points with betatron tunes $\nu_x = 13.08$, $\nu_y = 7.075$ and $\nu_x = 13.08$, $\nu_y = 7.10$. From the plots in Fig. 19 these working points appeared promising for the case of large λ.

Numerical simulations were done with a model including longitudinal motion, quantum fluctuation noise, and radiation damping, with the damping time expressed as the numbers of beam-beam collisions N=9090 for longitudinal and vertical oscillations and N=4545 for horizontal oscillations. Four sets of calculations were done at each working point: one for zero dispersion (λ=0), one for small

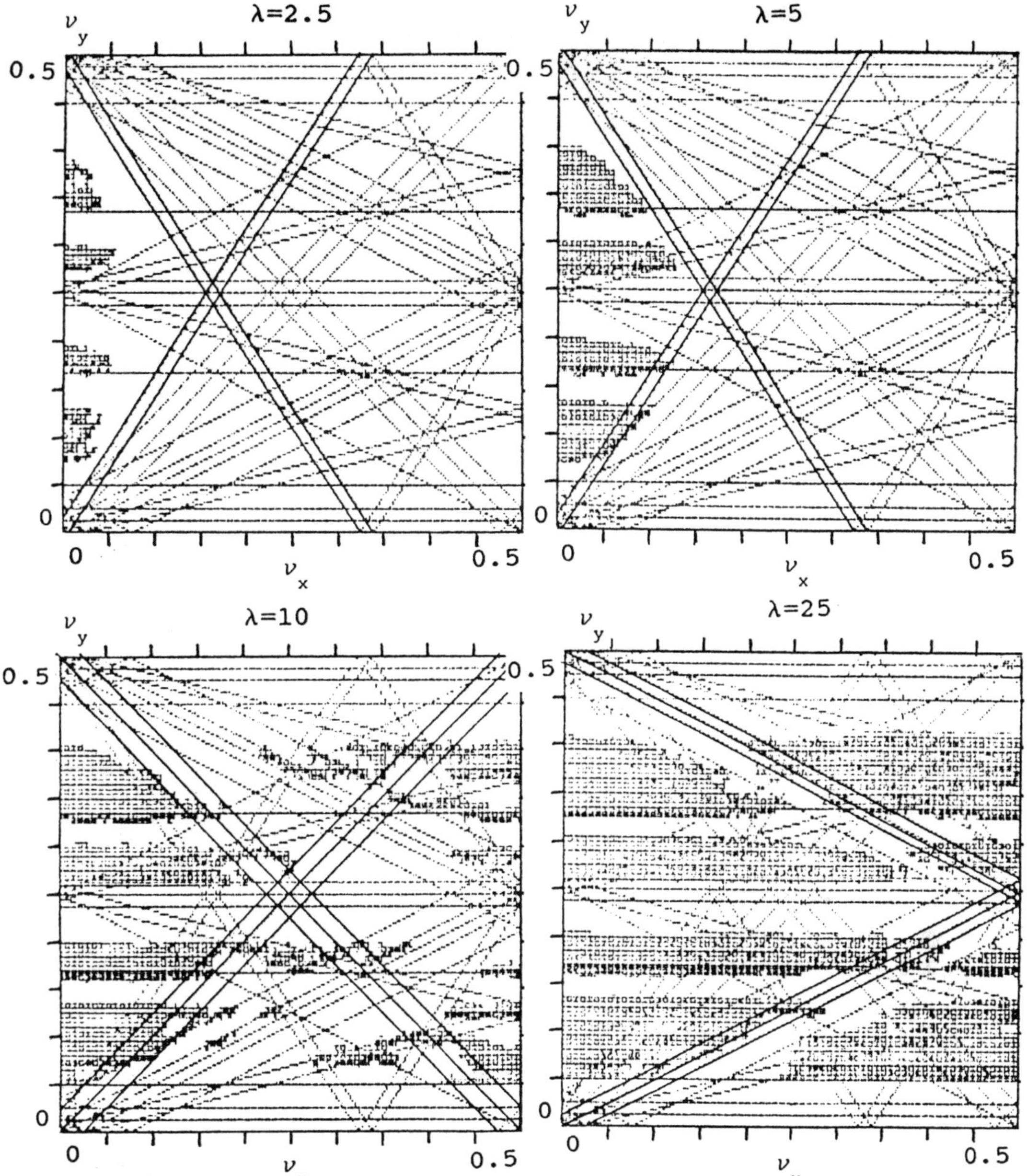

Fig. 19. Contour plot of the areas of tunes, where blowup of the weak beam vertical size does not exceed 15%. These areas are marked with shading. Four cases, with $\lambda = 2.5, 5, 10$, and 25, are shown. On each plot all resonances up to sixth order with $|\ell| \leq 3$ are presented.

dispersion (λ=0.7), and two for large dispersion (λ=5 and λ=25). The results (Fig. 20) give clear evidence of the effectiveness of large dispersion in suppressing the resonance $2\nu_x-2\nu_y=k$ and its synchrobetatron satellites. In our simulation the longitudinal beam size was $\sigma_z = 0.75$ cm and the vertical β-function at the IP was 1 cm.

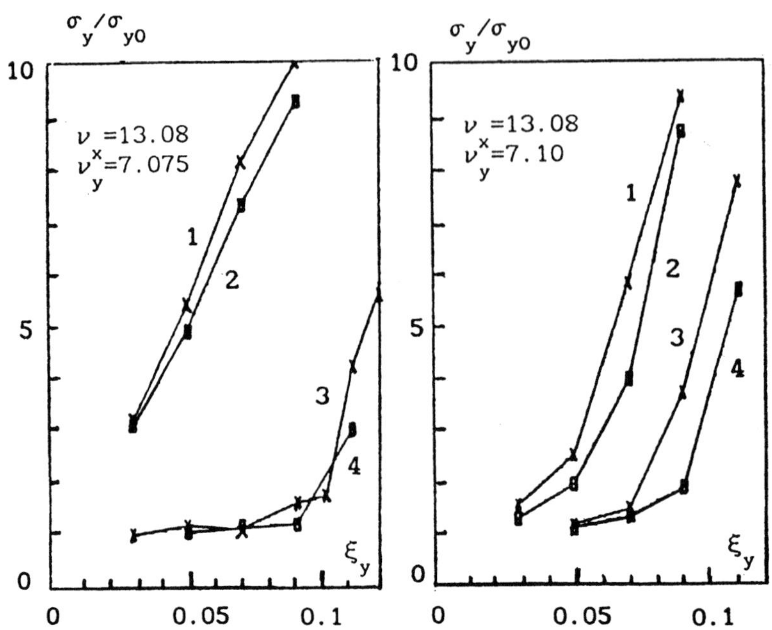

Fig. 20. Dependence of the weak beam vertical size on the vertical linear tune shift ξ_y. The horizontal tune shift ξ_x satisfies the condition $\xi_x = \xi_y/5$. (1) $\lambda = 0$; (2) λ=0.7; (3) λ=5; (4) λ=25.

8 COLLISION WITH CROSSING ANGLE

The idea of using crossed beams in e^+e^- storage rings has a long history. Its main purpose is to achieve fast separation of the colliding beam orbits. It allows an increase in collision frequency (number of bunches) and avoids beam-beam interactions in the parastic crossings.

Unfortunately, this method is not free of the problem shown in Fig. 21 (see also Ref. 22). The particle at a distance s from the center of its own bunch passes the center of the opposing bunch at a transverse position x + sφ, where x is the transverse displacement of a particle inside the bunch and φ is half the crossing angle.

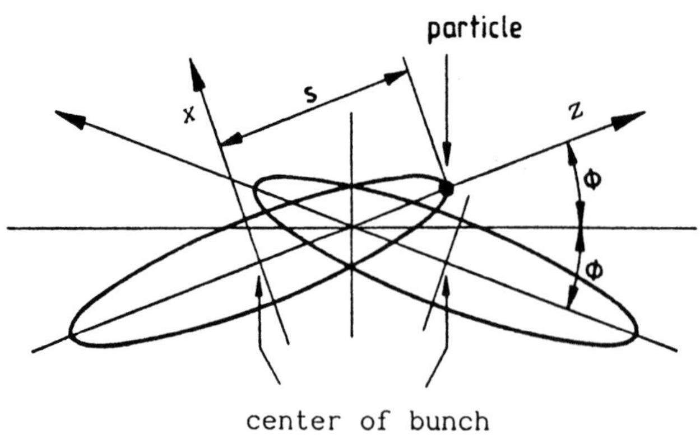

Fig. 21. Beam-beam interaction at a crossing angle.

Thus, the crossing angle causes modulation of the transverse particle coordinate with synchrotron oscillations as in the case of nonzero dispersion at the IP considered above. The value of the effective dispersion, ψ_{eff}, is

$$\psi_{eff} = \varphi\, \alpha R/\nu_s \tag{8.1}$$

where R is the average radius of the ring and α is the momentum compaction factor.

We consider two cases: zero normal dispersion at the IP, and large normal dispersion at the IP.

It is clear that, for a small geometrical loss in luminosity, the condition

$$\varphi\sigma_z/\sigma_x \lesssim 1 \tag{8.2}$$

must be realized. Here σ_x is the total transverse beam size at the IP.

In the first case this is only betatron size, and in terms of the parameter λ and effective dispersion we deal with

$$\lambda = \varphi\, \alpha R \sigma_\varepsilon/\nu_s\sigma_{x\beta} = \varphi\sigma_z/\sigma_{x\beta} \lesssim 1\ . \tag{8.3}$$

This is the worst case in our opinion, and the only recommendation is to have φ as small as possible to get λ<<1.

On the contrary, in the second case, with large nonzero dispersion at the IP, the angle φ can be large. The main difference consists of a general contribution of a synchrotron component σ_{xs} to the total transverse beam size σ_x at the IP. For $\sigma_{xs} >> \sigma_{x\beta}$, the condition (8.2) can be written as

$$\psi_{eff}/\psi \lesssim 1 \tag{8.4}$$

or

$$\varphi \cong \nu_s\, \psi/\alpha\bar{R}\ . \tag{8.5}$$

In terms of λ, the equality of ψ_{eff} and ψ gives an additional improving factor of $\sqrt{2}$,

$$\lambda = |\psi + \psi_{eff}|\sigma_\varepsilon/\sigma_{x\beta} \cong \sqrt{2}\,\lambda_0 \tag{8.6}$$

where λ_0 is the value of λ without crossing angle.

Thus, in the second case the crossing angle does not lead to new phenomena and moreover leads to better conditions in the beam-beam interaction. We can therefore expect the sum of the effects of large dispersion and collision at a crossing angle to open a new path towards high luminosity.

To be accurate, we should mention that the synchrotron oscillation is also influenced by the betatron one. The energy shift $\delta E/E$ after the interaction is

$$\delta E/E = \varphi\, \Delta x'\ . \tag{8.7}$$

Because of the big difference between the longitudinal and transverse emittances, however, this does not result in any destructive effects.

9 INTERFERENCE OF MACHINE IMPERFECTIONS WITH BEAM-BEAM EFFECTS

We consider below several types of machine imperfections that can cause reduction of the maximal ξ.

The first is nonlinear components of the guiding magnetic field. These include the sextupole and octupole lenses and other uncontrolled cubic and high-order machine nonlinearities. Even in the single-beam case, these may reduce the aperture from the free mechanical aperture to the so-called dynamic aperture. In the two-beam case with beam-beam effects, small size of the dynamic aperture might be a case of ξ limitation. This was observed on the VEPP-2M storage ring,[23] where the dependence of the ultimate ξ_y value on the position of the aperture scraper was studied. The result (Fig. 22) shows that ξ_y is independent of scraper position with an aperture above $35\sigma_y$; below this, ξ_y decreases. The existance of this threshold is clear evidence for a dynamic aperture limit at $35\sigma_y$.

Recently, it was noted on an example of cubic nonlinearity[24] that, even in the case of betatron tunes chosen far from the destructive machine resonances, the zero harmonic of the nonlinearity can lead to significant effects. A dangerous situation arises at the oscillation amplitudes where beam-beam detuning is partly or totally compensated by cubic machine nonlinearity. This means that resonance self-stabilization is absent here, and the resonance width can extend to large amplitudes. The same follows from formula (3.24) for resonance width. At constant harmonic amplitude, the resonance width can be sufficiently increased by diminishing the nonlinearity α.

According to conventional notation we define the average cubic machine nonlinearity as

$$R_x = \frac{1}{\varepsilon_x}\frac{d\nu_x}{dA_x^2}, \qquad R_y = \frac{1}{\varepsilon_y}\frac{d\nu_y}{dA_y^2}, \qquad R_{xy} = \frac{1}{\varepsilon_y}\frac{d\nu_x}{dA_y^2} = \frac{1}{\varepsilon_x}\frac{d\nu_y}{dA_x^2} \tag{9.1}$$

where $\varepsilon_x,\varepsilon_y$ are horizontal and vertical beam emittances and A_x,A_y are normalized betatron amplitudes. These nonlinearities contribute to the Hamiltonian (3.10) with the term[24]

$$W_{00} = R_x I_x^2 + R_y I_y^2 + 2\,R_{xy} I_x I_y\,. \tag{9.2}$$

All the formulae of Section 3 are valid after the substitution

$$V_{00} + W_{00} \to V_{00}\,. \tag{9.3}$$

It is not surprising, therefore, that the positions of the resonance lines in the amplitude plane are charged and under certain conditions the new nonlinearity α can be very small.

Gerasimov et al.[11] studied numerically the effect of cubic machine nonlinearity on an example with resonance $10\nu_y = 96$ (see Fig. 23). They calculated three cases, with (a) zero, (b) positive, and (c) negative nonlinearity, all at $\nu_y = 9.599$, $\nu_x = 8.555$, $\xi_y = 0.03$, $\xi_x = 0.01$. The magnitudes of the machine-induced tune shifts were $\Delta\nu_x = \pm 2.5\cdot 10^{-5}$ at $A_y = 0$, $A_x = 1$ and $\Delta\nu_y = \pm 2.5\cdot 10^{-7}$ at $A_y = 1$, $A_x = 0$.

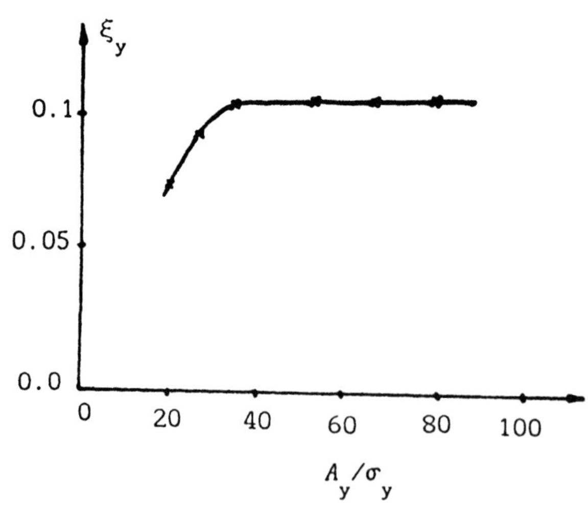

Fig. 22. Maximum ξ_y versus aperture (VEPP-2M, 0.5 GeV, weak-strong beams, $\sigma_y \ll \sigma_x$, $\beta_x = 40$ cm, $\beta_y = 3$ cm).

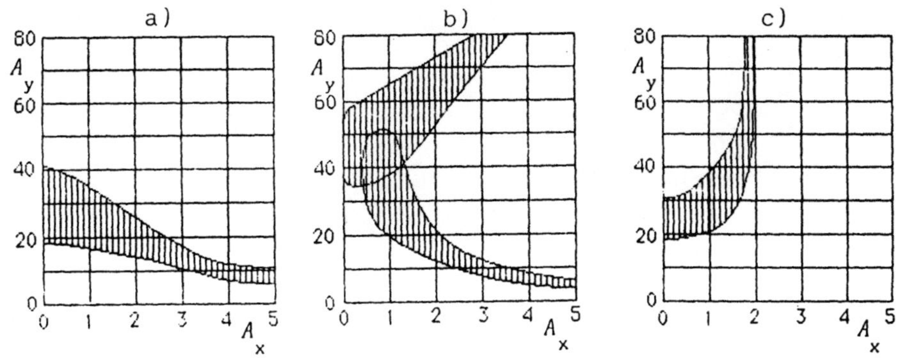

Fig. 23. The resonance $10\nu_y = 96$ in the amplitude plane for the cases of machine nonlinearity (a) zero, (b) positive, and (c) negative.

In all three cases the direction of resonant oscillations is vertical. In case (a) the resonance spans the amplitudes $A_y = 20$ to 40 at $A_x = 0$. However, the aperture is at about 80, and without additional resonances it is difficult for a particle to overcome the strong damping between 40 and 80. The situation changes dramatically when positive nonlinearity is added, case (b). In a certain region in amplitude space the beam-beam-induced nonlinearity is canceled by the machine nonlinearity and a region of very wide oscillations appears. At $A_x = 1.5$ the resonance spans the entire distance from $A_y = 15$ to $A_y = 70$. In this case a lower edge of the resonance can be considered as a nonlinear dynamic aperture. In case (c) with negative machine nonlinearity, such a region is absent and the machine nonlinearity simply bends the resonance upwards as A_x increases.

Temnykh[25] made experimental observations on the effect of machine nonlinearity on the resonances induced by beam-beam effects, at the VEPP-4 storage rings. The method was to observe the particle loss rate during the betatron tune scan. In Fig. 24 all regions with high particle loss rage are easily identified with certain resonances.

Several scans were made across the resonance $7\nu_x = 60$ with different values of cubic nonlinearity. Figure 25 shows the results of two scans with $R_x \cong \pm 6.9\ \text{cm}^{-1}$. It is interesting that the specific luminosity, $L_{sp} = L/i^+i^-$, does not change while the loss rate changes significantly.

A second example of a machine imperfection is a small $(\lesssim 0.1\sigma_{x,y})$ separation of the colliding beam orbits. Its effect is especially important for two-ring colliding beam facilities. The orbit separation breaks the symmetry of the beam-beam interaction potential, and this results in the appearance of odd resonances in addition to the even ones. The set of even and odd resonances can overlap much more easily than the set of only even resonances.

The appearance of odd resonances at a small orbit separation was observed on the VEPP-4 storage ring.[25] Figure 26 shows the loss rate and specific luminosity during scans across resonance $7\nu_x = 60$ at different orbit separations. At zero separation (b) the resonance $7\nu_x = 60$ disappears and the resonance $14\nu_x = 120$ is not observed.

It is interesting that observing the loss rate on the anomalous high-order beam-beam resonances appears to be more effective for accurate tuning of beam collisions than observing the specific luminosity. In practice, static orbit separation can result from orbit separation ripple, which only worsens the situation.

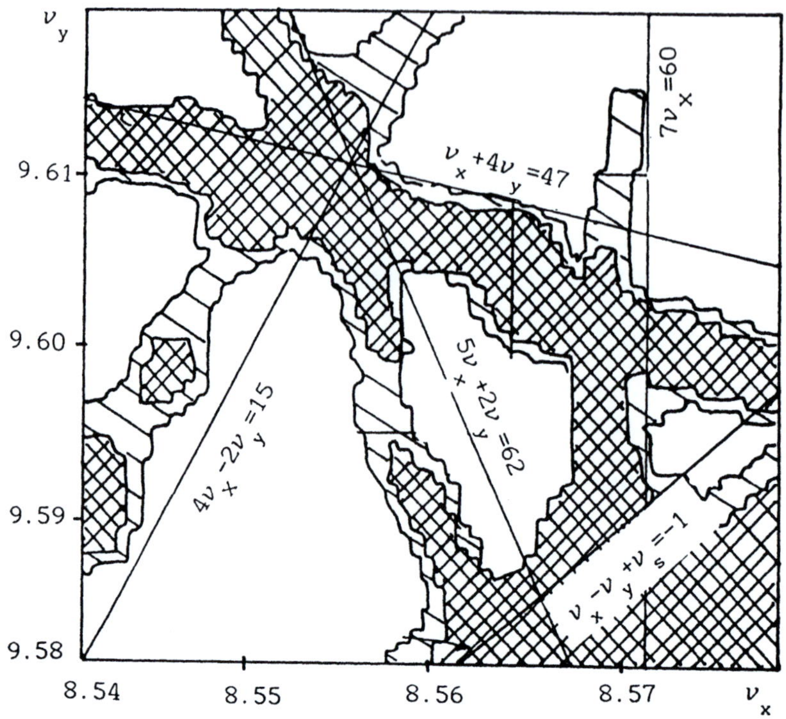

Fig. 24. Positron loss rate versus position on tune plane near the main working point of the VEPP-4 storage ring. The following resonances are identified: $7\nu_x = 60$; $\nu_x + 4\nu_y = 47$; $5\nu_x + 2\nu_y = 62$; $4\nu_x - 2\nu_y = 15$; $\nu_x - \nu_y + \nu_s = -1$. The orbit separation was $\sim 0.1\sigma_x$ in the horizontal direction during the experiment.

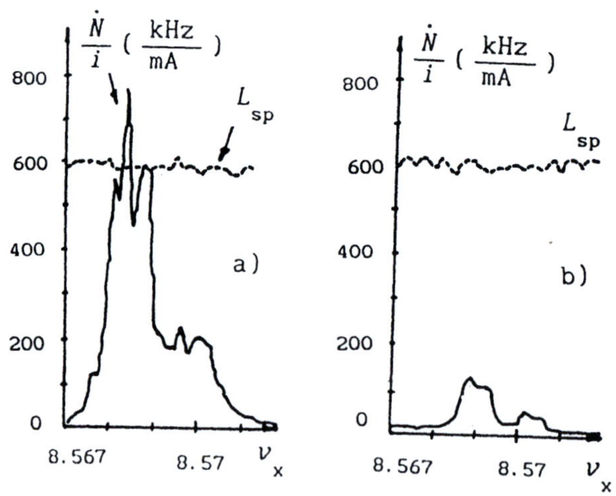

Fig. 25. Dependence of positron loss rate and specific luminosity, L_{sp}, on the horizontal betatron tune during the scan across the resonance $7\nu_x = 60$ on VEPP-4 with opposite values of the cubic machine nonlinearity: (a) $R_x \cong 6.9\ cm^{-1}$; (b) $R_x = -6.9\ cm^{-1}$.

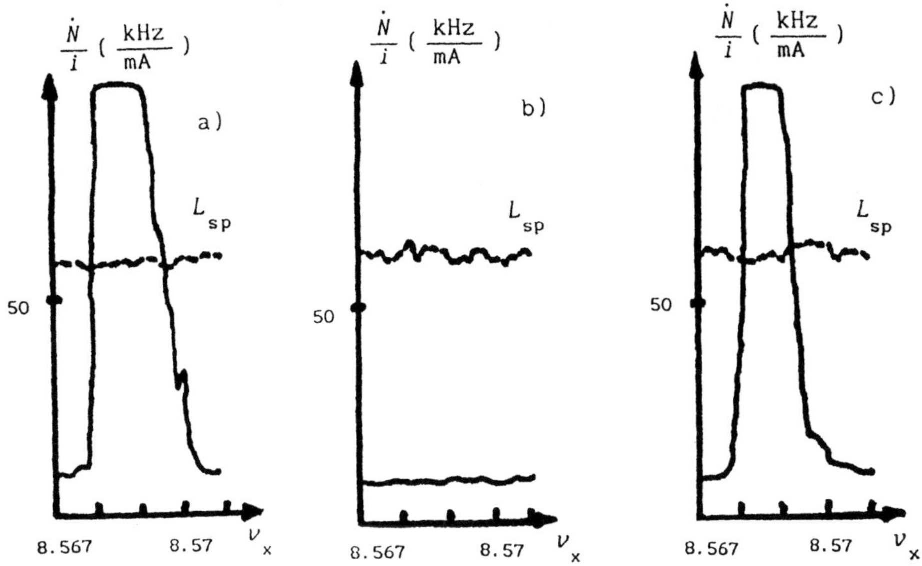

Fig. 26. Dependence of the positron loss rate and specific luminosity, L_{sp}, on the horizontal betatron tune during the scan across the resonance $7\nu_x = 60$ on VEPP-4. Orbit separation: (a) $-0.1\sigma_x$; (b) zero; (c) $0.1\sigma_x$.

A third source of machine effects on the beam-beam interaction is asymmetry in the betatron phase advance, $\mu_{x,y}$, between interaction points, which also leads to anomalous resonances. Ivanov et al.[26] observed two effects on the VEPP-2M storage ring with the variation of phase advance of the vertical betatron oscillations, $\Delta\mu_y$: blowup of the vertical beam size (Fig. 27a) and degradation of the maximal ξ_y (Fig. 27b).

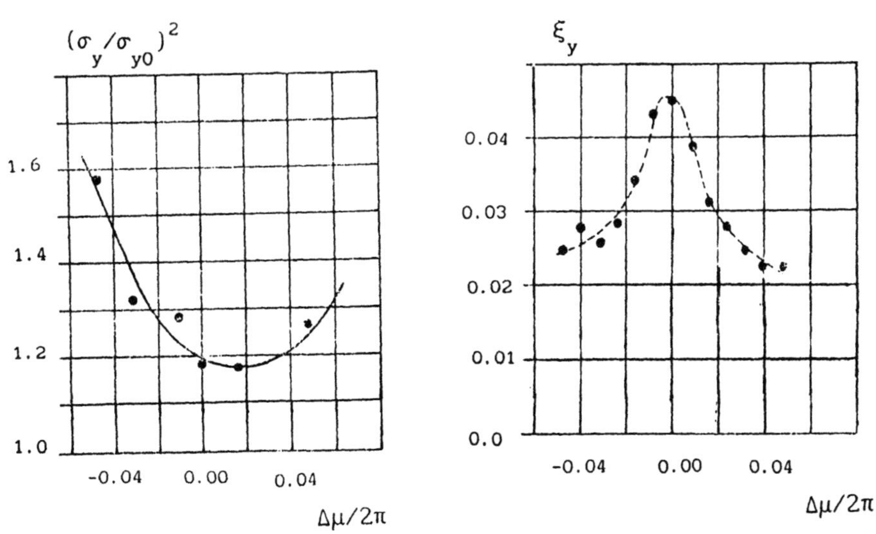

Fig. 27. Vertical emittance and maximal ξ_y versus the asymmetry of the vertical betatron phase advance between the interaction points (VEPP-2M, 510 MeV, $\nu_x = 3.059$, $\nu_y = 3.097$).

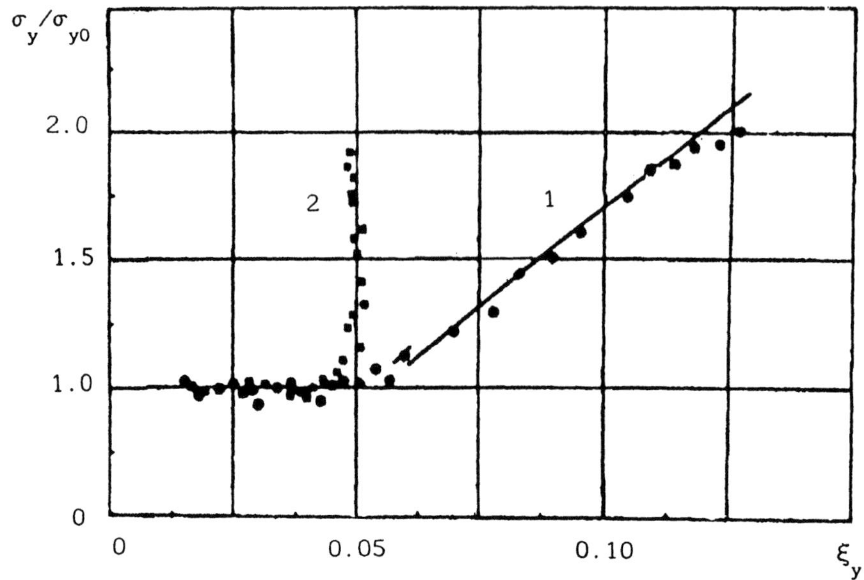

Fig. 28. Vertical beam size blowup versus the attained value of ξ_y. (1) $i^+ << i^-$; (2) $i^+ = i^-$. (VEPP-2M, E = 510 MeV, $\beta_y/\sigma_z = 2.5$, $\nu_x = 3.054$, $\nu_y = 3.087$.)

10 STRONG-STRONG INTERACTION AND FLIP-FLOP EFFECT

Consideration of the beam-beam effects in the weak-strong approximation can provide some understanding of the phenomena leading to beam intensity limitation, but it can hardly pretend to provide an exact value of the limit in a real machine, where we are dealing with two beams of almost equal intensity. This is clearly demonstrated by observations on the VEPP-2M storage ring[26] (Fig. 28), which show a big difference between weak-strong and strong-strong cases in vertical beam size behavior versus strong beam intensity.

This means that an ideal solution of the beam-beam problem must be self-consistent. Temnykh[27] attempted a phenomenological explanation of the flip-flop effect. Irwin et al.[28] and Tennyson[29] carried out similar studies.

A flip-flop effect between the sizes of the two beams has been observed. With increased intensity of both beams, one beam gains dominance over the other, blowing it up while reducing its own size. The collapsed beam effectively becomes stronger while the blown-up beam effectively becomes weaker. One can reverse the roles of the beams by exerting some asymmetric external influence.

Temnykh[27] proposed the following system of algebraic equations to describe the internal links between the beam dimensions and currents:

$$\sigma_+^2 = \sigma_0^2 + \left(ai^-/\sigma_-\right)^p, \qquad \sigma_-^2 = \sigma_0^2 + \left(ai^+/\sigma_+\right)^p. \tag{10.1}$$

Here σ_0 is the natural size, assumed to be the same for the two beams; σ_+, σ_- are the blown-up beam dimensions; i^+, i^- are beam currents; and a, p are parameters to be chosen, which are supposedly related to the constants affecting beam size, but the nature of whose effects has not been determined.

At constant currents each equation in (10.1) describes the dependence of one beam size on the opposite beam size. This is shown in Fig. 29 for three values of i^+.

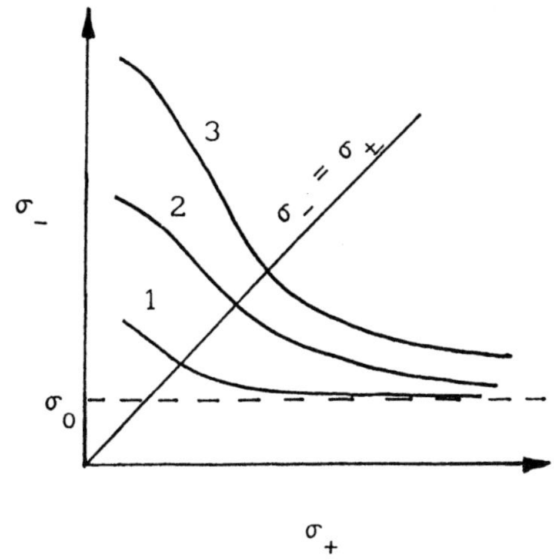

Fig. 29. Behavior of one beam size as a function of the opposite beam size. (1) $i^+ = i_0$; (2) $i^+ = 2i_0$; (3) $i^+ = 3i_0$.

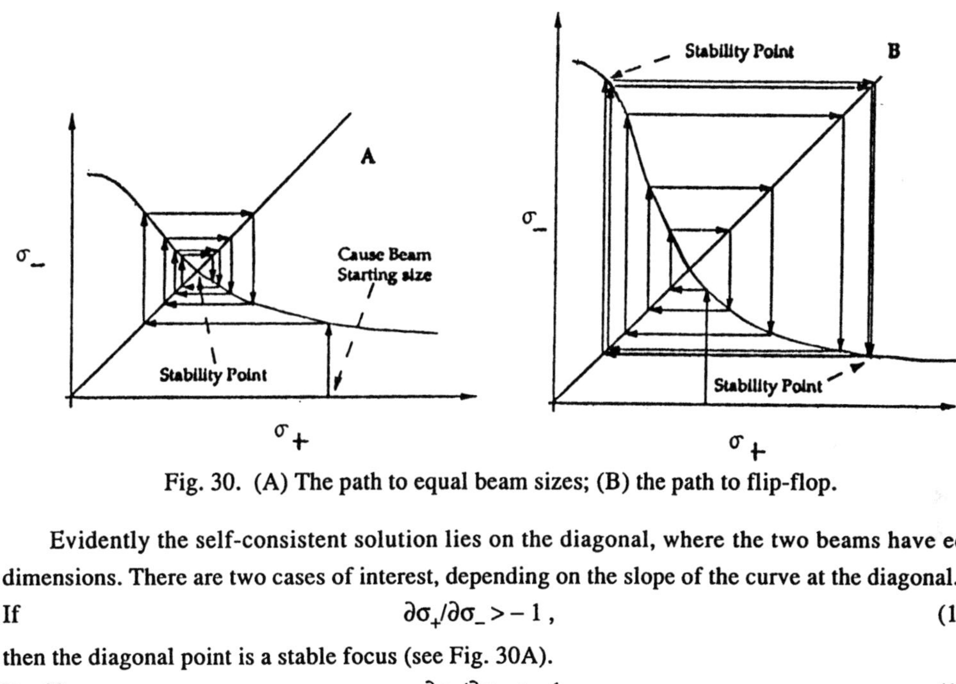

Fig. 30. (A) The path to equal beam sizes; (B) the path to flip-flop.

Evidently the self-consistent solution lies on the diagonal, where the two beams have equal dimensions. There are two cases of interest, depending on the slope of the curve at the diagonal.

If
$$\partial\sigma_+/\partial\sigma_- > -1\,, \tag{10.2}$$
then the diagonal point is a stable focus (see Fig. 30A).

But if
$$\partial\sigma_+/\partial\sigma_- < -1\,, \tag{10.3}$$
the diagonal point becomes an unstable saddle (see Fig. 30B) and two new stable points appear (see Fig. 31).

At
$$\partial\sigma_+/\partial\sigma_- = -1\,, \tag{10.4}$$
the solution of Eq. (10.1) is changing from equal-dimension beams to flip-flop beams. This condition corresponds to a certain beam current, which we define as the critical one. At the natural assumption used for simplification, $i^+ = i^- = i$, the critical current i^* is[27]
$$i^* = \frac{\sigma_0^{(1+2/p)}}{a}\left(\frac{p}{p-2}\right)^{1/2}\left(\frac{2}{p-2}\right)^{1/p}, \tag{10.5}$$

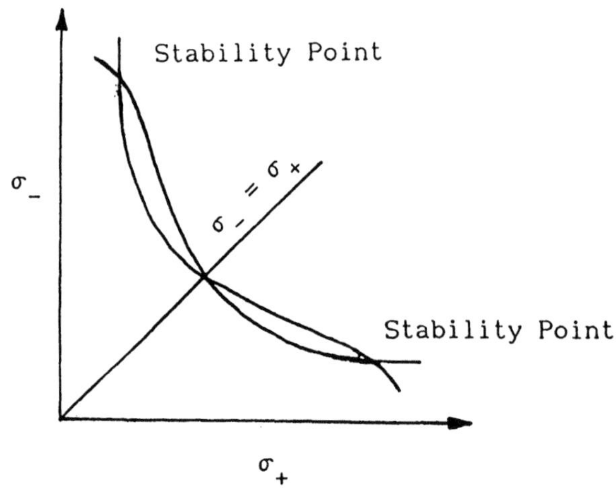

Fig. 31. The system of equations (10.1) above the critical condition, Eq. (10.2).

and the critical size σ^* is

$$\sigma^* = \sigma_0\left(\frac{p}{p-2}\right)^{1/2}. \tag{10.6}$$

It is interesting that i^* and σ^* increase as p approaches 2. At $p = 2$, i^* and σ^* are infinitely large, which means absence of the flip-flop effect.

The above discussion of the flip-flop effect provided an example where strong-strong behavior of the beams can be derived from weak-strong behavior. This is advantageous because it allows reduction of the complex strong-strong model of beam-beam interactions to an easier weak-strong model.

11 ADDING RADIATION DAMPING AND QUANTUM FLUCTUATION NOISE

It is well known that the electrons and positrons in a storage ring undergo radiation damping accompanied by quantum fluctuation noise; therefore, a correct model of beam-beam effects should include these. Below we show that damping and noise in the particle motion lead to a new phenomenon which is absent in the conservative case.

We begin with experimental observations. Figure 32[6] shows data on the dependence of ξ on damping decrements, from various machines. The maximal ξ values, measured under almost identical experimental conditions during an energy scan on VEPP-4, show a weak dependence of ξ on the damping decrements (Fig. 33). In both plots the spread at any particular decrement value is quite large, and any correlations in which the maximal values of ξ are calculated from the decrements appear quite speculative.

During the last decade several theories of beam-beam effects including noise and damping appeared.[30–32] The goal of all of them was calculation of the particle distribution in the presence of nonlinear resonances, which would allow finding the beam blowup and particle losses at a certain aperture. A common method is to solve the Fokker-Planck equation for the evolution of the particle distribution. Recently, these studies were reviewed and several new results were obtained.[33–35] Here, we review briefly Gerasimov's main ideas[33,34] on the application of this method to beam-beam effects.

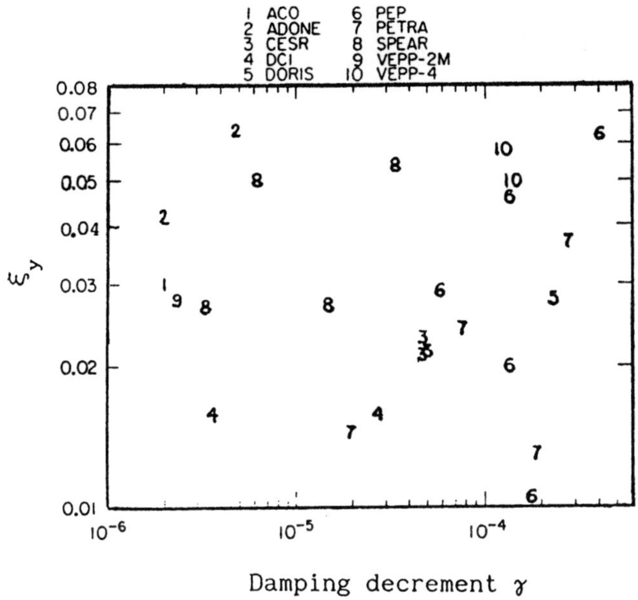

Fig. 32. Maximal ξ_y versus damping decrements observed on various machines.

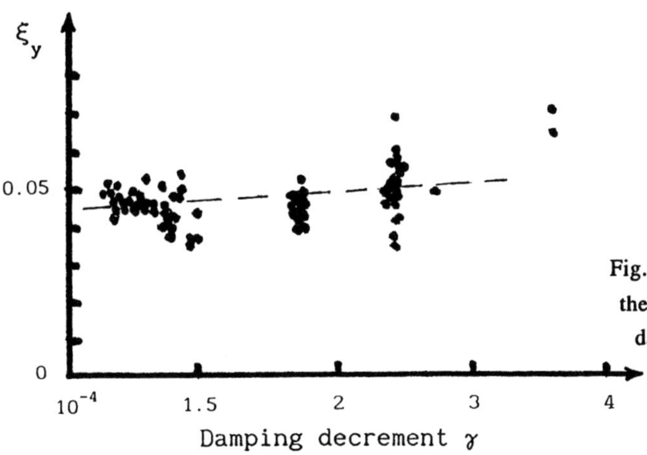

Fig. 33. Dependence of maximal ξ_y on the damping decrement. Experimental data from VEPP-4 during the energy scan in the physic runs.

The influence of radiation damping and quantum fluctuation noise can be considered by adding to the Hamiltonian equations the dissipative and random terms:

$$\dot{\mathbf{X}} = \partial H/\partial \mathbf{P}\,, \qquad \dot{\mathbf{P}} = (\partial H/\partial X) - \gamma \mathbf{P} + \sqrt{2\eta}\,\zeta(t)\,, \tag{11.1}$$

where H is the Hamiltonian (3.1), γ is the damping decrement, and η is the diffusion coefficient. The random process $\zeta(t)$ is a white noise process,

$$\langle \zeta_i(t)\, \zeta_j(t+\tau) \rangle = \delta_{ij}\delta(\tau)\,, \tag{11.2}$$

and the symbol $\langle \quad \rangle$ denotes averaging over time.

The Fokker-Plank equation describing the evolution of the particle distribution function has the form[36]

$$\frac{\partial\rho}{\partial t}+P\frac{\partial\rho}{\partial X}-\omega X-\frac{\partial\left[V-\delta_T(t)\right]}{\partial X}\frac{\partial\rho}{\partial P}=\frac{\partial}{\partial P}\left(\gamma P\rho+\eta\frac{\partial\rho}{\partial P}\right), \tag{11.3}$$

where V is a perturbation potential from the Hamiltonian (3.1).

In the absence of perturbation, Eq. (11.3) yields the stationary Hibbs distribution,

$$\rho=N\exp\left\{-\frac{\gamma}{\eta}\left[\frac{1}{2}\left(P^2+\omega_\ell^2X_\ell^2\right)\right]\right\}, \tag{11.4}$$

with the temperature $kT=\eta/\gamma$ and ω_ℓ being the betatron frequencies (summation over repeated indices is implied).

For solving (11.3) in the case of time-dependent perturbation, the following assumptions are useful: damping and noise are small relative to unperturbed dynamics; ρ is averaged in oscillation phases on the large interval of betatron oscillations; the slow variation of ρ depends only on the action variables, $\rho(J,t)$. The resulting so-called thermal-averaged Fokker-Plank equation is

$$\frac{\partial\rho}{\partial\tau}=\frac{\partial}{\partial J_k}\left(-\gamma_k(J)+\eta G_{k\ell}(J)\frac{\partial\rho}{\partial J_\ell}\right), \tag{11.5}$$

where the quantities F_k, $G_{k\ell}$,

$$F_k=-\left\langle p_1\frac{\partial J_k(X,P)}{\partial P_1}\right\rangle, \qquad G_{k\ell}=\left\langle\frac{\partial J_k(X,P)}{\partial P_1}\frac{\partial J_\ell(X,P)}{\partial P_1}\right\rangle, \tag{11.6}$$

are the averages along the trajectories of the unperturbed Hamiltonian and time.

The relaxed distribution function is the solution of the stationary form of Eq. (11.5):

$$\frac{\partial}{\partial J_k}\left(-\gamma F_k(J)+\eta G_{k\ell}(J)\frac{\partial\rho}{\partial J_\ell}\right)=0\,. \tag{11.7}$$

The divergence form of Eq. (11.7) means the existence of fluxes in phase space, which are circulated in closed loops. This process was called phase convection.

Of most interest to us are the probabilities of large fluctuations, defined by the tails of the relaxed distribution function at $|J|>>\eta/\gamma$. Therefore, we can apply the weak-noise asymptotic $\nu\to 0$, corresponding to the low-temperature limit $kT\ \to\ 0$, for a description of the relaxed distribution function

$$\rho(J,T)=z\exp\left\{-\frac{\Phi(J)}{kT}+O(kT)\right\}, \tag{11.8}$$

and reduce the complexity of the second-order Fokker-Planck equation for ρ to the first-order equation for function Φ.[37]

The influence of isolated nonlinear resonances on the relaxed distribution function and the associated speed-up of the escape rate can be best described in the weak-noise asymptotic through the concept of the extreme trajectory, i.e. the most probable trajectory of a particle in reaching a point in space, **J**, starting from the center **J** = 0.

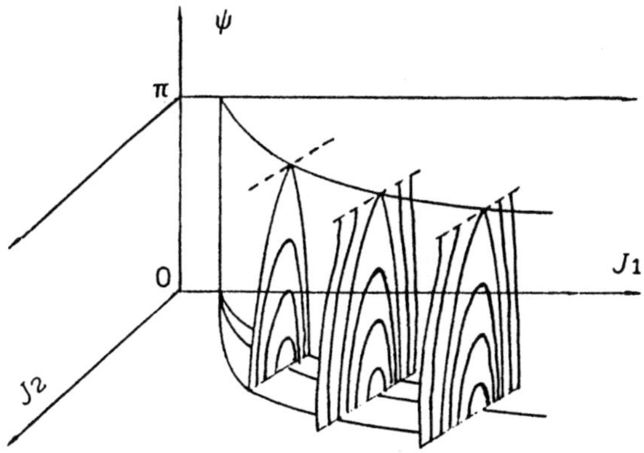

Fig. 34. Phase trajectories in the space of unperturbed actions J_1, J_2 and resonant phase $\psi = \ell\Phi_x + m\Phi_z - k\theta$ in the vicinity of an isolated nonlinear resonance.

For the initial conditions outside the separatrix (the "tube") of Fig. 34, the trajectories of damping particles differ only a little, $\sim V^{1/2}$, from those of unperturbed ones. For initial conditions inside the separatrix, the trajectories go by the contracting spiral along the central line of the tube, if the damping is small enough, $\gamma << V$. This phenomenon was first identified by Tennyson[38] and named resonance streaming. The speed $\dot{\mathbf{J}}_r$ of the motion along the resonance line can be found by decomposing the vector of the damping force $\mathbf{F}_0$ into two components: $\mathbf{F}_\ell$, parallel to the vector of the resonance oscillations $\bar{\ell} = (\ell, m)$, and $\mathbf{F}_r$, tangent to the resonance line, so that $\dot{\mathbf{J}}_r = \gamma\mathbf{F}_r$. Thus, neglecting the motion on the small scale of the order of the resonance width $\sim V_{1/2}$, one can say that the resonance cancels the component of the damping force $\mathbf{F}_\ell$ along the direction of resonance oscillations and the particle moves along the resonance line under the influence of the component $\mathbf{F}_r$.

The particle motions driven by noise and damping along the narrow $\sim V^{1/2}$ resonance tube can be effectively considered one-dimensional. With this additional simplification the variation of Φ along the resonance line is described by a one-dimensional equation,

$$F_r \frac{d\Phi}{dJ_r} + G_r \left(\frac{dF}{dJ_r}\right)^2 = 0\,, \tag{11.9}$$

where J_r is the distance $|\mathbf{J}|$ along the resonance line, and F_r and G_r are the components of damping force and diffusion taken along the resonance.

Two possible solutions of Eq. (11.9), $d\Phi/dJ_r = 0$ and $d\Phi/dJ_r = -F_r/G_r$, correspond to two possible directions of the extreme trajectories in the resonance: along and against the component $\mathbf{F}_r$. It is clear that the function Φ measuring the difficulty of attaining given points in the phase space stays constant on a certain section of the extreme trajectory if this section coincides with the motion under the influence of damping. An example of this case is shown in Fig. 35A. The motion against damping is difficult, so that the function Φ grows along the extreme trajectory $(d\Phi/dJ_r \neq 0)$. Nevertheless, the easiest path to some point $\mathbf{J}_0$ from the center $\mathbf{J} = 0$ may include large distances inside the resonance tube. An example of this case is shown in Fig. 35B.

Because of the existence of extreme paths, new attractor phase-space regions other than the center $\mathbf{J} = 0$ are possible, and the circulating particle fluxes are formed as shown in Fig. 36.

Similar effects were observed in the VEP-1 storage ring after excitation of a single beam by an external periodically driven force.[39] In Fig. 37, a TV representation of the beam transverse cross-section, new attractor regions other than the beam center are clearly observed.

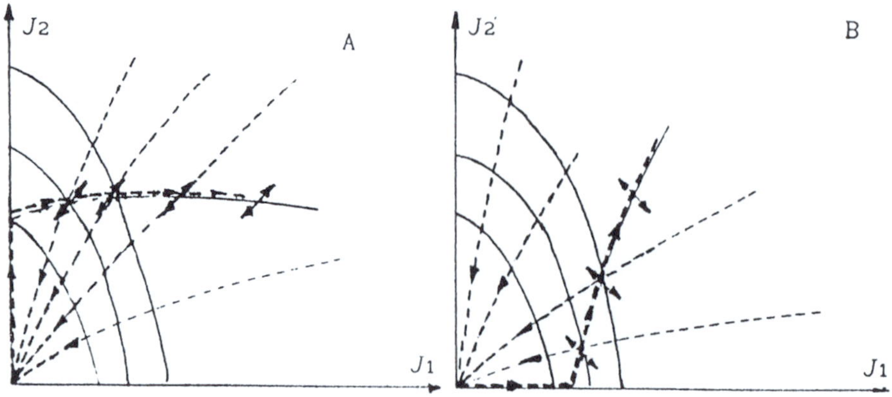

Fig. 35. (A) The extreme path along the resonance "along damping."
(B) The extreme path along the resonance "against damping."
Dashed lines: unperturbed motion with damping; heavy dashed line: extreme path.
Heavy solid line: resonance line. Small arrows: oscillations at the resonance.

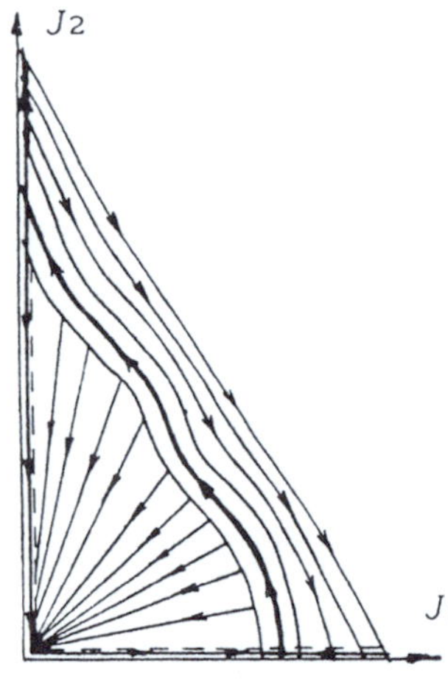

Fig. 36. Distributions of fluxes in the plane J_1, J_2.

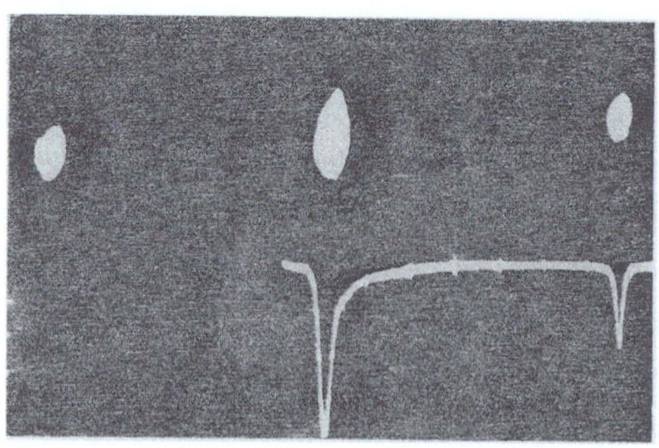

Fig. 37. TV representation of the beam transverse cross-section and the oscillogram of the particle density distribution. The vertical betatron tune is chosen near the fourth-order resonance $4\nu_y = 3$.

Inside the resonance tube the particle may deviate from the central line and reach the separatrix surface with an exponential small probability

$$\Delta W \sim \exp\,(-\gamma B/\eta) \tag{11.10}$$

where B is proportional to the resonance width. The weaker the resonance, the higher the probability of escape.

With the technique discussed above, one can calculate the relaxed distribution function for the case of particle motion perturbed by several isolated (non-overlapping) nonlinear resonances. An example of two-dimensional calculations, for the regime with betatron tunes $\nu_x = 0.025$, $\nu_y = 0.15$, and for radiation damping time (in numbers of beam-beam interactions) $N = 3 \cdot 10^3$, is shown in Fig. 38a. Here, the constant value lines of a function $\rho(A_x, A_y)/(A_x A_y)$ are presented. Each line corresponds to expontential reduction of particle density with factor e. A similar distribution of particle density was constructed by a special tracking technique in the weak-strong simulation of beam-beam interactions for the case of large dispersion at the IP. The results with the model not including the effect of tune modulation are shown in Fig. 38b. The coincidence of anaytical and numerical results is quite satisfactory here, but it is lower in the fully three-dimensional case with tune modulation included (see Fig. 39).

A similar tracking technique with step-by-step determination of the constant value lines of particle density was developed for lifetime calculations in beam-beam interactions. Then it was applied to analysis of the regime with large dispersion of the IP.[21] The case of zero dispersion ($\lambda = 0$) and moderate dispersion ($\lambda = 0.7$) and two cases of large dispersion ($\lambda = 5$, $\lambda = 25$) were studied at two working points of the betatron tunes near the main coupling resonance. The results (Fig. 40) show that the best lifetimes are for the large dispersion cases. This provides additional evidence for the effectiveness of large dispersion in suppressing all coupling resonances.

ACKNOWLEDGEMENTS

I am grateful to my wife Lyuba for her patience and encouragement. I also with to thank V. Lebedev, D. Rice and D. Shatilov for reviewing the manuscript and giving useful suggestions.

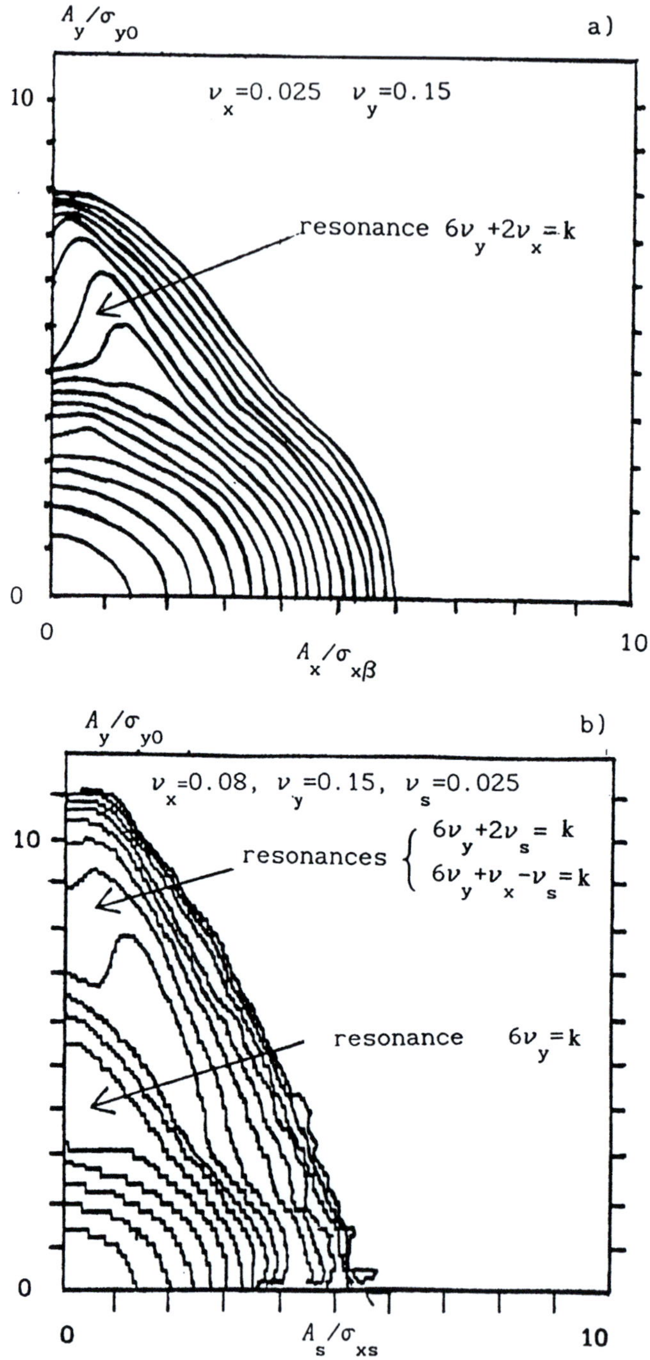

Fig. 38. Contour plot of the particle density distribution function $\rho(A_x,A_y)/a_x,A_y$. The constant value lines show the exponential reduction of particle density with factor e. (a) Model calculations; (b) results of simulation, not including the effect of tune modulation.

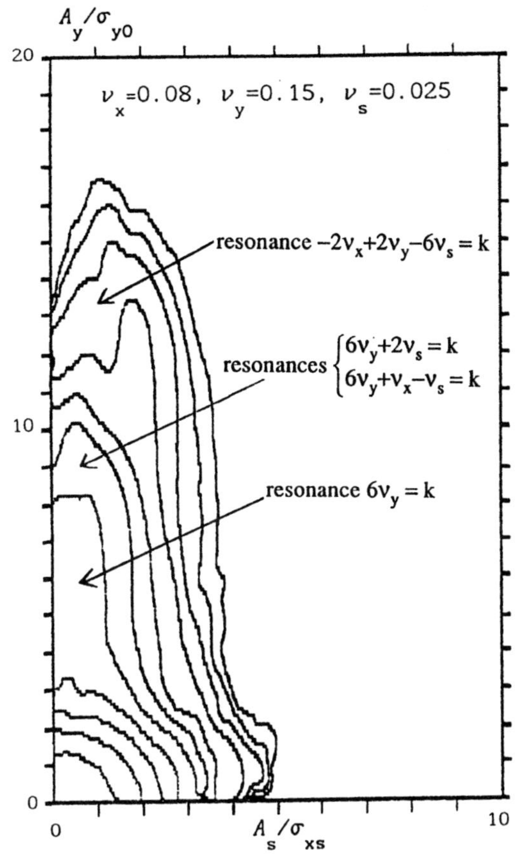

Fig. 39. The same as Fig. 38b but with the effect of tune modulation included in the simulation.

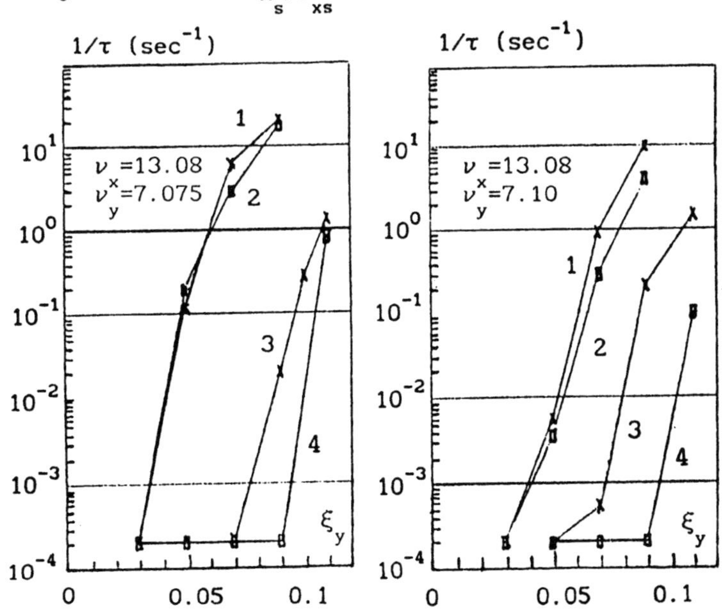

Fig. 40. Dependence of the inverse lifetime of the weak beam on the vertical tune shift ξ_y. The horizontal tune shift satisfies the condition $\xi_x = \xi_y/5$. (1) λ=0; (2) λ=0.7; (3) λ=5; (4) λ=25.

APPENDIX: POTENTIAL OF A 3-DIMENSIONAL GAUSSIAN CHARGE DISTRIBUTION[1]

Consider the Poisson equation

$$\nabla^2 U(\mathbf{r}) = -4\pi\rho(\mathbf{r}) \tag{A.1}$$

with charge distribution $\rho(\mathbf{r})$, where $\mathbf{r}$ denotes $(\bar{x},\bar{y},\bar{z})$.

The Green's function corresponding to Eq. (A.1) is well known in the form

$$G(\mathbf{r},\xi) = \frac{1}{4\pi|\mathbf{r}-\xi|}, \tag{A.2}$$

which satisfies the equation

$$\nabla^2 G(\mathbf{r},\xi) = -\delta(\mathbf{r}-\xi) \tag{A.3}$$

where ξ denotes (ξ_1,ξ_2,ξ_3). Now we rewrite the Green's function (A.2) as an integral representation

$$\frac{1}{4\pi|\mathbf{r}-\xi|} = \frac{1}{2\pi^{3/2}} \int_0^\infty \exp\left[-|\mathbf{r}-\xi|^2 q^2\right] dq\,, \tag{A.4}$$

which is well known as the integration formula. Note that the couplings among the three coordinates $(\bar{x},\bar{y},\bar{z})$ appearing on the left-hand side of (A.4) are separated in the integral. For later application, we change the integration variable

$$t = 1/q^2\,, \tag{A.5}$$

then

$$G(\mathbf{r},\xi) = \frac{1}{4\pi^{3/2}} \int_0^\infty \frac{\exp\left[-|\mathbf{r}-\xi|^2/t\right]}{t^{3/2}}\, dt\,. \tag{A.6}$$

Using Eq. (A.6), we obtain a formal solution of the Polsson equation

$$U(\mathbf{r}) = \frac{1}{\pi^{1/2}} \int_0^\infty \frac{dt}{t^{3/2}} \iiint_{-\infty}^{\infty} d\xi\rho\,(\xi) \exp\left[\frac{|\mathbf{r}-\xi|^2}{t}\right]. \tag{A.7}$$

Now we consider a three-dimensional Gaussian charge distribution

$$\rho(\mathbf{r}) = \frac{Ne}{(2\pi)^{3/2}abc} \exp\left[-\frac{\bar{x}^2}{2a^2} - \frac{\bar{y}^2}{2b^2} - \frac{\bar{z}^2}{2c^2}\right], \tag{A.8}$$

where a, b, c are the standard deviations.

Substitution of (A.8) into (A.7) yields

$$U(\mathbf{r}) = \frac{Ne}{(2\pi)^{3/2}abc\pi^{1/2}} \int_0^\infty \frac{dt}{t^{3/2}} \iiint_{-\infty}^{\infty} d\xi \exp\left[-\frac{|\mathbf{r}-\xi|^2}{t} - \frac{\xi_1^2}{2a^2} - \frac{\xi_2^2}{2b^2} - \frac{\xi_e^2}{2c^2}\right] \tag{A.9}$$

and integration results in

$$U(\mathbf{r}) = \frac{Ne}{\sqrt{\pi}} \int_0^\infty \frac{\exp\left[-\frac{x^2}{(2a^2+t)} - \frac{y^2}{(2b^2+t)} - \frac{z^2}{(2c^2+t)}\right]}{\sqrt{(2a^2+t)(2b^2+t)(2c^2+t)}} dt . \tag{A.10}$$

REFERENCES

1. K. Takayama. A New Method for the Potential of a 3-Dimensional Nonuniform Charge Distribution. Lett. al Nouvo Cimento **34** (N7), 190 (1982).
2. M. Bassetti and G. Erskine. CERN ISR-TH/80-06 (1980).
3. S. Myers. Review of Beam-Beam Simulations, in Nonlinear Dynamics Aspects of Particle Accelerators, Lecture Notes in Physics Vol. **247**, p. 176, Springer (1985).
4. A. Chao. Bream-Beam Instability, in Physics of High Energy Particle Accelerators, AIP Conf. Proc. **127**, p. 201 (1985).
5. J.M. Greene, in Proc. Beam-Beam Interaction Seminar, SLAC, SLAC-PUB-2624 (1980).
6. J.T. Seeman. Observations of the Beam-Beam Interaction, in Ref. 3, p 121.
7. B.V. Chirikov. A Universal Instability of Multi-Dimensional Oscillator Systems. Phys. Rep. **52**, 263 (1979).
8. A. Gerasimov et al. The Dynamics of the Beam-Beam interaction, in Ref. 3, p.154.
9. S. Peggs. Some Aspects of Machine Physics at CESR, Ph.D. Thesis, Cornell U. (1981).
10. S.V. Milton. The Beam-Beam Interaction in Electron Storage Rings: A Study of the Weak/Strong Case, Ph.D. Thesis, Cornell U. (1990).
11. A. Gerasimov et al. Nonlinear Resonances and Beam-Beam Effects for Elliptical Beam, in Proc. 13th Int. Conf. on High Energy, Accelerators, Novosibirsk, Vol. **2**, p. 97 (1986).
12. R.D. Ruth. Single Particle Dynamics and Nonlinear Resonances in Circular Accelerators, in Ref. 3, p. 38.
13. F.M. Izrailev et al. Numerical Experiments for Determination of Stochasticity Criterion in Beam-Beam Interaction. Preprint INP 77-43, Novosibirsk (1977).
14. E.D. Courant. Beam-Beam Interaction for Bunched and Unbunched Beams, in Proc. 11th Int. Conf. on High Energy Accelerators, Geneva, pp. 763-6 (1970).
15. F.M. Izrailev. Nearly Linear Mappings and their Applications. Physica 1D, 243 (1980).
16. J.L. Tennyson, The Instability Treshold for Bunched Beams in ISABELLA, in Nonlinear Dynamics and the Beam-Beam Interaction, Eds. Month, Herrera, AIP Conf. Proc. **57**, p. 158 (1979).
17. F.M. Izrailev and I.B. Vasserman, The Influence of Different Types of Modulations on a Decrease of the Stochasticity Limit in Beam-Beam Effects, in Proc. 7th All-Union Conf. on Charged Particle Accelerators, Dubna, p. 288 (1981).
18. S.G. Peggs. Beam-Beam Synchrobetatron Resonances, Particle Accelerator Conf., Santa Fe, 1983, IEEE Trans. Nucl. Sci. **N5-30**, 2457 (1983).

19. S. Krishnagopal and R. Siemann, Bunch Length Effects in the Beam-Beam Interaction. Phys. Rev. D **41**, 2312 (1990).
20. N.S. Dikansky et al. Simulation of Beam-Beam Effects of Round Beams in Φ-meson Project. Presented at 12th All-Union Conf. on High Energy Accelerators (1990).
21. A.L. Gerasimov et al. Beam-Beam Effects with Big Dispersion Function at the Interaction Point, submitted to Nucl. Instr. Meth., Sept. 1990.
22. A. Pivinski, Synchrobetatron Resonances, in Ref. 3, p. 104.
23. I.B. Vasserman et al. Proc. 7th All-Union Conf. on Charged Particle Accelerators, Dubna, Vol. **1**, p. 184 (1981).
24. A.B. Temnykh. Influence of Cubic Nonlinearity of Storage Ring Magnetic Field on Beam-Beam Effects on VEPP-4, in Proc. 13th Int. Conf. on High Energy Accelerators, Novosibirsk, Vol. **1**, p. 78 (1986).
25. A.B. Temnykh, Observation of Beam-Beam Effects on VEPP-4, in Proc. 3rd Advanced ICFA Beam Dynamics Workshop on Beam-Beam Effects in Circular Colliders, Novosibirsk, p. 5 (1989).
26. P.M. Ivanov et al. Luminosity and Beam-Beam Effects on VEPP-2M with Superconducting Wiggler Magnet, in Ref. 25, p. 26.
27. A.B. Temnykh. Phenomenological Model of Flip-Flop in Colliding Beams, Preprint INP 82-148, Novosibirsk, 1982.
28. J. Irwin et al. The Flip-Flop Effect in Beam-Beam Interaction. Cornell Internal Note CON 87-15 (1987).
29. J.L. Tennyson. Flip-Flop Modes in Symmetric and Asymmetric Colliding-Beam Storage Rings. Preprint LBL-28013 (1989).
30. F. Ruggiero. Renormalized Fokker-Planck Equation for the Problem of the Beam-Beam Interaction in Election Storage Rings. Ann. Phys. **153**, 122 (1984).
31. S. Kheifets. Study of a Weak Beam Interaction with a Flat Strong Beam. Part. Accel. **15**, 153 (1984).
32. J.F. Schonfeld. Statistical Mechanics of Colliding Beams. Ann. Phys. **160**, 149 (1985).
33. A.L. Gerasimov. Phase Convection and Nonequilibrium Fluctuations in Nearly-Integrable Systems. Phys. Lett. A **135**, 92 (1989).
34. A.L. Gerasimov. Phase Convection and Distribution Tails in Periodically Driven Brownian Motion. Physica D **41**, 89 (1990).
35. Y.H. Chin. Renormalization Theory of Beam-Beam Interaction in Electron-Positron Colliders, in Ref. 25, p. 69.
36. H. Risken. The Fokker-Planck Equation. Springer (1983).
37. A. Ventzel and M. Freidlin. Random Perturbations of Dynamical Systems. Springer (1984).
38. J.L. Tennyson. Resonance Transport in Near Integrable System with Many Degrees of Freedom. Physica D 5, 123 (1982).
39. G.N. Kulipanov et al. Influence of Nonlinearities on Betatron Oscillations in Storage Rings. Preprint INP-251, Norosibirsk, 1968.

COHERENT BEAM-BEAM EFFECTS

Alexander W. Chao

Accelerator Division
Superconducting Super Collider Laboratory*
2550 Beckleymeade Avenue
Dallas, TX 75237

1 INTRODUCTION

There are two physical pictures that describe the beam-beam interaction in a storage ring collider: the weak-strong and the strong-strong pictures as shown in Figure 1. Both pictures play a role in determining the beam-beam behavior. This review addresses only the strong-strong picture. The corresponding beam dynamical effects are referred to as the coherent beam-beam effects. Some basic knowledge of the weak-strong picture is assumed.[1] To be specific, two beams of opposite charges are considered.

In the weak-strong picture, the strong beam is unperturbed by the weak beam. It is regarded as a rigid nonlinear lens at the beam-beam interaction points (IP). Beam dynamics in this picture is that of an incoherent single-particle motion

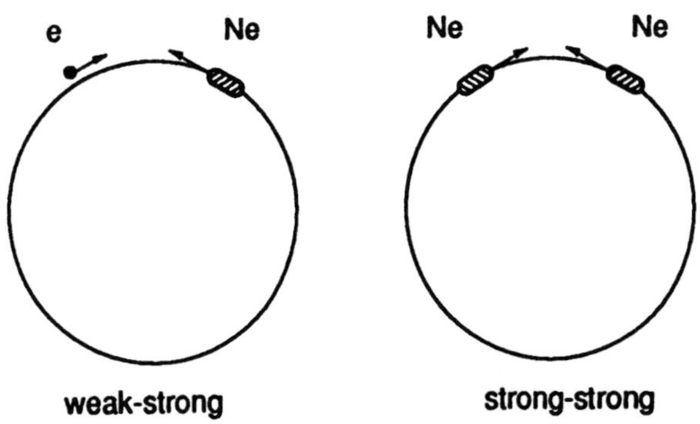

Figure 1. Weak-strong and strong-strong beam-beam interaction.

in the presence of a static nonlinear lens. The beam-beam problem is reduced to a nonlinear mapping problem.

In the strong-strong picture, both beams are perturbed. To study the coherent beam-beam effects, the procedure in principle is as follows:

(a) Let the steady-state distribution be Ψ_0. For example, Ψ_0 may be gaussian in the absence of the beam-beam interaction. With strong-strong perturbation, Ψ_0 must be solved in a self-consistent manner and in general is no longer gaussian.

(b) Given the perturbed, generally non-gaussian Ψ_0, consider distributions of the $\pm$ beams that deviate from Ψ_0, i.e.,

$$\begin{aligned} \Psi_+ &= \Psi_0 + \Delta\Psi_+ \\ \Psi_- &= \Psi_0 + \Delta\Psi_- \, . \end{aligned} \tag{1}$$

The $+$ beam sees the beam-beam force from the $-$ beam, which is related to Ψ_-, while the $-$ beam sees a beam-beam force similarly related to Ψ_+. The beam-beam force thus provides the coupling between $\Delta\Psi_\pm$. The question is then: are infinitesimal perturbations $\Delta\Psi_\pm$ stable under their mutual interaction? Stability of $\Delta\Psi_\pm$ determines the stability of the beam-beam system.

However, neither of the above steps is straightforward. Although much progress has been made, studies on coherent beam-beam effects so far have covered only pieces of the complex problem. This paper is a summary of some of these studies.

2 THE DYNAMIC-BETA MODEL

This is the simplest model in the strong-strong picture.[2] To proceed, consider a linearized beam-beam force (linearized with respect to the coordinates x and y of the particle which is experiencing the force). In this case, the perturbed steady-state Ψ_0 is gaussian if the unperturbed one is gaussian. The linear beam-beam force of one beam perturbs the other beam in such a way that the betatron phase advance between IPs $\mu_{x0,y0}$, the beta-functions at the IPs $\beta^*{}_{x0,y0}$, and the rms beam sizes at the IPs $\sigma_{x0,y0}$ are perturbed to become $\mu_{x,y}$, $\beta^*{}_{x,y}$, and $\sigma_{x,y}$ respectively. The perturbed quantities are given by

$$\cos\mu_{x,y} = \cos\mu_{x0,y0} - 2\pi\zeta_{x,y}\sin\mu_{x0,y0} \tag{2a}$$

$$\beta^*{}_{x0,y0}/\beta^*{}_{x,y} = \sin\mu_{x,y}/\sin\mu_{x0,y0} \tag{2b}$$

where $\zeta_{x,y}$ is the beam-beam strength parameter

$$\zeta_{x,y} = N\,r_0\,\beta^*{}_{x0,y0}/2\pi\gamma\sigma_{x,y}\,(\sigma_x+\sigma_y) \tag{3}$$

with γ the relativistic factor, N the number of particles per beam bunch, and $r_0 = e^2/mc^2$ the classical radius of the particle. Note that it is the unperturbed beta-functions and the perturbed beam sizes that appear in the expression of $\zeta_{x,y}$, Eq. (3).

Stability of the beams under linearized perturbations requires $|\cos\mu_{x,y}| \le 1$. This means, using Eq. (2a) and dropping subscripts x and y for abbreviation,

$$\zeta \le \cot(\mu_0/2)/2\pi\,. \tag{4}$$

Figure 2 shows the region of stability in the (ζ, ν_0) space, where ν_0, defined as $\mu_0/2\pi$, is the tune advance between IPs. The diagram repeats with a period of $\nu_0 = 1/2$. Sometimes, it is referred to as the sawtooth diagram because of its appearance when illustrated with a wider range of ν_0. Note that when ν_0 is close to but slightly below a multiple of 1/2, the stability region shrinks to zero. In other words, near 1/2-integer resonances, when $\nu_0 = m/2$, even a weak beam-beam perturbation makes the beam unstable.

In the incoherent case, the weak beam is perturbed by the strong beam. If the beam-beam force is linearized, the perturbed quantities of the weak beam are also given by Eq. (2), except that the beam-beam strength parameter $\zeta_{x,y}$ is

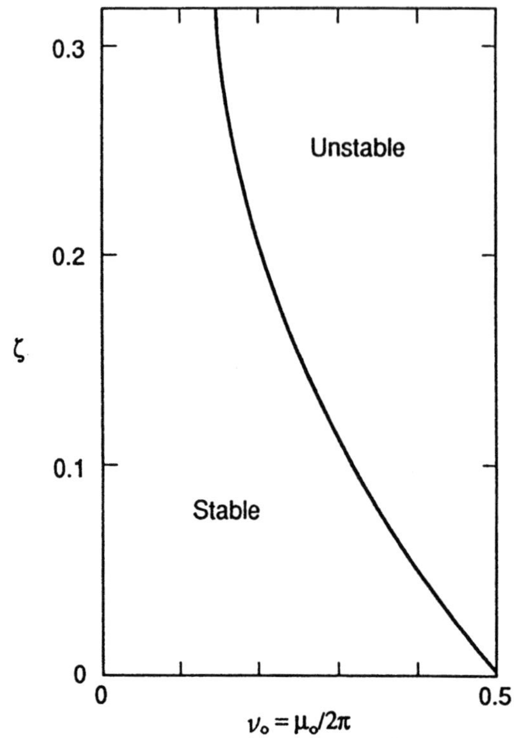

Figure 2. Stability diagram of the dynamic-beta effect. The incoherent weak-strong case has the same diagram if ζ is replaced by ζ_0. The behavior repeats with period $\nu_0 = 1/2$.

given by the strong-beam values which are always unperturbed. The incoherent stability condition is

$$\zeta_0 \leq \cot(\mu_0/2)/2\pi. \tag{5}$$

Although similar, Eqs. (4) and (5) have very different contents. In particular, Eq. (5) can be violated if one increases the beam intensity N, but it will be shown later that Eq. (4) is always satisfied for any value of N.

In solving Eqs. (2) and (3) for the strong-strong case, we note that the perturbed beam sizes scale with the perturbed beta-functions (the dynamic-beta), i.e., $\sigma_{x,y} \propto \sqrt{\beta^*{}_{x,y}}$, not $\sqrt{\beta^*{}_{x0,y0}}$, assuming beam emittances are unchanged due to beam-beam interaction. Further noting that $\beta^*{}_{x,y}$ depend on the perturbed beam-beam parameters $\zeta_{x,y}$, which in turn depend on $\sigma_{x,y}$, we conclude that $\sigma_{x,y}$, $\beta^*{}_{x,y}$, $\mu_{x,y}$, $\zeta_{x,y}$ and luminosity L need to be found self-consistently for a given value of N. The luminosity is given by

$$L = N^2 f M / 4\pi\sigma_x\sigma_y \tag{6}$$

where f is the revolution frequency of the beams around the storage ring, and M is the number of bunches in each beam.

As an example, consider the case of round beams $\sigma_{x0} = \sigma_{y0} = \sigma_0$ and $\mu_{x0} = \mu_{y0} = \mu_0$. Squaring both sides of Eq. (2b) and eliminating $\sin^2 \mu_{x,y}$ using Eq. (2a) give a self-consistency expression for $\beta^*/\beta^*{}_0$:

$$\left(\frac{\beta^*{}_0}{\beta^*}\right)^2 = 1 + 2\rho_0 \cot\mu_0 \frac{\beta^*{}_0}{\beta^*} - \rho_0{}^2 \left(\frac{\beta^*{}_0}{\beta^*}\right)^2 \tag{7}$$

where $\rho_0 = 2\pi\zeta_0$ has been defined. Beam intensity N is contained in ρ_0.

Equation (7) can be solved to give $\beta^*/\beta^*{}_0$ in terms of the unperturbed quantities:

$$\frac{\beta^*}{\beta^*{}_0} = \sqrt{1 + \left(\frac{\rho_0}{\sin\mu_0}\right)^2} - \rho_0 \cot\mu_0 \ . \tag{8}$$

Ratios of other quantities follow from Eq. (8):

$$\frac{\beta^*{}_0}{\beta^*} = \frac{\zeta}{\zeta_0} = \frac{\sigma_0{}^2}{\sigma^2} = \frac{L}{L_0} \ . \tag{9}$$

Figure 3 shows the dependences of the dynamic-beta, the perturbed beam-beam parameter, and the luminosity on beam intensity N for various values of ν_0. As far as the dynamic-beta effect is concerned, the luminosity would benefit from having ν_0 slightly above a 1/2-integer and would suffer if it is slightly below a 1/2-integer.

As N increases, the beam-beam parameter approaches a constant value from below as shown in Figure 3(b). The constant value is that of the stability condition, Eq. (4). This means that for round beams, $\zeta \to$ constant, $\sigma \to \sqrt{N}$, and $L \to N$ as $N \to \infty$, to be compared with the weak-strong case which pre-

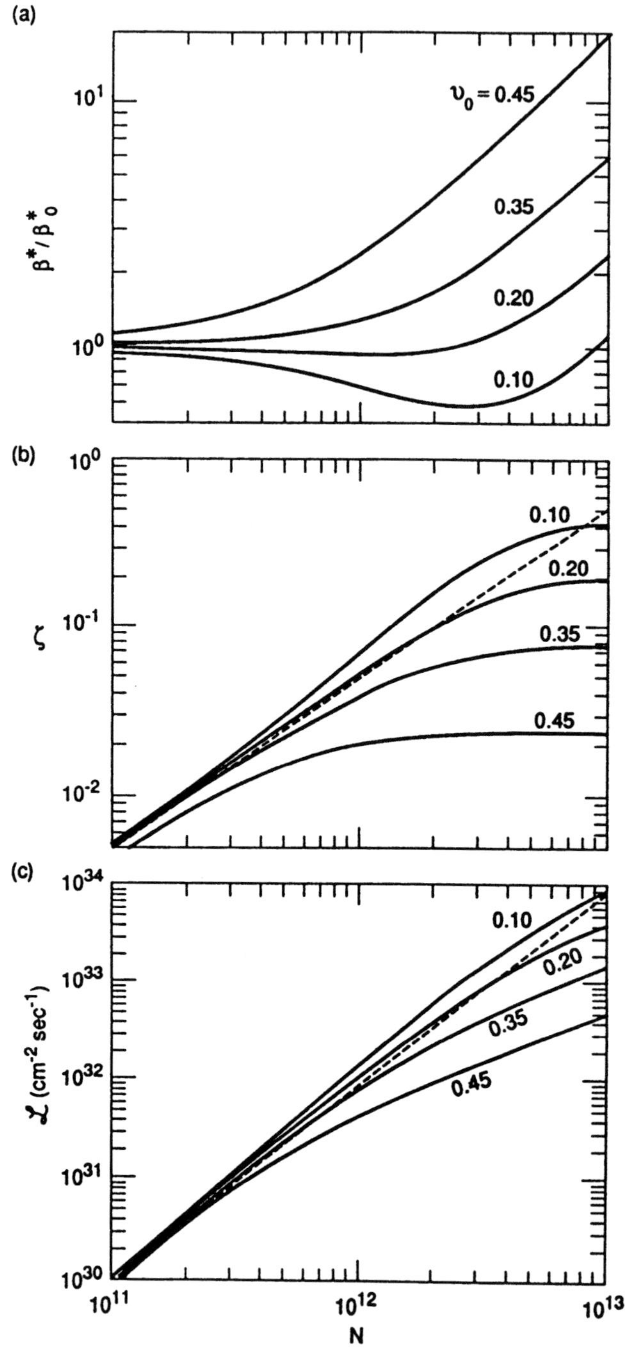

Figure 3. Dynamic-beta behavior of $\beta^*/\beta^*{}_0$, ζ and L as functions of N for various values of ν_0 (the unperturbed tune advance between IPs, modulus 1/2). Dashed curves in (b) and (c) are the unperturbed case. Numbers are normalized so that $L_0 = 10^{32}\,\mathrm{cm}^{-2}\mathrm{s}^{-1}$, $\rho_0 = 0.05$ when $N = 10^{12}$.

dicts $\zeta \to N$, $\sigma \to$ constant, and $L \to N^2$ as $N \to \infty$. More significantly, this means that as N is increased, the beam size always expands sufficiently to avoid violating the instability condition (4). The two beams are always stable.

Although the system is always stable, it still suffers from resonances. The beams blow up and L levels off faster when ν_0 is just below a 1/2-integer.

Figure 3 is true for a round beam. In case of a flat beam, a different scaling is obtained with N. As $N \to \infty$, dynamic-beta model gives $\zeta_x \to N$ (provided it stays well below the instability condition), $\sigma_x \to$ constant, $\zeta_y \to$ constant, $\sigma_y \to N$, and $L \to N$. Compare with Figure 26 later.

3 FLIP-FLOP EFFECT

The dynamic-beta model discussed in the previous section can be generalized. Still consider round beams and linearized beam-beam forces, but now allow the two beams to behave differently. Using a similar derivation to Eq. (7) but keeping track of the fact that the perturbation of one beam is due to the beam-beam force from the other beam, we obtain (let $\rho_{o+} = \rho_{o-} = \rho_0$)

$$\begin{aligned} \left(\frac{\beta^*{}_0}{\beta^*{}_+}\right)^2 &= 1 + 2\rho_0 \cot\mu_0 \frac{\beta^*{}_0}{\beta^*{}_-} - \rho_0{}^2 \left(\frac{\beta^*{}_0}{\beta^*{}_-}\right)^2 \\ \left(\frac{\beta^*{}_0}{\beta^*{}_-}\right)^2 &= 1 + 2\rho_0 \cot\mu_0 \frac{\beta^*{}_0}{\beta^*{}_+} - \rho_0{}^2 \left(\frac{\beta^*{}_0}{\beta^*{}_+}\right)^2 . \end{aligned} \tag{10}$$

One obvious solution is $\beta^*{}_+ = \beta^*{}_-$, i.e., Eq. (7). The other set of solutions with $\beta^*{}_+ \neq \beta^*{}_-$ is found to be

$$\frac{\beta^*{}_0}{\beta^*{}_\pm} = \frac{1}{\rho_0{}^2 - 1} \left\{ \rho_0 \cot\mu_0 \pm \sqrt{\frac{\rho_0{}^2 \cot^2\mu_0 \,(\rho_0{}^2 - 3) + (\rho_0{}^2 - 1)^2}{\rho_0{}^2 + 1}} \right\} . \tag{11}$$

In case the beams remain in this solution, one beam is statically blown up ($\beta^*{}_+$), while the other pinches ($\beta^*{}_-$). Equation (11) constitutes a bi-stable state of the beams because switching the roles of the blown-up beam and the pinched beam is also a solution. This situation can be a model for the "flip-flop" phenomenon[3-6] observed in storage ring colliders. Note that this is only the simplest possible explanation of the flip-flop effect. Another possibility will be mentioned later.

Figure 4 shows the steady-state dynamic-beta solutions as functions of ρ_0. The equal-beam solution (8) always exists for all values of ρ_0. The flip-flop solution (11) exists only over a finite range of ρ_0:

$$\cot\frac{\mu_0}{2} > \rho_0 > \sqrt{1+\frac{\cos^2\mu_0+\sqrt{\cos^4\mu_0+8\cos^2\mu_0}}{2}}\ . \tag{12}$$

At the lower end of the range, the equal-beam and the flip-flop solutions coincide. At the higher end, the blown-up beam in the flip-flop solution has infinite beam size. The pinched beam in this case sees no beam-beam force from the blown-up beam, and its beam size is unperturbed, i.e., $\sigma_- = \sigma_0$.

Note that not all steady-states are stable. In particular, in region (12), it will be shown later that the equal-beam steady-state is in fact unstable; any

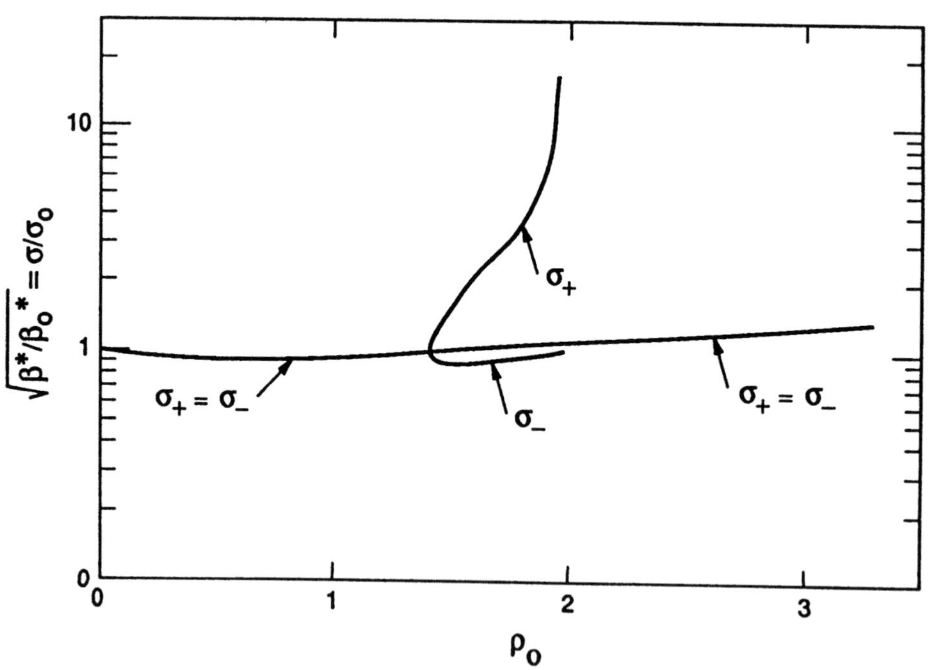

Figure 4. The dynamic-beta steady-states.

infinitesimal perturbation will force the beam either to converge into the flip-flop state or to become unstable.

4. STEADY-STATE WITH NONLINEAR BEAM-BEAM FORCE

In the previous section, we assumed a linearized beam-beam force. With this assumption, the perturbed steady-state remains gaussian, and only the rms beam size has changed. However, if the beam distribution is indeed gaussian, the beam-beam force cannot be strictly linear. (Only a uniform elliptical distribution gives strictly linear beam-beam force, and even then, only inside the elliptical region of the distribution.) Therefore an inconsistency exists here. One could consider two ways to proceed:

(a) Include nonlinear beam-beam force for gaussian beams, but insist that the beam-beam force acts only on the second moments of the beam distributions in such a way that the distributions are describable by gaussians. This is still not self-consistent, but hopefully provides useful insights. Additional information on this approach will be provided later.

(b) In general, beam-beam forces and beam distributions are to be found self-consistently. Assuming flat beams for simplicity, the two beam distributions are coupled through the Vlasov equation

$$\begin{aligned} \frac{\partial \Psi_+}{\partial s} + y' \frac{\partial \Psi_+}{\partial y} - F_-(y,s) \frac{\partial \Psi_+}{\partial y'} &= 0 \\ \frac{\partial \Psi_-}{\partial s} + y' \frac{\partial \Psi_-}{\partial y} - F_+(y,s) \frac{\partial \Psi_-}{\partial y'} &= 0 \end{aligned} \tag{13}$$

where

$$\begin{aligned} F_{-,+}(y,s) &= K(s)y + \text{BB force} \\ \text{BB force} &= \frac{2\pi N r_0}{L_x \gamma} \delta_p(s) \int_{-\infty}^{+\infty} d\bar{y}\, H(y-\bar{y})\, \rho_{-,+}(\bar{y}) \\ &= \frac{2\pi N r_0}{L_x \gamma} \delta_p(s) \int_{-\infty}^{+\infty} d\bar{y}\, H(y-\bar{y}) \int_{-\infty}^{+\infty} d\overline{y'}\; \Psi_{-,+}(\bar{y},\overline{y'},s)\,. \end{aligned} \tag{14}$$

The external focussing force provided by the accelerator is given by the term $K(s)y$ in Eq. (14). The beam-beam force occurs only at the IP as described by $\delta_p(s)$, the periodic δ-function with period given by the distance between IPs for equally spaced IPs. In a flat beam case, the beam-beam force is obtained by folding the beam distribution $\rho(y)$ with the function $H(x) = 1$ if $x > 0$, and -1 if $x < 0$, as Eq. (14) describes.

The equal-beam, steady-state self-consistent distribution satisfies

$$\partial\Psi_0/\partial s + y'\partial\Psi_0/\partial y - F_0(y,s)\partial\Psi_0/\partial y' = 0 \tag{15}$$

where

$$\begin{aligned} F_0(y,s) &= K(s)y + \text{BB force} \\ \text{BB force} &= \frac{2\pi N r_0}{L_x\gamma}\,\delta_p(s)\int_{-\infty}^{+\infty} d\bar{y}\, H(y-\bar{y})\int_{-\infty}^{+\infty} d\overline{y'}\,\Psi_0(\bar{y},\overline{y'})\,. \end{aligned} \tag{16}$$

In the dynamic-beta analysis, it is assumed

$$\text{BB force} \propto \delta_p(s)y\ . \tag{17}$$

With this assumption, the beam-beam problem is soluble, as discussed in the previous section.

It turns out that there exists another soluble case. Consider two relativistic, unbunched, round beams of opposite charges penetrating each other. Let both beams have a distribution that is gaussian in the transverse momenta x' and y', and algebraic in coordinates x and y as follows:

$$\Psi_0 = \frac{e\lambda_0 \exp\left[-\dfrac{x'^2+y'^2}{2\sigma'^2}\right]}{2\pi^2\sigma'^2 A^2\left(1+\dfrac{x^2+y^2}{A^2}\right)^2} \tag{18}$$

where λ_0 is the line density of the two beams, and $\sigma'^2 = \lambda_0 r_0/\gamma$. It can be shown[7] using the Vlasov equation (15) that if the two beams have the distribution (18) with an arbitrary beam size A, they will pinch each other with their beam-beam forces in such a way that they both maintain the steady-state (18).

Example:

(*a*) Generalize Eq. (18) by including a uniform, linear external focussing in the derivation.

(*b*) Find the counterpart of Eq. (18) for the case of flat beams.

In the dynamic-beta model, it was noted that a localized linear beam-beam force (17) drives half-integer resonances $\nu_0 = m/2$. Now note that in the case where the beam-beam force is nonlinear but smooth (nonlocalized, independent of s), as in the case of two self-pinching unbunched beams, a steady-state can be found without indication of resonances, i.e., a nonlocalized nonlinear beam-beam force does not drive resonances.

In general, unfortunately, the beam-beam force is both nonlinear and localized, i.e.,

$$\text{BB force} \propto \delta_p(s) \times (\text{nonlinear function of } y)\,. \tag{19}$$

In this case, like the case in single-particle nonlinear dynamics, the problem becomes very difficult to solve. In particular, resonances are driven when ν_0 is close to a rational number m/n. A self-consistent strong-strong distribution is yet to be found in this case. The closest soluble problem is that of the weak-strong case when there is only one prominent resonance playing a role.

Drawing analogy with the longitudinal potential-well distortion due to coherent wake-fields in a storage ring, it may be stated that in the present case the steady-state solution Ψ_0 is "transverse potential well distorted" by the beam-beam force at the IP.

Assuming the problem is solved, the preceding analysis provides only the equal-beam solution. One could also consider the nonlocal, nonlinear beam-beam forces, and unequal beams, just as we did in the dynamic-beta consideration.

This will then give other flip-flop steady-state solutions with nonlinear beam-beam forces. These solutions will be sensitive to the proximity of the tune ν_0 to rational values m/n just as the flip-flop solution in the dynamic-beta model is sensitive to $\nu_0 = m/2$.

5 DYNAMICS ABOUT THE STEADY-STATE

Assuming the steady-state, whether the equal-beam or flip-flop solution, is known, the next question to ask is whether the beam motion is stable against small perturbations from it. This is another difficult problem. To approach this problem, consider the steps, arranged in order of complexity, given in Table 1.

Table 1. Steps taken in studying the coherent beam-beam dynamics

Step	Steady-state	Perturbation
1	Unperturbed	center-of-mass motion
2	Unperturbed	rms beam-sizes
3	Dynamic-beta (equal-beam)	rms beam-sizes
4	Unperturbed	higher-order modes
5	Dynamic-beta (equal beam)	higher-order moments
6	Potential-well distortion (equal beam)	repeat 1–5
7	Flip-flop states	repeat 1–5

Steps 1–4 have been addressed, at least partially, in the literature; some of them are discussed below. Steps 5–7 are yet to be studied.

6 RIGID DIPOLE MODEL

As mentioned before, the dynamic-beta effect is the simplest strong-strong model, which involves static beam distortions. The simplest coherent beam-beam model that involves dynamics (moving beams) is the rigid dipole model.[8–15] In this model (step 1 of Table 1), each beam is assumed to have the unperturbed distribution rigidly except that their centers-of-mass are allowed to move in accordance with the beam-beam forces. To illustrate the model, assume the unperturbed beam distribution Ψ_0 is round at the IPs. Let the center-of-mass coordinates of the two beams be designated by $Y_\pm$. For $Y_\pm$ smaller than the beam size so that the beam-beam force can be linearized, the centers-of-mass exert beam-beam kicks on each other according to[16–19]

$$\begin{aligned} \Delta Y'_+ &= -\frac{G}{f}(Y_+ - Y_-) \\ \Delta Y'_- &= -\frac{G}{f}(Y_- - Y_+) . \end{aligned} \tag{20}$$

In Eq. (20), f/G is the focal length of the center-of-mass motions, where $f = \beta^*{}_0/4\pi\zeta_0$ is the focal length for incoherent motion and G is a form factor obtained approximately by averaging the single-particle beam-beam kicks over the rigid distribution $\Psi_0(r)$:

$$G = \frac{\int_0^\infty r\, dr\, \Psi_0{}^2(r)}{\Psi_0(0) \int_0^\infty r\, dr\, \Psi_0(r)} = \begin{cases} 1/2 & \text{for round gaussian.} \\ 1 & \text{for round uniform disk.} \end{cases} \tag{21}$$

Equation (21) applies only for round distributions, but it has been shown[18] that $G = 1/2$ even for an elliptical gaussian.

Consider the case of one bunch per beam. The transformation for the vector $[Y_+, Y'_+, Y_-, Y'_-]$ from IP to IP is

$$\begin{bmatrix} 1 & 0 & 0 & 0 \\ -\frac{G}{f} & 1 & \frac{G}{f} & 0 \\ 0 & 0 & 1 & 0 \\ \frac{G}{f} & 0 & -\frac{G}{f} & 1 \end{bmatrix} \times \mathbf{R} , \tag{22}$$

where **R** is the transformation through the storage ring arcs,

$$\mathbf{R} = \begin{bmatrix} \cos\mu_0 & \beta^*{}_0 \sin\mu_0 & 0 & 0 \\ -\frac{1}{\beta^*{}_0}\sin\mu_0 & \cos\mu_0 & 0 & 0 \\ 0 & 0 & \cos\mu_0 & \beta^*{}_0 \sin\mu_0 \\ 0 & 0 & -\frac{1}{\beta^*{}_0}\sin\mu_0 & \cos\mu_0 \end{bmatrix} . \tag{23}$$

The transformation matrix (22) can be analyzed for eigenvalues. The four eigenvalues form two complex conjugate pairs: each corresponds to a coherent beam-beam dipole mode: the 0- and the π-mode.

In the 0-mode, the two beams move up and down (or left and right) together, i.e., $Y_+ = Y_-$, at the IP. The coherent beam-beam forces on the centers-of-mass vanish [see Eq. (20)]. The motion is as if unperturbed and is therefore always stable regardless of the beam intensity. The 0-mode coherent frequency μ is equal to the unperturbed frequency μ_0.

In the π-mode, the bunches move out of phase with $Y_+ = -Y_-$ at the IPs. The effective coherent beam separation is therefore $2G$ times that in the incoherent case (G from the form factor, 2 due to both beams moving). The π-mode frequency μ therefore satisfies Eq. (2a) with the replacement $\zeta \to 2G\zeta_0$. This beam-beam system is stable if the π-mode is stable, i.e.,

$$\zeta_0 < \cot(\mu_0/2)/4\pi G\,. \tag{24}$$

In the unstable region, except when ζ is very close to the stability boundary, the instability growth rate obtained from the imaginary part of the eigenvalues of the π-mode is fast, which means it would be difficult to fight this dipole instability by a typical feedback system.[12,20]

According to Eq. (24), resonance occurs when $\mu_0 = \pi$, or $\nu_0 = 1/2$, just like the incoherent and the dynamic-beta cases. Furthermore, for the case of one bunch per beam assumed so far, the stability conditions (4), (5) and (24) all look similar, especially for gaussian beams for which $G = 1/2$. Figure 5, however, illustrates one important difference among them.

Similar analysis can be carried out for M bunches per beam in a storage ring with $2M$ IPs. In this case, the transformation matrix will be $4M \times 4M$ and there are $2M$ coherent dipole modes. Stability of the system requires all modes be stable, i.e., all eigenvalues have $|\lambda| = 1$.

For multiple bunches per beam, the coherent dipole instability distinguishes itself from the incoherent and the dynamic-beta effects in the resonances it excites. In the $M = 1$ case, all three cases excite only the $\nu_0 = m/2$ resonances. For multiple bunches per beam, the difference comes from the fact that the superperiodicity suppresses incoherent effects, but not coherent effects. Let the total tune of the storage ring be $\nu = 2M\nu_0$. (Throughout this paper, the notations $\mu_0 =$ phase advance between IPs, $\nu_0 =$ tune advance between IPs, and $\nu =$ total tune of the storage ring are used.) Coherent dipole instability occurs when $\nu = m$, while the incoherent and dynamic-beta resonances occur at a much sparser density, i.e., when $\nu = mM$. Coherent effects thus impose stricter limit on beam stability for $M > 1$, and the stability condition no longer look the same as the incoherent and dynamic-beta effects. Figure 6 shows the sawtooth diagrams for the cases of $M = 1, 2, 3$ and 4.

Coherent beam-beam dipole modes have been observed in storage rings such as SPEAR,[21] PETRA[22,23] and TRISTAN.[24–26] Figure 7 shows the observation of the 0- and the π-modes in PETRA with $M = 1$.[22] The peak at the lower frequency corresponds to the 0-mode, the one at higher frequency is the π-mode. In Figure 7(b), the π-mode frequency is slightly shifted because, although separated, the beams still couple slightly. Figure 8 shows the observation in TRISTAN with $M = 2$.[24] The parameters are such that all four modes are stable and these modes are observed by external driving.[18] With proper filtering, selected modes are observed.

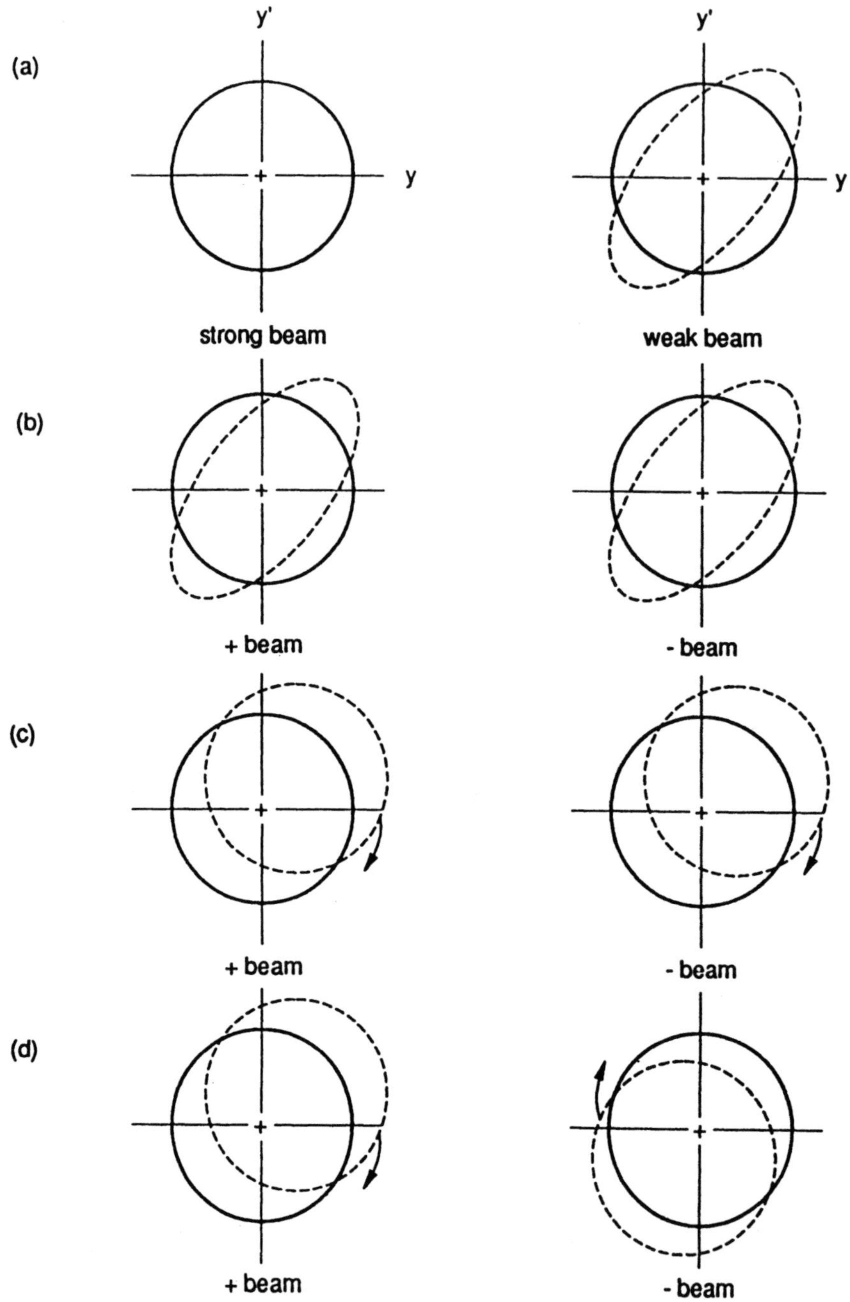

Figure 5. Distributions in phase space of two colliding beams for four cases: (a) incoherent weak-strong case, (b) dynamic-beta effect, (c) coherent dipole 0-mode, and (d) coherent dipole π-mode. (b), (c) and (d) are strong-strong cases. The solid and dashed curves indicate the unperturbed and the beam-beam perturbed distributions, respectively. The perturbed distribution moves with time in (c) and (d) but is static in (a) and (b). The beam dynamics behavior is very distinct in all cases, even though the stability criteria look similar.

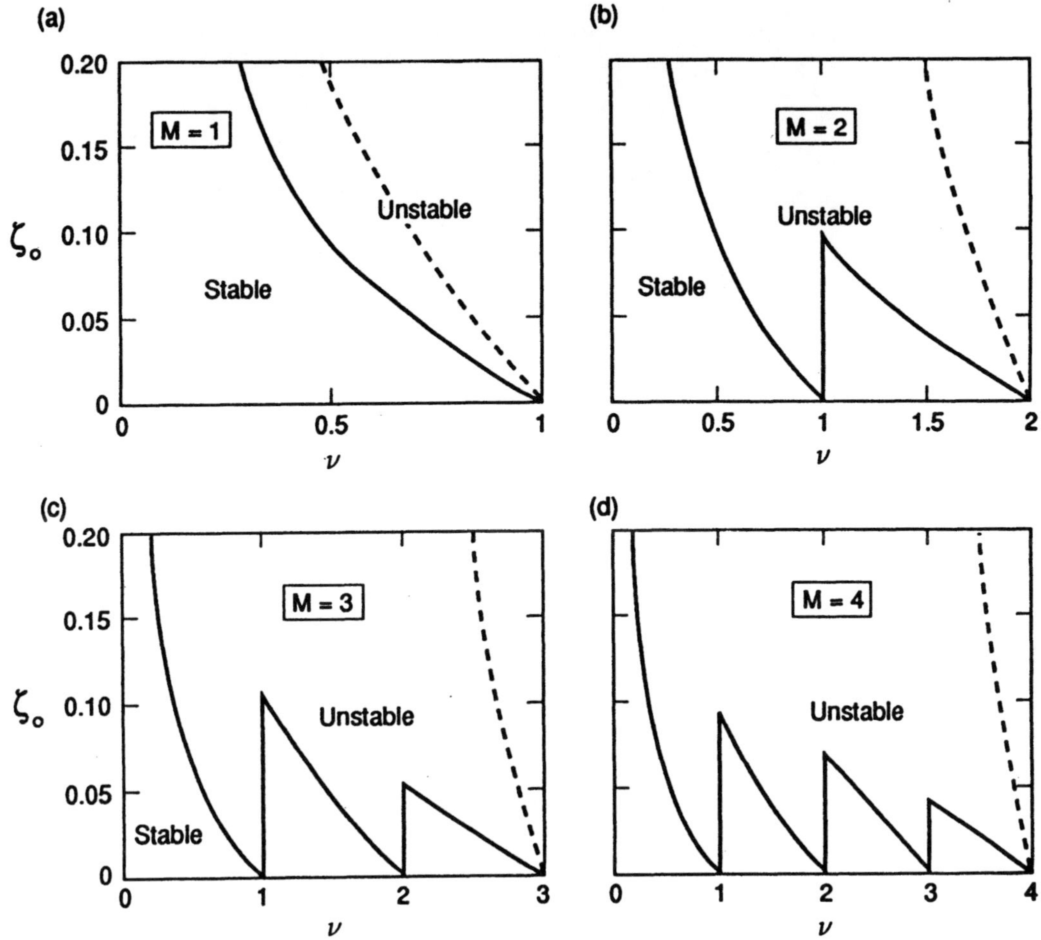

Figure 6. Stability diagrams (when $G = 1/\sqrt{2}$) for $M = 1, 2, 3, 4$ bunches per beam. $\nu = 2M\nu_0$ is the total tune. The figure repeats with a period of $\nu = M$. The dashed curves give the stability region for the weak-strong case for comparison.

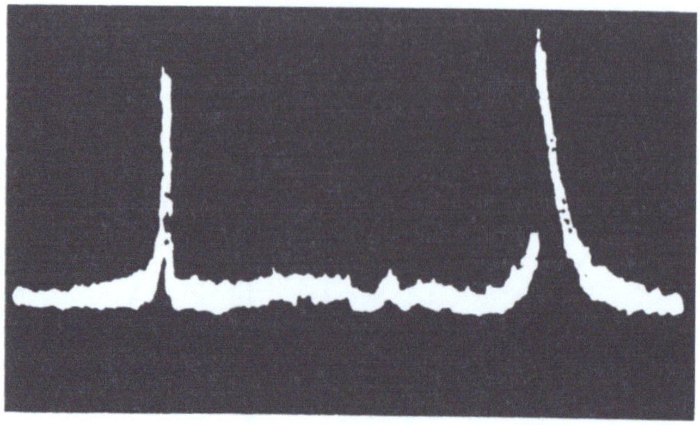

Vertical eigenfrequencies of two colliding bunches.

(a)

Vertical eigenfrequencies of two separated bunches.

(b)

Figure 7. Observation of the 0- and π-modes in PETRA with one bunch per beam: (a) beams colliding and (b) separated.

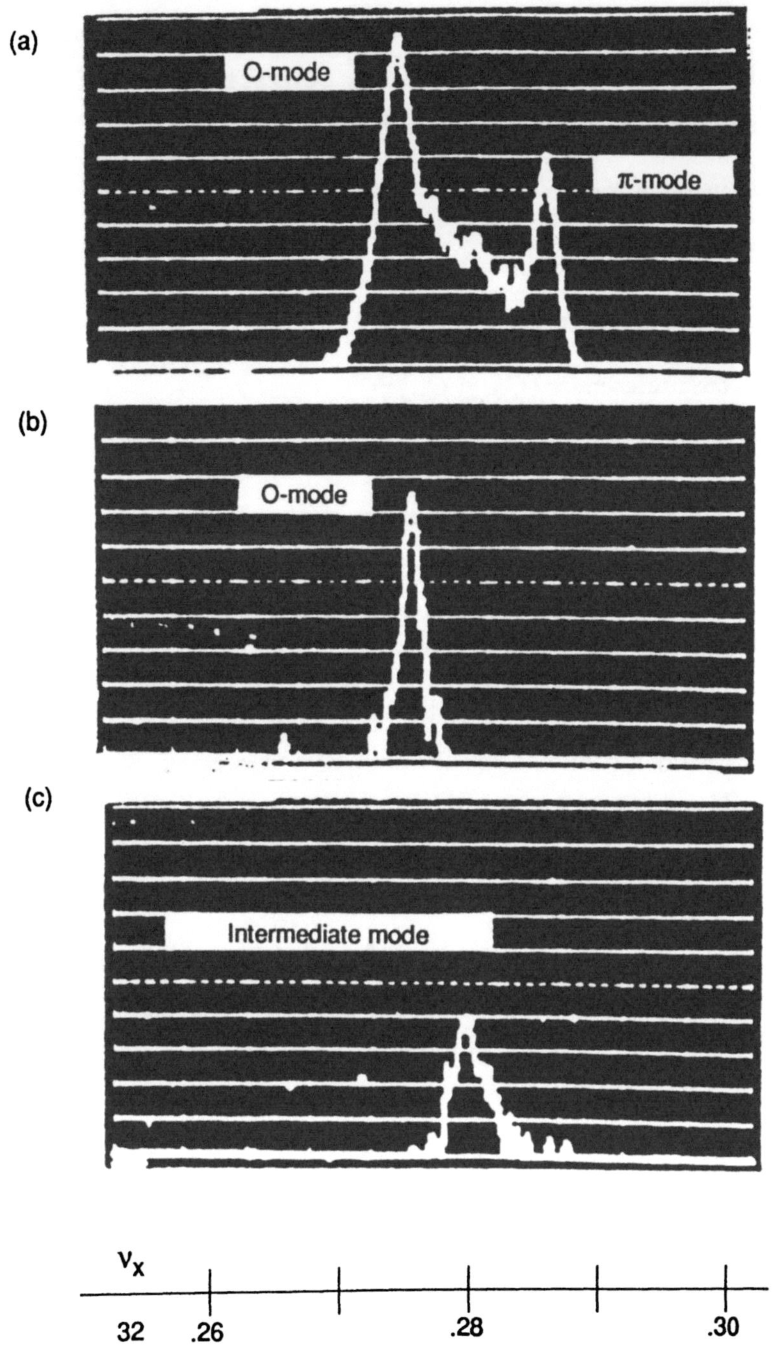

Figure 8. Observation of coherent dipole modes at TRISTAN with 2 bunches per beam. (a) Beam spectrum with four modes superimposed. With proper filtering, (b) the 0-mode, and (c) an intermediate mode could be observed.

The dipole instabilities have been studied in the literature more extensively than the other coherent beam-beam effects. Several extensions of the basic model described so far have been worked out.[27] These include:

unequal beams ($N_+ \neq N_-$)[27]

unequal tunes ($\nu_+ \neq \nu_-$)[27]

error in phase advances between IPs[28]

unequal circumferences ($C_+ \neq C_-$)[29,30]

four compensating beams[20]

coherent beam-beam synchro-betatron resonances[31]

spontaneous beam separation[29,32]

long-range beam-beam interactions.[33–37]

One in fact makes the observation that for each incoherent beam-beam effect, there is most likely a coherent counterpart, and vice versa. Some of these extensions are reviewed below.

6.1 Unequal Beams

Consider the case when the two beam bunches have different intensities, $\rho_+ \neq \rho_-$, and the two beams are in separate storage rings with different phase advances between IPs, $\mu_+ \neq \mu_-$. The eigenmode frequencies $\mu_{1,2}$ are determined from[27]

$$2\cos\mu_{1,2} = \cos\mu_+ + \cos\mu_- - G\rho_+ \sin\mu_+ - G\rho_- \sin\mu_- \pm \sqrt{Q} \qquad (25)$$

where

$$Q = (\cos\mu_+ - \cos\mu_- - G\rho_+ \sin\mu_+ + G\rho_- \sin\mu_-)^2 + 4G^2\rho_+\rho_- \sin\mu_+ \sin\mu_- \ .$$

Note that this analysis allows other differences in the two beams, e.g. differences in $N_\pm$, $\gamma_\pm$, $\beta^*{}_\pm$ and $\sigma_\pm$ can all be incorporated into $\rho_\pm$.

Example: Show that Eq. (25) reduces to Eq. (2a) when $\mu_+ = \mu_- = \mu_0$ and $\rho_+ = \rho_- = \rho_0$.

Example: When $\rho_+ = 0$, show that the $-$ beam is unperturbed, while the $+$ beam frequency shift $= G \times$ (the incoherent shift). This is the coherent dipole mode in the weak-strong limit.

Example: When $\mu_+ = \mu_- = \mu_0$ but $\rho_+ \neq \rho_-$, show that the two eigenmodes are still described by a 0-mode and a π-mode, and that the mode frequencies are given by

$$\cos\mu = \cos\mu_0 - \begin{bmatrix} 0 \\ 1 \end{bmatrix} G(\rho_+ + \rho_-)\sin\mu_0 \,. \tag{26}$$

6.2 Unequal Tunes

Equation (25) also includes the effect of unequal tunes. Figure 9 shows the stability region in the (ν_{o+}, ν_{o-}) plane.[27] A new set of sum resonances $\nu_{o+} + \nu_{o-} = m$ is introduced, where $\nu_{o\pm} = \mu_{o\pm}/2\pi$ are the tune advances between IPs in the two storage rings. Strictly speaking, only these sum resonances are true coupled-beam modes. Coherent beam-beam modes are strongly excited by the sum resonances.

In passing, it can be noted that splitting ν_{o+} and ν_{o-} did not improve the instability. The possible expectation that splitting the tunes introduces Landau damping, and therefore stability, to the system is incorrect.

6.3 Phase Advance Errors

Consider again the case with one bunch per beam. In the previous analysis, the two IPs in the storage ring are separated by two arcs with equal betatron phase advances. The analysis can be extended to unequal phase advances

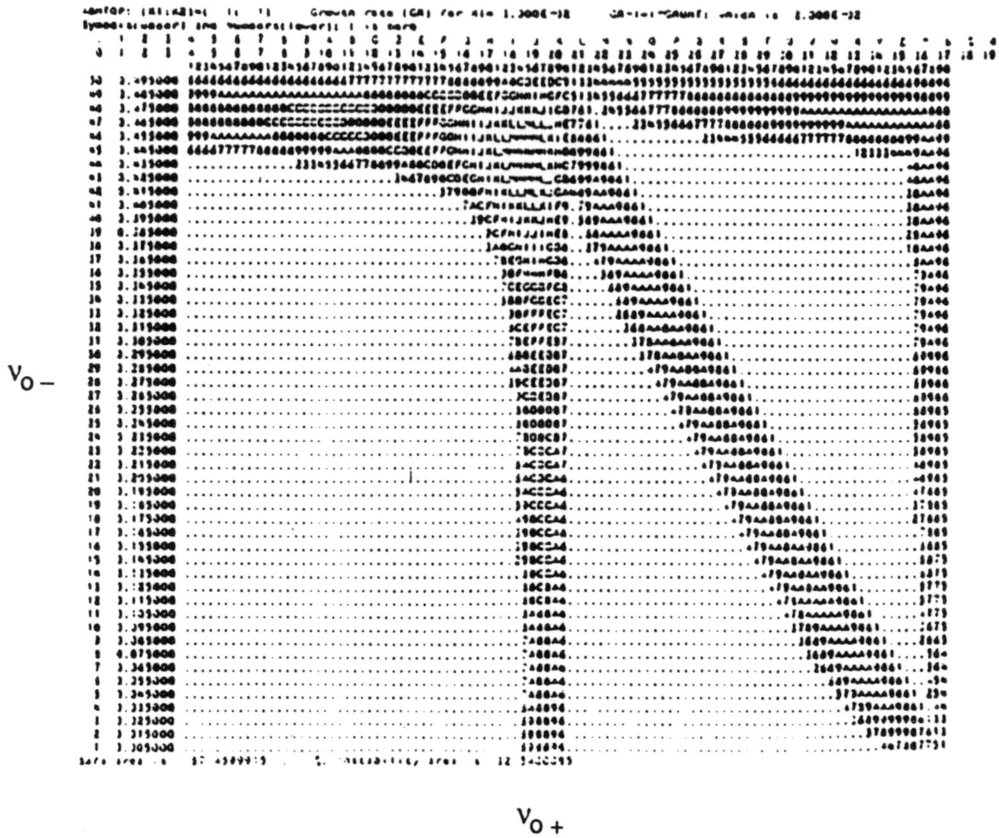

Figure 9. Stability diagram in (ν_{o+}, ν_{o-}) space, which shows the excitation of sum resonances $\nu_{o+} + \nu_{o-} = m$. The horizontal ν_{o+} axis is from 0 to 1. The vertical ν_{o-} axis is from 0 to 1/2. The pattern repeats if $\nu_{o\pm}$ are replaced by $\nu_{o\pm} + 1/2$.

parametrized as $\mu_0 \pm \delta\mu$.[9] The result is shown in Figure 10(a). The most pronounced new feature is the introduction of a stopband when the total tune $\nu = 2\nu_0$ is close to 1/2. The stopband edges are found to be

$$\zeta_0 = \frac{\cot \mu_0}{2G\pi} \frac{1 \pm \sin \delta\mu}{\cos^2 \delta\mu}. \tag{27}$$

The sawtooth part of the stability boundary becomes [cf. Eq. (24)]

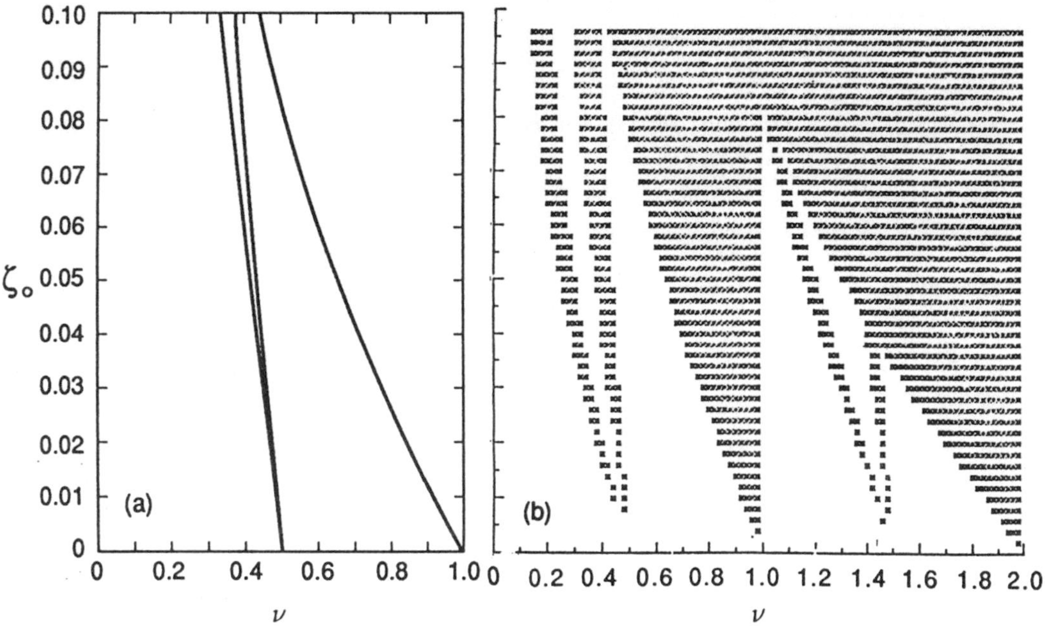

Figure 10. Effect of phase advance errors in the storage ring arcs on the sawtooth diagram. (a) $M = 1$, $\delta\mu/2\pi = 0.02$. (b) $M = 2$, $\delta\mu/2\pi = 0.03$. The form factor is taken to be $G = 1$.

$$\zeta_0 = \frac{\sin\mu_0}{4G\pi(\cos\delta\mu - \cos\mu_0)}. \tag{28}$$

Equation (28) reduces to (24) if $\delta\mu = 0$. For multiple bunches per beam, the results resemble Figure 6, except that new stopbands of complicated structure are created around half-integer resonances $\nu = m/2$. Figure 10(b) shows the result obtained numerically for the case of $M = 2$.[28]

6.4 Unequal Circumferences

Consider the case when the two beams are stored in separate storage rings of different circumferences $C_\pm$. The number of bunches $K_\pm$ is chosen so as to have a fixed bunch spacing $C_+/K_+ = C_-/K_-$. The system has a superperiod $= K_+C_- = K_-C_+$. The total tune advances per superperiod of the two storage rings are $K_+\nu_{o-}$ and $K_-\nu_{o+}$. The coherent dipole beam-beam resonances in this system are given by

$$K_+\nu_{o-} = m/2 \tag{29a}$$

$$K_-\nu_{o+} = m/2 \tag{29b}$$

$$K_+\nu_{o-} + K_-\nu_{o+} = m\,. \tag{29c}$$

The consequence is that there are now many more resonances, which makes it undesirable to have unequal circumferences for the two beams.[30]

Figure 11 shows the result for the case $K_+ = 2$, $K_- = 3$.[29,30] Note particularly the three coupling resonances of the type (29c), which are strongly driven by the beam-beam coupling. In comparison, the single-beam resonances of types (29a) and (29b) are driven only when $\nu_{o\pm} = m/2$.

6.5 Four-Beam Compensation

One method to compensate for the beam-beam effect—therefore greatly increasing the luminosity—is the four-beam idea of DCI.[38,39] In this idea, two beam bunches +Ne and −Ne are made to collide with two other beam bunches +Ne and −Ne, as shown in Figure 12(a). The colliding bunches have neutral net charges and thus presumably produce no beam-beam forces. Unfortunately, this idea had not worked at DCI.

The problem is that what has been eliminated is the incoherent beam-beam force. Coherent beam-beam forces are not dealt with, because infinitesimal de-

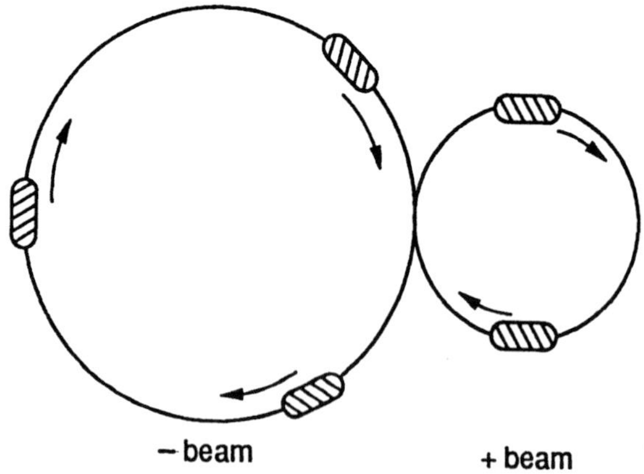

Figure 11. (a) Unequal circumferences for the case $K_+ = 2$, $K_- = 3$.
(b) See next page.

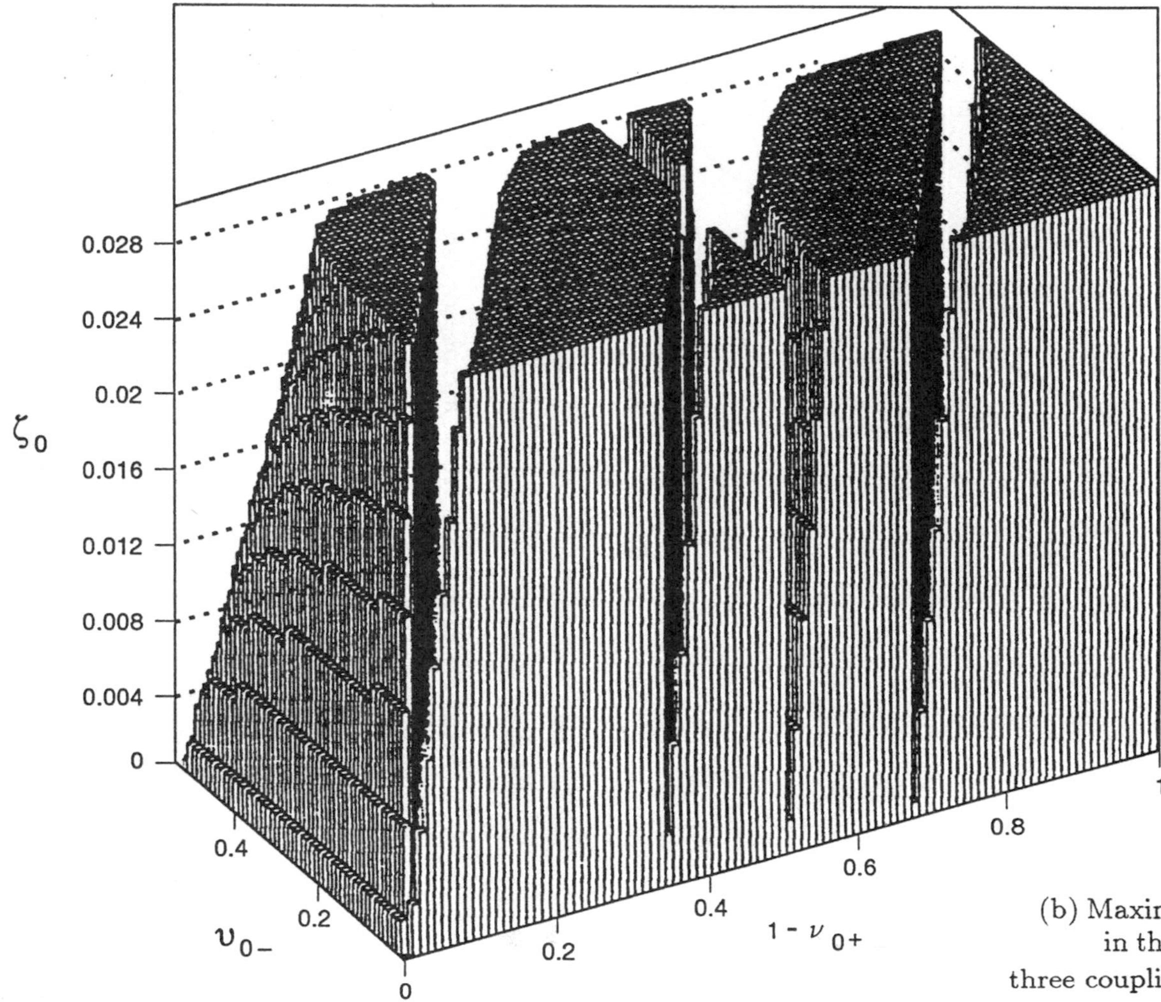

Figure 11.
(b) Maximum stable ζ_0 value plotted in the (ν_{o+}, ν_{o-}) plane, showing three coupling resonances of type (29c).

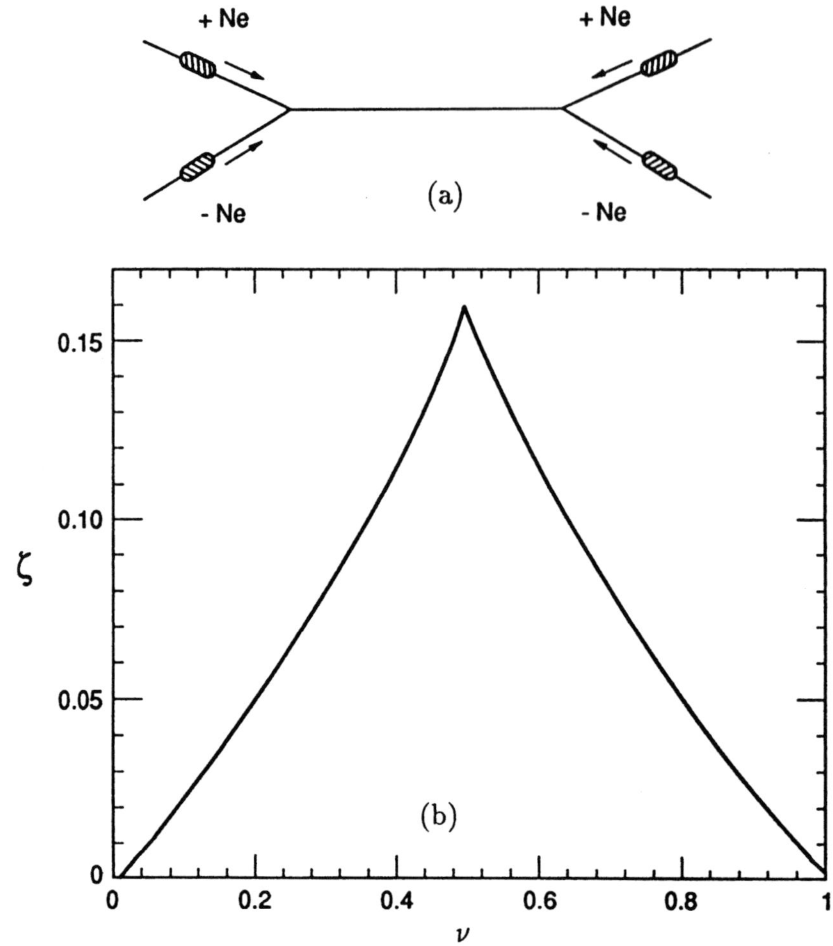

Figure 12. (a) Schematic of the four-beam compensation arrangement.
(b) Stability diagram for the 4-beam scheme.

viations from the steady-state can still grow as coherent modes. In fact, this four-beam combination has worsened the coherent beam-beam effect. The net result is that one problem has been traded for another—better incoherent effects for worse coherent effects.

Figure 12(b) illustrates the sawtooth diagram for the four-beam case.[20] The stability boundary consists of two segments, separated at $\nu = 1/2$. The segment below $\nu = 1/2$ turns out to be close to what would be obtained with colliding $-Ne$ and $-Ne$ (or $+Ne$ and $+Ne$), while the segment above $\nu = 1/2$ is close to that for colliding $+Ne$ and $-Ne$. The coherent dipole instability therefore is a result of the worse of the two possible combinations.

The situation will change if the four colliding beams are provided by linac colliders instead of storage rings, in which case the beams are discarded after collision. There will be no coherent beam-beam effects (at least not the multi-turn resonant type of interest here), and the four-beam idea could still be applied to avoid beamstrahlung, which is one of the main limiting effects of high energy linac colliders. In fact, only three beams are needed to come from linac colliders; one beam can still come from a storage ring.

Also, analyses have been carried out for the four-beam scheme for higher-order modes using Vlasov techniques.[40–45] One such analysis[41] yields the result shown in Figure 13. Additional information on the Vlasov technique for higher-order modes is provided later.

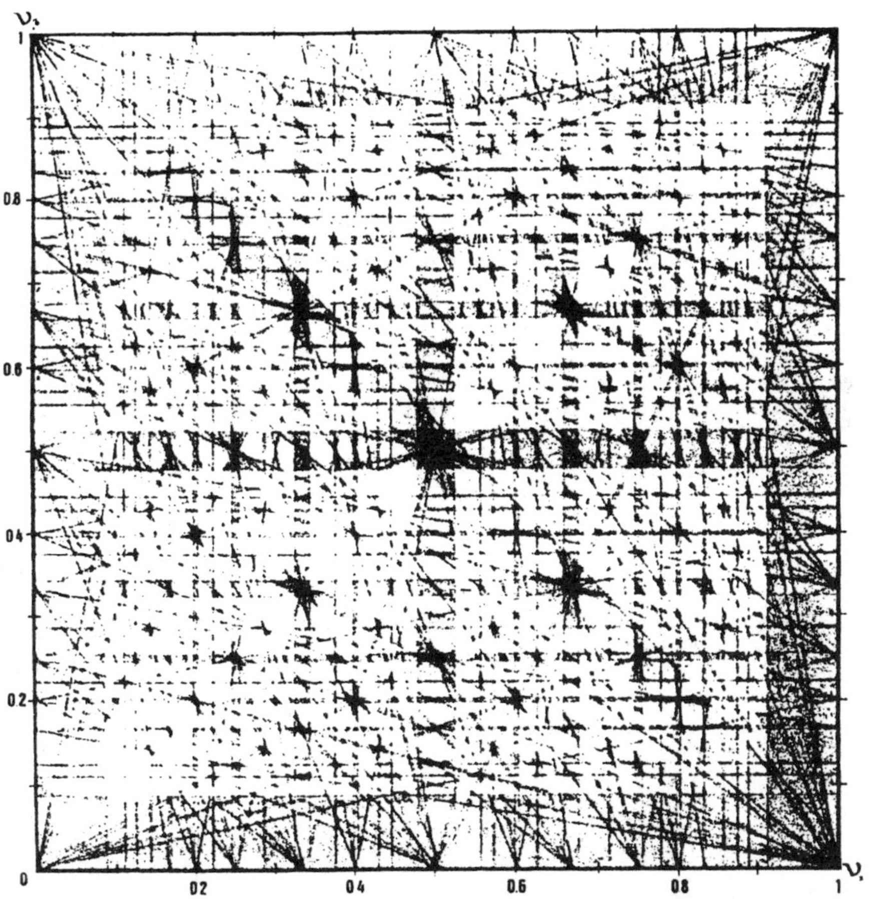

Figure 13. Stability region in (ν_x, ν_y) space for DCI with $\zeta_0 = 0.05$, according to a mode analysis.

6.6 Synchro-betatron Resonances

Finite dispersions at the IP or the rf cavity excite coherent beam-beam synchro-betatron difference resonances $\nu_\beta - n\nu_s = m$. Similar to the single-particle case, the sum resonances are stable above transition.

The analysis has been done when the beam-beam force is linearized,[31] in which case only the linear resonances ($n = 1$) are excited. The dynamics can be described by 8×8 transformation matrices. Eigen-analysis of the system gives the stability region in the (ν_β, ν_s) space as shown in Figure 14, where ν_β and ν_s are the total betatron and synchrotron tunes. The synchro-betatron resonances occupy small regions in the tune diagram even with exaggerated values of dispersion at the IP and the rf cavity as used in Figure 14, and are therefore not a very significant concern in practice.

6.7 Spontaneous Beam Separation

So far in analyzing the coherent dipole effects, we have linearized the beam-beam force. This is valid only for center-of-mass motions small compared with the beam size σ. It suffices if one is interested only in the beam-beam effect on luminosity because even small beam motion must be avoided for luminosity purposes. The rigid dipole motion in the presence of nonlinear beam-beam forces has been simulated numerically. Figure 15 shows one such result.[14]

With a nonlinear beam-beam force, the magnitude of center-of-mass motion saturates when beam separation becomes comparable to σ. In fact, as beams separate far enough, they can find a steady-state separation to stay in. For example, in case of 1 bunch per beam with round, gaussian, equal beams, the steady-state beam separation is zero in the linearly stable region $2\pi\zeta_0 < \cot(\mu_0/2)$, while in the unstable region, the steady-state beam separation Δ is determined by

$$\frac{2\sigma^2}{\Delta^2}(1 - e^{-\Delta^2/2\sigma^2}) = \frac{\cot\frac{\mu_0}{2}}{2\pi\zeta_0} < 1\,. \tag{30}$$

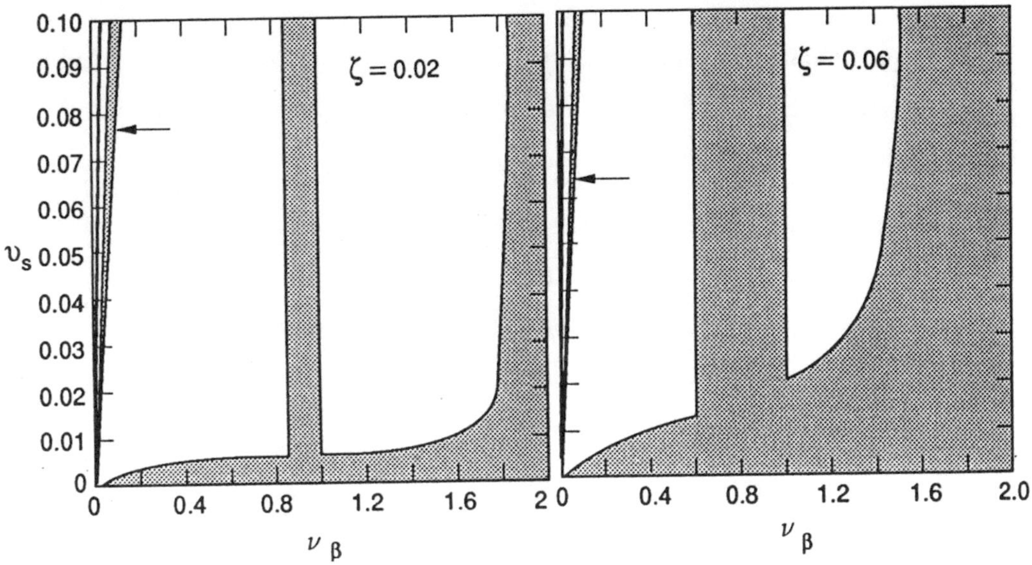

Figure 14. Stability region in the (ν_β, ν_s) space for two values of $\zeta_0 = 0.02$ and 0.06. The instability regions due to synchro-betatron resonances are indicated by arrows.

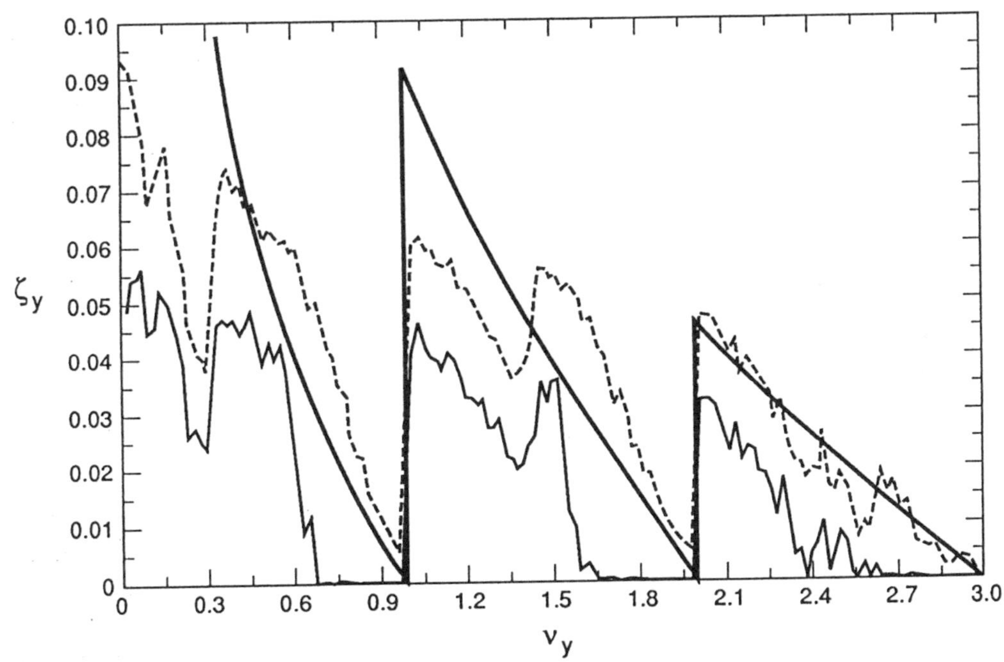

Figure 15. The value of ζ_y as a function of the total vertical tune ν_y and fixed $\nu_x = 0.2$ for the case of $M = 3$. Rigid center-of-mass motions are simulated with nonlinear beam-beam forces. Solid and dotted curves represent the minimum and average values of ζ_y over 200 turns simulated. Sawtooth curve is that predicted by a linear theory.

With beams separated, the luminosity is reduced by a factor $\exp(-\Delta^2/4\sigma^2)$. Figure 16 shows the behavior of Δ/σ and luminosity reduction factor as functions of the stability parameter $x = 2\pi\zeta_0/\cot(\mu_0/2)$.

It may be desirable to study the behavior of small perturbations around a separated-beam configuration. The analysis will yield the corresponding mode frequencies and the stability conditions. The case of unequal tunes has also been studied. It was found that the behavior of beam separation becomes complicated near a sum resonance $\nu_{o-} + \nu_{o+} = m$.[29,32]

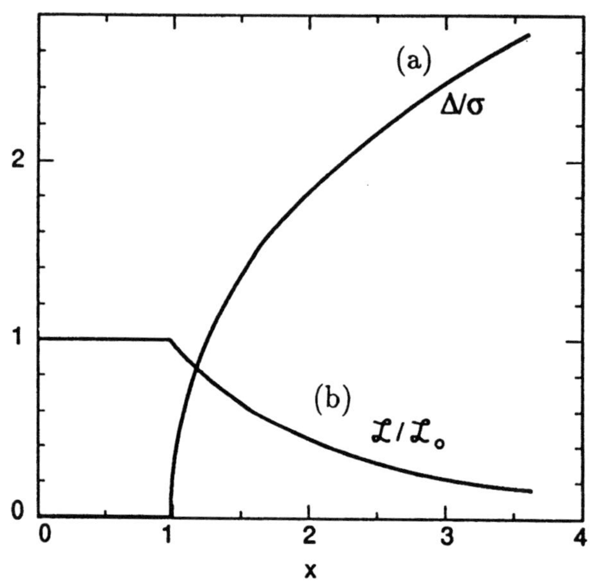

Figure 16. Steady-state beam separation due to nonlinear beam-beam forces. (a) Δ/σ as a function of $x = 2\pi\zeta_0/\cot(\mu_0/2)$. (b) Luminosity reduction as function of x.

7 QUADRUPOLE MODES

In the dipole model, the beam distribution is assumed to be rigid; only center-of-mass motions are allowed. In general, coherent beam motion is to be described as a superposition of all modes, not only dipole modes. The dynamics of these modes altogether will determine the stability of the beam. When proceeding to higher-order modes, more resonances will play a role, as shown in Table 2.

Table 2. Coherent beam-beam modes and their driving resonances

Mode	Driving Resonances
rigid dipole	total tune $\nu = m$
quadrupole	$\nu = m/2$
sextupole	$\nu = m/3$
etc.	

Consider two bunches colliding head-on (i.e., no dipole motion), but their beam sizes oscillate (i.e., they execute quadrupole oscillation). Define the Σ-matrix by the second moments as

$$\Sigma = \begin{bmatrix} \langle x^2\rangle & \langle xx'\rangle & 0 & 0 \\ \langle xx'\rangle & \langle x'^2\rangle & 0 & 0 \\ 0 & 0 & \langle y^2\rangle & \langle yy'\rangle \\ 0 & 0 & \langle yy'\rangle & \langle y'^2\rangle \end{bmatrix} . \tag{31}$$

There is a Σ-matrix for each beam. Designate the two Σ-matrices by $\Sigma_\pm$. The transformations through the arcs are

$$(\Sigma_\pm)_{\text{out}} = \mathbf{R}\,(\Sigma_\pm)_{\text{in}}\,\widetilde{\mathbf{R}} \tag{32}$$

where the matrix $\mathbf{R}$ is the same as that defined in Eq. (23), if $\beta^*{}_{x0} = \beta^*{}_{y0} = \beta^*{}_0$ and $\mu_{x0} = \mu_{y0} = \mu_0$.

As indicated by Table 2, nonlinear dynamics plays a fundamental role in the coherent beam-beam dynamics. Unlike the single-particle nonlinear dynamics, however, even a linearlized beam-beam force is capable of driving nonlinear resonances because the focal lengths depend on the beam sizes. The transformation of the Σ-matrices through the IP due to a linearized beam-beam force is given by

$$\begin{aligned} (\Sigma_+)_{\text{out}} &= \mathbf{T}_{\text{BB}-}(\Sigma_+)_{\text{in}}\,\widetilde{\mathbf{T}_{\text{BB}-}} \\ (\Sigma_-)_{\text{out}} &= \mathbf{T}_{\text{BB}+}(\Sigma_-)_{\text{in}}\,\widetilde{\mathbf{T}_{\text{BB}+}} \;. \end{aligned} \tag{33}$$

$\mathbf{T}_{\mathrm{BB}\pm}$ are the beam-beam matrices

$$\mathbf{T}_{\mathrm{BB}\pm} = \begin{bmatrix} 1 & 0 & 0 & 0 \\ -\frac{1}{f_{x\pm}} & 1 & 0 & 0 \\ 0 & 0 & 1 & 0 \\ 0 & 0 & -\frac{1}{f_{y\pm}} & 1 \end{bmatrix} \tag{34}$$

which give the effect on the beam size of one beam at IP due to the oscillating size of the other beam. The focal lengths $f_{x\pm}$ and $f_{y\pm}$ depend on $\sigma_{x\pm} = \langle x^2 \rangle^{1/2}$ and $\sigma_{y\pm} = \langle y^2 \rangle^{1/2}$ which are the instantaneous, oscillating beam sizes of the two beams.

Consider small perturbations around a steady-state Σ_0:

$$\Sigma_- = \Sigma_0 + \Delta\Sigma_-, \qquad \Sigma_+ = \Sigma_0 + \Delta\Sigma_+. \tag{35}$$

To simplify the analysis, assume the unperturbed beams are round. Next linearize the system of equations (32) – (34) with respect to the infinitesimal elements in the matrices $\Delta\Sigma_\pm$. Since each beam has six elements $\Delta\langle x^2\rangle$, $\Delta\langle xx'\rangle$, $\Delta\langle x'^2\rangle$, $\Delta\langle y^2\rangle$, $\Delta\langle yy'\rangle$, and $\Delta\langle y'^2\rangle$, the dynamics of the system can be described by 12×12 transformation matrices. Eigen-analysis of the system then gives 6 modes. The result[46] is that four of the twelve eigenvalues are identically unity. They correspond to the two constants of the motion $\langle x^2\rangle\langle x'^2\rangle - \langle xx'\rangle^2$ and $\langle y^2\rangle\langle y'^2\rangle - \langle yy'\rangle^2$. The remaining four dynamic modes have frequencies[47]

$$\cos\mu = \cos 2\bar{\mu} + \begin{bmatrix} 2 \\ 1 \\ -1 \\ -2 \end{bmatrix} \rho_0 \sin\mu_0 \cos\bar{\mu}$$

$$\cos\bar{\mu} = \cos\mu_0 - \rho_0 \sin\mu_0 \,. \tag{36}$$

This result resembles Eq. (2a), although the dynamics is very different. The sawtooth diagram of the system is shown in Figure 17(a). Quadrupole instabilities are most pronounced near resonances $\nu = m/2$.

Similar analysis has been carried out for multiple bunches per beam, in which case there are $4M$ dynamic modes. Figure 17(b) gives the sawtooth diagram for the case of $M = 3$. Again, resonances occur when the total tune is half-integral.

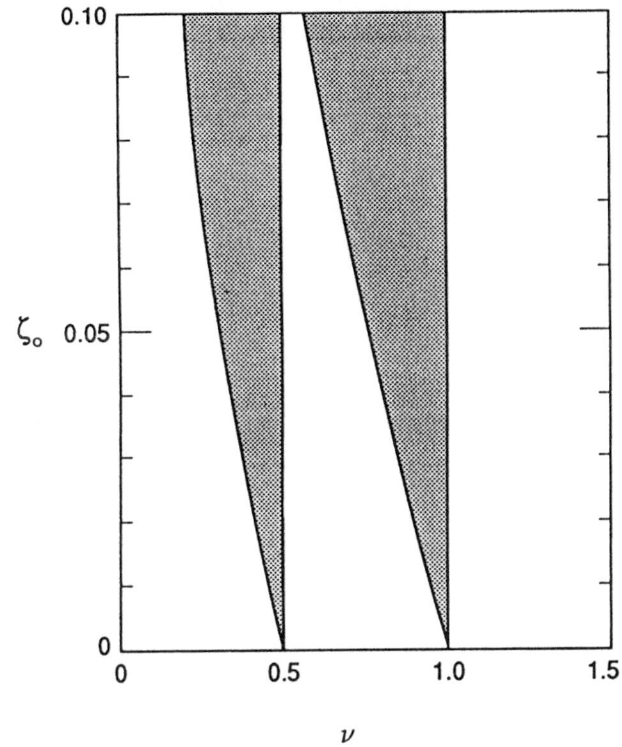

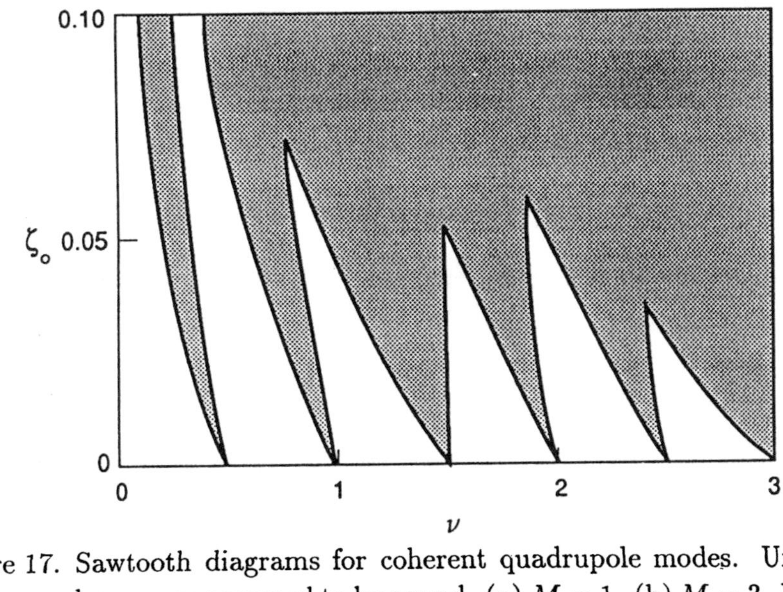

Figure 17. Sawtooth diagrams for coherent quadrupole modes. Unperturbed beams are assumed to be round. (a) $M = 1$. (b) $M = 3$. Resonances occur around $v = m/2$.

Just like the dipole case, it is possible to extend the basic model described above for the quadrupole modes.[47] For example, one could include in the analysis the phase advance errors to obtain Figure 18. One observes that the phase errors introduce new resonances at $\nu = m/4$. Similar procedure can be extended to the case $\nu_x \neq \nu_y$ to yield Figure 19, which shows structures near the coupling sum resonances $\nu_x + \nu_y = m$.

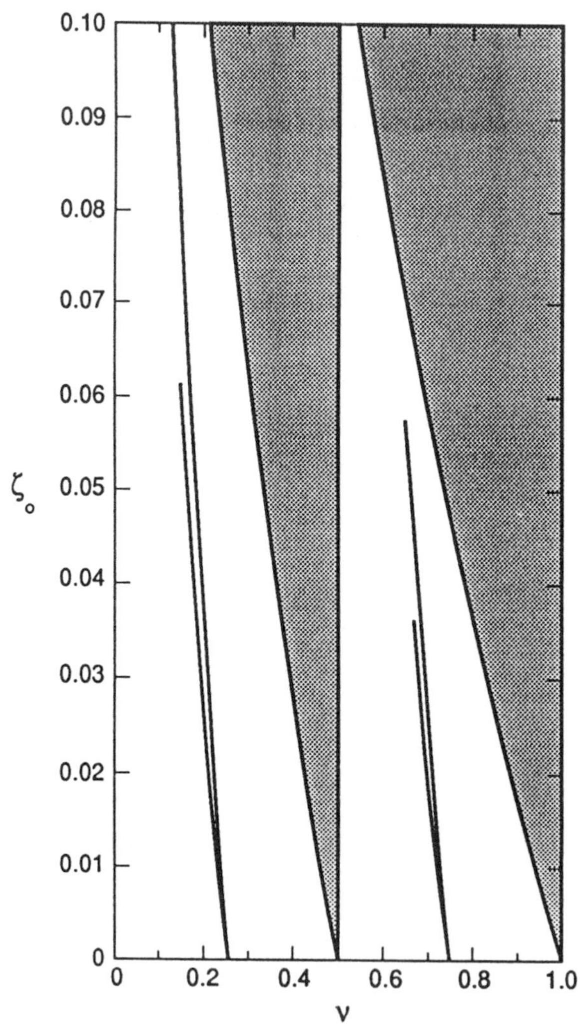

Figure 18. Quadrupole mode sawtooth diagram when phase advance errors $\pm 2\pi \times 0.05$ are introduced in the storage ring arcs. Stopbands are created near $v = m/4$.

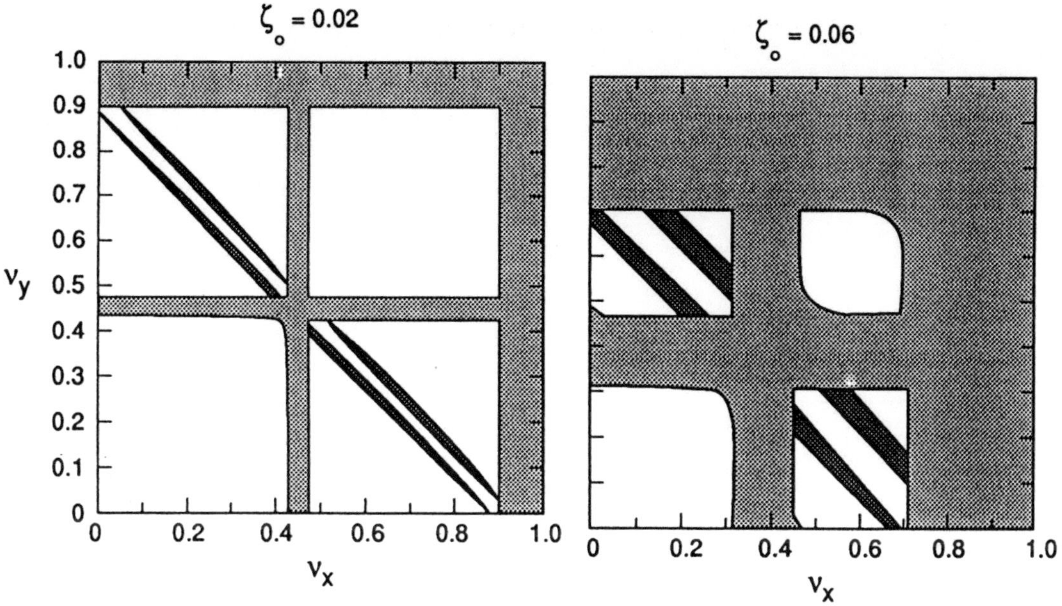

Figure 19. Quadrupole mode stability region in the (ν_x, ν_y) space for the case $M = 1$ and round beams, for two values of $\zeta_0 = 0.02$ and 0.06. Resonances occur when $\nu_x = m/2$, $\nu_y = m/2$, and $\nu_x + \nu_y = m$. No phase errors in arcs are included.

8 QUADRUPOLE MODES AROUND DYNAMIC-BETA

The previous section addresses the quadrupole oscillations around the unperturbed state. One may also investigate quadrupole oscillations around the dynamic-beta steady-state, i.e., that shown in Figure 4. This will give information on whether the beams will choose to stay in the equal-beam, the flip-flop, or neither state. There have been two investigations in this direction using two models, as discussed next.[48]

8.1 The First Model

In this model,[49–51] the dynamics of the second moments is still given by Eqs. (32)–(34) and the beam-beam force is still linearized, but this time the infinitesimal quantities $\Delta\Sigma_\pm$ are not introduced and the problem is not linearized with respect to them. The problem is then studied by numerical simulation. Figure 20(a)[49] shows the equal-beam and the flip-flop solutions, like Figure 4

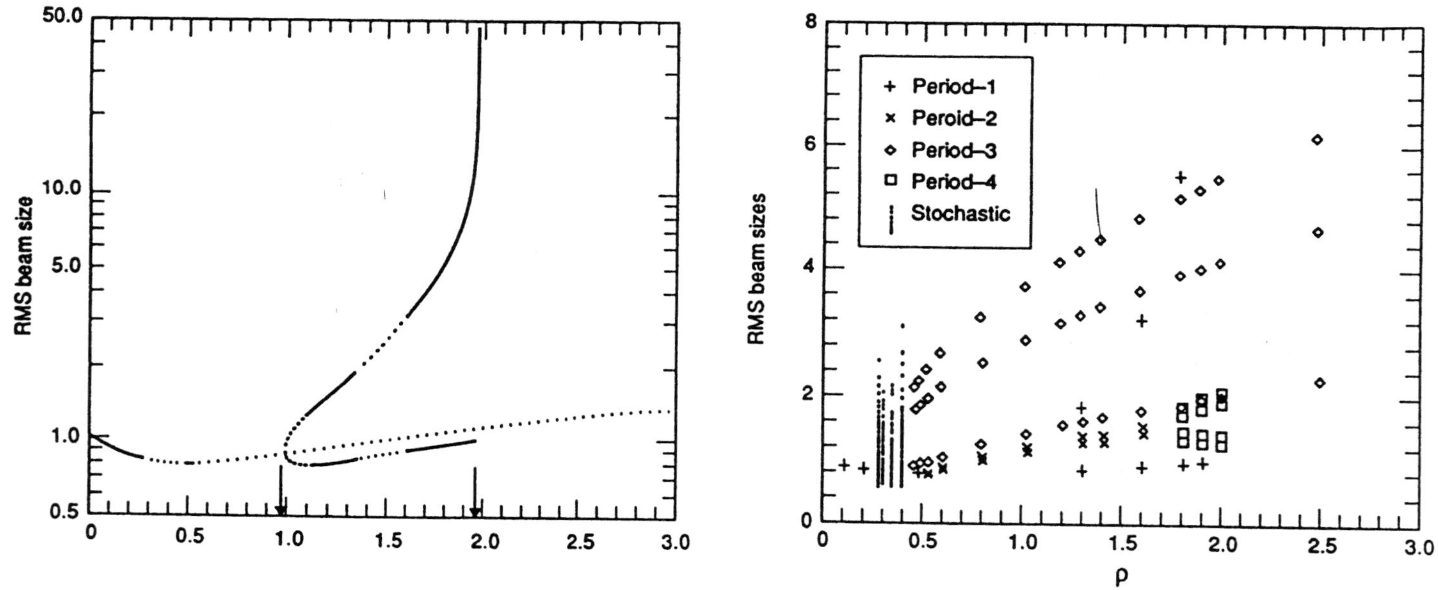

(a) (b)

Figure 20. Stability of the dynamic-beta steady-state solutions for the case of linearized beam-beam force. (a) Period-1 fixed points. Regions stable and unstable against small perturbations are plotted as solid and dotted curves respectively. (b) More complicated motions of second moments are indicated by higher-order periods and chaos.

but slightly different (e.g. a small radiation damping has been included). Solid and dotted sections in Figure 20(a) indicate stable and unstable regions of the steady-states. Note that the equal-beam steady-state is unstable when the flip-flop state exists. In the region ρ less than approximately 0.25, the equal-beam solution is stable. In the present model, $\rho = 0.25$, which corresponds to a beam-beam strength of $\zeta = \rho/2\pi = 0.04$, may be associated with the beam-beam limit observed in colliding beam storage rings. Beyond this region, the equal-beam solution is unstable except for a very small region around $\rho = 0.5$. Between approximately $\rho = 1$ and $\rho = 2$, there are regions where the flip-flop state is stable, meaning the beams will choose to have one beam statically blown-up and the other beam pinched. In the rest of the regions, both the equal-beam and flip-flop steady-states are unstable.

Figure 20(b) shows more details of what is happening to the beams. First, the steady-states are found when they are stable; these agree with Figure 20(a) and are indicated as period-1 states in Figure 20(b). In addition, states with higher periods are found. In a period-2 state, for example, the beams have two states (two sizes) to stay in, and they occupy these two states alternately from one IP to the next. Similarly there are period-3 and -4 states as indicated. In the region approximately between $\rho = 0.25$ to 0.4, the motion of the second moments is apparently chaotic. Figure 20(b) thus shows a rather complex dynamics of the quadrupole modes.

8.2 The Second Model

As mentioned before, linear beam-beam force is inconsistent with gaussian distribution. Insisting on linear beam-beam force in the previous model has violated the Maxwell equations.

Below we introduce another simplified quadrupole mode model[52–54] in which (a) beam distributions are gaussian, (b) the beam-beam force is that for gaussian distributions, therefore nonlinear, but (c) the beam-beam force affects only the second moments in such a way that the beam distribution remains always gaussian. Effects of the nonlinear beam-beam force on higher moments are ignored.

Obviously this model is still not self-consistent because of step (c). Although step (b) has rescued the Maxwell equations, step (c) makes the system nonsymplectic, thus violating the Hamiltonian feature which might be important when studying stability of the dynamical system.

One then proceeds as before, i.e., to obtain the dynamic-beta and the flip-flop solutions, and the quadrupole mode dynamics around them. This model has been extensively studied. One interesting result of the study is Figure 21 which shows the steady-state solutions and their stability.[54] Note that although the qualitative behavior of the equal-beam and flip-flop states looks similar to that of Figure 20(a), their stability behavior is quite different. In Figure 21, there is always a stable steady-state for the beams to stay in. Below $\zeta \approx 0.05$, the beams stay in the equal-beam state, while they stay in the flip-flop state above $\zeta \approx 0.05$. The behavior is much simpler than that of Figure 20(b). Table 3 is a comparison between the two models.

It is not very useful to ask which of the two models is more correct because neither is, each having violated something fundamental. However, they do provide information on what to expect in the ultimate analysis.

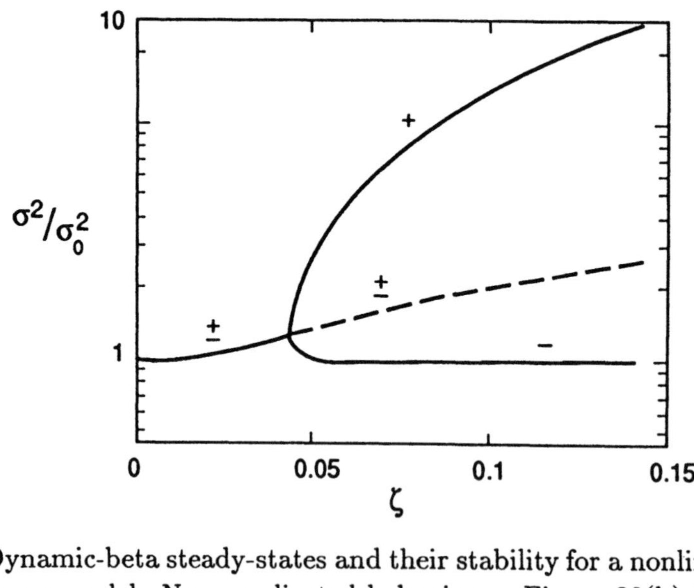

Figure 21. Dynamic-beta steady-states and their stability for a nonlinear beam-beam model. No complicated behavior as Figure 20(b) is observed in this model.

Table 3. Comparison between the two simplified quadrupole mode models

	First Model	Second Model
Self-consistency		
Maxwell equations	no	yes
Nonlinear dynamics	yes	no
Dynamic-beta	yes	yes
Flip-flop	yes	yes
Chaos	yes	no
First instability encountered	chaos	flip-flop

9 HIGHER-ORDER MOMENTS

To improve beyond the two models discussed in the previous section, it becomes necessary to deal with nonlinear beam-beam force, and allow this beam-beam force to distort the distribution from gaussian. It is clear that a self-consistent treatment that takes into account all moments is needed here.[40–45,55–62] In other words, we have to consider, in addition to the dipole and quadrupole modes, also the sextupole and octupole modes, etc. A proper treatment will simultaneously be symplectic, and satisfy the Maxwell equations, at least in principle, at the cost of a more complicated analysis.

Use the octupole mode as an example. It can be unstable due to the beam-beam interaction according to the mechanism shown in Figure 22. Let Figure 22(a) be the unperturbed beam distribution in the phase space (y, y') for the two beams. To investigate the coherent octupole mode stability, perturb one of the beams slightly in an octupolar manner as shown in Figure 22(b). This perturbation will couple to the other beam through the beam-beam interaction. For example, the octupolar modulations of the two beams may interlace each other at the IPs as one possible mode pattern. If the motion is stable, the coupled

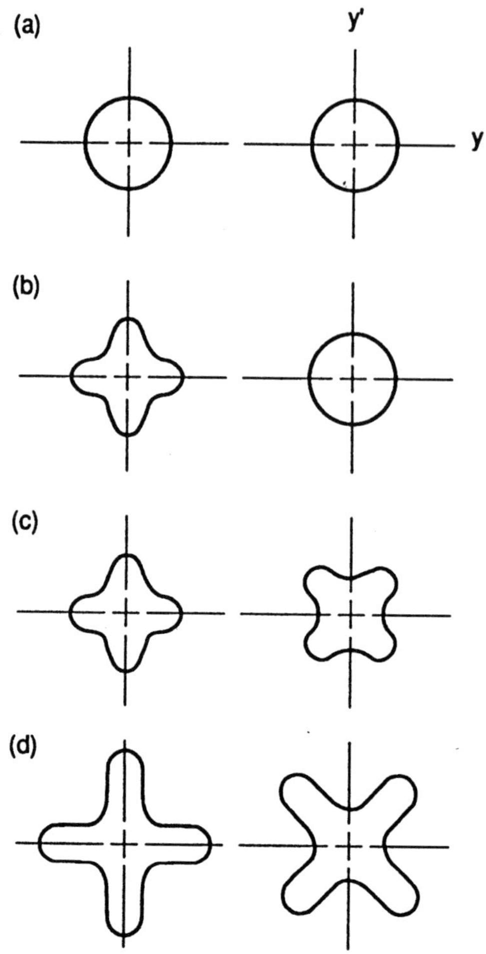

Figure 22. Coherent beam-beam octupole mode instability.

octupolar motion will remain small, as shown in Figure 22(c). However, if the total betatron tune is sufficiently close to a quarter-integer, i.e., $\nu \approx m/4$, the small perturbations will grow exponentially as shown in Figure 22(d), resulting in an octupole mode instability.

In general, instability occurs near resonances $\nu \approx m/n$, as in the incoherent case. However, analogous to the situation shown in Figure 5, there is one important difference in phase space. The cases of incoherent effect, the dynamic-beta effect and the octupole mode effect when $\nu \approx m/4$ (tune per superperiod $\approx m/8$) are illustrated in Figure 23 for one bunch per beam. Note that the dynamic-beta state mentioned in Figure 23 is not that of Figure 4 with linearized beam-beam force, but that mentioned in the discussion following Eq. (19).

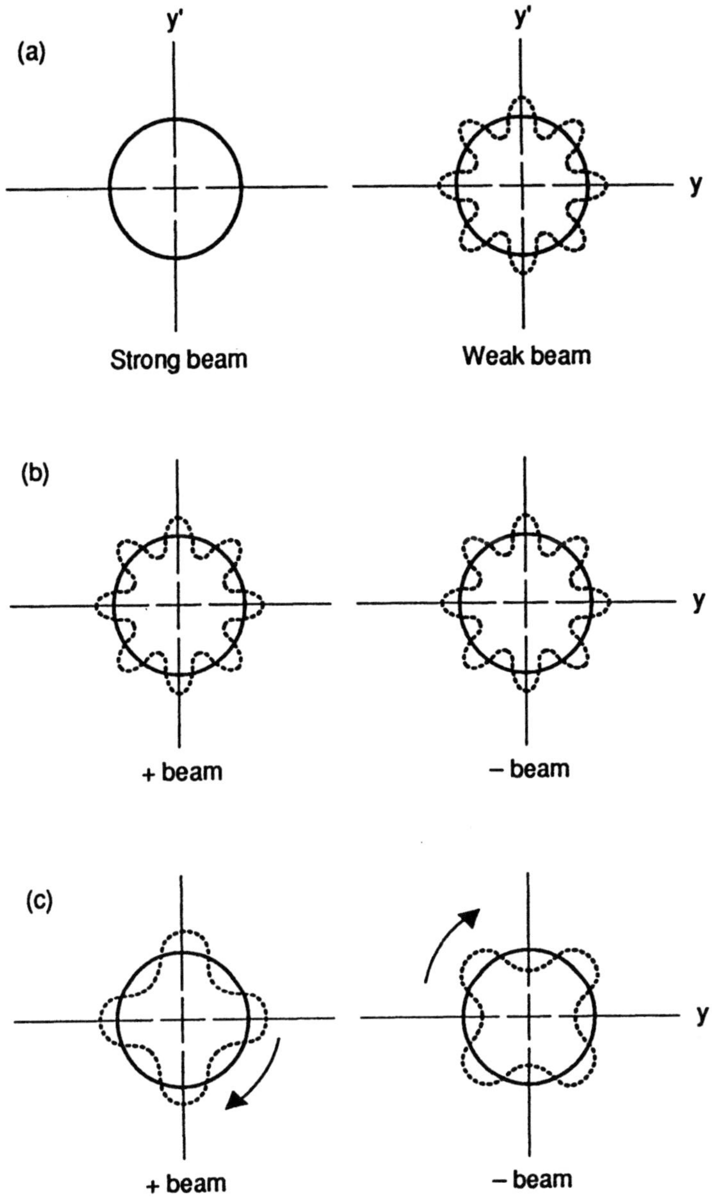

Figure 23. Beam behavior in phase space of two beams for (a) incoherent strong-weak case, (b) dynamic-beta effect, (c) one of the coherent octupole modes; (b) and (c) are strong-strong cases. The solid and dashed curves indicate the unperturbed and the beam-beam perturbed distributions, respectively. The perturbed distribution moves with time in (c) but is static in (a) and (b).

The Vlasov equation (13–14) provides a tool for the general treatment of higher-order modes (assuming flat beams). Below, we will mention the analysis of Ref. 58 as an illustration of the technique. To proceed, let the beam distributions be written as infinitesimal perturbations on top of a steady-state Ψ_0 as Eq. (1). The Vlasov equation is then linearized in $\Delta\Psi_\pm$ and the resultant equations are transformed from the Cartesian variables (y, y') to the action-angle variables (J, ϕ). The beam-beam force is localized in this analysis.

The problem becomes soluble if we assume a water-bag model of the unperturbed steady-state distribution

$$\Psi_0 = (1/\pi\varepsilon)\, H(\varepsilon/2 - J) \tag{37}$$

where ε is the unperturbed beam emittance; $H(x) = 1$ if $x > 0$ and 0 if $x < 0$.

The choice of water-bag model makes the solution of the problem particularly simple, but this is not without cost. The imposed distribution Eq. (37) is inconsistent with the steady-state condition, Eqs. (15)–(16). The problem is therefore solved but not self-consistently.

Coherent beam-beam instability is pronounced near resonances

$$\nu = 2\nu_0 \approx m/n \tag{38}$$

in which case the most perturbed mode is the n-th mode ($n = 1, 2, 3\ldots$ for dipole, quadrupole, sextupole... modes, respectively),

$$\Delta\Psi \sim \exp(\mathrm{i}n\phi)\,. \tag{39}$$

For the water-bag model, the perturbed mode frequency is given by[58]

$$\cos\mu = \cos(2\pi n\nu) \pm \frac{8n}{4n^2 - 1}\,\zeta \sin(2\pi n\nu) \tag{40}$$

where $\pm$ refer to the in-phase or out-of-phase mode patterns (the 0- and the π-modes). Instability results from having $|\cos\mu| > 1$, which occurs when $|\cos(2\pi n\nu)|$ is close to unity, or when Eq. (38) is satisfied. Around the resonance (38), there is a stopband within which the beams are unstable. From Eq. (40), one obtains the stopband width around the resonance:

$$\delta\nu_n = \frac{\zeta}{2\pi}\frac{16}{4n^2-1}. \tag{41}$$

Figure 24 is the sawtooth diagram for 2 bunches per beam. As more and more modes are included, higher-and-higher-order resonances become involved. When the maximum mode number is n, resonances up to order n near $\nu = m/n$ are excited. Figure 25 shows the dependence of the sawtooth diagram on the number of bunches. The higher-order resonance behavior can be rather complex.

In analogy to single-particle nonlinear dynamics, the Chirikov criterion may be applied by summing the stopband widths of all resonances. The total width is $\Sigma n\delta\nu_n$ which diverges if Eq. (41) is used. This means that, for the water-bag model, no region in tune space is free of resonance effects. However, the water-bag model does not provide any tune dependence with amplitude because all dynamics occur at one amplitude only, namely the amplitude corresponding to the edge of the bag. This means there is no consideration of Landau damping effects, which is a deficiency of the water-bag model.

It is important to keep in mind one special feature of the beam-beam interaction. Unlike nonlinear fields due to magnet imperfections, the beam-beam effect is intrinsically stable in the sense that it perturbs only particles in the beam core and not particles with large amplitudes. As a core particle becomes unstable, it necessarily grows out of the instability as its amplitude increases. This detuning effect plays an important role in determining the characteristics of the incoherent beam-beam effects. The same physical feature will also play an important role in the coherent beam-beam effects by manifesting itself through Landau damping. A complete analysis of coherent beam-beam effects will have to consider

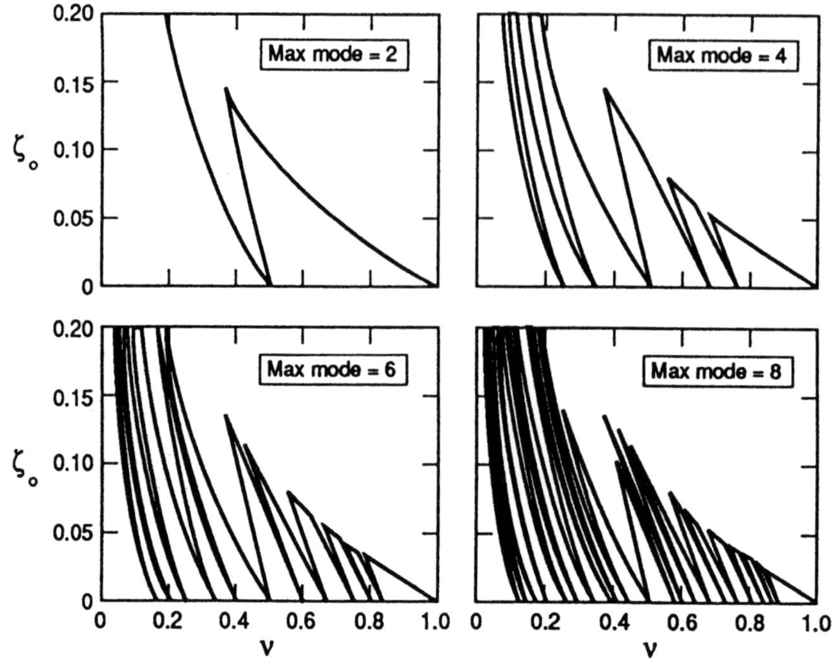

Figure 24. Sawtooth diagrams for the case of $M = 2$ and varying maximum mode number $= 2, 4, 6$, and 8.

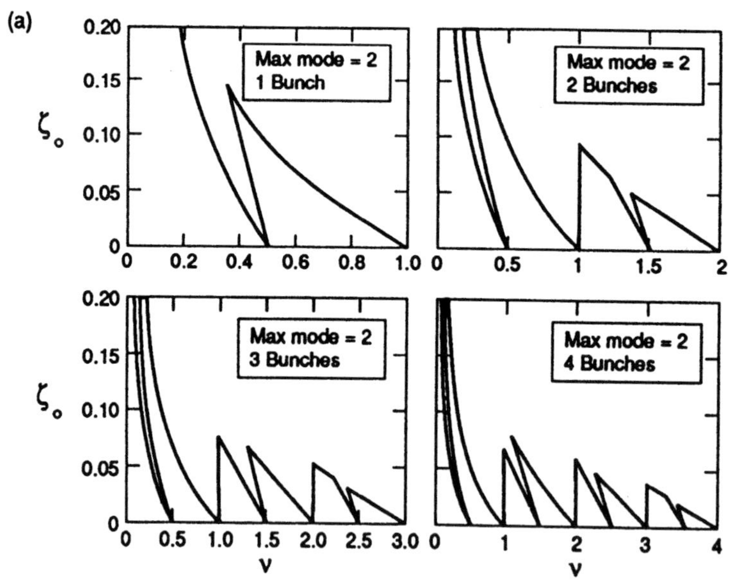

Figure 25. Sawtooth diagrams for different number of bunches $M = 1, 2, 3, 4$; (a), (b) and (c) are three sets for varying maximum mode number $= 2, 3$ and 4, respectively. (See next page)

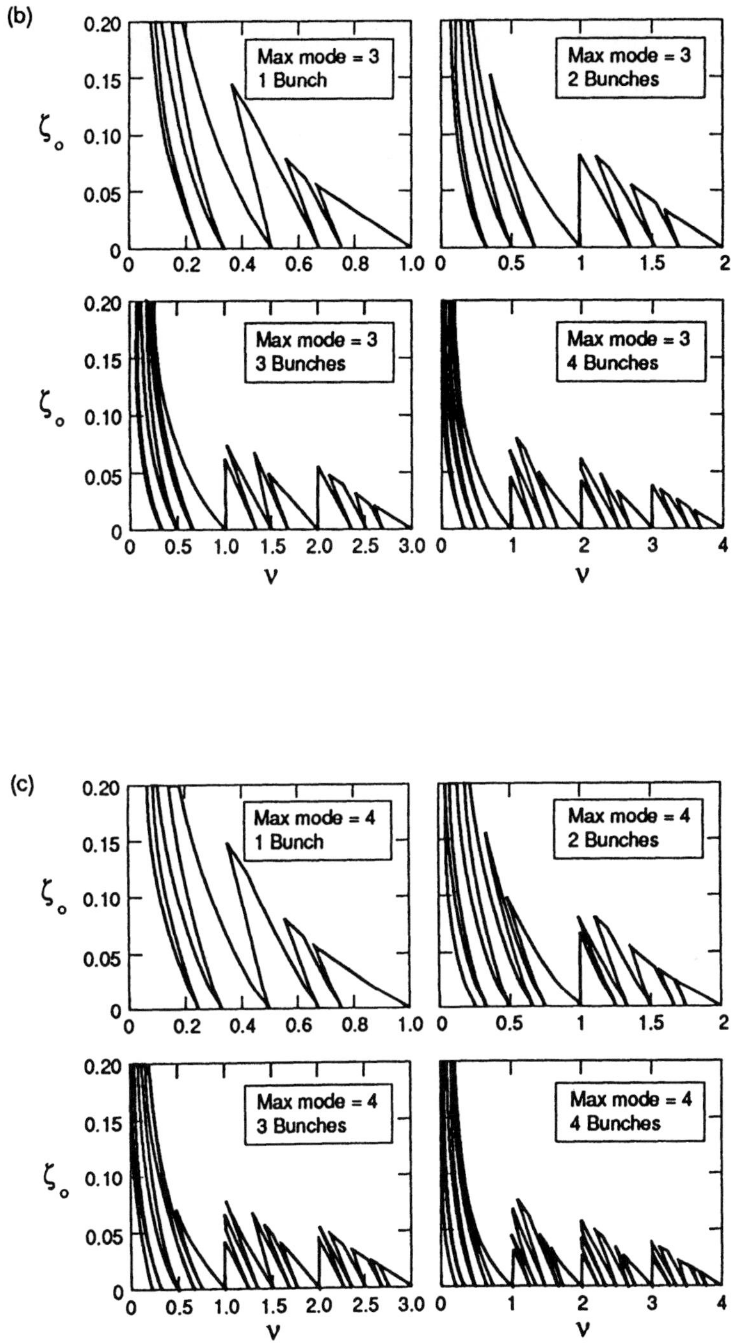

Figure 25 (Continued)

Landau damping effects. When Landau damping is included, the difference between coherent and incoherent effects blurs.

Higher-order modes can also be studied by numerical simulations. Although not emphasized in this review, a remark is to be made consistent with previous discussions. To study the coherent beam-beam effects, it is important in the simulation to allow the beam distribution to deviate from gaussian so that higher-order modes are allowed to grow. Restricting the beam distribution to gaussian shape is equivalent to simulating each beam by two macro-particles even when a thousand particles are used. The approach of allowing for nongaussian distortions has been taken by the more recent simulation codes.[63–66]

10 COHERENT BEAM BLOW-UP MODEL

The coherent beam-beam instability provides a possible description of the observed beam-beam limit.[67] For a given choice of tune away from resonances, consider the situation as the beam intensity N is increased from zero. At first, the beams are stable, beam sizes are given by their unperturbed values, beam-beam parameter is proportional to N, and luminosity is proportional to N^2. As N is increased to reach the boundary of the sawtooth curve, the beams become unstable, the beam sizes grow. However, they grow only by so much as to make ζ stay just under the sawtooth curve. In other words, as N is increased beyond a certain threshold, ζ is saturated at a certain critical value (the beam-beam limit). The beam size and the luminosity then behave as shown in Figure 26 for the flat- and round-beam cases.

This coherent blow-up model is analogous to the bunch lengthening phenomenon observed in electron storage rings. Observe that a similar situation has been encountered when the dynamic-beta effect was discussed, in which the beams continue to blow up as N is increased in such a way that instability (4) is always avoided.

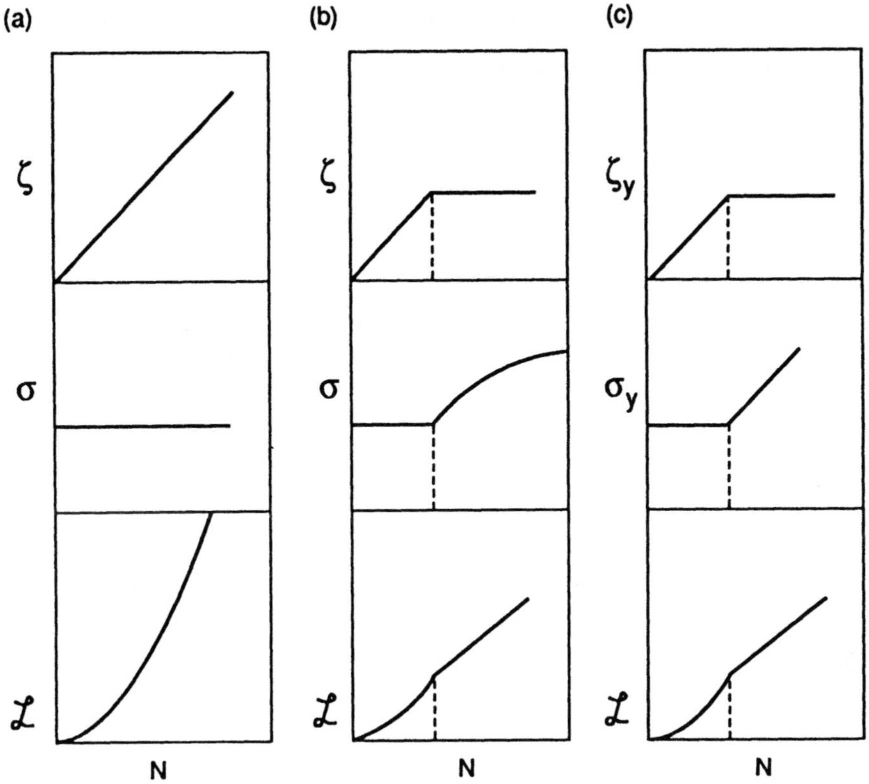

Figure 26. Behavior of beam-beam parameter, beam size and luminosity as functions of beam intensity for three cases: (a) unperturbed, (b) round beams, (c) flat beams. Dotted lines indicated a beam-beam threshold.

REFERENCES

1. See, for example, *Nonlinear Dynamics and the Beam-beam Interaction, BNL*, AIP Proc. No. 57, M. Month and J.C. Herrera, eds., 1979.

2. B. Richter, Proc. Int. Symp. Electron and Positron Storage Rings, Saclay, 1966, p. I-1-1.

3. M.H.R. Donald and J.M. Paterson, *An Investigation of the Flip-flop Beam-beam Effect in SPEAR*, IEEE Trans. Nuc. Sci. **NS-26**, 3580 (1979).

4. H. Wiedemann, *Experiments on the Beam-beam Effect in e+e− Storage Rings*, AIP Proc. No. 57, M. Month and J.C. Herrera, eds., 84 (1979).

5. J.F. Tennyson, *Flip-flop Modes in Symmetric and Asymmetric Colliding-beam Storage Rings*, LBL-28013 (1989).

6. J.F. Tennyson, lectures at this school. Personal communication.

7. S. Kheifets and A.W. Chao, *Steady-state Distribution for Unbunched Beams Colliding in a Drift Space*, SLAC report PEP-325 (1979).

8. A. Piwinski, *Coherent Beam Break-Up Due to Space Charge*, Proc. 8-th Int. Conf. High Energy Accel., CERN, 357 (1971).

9. A. Chao and E. Keil, *Coherent Beam-beam Effect*, CERN/ISR-TH/79-31 or SLAC report PEP-300 (1979).

10. E. Keil, *Coherent Beam-beam Effect in LEP*, CERN report LEP-178, (1979).

11. E. Keil, *Visible Frequencies in the Coherent Beam-beam Effect With More Than One Bunch in Each Beam*, CERN report LEP-268 (1980).

12. E. Keil, *Theoretical Aspects of the Beam-beam Interaction*, Proc. 11-th Int. Conf. High Energy Accel., CERN, 759 (1980).

13. E. Keil, *Visible Frequencies in the Nonlinear Coherent Beam-beam Effect*, CERN report LEP-294 (1981).

14. E. Keil, *Nonlinear Coherent Effects Between Two Strong Bunched Beams Colliding in a storage Ring*, Nucl. Instr. Meth. **188**, 9 (1981).

15. J.D. Bjorken, Workshop on p-pbar Options for the Super Collider, U. Chicago, Fermilab Conf.-84/29-THY (1984).

16. R. Talman, *Multiparticle Phenomena and Landau Damping*, AIP Proc. 153, Accelerator Summer School, SLAC and Fermilab, 789 (1987).

17. A. Hofmann and S. Myers, *Evaluation of the Absolute LEP Luminosity from Measurement of the Coherent Beam-beam Tune Split*, CERN report LEP-604, (1988).

18. K. Hirata, *Coherent Betatron Oscillation Modes Due to Beam-beam Interaction*, Nucl. Instr. Meth. Phys. Res. **A269**, 7 (1988).

19. K. Yokoya, Y. Funakoshi, E. Kikutani, H. Koiso and J. Urakawa, KEK Preprint 89-14 (1989).

20. E. Keil, *Four Beam Compensation Schemes*, CERN LEP-TH/89-37, 3rd Advanced ICFA Beam Dynamics Workshop on Beam-beam Effects in Circular Colliders, Novosibirsk, 85 (1989).

21. SPEAR Group, *Beam-Beam Coupling in SPEAR*, Proc. 9th Int. Conf. High Energy Accel., Stanford, 1974, p.66.

22. A. Piwinski, *Recent Results from DORIS and PETRA*, AIP Proc. No. 57, M. Month and J.C. Herrera, eds., 1979.

23. A. Piwinski, *Observation of Beam-beam Effects in PETRA*, IEEE Trans. Nucl. Sci. **NS-26**, 4268 (1979).

24. T. Ieiri, T. Kawamoto and K. Hirata, Nucl. Instru. Meth. Phys. Res. **A265**, 364 (1988).

25. K. Satoh, Proc. 6th Symp. Accel. Tech., Tokyo, 1987, p.18.

26. T. Ieiri and K. Hirata, *Observation and Simulation of Nonlinear Behavior of Betatron Oscillations During the Beam-beam Collision*, Proc. IEEE Part. Accel. Conf., Chicago, 709 (1989).

27. For a review of coherent dipole and quadrupole effects, see K. Hirata, *The Beam-beam Interaction: Coherent Effects*, AIP Proc. 214, Workshop on Beam Dynamics Issues of High Luminosity Asymmetric Colliders, LBL, 175 (1990).

28. E. Keil, *Coherent Beam-beam Effects in Machines with Unequal Betatron Phase Advances Between Crossing Points*, CERN report LEP-226 (1980).

29. K. Hirata and E. Keil, *Barycentre Motion of Beams Due to Beam-beam Interaction in Asymmetric Ring Colliders*, CERN/LEP-TH/89-76 (1989).

30. K. Hirata and E. Keil, *Coherent Beam-beam Interaction Limit in Asymmetric Ring Colliders*, Phys. Lett. **B232**, 413 (1989).

31. Y. Kamiya and A.W. Chao, *Coherent Synchro-betatron Oscillations Excited by the Beam-beam Interaction*, SLAC/AP-10 (1983).

32. K. Hirata, *Barycenter Motion of Beams Under the Beam-beam Interaction*, KEK report TN-880010 (1989).

33. M. Furman and A. Chao, *Effect of Long Range Beam-beam Interaction on the Stability of Coherent Dipole Motion*, IEEE Trans. Nucl. Sci. **NS-32**, 2297 (1985).

34. E. Forest, *Treatment of the Long Range Coherent Beam-beam with the Vlasov Equation*, SSC-51 (1986).

35. K. Hirata, *The Long Range Beam-beam Force: Effects on the B Factory*, AIP Proc. 214, Workshop on Beam Dynamics Issues of High Luminosity Asymmetric Colliders, LBL, 441 (1990).

36. E. Forest and M. Furman, *Algebraic Theory of Beam-beam Interaction in the Thin Lens Model*, SSC-32, (1985).

37. M.A. Furman, *Results of Coherent Dipole Beam-beam Interaction Studies for SSC Lattices*, SSC-62, (1986).

38. J.E. Augustin et al., Proc. 7-th Int. Conf. High Energy Accel., Yerevan, Vol. 2, 113 (1970).

39. G. Arzelier et al., Proc. 8-th Int. Conf. High Energy Accel., CERN, 150 (1971).

40. N.N. Chau and D. Potaux, *Stabilite des Oscillations Transverses dans un Anneau a Charge D'Espace Compensee, I. Methode de Calcul*, Orsay Tech. Report 5-74 (1974).

41. N.N. Chau and D. Potaux, *Stabilite des Oscillations Transverses dans un Anneau a Charge D'Espace Compensee, II. Application au Choix d'un Point de Fonctionnement*, Orsay Tech. Report 2-75 (1975).

42. Ya.S. Derbenev, Proc. III All Union Part. Accel. Conf., Moscow, 382 (1973).

43. Ya.S. Derbenev, *Collective Instability of Compensated Colliding Beams*, SLAC TRANS-151 (1973).

44. N.S. Dikansky and D.V. Pestrikov, *Colliding Beams Coherent Instability*, Part. Accel. **12**, 27 (1982).

45. Ya.S. Derbenev, N.S. Dikansky, D.V. Pestrikov, Preprint 7-75, INPh, Siberian Division, USSR Academy of Sciences, 1972.

46. A.W. Chao, *The Beam-beam Instability*, AIP Proc. 127, Accel. Summer School, BNL/SUNY, M. Month, P. Dahl, and M. Dienes, eds., 201 (1983).

47. Y. Kamiya and A. Chao, *Stability of the Coherent Quadrupole Oscillations Excited by the Beam-beam Interaction*, SLAC/AP-8 (1983).

48. Historically the second model was introduced before the first model. The order has been reversed below to simplify the discussions.

49. M.A. Furman, K.Y. Ng and A.W. Chao, *A Symplectic Model of Coherent Beam-beam Quadrupole Modes*, SSC-174 (1988).

50. M.A. Furman, *A Symplectic Coherent Beam-beam Model*, 3rd Advanced Beam Dynamics Workshop on Beam-beam Effects in Circular Colliders, Novosibirsk, 52 (1989).

51. K.Y. Ng, *Linear Beam-beam Effects for Round Beams*, SSC-161, (1988).

52. K. Hirata, *Solvable Model of the Beam-beam Limit in e+e− Colliding Rings*, Phys. Rev. Lett. **58**, 25 (1987); **58**, 1798(E) (1987).

53. K. Hirata, *Solvable Model, Flip-flop Hysteresis and Catastrophe in e+e− Colliding Storage Rings*, KEK-Preprint 86-102 (1987).

54. K. Hirata, *Solvable Model of Beam-beam Effects in e+e− Colliding Storage Rings*, Phys. Rev. **D37**, 1307 (1988).

55. H.S. Uhm and C.S. Liu, *Filamentation Instability of Electron and Positron Colliding Beams in Storage Ring*, Phys. Rev. Lett. **43**, 914 (1979).

56. B. Zotter, *Filamentation Instability in LEP*, CERN Report LEP-194, (1979).

57. R.E. Meller and R.H. Siemann, *Coherent Normal Modes of Colliding Beams*, IEEE Trans. Nucl. Sci., **NS-28**, 2431 (1981).

58. A.W. Chao and R.D. Ruth, *Coherent Beam-beam Instability in Colliding-beam Storage Rings*, Part. Accel. **16**, 201 (1985).

59. E. Forest, *A Theory of Coherent Beam-beam Effects with Long-range Interactions*, Part. Accel. **21**, 133 (1987).

60. K. Hirata, *Beyond Gaussian Approximation for Beam-beam Interaction—an Attempt*, CERN/LEP-TH/88-56 (1988).

61. N.S. Dikansky and D.V. Pestrikov, *Collective Beam-beam Phenomena*, 3rd Advanced Beam Dynamics Workshop on Beam-beam Effects in Circular Colliders, Novosibirsk, 76 (1989).

62. K. Hirata, *Stratonovich Expansion and Beam-beam Interaction*, CERN/LEP-TH/89-14 (1989).

63. K. Hirata and E. Keil, *A Program for Computing Beam-beam Modes*, CERN/LEP-TH/89-57 (1989).

64. S. Krishnagopal, Ph.D. thesis, Cornell University, 1990.

65. J. Koga, Ph.D. thesis, University of Texas at Austin, 1990.

66. A. Chao, M. Furman, N.K. Mahale, R. Siemann, S. Krishnagopal, T. Tajima, J. Koga, *Summary Report of the Mini-workshop on Beam-beam Simulations*, SSC-N-702, (1990).

67. A.W. Chao, P. Bambade and W.T. Weng, *Nonlinear Beam-beam Resonances*, Lecture Notes in Phys. 247, Springer-Verlag, J.M. Jowett, M. Month and S. Turner, eds., 77 (1985).

BEAM-BEAM PHENOMENA IN LINEAR COLLIDERS

Kaoru Yokoya and Pisin Chen[†]

Natinal Laboratory for High Energy Physics, Oho, Tsukuba-shi, Ibaraki, 305, Japan
[†]Stanford Linear Accelerator Center, Stanford University, Stanford, CA94309, USA

1 Introduction

A linear collider is a huge complex consisting of several components each of which calls for new technologies and theoretical understandings. Among these, the problem of the beam-beam interaction seems to be the "cleanest" one in the sense that it contains only a few parameters such as the beam size and the number of particles in a bunch and that it involves almost no technological developments.

Although the basic physics of the beam-beam interaction in linear colliders is the same as that in circular colliders, the actual theory is totally different because of the very strong interaction within one collision and the single-pass nature. The phenomenon consists of two aspects, the classical and the quantum mechanical. During the collision the bunches are deformed by the electromagnetic attraction between the electron and positron beams, giving rise to enhancement of the luminosity. Because of the high energy and high beam-beam field we expect a large amount of energy to be deposited in the form of the synchrotron radiation, which, in the case of beam-beam interaction, is called beamstrahlung. Also, for a couple of years, it has been recognized that the e^+e^- pair creation process is a significant source of background for the experiments.

The change in particle energy and population due to the quantum processes can affect the classical phenomenon in principle. Nevertheless, one normally chooses the design parameters such that the energy loss by beamstrahlung is only a small fraction of the initial energy and that the pair-created particles are much fewer in number than the incident particles. Therefore, to a good approximation we can discuss the classical and quantum phenomena separately. If necessary, we can take into account effects such as beam deformation on the quantum processes by using effective beam size.

In this report we shall mostly discuss linear colliders of the next generation in which the beam energy is up to about 1 TeV.

2 Classical Phenomena

The major issues regarding the classical effects are the enhancement of the luminosity due to the electromagnetic attraction, and the deflection angles of the individual particles and the bunch center. These issues have been studied in detail by many authors [1,2,3,4,5], although many topics remain to be studied such as asymmteric collisions. We shall describe the results so far obtained with emphasis on symmetric collisions of flat beams, which have been commonly invoked for the purpose of suppressing the beamstrahlung.

The fact that the beam energy is much higher than the electron rest mass simplifies the problem of classical particle dynamics greatly:

(1) The acceleration by the longitudinal field is negligible.

(2) The transverse force between e^+ and e^- acts only when their longitudinal coordinates nearly coicide because of the Lorentz contraction.

(3) The interaction within the same bunch can be ignored because $\mathbf{E}+\mathbf{v}\times\mathbf{B}=O(\mathbf{E}/\gamma^2)$.

(4) Only the electrostatic field is needed: $\mathbf{E}+\mathbf{v}\times\mathbf{B} \sim 2\mathbf{E}$, even in the presence of a (reasonable) crossing angle.

Coordinate system

We define the coordinates (x,y,s,t) for a head-on collision such that the longitudinal coordinate s is along the direction of motion of the electron beam and that s=t=0 at the moment when the two bunch centroids overlap. We also define comoving longitudinal coordinate z_1 (z_2) whose origin is the center of the electron (positron) bunch. [Therefore the coordinate (x,y,z_2,t) is left-handed.] Since particles travel almost at the speed of light, z_1 (z_2) is a constant for each particle. The s coordinate of a particle in an electron (positron) bunch satisfies the relation $s = z_1 + t$ ($s = -z_2 - t$). Unless specified otherwise, we use the convention c=$\hbar$=1 throughout the discussions.

2.1 Equation of Motion and the Disruption Parameters

The equation of motion of an electron is given by

$$\frac{d^2x}{dt^2} + \frac{4Nr_e}{\gamma} n_L(z_2) \frac{\partial \Phi}{\partial x} = 0 \qquad (z_2 = -z_1 - 2t) \tag{2.1}$$

(similarly for y). Here, N is the number of particles in a bunch, r_e the classical electron radius, γ the particle energy in units of rest mass and n_L the longitudinal density, which is independent of t, defined by

$$n_L(z) = \int n(x, y, z, t) dx dy \tag{2.2}$$

where the density n is normalized such that $\int n dx dy dz = 1$. The electrostatic potential Φ satisfies the Poisson equation

$$\Delta\Phi \equiv \frac{\partial^2 \Phi}{\partial x^2} + \frac{\partial^2 \Phi}{\partial y^2} = 2\pi n_T(x, y, z_2, t) \tag{2.3}$$

where n_T is the transverse distribution defined by $n(x,y,z,t)/n_L(z)$.

A formal solution to the Poisson equation is given by

$$\Phi(x, y) = \frac{1}{2} \int \log[(x - X)^2 + (y - Y)^2] n_T(X, Y) dX dY. \tag{2.4}$$

If the transverse particle distribution is axisymmetric (i.e., "round" beam) and Gaussian with r.m.s. radius $\sigma = \sigma_x = \sigma_y$, then Φ is given by

$$\Phi(x, y) = \int_0^r \frac{1 - e^{-r^2/2\sigma^2}}{r} dr \qquad (r^2 = x^2 + y^2). \tag{2.5}$$

If the beam is very "flat," i.e., $\sigma_x \gg \sigma_y$,

$$\frac{\partial \Phi}{\partial x} = \frac{1}{\sigma_x} e^{-x^2/2\sigma_x^2} \int_0^{x/\sigma_x} e^{\tau^2/2} d\tau \tag{2.6}$$

$$\frac{\partial \Phi}{\partial y} = \frac{1}{\sigma_x} e^{-x^2/2\sigma_x^2} \int_0^{y/\sigma_y} e^{-\tau^2/2} d\tau \qquad (|y| \ll \sigma_x). \tag{2.7}$$

(These expressions do not exactly satisfy $\partial^2\Phi/\partial x \partial y = \partial^2\Phi/\partial y \partial x$.)

When the transverse distribution is a uniform elliptic cylinder with radii a_x=$2\sigma_x$ and a_y=$2\sigma_y$, Φ can be exactly expressed as

$$\Phi = \begin{cases} \frac{1}{a_x + a_y}\left(\frac{x^2}{a_x} + \frac{y^2}{a_y}\right) & \left(\frac{x^2}{a_x^2} + \frac{y^2}{a_y^2} < 1\right) \\ \log \frac{\sqrt{a_x^2+q}+\sqrt{a_y^2+q}}{a_x+a_y} + \frac{x^2/\sqrt{a_x^2+q}+y^2/\sqrt{a_y^2+q}}{\sqrt{a_x^2+q}+\sqrt{a_y^2+q}} & \left(\frac{x^2}{a_x^2} + \frac{y^2}{a_y^2} > 1\right) \end{cases} \tag{2.8}$$

where $q(x, y)$ is the positive solution to the equation $x^2/(a_x^2 + q) + y^2/(a_y^2 + q) = 1$.

At a distance far from the axis for any distribution, Φ can always be given by

$$\Phi = \log r + \text{const.} \qquad (r \gg \sigma_x, \sigma_y). \tag{2.9}$$

Now let us define the so-called disruption parameter. Near the center of a (transversely) Gaussian beam, we have

$$\Phi = \frac{x^2}{2\sigma_x(\sigma_x+\sigma_y)} + \frac{y^2}{2\sigma_y(\sigma_x+\sigma_y)} \qquad (|x| \ll \sigma_x, \quad |y| \ll \sigma_y) \tag{2.10}$$

and the equation of motion becomes

$$\frac{d^2x}{dt^2} + \frac{4Nr_e}{\gamma} n_L(z_2) \frac{x}{\sigma_x(\sigma_x+\sigma_y)} = 0. \tag{2.11}$$

If the initial motion of a particle is paraxial ($x = x_0, \dot{x} = 0$) and if the beam-beam force is so weak that the change of x during the collision is small, the final deflection angle is given by

$$\dot{x}_{fin} = -\frac{2Nr_e}{\gamma} \frac{x_0}{\sigma_x(\sigma_x+\sigma_y)}. \tag{2.12}$$

[Note that $\int n_L(z_2)dt = \int n_L(-z_1 - 2t)dt = 1/2$.] This is equivalent to a thin focusing lens with focal length $\sigma_x(\sigma_x+\sigma_y)\gamma/(2Nr_e)$. The disruption parameter D is defined as the ratio of the r.m.s. bunch length σ_z to the focal length. Thus,

$$D_{x(y)} \equiv \frac{2Nr_e}{\gamma} \frac{\sigma_z}{\sigma_{x(y)}(\sigma_x+\sigma_y)}. \tag{2.13}$$

We define the disruption parameter by this expression for any initial distribution although the expression for the focal length may be different for other distributions. Therefore, the expression $D = \sigma_z/(\text{focal length})$ is only approximate. Note that the horizontal-to-vertical beam size ratio is the inverse of the ratio of the disruption parameters: $R \equiv \sigma_x/\sigma_y = D_y/D_x$.

If the longitudinal distribution is uniform, the equation of motion near the axis is

$$\frac{d^2x}{dt^2} + \frac{D_x}{\sqrt{3}{\sigma_z}^2} x = 0 \qquad (-\sqrt{3}\sigma_z/2 < t < +\sqrt{3}\sigma_z/2), \tag{2.14}$$

which gives rise to sinusoidal oscillations. The number of oscillations is given by $\sqrt{\sqrt{3}D}/2\pi$. Russian papers often invoke the parameter $n = \sqrt{D}/2\pi$ instead of D.

The disruption parameter is the most important parameter characterizing the classical effects of the beam-beam interaction in linear colliders.

2.2 Computer Simulations

Although analytic calculations are important in understanding the physics of the collision, computer simulations are indispensable for obtaining quantitative conclusions when the dynamics involved is highly nonlinear. This is especially true for strong, single-pass collisions.

In computer simulations the bunches are usually described by ensembles of macroparticles. The number of macroparticles is typically 10^3 to 10^5. The whole process is divided into time slices. At each time step the bunches are divided into longitudinal slices and the Poisson equation is solved in each slice. The central issue of the computing technique is how to solve this equation efficiently. We can compute the beam-beam force by using the direct Coulomb force without invoking the Poisson equation. But the computing time is proportional to the number of macroparticles squared, which is extremely large when we use more than a few thousand macroparticles.

Round beams. If the problem is exactly axisymmetric (head-on collision of round beams), the potential Φ is a function of r only and is given by

$$\frac{\partial\Phi}{\partial r} = \frac{1}{r}\left[2\pi\int_0^r n_T(r)rdr\right], \tag{2.15}$$

where the quantity in brackets is the total charge within radius r and can be computed easily by counting the number of macroparticles enclosed.

Very flat beams. If the beam is very flat ($\sigma_x \gg \sigma_y$), from Eq. (2.4) we obtain for $|y| \ll \sigma_x$

$$\frac{\partial\Phi}{\partial x} \sim \text{p.v.}\int \frac{n_T(X)}{x-X}dX, \qquad \frac{\partial\Phi}{\partial y} \sim \pi\int n_T(x,Y)\,\text{sgn}(y-Y)dY \tag{2.16}$$

where p.v. is Cauchy's principal value and $n_T(x) = \int n_T(x,Y)dY$. The horizontal force is independent of y. In the limit of very flat beams, we can ignore the horizontal motion and can replace the vertical force with

$$\frac{\partial\Phi}{\partial y} \sim \pi n(x)\int n_T(Y)\,\text{sgn}(y-Y)dY. \tag{2.17}$$

Note, however, that the horizontal force is always the same order of magnitude as the vertical. The former can be ignored only because its effect is small compared with the horizontal dimension of the beam. Therefore, when we need the magnitude of the beam-beam force, e.g., in estimating the beamstrahlung, the horizontal force cannot be ignored.

PIC: particles-in-cells. For round or very flat beams, the problem is reduced to a one-dimensional equation and the computation is very fast. In general cases where σ_x and σ_y are comparable, we need to solve real 2-D equations. For that, we use a 2-D mesh in each longitudinal slice. The particle distribution is represented by the number of macroparticles in each cell $Q_{i,j}$ and the Laplacian is approximated by a difference

$$\Delta\Phi = \frac{\Phi_{i+1,j} - 2\Phi_{i,j} + \Phi_{i-1,j}}{(\delta x)^2} + \frac{\Phi_{i,j+1} - 2\Phi_{i,j} + \Phi_{i,j-1}}{(\delta y)^2} \tag{2.18}$$

where δx and δy are the mesh sizes. The boundary condition is given at infinity by Eq. (2.9), which requires an infinite mesh region. However, the potential on the outer boundary of the mesh region can be computed explicitly by using Eq. (2.4):

$$\Phi_{i,j} = \frac{1}{2}\sum_{i',j'} \log[(x_i - x_{i'})^2 + (y_j - y_{j'})^2]Q_{i',j'}. \tag{2.19}$$

(Actually, it is better to take an average over each cell.) To use this formula for all the grid points is much too time consuming. The Poisson equation can then be reduced to a matrix equation $M\Phi = Q$, which can in turn be solved by the method of LU-decomposition [6]. The decomposition is time consuming (its computing time is proportional to m^6, m being the number of mesh in x and y dimensions), but it is needed only once if we use the same value of $\delta x/\delta y$ throughout the process. In each longitudinal slice the computing time is proportional to m^3 for solving Eq. (2.19) and to m^4 for the backward substitution in the LU-method.

Instead of using the LU-decomposition, we can use Fast Fourier Transformation. If we employ the 2-D Fourier transform $\tilde{\Phi}(k_x,k_y)$ and $\tilde{n}_T(k_x,k_y)$, the Poisson equation becomes $(k_x^2 + k_y^2)\tilde{\Phi} = \tilde{n}$ and is easy to solve. (In this approach special care must be taken in constructing a Fourier transformed equation, because the Fourier transformation implicitly assumes periodic distribution.) The computing time is proportional to $m^2 \log m$. This method is advantageous if m is very large.

2.3 Luminosity Enhancement

Our primary concern in the classical effects is the enhancement of luminosity due to the Coulomb attraction between the electron and the positron beams. The luminosity per collision ($1/\text{cm}^2$) is given by

$$L = 2N^2 \int dx\, dy\, ds\, dt\, n_1(x,y,z_1,t)\, n_2(x,y,z_2,t) \qquad (s = z_1 + t = -z_2 - t), \tag{2.20}$$

which is to be multiplied by the repetition frequency of collisions to obtain the usually cited luminosity ($1/\text{cm}^2/\text{sec}$). In the case of a head-on collision of (transverse) Gaussian beams without disruption we have

$$L_{00} = \frac{N^2}{4\pi\sigma_x\sigma_y}. \tag{2.21}$$

This needs to be modified if there is a finite crossing angle and/or finite emittance. The latter is characterized by the parameter $A_{x(y)} = \sigma_z/\beta^*_{x(y)}$ where β^* is the beta function at the collision point. The geometrical luminosity is then given by

$$L_0 = L_{00} \times \eta(\phi_c, A) \tag{2.22}$$

where ϕ_c is the full crossing angle (we assume horizontal crossing). If the beam is very flat and Gaussian, the correction factor η can be expressed by using the modified Bessel function K_0 as

$$\eta = \frac{1}{\sqrt{\pi}A_y} \exp\left[-\frac{1+c_\phi^2/4}{2A_y^2}\right] K_0\left(\frac{1+c_\phi^2/4}{2A_y^2}\right) \qquad (c_\phi = \frac{\phi_c}{\sigma_x/\sigma_z}). \tag{2.23}$$

(We shall not include the effect of initial displacement in the definition of L_0.) The luminosity enhancement factor H_D is defined as the effective luminosity under disruption over the geometrical one:

$$H_D = L/L_0. \tag{2.24}$$

Analytic derivation of $H_D(D)$ is very difficult. In the case of round beams with very small D, we have [4]

$$H_D = 1 + D\frac{\sigma_x{}^2 \int r dr\, n_T^3}{\sigma_z \int r dr\, n_T^2} \int_{-\infty}^{\infty} d\tau \int_0^{\infty} \tau' d\tau' n_L(\tau) n_L(\tau + \tau') \tag{2.25}$$

which gives $H_D = 1 + 2D/(3\sqrt{\pi})$ for round Gaussian beams. But this does not help very much in practice when $D \gtrsim O(1)$. We need computer simulations.

In computer simulations we must be careful with the following points. First, $H_D(D)$ is not a well-defined function if the initial motions of all the particles are parallel (i.e., zero emittance). In such a case, the particles near the axis are focused to a point, causing infinite luminosity. The nonlinearity of the force relaxes the singularity to some extent but a weak infinity still remains. In the simulations with A=0, of course, an infinite luminosity does not actually appear, but the result diverges logarithmically as the mesh size decreases. When the emittance is finite ($A \neq 0$) or the depth-of-focus is non-zero, the inherent divergence of the beam tends to smear out the singularity and affect the eventual luminosity. In many recent studies of next-generation linear colliders, A_y is often as big as ~ 1.

Another point is that the problem seems to be rather singular for large D even if $A \neq 0$. We normally expect H_D to increase with D up to a certain point and then saturate before it starts to decrease. However, it increases monotonically in some of the simulation results [3,4], which assume either exact axial symmetry for round beams or exact up-down ($+y$ and $-y$) symmetry for flat beams. This problem is still contraversial but seems to be of only academic interest because actual beams are more or less asymmetric in any case. To avoid this problem, we introduce an initial beam offset (displacement of the whole bunch) Δ. Thus, H_D is a function of D, A and Δ.

Fig. 1. Luminosity enhancement factor for round Gaussian beams for various values of the offset. A=0.4.

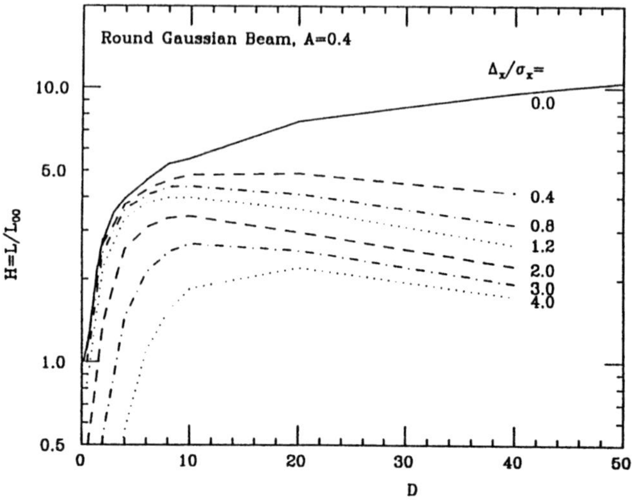

Fig. 2. Enhancement factor for flat Gaussian beams for various values of the offset Δ_y. A_y=0.2 in Fig. 2a and 1.0 in Fig. 2b.

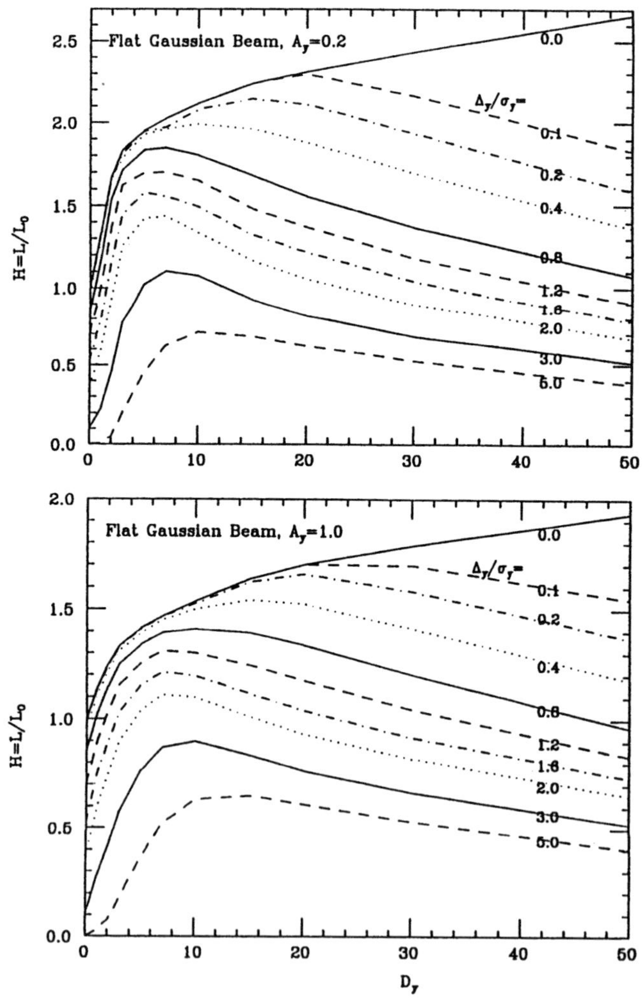

Fig. 1 shows H_D for round Gaussian beams as a function of D with various values of Δ for $A = 0.4$. The curve for Δ=0 (solid line) was computed assuming exact axial symmetry.

The region of large D and small but non-zero Δ is sensitive to computing errors. We can see in this figure that, when D is large, a tiny offset is enough to degrade the luminosity from the line of vanishing Δ. Therefore, in practice, the enhancement rapidly increases up to ~5 with D and then falls off if Δ is sufficiently small but non-zero. This behavior is common to all the simulations so far reported.

Fig. 2a and 2b show results for flat beams with A_y=0.2 and 1.0 respectively. We find behavior similar to that for round beams. The dependence on Δ for large D_y is clearer since we can simulate down to as small a value as $\Delta = 0.1\sigma_y$ because the problem is nearly one-dimensional. A quantitative difference from the round beam case is that we can expect enhancement by at most a factor of 2 for flat beams. Therefore, luminosity enhancement is not a big issue for colliders of the next generation, although some further enhancement might be expected from careful shaping of the longitudinal distribution. More important is the dependence on the offset Δ, which will be discussed later.

2.4 Kink Instability

The saturation of H_D as a function of D and its subsequent decrease are explained by the effect called kink (or two-stream) instability. Fig. 3 shows typical behavior of the bunches during collision. In each plot corresponding to a different time stage t/σ_z, the vertical center-of-mass coordinate y/σ_y of longitudinal slices is plotted as a function of the longitudinal coordinate s/σ_z. The two beams are initially displaced by $\Delta = 0.2\sigma_y$. (D_y=20 is used.) The displacement between the two beams grows in time, causing a loss of luminosity.

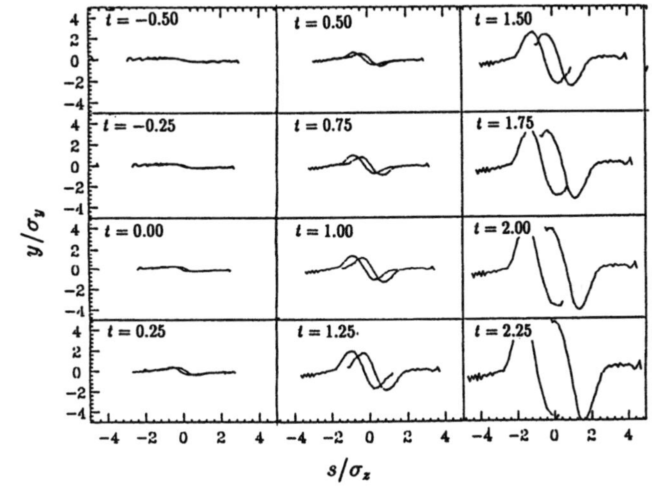

Fig. 3. Evolution of the kink instability. The beams collide with initial vertical offset Δ_y=$0.2\sigma_y$. The time range, $-0.5 < t < 2.25$ (t in units of σ_z), is shown with time mesh 0.25. In each plot the vertical center-of-mass y/σ_y is shown as a function of the longitudinal position s/σ_z. Flat beam with D_y=20.

This phenomenon can be explained by a simple model [7]. Let us model the bunch by a sheet of charge which is uniform in x and z and Gaussian in y. The particle motion for small y can be approximated by

$$\frac{d^2y_1}{dt^2} = -\omega_0^2(y_1 - y_2) \qquad (\omega_0^2 = \frac{\sqrt{2\pi}}{6}\frac{D_y}{{\sigma_z}^2}). \tag{2.26}$$

If we transcribe it to the Euler picture $y(s,t)$, we have

$$\left(\frac{\partial}{\partial t} + \frac{\partial}{\partial s}\right)^2 y_1 = -\omega_0^2(y_1 - y_2), \qquad \left(\frac{\partial}{\partial t} - \frac{\partial}{\partial s}\right)^2 y_2 = \omega_0^2(y_1 - y_2). \tag{2.27}$$

We easily get a solution of the form

$$y_j = a_j e^{i(ks-\omega t)} \qquad (j = 1,2) \tag{2.28}$$

with

$$\begin{pmatrix} (\omega-k)^2-\omega_0^2 & \omega_0^2 \\ \omega_0^2 & (\omega+k)^2-\omega_0^2 \end{pmatrix} \begin{pmatrix} a_1 \\ a_2 \end{pmatrix} = 0, \tag{2.29}$$

which leads to a dispersion relation

$$\omega^2 = k^2 + \omega_0^2 \pm \sqrt{4\omega_0^2 k^2 + \omega_0^4}. \tag{2.30}$$

The solution is unstable if $|k| < \sqrt{2}\omega_0$, and the growth rate is given by

$$|\Im\omega| = [\sqrt{4\omega_0^2 k^2 + \omega_0^4} - (\omega_0^2 + k^2)]^{1/2}. \tag{2.31}$$

The most unstable solution (the largest $\Im\omega$) is

$$\begin{aligned} k &= \pm\frac{\sqrt{3}}{2}\omega_0, \qquad \omega = \frac{i}{2}\omega_0 \\ y_{1(2)} &= a_0 \exp\left[\frac{1}{2}\omega_0 t \pm i(\frac{\sqrt{3}}{2}\omega_0 s - \frac{\pi}{6})\right] \qquad (0 < t < \sqrt{3}\sigma_z). \end{aligned} \tag{2.32}$$

The growth factor of the amplitude is $\exp(\sqrt{3}/2\omega_0\sigma_z) = \exp(\sqrt{2\pi}D_y/8)$. This solution can qualitatively explain the phenomena seen in the simulation. For example, Eq. (2.32) shows a wave pattern which does not oscillate in time t. This is clearly seen in Fig. 3. Also, the phase difference of 60 degrees between the electron and positron oscillations agrees well with the simulation.

The exponential dependence of the growth factor on the disruption parameter might seem disastrous, but in practice the kink phenomenon is actually rather beneficial. For a large enough D_y, even when the initial displacement Δ_y is much larger than σ_y so that the luminosity is expected to be exponentially small in the absence of the interaction, the initial stage of the kink instability helps to bring the beam centers closer to give significant luminosity.

If $\sigma_y \ll \Delta_y \ll \sigma_x$, the equation of motion is approximately given by

$$\frac{d^2y}{dt^2} + \frac{\pi}{6}\frac{\sigma_y D_y}{\sigma_z{}^2} = 0. \tag{2.33}$$

The solution is a parabolic function of t, and the two beam centers can overlap in time if

$$\Delta_y \lesssim \frac{\pi}{4}\sigma_y D_y. \tag{2.34}$$

Thus, so long as $\Delta_y/\sigma_y \lesssim D_y$, there is no significant loss of luminosity.

The enhancement factor H_D is plotted in Fig. 4 as a function of the initial displacement for various values of D_y. (A_y=0.2 and 1.0 in Figs. 4a and 4b, respectively.) The curve for $D_y = 0$ shows the analytic formula $\exp(-\Delta_y^2/4\sigma_y{}^2)$. H_D falls off rapidly when D_y is small. But when D_y is large, H_D sustains within a large range of Δ_y except for a sharp decrease when the situation departs from perfect alignment. In particular, H_D tends to be most insensitive to the displacement for $5 \lesssim D_y \lesssim 20$. Above this region, the initial drop is serious.

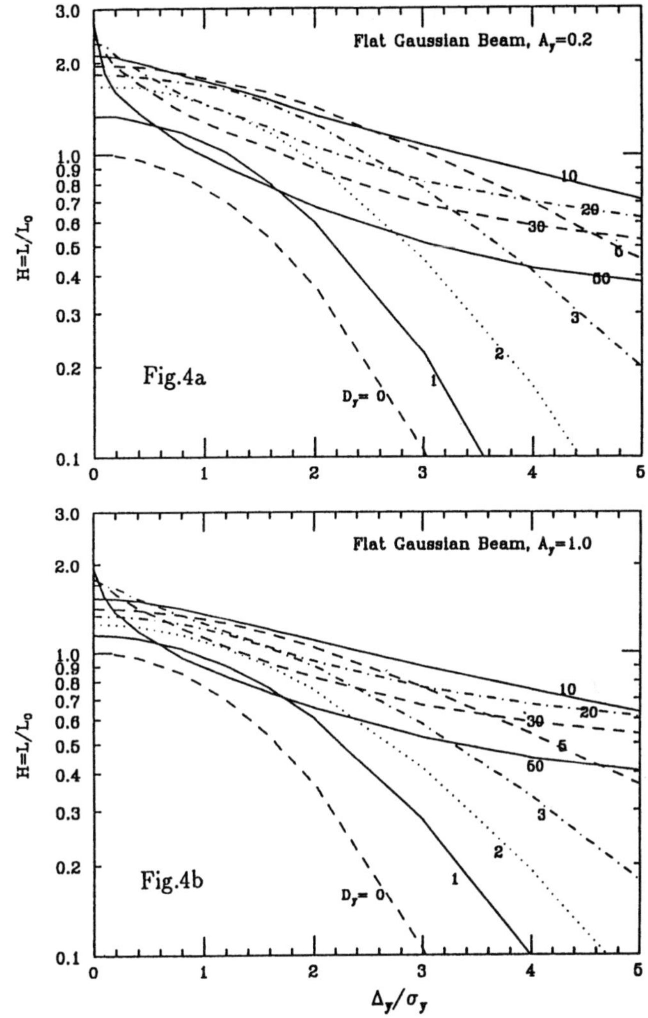

Fig. 4. Enhancement factor H_D vs. initial offset Δ_y for various values of the disruption parameter D_y. Flat Gaussian beam. A_y=0.2 in Fig. 4a and 1.0 in Fig. 4b. An analytic expression is used for the curve D_y=0.

2.5 Disruption Angle

The distribution of the deflection angle is another important issue among the classical effects. It was a major factor in determining the aperture of the final quadrupole magnet in the SLC.

First, let us consider the outgoing angles of the full energy particles. The angle is characterized by the parameter

$$\theta_0 \equiv \frac{2Nr_e}{\gamma(\sigma_x+\sigma_y)} = \frac{D_x\sigma_x}{\sigma_z} = \frac{D_y\sigma_y}{\sigma_z}. \tag{2.35}$$

The horizontal and vertical angles are always comparable but their distributions are very different if the beams are very flat. The outgoing angles as functions of the initial particle position (x_0,y_0) and their distribution are schematically plotted in Fig. 5 for flat beams with very small D_y. The horizontal angle θ_x as a function of x_0 has a turning point near $x_0 \sim 1.3\sigma_x$, which gives a cusp in the distribution of θ_x at its maximum $\theta_{x,max} = 0.765\theta_0$. On the other hand the turning point of θ_y is located at large y_0 (infinity if $\sigma_x/\sigma_y \to \infty$) where the particle density is exponentially small. Therefore, the distribution function of θ_y is small near its maximum $\theta_{y,max} = 1.25\theta_0$. The r.m.s. angles are the same: $\theta_{x,rms}$=$\theta_{y,rms}$=$0.550\theta_0$.

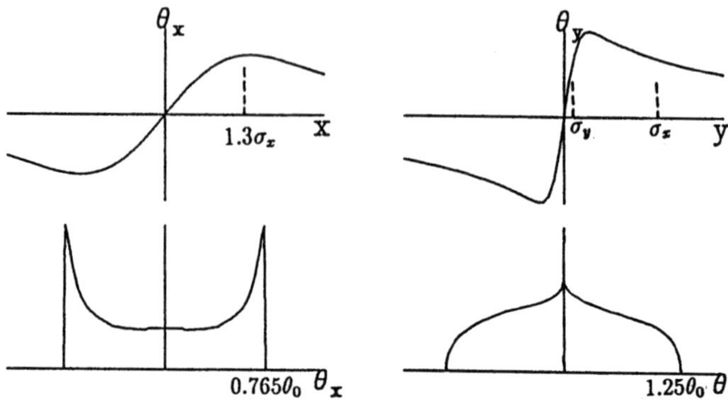

Fig. 5. Horizontal and Vertical deflection angles as functions of x and y (top) and their distribution functions (bottom). Flat beams with small disruption parameters.

A simulation result for finite D_y is shown in Fig. 6, where $\theta_{y,max}/\theta_0$ (solid curve) and $\theta_{y,rms}/\theta_0$ (dashed curve) are plotted vs. D_y for A_y=0.2. The dependence on A_y is weak except for small D_y, where the initial angular spread (truncated at 2.5 sigmas) dominates. The curve for $A_y \to 0$ can be fitted by

$$\theta_{y,rms} \sim \frac{0.55\theta_0}{[1+(0.5D_y)^5]^{1/6}}$$
$$\theta_{y,max} \sim 2.5\theta_{y,rms}. \qquad (2.36)$$

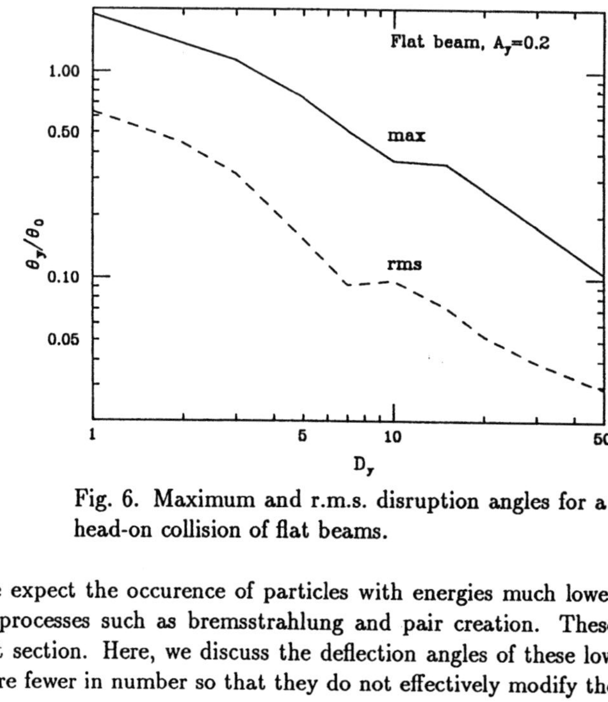

Fig. 6. Maximum and r.m.s. disruption angles for a head-on collision of flat beams.

In the next-generation colliders we expect the occurence of particles with energies much lower than the initial beam energy due to processes such as bremsstrahlung and pair creation. These processes will be discussed in the next section. Here, we discuss the deflection angles of these low energy particles, assuming that they are fewer in number so that they do not effectively modify the nature of disruption.

Consider a particle with energy ϵE_0 ($\epsilon \ll 1$). The effective disruption parameters D_x/ϵ and D_y/ϵ can be very large so that the outgoing angle is much larger than that of the full energy particles. The pair creation process always generates oppositely charged particles (e.g., positrons traveling along the same direction as the initial electrons). Their trajectories are very different from those of particles with the same sign of charge.

Let us first consider the same charge particles. They are focused by the oncoming beam and oscillate within the beam. The equation of motion is approximately

$$\frac{d^2x}{dt^2} + \frac{D_x}{\epsilon}\frac{x}{\sqrt{3}{\sigma_z}^2} = 0, \qquad (2.37)$$

where, for simplicity, we assume a uniform longitudinal distribution. If the particle is created at the beginning of the collision at $x=x_0$ (we ignore the initial angle), the angle at exit $t=\sqrt{3}\sigma_z$ is

$$\theta_x = -x_0\sqrt{\frac{D_x}{\sqrt{3}\epsilon}\frac{1}{\sigma_z}}\sin\sqrt{\frac{\sqrt{3}D_x}{\epsilon}}. \tag{2.38}$$

Assuming $|x_0|\lesssim\sigma_x$ (the deflection is smaller for $|x_0|\gtrsim\sigma_x$ because of the nonlinearity), we get the maximum horizontal deflection angle

$$\theta_{x,max} \sim \begin{cases} \theta_0/\epsilon & (D_x/\epsilon\lesssim 1/\sqrt{3}) \\ \theta_0/\sqrt{\sqrt{3}\epsilon D_x} & (D_x/\epsilon\gtrsim 1/\sqrt{3}). \end{cases} \tag{2.39}$$

A similar expression holds for the vertical angle but, since D_y is usually larger than 1, we have only the region $D_y/\epsilon \gg 1$:

$$\theta_{y,max} \sim \sqrt{\frac{1}{\sqrt{3}\epsilon D_y}}\theta_0. \tag{2.40}$$

Since $D_y > D_x$, the maximum horizontal angle is larger than the vertical by a factor $\sqrt{D_y/D_x}$ for the same charge particles. The reason is that the vertical oscillation number is larger than the horizontal so that the vertical effect does not accumulate.

Next, consider an oppositely charged particle. In this case the beam-beam force is defocusing and, if D/ϵ is large, the particle can be quickly deflected out of the beam and the effect of the far region $\Phi \sim \log r$ dominates eventually.

Assume a uniform elliptic cylinder distribution. The potential Φ on the line $y=0$ is given by

$$\Phi(x) = \begin{cases} x^2/a_x^2 & (|x| < a_x) \\ \log\frac{|x|+\sqrt{x^2-a_x^2}}{a_x} + \frac{|x|}{|x|+\sqrt{x^2-a_x^2}} & (|x| > a_x) \end{cases} \tag{2.41}$$

where $a_x=2\sigma_x$, $a_y=2\sigma_y$ and we assumed $a_x \gg a_y$. The equation of motion can be integrated once and gives

$$\frac{1}{2}\dot{x}^2 = \frac{D_x{\sigma_x}^2}{\sqrt{3}\epsilon{\sigma_z}^2}[\Phi(x)-\Phi(x_0)] \qquad (0 < t < \sqrt{3}\sigma_z). \tag{2.42}$$

By approximately solving this equation, we find the maximum deflection angle to be

$$\theta_{x,max} \sim \begin{cases} \theta_0/\epsilon & (D_x/\epsilon\lesssim 1) \\ \left[\log(4\sqrt{3}D_x/\epsilon)/(\sqrt{3}\epsilon D_x)\right]^{1/2}\theta_0 & (D_x/\epsilon\gtrsim 1). \end{cases} \tag{2.43}$$

The potential Φ on the line $x=0$ is given by

$$\Phi(y) = \begin{cases} y^2/a_x a_y & (|y| < a_y) \\ \log\frac{|y|+\sqrt{y^2+a_x^2}}{a_x} + \frac{|y|}{|y|+\sqrt{y^2+a_x^2}} & (|y| > a_y). \end{cases} \tag{2.44}$$

If $D_y/\epsilon\gtrsim 1$, the contribution of the region $|y| < a_y$ is negligible, and we get the same approximate expression for θ_y as for θ_x regardless of D_x/ϵ. Thus, so long as $D_y/\epsilon \gg 1$, three angles $\theta_{x,same}$, $\theta_{x,opp}$ and $\theta_{y,opp}$ are comparable apart from the logarithmic factor $\log(D_x/\epsilon)$ and $\theta_{y,same}$ are much smaller than these.

2.6 Center-of-Mass Deflection

If the bunches are displaced transversely before collision, the center-of-mass of the bunches is deflected by the beam-beam force. The relation between the initial displacement $\Delta_{x,y}$ and the center-of-mass deflection angle $\Theta_{x,y}$ is useful for monitoring the beam position. For small displacement and small disruption, $\Theta_{x(y)}$ is given by $\frac{1}{2}\theta_0\Delta_{x(y)}/\sigma_{x(y)}$. The relation between Δ and Θ is plotted schematically in Fig. 7 for very flat beams. Θ_x has a maximum near $\sim\sigma_x$ whereas Θ_y is almost flat in a wide range $\sigma_y \lesssim \Delta_y \lesssim \sigma_x$. In general, we define the form factor F for the vertical deflection by

$$\Theta_y = \frac{1}{2}\theta_0 F(\Delta_y/\sigma_y). \qquad (2.45)$$

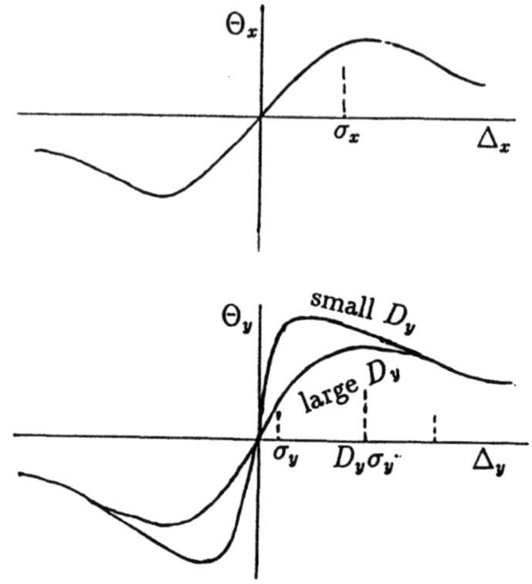

Fig. 7. Center-of-mass deflection as a function of the displacement.

For small D_y, F is given by

$$\begin{aligned} F(\delta) &= \frac{1}{\sqrt{\pi}}\int_0^\delta d\tau \int_{-\infty}^{\infty} d\xi \frac{1}{\sqrt{1+A_y^2\xi^2}} \exp\left[-\xi^2 - \frac{\tau^2}{4(1+A_y^2\xi^2)}\right] \\ &= \int_0^\delta d\tau e^{-\tau^2/4} \qquad (A_y = 0) \\ &= \delta/(1+A_y^2/4) \qquad (\delta \to 0,\ A_y \to 0), \qquad = \sqrt{\pi} \qquad (\delta \to \infty). \end{aligned} \qquad (2.46)$$

The dependence on A_y is small as long as $A_y \lesssim 1$.

The form factor F for finite D_y obtained by simulations is plotted in Fig. 8. When D_y is small, Θ_y saturates at a few σ_y. If, therefore, Δ_y is large, the deflection does not give information about Δ_y and cannot be used as a sensitive position monitor. However, when D_y is large, say 10, the deflection measurement can be useful even for $\Delta_y \gtrsim 5\sigma_y$.

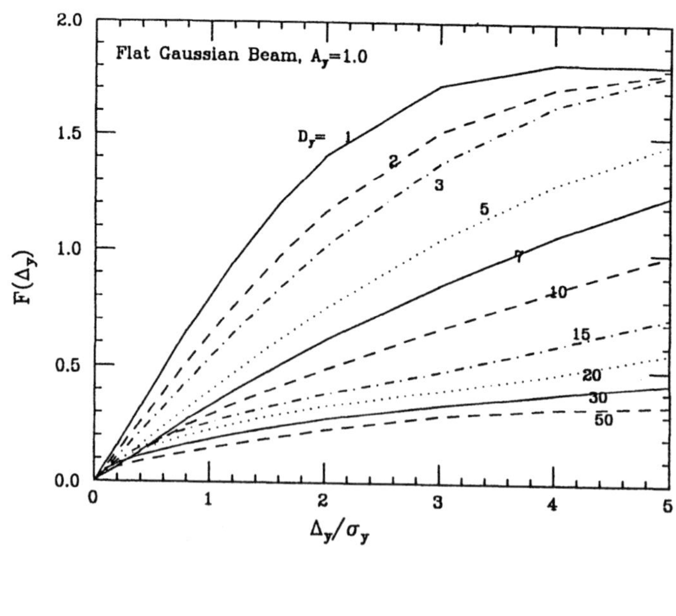

Fig. 8. The form factor F of the center-of-mass deflection angle as a function of the initlal offset Δ_y and the disruption parameter D_y. Flat beam with A_y=1.0.

Multibunch crossing instability

When a beam pulse consists of several bunches, there can be undesired close encounters of bunches before and after the central collision point. Let us consider this peripheral interaction in the case of flat beams with a horizontal crossing angle. (The multi-bunch head-on scheme seems to be impractical.)

The second electron bunch before collision is attracted by the first positron bunch at a distance after the first collision. As a result the former arrives at the collision point with a horizontal displacement. This effect gives the same horizontal displacement for the second positron bunch. Thus the second collision will take place at a collision point which is slightly shifted horizontally but with no relative offset.

However, this is not true if there is a vertical displacement due to errors somewhere upstream [8]. If the vertical shift of the first electron bunch is positive, for example, it will kick the first positron upwards at the central collision point and itself will be kicked downwards. Consequently, this electron bunch will, at peripheral encounters, kick downwards all the following positron bunches, which, in turn, will arrive at the collision point with negative displacements. Thus, all the effects add up with opposite signatures between electrons and positrons to cause an instability. This phenomenon is physically the same mechanism as the kink instability discussed before, albeit that it is magnifested in a discrete form.

Let us denote the vertical offset of the k-th $e^{\pm}$ bunch in units of σ_y by $\Delta_k^{\pm}$. Then, the l-th $e^{\mp}$ bunch is kicked by the l-th $e^{\pm}$ bunch at the collision point by an angle $y_l' = \frac{1}{2}\theta_0 F(\Delta_l^{\pm} - \Delta_l^{\mp})$. When this bunch encounters the k-th ($k > l$) $e^{\pm}$ bunch, the latter is kicked vertically by an angle $2Nr_e/\gamma \times y_l' L_{k,l}/(L_{k,l}\phi_c)^2$. Here, $L_{k,l}$ is the distance between the central collision point and the encounter point of the k-th and l-th bunches and ϕ_c is the crossing angle. Therefore, $L_{k,l}\phi_c$ is the distance of encounter. (We assume $L_{k,l}\phi_c \gg \sigma_x$.) After this encounter, the k-th $e^{\pm}$ bunch travels a distance $L_{k,l}$ till the collision. Therefore, the vertical shift of the k-th bunch due to this encounter is given by $Nr_e\theta_0/(\gamma\phi_c^2)F(\Delta_l^{\pm} - \Delta_l^{\mp})$, which is independent of $L_{k,l}$. Thus, denoting the relative displacement by $\Delta_k \equiv \Delta_k^+ - \Delta_k^-$, we obtain

$$\Delta_k = C\sum_{l=1}^{k-1} F(\Delta_l) + \Delta_{k,0}, \qquad C = D_x D_y \left[\frac{\sigma_x/\sigma_z}{\phi_c}\right]^2 = \frac{D_x D_y}{c_\phi^2} \tag{2.47}$$

where $\Delta_{k,0}$ is the displacement due to errors upstream.

If $\Delta_{k,0}=\Delta_0$ for all the bunches and if D_y and Δ are small so that $F(\Delta) \sim \Delta$, we get

$$\Delta_k = (1 + C)^{k-1}\Delta_0. \tag{2.48}$$

Therefore, this effect is serious if $C(m_b - 1)\gtrsim 1$ where m_b is the number of bunches which encounter in the magnet-free region near the IP. However, this criterion is too pessimistic because, as we can see in Fig. 8, $F(\Delta)$ is considerably smaller than Δ if D_y is large.

By converting Eq. (2.47) into a differential equation with regard to k, we can approximately solve the equation. Then the criterion of tolerable instability becomes

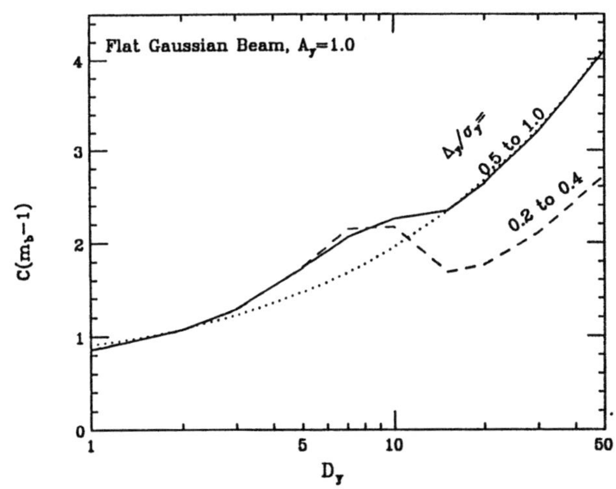

Fig. 9. Criterion for the multibunch crossing instability.

$$C(m_b-1) \lesssim \int_{\Delta_0}^{\Delta_{fin}} \frac{d\Delta}{F(\Delta)} \tag{2.49}$$

where Δ_{fin} is the maximum tolerable displacement of the last bunch. Because $F(\Delta)$ is not linear in Δ, the criterion depends both on Δ_0 and Δ_{fin}. The function on the right-hand side is plotted in Fig. 9 as a function of D_y for A_y=1.0. Two curves correspond to (Δ_0,Δ_{fin})= (0.2,0.4) (dashed line) and (0.5,1.0) (solid line), respectively.

The latter range of Δ seems to be more practical. Thus, the criterion that the blow-up factor is to be less than two is given by

$$C(m_b-1) \lesssim \sqrt{\frac{1}{2}+\frac{D_y}{3}}, \tag{2.50}$$

which is shown by the dotted line in Fig. 9.

3 Quantum Beamstrahlung

The magnetic field in the bunch during collision is typically of the order of kilo-Teslas in the linear colliders of the next generation. Therefore, even though the longitudinal length of the field is small $\sim \sigma_z \sim 10^{-4}$ m, the synchrotron radiation, called beamstrahlung in this case, plays an important role. As we shall see later, the so-called radiation coherence length $l_R \approx \rho/\gamma$, ρ being the orbit radius of curvature, is typically $\sim 10^{-6}$m, which is considerably smaller than the bunch length. (When the disruption parameter is large, the field seen by a particle changes within a longitudinal distance $\sim \sigma_z/\sqrt{D}$. But l_R is usually still smaller than this length though a bit marginally.) Therefore, we can invoke the formula of the synchrotron radiation in a constant field.

The radiation is characterized by the critical energy ω_c. Let us introduce the dimensionless Lorentz invariant parameter Υ, defined by

$$\Upsilon \equiv \frac{2}{3}\frac{\hbar\omega_c}{E} = \frac{\lambda_e\gamma^2}{\rho} = \gamma\frac{B}{B_c} = \frac{e}{m^3}\sqrt{|(F_{\mu\nu}p^\nu)^2|} \tag{3.51}$$

where E is the electron energy before radiation, λ_e the Compton wavelength, $F_{\mu\nu}$ the energy-momentum tensor of the field, p^ν the 4-momentum of the electron, B the magnetic field (actually, $|\mathbf{E}|+|\mathbf{B}|$) and B_c the critical magnetic field defined as $B_c = m^2/e \approx$4.4 GTeslas. (The parameter $\xi \equiv \omega_c/E = 3\Upsilon/2$ is also often adopted instead of Υ.)

This parameter Υ is not constant during the collision. In the case of Gaussian beams, the maximum and average of its value can be estimated by

$$\Upsilon_{max} \approx \frac{2Nr_e^2\gamma}{\alpha\sigma_z(\sigma_x+\sigma_y)}, \qquad \Upsilon_{av} \approx \frac{5}{12}\Upsilon_{max} \tag{3.52}$$

where α is the fine structure constant. The SLC design parameters give $\Upsilon_{av} \sim 0.004$. In the linear colliders of the next generation we will have Υ_{av} ~0.1 to 1.0, i.e., the typical photon energy is comparable to the initial electron energy. Therefore, we need to take into account the recoil of the electron. We cannot use the standard radiation formula used in storage ring theories, and the quantum electrodynamic (QED) formulation is necessary.

3.1 Sokolov-Ternov Formula

The radiation formula for arbitrary Υ was first derived by Sokolov and Ternov [9]. They used the exact solution of the Dirac equation in a uniform magnetic field and computed the transition rate. Baier and Katkov [10] rederived the same formula by a method that is very powerful for general semi-classical problems such as the radiation in varying fields. Here, we shall follow the latter.

The matrix element for one photon emission with momentum $\mathbf{k}$ and energy $\omega = |\mathbf{k}|$ is written in the form

$$\left\langle f \left| \frac{e}{\sqrt{2\omega}} \int dt\, e^{i\omega t} M(t) \right| i \right\rangle, \qquad M(t) = u_{s'}^{\dagger} \boldsymbol{\alpha} \cdot \boldsymbol{\varepsilon} e^{-i\mathbf{k}\cdot\mathbf{x}(t)} u_s$$

where s (s') is the initial (final) electron spin state, $\boldsymbol{\varepsilon}$ the photon polarization vector, $\boldsymbol{\alpha}$ the Dirac matrix, and $\mathbf{x}(t)$ the Heisenberg position operator. The probabality of photon emission is given by

$$dw_\gamma = \frac{e^2}{4\pi} \frac{d\mathbf{k}}{(2\pi)^2 \omega} \left\langle i \left| \int dt_1\, dt_2\, e^{i\omega(t_1 - t_2)} M^\dagger(t_2) M(t_1) \right| i \right\rangle. \tag{3.53}$$

Baier and Katkov reduced this expression into the following form assuming that the electron's dynamical variables commute (treating the electron orbit classically) but retaining the commutator between electron and photon variables (taking into account the recoil):

$$dw_\gamma = \frac{e^2}{4\pi} \frac{d\mathbf{k}}{(2\pi)^2 \omega} \int dt_1\, dt_2\, \exp i\{\omega(t_1 - t_2) - \frac{E}{E - \omega} \mathbf{k} \cdot [\mathbf{x}(t_1) - \mathbf{x}(t_2)]\} R^*(t_2) R(t_1) \tag{3.54}$$

where

$$R(t) = \varphi_f^* \left[\frac{1}{2} \boldsymbol{\varepsilon} \cdot \mathbf{p} \left(\frac{1}{E_{\mathbf{p}} + m} + \frac{1}{E_{\mathbf{p}'} + m} \right) + \frac{1}{2} \left(\frac{\boldsymbol{\varepsilon} \times \mathbf{p}}{E_{\mathbf{p}} + m} - \frac{\boldsymbol{\varepsilon} \times \mathbf{p}'}{E_{\mathbf{p}'} + m} \right) \right] \varphi_i.$$

Here, φ's are the 2-component spinors and $\mathbf{p} = E\mathbf{v}(t)$, $\mathbf{p}' = \mathbf{p} - \mathbf{k}$, $E_{\mathbf{p}} = \sqrt{m^2 + \mathbf{p}^2}$, $E_{\mathbf{p}'} = \sqrt{m^2 + \mathbf{p}'^2}$. In these expressions all the variables are already c-numbers (classical numbers). Thus, $\mathbf{x}(t)$ is the classical electron trajectory and $\mathbf{v}(t) = d\mathbf{x}/dt$. Summing over the final spins and photon polarizations and averaging over the initial spins, one can replace $R^* R$ with

$$\left(1 + \frac{\omega}{E'} + \frac{1}{2} \frac{\omega^2}{E'^2}\right) [\mathbf{v}(t_1) \cdot \mathbf{v}(t_2) - 1] + \frac{1}{2\gamma^2} \frac{\omega^2}{E'^2}, \qquad E' = E - \omega.$$

Let us define $t_{1(2)} = t \mp \tau/2$. When the field variation length [$\sim \min(\sigma_z, \sigma_z/\sqrt{D})$] is small compared with l_R, we can expand the integrand in terms of τ. To get the constant field formula, the following terms are enough:

$$\mathbf{v}(t_1) \cdot \mathbf{v}(t_2) - 1 = -\frac{1}{\gamma^2} - \frac{1}{2} |\dot{\mathbf{v}}(t)|^2 \tau^2, \qquad \mathbf{x}(t_2) - \mathbf{x}(t_1) = \mathbf{v}(t)\tau + \frac{1}{24} \ddot{\mathbf{v}}(t) \tau^3.$$

By integrating Eq. (3.54) over the photon angle $\mathbf{k}/|\mathbf{k}|$ we get

$$\frac{dW_\gamma}{d\omega} = \frac{ie^2}{8\pi^2} \frac{E'}{E} \int_{-\infty}^{\infty} \frac{d\tau}{\tau - i0} \left[\frac{1 + (\omega/E')}{\gamma^2} + (1 + \frac{\omega}{E'} + \frac{1}{2} \frac{\omega^2}{E'^2}) \frac{\tau^2}{2\rho^2} \right] \exp\left[-i \frac{\omega}{E'\Upsilon} \frac{\gamma\tau}{2\rho} [1 + \frac{1}{3} (\frac{\gamma\tau}{2\rho})^2] \right] \tag{3.55}$$

where $1/\rho = |\dot{\mathbf{v}}(t)|$ and $W_\gamma = dw_\gamma/dt$.

The range of τ that gives a significant contribution to the integral is called the radiation coherence length l_R. Note that, in an integration like $\int dx \exp[-ic(x + x^3/3)]$, the significant contribution comes from the region $|x| \lesssim c^{-1/3}$ when $c \ll 1$ (x^3 term) and from $|x| \lesssim c^{-1/2}$ when $c \gg 1$ (saddle point of the phase). Therefore, the radiation coherence length is

$$l_R \approx \frac{\rho}{\gamma} \left[\frac{\omega}{E - \omega} \frac{1}{\Upsilon} \right]^{-1/3}, \ (\frac{\omega}{E - \omega} \frac{1}{\Upsilon} \ll 1)\ ; \quad \approx \frac{\rho}{\gamma} \left[\frac{\omega}{E - \omega} \frac{1}{\Upsilon} \right]^{-1/2}, \ (\frac{\omega}{E - \omega} \frac{1}{\Upsilon} \gg 1). \tag{3.56}$$

In the case $\Upsilon \ll 1$, as in storage rings, for most photons $\omega/\Upsilon E$ is $O(1)$ which gives $l_R \approx \rho/\gamma$. [This is not true for photons with energies much lower than the critical energy. For such radiation we need to use $l_R \approx (\rho/\gamma)(\omega_c/\omega)^{1/3}$.] When Υ is large, the coherence length is $l_R \approx (\rho/\gamma)\Upsilon^{1/3}$ because $\omega/(E - \omega)$ is $O(1)$. For the next generation linear colliders, in which Υ is $O(1)$, we can ignore the factor $\Upsilon^{1/3}$.

Now, by using the formulas

$$\int_{-\infty}^{\infty} \frac{1}{x-i0} e^{-ic(x+x^3/3)} dx = \frac{2i}{\sqrt{3}} \int_{2c/3}^{\infty} K_{1/3}(x) dx$$

$$\int_{-\infty}^{\infty} x \sin c(x+\frac{x^3}{3}) dx = \frac{1}{\sqrt{3}} K_{2/3}(\frac{2c}{3})$$

where K_ν is the modified Bessel function and $i0$ defines the integration contour to pass below the pole at the origin, we obtain the Sokolov-Ternov spectrum formula

$$\frac{dW_\gamma}{d\omega} = \frac{\alpha}{\sqrt{3}\pi\gamma^2} F_{BS} \tag{3.57}$$

with

$$F_{BS} = -\int_\xi^\infty K_{1/3}(\xi')d\xi' + \left(\frac{E-\omega}{E} + \frac{E}{E-\omega}\right) K_{2/3}(\xi) \tag{3.58}$$

$$= \int_\xi^\infty K_{5/3}(\xi')d\xi' + \frac{y^2}{1-y} K_{2/3}(\xi), \tag{3.59}$$

where

$$\xi = \frac{2\omega}{3\Upsilon(E-\omega)} = \frac{\omega}{\omega_c}\frac{1}{1-\omega/E} = \frac{2}{3\Upsilon}\frac{y}{1-y} \qquad (y=\frac{\omega}{E}). \tag{3.60}$$

In the classical limit $\Upsilon \to 0$, the second term in Eq. (3.59) can be ignored and we come to the classical spectrum formula.

The Sokolov-Ternov power spectrum is schematically shown in Fig. 10. The low energy behavior $P(\omega) \propto \omega^{1/3}$ is the same as the classical formula. The high energy part extends to infinite (unphysical) energy as $\exp(-\omega/\omega_c)$ in the classical formula but it is truncated at $\omega = E$ in the Sokolov-Ternov formula. The high energy tail is expressed as $e^{-\xi}$. In the log-log plot, the spectrum approaches nearly a triangular form in the limit $\Upsilon \to \infty$.

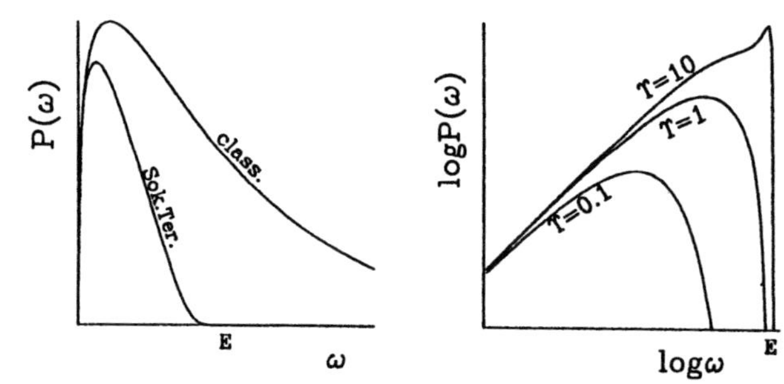

Fig. 10. Sokolov-Ternov Spectrum Function

3.2 Number of Photons and Average Energy Loss

By integrating Eq. (3.58) over the photon energy, we obtain the expected number of photons per unit time:

$$\frac{dN_\gamma}{dt} = \int_0^E \frac{dW_\gamma}{d\omega} d\omega = \frac{5}{2\sqrt{3}}\frac{\alpha\Upsilon}{\lambda_e\gamma} \cdot U_0(\Upsilon) \tag{3.61}$$

with

$$U_0(\Upsilon) = \left\{ \begin{array}{ll} 1 & (\Upsilon \to 0) \\ (28\sqrt{3}/45)\Gamma(2/3)(3\Upsilon)^{-1/3} = 1.012\Upsilon^{-1/3} & (\Upsilon \to \infty) \end{array} \right\} \approx \frac{1}{(1+\Upsilon^{2/3})^{1/2}}. \tag{3.62}$$

The average energy loss per unit time is

$$\left\langle -\frac{1}{E}\frac{dE}{dt} \right\rangle = \int_0^\infty \frac{\omega}{E}\frac{dW_\gamma}{d\omega} d\omega = \frac{2}{3}\frac{\alpha\Upsilon^2}{\lambda_e\gamma} \cdot U_1(\Upsilon) \tag{3.63}$$

with

$$U_1(\Upsilon) = \left\{ \begin{array}{ll} 1 & (\Upsilon \to 0) \\ (16/9)\Gamma(2/3)(3\Upsilon)^{-4/3} = 0.556\Upsilon^{-4/3} & (\Upsilon \to \infty) \end{array} \right\} \approx \frac{1}{[1+(1.5\Upsilon)^{2/3}]^2}. \tag{3.64}$$

The functions U_0 and U_1 are plotted in Fig. 11.

The average photon energy is

$$\frac{\langle\omega\rangle}{E} = \frac{4\sqrt{3}}{15}\Upsilon\frac{U_1(\Upsilon)}{U_0(\Upsilon)} \tag{3.65}$$

$$= \left\{ \begin{array}{ll} 0.462\Upsilon & (\Upsilon \to 0) \\ 16/63 = 0.254 & (\Upsilon \to \infty). \end{array} \right.$$

Note that the average photon energy is finite (about one quarter of the initial energy) in the limit $\Upsilon \to \infty$.

In the case of a collision of Gaussian bunches, the average number of emitted photons per electron n_γ and the relative energy loss δ_E can be given approximately by

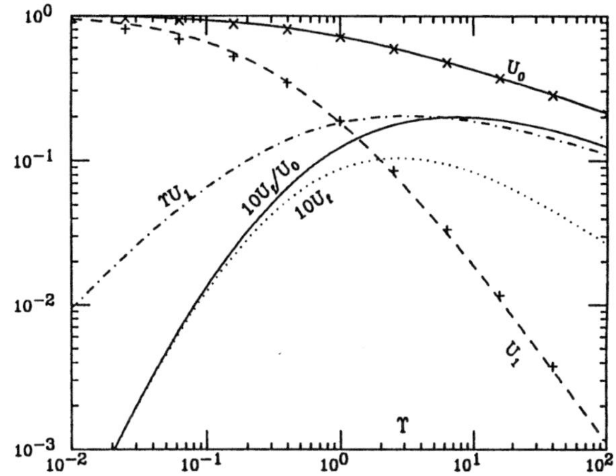

Fig. 11. Functions U_0, U_1 and U_f. The crosses are the approximate formulas in Eqs. (3.62) and (3.64).

$$n_\gamma \approx 1.06\alpha r_e N\frac{2}{\sigma_x+\sigma_y}U_0(\Upsilon_{av}) \approx 2.54\left[\frac{\alpha\sigma_z\Upsilon_{av}}{\lambda_e\gamma}\right]U_0(\Upsilon_{av}) \tag{3.66}$$

$$\delta_E = \left\langle -\frac{\Delta E}{E}\right\rangle \approx 0.216\frac{r_e^3N^2\gamma}{\sigma_z}\left(\frac{2}{\sigma_x+\sigma_y}\right)^2 U_1(\Upsilon_{av}) \approx 1.24\left[\frac{\alpha\sigma_z\Upsilon_{av}}{\lambda_e\gamma}\right]\Upsilon_{av}U_1(\Upsilon_{av}). \tag{3.67}$$

The beamstrahlung causes a spread in the center-of-mass energy of e^+ and e^-. This effect is characterized by the parameter δ_E. Although it is smaller than 0.1% in the SLC, it is a severe limiting factor for the performance of colliders in the future. As seen in Fig. 11, $\Upsilon U_1(\Upsilon)$ is almost constant (0.1 to 0.2) in the wide range $0.2 \lesssim \Upsilon \lesssim 200$. Therefore, in order to keep δ_E below 10 to 20%, most designs choose $\alpha\sigma_z\Upsilon/\lambda_e\gamma$ around unity. Since U_0 is not far from unity unless Υ is very large, n_γ is also of order unity. It should be noted that the ratio of l_R to σ_z is approximately equal to α/n_γ, which means l_R/σ_z is about 10^{-2} if $n_\gamma \sim 1$.

3.3 Energy Spectrum of Electrons

The beamstrahlung alters the energy spectrum of electrons. The final spectrum is complicated in general because of the possibility of emitting more than one photon. If $n_\gamma \ll 1$, however, we may ignore the multi-photon emission. The final electron spectrum is then given approximately by

$$\psi_{fin}(x) = e^{-n_\gamma}\delta(x-1) + (1-e^{-n_\gamma})f(1-x) \tag{3.68}$$

where δ is the Dirac delta function (the energy spread before collision is negligible) and $f(y)$ is the Sokolov-Ternov spectrum function normalized as $\int f(y)dy = 1$. Υ in $f(y)$ may be replaced by the average value. A more detailed calculation is given in Ref [11]. But Eq. (3.68) is sufficient as a coarse approximation to be used in background estimations, if $n_\gamma \lesssim 1$.

For the average spectrum we may replace n_γ with $n_\gamma\tau$ $(0 < \tau < 1)$ and average over τ:

$$\psi_{av}(x) \sim \int_0^1 \psi_{fin}(x, n_\gamma\tau)d\tau = \frac{1-e^{-n_\gamma}}{n_\gamma}\delta(x-1) + (1-\frac{1-e^{-n_\gamma}}{n_\gamma})f(1-x). \tag{3.69}$$

Simulation results show that the 2-D differential luminosity spectrum $dL/dx_1dx_2 = L \times \psi_2(x_1, x_2)$ factorizes quite well as $\psi_{avr}(x_1)\psi_{avr}(x_2)$. Therefore, the center-of-mass spectrum is approximately given by

$$\psi_s(s) \sim \int dx_1\, dx_2\, \delta(s/4E_0^2 - x_1x_2)\psi_{av}(x_1)\psi_{av}(x_2). \tag{3.70}$$

Actually, the 2-D spectrum is dominated by the two edges $x_1 = 1$ and $x_2 = 1$, i.e., the case where one of the colliding particles is at full energy. Therefore, the events with low s ($\ll 4E_0^2$) go mostly forward along with one of the two beams.

3.4 How to Reduce Beamstrahlung

Since the energy loss due to the beamstrahlung is a big limitation for high energy linear colliders, several methods of reducing it have been investigated.

Short Bunches

As we can see in Eq. (3.67), we can reduce δ_E by using a longer bunch in the classical regime $\Upsilon \ll 1$. This is limited, however, by other constraints in colliders such as control of the energy spread in the linac. In the extreme quantum regime $\Upsilon \gg 1$, δ_E is proportional to $\sigma_z{}^{-1/3}$ owing to the Υ-dependence of U_1. This does not help unless Υ is extremely large. The Υ-dependence of δ_E is like $\Upsilon U_1(\Upsilon)$ which is plotted in Fig. 11. It is almost flat in the region $0.2 \lesssim \Upsilon \lesssim 100$, which is the region for linear colliders of the next and, probably, the next-to-next generation.

Another possibility is that, when we go to very short bunch lengths comparable to l_R, the radiation might be suppressed. Radiation by very short bunches has been discussed by many authors [12]. It is concluded, however, that the correction to the constant field formula is small for Gaussian bunches.

Flat Beams

If we go to very flat beams $R = \sigma_x/\sigma_y \to$ large with fixed area $\sigma_x\sigma_y$, we can reduce $\delta_E \propto R/(1+R)^2$ without sacrificing the luminosity. This is the most practical cure for the next generation colliders. Recent designs mostly adopt flat beams with $R{\sim}10$ for CLIC, $R{\sim}100$ for NLC and JLC, or even 3000 for VLEPP. It has been demonstrated that flat beams have many other advantages [13], but the price to pay is that the vertical tolerances in the damping ring, the linac and the final focus system become tighter.

Charge Compensation

If each beam is a charge-neutral mixture of electrons and positrons, then the beamstrahlung can in principle be completely suppressed. Apart from the technical difficulties of mixing beams, it has been shown [15,16] that the system of four beams is unstable when mixing is not perfect. Fig. 12a shows the luminosity enhancement factor H_D as a function of the residual charge $\Delta Q/Q$. When n_y ($= \sqrt{D_y/2\pi}$) exceeds $\sim$1, imperfection of 0.1% is enough to degrade the luminosity. Fig. 12b shows $H_D(n_y)$ for two values of the center-of-mass shift Δ_y/σ_y between e^+ and e^- in the same beam. Even a small shift of 1% has a serious effect. Thus, the charge compensation scheme is very hard to realize.

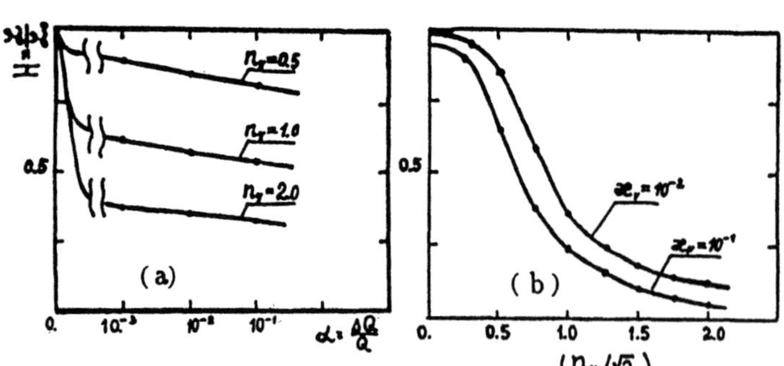

Fig. 12. Luminosity enhancement factor in charge compensation scheme. The effect of inaccurate charge compensation (Fig. 12a) and that of the shift of one of the beams (Fig. 12b). From Solyak [15].

Plasma Return Current

During the collison the charges of e^+ and e^- cancel each other but the currents add. If there is a plasma at the collision point, this current may be neutralized by the plasma return current [14]. However, there is a problem of hadronic backgrounds due to the collision with ions.

γ-γ Colliders

If everything fails, we can give up e^+e^- colliders. A promising proposal has been made for converting an electron collider into a γ-γ collider: by irradiating electrons with an intense laser beam just before the collision point, we can convert the electrons into photons by the Compton back-scattering. High energy, highly focused photon beams can be produced this way. There is, however, a different constraint on the γ-γ colliders, associated with "coherent pair creation" (discussed in the next section) [17]. The present status of study has been summarized by Telnov [18].

3.5 Spin-Flip Radiation

It is easier to have longitudinally polarized electrons in linear colliders than in storage rings. It is also possible to have polarized positrons by using polarized photons generated in helical wigglers [19]. There is no major difficulty in preserving the polarization from the source to the collision point. Here, we mention possible depolarization due to the beam-beam interaction.

In storage rings electron/positron beams tend to polarize (transversely) because of the spin-flip radiation. The ratio of spin-flip radiation to total radiation is very small: $\sim \Upsilon^2 \sim 10^{-11}$ in storage rings. This ratio is much larger in the beam-beam interaction in linear colliders. It is given by

$$\frac{dN_{\gamma,flip}}{dt} = \frac{dN_\gamma}{dt} U_{f0}(\Upsilon) \tag{3.71}$$

where the function $U_{f0} = U_f/U_0$ is plotted in Fig. 11. The depolarization is given by $\Delta P = 2n_\gamma U_{f0}$. As seen in Fig. 11, U_{f0} is almost constant in the region $1 \lesssim \Upsilon \lesssim 100$ and does not exceed 0.02. Therefore, if $n_\gamma \sim 1$ the depolarization is less than a few percent.

There is another mechanism of depolarization, i.e., due to the classical precession in the magnetic field. The depolarization by this effect was estimated [20] to be $\sim 0.006(n_\gamma/U_0)^2$, which is also less than a few percent if $n_\gamma \lesssim 3$.

3.6 Bremsstrahlung

Beamstrahlung is produced by the interaction between $e^\pm$ and the collective beam-beam field. There are of course photons emitted by the collisions of individual particles, namely the bremsstrahlung $e^+e^- \to e^+e^-\gamma$. The total cross section is

$$\sigma_{brems} \approx \frac{1}{2}\frac{16}{3}\alpha r_e^2 [\log 4\gamma^2]^2 \times \left\{ \frac{\log \sigma_y/\lambda_e}{\log 4\gamma^2} \right\} \sim 10^{-24}\ \mathrm{cm}^2 \tag{3.72}$$

where the factor in braces is the correction due to the finite beam size [21](see Sec. 4.2.4 for more details). We count the photons parallel to one of the incident beams only, which gives the factor 1/2. With the typical luminosity per collision 10^{30} cm^{-2}, we have $\sim 10^6$ photons per collision, which is much smaller than the beamstrahlung.

The energy spectrum is

$$\frac{d\sigma_{brems}}{dy} = \frac{1}{2}\frac{16}{3}\frac{\alpha r_e^2}{y}\left[(1 - y + \frac{3}{4}y^2)(\log\frac{\sqrt{2}\sigma_y}{\lambda_e} - \frac{1+\gamma_E}{2}) - \frac{3}{4}(\frac{13}{9} - \frac{13}{9}y + y^2)\right] \tag{3.73}$$

where $y = \omega/E$ and γ_E=0.577 is Euler's constant. This can exceed the beamstrahlung only at the high energy end $y \sim 1$. The number of low energy final electrons with energy below xE ($x \ll 1$) is $\sim 10^4 x$ per collision if the luminosity per collision is 10^{30} cm^{-2}.

4 Electron-Positron Pair Creation

In general, low energy electrons and positrons comoving with the e^+e^- beams during collision are a source of deleterious backgrounds. Being lower in energy, these particles can either be deflected more severely by the beam-beam field, or even carry an inherently large angle and transverse momentum, such that they fail in exiting through the exhaust port and pollute the detector. One such source of low energy particles is a direct result of beamstrahlung or bremsstrahlung, as described in the last section. A more severe source, however, is the second-order quantum electrodynamic (QED) process of e^+e^- pair creation during the collision. In this case, there is always one particle in the pair that carries the same sign of charge as the oncoming beam, and will be severely deflected unbound. At lower energies, the pairs are made essentially by an *incoherent* process, in which a pair is created by the interaction of particles with individual particles in the oncoming beam, e.g., the interaction of beamstrahlung photons with oncoming electrons. This problem was first identified by Zolotarev *et al.* [22].

At energies where the beamstrahlung parameter Υ lies approximately in the range $0.5 \lesssim \Upsilon \lesssim 100$, pair creation from the beamstrahlung photons is dominated by a *coherent* process, first noted by Chen [23]. In this case the photon turns into an e^+e^- pair by interacting with the collective field of the oncoming beam, not with the individual particles. For the next generation of linear colliders, the beamstrahlung parameter happens to lie in the range between these two regimes. In this section, we shall describe these processes.

4.1 Coherent Pair Creation

4.1.1 Pair Creation in External Fields

A photon in vacuum is always accompanied by virtual electron-positron pairs (vacuum polarization). These pairs cannot emerge as real particles because of energy-momentum conservation. In a strong electromagnetic field, however, the energy-momentum can be carried by the field and the pair can be kicked "on-shell." Consider the boosted frame where the e^+e^- pair is created at rest. In this frame there is an electric field which is $E' = (\hbar\omega/2mc^2)B$, where B is the magnetic field in the lab frame. At the threshold, the created particle with unit charge e should aquire enough energy within one Compton wavelength to supply its rest mass. Thus the threshold condition is $eE'\lambda_e \sim mc^2$. Let us define a parameter χ as

$$\chi \equiv \frac{\omega}{m}\frac{B}{B_c} \equiv \frac{\omega}{E}\Upsilon \equiv y\Upsilon. \tag{4.74}$$

The last two definitions are convenient when the photon is induced by a primary particle with energy E in the lab frame. Thus $\chi \sim 1$ corresponds to the threshold condition for coherent pair creation. It plays a role similar to that of Υ in beamstrahlung.

The problem of pair creation in magnetic fields is not new. Klepikov [24] first calculated this problem in a uniform field. Several others [25] have re-examined the problem with different formalisms. In the lowest order of the perturbation theory the matrix element for the pair creation process is essentially the same as that for beamstrahlung, cf. Eq. (3.58), except that now the initial electron momentum k_μ has to be replaced by $-k_\mu$ in the cross channel.

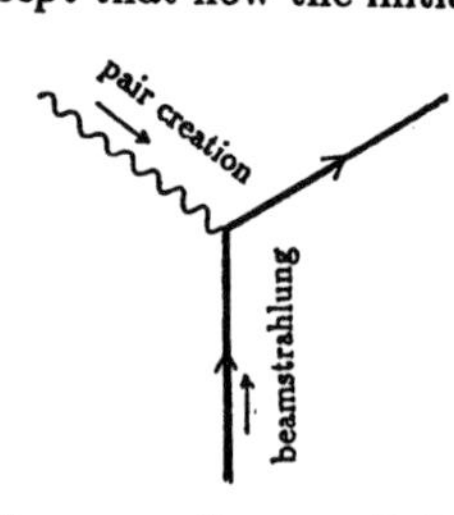

The number spectrum as a function of energy of pair creation per unit time can then be obtained directly from Eq. (3.58):

$$\frac{dW_{CP}}{dE_+} = \frac{\alpha}{\sqrt{3}\pi}\frac{m^2}{\omega^2}F_{CP} \tag{4.75}$$

with

$$F_{CP} = \int_\eta^\infty K_{1/3}(\eta')d\eta' + \left(\frac{E_-}{E_+}+\frac{E_+}{E_-}\right)K_{2/3}(\eta), \qquad \eta = \frac{2}{3\chi}\frac{\omega^2}{E_+E_-}, \qquad E_- = \omega - E_+, \tag{4.76}$$

where $E_\pm$ is the energy of the created $e^\pm$.

The total probability of pair creation per unit time is obtained by integrating this expression over ω. We find

$$W_{CP} = \int_0^\omega \frac{dW_{CP}}{dE_+} dE_+ = \begin{cases} 0.23(\alpha m^2/\omega)\chi e^{-8/3\chi} & (\chi \ll 1) \\ 0.38(\alpha m^2/\omega)\chi^{2/3} & (\chi \gg 1). \end{cases} \tag{4.77}$$

We see that $\chi \sim 1$ indeed corresponds to a thereshold below which the probability is exponentially small. For the entire range of χ, W_{CP} can be well approximated by the following expression [26]:

$$W_{CP} \approx \frac{4}{25} \frac{\alpha m^2}{\omega} K_{1/3}^2\left(\frac{4}{3\chi}\right), \tag{4.78}$$

or simply by

$$W_{CP} \approx \frac{\alpha m^2}{\omega} \frac{0.23\chi}{(1+0.22\chi)^{1/3}} e^{-8/3\chi}, \tag{4.79}$$

where the coefficients are chosen so as to satisfy Eq. (4.77).

The energy spectrum (4.76) (normalized to unity) is plotted in Fig. 13 for various values of χ. We find that, when χ is small, the spectrum has a peak at E_+/ω=0.5, which means the final electron and positron equally share the energy of the initial photon. On the other hand when $\chi \gg 1$, the maximum is around $E_\pm/\omega \sim 1.6/\chi$.

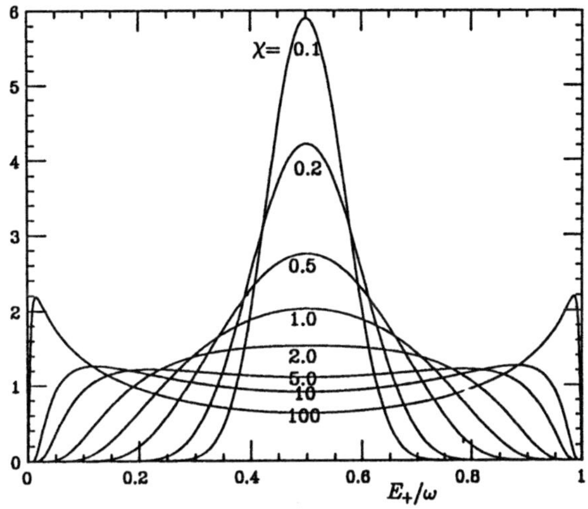

Fig. 13. Energy distribution of e^+ from coherent pair creation for various values of χ.

4.1.2 Coherent Beamstrahlung Pair Creation

Beamstrahlung is the most important source of photons which act as initial photon for the coherent pair creation process in linear colliders. For this purpose, it is useful to express the energy spectrum (4.76) in terms of Υ of the primary particle instead of χ of the intermediate photon. Let $x \equiv E_+/E$ be the fractional energy of the pair particle and $y \equiv \omega/E$ be that of the photon, then $(E_+/\omega)\chi = x\Upsilon$. Then Eq. (4.76) becomes $dW_{CP}/dx = (\alpha m/\sqrt{3}\pi\gamma y^2)F_{CP}$ with

$$F_{CP}(x,y,\Upsilon) = \int_\eta^\infty K_{1/3}(\eta')d\eta' + \left(\frac{y-x}{x} + \frac{x}{y-x}\right) K_{2/3}(\eta), \qquad \eta = \frac{2}{3\Upsilon}\frac{y}{x(y-x)}. \tag{4.80}$$

Then the energy spectrum of the particle created through the cascade process of beamstrahlung and coherent pair creation is obtained by the convolution of Eq. (3.59) and Eq. (4.80) as

$$\begin{aligned} \frac{dn_b}{dx} &= \iint_{t_2>t_1} dt_2\, dt_1\, d\omega \frac{dW_\gamma}{d\omega}\frac{dW_{CP}}{dx} \\ &= \frac{1}{3\pi^2}\left(\frac{\alpha}{\gamma\lambda_e}\right)^2 \iint dt_2\, dt_1 \int_x^1 \frac{dy}{y^2} F_{BS}(y,\Upsilon(t_1))\, F_{CP}(x,y,\Upsilon(t_2)). \end{aligned} \tag{4.81}$$

Here, t_1 and t_2 are the time of photon emission and pair creation, respectively. In the case of longitudinally uniform beams with small disruption, Υ is almost constant along the photon path and we can replace $\int\int dt_1\, dt_2$ by $3\sigma_z{}^2/2$. Then the total number of pairs per primary electron is given by integrating this expression over x:

$$n_b = \frac{3\sigma_z{}^2}{2} \int dy \frac{dW_\gamma}{dy} W_{CP} = \left[\frac{\alpha\sigma_z}{\gamma\lambda_e}\Upsilon\right]^2 \Xi(\Upsilon). \tag{4.82}$$

(This definition of Ξ is different from that in Ref [30] by a numerical factor.) The function Ξ is plotted in Fig. 14. Its asymptotic form is

$$\Xi(\Upsilon) = \begin{cases} (7/128)\exp(-16/3\Upsilon) & (\Upsilon \lesssim 1), \\ 0.295\Upsilon^{-2/3}(\log\Upsilon - 2.488) & (\Upsilon \gg 1). \end{cases} \tag{4.83}$$

As pointed out in Sec. 3, the quantity $\alpha\sigma_z\Upsilon/\gamma\lambda_e$ is order unity for next-generation colliders. Thus $n_b \sim O(10^{-2})$ for $\Upsilon \gtrsim 1$. Since the typical number of particles in a bunch is $\sim O(10^{10})$, we expect to have $\sim O(10^8)$ e^+e^- pairs per collision. On the other hand, Ξ is exponentially small for $\Upsilon \ll 1$. If we are to suppress coherent creation below ~ 1 per collision we need to choose $\Upsilon \lesssim 0.3$, which gives $\Xi \lesssim 10^{-9}$. (In this context, it is better to use Υ_{max} instead of Υ_{av}.)

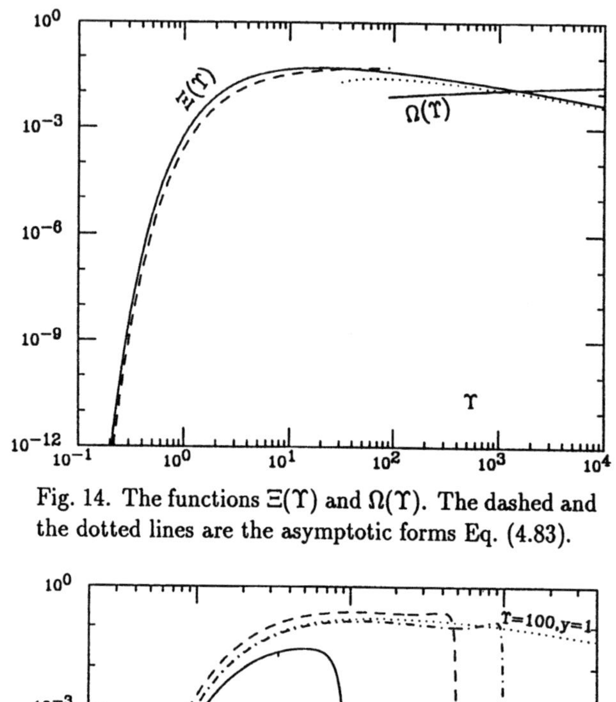

Fig. 14. The functions $\Xi(\Upsilon)$ and $\Omega(\Upsilon)$. The dashed and the dotted lines are the asymptotic forms Eq. (4.83).

Next, let us discuss the energy spectrum. As seen in Fig. 13 low energy pairs are strongly suppressed in coherent pair creation. To see this in more detail we plotted $dW_{CP}/d(\Upsilon x)$ as a function of $x\Upsilon$ in Fig. 15 for various combinations of Υ and the intermediate photon energy y. We find that, for a given value of Υ, the minimum energy x_{min} is independent of the intermediate photon energy. In addition, for different values of Υ, x_{min} scales as $1/\Upsilon$. This can be explained by the following qualitative arguments.

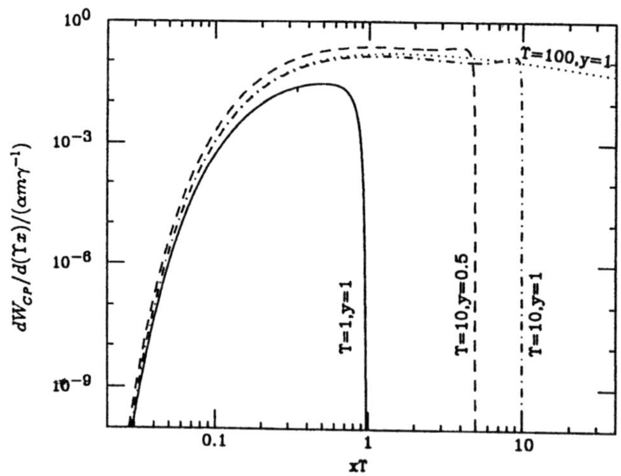

Fig. 15. $dW_{CP}/d(\Upsilon x)/(\alpha m\gamma^{-1})$ vs. $x\Upsilon$.

In the Lorentz frame where the pair is created at rest, the invariant mass of the system is $W = 2eE'\lambda_e$. The Lorentz factor for the boost is obviously the photon energy ω divided by the invariant mass. Thus we have $W^2 = 2eB\omega\lambda_e$. On the other hand, from the final state we have $W^2 = \omega^2 m^2/E_+E_-$. In the case where one particle is at very low energy, e.g., $E_+ \ll E_- \sim \omega$, we have $W^2 \sim \omega m^2/E_+$. Thus $E_{min} \sim \gamma m/2\Upsilon$, or $x_{min} \sim 1/2\Upsilon$, and is independent of the photon energy. The actual value of x_{min} is somewhat different from this naive picture, depending on how small a fraction one allows. One can take $x_{min} \sim 1/20\Upsilon$ if the conversion rate from a photon to a pair is to be less than $\sim 10^{-8}$.

The entire spectrum from the coherent beamstrahlung pair creation is plotted in Fig. 16 (normalized to unity). As pointed out by Baier et al. [27], the maximum is near x=1/3 when Υ is small. The peak moves towards small x for large Υ. If Υ is not very large ($\Upsilon \lesssim 10$), the following formula gives a good approximation for any x:

$$\frac{dn_b}{dx} = 0.2 \left[\frac{\alpha\sigma_z}{\gamma\lambda_e}\Upsilon\right]^2 \sqrt{\frac{1-x}{\Upsilon x}} \times \exp\left[-\frac{2}{3\Upsilon}\left(\frac{1-x}{x} + \frac{4}{1-x}\right)\right] \quad (4.84)$$

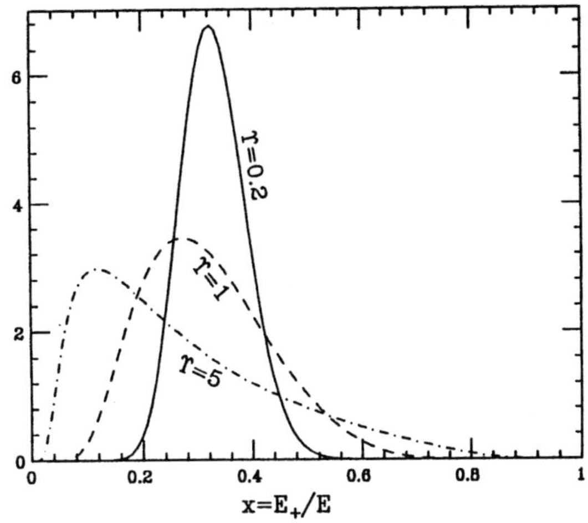

Fig. 16. Energy spectrum of the cascade of beamstrahlung and coherent pair creation.

The beamstrahlung photon is definitely polarized in the orbital plane of the electron. Baier et al. [27] pointed out that the polarization effect can be taken into account in Eq. (4.81) by replacing $F_{BS}F_{CP}$ with $F_{BS}F_{CP} - K_{2/3}(\xi)K_{2/3}(\eta)$ where ξ and η are defined in Eqs. (3.60) and (4.80). Then, the factor 7/128 in Eq. (4.83) becomes 3/64. There is no considerable effect on the spectrum.

4.1.3 The Trident Cascade

In addition to coherent beamstrahlung pair creation, e^+e^- pairs can also be coherently created through virtual photons carried by the primary charged particles, i.e., $e \to ee^+e^-$, in an electromagnetic field. This process is sometimes called the *trident cascade*. It has been studied in detail [28]. For comparison, we express it as

$$n_v = \left(\frac{\alpha\sigma_z}{\gamma\lambda_e}\Upsilon\right)\Omega(\Upsilon), \quad (4.85)$$

where, according to Ritus [29],

$$\Omega(\Upsilon) \sim 0.23\,\alpha \log \Upsilon, \qquad \Upsilon \gg 1. \quad (4.86)$$

The auxiliary function $\Omega(\Upsilon)$ is also shown in Fig. 14. Besides the different scalings between Ξ and Ω, beamstrahlung pair creation increases quadratically with the quantity $(\alpha\sigma_z/\gamma\lambda_e)\Upsilon$ whereas the trident cascade scales linearly. This is simply because the former necessarily involves a real intermediate process and thus a double integration in time. Note that the crossover between $\Xi(\Upsilon)$ and $\Omega(\Upsilon)$ occurs at $\Upsilon \sim 1000$. Since the quantity $(\alpha\sigma_z/\gamma\lambda_e)\Upsilon$ is usually of order unity, we see that the contribution from the trident cascade is relatively unimportant unless $\Upsilon \gg 1$.

4.1.4 Deflection Angle and Maximum Υ

If e^+e^- pairs are created in a field-free space, their outgoing angles will be simply the final angles from the creation process. In the case of beam-beam interaction, however, these low energy secondary particles will be strongly deflected by the macroscopic collective field. The deflection angle of low energy particle is discussed in Sec. 2.5.

If an e^+e^- collider is designed such that the beams collide head-on, then the above consideration imposes a severe design constraint. The typical distance between the final focusing magnet and the IP is $\sim 10^2$ cm, while the aperture of the magnet is $\sim 10^{-1}$ cm. This means that any particle with outcoming angle larger than $\sim O(1)$ millirad will necessarily hit the magnet and generate backgrounds.

One obvious means of alleviating the problem is sufficient suppression of the coherent process. This can in principle be achieved by imposing a constraint on the value of Υ. As we have seen, if $\Upsilon \lesssim 0.3$, the pair creation rates are exponentially suppressed to $\sim O(10^{-9})$. To achieve this goal, the charge density of the beams must be reduced by either lengthening the bunch or reducing the number of particles per bunch, or both, with a corresponding increase in the number of bunches in order not to sacrifice luminosity. In practice, however, these leverages can be handled only within certain limits because of various other accelerator constraints.

It happens that total suppression of the coherent pairs is not necessary [30]. Remember that for any given Υ there is a minimum x_{min}. Assume that the beams are colliding at an angle ϕ_c. Then by inserting x_{min} into Eq. (2.43), one finds the maximum allowable value of Υ to suppress almost perfectly the large angle pairs $\theta \gtrsim \phi_c$:

$$\Upsilon \lesssim \max\left(0.3\,,\ \frac{\sqrt{3}}{20}\frac{(\phi_c\sigma_z/\sigma_x)^2}{\log[5(\phi_c\sigma_z/\sigma_x)^2]}\right). \tag{4.87}$$

(The factor 5 in the log is somewhat arbitrary.) When the crossing angle is much larger than the beam diagonal angle σ_x/σ_z, this limit is much less stringent than the constraint for total suppression of the coherent pairs.

4.2 Incoherent Pair Creation

The seriousness of the pair creation problem lies in the transverse momenta that the pair particles carry when leaving the interaction point (IP) with large angles. In the next generation of linear colliders, as it occurs, the coherent pairs can be exponentially suppressed by properly choosing the Υ ($\lesssim 0.3$). When this is achieved, the incoherent pair creation processes, where the e^+e^- pairs are created through individual scatterings, become dominant. Furthermore, these incoherent pairs tend to have more uniform energy spectra than the pairs from the coherent process, which results in a greater abundance of deleterious low energy particles.

Note that both real (beamstrahlung) and virtual photons can turn into e^+e^- pairs through this channel. These are the so-called Breit-Wheeler ($\gamma\gamma \to e^+e^-$), Bethe-Heitler ($e^\pm\gamma \to e^\pm e^+e^-$), and Landau-Lifshitz ($e^+e^- \to e^+e^-e^+e^-$) processes. In this section, we will describe the spectrum, the cross section, and the large inherent angle events of the incoherent pair creation processes using the *equivalent photon approximation*. Numerical examples are given wherever suitable.

4.2.1 The Equivalent Photon Approximation

The Breit-Wheeeler, Bethe-Heitler, and Landau-Lifshitz processes are well-known phenomena in QED, and are well described in any standard QED textbook [31]. In our particular setting, the BW and BH processes occur as a result of beamstrahlung, which has a wide range of spectra. Thus for our purpose it is most convenient to describe these processes in the laboratory (lab) frame (which is also the center-of-mass frame for the colliding e^+e^- beams.) Moreover, we note that the basic kernel of all these processes is the same. For the BW process both photons are real beamstrahlung photons; for the BH process one is real and one is virtual; and for the LL process both photons are virtual.

We shall thus invoke the so-called equivalent photon picture in our discussion. This is an approximation in which the two photons involved in each process are treated as being real. Then the relativistic kinematics relates the fractional energy x of the outcoming positron (or electron) to its angle θ (relative to the initial particle trajectory) and the fractional energies of the two photons, y_1, y_2, as

$$x = \frac{2y_1y_2}{y_1(1-\beta c) + y_2(1+\beta c)}, \tag{4.88}$$

where $c \equiv \cos\theta$ and β is the speed of the particle. In the following, we shall assume $\beta \sim 1$, as the pair particles of concern are still relativistic.

The differential cross section for the kernel $\gamma\gamma \to e^+e^-$ is

$$d\sigma_{\gamma\gamma} = \frac{8\pi r_e^2 m^2 dt}{s^2}\left[\frac{1}{4}\left(\frac{t-m^2}{u-m^2}+\frac{u-m^2}{t-m^2}\right) - \left(\frac{m^2}{t-m^2}+\frac{m^2}{u-m^2}\right) - \left(\frac{m^2}{t-m^2}+\frac{m^2}{u-m^2}\right)^2\right], \tag{4.89}$$

where s,t,u are the Mandelstam variables. In terms of our variables, we have

$$s = 4\gamma^2 m^2 y_1 y_2, \qquad t - m^2 = -2\gamma^2 m^2 y_1 x(1-c), \qquad u - m^2 = -2\gamma^2 m^2 y_2 x(1+c). \tag{4.90}$$

With the help of Eq. (4.88), we find

$$\begin{aligned} d\sigma_{\gamma\gamma} = & \frac{\pi}{4}\frac{r_e^2}{\gamma^2 y_1 y_2}\frac{dc}{y_2}\left\{2\frac{y_1^2(1-c)^2 + y_2^2(1+c)^2}{(1-c^2)[y_1(1-c)+y_2(1+c)]}\right. \\ & \left. +\frac{2}{\gamma^2}\frac{y_1(1-c)+y_2(1+c)}{y_1 y_2(1-c^2)} - \frac{1}{2\gamma^4}\frac{[y_1(1-c)+y_2(1+c)]^3}{y_1 y_2[y_1 y_2(1-c^2)]^2}\right\}. \end{aligned} \tag{4.91}$$

The first term obviously dominates. Furthermore, we expect that the major contribution comes from $c \lesssim c_0 \lesssim 1$. Thus, $y_1(1-c) + y_2(1+c) \sim 2y_2$. With $\gamma^2 y_1 y_2 \gg 1$, we arrive at

$$\sigma_{\gamma\gamma}(y_1, y_2) = \frac{d\sigma_{\gamma\gamma}}{dc} \sim \frac{\pi r_e^2}{\gamma^2 y_1 y_2}\frac{1}{1-c^2}\left\{\frac{y_1^2(1-c)^2 + y_2^2(1+c)^2}{[y_1(1-c)+y_2(1+c)]^2}\right\} \sim \frac{\pi r_e^2}{\gamma^2 y_1 y_2}\frac{1}{1-c^2}. \tag{4.92}$$

The last approximation is made due to the fact that the factor in braces is a slowly varying function ranging from 1/2 to 1. Thus our estimates are upper bounds that are too high by less than a factor of 2.

Next we need to find the spectrum for the virtual photons that comove with the primary charged particle. From the virtual photon propagator, we can establish the spectrum of the equivalent photons:

$$dn_v = \frac{\alpha}{\pi}\frac{dy}{y}\frac{y_\perp^2 dy_\perp^2}{(y_\perp^2 + y^2/\gamma^2)^2}, \tag{4.93}$$

where $y_\perp$ is the normalized transverse momentum q of the photon: $y_\perp \equiv q/\gamma m$. Integrating over $y_\perp^2$, we get

$$n_v(y) = \frac{\alpha}{\pi}\frac{1}{y}\log\left(\frac{y_{\perp max}^2 + y^2/\gamma^2}{y_{\perp min}^2 + y^2/\gamma^2}\right), \tag{4.94}$$

where $y_{\perp max}$ and $y_{\perp min}$ are the maximum and minimum momentum transfer, respectively. In the equivalent photon approximation, it is assumed that the transverse momentum of the virtual photon satisfies the condition

$$y/\gamma^2 \ll y_\perp \ll 1/\gamma. \tag{4.95}$$

Since n_v depends on $y_{\perp max}$ and $y_{\perp min}$ only logarithmically, it is customary to take the bounds in Eq. (4.95) as their values. We then have

$$n_v(y) = \frac{2\alpha}{y}\frac{1}{y}\log\left(\frac{1}{y}\right). \tag{4.96}$$

This expression is indeterminate up to a numerical coefficient in the argument of the logarithm because of the nature of the approximation [22]. In the equivalent photon approximation, one combines this virtual photon spectrum with the above differential cross section to obtain the desired expression for the process that involves virtual photon exchange.

4.2.2 The Total Cross Sections

We now want to find the total incoherent e^+e^- pair creation events:

$$\sigma = g\int dc \int dy_2 \int dy_1\, n_a(y_1)\, n_b(y_2)\, \sigma_{\gamma\gamma}(y_1, y_2), \tag{4.97}$$

where g=1/4 for the BW process and 1 for both the BH and LL processes. For the BW process, the effective collision time for the beamstrahlung photons, which are emitted during the collision itself, is only 1/4 of that for the primary beam particles. For the BH process, the factor 1/2 that arises from the effective collision time is compensated by matching between the real and the virtual photons from the two beams. The photon spectra n_a and n_b depend on whether the photons are real or virtual. For virtual photons, the spectrum is the one derived in Sec. 4.2.1. For real beamstrahlung photons, the spectrum is given in Eq. (3.58). For our purpose here, we integrate it over the collision time and approximate it in the form

$$n_b(y) = \frac{\Gamma(2/3)}{\pi}\left(\frac{\alpha\sigma_z}{\gamma\lambda_e}\right)(3\Upsilon)^{2/3}y^{-2/3}, \tag{4.98}$$

where $\Gamma(2/3)$=1.3541.

First we carry out the angular integration. The dominant contribution to the cross section is associated with the typical transverse momentum m versus half the center-of-mass energy $\sqrt{\omega_1\omega_2} = m\gamma\sqrt{y_1y_2}$, or $\theta_0 \sim 1/\gamma\sqrt{y_1y_2}$. Thus for the BW process with fixed initial photon energies in the lab frame, we have

$$\sigma_{BW}(y_1, y_2) = \int_{-c_0}^{c_0} \sigma_{\gamma\gamma}(y_1, y_2)\, dc = \frac{\pi r_e^2}{2\gamma^2 y_1 y_2}\log(4\gamma^2 y_1 y_2). \tag{4.99}$$

For the BH and LL processes we need to integrate further over the virtual photon spectrum. The threshold condition for pair creation is that $m = \sqrt{\omega_1\omega_2}$. Thus the lower bound for the virtual photon energy with a fixed real photon energy y_1 is $y_2 = 1/(\gamma^2 y_1)$. We then have, for the BH process,

$$\sigma_{BH}(y_1) = \int_{1/\gamma^2 y_1}^{1} n_v(y_2)\sigma_{BW}(y_1, y_2)dy_2 = \alpha r_e^2\{(\log 4+1)\log 4\gamma^2 y_1 - (\log 4+2)\log 4 - 1\}. \tag{4.100}$$

For the LL process, we need to integrate once more over the other virtual photon. Again from the threshold condition, the lower bound for the second photon is $1/\gamma^2$. Thus

$$\sigma_{LL} = \int_{1/\gamma^2}^{1} dy_1 \int_{1/\gamma^2 y_1}^{1} dy_2\, n_v(y_1)n_v(y_2)\sigma_{BW}(y_1, y_2) = \frac{\alpha r_e^2}{\pi}(\log 4+1)(\frac{1}{3}\log^3\gamma^2 - 2\log^2\gamma^2). \tag{4.101}$$

When compared with the more accurate calculations [31,32]

$$\sigma_{BH}(y_1) = (28/9)\alpha r_e^2 \log(4\gamma^2 y_1) \tag{4.102}$$

$$\sigma_{LL} = (\alpha^2 r_e^2/\pi)[(28/27)\log^3(4\gamma^2) - 6.59\log^2(4\gamma^2) - 11.8\log^2(4\gamma^2) + 104], \tag{4.103}$$

we find that σ_{BW} has the precise logarithmic term and, while both σ_{BH} and σ_{LL} indeed reproduce the correct logarithmic behavior, our expressions tend to have somewhat smaller coefficients. This, however, is consistent with the accuracy of our approximations.

The above intermediate expressions serve to compare with the well-known QED results. In our specific context, the real photons are induced from beamstrahlung. Thus for the BW and BH processes we further combine with the beamstrahlung spectrum, with the proper g factor in Eq. (4.97) included, and get

$$\sigma_{BW} = \frac{3}{16\pi}\Gamma^2(2/3)\frac{r_e^2}{\gamma^{2/3}}\left(\frac{\alpha\sigma_z}{\gamma\lambda_e}\right)^2(3\Upsilon)^{4/3}\left[\log 4 + \frac{3}{2}\right]\log\gamma^2, \tag{4.104}$$

$$\sigma_{BH} = \frac{3}{\pi}\Gamma(2/3)\alpha r_e^2\left(\frac{\alpha\sigma_z}{\gamma\lambda_e}\right)(3\Upsilon)^{2/3}[(\log 4+1)\log 4\gamma^2 - (\log 4+5)\log 4 - 4]. \tag{4.105}$$

Consider a 1/2-TeV Intermediate Linear Collider (ILC) designed by Palmer [34] (Palmer's Machine I), where $\gamma = 5\times10^5$, σ_z=0.11 mm, Υ=0.17 and luminosity L=1.95$\times10^{31}$ /cm^2 per bunch train (10 bunches per train), and 130 Hz collision repetition rate. The total cross sections in Eqs. (4.101) to (4.105) are

$$\sigma_{BW} \approx 8\times10^{-28}\ \text{cm}^2 \quad , \quad \sigma_{BH} \approx 1\times10^{-25}\ \text{cm}^2 \quad , \quad \sigma_{LL} \approx 2\times10^{-28}\ \text{cm}^2. \tag{4.106}$$

The total numbers of pairs are then $N_{BW} \approx 2\times10^4$, $N_{BH} \approx 2\times10^6$, and $N_{LL} \approx 3\times10^5$, per bunch train.

For a 1-TeV linear collider (TLC) (Palmer's Machine J), where γ=1$\times10^6$, σ_z=0.12 mm, Υ=0.60, one finds that the cross sections are essentially the same. With a luminosity L=8.04$\times10^{31}$ /cm^2 per bunch train (17 bunches per train), the number of incoherent pairs is about 4 times that of ILC. These are about two to three orders of magnitude smaller than the corresponding coherent yields.

4.2.3 Pair Creation at Large Inherent Angles

Since the transverse momentum is the central issue for particles with large angles, we note that it has an additional source besides the deflections due to the collective fields of the oncoming beam. Namely, when the pair particles are created at low energies, the intrinsic angles of these pairs when produced may already be large. This issue was first studied by Zolotarev et al. [22], and more recently by Chen et al. [33]

To obtain the partial cross sections for all events with transverse momentum (divided by γm) $p_\perp \ge p_{\perp 0}$ and outcoming angle $\theta_0 \le \theta \le \pi - \theta_0$ (or $-c_0 \le c \le c_0$), it is more convenient first to carry out the integrations over y_1 and y_2 in Eq. (4.97). Since the transverse momentum $p_\perp = x\sqrt{1-c^2}$, with any choice of $p_\perp$ and c, the initial photon energies are limited by

$$y_\pm = \frac{x}{2}(1\pm c) = \frac{p_\perp}{2}\sqrt{\frac{1\pm c}{1\mp c}}. \tag{4.107}$$

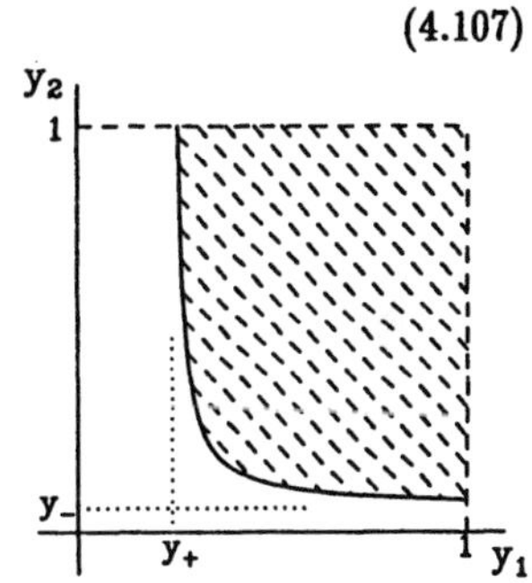

Furthermore, for any given value of $y_2 \ge y_1$, the lower bound for y_1 is

$$y_b = \frac{y_2 y_+}{y_2 - y_-}. \tag{4.108}$$

Equation (4.97) can then be specified as

$$\sigma_{ab\to e^+}(p_{\perp 0}, c_0) = g\int_{-c_0}^{c_0} dc \int_{y_-}^{\infty} dy_2 \int_{y_b}^{\infty} dy_1\, n_a(y_1)\, n_b(y_2)\, \sigma_{\gamma\gamma}(y_1, y_2). \tag{4.109}$$

The upper limits of the spectral integrals are set at infinity since the contributions are essentially dominated by the lower bounds.

Inserting the photon spectra and Eq. (4.92) into Eq. (4.109) and invoking the boundary conditions in Eqs. (4.107) and (4.108), we can calculate the partial cross section for positrons with momentum larger than $p_{\perp 0}$ in all 4π solid angles, excluding the forward and backward cones of half-angle θ_0. The results are [33]

$$\sigma_{BW} = \frac{9}{8\pi}\frac{\Gamma^4(2/3)}{\Gamma(1/3)}\frac{r_e^2}{\gamma^2}\left[\frac{\alpha\sigma_z}{\gamma\lambda_e}\right]^2\left[\frac{6\Upsilon}{p_{\perp 0}}\right]^{4/3}\log\frac{1}{\tau_0}\ , \tag{4.110}$$

$$\sigma_{BH} = \frac{54}{5\pi}\Gamma(\frac{2}{3})\frac{\alpha r_e^2}{\gamma^2}\frac{\alpha\sigma_z}{\gamma\lambda_e}\left[\frac{36\Upsilon^2}{{p_{\perp 0}}^5}\right]^{1/3}[\tau_0^{-1/3} - \tau_0^{1/3}][-\log\frac{p_{\perp 0}\tau_0}{2} - \psi(\frac{8}{3}) + \psi(1)], \tag{4.111}$$

$$\sigma_{LL} = \frac{16}{\pi}\frac{\alpha^2 r_e^2}{\gamma^2}\frac{1}{{p_{\perp 0}}^2}\log\frac{1}{\tau_0}\left[\log\frac{p_{\perp 0}}{2\tau_0}\log\frac{p_{\perp 0}\tau_0}{2} + 3\log(p_{\perp 0}/2) + \frac{73}{12} - \frac{\pi^2}{6}\right]. \tag{4.112}$$

Here $\Gamma(1/3)$=2.6789, $\psi(8/3)$=0.7818, $\psi(1)$=0.5772, and $\tau_0 = \tan(\theta_0/2)$. We note that both BW and LL processes are forward-backward symmetric, whereas BH is asymmetric in c. This is the result of the matching between the two different photon spectra in the BH case.

The above expressions account for only one of the two particles (say positron) in the e^+e^- pairs. In the case where the inherent transverse momentum and angle are already large, further deflections on either species in the pair by the beam should not effectively alter the ultimate outcoming transverse momenta and angles. Therefore both electrons and positrons, regardless of the directions of flight, should be counted. Thus the contribution should be twice what would be given from the partial cross sections in Eqs. (4.110) to (4.112).

Let us again estimate the yields from ILC and TLC. First we plot the partial cross sections in Eqs. (4.110) to (4.112) and their sum as a function of $p_{\perp 0}$ in Fig. 17, with the cut-off angle fixed at θ_0=0.1. The dominant scalings of $p_{\perp 0}{}^{-4/3}$, $p_{\perp 0}{}^{-5/3}$, and $p_{\perp 0}{}^{-2}$, for the BW, BH, and LL processes, respectively, are clearly seen. On the other hand, the dependence of the partial cross sections on the cut-off angles is much milder, as expected. Fig. 18 shows such a plot, again with the ILC parameters and $p_{\perp 0}$=10 MeV/c. The choice of the cut-offs depends on practical considerations in the design of the detector masking [35]. From Ref [30] it seems reasonable to assume θ_0 =100 mrad=0.1 and the transverse momentum cut-off at 10 MeV/c, or $p_{\perp 0} = 4 \times 10^{-5}$. The total number of large inherent angle e^+ and e^- is obtained simply by doubling the partial cross sections in Eqs. (4.110) to (4.112) and multiplying by the luminosity. The corresponding events per bunch train are $N_{BW} \approx 220$, $N_{BH} \approx 1230$, and $N_{LL} \approx 500$. Thus the total yield per bunch is $\approx 2 \times 10^3$.

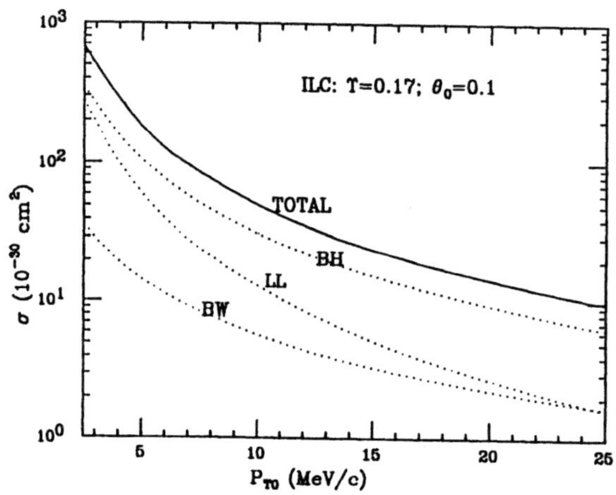

Fig. 17. The partial cross section as a function of the cut-off transverse momentum, at a fixed cut-off angle θ_0=0.1, for the BW, BH and LL processes, shown in dotted curves. The sum of these three processes is shown in solid curve. The parameters are based on the ILC in Ref [34].

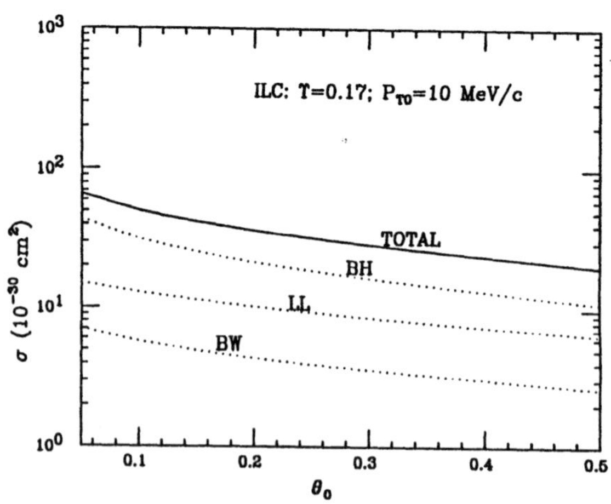

Fig. 18. The partial cross section as a function of the cut-off angle. The transverse momentum cut-off is fixed at $p_{\perp 0}$=10 MeV/c. The same ILC parameters are assumed.

For TLC we find for every bunch train $N_{BW} \approx 900$, $N_{BH} \approx 5300$, and $N_{LL} \approx 2400$. The total yield is $\approx 9 \times 10^3$. These numbers are not small.

To suppress the yield further, it is most effective to increase the longitudinal solenoidal magnetic field in the detector so as to raise the cut-off transverse momentum. The dependence of the total events on the transverse momentum cut-off for the ILC and TLC is shown in Fig. 19, where the angular cut-off is fixed at θ_0=0.1.

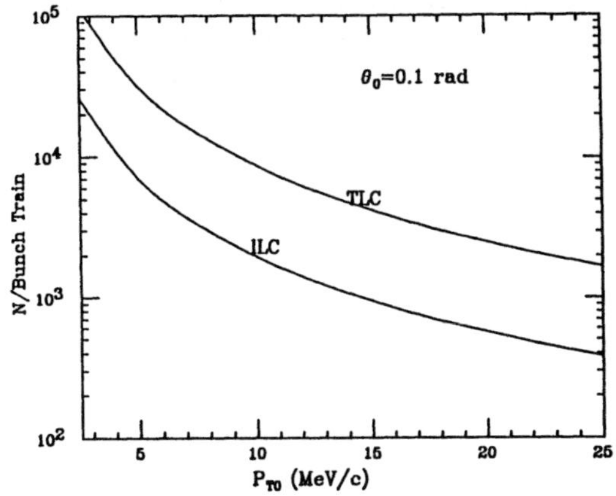

Fig. 19. The total yield of pair particles in ILC and TLC per bunch train as a function of the transverse momentum cut-off.

4.2.4 Geometric Reduction

In Secs. 4.2.2 and 4.2.3 we saw that for a given equivalent photon energy y_1, the dominant contribution to the cross section comes from the region of small transverse momentum y_{bot}. Quantum mechanically, this corresponds to the region of large *impact parameters*: $b \sim 1/q_\perp = 1/(\gamma m y_\perp)$, up to a typical value of $b_m \sim 1/my = \lambda_e/y$. If b_m turns out to be larger than the beam transverse size, these equivalent photons will extend physically to the outside of the oncoming beam. As a result, the part of the equivalent photons having impact parameters larger than the beam size cannot participate in the interaction. Therefore the effective cross section will be smaller than the standard calculations. This *geometric reduction* effect was first observed at Novosibirsk [36,37], and subsequently developed theoretically by several authors [38,39,21].

For the next-generation linear colliders, such as ILC and TLC that we discussed above, the e^+e^- colliding beams are typically very *flat*, i.e., $\sigma_x \gg \sigma_y$. Thus the geometric reduction is dominated by the minor dimension. Let the typical beam height be $2\sigma_y$, then the corresponding "cut-off" impact parameter is $b_c \sim 2\sigma_y$. Thus the region of transverse momentum transfer, $1/b_c \gtrsim q_\perp \gtrsim 1/b_m$, is suppressed. Let us denote the effective cross section by $\bar{\sigma} = \sigma - \sigma'$, then the cut-off cross section σ' is associated with the equivalent photon spectrum in Eq. (4.94), where $y_{\perp max}$ and $y_{\perp min}$ are related to b_m and b_c, respectively, i.e.,

$$n'_v(y) \simeq \frac{2\alpha}{\pi}\frac{dy}{y}\log\left(\frac{\lambda_e/2\sigma_y}{y}\right). \tag{4.113}$$

The cut-off cross section σ' for the BH and LL processes can be derived by inserting n'_v into Eqs. (4.100) and (4.101). By construction, the above spectrum is applicable for $y \lesssim \lambda_e/2\sigma_y$. Thus the upper bounds of the y-integrations have to be replaced by $\lambda_e/2\sigma_y$. But since the dominant contribution comes from the lower bounds, this change does not affect the leading logarithmic behavior. We find the reduced effective cross section to be

$$\bar{\sigma}_{BH} \sim \alpha r_e^2(\log 4 + 1)\log(2\sigma_y/\lambda_e), \tag{4.114}$$

$$\bar{\sigma}_{LL} \sim \frac{1}{3\pi}\alpha^2 r_e^2(\log 4 + 1)[\log^3\gamma^2 - \log^3(\gamma^2\lambda_e/2\sigma_y)]. \tag{4.115}$$

For ILC, the beam height is as miniscule as $2\sigma_y$=8 nm. The geometric reduction turns out to be $\bar{\sigma}_{BH}/\sigma_{BH} \sim 0.36$ and $\bar{\sigma}_{LL}/\sigma_{LL} \sim 0.79$. For TLC, $2\sigma_y$=6.2 nm. The corresponding reductions are 0.33 and 0.73 for the BH and LL processes, respectively. One could in principle also repeat the calculations of large inherent angles for the geometric reductions. As a rough estimate, the corresponding reductions should not be too different from the above values. The geometric reduction is therefore a welcome effect with regard to the e^+e^- backgrounds.

References

[1] R. Holebeek, Nucl. Instr. Meth. **184** (1981) 333.

[2] N. Solyak, *Flat Beam Disruption*, in ICFA Workshop on Physics of Linear Colliders, Capri, Italy, June 1988.

[3] W. M. Fawley and E. P. Lee, *Particle in Cell Simulations of Disruption*, in New Developments in Particle Acceleration Techniques, Orsay, June 1987, CERN 87-11, ECFA 87/110.

[4] P. Chen and K. Yokoya, *Disruption Effects from the Interaction of Round e^+e^- Beams*, Phys. Rev. **D38** (1988) 987; K. Yokoya, Proc. 1988 Linear Acc. Conf. Oct. 3-7, 1988, Williamsburg, VA (CEBAF-Report 89-001, June 1989), page 494.

[5] L. Wood, private communication. Some results appear in O. Napoly and B. Zotter, Proc. 2nd European Accelerator Conf., Nice, June 1990, page 1408.

[6] K. Yokoya, *A Computer Simulation Code for the Beam-Beam Interaction in Linear Colliders*, KEK-report 85-9, Oct. 1985.

[7] Y. H. Chin, *Stability of a Colliding Beam in a Linear Collider*, DESY 87-011, Jan. 1987.

[8] P. Chen and K. Yokoya, *Multibunch Crossing Instability*, SLAC-PUB-4653, June 1988.

[9] A. A. Sokolov and I. M. Ternov, *Radiation from Relativistic Electrons*, AIP, Translation Series, New York, 1986.

[10] V. N. Baier and V. M. Katkov, Sov. Phys. JETP **26** (1968) 854; **28** (1969) 807.

[11] K. Yokoya and P. Chen, *Electron Energy Spectrum and Maximum Disruption Angle under Multi-Photon Beamstrahlung*, Proc. IEEE Part. Acc. Conf., Chicago, March 1989 (SLAC-PUB-4935, Mar.1989).

[12] M. Jacob and T. T. Wu, Phys. Lett. **197B** (1987) 253, Nucl. Phys. **B303** (1988) 373, 389.
P. Chen and K. Yokoya, Phys. Rev. Lett. **61** (1988) 1101.
M. Bell and J. S. Bell, Part. Acc. **22** (1988) 301.
V. N. Baier, V. M. Katkov and V. M. Strakhovenko, *Quantum Radiation Theory in Inhomogeneous External Field*, INP-preprint 88-168, Novosibirsk, 1988.

[13] R. B. Palmer, Workshop on New Developments in Particle Accelerator Techniques, Orsay, June 1987 (SLAC-PUB-4295, Apr. 1987).

[14] D. H. Whittum, A. M. Sessler, J. J. Stewart and S. S. Yu, *Plasma Suppression of Beamstrahlung*, LBL-25759, June 1989.

[15] N. Solyak, *Collision Effects in Compensated Bunches of Linear Colliders*, preprint 88-44, INP Novosibirsk, 1988.

[16] J. B. Rosenzweig, B. Autin and P. Chen, *Instability of Compensated Beam-Beam Collisions*, in Advanced Accelerator Concepts, C. Joshi ed., AIP Conf. Proc. **193** (1989).

[17] V. Telnov and P. Chen, Phys. Rev. Lett. **63** (1989) 1796.

[18] V. Telnov, Nucl. Instr. Meth. **A294** (1990) 72.

[19] V. E. Balakin and A. A. Michailichenko, *The Conversion System for Obtaining Highly Polarized Electrons and Positrons*, INP-Preprint, 79-85, Novosibirsk, 1979; A. A. Michailichenko, in ICFA Workshop on Physics of Linear Colliders, Capri, June 1988, page 391.

[20] K. Yokoya and P. Chen, *Depolarization due to Beam-Beam Interaction in Electron-Positron Linear Colliders*, 8th Int. Symp. on High Energy Physics, Minneapolis, Sept. 1988. (SLAC-PUB-4692, Sept.1988).

[21] G. L. Kotkin, S. I. Polityko and V. G. Serbo, Sov. J. Nucl. Phys. **42** (1985) 440.

[22] M. S. Zolotarev, E. A. Kuraev and V. G. Serbo, INP-Preprint 81-63 (1981), Novosibirsk; SLAC TRANS-227 (1987).

[23] P. Chen, Intl. Workshop on the Next Generation Linear Colliders, SLAC Report 335, 1989; SLAC-PUB-4822 (1988), appeared in Proc. DPF Summer Study, SNOWMASS '88, World Scientific (1989).

[24] N. P. Klepikov, Sov. Phys. JETP **19** (1954) 19.

[25] V. N. Baier and V. M. Katkov, Sov. Phys. JETP **26** (1968) 854; W. Y. Tsai and T. Erber, Phys. Rev. **D10** (1974) 492.

[26] T. Erber, Rev. Mod. Phys. **38** (1966) 626.

[27] V. N. Baier, V. M. Katkov and V. M. Strakhovenko, Part. Acc. **30** (1990) 43.

[28] V. N. Baier, V. M. Katkov and V. M. Strakhovenko, Soviet J. Nucl. Phys. **14** (1972) 572.

[29] V. I. Ritus, Nucl. Phys. **B44** (1972) 236.

[30] P. Chen, Part. Acc. **30** (1990) 1013.

[31] See, for example, V. B. Berestetskii, E. M. Lifshitz and L. P. Pitaevskii, *Relativistic Quantum Theory*, Part 1, Pergamon Press (1971).

[32] V. N. Baier, V. S. Fadin, V. A. Khoze and E. A. Kuraev, Phys. Reports **78** (1981) 293.

[33] P. Chen, T. Tauchi and D. V. Schroeder, *Pair Creation at Large Inherent Angles* to appear in Snowmass Proceedings, 1991.

[34] R. B. Palmer, SLAC-PUB-5190 (1990); submitted to Ann. Rev. Nucl. Part. Sci.

[35] T. Tauchi et al., *Background Problem at Interaction Point for an e^+e^- Linear Collider*, submitted to IEEE Conf.

[36] A. E. Blinov, A. E. Bondar, Yu. I. Eidelman et al., Phys. Lett. **113B** (1982) 423.

[37] Yu. A. Tikhonov, *Candidates's Dissertation*, Inst. Nucl. Phys., Novosibirsk (1982).

[38] V. N. Baier, V. M. Katkov and V. M. Strakhovenko, Sov. J. Nucl. Phys. **36** (1982) 95.

[39] A. I. Burov and Ya. S. Derbenev, INP Preprint 82-07, Novosibirsk (1982).

Engineering Limitations of Proton Colliders

D. A. Edwards and M. J. Syphers
*SSC Laboratory**
2550 Beckleymeade Ave.
Dallas, Texas 75237, USA

Abstract

The higher collision energies of colliders planned and contemplated lead to a demand for higher luminosity as well. The dynamical limitations are usually expressed in terms of beam-beam tune shifts. There are other limitations that can be characterized as technological or engineering concerns. These inevitably involve a balance between cost and other considerations. The latter include materials properties, especially superconducting materials. Higher luminosity means higher beam intensity which leads to higher levels of radiation damage and synchrotron radiation with consequences for equipment lifetime and vacuum stability.

1 Introduction

The present luminositv record for both electron-positron and proton-proton colliders is 2×10^{32} $\mathrm{cm}^{-2}\ \mathrm{s}^{-1}$ achieved at CESR and the ISR. The former is an electron-positron storage ring operating at center-of-mass energies up to 16 GeV; the latter was a pair of proton-proton storage rings which produced collisions up to center-of-mass energy of 64 GeV. The Tevatron and Sp$\bar{\mathrm{p}}$S, in their role as proton-antiproton colliders, operate at the 10^{31} level at center-of-mass energies up to 1.8 TeV and 0.9 TeV respectively. The cross sections for particle production vary roughly as the inverse square of the mass state of interest, so progress toward higher mass physics leads to the desire for higher luminosity. For example, the nominal design luminosity selected for the SSC is 10^{33} while that for the LHC is an order of magnitude higher.

Insofar as beam dynamics is concerned, the leading limiting effect is the beam-beam interaction and in the next section we summarize the guidance of the beam dynamicists. The real limitations, however, are imposed by cost and technology. Neither necessarily presents firm impassable barriers. The cost of today's accelerator projects was virtually inconceivable 40 years ago, nor had the promise

*Operated by the Universities Research Association, Inc., for the U.S. Department of Energy.

of superconductivity been fulfilled. Nevertheless, the next generation of hadron colliders will be constructed in today's cost and materials environment.

This next generation will employ superconducting magnets and the technologies associated with them. In this paper, we comment on that technology, the implications of energy deposition due to beam loss, the increasing impact of synchrotron radiation, and the as yet unresolved questions regarding the vacuum behavior in superconducting magnets exposed to significant synchrotron radiation. We will not discuss the ever present issue of reliability which, though a major concern, has been successfully confronted by each succeeding generation of accelerators.

2 What the Dynamicists Tell Us

The maximum achievable luminosity in a hadron collider was the subject of a study at a recent Snowmass workshop.[1] If the only limiting effect is the beam-beam interaction the luminosity expression, since it contains the ratio of particles per bunch to normalized emittance, can be recast in terms of the head-on beam-beam tune shift parameter. The total tune spread is the sum of the head-on and long-range contributions. If it is assumed that it is the total tune spread that is the limiting parameter, then the maximum luminosity at a single interaction point is given by

$$\mathcal{L} \leq 1.7 \times 10^{34}\ \mathrm{cm^{-2}sec^{-1}}\ \frac{\gamma^{7/2}\alpha^4}{\epsilon L^*}. \tag{1}$$

In this expression the normalized emittance is defined by $\epsilon = \gamma\sigma^2/\beta$, where σ is the rms beam size at a location in the ring where the amplitude function has the value β and the maximum tune spread is taken to be 0.024. This latter figure is inferred from experience at the Sp$\bar{\text{p}}$S and the Tevatron.

This formula sets a challenging luminosity scale indeed. For typical SSC parameters, the maximum achievable luminosity under these ground rules is $\sim 7 \times 10^{35}$. (When one of the authors of the Snowmass article was asked why this figure wasn't simply rounded off to 10^{36}, the reply was that the answer might seem beyond the limits of credibility.) Why do both the SSC and LHC fall short of this ultimate goal? The limitations of the following sections come into play.

3 The Technological Environment

Whatever long-range promise high temperature superconductivity may hold, the magnets for the next generation of hadron colliders will be based on NbTi or Nb_3Sn materials. Magnetic fields in the 10-Tesla range imply stresses due to magnetic pressure in the neighborhood of 400 atmospheres; significantly higher fields are unlikely due to strength limits of materials.

Coil cooling will be provided by liquid helium at a temperature at or near 4 K. If the refrigeration system operated at the ideal Carnot efficiency, 75 W input at room temperature would be required to remove 1 W at 4 K. In reality, the best one can do is worse by a factor of about four.

To set a scale for refrigeration load, suppose one could construct a cryostat with 0.1 W/m static heat loss. Then, for instance, two synchrotrons each of circumference 87 km would represent a total refrigeration load of 17.4 kW, or 5 MW at room temperature. Already we see that 0.1 W/m is a significant amount. For comparison, the static heat load of the Tevatron is somewhat in excess of 1 W/m. So, as we turn to look at processes that add to the static heat load, we will look cautiously at levels much in excess of 0.1 W/m.

3.1 Energy Deposition from Beam Loss at the IP

Most any process which leads to beam loss will deposit energy in the superconducting magnets and add to the static heat load. Rather than attempt to catalogue all beam loss processes, we comment only on the one process which is inescapable in a collider — the energy streaming out in reaction products from the interaction point. For example, for SSC parameters the power generated at the interaction point is

$$\begin{aligned} P &= \mathcal{L} \cdot \sigma_{int} \cdot E \\ &= \left(\frac{10^{33}}{\mathrm{cm}^2\ \mathrm{sec}} \right) \left(0.1 \times 10^{-24}\ \mathrm{cm}^2 \right) \left(20 \times 10^{12}\ \mathrm{eV} \right) \\ &= 320\ \ \mathrm{W} \end{aligned} \tag{2}$$

in each direction. Despite all protective measures, much of this energy will be deposited in the focussing elements near the interaction point. These elements are apt to have very high gradients and are among the most technologically challenging pieces of hardware in the accelerator.

Figure 1 illustrates the initiation of a shower within a quadrupole magnet near the interaction point. The energy deposition is calculated with the aid of a Monte Carlo program which contains an accurate model of the superconducting magnet geometry and fields as well as the shower process. An example of the results of such calculations is shown in Table 1 for the nominal SSC luminosity and the optics associated with a high luminosity interaction region.[2] The energy deposition rates are listed in the third column. In the Tevatron, measurements have shown that an energy deposition rate of 8 mW/gram could be tolerated. Though it is not known what dose rate can be tolerated in SSC magnets, it is

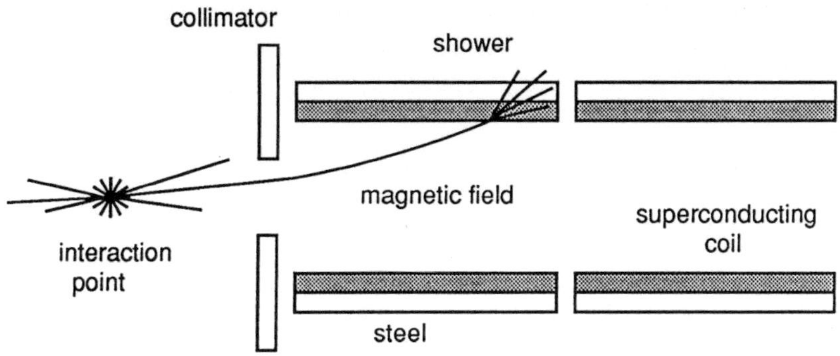

Figure 1: Schematic geometrical configuration used in energy deposition calculations.

Table 1: Maximum energy deposition dose rate D' and annual dose D in the superconducting coils of the SSC low-β IR beam elements. Interaction rate is 10^8/sec at $\mathcal{L} = 10^{33}$ cm^{-2} sec^{-1}. Here, the operational year is taken to be 10^7 sec. From Baishev, et al.

Name	Distance from IP (m)	D' (mW/gm)	D (MGy/year)
IP	0		
QL1	35	0.32	3.20
QL2a	47	0.19	1.92
QL2b	59	0.22	2.22
QL3	73	0.10	0.96
BV1	85	0.01	0.064
BV1	91	0.01	0.11
BV1	97	0.02	0.21

Table 2: Radiation resistance of selected materials. From Baishev, et al.

Material	Tolerable Dose (MGy)
Kapton, polyimide Kapton film Carbon-fiber reinforced tube Carbon-fiber-filled epoxy rods	50
G11 CR tube	20
PK102 (epoxy) Crest 7450 epoxy Fiberglass (epoxy impregnated)	10
Fiberglass rein. polyester resin Aluminum Mylar Superinsulation	5 2 2
Electrical insulation	0.1-10
Tefzel adhesive Cerex spunbonded polyester Teflon	0.5 0.06 0.01

likely that the situation is reasonably safe at the design luminosity. Upgrades to higher luminosity could pose more severe problems. The annual radiation dose is shown in the next column. Here, the situation is perhaps less comfortable. Table 2 indicates that a careful selection of materials will be necessary to achieve a ten-year lifetime of the cold mass of the interaction region quadrupoles. At a luminosity of 10^{34}, the lifetime would shrink to one year, while at the challenging number set forth by the dynamicists the lifetime would shrivel to less than a week.

3.2 Synchrotron Radiation

The formula for synchrotron radiation loss per turn from protons circulating in a synchrotron is

$$U = (0.0078\ \text{keV/turn})\ \frac{E_{TeV}^4}{\rho_{km}} \tag{3}$$

where E_{TeV} is the proton energy in TeV and ρ_{km} is the radius of curvature in kilometers. Table 3 shows the loss per turn and power dissipated due to synchrotron radiation for a selection of large hadron colliders. The last column indicates the operating circumstance. For the Tevatron, the situation is stated for the 1988-89 collider run. The SSC and LHC figures pertain to their nominal luminosities and operational parameters. For the Tevatron, synchrotron radiation is just not a

Table 3: Synchrotron radiation parameters from selected hadron colliders.

Device	U (keV/turn)	P (kW)	$P/2\pi\rho$ (W/m)	
Tevatron	0.006	1.6×10^{-5}	3.5×10^{-6}	$p\bar{p}$ Collider, 88-89
SSC	125	8.8	0.14	@ 10^{33}
LHC	2×12	2×10	2×0.63	@ $4 \cdot 10^{34}$

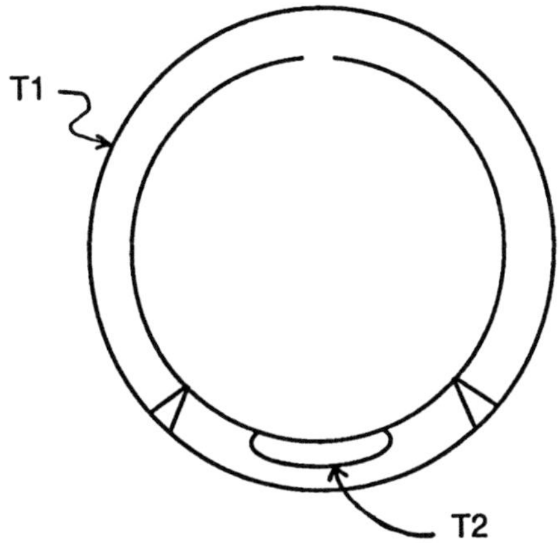

Figure 2: Conceptual design of a synchrotron radiation intercept.

consideration. It is amusing that at 900 GeV particles lost from a bucket take over an hour to spiral inward and strike the walls of the vacuum chamber. For the SSC, a particle lost from a bucket would strike the vacuum chamber in less than a minute, but more to the point, the synchrotron radiation load is the same as our static heat load goal of 0.1 W/m and so is significant. An increase in luminosity to 10^{34} by increasing the beam current by a factor of three would require a doubling of the cryogenic plant capacity. Further increase by this route is not attractive, but the SSC would still lag the LHC in luminosity!

An attractive proposal is to remove the synchrotron radiation load at a higher temperature. Because of the increased Carnot efficiency, the refrigeration power would be lowered. The magnet design is, of course, somewhat more complicated due to the introduction of the higher temperature intercept. A cartoon sketch of such a device appears in Figure 2. The superconducting coils are maintained at temperature T_1 outside the bore tube of the magnet. The liner is held at a temperature $T_2 > T_1$ by coolant flowing through the passage attached to the bottom of the liner. While the coils are at $T_1 = 2\text{-}4$ K, the temperature T_2 could be in the range of 20-80 K implying an improvement in Carnot efficiency of an order of magnitude or so. The slot at the top of our cartoon liner provides an escape route for hydrogen which is desorbed from the inner surface of the liner so that it can be trapped at the lower temperature of the bore tube. The LHC has incorporated a bore tube liner into its magnet design and a similar step is under consideration for the SSC.[3],[4]

4 Vacuum Stability

Even at 10^{33} luminosity there is a potentially unsavory vacuum situation in the SSC. To achieve the desired 150 hr partial lifetime of the beam due to interactions with the residual gas implies a gas density of $6 \times 10^8 H_2$ molecules/cm^3 or less. The saturated vapor pressure of hydrogen gas at 4.3 K corresponds to a gas density of 6×10^{11} molecules/cm^3. Therefore, a significant fraction of a monolayer of hydrogen must not be allowed to condense on the inner surface of the beam tube.

But we know that hydrogen is released even from clean baked surfaces that are exposed to synchrotron radiation. Experiments indicate that the time to build up the first monolayer at a luminosity of 10^{33} will be approximately 50 hours.[5] Then, a warm-up to 20 K would be necessary to pump out the hydrogen following which the system would be cooled down and operation resumed. After a number of such cycles this situation would eventually improve.

This is a fine picture if the entire ring cleaned up simultaneously. Unfortunately, the ring is made up of many isolated pieces. Each one is a ticking clock and clean-ups would need to be synchronized.

Further, at the saturated vapor pressure at 4.3 K, the energy deposition due to the beam-gas interaction is about 8 W/m — unsustainable by the cryogenic system. So, a conservative approach would be to have a liner as sketched above with T_1 at about 3 K consistent with a saturated vapor pressure appropriate to a beam lifetime of 150 hours. Obviously, further study is needed on these matters.

5 Concluding Remarks

A few years ago, some of us thought that reliability was the only major issue to be addressed in the next generation of hadron colliders. We still believe that it is the most significant single problem.

There are other considerations of an engineering nature that we have not mentioned in this short paper. The stored energy in the beam of these devices can be considerable; for the SSC, the 400 MJ in the beam of one ring represents a potential threat to the hardware. The nominal parameters of these collider designs involve many closely spaced bunches; the suppresion of multi-bunch instabilities will surely be a problem. Only passing mention has been made of the interaction region optics; the numbers advanced for β^* at a high luminosity interaction point may be most difficult to realize.

But the upgrade potential cannot be ignored. Some upgrade capability has to be addressed at the outset, within cost and schedule bounds. As a result, these synchrotrons are becoming a good deal more interesting as design challenges as they move out of the paper stages.

References

[1] Proc. of Snowmass 1990 SSC Summer Study. To be published.

[2] Baishev, I. S., A. I. Drozhdin, and N. V. Mokhov, *Beam Loss and Radiation Effects in the SSC Lattice Elements*, SSC Report SSCL-306, 1990.

[3] *Design Study of the Large Hadron Collider*, CERN Report 91-03, Geneva, May 1991.

[4] Edwards, H. T., *Study on Beam Tube Vacuum with Consideration of Synchrotron Light, Potential Liner Intercept, and Collider Quad/Spool Coil Diameter*, SSC Report SSCL-N-771, 1991.

[5] *Site-Specific Conceptual Design Report of the Superconducting Super Collider*, SSCL, ed. J. R. Sanford, SSCL-SR-1056, July, 1990.

Engineering Limitations with Electrons

O. Gröbner

CERN, Geneva, Switzerland

ABSTRACT

The vacuum required for electron machines is determined by the beam lifetime and must be in the ultrahigh vacuum range. Since the emission of intense synchrotron radiation by the electron beam produces strong dynamic outgassing from the vacuum chamber, this gas load must be removed by a powerful, distributed pumping system. To achieve good beam lifetimes an extended in-situ cleaning (beam conditioning) by synchrotron radiation must be applied. High pumping speeds can be obtained using a distributed pumping system, e.g. based on linear, non-evaporable getter pumps. The absorption of the synchrotron radiation power, either by the vacuum chamber directly or on local aborbers,is closely linked to the design of the global system and to the choice of the vacuum chamber material.

1 VACUUM REQUIREMENTS

The level of vacuum that is required in a high energy particle accelerator is determined by the tolerable amount of degradation of the beam during its passage through the machine. The loss of particles and the reduction of the particle density in the beam are directly proportional to the product of the residual gas pressure and the path length of the particles and therefore to the time the beam is maintained circulating in the machine. The necessary beam lifetimes are typically several hours or more and therefore require an ultrahigh vacuum (uhv) system.

1.1 Beam lifetime

The dominant loss mechanism for relativistic electrons by interaction with the residual gas is energy loss through the emission of photons due to bremsstrahlung on the nucleus and on the electrons of the residual gas molecules [1] The rate of energy loss for an electron of energy E is given by

$$-\frac{dE}{dt} = \frac{c\rho E}{X_0}$$

where c is the velocity of light, ρ (kg m^{-3}) the density of the residual gas and X_0 (kg m^{-2}) its radiation length. Particles are lost from the circulating beam only if their energy is outside the acceptance δE of the machine. This is accounted for by the factor $W = \ln(E/\delta E)$. The resulting time constant of the beam decay is

$$\frac{1}{t} = \frac{cW\rho}{X_0}.$$

The only machine-dependent parameter is the energy acceptance, giving W typically in the range from 6 to 10. The radiation length X_0 is defined as

$$X_0^{-1} = 4\,\alpha\, r_e^2\, Z(Z+1)\, N_0\, A^{-1}\, \ln(183/Z^{1/3})$$

where α is the fine structure constant, r_e the classical electron radius, Z the atomic number, N_o Avogadro's number and A the atomic mass of the gas molecules. For nitrogen $X_o = 393$ kg m^{-2} and converting from density to nitrogen equivalent pressure P_{N2} (Torr), the beam lifetime t (h) can be expressed approximately as

$$t = \frac{3\ 10^{-8}}{P_{N2}} .$$

This result shows that a residual pressure below 10^{-9} Torr is required in a storage ring. The dependence of the radiation length on the square of the nuclear charge points to the importance of avoiding heavy molecules in the residual gas, i.e. of producing a 'clean' uhv. The following table gives the average pressure, in units of nTorr, for different gas species equivalent to 1 hour beam lifetime calculated for LEP.

H_2	CH_4	H_2O	CO	Ar	CO_2
460	55	42	28	9.5	17.5

The beam lifetime expression given above neglects elastic scattering which would give rise to beam loss on the vacuum chamber aperture and on the rf bucket. These effects can be significant at low beam energy and in the case of a very small vacuum chamber aperture.

1.2 Synchrotron-radiation-induced gas desorption

The design of the vacuum system for an electron or positron accelerator and storage ring is dominated by the intense synchrotron radiation emitted by the particles and impinging on the vacuum chamber wall. The synchrotron radiation produces strong gas desorption [2, 3] from the vacuum chamber and intense thermal heating. The total power, P (W), emitted as synchrotron radiation photons in a given machine is

$$P = 6.032\ 10^{-9}\ \gamma^4\ \frac{I}{r}$$

where γ is the ratio of the total energy to the rest energy of the relativistic particles, I (A) the circulating beam current, and r (m) the radius of curvature in the bending magnets. This energy loss, which has to be compensated continuously by a powerful accelerating system to maintain the beam on its stable orbit, is an unwanted effect in high energy accelerators. Yet this same synchrotron radiation is the justification for the steadily growing number of dedicated synchrotron light sources, which are either operating or under construction all over the world. The synchrotron radiation power increases with the fourth power of the energy and the only 'free' parameter remains the bending radius and, therefore, the circumference of the machine. These considerations explain the very large size of LEP with its radius of curvature of 3096 m and its circumference of nearly 27 km [4].

The photon spectrum extends from below the visible spectrum up to gamma rays and is characterized by an average ('critical') energy ε_c (eV)

$$\varepsilon_c = 2.959\ 10^{-7}\ \frac{\gamma^3}{r} .$$

The total number of photons emitted per second is

$$\frac{dN}{dt} = 4.128\ 10^{17}\,\gamma\, I$$

and the flux of photons per meter of the vacuum chamber is

$$\frac{dN}{dt\ ds} = 6.57\ 10^{16}\ \gamma \frac{I}{r}\,.$$

The pressure due to the additional desorption of molecules from the wall of the vacuum chamber increases from its static value p_0 proportionally with the total beam current I.

$$p(I) = 6.57\ 10^{16}\ \frac{\gamma\eta I}{krS} + p_0. \qquad (1)$$

Here η is the molecular desorption yield (molecules photon^{-1}), S the linear pumping speed in l sec^{-1} per meter of vacuum chamber and $k = 3.2\ 10^{19}$ molecules per Torr l at room temperature.

The desorption of gas molecules is considered to occur in a two-step process, by which first photoelectrons are produced that subsequently desorb gas molecules by electron-stimulated desorption. The flux of desorbed gas molecules is related to the photons incident on the vacuum chamber by the molecular desorption yield η - a molecular desorption yield in the range from 10^{-2} to 10^{-4} molecules per photon being typical for a vacuum chamber made from aluminium. On stainless steel and copper the yield can be one to two orders of magnitude smaller. Instead of quoting the yield, it is common practice to express the overall vacuum performance of the vacuum system by the 'specific pressure rise',

$$\frac{\Delta p}{I} = 2.05\ 10^{-3}\ \frac{\eta\gamma}{rS} \qquad (2)$$

which also includes the performance of the pumping system, i.e. the pumping speed S.

2 DYNAMIC VACUUM AND BEAM CLEANING

When electron or positron beams circulate for the first time in a new machine, the dynamic pressure rise normally exceeds the initial static pressure by several orders of magnitude and the beam lifetime is very short. The vacuum improves gradually with the machine operation as more gas molecules are desorbed from the vacuum chambers and pumped out. This in-situ conditioning is called 'beam cleaning' and can give an improvement of the specific pressure rise by several orders of magnitude. Because of this beam cleaning effect vacuum systems for electron machines can be designed with a pumping system that would be inadequate for the high initial desorption rate, but which meets the low pressure requirement of the machine after an extended conditioning period. The beam cleaning observed during the initial running-in phase of LEP is shown in the Fig. 1.

Starting from an initial value above 10^{-7} Torr mA^{-1}, the specific pressure rise in LEP has decreased gradually to below 10^{-10} Torr mA^{-1} during the first year of operation. The horizontal

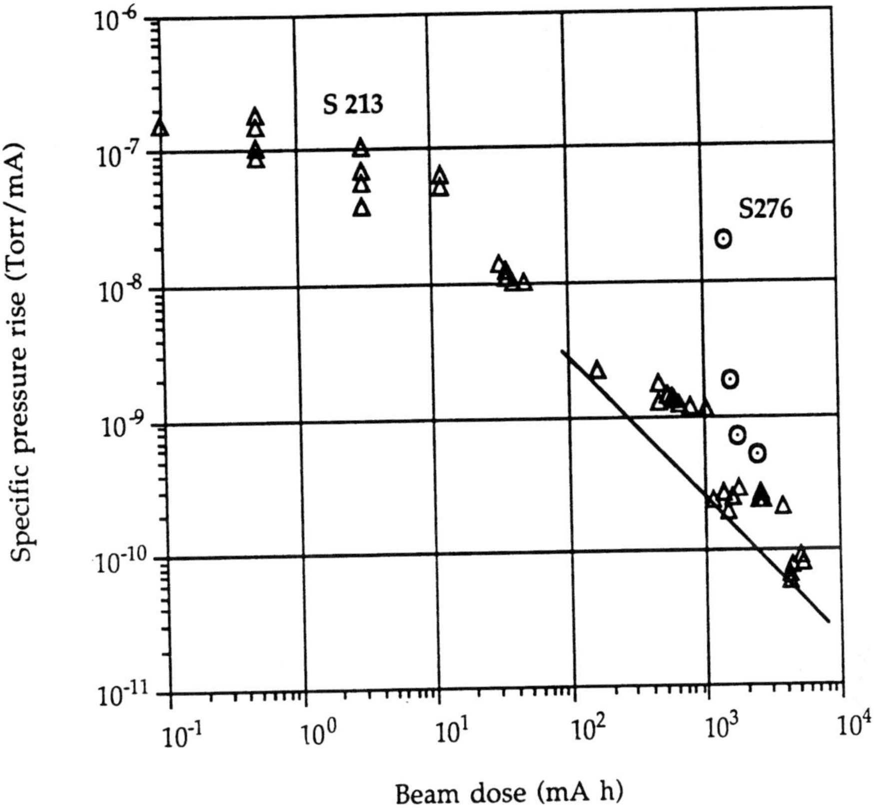

Fig. 1 Beam cleaning of the LEP vacuum system during the running-in phase. S213 represents the behaviour of the vacuum in the bending arcs. S276 shows the recontamination effect due to venting a section of the system to nitrogen without a subsequent bakeout. The line joins points with maximum NEG pumping speed after reconditioning, see section 3.2.

axis gives the 'beam dose' in mA h, i.e. the integrated product of beam current and time, which is a measure of the total exposure of the vacuum chamber to photons according to

$$\text{Beam dose (A h)} = 2.3\ 10^{20}\ \frac{\gamma}{r} \qquad \text{(photons m}^{-1}\text{)}.$$

In LEP at 45 GeV, 1 A h is equivalent to 6.5 10^{22} photons per meter of chamber.

It should be noted that the pumping speed for the LEP vacuum system is not constant, but decreases with the amount of gas sorbed by the non-evaporable (NEG) getter. The variable pumping speed, which is periodically restored by reconditioning of the getter, is the cause of the discontinuities of this cleaning curve. The points joined by the solid line correspond to the maximum pumping speed of the fully activated getter. This cleanup of the system continues as gas molecules escaping from the surface and from near surface oxide layers are pumped out. Experiments have shown that the rate at which gas enters the vacuum can be attributed to photon-induced diffusion in the surface oxide layer [5].

Unfortunately, whenever a section of a cleaned vacuum system has to be re-exposed to atmospheric pressure, recontamination of the surface occurs and a significant part of the cleaning effect is lost, see the circles marked S276 in Fig. 1. After atmospheric exposure, even under well controlled venting with dry nitrogen, the desorption yield increases to nearly its original value. Since in practice a large vacuum system is never opened entirely but only in 'small' sections, e.g. up to 474 m in LEP, the reconditioning has to be performed only in these subsections and, therefore, contributes proportionally less to the average pressure and to the lifetime. However, the cleanliness of the vacuum system can be restored almost completely if the vacuum section is re-baked in situ after the intervention.

From expressions (1) and (2) one would expect the specific pressure rise to increase proportionally with beam energy. This is only approximately correct since the molecular desorption yield also varies with photon energy, hence with the critical energy of the synchrotron radiation. Figure 2 combines results obtained at DCI - using a dedicated beam line for exposing an aluminium vacuum chamber of LEP - and from the LEP machine proper, which show that at low photon energies ($\varepsilon_c < 3$ keV) the yield increases, while at higher energies ($\varepsilon_c > 6$ keV) the yield decreases with energy.

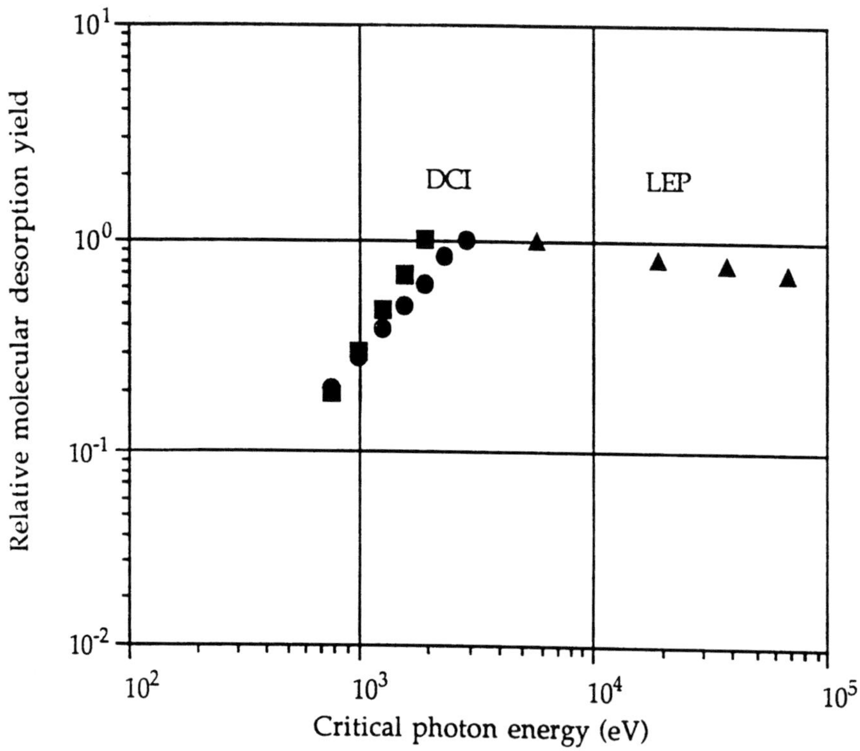

Fig. 2 The dependence of the molecular desorption yield of an aluminium vacuum chamber on the critical energy of the synchrotron radiation. The results are derived from two sets of measurements at DCI and at LEP and have been normalised to 1.

It should be noted that the molecular desorption yield depends in addition also on the angle of incidence - in practice the photons strike the vacuum chamber at grazing incidence which amplifies the desorption probability. The dependence of the pressure rise on the angle of incidence is shown in Fig. 3.

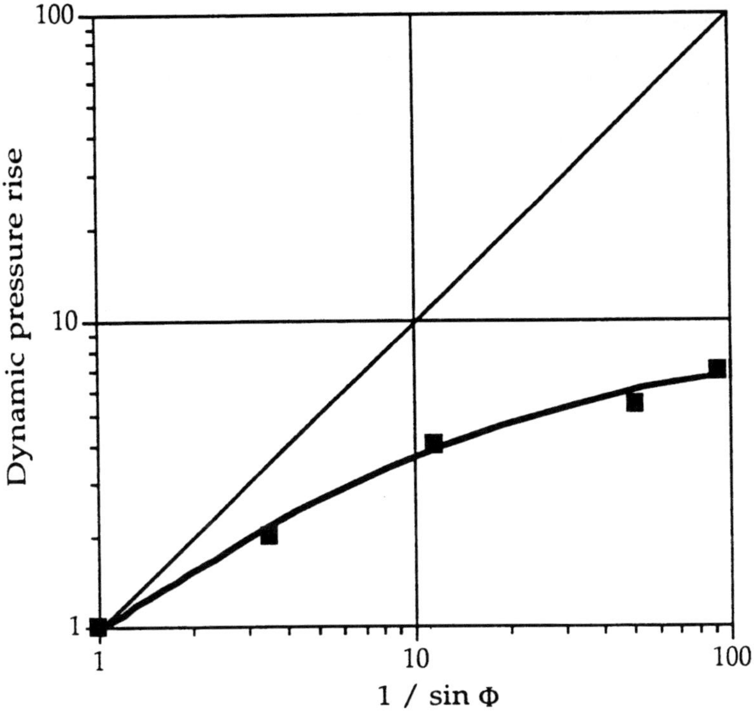

Fig. 3 The dependence of the dynamic pressure rise on the grazing angle of incidence. The straight line represents a 1/sinφ dependence expected on the basis of the photoelectron production.

Because of the angular dependence of the molecular desorption yield, it is preferable to arrange synchrotron radiation absorbers such that the photons are incident at perpendicular incidence. However, in this case the local power density is maximum and for any practical machine design it is necessary to make a compromise between the maximum tolerable power density and the molecular desorption. For this reason high energy machines like LEP, PETRA and PEP have adopted a design where the synchrotron radiation is evenly distributed around the bending magnet sections - giving a grazing incidence for the photons of about 7 to 20 mrad only, but with a reduced local power density.

3 PUMPING SYSTEM

3.1 Integrated ion pumps

Electron machines with small, low-conductance vacuum chambers and strong outgassing along all curved sections require a large and efficient pumping system which has to be distributed along the bending arcs. To achieve this goal, linear ion pumps [6] incorporated in the bending magnet vacuum chambers have been used extensively in the past. These integrated ion pumps operate in the stray magnetic field. The drawback is that their operation relies entirely on the bending magnet field and hence additional holding pumps must be provided to maintain the vacuum during the periods when the magnets are off. In addition, the pumping speed varies with the magnetic field and is therefore not constant between the injection and the operating energy of the machine. The achieved linear pumping speed is typically around 100 l s^{-1} m^{-1}.

3.2 NEG pumps

More recently, following the successful example of LEP, getter pumps have replaced integrated ion pumps in several new projects. The LEP vacuum system is pumped with a combination of a linear NEG [7] (Non-evaporable Getter) strip that has been inserted in the pump channel over more than 22 km of the circumference and, in addition, small (40 l s^{-1}) ion pumps placed at intervals of approximately 24 m. Figure 4 shows a section of the vacuum chamber with the elliptic beam channel and the adjacent pump duct.

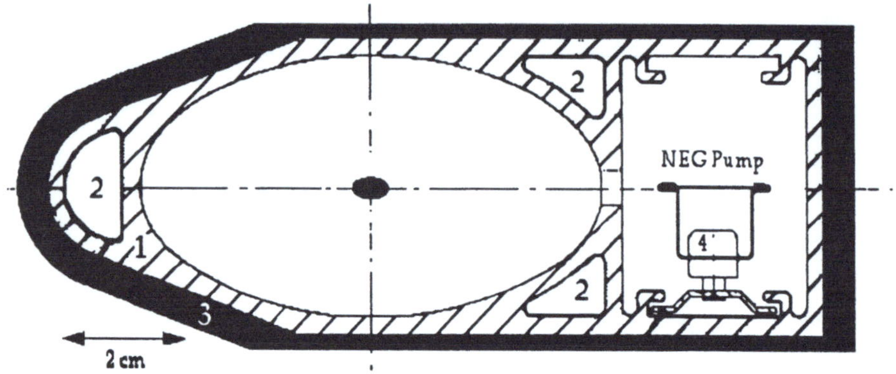

Fig. 4 Vacuum chamber of LEP made of extruded aluminium profile (1). The separation wall between the elliptic beam channel and the rectangular pump has longitudinal slots for pumping. Three water channels (2) are required for removing the heat generated by synchrotron radiation, but serve also for heating by hot water circulation during the vacuum bakeout. The chamber is surrounded by a 3 to 8-mm-thick lead shield (3). The NEG pump ribbon is supported by ceramic insulators (4).

The NEG, an Al Zr alloy [8] in the form of powder, is applied by cold sintering to a constantan ribbon and pumps chemically active gas molecules such as H_2, CO, CO_2 and N_2. Before it can pump, the getter must be activated under vacuum by heating to about 740°C for 1 hour, an operation that can easily be achieved by passing electric current through the metal support ribbon. Compared to linear ion pumps, the NEG [9] can give a substantially higher - more than a factor of 10 - linear pumping speed independent of magnetic field. The effective linear pumping speed of the getter ribbon in LEP as a function of the amount of gas pumped [10] is shown in Fig. 5.

Following the initial activation of the getter, which produces the high initial pumping speed, the NEG sorbes gas and gradually loses its pumping speed as the surface is progressively covered with gas molecules. Finally, when the pumping speed is below a minimum value required for the vacuum, the getter has to be 'reconditioned' by heating the NEG to about 450 °C for several minutes. During a typical pumping cycle, that is in between two conditionings, the NEG can sorb up to 0.1 Torr l m^{-1} with a speed exceeding 100 l s^{-1} m^{-1}. The lifetime capacity of the ribbon exceeds 30 Torr l m^{-1} for the gases CO and CO_2 and at least 10 times more for H_2. These quantities exceed by far what is expected during the lifetime of the machine. Inside the LEP vacuum chamber, the NEG strip is mounted on a stainless steel carrier and supported by insulating ceramics. At each end, the 10-m-long ribbon is attached to a current feedthrough to permit activation by resistive heating. For chemically inert gases and methane, ion pumps have been provided and these can also be used to measure the pressure in the system.

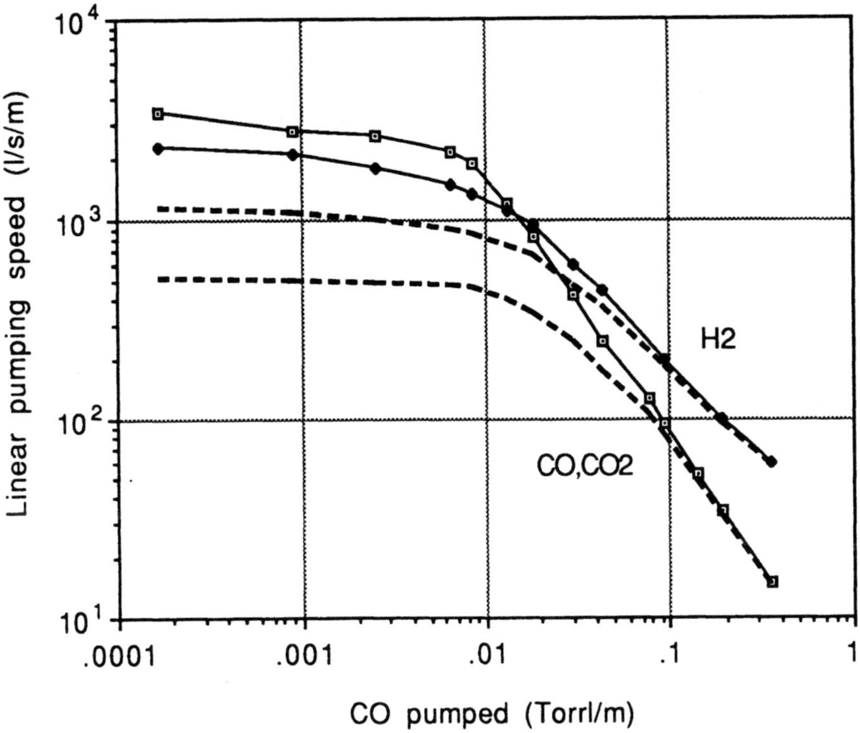

Fig. 5 Linear pumping speed of the NEG pump in the LEP vacuum chamber as a function of the amount of gas sorbed (full lines). The broken lines refer to the effective pumping speed when the conductance of the pumping slots between beam duct and pumping channel has been taken into account.

The operational experience in LEP has confirmed the results from the laboratory performance on small-scale tests. Due to the high effective pumping speed available, the pressure rise with beam can be kept low and the resulting beam-gas lifetime can be long. Furthermore, the frequency at which the NEG has to be reconditioned in a machine like LEP has not affected the operation. Between the start-up in July 1989 and March 1990, the NEG pumps have been reconditioned three times only. The first time at a beam dose of about 120 mA h, the second time at 950 mA h and the third time after 1.4 Ah. During the first two of these pumping cycles the effective linear pumping speed for CO and CO_2 has decreased from its maximum value of about 500 l s^{-1} m^{-1} to approximately 70 l s^{-1} m^{-1} while the third time the pumping speed was still well above 100 l s^{-1} m^{-1}. The total quantity of gas pumped by the NEG during this period is estimated to be less than 0.4 Torr l/m for CO and CO_2 and about 1 Torr l m^{-1} for H_2 - nevertheless equivalent to about three monolayers on the vacuum chamber. However, these quantities are still small compared to the total capacity of the getter which exceeds 30 Torr l m^{-1} for $CO+CO_2$ and 300 Torr l m^{-1} for H_2 [11].

The reconditioning of the NEG can be performed remotely from the control room. Because the heating to above 450°C produces a transient pressure rise, this operation is done with closed sector valves and with the ion pumps switched off to avoid deterioration of the ion pump performance. The whole vacuum system can be treated in less than 8 hours. For practical reasons, the reconditioning is performed towards the end of a long machine stop at intervals of several months.

Recent projects for B-factories require that the pumping system can cope with considerably larger gas loads than in LEP even after extended periods of beam conditioning. Consequently, the saturation time of the NEG pumps may become too short and the required reconditionings too frequent for machine operation. From the preceding expressions the dynamic gas flow, Q (Torr l s^{-1} m^{-1}), can be expressed as

$$Q = \frac{\Delta p}{I} \, I \, S \, .$$

The importance of the saturation time may be illustrated by the example of the BFI [12] Beauty Factory proposed at CERN. For this machine the tolerable pressure rise during operation is limited to 2 10^{-9} Torr A^{-1} with the nominal beam current of 1.3 A. The linear NEG pumps can provide an effective pumping speed of up to 500 l s^{-1} m^{-1}. These conditions correspond to a gas load of 1.3 10^{-6} Torr l s^{-1} m^{-1}. Therefore, it has been estimated that the NEG pumps, which may be designed with a capacity of 0.5 Torr l m^{-1}, will have to be reconditioned at intervals of less than 100 hours as compared to many months in LEP.

4 SYNCHROTRON RADIATION ABSORBERS

A limiting parameter for the design of the vacuum system of high energy and high intensity electron machines is the synchrotron radiation power which has to be absorbed inside the vacuum system. In LEP, at the maximum operating energy of 100 GeV, the average linear power density will be around 1.5 kW m^{-1} but locally this value may even double. The vacuum chamber requires, therefore, efficient water cooling. However, the maximum local temperature rise in the region where the power is deposited must stay within safe limits. In principle two solutions can be adopted:

i) a continuous absorption of the power along a smooth vacuum chamber at grazing incidence of the photons which distributes the total power over the circumference of the machine or

ii) local absorbers which intercept the radiation at near perpendicular incidence which shadow the vacuum chamber.

The advantage of local absorbers is that the radiation is absorbed in well defined locations which can be pumped efficiently with large, lumped pumps. In addition, the close to perpendicular incidence gives less outgassing [13]. With the distributed absorbers the very concentrated power deposition can be avoided. The particular choice depends largely on the detailed design parameters of the machine.

4.1 Synchrotron radiation absorption

The photon absorption at low energies is dominated by the photoelectric effect while at high energies, above a few tens of keV for aluminium, it is mainly due to Compton scattering. As a consequence, the local power deposition in the region of the primary impact of the radiation varies significantly with the photon energy. In LEP, the critical energy of the synchrotron radiation ranges from about 6 keV at injection to about 700 keV at 100 GeV beam energy. The photons penetrate into the vacuum chamber and deposit their energy either locally by the photoelectric effect or are scattered at sufficiently large angles as to penetrate the wall of the vacuum chamber without any significant further attenuation. This process makes the vacuum chamber transparent to photons above an energy which depends on the material. Figure 6 shows the result of an electromagnetic shower calculation [14] giving the fraction of the incident radiation which penetrates the 5-mm-thick LEP vacuum chamber as a function of the critical photon energy [15]. At the early design phase of LEP aluminium and copper have been considered as alternative materials. However, due to the very high photon energies, both materials become transparent to radiation and require the addition of a lead shield to reduce the radiation dose to machine elements, e.g. magnet coils and insulation of electric cables. At low photon energies such as in a B-factory, a copper vacuum chamber is self shielding.

From Fig. 6 it can be deduced that with aluminium, and because of the high photon energy, only about half of the synchrotron radiation is absorbed locally in the region of the direct impact while the rest is scattered and absorbed in the lead radiation shield surrounding the vacuum chamber. This has the consequence that the maximum local power density, causing high thermal stress, is reduced while the diffusely scattered radiation can easily be removed by the water cooling.

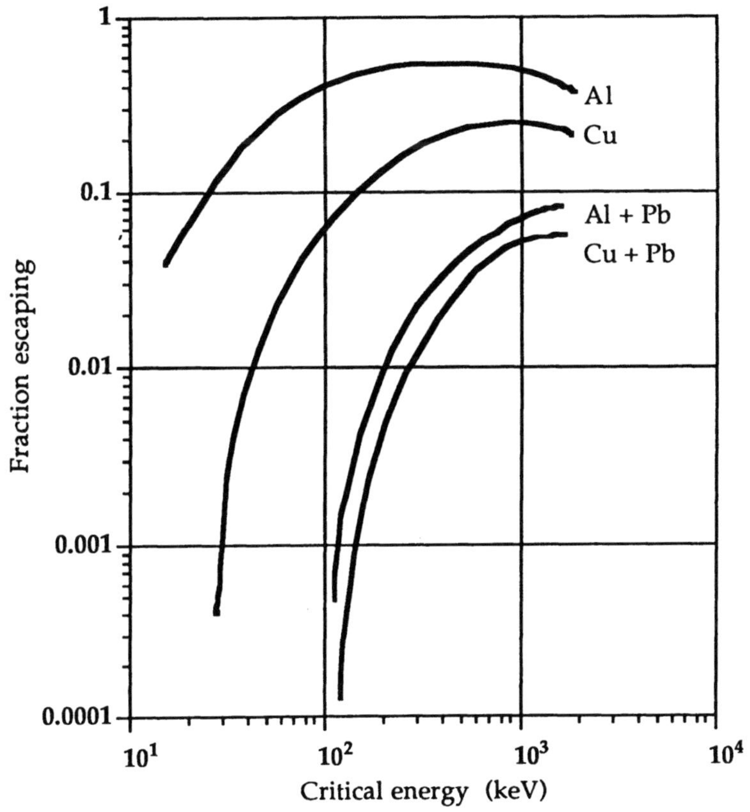

Fig. 6 The fraction of the synchrotron radiation power escaping from the vacuum chamber as a function of the critical energy of the synchrotron radiation in LEP. The photons are incident at about 7 mrad. The calculations have been made for aluminium and copper with an additional lead shield of 3 to 8 mm.

4.2 Width of the synchrotron radiation band

The thermal effects depend critically on the width of the band where the synchrotron radiation power is incident on the vacuum chamber. The width of the band of synchrotron radiation is related to the vertical size of the beam [16],

$$\sigma_y = \sqrt{\varepsilon_y \beta_y}$$

where ε_y is the vertical emittance of the beam and β_y the vertical beta function. The divergence of the synchrotron radiation for photons with the critical energy is proportional to $1/\gamma$, and depends also on the local vertical beam divergence given by

$$\sigma_y' = \sqrt{\frac{\varepsilon_y}{\beta_y}} .$$

In a bending magnet the synchrotron radiation travels typically a distance

$$d = \sqrt{2ra}$$

between the source point and the vacuum chamber, r being the bending radius and a the horizontal half aperture of the vacuum chamber. The region on the vacuum chamber exposed to the dense part of the radiation is approximately

$$w = 2(\sigma_y + d[1/\gamma + \sigma_y']).$$

For the arcs of LEP, the width of the synchrotron radiation band on the vacuum chamber is shown in Fig. 7.

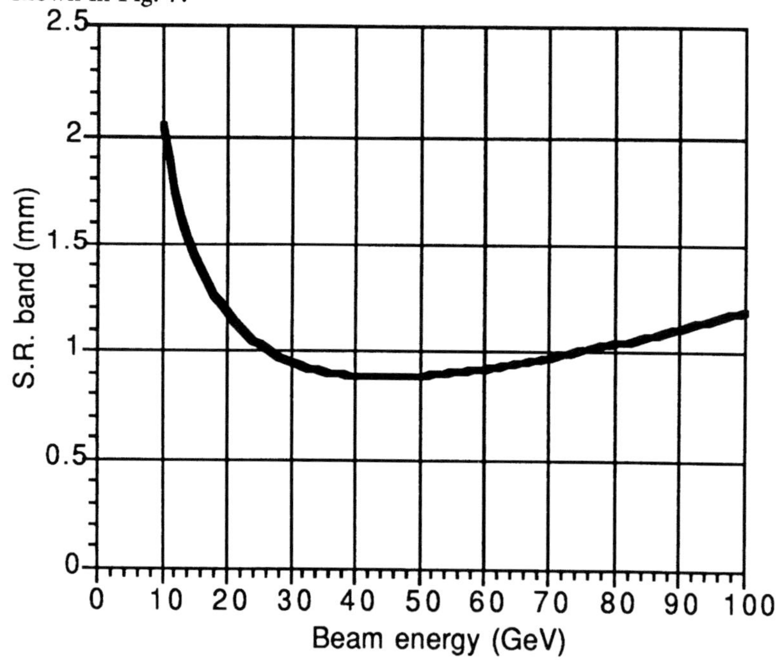

Fig. 7 Width of the vacuum chamber zone exposed to the dense part of the synchrotron radiation in LEP.

The initial decrease of the width is due to the $1/\gamma$ effect, while the following increase with beam energy comes from the growth of the vertical emittance. It can be seen that the dominant part of the synchrotron radiation power is concentrated in a very narrow region only 1 mm high on the vacuum chamber wall.

The maximum power which can be tolerated without risk may be estimated from the resulting thermal stress in the local zone of the vacuum chamber. Assuming that the thermal-induced stress ε (N mm^{-2}) should not exceed the elastic limit, one obtains from

$$\varepsilon = E \frac{\Delta l}{L}$$

and

$$\frac{\Delta l}{L} = \alpha \Delta T$$

the limiting condition

$$\varepsilon = E \alpha \Delta T.$$

Here α (°C^{-1}) is the coefficient of linear expansion and E (N mm^{-2}) is Young's modulus of the material. For a given vacuum chamber or synchrotron radiation absorber design, the temperature rise (temperature gradient) in the region of the direct synchrotron radiation impact can be calculated. For the design of the LEP vacuum chamber this method has been used to obtain the relative temperature rise as a function of the power, k (°C W^{-1} mm^{-2}). The maximum tolerable linear power density (W mm^{-1}) is then

$$P_{max} = \frac{\varepsilon}{E\alpha k} \,.$$

The following table compares the maximum linear power density for the vacuum chamber of LEP, which is made of extruded aluminium, with vacuum chambers made of different types of copper. A vacuum chamber profile made of hard, oxygen-free, high-conductivity (OFHC) copper has been proposed for the BFI project at CERN since an aluminium chamber would not meet the power requirement in excess of 10 kW m^{-1}.

Comparison of materials for vacuum chambers and synchrotron radiation absorbers

	AlMgSi0.5 (LEP)	OFHC (soft)	OHFC (hard)	"GlidCop"
Young's modulus (N mm^{-2})	7.5E+4	1.2E+5	1.2E+5	1.2E+5
Tensile strength (0.2) (N mm^{-2})	200	60	250	338
Coef. linear expansion (°C^{-1})	2.7E-5	1.7E-5	1.7E-5	1.7E-5
Thermal conductivity (W °C^{-1} mm^{-1})	23.7	38.5	38.5	38
Temperature rise (°C W^{-1} mm^{-1})	17.8	11.6	11.6	11.6
Max. temperature rise (°C)	98.77	29.41	122.55	165.69
Max. linear power (W mm^{-1})	5.55	2.54	10.56	14.28

The temperature rise shown in this table has been calculated for a 5-mm-thick vacuum chamber and assumes a uniformly distributed power (W mm^{-1}) along the chamber [17] The first three columns refer to common materials for vacuum systems. The data in the last column refer to "Glidcop", a special dispersion-strengthened copper which has not been used for vacuum chambers but only for special vacuum elements in high power applications [18].

It can be seen that the LEP vacuum chamber made of aluminium can absorb more than 5 kW/m. Since about half of the power is scattered at large angles and escapes the vacuum chamber, the total power radiated by the beam could even be a factor of 2 larger. However, when considering a copper vacuum chamber or the lower photon energy of the BFI-project, most of the power would be absorbed locally and this gain would not be available.

REFERENCES

[1] C. Benvenuti, R. Calder, O. Gröbner, Vacuum, **37**, 8/9, 699-707 (1987)

[2] G.E. Fischer and R.A. Mack, J Vac Sci Technol, **2**, 123-130 (1965)

[3] M. Bernardini, L. Malter, J Vac Sci Technol, **2**, 130-141 (1965)

[4] S. Myers, E. Picasso, Contemporary Physics, Vol 31, No. 6, 387 (1990)

[5] M. Andritschky, O. Gröbner, A.G. Mathewson, P. Strubin and R. Souchet, Vacuum, **38**, 8/10, 933-936 (1988)

[6] PEP, Conceptual design report, LBL, 4288/SLAC-189 (1976)

[7] C. Ferrario et al., Proc. 9th Symp. on Fusion Tech. (Pergamon), 51, (1976)

[8] C. Benvenuti, Nucl. Inst. and Methods **205**, 391, (1983)

[9] B. Ferrario, L. Rosai, P. della Porta, IEEE Trans. Nucl. Sci. Vol. NS-28, No 3, (1981)

10] C. Benvenuti, F. Francia, CERN-LEP-VA/89-61 (1989)

[11] C. Benvenuti, Nucl. Instr. Meth., 205, 391, (1983)

[12] T. Nakada, CERN 90-02 and PSI PR-90-08, March 1990

[13] R.J. Bennett, J.M. Lefebvre, B.A. Trickett, EPAC, European Particle Accelerator Conference, Rome, 1988, 1243, (1988)

[14] R.L. Ford, W.R. Nelson, SLAC-210, June 1978

[15] O. Gröbner, LEP-Note 245, June 1980

[16] M. Sands, SLAC-121, UC-28, November 1970

[17] O. Gröbner, H. Schuhbäck, R. Valbuena, T. Wikberg, CERN Technical Note AT VA/06/mpt-56 (1990)

[18] R. Digennaro et al., Nucl. Inst. and Methods A266, 498, (1988)

INTRABEAM SCATTERING

Allan H. Sørensen
Institute of Physics, University of Aarhus
DK-8000 Aarhus C, Denmark*
and
Department of Physics, Duke University
Durham, NC 27706, USA

1 PRELUDE

The individual particles of a charged particle beam circulating in a storage ring occasionally scatter on one another. The purpose of the present report is to give an introductory discussion of the influence of this *intrabeam scattering* on the evolution of the phase-space density of a charged particle beam. As we shall see below, intrabeam scattering leads to growth of the volume occupied in phase space by the beam. Growth rates become large at high densities and for high particle charge, and intrabeam scattering may thereby impose severe intensity limitations for highly charged ion beams.

To initiate our studies and obtain a background for understanding the specific case of a beam, we consider at first in general the rôle played by collisions in a plasma (an ionized gas): For a system containing a total of N particles we construct a $6N$-dimensional phase space. A continuity argument proves the Liouville theorem, which states that the density in this space remains a constant when measured along phase-space trajectories provided all forces acting are Hamiltonian. It is stressed that interparticle forces, which are indeed Hamiltonian, enter on equal footing with external forces. The multi-dimensional space is very impractical, however, and a reduction to everyday's 6-dimensional phase space is requested. In the course of this reduction, problems pop up with the interparticle interactions. They turn out to be separable into a Liouvillian "space-charge" force, associated with the collective action on a given "test particle" of the rest of the gas, and scattering events. The collisions, which add a term to the Liouvillian part of the equation governing the evolution of the distribution function in 6-dimensional phase space, are responsible for relaxation towards equilibrium. Assuming dominance of soft multiple collisions, the Fokker-Planck equation emerges, and further introduction of a naive model for the encounters leads to determination of the friction and diffusion coefficients. To complete the general discussion, we evaluate rates of diffusion for the specific cases of spatially homogeneous plasmas having isotropic and collapsed Maxwellian velocity distributions, respectively.

---

* Permanent address

When we turn to the case of a charged particle beam, the question immediately appears what may be different from the general plasma case? When the average longitudinal motion is transformed away—by working in the "particle frame" moving along the storage ring at the nominal beam velocity—the answer is: very little. Scattering events appear as in the plasma, only the distribution function is given in a set of generalized coordinates (x_β, x'_β,...) other than the usual rectangular ones ($x, p_x, \ldots$). In the beam case we encounter couplings, for instance between longitudinal and horizontal motion. We shall outline practical calculations which account for such couplings. Among these we mention the treatment of Piwinski[1] with later improvements.[2] However, instead of going into mathematical details at this point, we choose to show some theoretical results obtained by means of existing computer codes. The numerically obtained rates help to single out where the aforementioned couplings are essential for the scattering processes and where not. Furthermore, the data bring us in a natural way towards a discussion of the possible existence of a set of reduced variables, which would allow for a compact representation of results over a wide range of initial conditions. The question appears, whether we may create some reasonably general and simple curves, which could provide quick and reliable estimates of beam blow-up rates—without the computer. The study of the plasma gives some hints.

The reader who is familiar with the proceedings of the various general and advanced accelerator physics courses held by the CERN Accelerator School (CAS) will note that the present contribution very nearly reproduces one appearing in the proceedings of the 1986 general accelerator physics course.[3] An essential difference, however, is that whereas the discussion in the earlier paper focussed on low-energy machines, the present paper includes new results and discussion for high-energy storage rings.

2 GENERAL DESCRIPTION

2.1 Liouville in 6N dimensions

Consider a non-relativistic gas consisting of N interacting (identical) charged particles. Most generally, such a system possesses $3N$ degrees of freedom. A canonical representation therefore requires $3N$ generalized coordinates, q_i, and $3N$ conjugate momenta, p_i. In $6N$-dimensional phase space ($q_i, \ldots, q_{3N}$, $p_1, \ldots, p_{3N}$), the full system is represented by a single point at a given time, Fig. 1a. Often, for q_i we shall take the rectangular coordinates (x, y, z) of the positions of the individual particles and for p_i the corresponding linear momenta. In this case the coordinates $3n-2$, $3n-1$, $3n$ of the phase-space point define the position of particle number n whereas the coordinates $3(N+n)-2$, $3(N+n)-1$, $3(N+n)$ fix the momentum of that particle. In general, the conjugate momenta are obtained from the Lagrangian, L, according to the relation

$$p_i = \partial L / \partial \dot{q}_i, \qquad L(q, \dot{q}, t) \equiv T - U. \tag{1}$$

Here the dot represents the total time derivative d/dt and T denotes the kinetic energy of the system. The quantity U is the potential energy. As an example, if we choose the q's as the spherical coordinates (r, θ, φ) to the individual positions of the particles, one third of the p_i's correspond to the radial momenta of the particles, the remaining p_i's reflect the φ- and θ-components of the angular momenta.

To determine the path followed in time by the phase-space point, Fig. 1a, we need some equations of motion. These may be expressed as differential relations involving the Lagrangian. Alternatively, one may introduce a Hamiltonian,

$$H(q,p,t) \equiv \sum_i p_i \dot{q}_i - L. \tag{2}$$

We recall that in case only time-independent conservative forces are acting, H is a constant of motion, the total energy. In general, the total force divides into a conservative, or Hamiltonian, part and a non-conservative contribution. The former corresponds to the part of the force which is derivable from a potential U, where

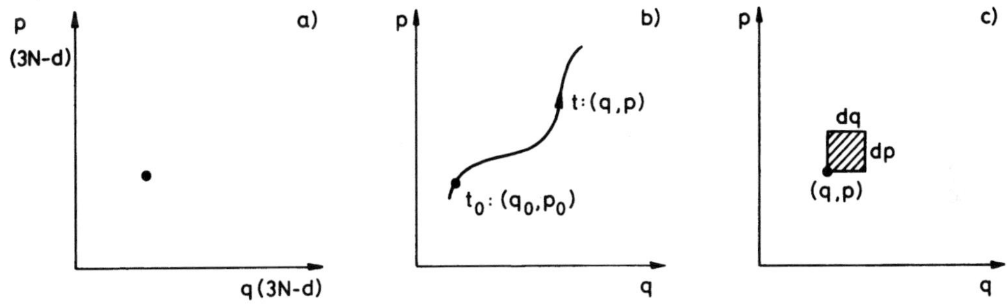

Figure 1: The $6N$-dimensional phase space.

$U = U(q,\dot{q},t)$ appears as a sum of the usual scalar potential and a term accounting for electromagnetic forces on moving charges.[4] The non-Hamiltonian part Q_i represents friction and inelastic processes. By means of the Hamiltonian, the $6N$ equations of motion then take the form

$$\dot{p}_i = -\frac{\partial H}{\partial q_i} + Q_i, \qquad \dot{q}_i = \frac{\partial H}{\partial p_i}. \tag{3}$$

Through specification of initial conditions, i.e., the $6N$ phase-space coordinates pertaining to the system at $t = 0$, the $6N$ first-order differential equations (3) fix in a unique manner the phase-space trajectory, Fig. 1b.

In practice, N is a large number and there is a lack of complete information on the initial conditions. Consequently, a statistical description is introduced. We define a phase-space distribution $\rho(q,p,t)$ such that

$$\rho(q,p,t)dqdp, \qquad dqdp \equiv \prod_i dq_i dp_i, \tag{4}$$

gives the probability for finding the particle system within a volume $dqdp$ near the point (q,p) at time t, Fig. 1c.

Liouville's theorem now states that: Provided solely Hamiltonian forces are acting, the phase-space density ρ remains unchanged along a phase-space trajectory, i.e., ρ is constant when measured at the variable position of a particle system moving through phase space, Fig. 2a. We may rephrase the theorem by stating that when $Q_i \equiv 0$, the volume enclosed by a given contour is conserved under the transformation (3), Fig. 2b.

In order to prove Liouville's theorem, consider a fixed domain Ω in phase space, Fig. 2c. Due to continuity, the speed, by which the total probability for finding the particle system within Ω changes, equals the total influx of probability per unit time through $S(\Omega)$, the surface of Ω,

$$\frac{d}{dt}\int_{\Omega}\rho\, dqdp = -\oint_{S(\Omega)}\vec{J}\cdot d\vec{S}, \qquad \vec{J} = \rho(\dot{q},\dot{p}). \tag{5}$$

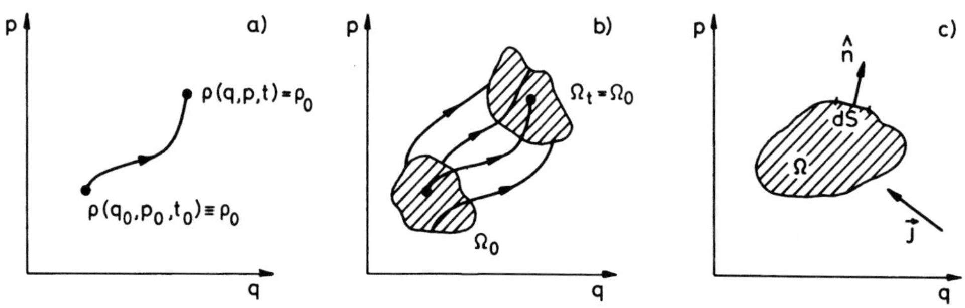

Figure 2: Liouville's theorem.

Here, $\vec{J}$ is the $6N$-dimensional probability current density and $d\vec{S}$ is an outward directed surface element, $d\vec{S} = \hat{n}dS$. Since Ω is fixed, the time derivative of ρ on the left-hand side may be taken directly as a partial derivative and by further means of the divergence theorem we may then rewrite Eq. (5) as

$$\int_{\Omega}\left[\frac{\partial\rho}{\partial t} + \sum_i\left(\frac{\partial}{\partial q_i}(\rho\dot{q}_i) + \frac{\partial}{\partial p_i}(\rho\dot{p}_i)\right)\right] dqdp = 0. \tag{6}$$

Since Eq. (6) holds for any choice of Ω, the quantity in square brackets itself vanishes identically. By differentiation of the products and collection of the terms which contain derivatives of ρ, we then obtain the relation

$$\frac{d\rho}{dt} = -\rho\sum_i\left(\frac{\partial\dot{q}_i}{\partial q_i} + \frac{\partial\dot{p}_i}{\partial p_i}\right). \tag{7}$$

When we now decide to follow the particle system in its motion through phase space, $\dot{p}_i$ and $\dot{q}_i$ are given by Eq. (3). Consequently, we arrive at the final result

$$\frac{d\rho}{dt} = -\rho\sum_i\frac{\partial Q_i}{\partial p_i}. \tag{8}$$

Evidently, if only Hamiltonian forces are acting, $Q_i \equiv 0$, the phase-space density remains constant, $d\rho/dt \equiv 0$. This is the Liouville theorem. The relation (8) provides a generalization to cases where also non-Hamiltonian forces are acting, $Q_i \neq 0$.

It is worthwhile stressing at this point that interparticle forces are Hamiltonian as they are derivable from potentials like

$$U_{jk} = \frac{(Ze)^2}{|\vec{r}_j - \vec{r}_k|}, \quad j \neq k, \tag{9}$$

where $\vec{r}_j$ denotes the position in direct space of particle number j (of charge Ze). Consequently, in the $6N$-dimensional phase space the introduction of interparticle interactions does not change the validity of the Liouville theorem as long as we do not consider inelastic processes like atomic excitation and ionization, charge transfer, or emission of radiation during collisions.

2.2 Reduction to 6 dimensions

The $6N$-dimensional phase space is clearly not very suitable for practical purposes. Therefore, it is customary to introduce an alternative description based on 6 dimensions only, three spatial ones and three momentum coordinates. Each single particle of the system is represented by a point, Fig. 3a, so instead of one point in a $6N$-dimensional space as in Fig. 1a we shall be working with N points in a 6-dimensional space. We introduce a particle density $f(\vec{q}, \vec{p})$ in such a way that

$$dN = f(\vec{q}, \vec{p}) d^3q d^3p \tag{10}$$

gives the number of particles expected within the volume $d^3q d^3p$ around the phase-space point $(\vec{q}, \vec{p})$. (Remember still that there is lack of complete information on initial conditions. The distribution function is chosen such as to reflect the best of our knowledge.) The basic question to be answered is then: What may be said about the evolution of f?

A formal reduction from $6N$ dimensions to 6 is described elsewhere.[5,6] We shall not embark on a discussion of this matter here. Instead, we shall be content by making a few simple observations. Consider first a situation where there is no inter-particle interaction. In this case, the single-particle Hamiltonian depends solely on $\vec{q}$ and $\vec{p}$—and not on the phase-space coordinates $\vec{q}\,'$ and $\vec{p}\,'$ of other particles. Hence, the individual particle trajectories are independent and determined by the single-particle Hamiltonian $H(\vec{q}, \vec{p})$ and, if active, by the non-conservative force $\vec{Q}(\vec{q}, \vec{p})$. We may therefore repeat the continuity argument, Eq. (5), with f substituted for ρ and a 6-dimensional current density of $\vec{j} = f \cdot (\dot{\vec{q}}, \dot{\vec{p}})$, Fig. 3b. In analogy to Eq. (8) we obtain as a result

$$\frac{df}{dt} = -f \operatorname{div}_{\vec{p}} \vec{Q} \tag{11}$$

i.e., a 6-dimensional version of Liouville's theorem holds for $\vec{Q} \equiv 0$.

Allow now for interactions between the N particles. In this situation the 6-d continuity argument goes wrong! Besides streaming smoothly through the surface $S(\Omega)$ of the phase-space volume according to the current density $\vec{j}$ (as determined by $H, \vec{Q}(\vec{q},\vec{p})$), Fig. 3b, there is the chance that a particle in its motion throughout phase space encounters closely another particle whereby it gets scattered, or kicked, into Ω, Fig. 3c. We may therefore write

$$\frac{df}{dt} = -f \operatorname{div}_{\vec{p}} \vec{Q} + \left(\frac{\partial f}{\partial t}\right)_c \tag{12}$$

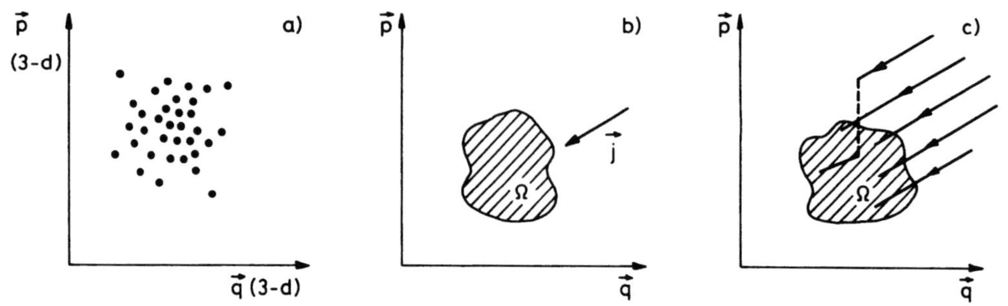

Figure 3: Distribution and motion in 6 dimensions.

where the last term on the right-hand side symbolizes the change in f due to collisions. The "soft" part of the interparticle interaction, the collective "space-charge" force due to all other particles, may be included in the single-particle Hamiltonian.[5,6,7]

In the special case $\vec{Q} \equiv 0$, Eq. (12) takes in $(\vec{r},\vec{v})$-space the form

$$\frac{\partial f}{\partial t} + \vec{v}\frac{\partial f}{\partial \vec{r}} + \frac{1}{m}\left(\vec{F}_{\text{ext}} + \vec{F}_{\text{sc}}\right)\frac{\partial f}{\partial \vec{v}} = \left(\frac{\partial f}{\partial t}\right)_c \tag{13}$$

with $\vec{v} \equiv \dot{\vec{r}}$, m being the particle mass and $\vec{F}_{\text{ext}}(\vec{r},\vec{v})$ denoting the external force. The space-charge force $\vec{F}_{\text{sc}}$ depends on the distribution function $f(\vec{r},\vec{v})$ and thereby the equation (13) is seen to be highly non-linear. However, we stress that a 6-d version of the Liouville theorem holds even with inclusion of this interaction term provided only that the collision term $(\partial f/\partial t)_c$ may be neglected.

To see the significance of the collision term, consider the simple example of a free, neutralized, spatially homogeneous gas: $\vec{F}_{\text{ext}} = \vec{F}_{\text{sc}} = \vec{Q} \equiv 0$ and $\partial f/\partial \vec{r} \equiv 0$. In this case Eq. (13) simply takes the form

$$\frac{\partial f}{\partial t} = \left(\frac{\partial f}{\partial t}\right)_c . \tag{14}$$

If collisions are absent, f clearly remains unchanged relative to its initial value—which could be anything that does not depend explicitly on time. On the other hand, if collisions are present, the collision term assures that the distribution function relaxes towards thermal equilibrium. Natural questions, which appear in this context, are: How fast is the relaxation and what is the final equilibrium?

2.3 Evaluation of collision term

In order to answer questions on the trend towards equilibrium, we need an expression for the collision term. Our strategy is to adopt some physical model for collisions which links $(\partial f/\partial t)_c$ to $f(\vec{r}, \vec{v})$.

Define $\psi(\vec{v}, \Delta\vec{v})$ as the probability that a particle of velocity $\vec{v}$ acquires an increment $\Delta\vec{v}$ (due to collisions) within a time interval Δt. The value of the distribution function f in the phase-space point $(\vec{r}, \vec{v})$ at time t may then be expressed in terms of the value of f in all points $(\vec{r}, \vec{v} - \Delta\vec{v})$ at the slightly earlier time $t - \Delta t$ as

$$f(\vec{r}, \vec{v}, t) = \int f(\vec{r}, \vec{v} - \Delta\vec{v},\, t - \Delta t)\psi(\vec{v} - \Delta\vec{v},\, \Delta\vec{v})d^3\Delta v. \tag{15}$$

We note that the function ψ is assumed to be time-independent, the process is assumed to be a so-called Markoff process. Upon expansion of Eq. (15) to second order, the collision term takes the form[6]

$$\begin{aligned}\left(\frac{\partial f}{\partial t}\right)_c &\equiv \frac{1}{\Delta t}\left(f(\vec{r}, \vec{v}, t) - f(\vec{r}, \vec{v}, t - \Delta t)\right) \\ &= \frac{1}{\Delta t}\left[-\sum_i \frac{\partial}{\partial v_i}(f\langle\Delta v_i\rangle) + \frac{1}{2}\sum_{ij}\frac{\partial^2}{\partial v_i \partial v_j}(f\langle\Delta v_i \Delta v_j\rangle)\right],\end{aligned} \tag{16}$$

with f evaluated in the phase-space point $(\vec{r}, \vec{v})$. The symbol $\langle\ \rangle$ indicates an average over increments $\Delta\vec{v}$ with the weight function $\psi(\vec{v}, \Delta\vec{v})$,

$$\langle\xi(\Delta\vec{v})\rangle \equiv \int \xi(\Delta\vec{v})\psi(\vec{v}, \Delta\vec{v})d^3\Delta v. \tag{17}$$

Consequently, $\langle\Delta\vec{v}\rangle$ denotes the average velocity increment due to collisions within the time interval Δt, i.e., the quantity

$$\frac{m}{\Delta t}\langle\Delta\vec{v}\rangle \equiv \vec{F}(\vec{v}) \tag{18a}$$

may be interpreted as a dynamical friction force experienced by a particle penetrating the gas (of other particles) at velocity $\vec{v}$. Similarly,

$$\frac{1}{\Delta t}\langle\Delta v_i \Delta v_j\rangle \equiv D_{ij}(\vec{v}) \tag{18b}$$

are diffusion coefficients. Substitution of Eqs. (16) and (18) into Eq. (13) leads to the Fokker-Planck equation.

In a plasma, collisions involve in general many particles. However, in the following we shall assume that such multiple collisions may simply be treated as sequences of binary encounters. We sometimes refer to this picture as a binary-encounter model. Consider therefore two identical particles of velocities $\vec{v}_1$ and $\vec{v}_2$, charge Ze and mass m. We may describe their collision in the centre-of-mass system as the scattering through some angle θ of a particle (number 2) of velocity $\vec{w} \equiv \vec{v}_2 - \vec{v}_1$ and mass $m/2$ on a fixed centre, Fig. 4. To determine the averages fixing the friction and diffusion coefficients,

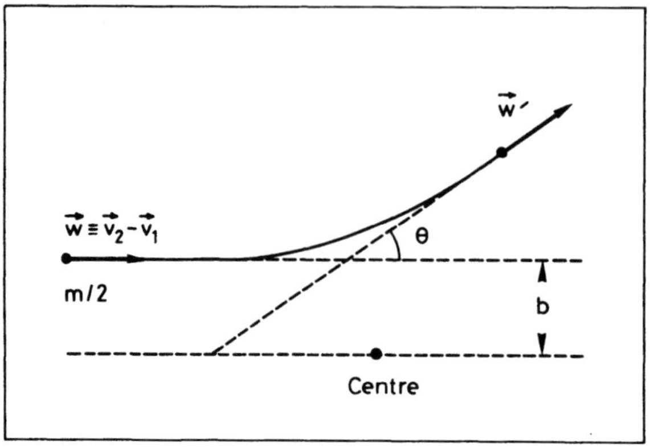

Figure 4: Diagram of binary collision.

Eq. (18), we have to average over all scattering events involving a given 'test-particle' velocity, $\vec{v} \equiv \vec{v}_1$. The influx of particles (number 2's) amounts to $f(\vec{v}_2) w d^3 v_2$ $(\mathrm{cm}^{-2}\mathrm{s}^{-1})$ and with a cross section $d\sigma(w,\theta)$ for scattering into an element $d\Omega$ of solid angle we may write

$$\frac{1}{\Delta t}\langle\xi\rangle = \int d^3v_2 d\Omega\, f(\vec{v}_2)\, w\, \frac{d\sigma(w,\theta)}{d\Omega}\,\xi. \tag{19}$$

Equations (18) then read

$$\begin{aligned} \vec{F}(\vec{v}) &= m \int d^3v_2 d\Omega\, f(\vec{v}_2)|\vec{v}_2 - \vec{v}|\frac{d\sigma(|\vec{v}_2 - \vec{v}|,\theta)}{d\Omega}\,\Delta\vec{v} \\ D_{ij}(\vec{v}) &= \int d^3v_2 d\Omega\, f(\vec{v}_2)|\vec{v}_2 - \vec{v}|\frac{d\sigma(|\vec{v}_2 - \vec{v}|,\theta)}{d\Omega}\,\Delta v_i \Delta v_j, \end{aligned} \tag{20}$$

where the change in test-particle velocity depends on scattering angle and initial velocities, $\Delta\vec{v} = \Delta\vec{v}(\theta, \vec{v}, \vec{v}_2)$.

To proceed, we need an expression for the differential scattering cross section. For pure Coulomb interaction, the Rutherford cross section applies, i.e.,

$$\frac{d\sigma}{d\Omega} = \frac{(Ze)^4}{m^2 w^4 \sin^4(\theta/2)} = \frac{d^2}{16\sin^4(\theta/2)}, \tag{21}$$

where $d \equiv 4(Ze)^2/mw^2$ denotes the collision diameter. Upon insertion into Eq. (20) and integration over scattering angles, the friction and diffusion coefficients take the form

$$\begin{aligned} \vec{F}(\vec{v}) &= \frac{8\pi(Ze)^4}{m} L \int d^3v_2\, f(\vec{v}_2)\,\frac{\vec{w}}{w^3}, \\ D_{ij}(\vec{v}) &= \frac{4\pi(Ze)^4}{m^2} L \int d^3v_2\, f(\vec{v}_2)\,\frac{w^2\delta_{ij} - w_i w_j}{w^3}. \end{aligned} \tag{22}$$

The "Coulomb logarithm" L is approximately given as $L \simeq \ln(2/\theta_{\min}) = \ln(2b_{\max}/d)$, where $\theta_{\min}$ denotes an effective minimum scattering angle corresponding to an effective maximum impact parameter $b_{\max}$, which may result from, for example, screening by other ions in the gas or from the spatial extension of the gas container. Typically, L attains values of the order of 10 and its variation with w is thereby very slow. In consequence, L has been taken outside the integrals in (22). We leave it as an exercise for the reader to prove the relations (22). Reference 6 or a similar textbook on plasma dynamics may be helpful.

2.4 Rates of diffusion

To estimate the rate of change of a non-equilibrium distribution $f(\vec{v})$ in its trend towards equilibrium, consider Eq. (16), which takes the form

$$\left(\frac{\partial f}{\partial t}\right)_c = -\frac{1}{m}\sum_i \frac{\partial}{\partial v_i}\left(f(\vec{v})F_i(\vec{v})\right) + \frac{1}{2}\sum_{ij}\frac{\partial^2}{\partial v_i \partial v_j}\left(f(\vec{v})D_{ij}(\vec{v})\right) \tag{23}$$

with the definitions (18). To characterize the distribution, let us compute various moments. The first moment defines the average value of the velocity $\vec{v}$, the second is related to the width of the distribution. To determine the n th moment, multiply Eq. (23) through by v_x^n, where x signifies a fixed but otherwise arbitrary direction. Under the assumption that f vanishes rapidly at infinity, integration over all velocities (partial over v_x) yields for $n = 1, 2$ the results

$$\frac{\partial}{\partial t}\overline{v_x} = \frac{1}{m}\overline{F_x},$$

$$\frac{\partial}{\partial t}\overline{v_x^2} = \frac{2}{m}\overline{F_x v_x} + \overline{D_{xx}}, \tag{24}$$

where the bar indicates an average over the distribution function f,

$$\overline{\zeta(\vec{v})} \equiv \int d^3v \, f(\vec{v})\, \zeta(\vec{v}). \tag{25}$$

Note that f enters both in the averaging in Eqs. (24) and in the determination of the friction and diffusion coefficients.

Let us now pose the question: How quickly do particles immersed into the gas randomize? Or, in other words, how fast may we expect such particles to lose orientation? If we choose particles of the same type as the gas particles and with comparable speed, the answer clearly will provide information on the trend of the distribution $f(\vec{v})$ towards isotropy. Therefore, consider a single "test particle" of velocity $\vec{u}$, i.e., with a distribution function

$$f_t(\vec{v}) = \delta(\vec{v} - \vec{u}), \tag{26}$$

penetrating a "background" gas characterized by the usual distribution $f(\vec{r}_2, \vec{v}_2)$. The development with time of the test-particle distribution is governed by Eqs. (24), where the averages now are taken over f_t, whereas $\vec{F}$ and D_{ij} are determined by f. The first of Eqs. (24) determines the slowing-down on average of the test particle towards the "centre-of-gravity" of the gas distribution f. The second determines the noise or fluctuations in test-particle velocity acquired during slowing down. Randomization is inherent in this diffusion equation. To estimate the rate of loss of test-particle orientation, consider the blow-up of the square of the test-particle velocity in directions transverse to $\vec{u}$. With the distribution (26) inserted into the second of Eqs. (24), we obtain

$$\frac{\partial}{\partial t}\overline{v_\perp^2} = D_{\alpha\alpha}(\vec{u}) + D_{\beta\beta}(\vec{u}), \tag{27}$$

where α and β denote mutually perpendicular directions transverse to $\vec{u}$.

By means of the result (27), let us now estimate the time constant τ for relaxation towards isotropy for a few specific distributions $f(\vec{v})$. For a spatially uniform gas of density n and isotropical Maxwellian velocity distribution, the rate at which a test particle randomizes assumes the value

$$\tau^{-1} \equiv u^{-2}\frac{\partial}{\partial t}\overline{v_\perp^2} \simeq \frac{4\pi n(Ze)^4}{m^2} L\,\Delta^{-3} \tag{28}$$

when u is chosen as the root-mean-square velocity Δ of the gas (in one dimension) and the diffusion coefficients (22) are used, Ref. 6 or a similar textbook may be consulted during derivation of the relation (28). Alternatively, through introduction of a measure μ for the phase-space density of the gas, we may express the relaxation rate as

$$\tau^{-1} = 4\pi(Ze)^4 m\, L\,\mu, \quad \mu \equiv \frac{n}{m^3\Delta^3}. \tag{29}$$

We prefer this representation when we later turn to the case of a beam. (Note that the actual phase-space density of the gas varies from point to point according to $f(\vec{v})$—the quantity μ introduced in Eq. (29) only reflects the scale.)

Another example is that of a spatially uniform gas with an anisotropical Maxwellian velocity distribution,

$$f(\vec{r},\vec{v}) = nf(\vec{v}) = \frac{n}{\pi\sqrt{2\pi}\Delta_\perp^2\Delta_\parallel}\, e^{-(v_x^2+v_y^2)/\Delta_\perp^2}\, e^{-v_z^2/2\Delta_\parallel^2}, \tag{30}$$

where we assume $\Delta_\parallel \ll \Delta_\perp$. To estimate the rate of relaxation towards isotropy of this distribution, take a test particle with a velocity in the disc of gas velocities, $\vec{u} = u\hat{x}$ with $u \simeq \Delta_\perp$, and consider the diffusion to z-velocity space. This is determined by the coefficient $D_{zz}(\vec{u}) = D_{zz}(u\hat{x})$. Clearly, the distribution (30) factorizes into a transverse and a longitudinal velocity distribution, $f(\vec{v}) = f_z(v_z)f_\perp(v_\perp)$. A relevant quantity to compute would therefore be the blow-up rate for the longitudinal distribution,

$$\tau^{-1} \equiv \Delta_\parallel^{-2}\frac{\partial}{\partial t}\overline{v_z^2} = \Delta_\parallel^{-2}D_{zz}(u\hat{x}). \tag{31}$$

With the distribution (30) inserted into Eq. (22) we obtain, upon integration over all gas velocities, the result

$$\tau^{-1} \simeq 4\pi(Ze)^4 m\, L\, \mu\, \frac{\Delta_\perp}{\Delta_\parallel}, \qquad \mu = \frac{n}{m^3 \Delta_\perp^2 \Delta_\parallel}, \tag{32}$$

for $u \sim \Delta_\perp$, cf. the Appendix. Again, μ provides a measure of the phase-space density of the gas.

With this example, we close the discussion of the general description of relaxation towards isotropy of a non-equilibrium plasma. We are ready for the beam case!

3 SCATTERING WITHIN BEAM

3.1 What is special?

At first sight, the beam looks very different from the gas discussed above. However, if we transform away the drift $\vec{v}_0$ by working in a reference frame, the so-called particle frame (PF), which moves along with the average beam particle, scattering processes appear exactly as in the gas case. In particular, the motion in PF is non-relativistic and the scattering processes classical (i.e., non-quantal). The only complication to be encountered is that while the scattering naturally is described in physical space, the distribution functions, whose change we are interested in estimating (emittances, etc.), are given in a different set of generalized coordinates. In the accelerator we encounter couplings between the motion in x, y and z directions.

To illustrate the importance of such couplings, let us consider the following simple example, which is due to Derbenev.[8] At a given instant, two beam particles move transversely to the beam axis in PF in the horizontal plane at equal but opposite momenta, $\pm p_x$. Then, they perform 90° collisions whereby they are turned parallel

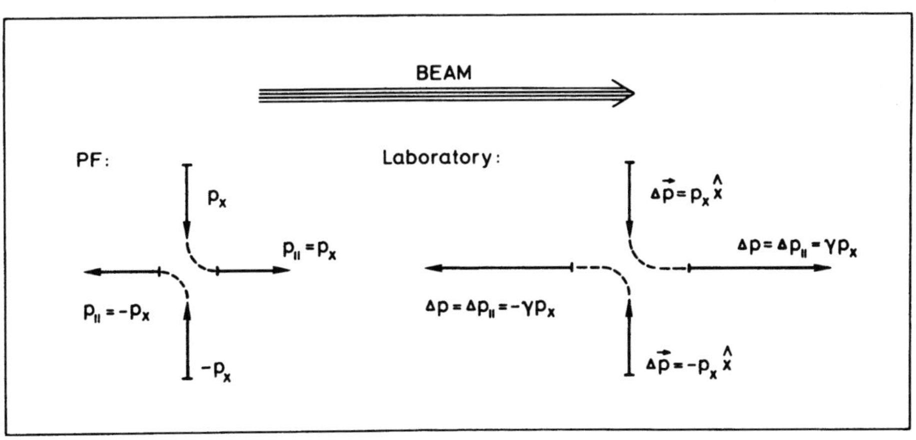

Figure 5: Sketch of Derbenev's collision of two beam particles.

to the beam axis, the momenta being conserved in magnitude. In the laboratory frame, the transverse components of the momenta of the two particles remain unchanged from their PF values. However, the gain in longitudinal momentum during the collision, $\Delta p_{\|}$, is magnified by a factor of γ, where $\gamma \equiv (1-v_0^2/c^2)^{-1/2}$ is related to the total relativistic energy E of the average beam particle as $E = \gamma mc^2$. The situation is sketched in Fig. 5, for the laboratory we have only shown the deviation in momentum from the average $\vec{p} \equiv \gamma m \vec{v}_0$. In PF, the sum of the momentum squares clearly is unchanged by the collision—the longitudinal blow-up is compensated by a transverse collapse. However, what happens to the single-particle emittances and $\Delta p/p$?

The collision illustrated in Fig. 5 leads to a growth in longitudinal phase space because, initially, the longitudinal momentum components vanished. Consider now the transverse phase space. For simplicity, we shall neglect derivatives of the lattice functions, $\beta' = D' \equiv 0$, whereby the horizontal (single-particle) emittance takes the form

$$\varepsilon_x = \frac{\pi}{\beta_x}\left[x_\beta^2 + \beta_x^2 x_\beta'^2\right], \quad x_\beta = x - D\frac{\Delta p}{p}. \tag{33}$$

The physical point of scattering x remains unchanged by the collision. In terms of the betatron amplitude x_β and its derivative x'_β the scattering event is mapped by Table 1. Here the sign on terms containing p_x varies according to the particle considered. Insertion into Eq. (33) leads to a change of the sum of emittances for the two particles of

$$\Delta(\varepsilon_{x,1} + \varepsilon_{x,2}) = 2\frac{\pi}{\beta_x}\left(\frac{p_x}{p}\right)^2 \left[(D\gamma)^2 - \beta_x^2\right]. \tag{34}$$

Evidently, the sign of the square-bracket factor in Eq. (34) decides whether the total emittance de- or increases. For $(D\gamma/\beta_x)^2 > 1$, we encounter simultaneous longitudinal and transverse growth!

Table 1: Map of x_β and x'_β in Derbenev's collision

Time	x_β	x'_β
Before	x	$\pm p_x/p$
After	$x \mp D\gamma p_x/p$	0

3.2 Practical IBS calculation

Descriptions of practical calculations of intrabeam scattering (IBS) may be found elsewhere.[1–2,9–11] The classical reference is Piwinski's paper.[1] Here, derivatives of the lattice functions are neglected, which strongly limits the applicability. Improvements of Piwinski's treatment to account for derivatives have been reported, for instance by Martini.[2]

In general, the procedure used in the IBS-calculations is quite similar to that for the plasma case, Section 2, only the formulas are more complicated. We may sketch a typical calculation as follows:

i) go to PF; here we have classical Rutherford scattering between pairs of identical particles;
ii) compute the change of single-particle emittances etc. (as defined in the laboratory frame) in a given binary (PF) collision;
iii) assume simple phase-space distribution functions (Gaussians);
iv) average over all collisions (scattering angles, incidence and exit momenta, place of scattering).

These are the ingredients—practical performance is 'only' a question of technique!

Various IBS computer codes are available on the market. Inputs are storage-ring lattice and beam (type and number of particles, momentum, emittances and $\Delta p/p$) and the output consists of heating rates for vertical and horizontal emittances and longitudinal momentum spread. We advocate Martini's version since this is written in a transparent way and appears well documented.[2,12]

3.3 Examples of IBS rates

Rather than dwell on technical details of the IBS calculations, let us play around a bit with some of the available computer codes. In doing so, the question immediately appears how to represent the data in a reasonably compact manner?

Let us for a moment assume that the velocity distribution of the beam in PF is a Maxwellian and the spatial density a constant. In this case the diffusion rate in PF is defined by Eq. (29). To transform to the laboratory frame, we note that the spatial density n and the longitudinal momentum spread $\Delta p_{\|}$ are magnified by a factor of γ, whereas the transverse component of the momentum $p_{\perp}$ is unaffected upon transformation. This leads to a conserved phase-space density,

$$\left.\begin{aligned} n^L &= \gamma n^{PF} \\ \Delta p_{\|}^L &= \gamma p_{\|}^{PF} \\ p_{\perp}^L &= p_{\perp}^{PF} \end{aligned}\right] \Longrightarrow \mu^L = \mu^{PF}. \tag{35a}$$

Accounting for time dilation

$$\tau^L = \gamma \tau^{PF}, \tag{35b}$$

we obtain for the Maxwellian PF distribution a rate of

$$\tau_L^{-1} = \gamma^{-1}\, 4\pi (Ze)^4 m\, \mu\, L. \tag{36}$$

To represent IBS data for a realistic beam, our first suggestion could therefore be to plot $\tau^{-1}\gamma/\mu$ or, rather, $(\sum \tau^{-1})\gamma/\mu$ as a function of emittance and longitudinal momentum spread. Here, the summation sign indicates the sum over blow-up rates for horizontal and vertical emittances and longitudinal momentum spread, $\sum \tau^{-1} \equiv \tau_H^{-1} + \tau_V^{-1} + \tau_p^{-1}$.

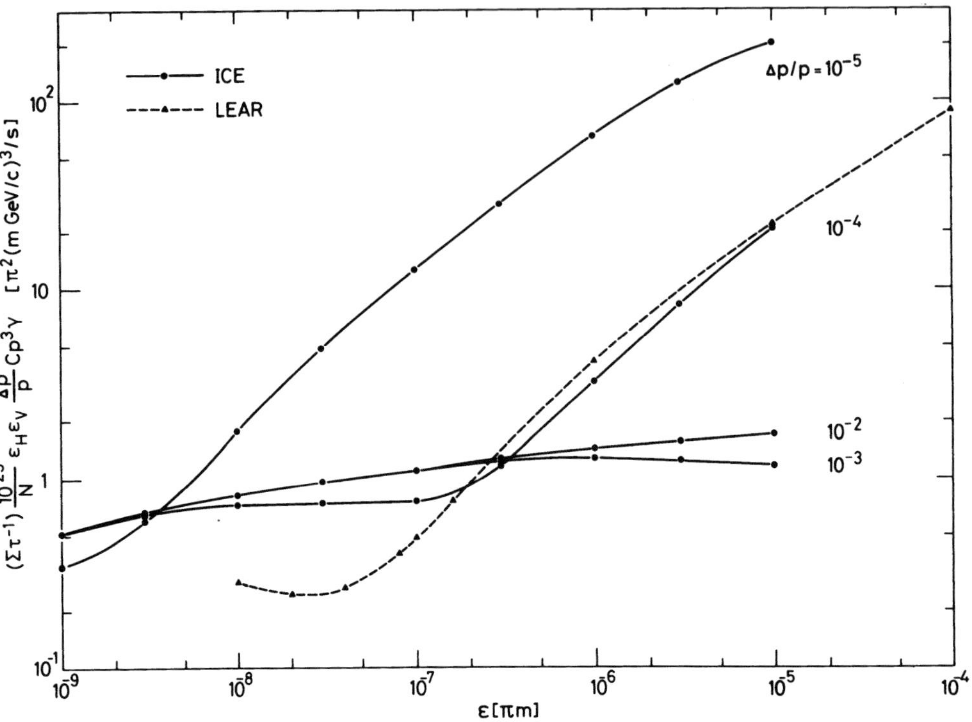

Figure 6: Intrabeam scattering for coasting proton beams in ICE and LEAR.

Figure 6 shows examples of such plots for coasting proton beams of equal horizontal and vertical emittances circulating in two different CERN storage rings, the former ICE-ring ($\gamma_t = 1.3$) at p=0.3GeV/c and LEAR ($\gamma_t^2 = -(14.5)^2$) at p=0.6GeV/c. For ICE we have used Martini's code, for LEAR a code written by Möhl.[13] Corrections have not been made for the slightly different definitions of beam emittances and $\Delta p/p$ (relations to widths of Gaussians) applied in the two programs. For the measure μ of the phase-space density entering the ordinate in Fig. 6, we have substituted the expression

$$\mu^{-1} = N^{-1}p^3\varepsilon_H\varepsilon_V C\Delta p/p, \tag{37}$$

where N denotes the total number of beam particles and C is the circumference of the ring in question; our unit of emittance is πmrad, of momentum GeV/c and of length and time m and sec, respectively. Note that only the points are calculated—curves are drawn to 'guide the eye'.

Let us stress some of the main observations to be made from Fig. 6: First of all, we note that there is always growth, i.e., the ordinate in the figure is always positive (the individual blow-up or 'heating' rates may be negative as well as positive). This implies that, in general, the beam never reaches an equilibrium. Secondly, we encounter for any fixed value of $\Delta p/p$ a smooth behaviour, that is, a slow increase in the normalized rate with increasing emittance up to a certain point where a more rapid dependence sets in. To the left of the point (the flat portion of the curve) the blow-up rate is dominated by

τ_H^{-1}, to the right (steep portion of the curve) τ_p^{-1} dominates. Throughout, τ_V^{-1} plays a minor rôle. Thirdly, the break-up point is positioned exactly in the region where coupling in the horizontal plane becomes important. Since the (Gaussian) width of the beam in this direction is composed of two contributions, one, $\sigma_{H\beta}$, which is due to betatron oscillation and another, $D\sigma_p$, which appears because of the coupling via the dispersion to the momentum spread σ_p, that is, since

$$\sigma_H^2 = \sigma_{H\beta}^2 + D^2\sigma_p^2, \tag{38}$$

coupling becomes important for $\sigma_{H\beta}^2 \sim D^2\sigma_p^2$ or, equivalently, for

$$\varepsilon_H \sim \left\langle \frac{D^2}{\beta_H} \right\rangle \left(\frac{\Delta p}{p}\right)^2 . \tag{39}$$

A typical value for the average $\langle\ \rangle$ around the storage ring of D^2/β_H could be in the range 1-10m. Finally, the patterns observed for different machines are similar.

3.4 Representation of IBS-data by reduced variables

The study of Fig. 6 immediately suggests further condensation of the IBS data by plotting the curves as functions of the reduced variable $\varepsilon/(\Delta p/p)^2$, cf. Eq. (39). The only question which remains is then what the optimal choice for the second axis may be. To the right of the break-up (or, perhaps, "break-even") point, couplings between the three planes of motion may be neglected. In the same region, the beam has collapsed longitudinally in momentum space, in PF even with an additional factor of γ. Consequently, the steep portion of the curve corresponds to a PF-rate similar to (32). Transformation to the laboratory frame proceeds via Eqs. (35). Since the phase-space density is conserved, while the γ-factors from time dilation and longitudinal momentum spread cancel, we obtain

$$\tau_L^{-1} \simeq 4\pi(Ze)^4 m\mu L \left[\frac{\langle p_\perp^2\rangle}{\langle \Delta p_\parallel^2\rangle}\right]_L^{1/2} \propto Z^4 m\mu L \frac{\sqrt{\varepsilon}}{\Delta p/p} \quad , \tag{40}$$

which is independent of γ.

Biased by the result (40), we show in Fig. 7 a plot of $(\sum \tau^{-1})\mu^{-1}\langle L\rangle^{-1}$ as a function of $\varepsilon/(\Delta p/p)^2$ for a coasting proton beam circulating in ICE. The conditions are the same as in Fig. 6 and for μ we have again substituted the measure (37). Through the points computed for $\Delta p/p = 10^{-4}$ we have drawn a curve to guide the eye. To our great satisfaction we observe that the normalized rates corresponding to the three remaining ICE curves of Fig. 6 fall very close to the 10^{-4} curve. For the storage ring (and beam energy) given, the intrabeam scattering may to a very good approximation be represented by a single universal curve! It may be noted that to obtain a single curve with as high accuracy as in Fig. 7, it is essential to include in the ordinate the average $\langle L\rangle$ around the ring of the Coulomb logarithm.

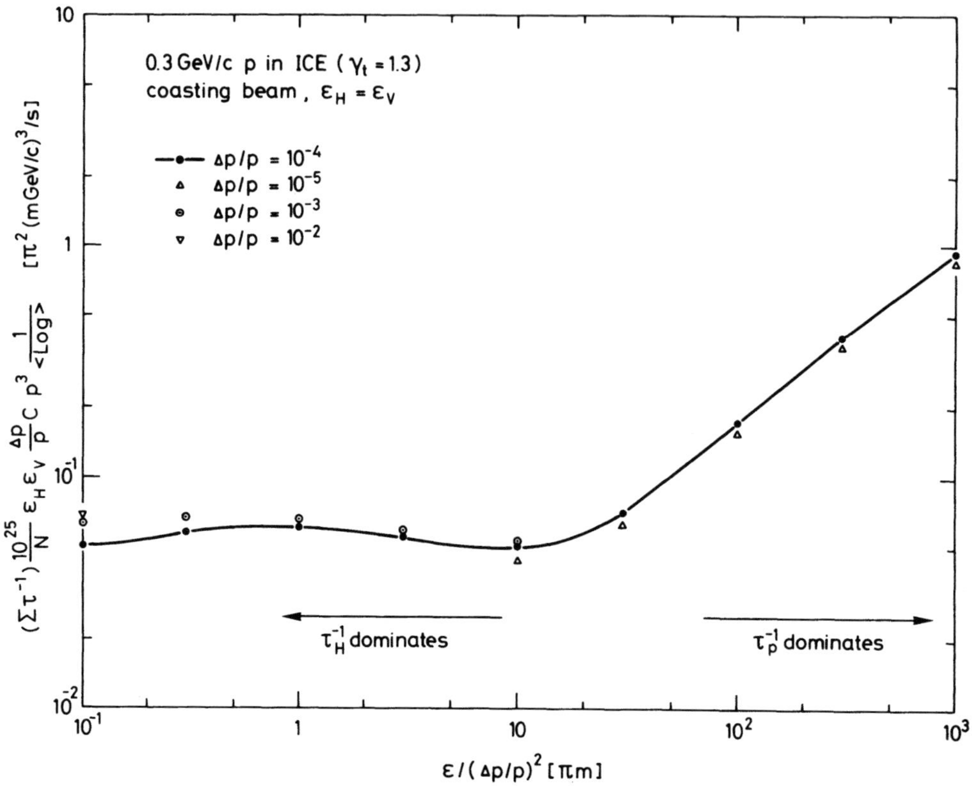

Figure 7: IBS rates for ICE in reduced variables.

Scaling to particles of different charge and mass is straightforward, cf. expressions like (36) and (40). By inclusion of an extra factor $Z^{-4}m_p/m$ in the legend for the ordinate, Fig. 7 is valid as it stands for beams of (bare) ions of charge Ze and mass m (m_p denotes here the proton mass).

3.5 High energies

The numerical results presented in Figs. 6 and 7 were obtained for non-relativistic coasting beams. What happens if the energy is raised into the relativistic regime? Figure 8 displays results of IBS calculations (with Martini's code) for coasting proton beams circulating in RHIC, the Relativistic Heavy Ion Collider, at $\gamma = 5$ and $\gamma = 100$. Reduced variables are used in the plot exactly as in Fig. 7. The curves correspond to a momentum spread of $\Delta p/p = 10^{-3}$. A few points corresponding to $\Delta p/p = 10^{-4}$ have also been included and we immediately note that for given value of γ, IBS may be represented by a single curve—quite like the nonrelativistic results, Fig. 7, all fell on a single curve. (By inclusion of the factor $Z^{-4}m_p/m$ mentioned at the end of section 3.4, gold ions have proven to yield reduced IBS rates reproducing the proton rates to within a few per cent. The curves may hence be considered valid for any beam of bare ions.) However, it is also very apparent that the two curves in Fig. 8 are somewhat different:

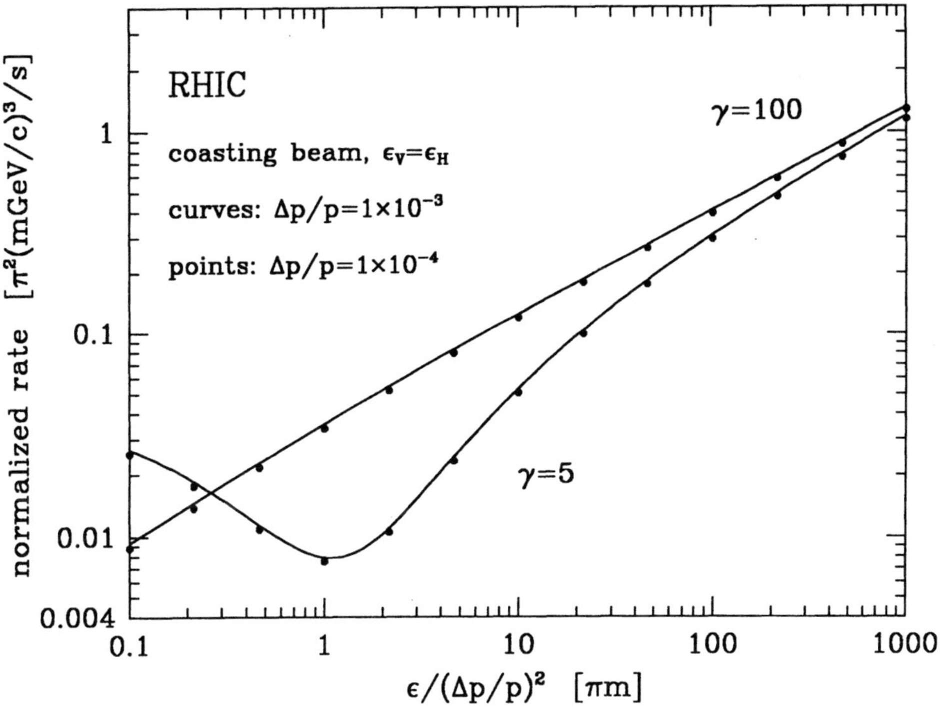

Figure 8: IBS rates for RHIC in reduced variables.

The $\gamma = 5$ curve qualitatively has the same shape as the curve in Fig. 7 (only the break-up point now appears as a minimum and at a lower value of $\varepsilon/(\Delta p/p)^2$ than before). The $\gamma = 100$ curve, on the other hand, is essentially a straight line corresponding everywhere to the asymptote which the low-energy curve approaches to the right, that is, at high values of $\varepsilon/(\Delta p/p)^2$. For the high-energy curve, heating in the longitudinal direction dominates throughout. In the low-energy case longitudinal blow-up dominates to the right of the minimum point (in agreement with the non-relativistic results, Fig. 7) whereas transverse blow-up dominates to the left (and here horizontal and vertical heating rates are about equal).

The high-energy curve ($\gamma = 100$) in Fig. 8 is particularly simple; the reduced rate increases by two orders of magnitude upon an increase in $\varepsilon/(\Delta p/p)^2$ by four orders of magnitude. In other words, the reduced rate grows in proportion to $\sqrt{\varepsilon/(\Delta p/p)^2}$. This is exactly the result (40) corresponding to the longitudinal blow-up rate for a longitudinally collapsed velocity distribution. Hence, at very high energies IBS may be represented by a single universal curve valid for *any* machine, namely the curve labelled $\gamma = 100$ in Fig. 8.

At this point we are obviously curious to know why we observe a variation with γ in the reduced IBS rates. Under suspicion falls the γ-dependence of the degree of collapse of the PF-velocity distribution longitudinally (for fixed value of $\varepsilon/(\Delta p/p)^2$). With momentum transformations according to (35a), the longitudinal and transverse

PF velocity spreads amount to

$$\Delta v_{\parallel}^{PF} = \frac{1}{\gamma m}\Delta p, \qquad \Delta v_x^{PF} \simeq \frac{1}{m}\sqrt{\frac{\varepsilon}{\pi\beta_x}}p. \tag{41}$$

These spreads are equal for

$$\frac{\varepsilon}{(\Delta p/p)^2} \simeq \frac{\pi\beta_x}{\gamma^2} \tag{42}$$

(where the right-hand side should be understood as some proper average taken around the ring). The important observation to make here is that the value of the reduced variable $\varepsilon/(\Delta p/p)^2$ where isotropy in PF velocity space occurs scales as γ^{-2}. Note that a simple comparison of the rates (36) and (40) would yield the same result.

Figure 9a shows a plot of the position of the IBS minimum point for coasting proton beams in RHIC as a function of γ. The momentum spread is chosen as $\Delta p/p = 10^{-4}$, horizontal and vertical emittances are identical and the value of $\varepsilon/(\Delta p/p)^2$ at minimum has been multiplied by γ^2. For γ-values in excess of 40, the maximum in Fig. 9, minimum occurs at $\varepsilon/(\Delta p/p)^2 < 0.1\pi$m; high-energy beams essentially always appear collapsed in PF velocity space. As is obvious from Fig. 9a, the position of the

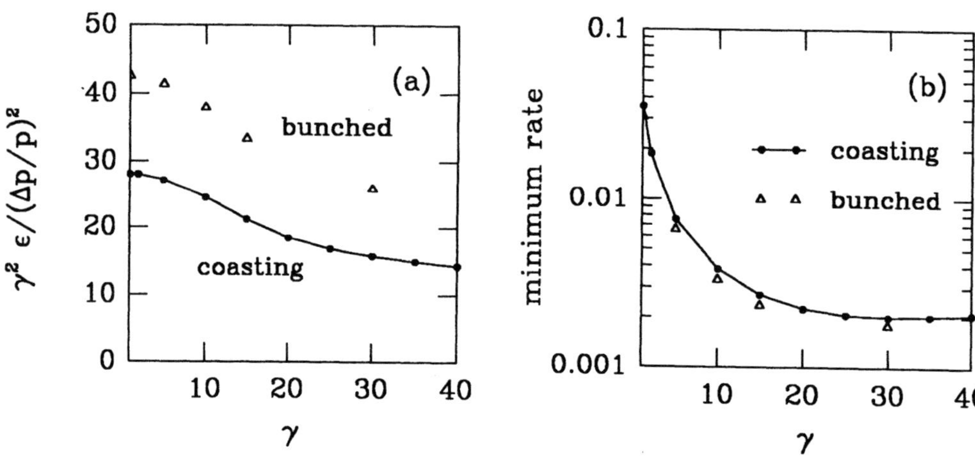

Figure 9: Coordinates of minimum in reduced-variable IBS for RHIC.

minimum scales roughly as γ^{-2}. Hence collapse of the velocity distribution, Eq. (42), is much more important than coupling, Eq. (39), in defining minimum (or "break-up"). To the right of the γ-dependent minimum (or "break-up" point), IBS follows roughly the machine-independent longitudinal blow-up formula (40), which defines the asymptote. To the left, transverse blow-up and couplings dominate and results are machine dependent (although the shape and absolute values of the IBS curves encountered for different storage rings do not differ wildly).

Figure 9b displays the actual minimum value of the normalized rate for RHIC as a function of γ for the same beams as in Fig. 9a.

3.6 Bunched beams

Intrabeam scattering in bunched beams appears much the same as IBS in coasting beams. Let us here briefly report on results obtained with Martini's code for bunched RHIC beams (with $\varepsilon_H = \varepsilon_V$): For given γ-value, the IBS rates are quite close to the coasting beam rates of Fig. 8 when presented in reduced variables. When the ring circumference used in the coasting beam case, Fig. 7, is replaced by the bunch length multiplied by $2\sqrt{\pi}$, there is exact agreement between coasting and bunched IBS to the (far) left of the minimum point. To the (far) right of the minimum, the reduced IBS rate for a bunched beam equals (with the same substitution) exactly half of that for the coasting beam. This difference is reflected in an appearance of the minimum at a higher value of $\varepsilon/(\Delta p/p)^2$ for bunched compared to coasting beams. Minimum bunched beam results are indicated in Fig. 9.

4. CONCLUDING REMARKS

Through the previous sections we have intended to give a transparent introduction to the phenomenon of intrabeam scattering. We have emphasized the close analogy between IBS and the trend towards isotropy in velocity space of a plasma and, at the same time, stressed the differences which are due to the couplings of the degrees of freedom in the accelerator. Rather than going through all the technical details of IBS calculations, we have chosen to show results of numerically computed rates. A study of such data, combined with the general considerations for the plasma case, led to the production of a strongly reduced representation of IBS rates: For a given lattice and given beam energy, a single curve may be created, which allows for a quick read-off of heating rates in any situation, i.e., for any type of ion beam and any number of beam particles, momentum spread, emittance and, if applicable, bunch length. Furthermore, the curves corresponding to even very different machines (as far as tested) are qualitatively the same at given energy—even the absolute values are within factors of, typically, 2 to 4. At ultra-relativistic energies, $\gamma \gtrsim 100$, IBS may be represented by a single γ-independent universal curve valid for any storage ring.

Finally, let us mention that a few experimental measurements of IBS heating rates exist. Experimental data for the CERN Antiproton Accumulator have been reported to agree with the results obtained by the computer codes, cf. Refs. 2 and 9, when integration is performed over finite (relatively long) time intervals. (Similar integrations are performed in Ref. 11 for RHIC.) For the low-energy ring ASTRID in Aarhus, comparison between experimental data on heating rates and the simple analytical expression obtained above for a collapsed Maxwellian velocity distribution has been successful, as far as I understand [14].

ACKNOWLEDGEMENT

The author wishes to thank G. Parzen for promptly providing a print-out of the RHIC lattice parameters shortly before my departure for Hilton Head Island.

REFERENCES

1) A. Piwinski, *Proc. CERN Accelerator School on General Accelerator Physics*, Gif-sur-Yvette, Paris, 1984 (CERN 85-19, CERN, Geneva, 1985) 451-462; this contribution is essentially identical to the original paper published in *Proc. 9th Int.Conf. on High Energy Accelerators*, 1974.

2) M. Martini, CERN PS/84-9(AA) (1984).

3) A. H. Sørensen, *Proc. CERN Accelerator School, Second General Accelerator Physics Course*, Aarhus, Denmark, 1986 (CERN 87-10, CERN, Geneva, 1987; Ed. S. Turner) 135.

4) H. Goldstein, *Classical Mechanics* (Addison-Wesley, Reading MA, 1957).

5) A.J. Lichtenberg, *Phase-Space Dynamics of Particles* (Wiley, New York, 1969).

6) T.J.M. Boyd and J.J. Sanderson, *Plasma Dynamics* (Nelson, London, 1969).

7) A. H. Sørensen, *Proc. CERN Accelerator School, Third General Accelerator Physics Course*, Salamanca, Spain, 1988 (CERN 89-05, CERN, Geneva, 1989; Ed. S. Turner) 18.

8) Ya.S. Derbenev, Fermilab $\bar{p}$-Note 176 (1981).

9) M. Conte and M. Martini, *Part.Acc.* **17** (1985) 1.

10) J.D. Bjorken and S.K. Mtingwa, *Part.Acc.* **13** (1983) 115.

11) G. Parzen, *Nucl. Instrm. Methods* **A251** (1986) 220; **A256** (1987) 231.

12) M. Martini, PS/AA/Note 84-7 (1984).

13) D. Möhl (unpublished).

14) M. Kristensen et al., private communication

15) M. Abramowitz and I.A. Stegun, *Handbook of Mathematical Functions* (Dover, New York, 1972).

APPENDIX

In order to derive the expression (32) for the blow-up rate in the axial direction of a flattened Maxwellian velocity distribution, Eq. (30), we need an expression for D_{zz}, cf. Eq. (31). With a transverse test-particle velocity, $\vec{u} = u\hat{x}$, this component of the diffusion tensor reads

$$D_{zz} = \frac{4\pi n(Ze)^4}{m^2} L \int d^3v\, f(\vec{v})\, \frac{w^2 - w_z^2}{w^3}, \quad \vec{w} = \vec{v} - u\hat{x}, \tag{A1}$$

according to Eq. (22). A change of integration variable gives

$$D_{zz} = \frac{4n(Ze)^4}{\sqrt{2\pi} m^2 \Delta_\perp^2 \Delta_\parallel} L \int d^2w_\perp dw_z \frac{w_\perp^2}{w^3}\, e^{-(\vec{w}_\perp + u\hat{x})^2/\Delta_\perp^2}\, e^{-w_z^2/2\Delta_\parallel^2}, \tag{A2}$$

where $\vec{w}_\perp$ and $w_z\hat{z}$ denote the components of the relative velocity transverse and parallel to the axis, respectively. When the velocity distribution is assumed excessively flattened, $\Delta_\parallel \ll \Delta_\perp$, the factor w^{-3} may be replaced by $w_\perp^{-3}$. The diffusion coefficient (A2) then reduces to

$$D_{zz} = \frac{4n(Ze)^4}{m^2\Delta_\perp^2} L \int_0^\infty dw_\perp\, e^{-(w_\perp^2 + u^2)/\Delta_\perp^2} \int_0^{2\pi} d\varphi\, e^{-2w_\perp u \cos\varphi/\Delta_\perp^2}. \tag{A3}$$

Integration over the azimuthal angle φ yields the modified Bessel function I_0,

$$D_{zz} = \frac{8\pi n(Ze)^4}{m^2\Delta_\perp^2} L \int_0^\infty dw_\perp\, e^{-(w_\perp^2 + u^2)/\Delta_\perp^2}\, I_0\left(\frac{2w_\perp u}{\Delta_\perp^2}\right), \tag{A4}$$

cf. Ref. 15. In turn, the $w_\perp$ integration may be performed to give[15]

$$D_{zz} = \frac{4\pi n(Ze)^4}{m^2\Delta_\perp} L \sqrt{\pi}\, e^{-u^2/2\Delta_\perp^2}\, I_0\left(\frac{u^2}{2\Delta_\perp^2}\right). \tag{A5}$$

The function $e^{-x}I_0(x)$ decreases slowly with x; at $x = 0$ it equals 1, for $x \gg 1$ it approaches $(2\pi x)^{-1/2}$ and for $x \approx \frac{1}{2}$ we have $\sqrt{\pi}\, e^{-x}I_0(x) \simeq 1$ (Ref. 15). In consequence, we may write

$$D_{zz}(u\hat{x}) \simeq \frac{4\pi n(Ze)^4}{m^2\Delta_\perp} L, \qquad u \simeq \Delta_\perp. \tag{A6}$$

Insertion of this result into Eq. (31) leads to the rate (32).

IONS AND NEUTRALIZATION

A. PONCET

CERN, Geneva, Switzerland

ABSTRACT

After a short presentation of intensity limitations examples due to trapped ions, the processes of ionization and neutralization build up in particle accelerators and storage rings are briefly reviewed. The tolerable limits in neutralization are then assessed at the light of current theories of incoherent and coherent effects driven by ions. Finally the usual antidotes such as clearing electrodes, missing bunch schemes and beam shaking are presented.

1. INTRODUCTION

Charged particles created by ionization of the residual gas (ions, electrons) or dust (charged matter) by the circulating particles may be trapped in the beam's space charge potential, where they can cause beam loss or simply emittance blow-up. Small electron storage rings (bunched) and antiproton accumulators (coasting) are extremely sensitive to this phenomenon (neutralization by ions). Proton storage rings with coasting beams (neutralization by electrons) such as the Bevatron and the ISR are also affected, but to a lesser extent. Large machines with bunched beams such as proton-antiproton (SPS-Tevatron) or electron-positron colliders (LEP) are not subject to ion trapping, as ions are either naturally unstable or have enough time to escape to the vacuum chamber wall in between bunch passages. Ions interact with the beam particles mainly via the Coulomb force : individually they add up to the residual gas density, thus enhancing single-particle loss through Bremsstrahlung and single Coulomb scattering, and emittance blow-up through multiple scattering at small angles. Macroscopically, due to their highly non-uniform longitudinal and transverse density, *their collective* space charge may represent non-linear optical defects which add to those of the machine magnet structure, thus enhancing the individual particle's incoherent motion (ion induced non-linear resonances), causing beam loss and emittance blow-up. Last but not least, transverse collective coherent motion between the ion stream (or neutralization pockets) and the beam may exist, leading to emittance blow up and eventually beam loss.

Figures 1 and 2 below illustrate ion-induced intensity-limiting phenomena, as experienced in the CERN EPA (Electron Positron Accumulator - 600 MeV) and AA (Antiproton Accumulator - 3.5 GeV/c) machines.

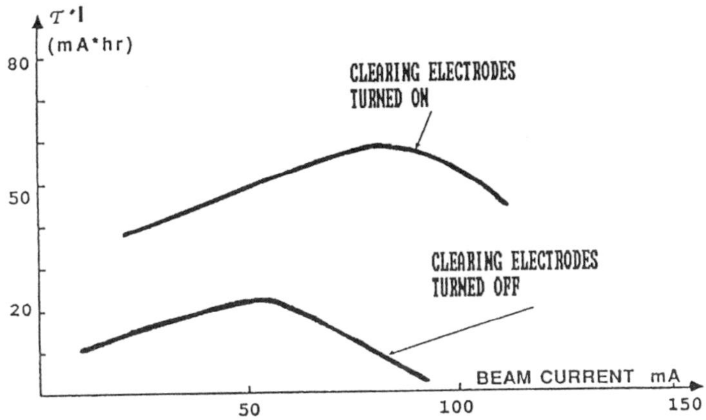

Figure 1 : Beam lifetime τ in EPA as a function of beam current, with and without ion trapping [1].

Ion neutralization of beams and its related effects have been the object of many papers since the late 1960's. A rather complete and useful list of references can be found in two previous CAS proceedings where general reviews of the subject are presented [2-3]. In this present paper the accent will be put on neutralization effects such as ion-induced coherent instabilities, and beam shaking as a clearing mechanism. But in order to be self-consistent the main aspects of ion trapping will first be briefly recalled.

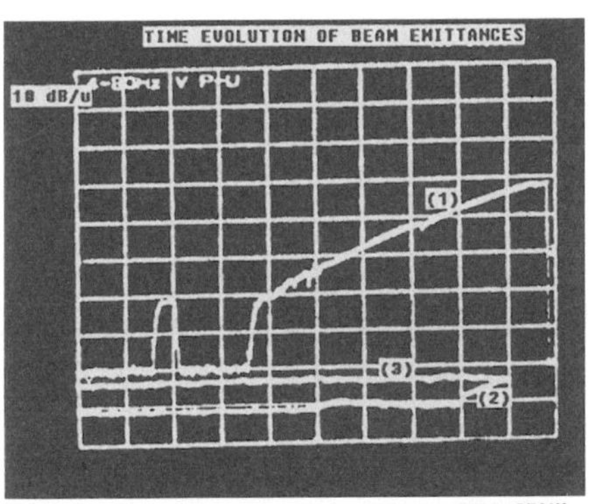

TRACE 1 : HORIZONTAL EMITTANCE (10 sec FULL SPAN)
TRACE 2 : " IDEM " (10 mn FULL SPAN)
TRACE 3 : VERTICAL EMITTANCE (10 mn FULL SPAN)

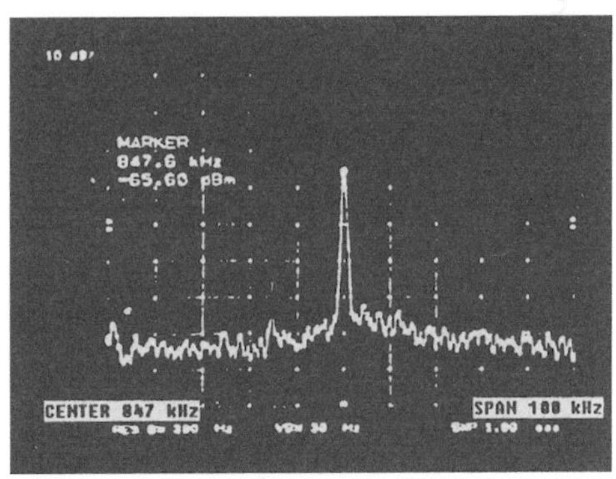

FREQUENCY SPECTRUM AROUND 847 kHz SHOWING THE COHERENT QUADRUPOLAR SIGNAL OF THE $(5-2Q_h)f_r$ MODE

Figure 2 : Left, emittance blow up in the CERN AA [4] caused by ion induced transverse coherent quadrupole instabilities. Right, the signal observed with a quadrupolar pickup electrode at the frequency of the unstable mode.

2. ION PRODUCTION AND CONDITIONS OF TRAPPING

Trapping occurs when electrons or positive ions resulting from a Coulomb collision between a beam particle and a residual gas molecule do not have enough kinetic energy to escape the beam space-charge potential.

Ionisation of a gas atom is the result of an energy transfer ΔE (b) larger than the binding energy (or ionisation potential) of one of its orbital electrons in an elastic collision event with a beam particle [3] :

$$\Delta E\,(b) = \frac{(Ze)^2\ (ze)^2}{8\,\pi^2\,\varepsilon_o{}^2\,m\,c^2\,\beta^2\,b^2} > E_I \tag{1}$$

where e is the elementary charge, Z and z are the charge numbers of the incident and of the target particles, respectively, c β is the velocity of the incident particle, b the impact parameter (the minimum distance of the two particles during the collision) and E_I the ionisation energy $\cong z\ 13.5$ (eV) [5]. The mass of the target particle m is ~ 2 m_p A for an atom of mass number A, m_e for an electron (m_p is the proton mass).

The above expression, only approximately valid for the useful range of impact parameters defined by the radii of the atom and its nucleus, can be used to show that the transfered energy to an electron and its mucleus is in the ratio :

$$\frac{2\,m_p}{z\,m_e} = \frac{3672}{z} >> 1 \tag{2}$$

thus establishing the facts that beam energy loss is mainly on the electrons of the gas atoms, and that positive ions are essentially created at rest, i.e. with no or negligible kinetic energy resulting from the ionizing collision [3].

The rate at which one individual particle in the beam ionizes an atom is given by :

$$R_p = c\,\beta\,\Sigma\,n_j\,\sigma_j \tag{3}$$

where n_j and σ_j are respectively the densities and ionization cross sections of molecules of gas species j. The cross sections depend on the molecule and on the beam relativistic speed by the Bethe formula [6] :

$$\sigma_j = 4\,\pi \left(\frac{\bar{h}}{mc}\right)^2 (M^2 x_1 + C\,\beta^2) \tag{4}$$

where $x_1 = \beta^2 \ln(\gamma^2\text{-}1)$ and M^2 and C are characteristics of the molecules and $\bar{h}$ is Planck's constant. As a typical example, the values for EPA are shown in Table 1.

Distant collisions with large impact parameters — much more probable than close ones leading to ionisation — are important since they feed energy to the ions. In some circumstances (neutralization pockets) this may be a natural clearing mechanism, i.e. when the trapped species receive enough energy to escape the beam potential U. This natural clearing rate can be calculated from the transfered energy [3] :

$$R_c = \frac{2\pi\, m_o\, c^3\, r_o^{\,2}}{\beta}\,\frac{N_p\, z}{e\,U}\,\log_n\left(3.10^4 z^{-2/3}\right) \tag{5}$$

with m_o and r_o being m_e, r_e if the trapped species is an electron, m_p, r_p for a proton and $2m_p$, r_p for an ion of charge number z. However the process is slow compared to typical ionization rates and Table 1 gives clearing rates for the EPA machine.

Table 1

EPA ionization cross sections and clearing rates (600 MeV, electrons with $\gamma = 1200$, $P = 10^{-9}$ mbar)

Gas	$\sigma_j(m^2)$	$R_p\ (s^{-1})$	Ionisation time (s)	$R_c(s^{-1})$	Clearing time (s)
H_2 (~50%)	0.4×10^{-22}	0.4	2.5	6.10^{-3}	166
CO (~50%)	1.5×10^{-22}	1.5	0.7	0.04	25

The machine average neutralization is usually defined as

$$\eta = \frac{N_i}{N_p} = \frac{R_p}{R_c} \leq 1 \tag{6}$$

where N_i and N_p are respectively the total number of ions and beam particles. A local value of the neutralization can be defined as

$$\hat{\eta} = \frac{n_i}{n_p}$$

where n relates to each species' density. An upper limit for η ($\hat{\eta}$) is obviously 1 since larger values would reverse the beam-plus-ion space-charge potential, thus repelling ions to the wall.

Only in places where an external magnetic field confines excess ions (i.e. if no clearing), which is possible in quadrupoles [3] for instance, could $\hat{\eta}$ be perhaps larger than one.

In the vast majority of cases (electron storage rings with typical pressures of 10^{-9} mbar and ionisation times of a second or less), ionization is the dominant effect in the absence of any clearing mechanism. Due to successive ionizations of already trapped ions, the neutralization builds up exponentially in time and reaches its maximum value in less than the primary ionization time. In the limit of full neutralization, two mechanisms contribute to light-ion filling of the beam :

1) multiple ionization which can be shown to limit an ion species to a density equal to that of the residual gas one [3]; 2) heating by the beam, more pronounced for high-z species, becoming more and more efficient as the beam potential smears out when neutralization builds up [6].

If ion clearing is applied via electrodes, the maximum ion lifetimes are of the order of a ms or less, such that single-charge ions predominate [H_2^+, CO^+, CO_2^+...]. This can be globally confirmed by measuring clearing currents on electrodes [7]. Figure 3 shows such a measurement for the AA. The total ion current is compatible with primary ionization events, using calculated cross sections (Eq. 4) and measured gas composition and pressure.

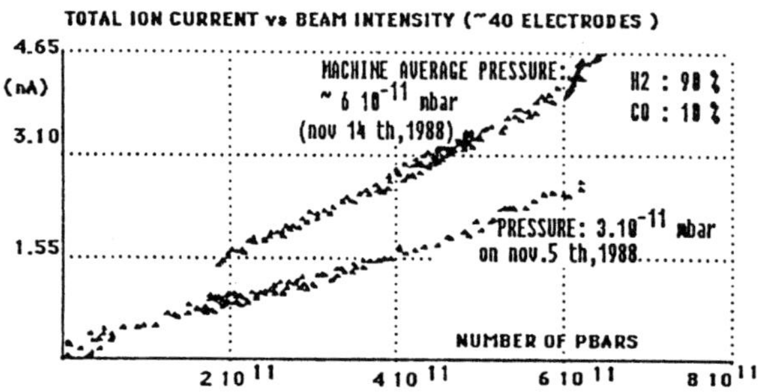

Figure 3 : Sum of ion current drawn on CERN AA clearing electrodes with $N_{\bar{p}} = 6 \times 10^{11}$, and 90% H_2, 10% CO. The pressure is $\cong 3.10^{-11}$ mbar.

3. CONDITIONS OF TRAPPING

3.1 Beam potential and ion pockets

The main trapping mechanism is the beam potential, since the ions at birth seldom have kinetic energies above thermal ones (i.e. < 1eV), and this is usually not sufficient to overcome the potential energy given by the beam potential at the birth location for a singly charged ion [2] :

$$U = \frac{\lambda}{2\pi\varepsilon_0} \begin{cases} \frac{r^2}{2a^2} - \frac{1}{2} + \log_n\left(\frac{r_c}{a}\right) & r \leq a \\ \log_n\left(\frac{r}{r_c}\right) & r \geq a \end{cases} \qquad (7)$$

this expression being valid for a round uniform beam of radius a, line charge density λ, into a round vacuum chamber of radius r_c.

The $\log_n$ dependence of U on the ratio of chamber to beam radii shows that U is modulated along the machine azimuth. Any chamber enlargment (bellows, flanges) represents a potential well into which ions are trapped (they cannot move longitudinally : this is an "ion pocket"). In addition to a transverse oscillation of the ion in the potential well, there is an azimuthal ion drift resulting from the longitudinal beam potential gradient which usually dominates the thermal motion and the beam induced electro-magnetic force [3]. For low values of the neutralization, typical longitudinal electric fields due to the variation of lattice functions alone are of the order of a fraction to a few V/m, resulting in ion azimuthal velocities of several km/s in field free regions.

3.2 External magnetic fields

The main effect of an external magnetic field is to induce a cycloïdal motion in the azimuthal direction, perpendicular to the field lines, resulting in drifts along the beam with velocities [2] :

for a dipole : $$v_D = \frac{E_x}{B} \qquad (8)$$

a quadrupole : $$v_D = W_{kin} \frac{1}{g\, e\, B^2} \frac{\partial B}{\partial x} \qquad (9)$$

where B is the magnetic field, E the beam space charge, x the transverse horizontal coordinate, W_{kin} the ion kinetic energy in the plane perpendicular to B, and g e the electric charge of the ion.

These drift velocities, of the order of several 100 m/s to km/s at the edge of the beam, fall to zero or very low values for ions born at the center of the beam (the drift along the beam is in opposite directions on either side of its axis). For this reason, the possibility of high ion neutralization levels in magnets exists, and it is important to place clearing electrodes in these locations. In uniform magnetic fields, the motion parallel to the field of the center of gyration of the cyclotron motion is not affected by the field.

Magnetic mirror effects occur for low energetic ions drifting from field-free regions towards magnet fringe fields (for some conditions of creation). The longitudinal gradient of the field may reverse the ion motion thus creating a barrier [8]; for this reason sections between magnets must incorporate clearing electrodes. Similar effects are suspected for electrons repelled by the beam potential in quadrupoles; the gradient along the field lines may provoke a containment effect, resulting in an enhanced local ionization rate (electrons from ionization are much more energetic than ions and may ionize the gas) [3].

Without clearing, coasting-beam machines with negative space charge such as antiproton accumulators can reach very high neutralization levels, close to one [9]. DC proton machines trapping electrons, such as the ISR [10], reached neutralization values of a few percent only, most electrons at birth receiving enough momentum to escape the beam potential.

3.3 Ion stability in bunched-beam machines

As we have seen the main force from the beam on an ion is radially focusing; it results from the space charge and leads to the equation of motion :

$$\ddot{x} + k\,x = 0 \tag{10}$$

where $k = \frac{ge}{m}\frac{\partial E}{\partial x}$, with E the beam electric field. For a continuous (DC) beam, E does not depend on time and the resulting motion is a transverse harmonic oscillation (to first order) of frequency ω_i (bounce frequency) and constant amplitude.

When the beam is bunched, the k term in (10) becomes a periodic function of time, yielding the possibility of parametric resonances and thus of open transverse motion for the ions (i.e. loss). It is for this reason that bunched beams reach lower neutralization levels than coasting ones, at least for some conditions of size, emittances, number of bunches and intensity thresholds, contained into a single criterium : the critical mass to charge ratio of the ion.

For uniformly spaced equal bunches around a machine such that an ion receives a succession of instantaneous kicks, similarly to particles travelling through a series of thin focusing lenses, Eq. (10) becomes a Hill equation and yields the following condition of stability for a singly-charged ion of mass-to-charge ratio A_i/g ($g = 1$) [11] :

$$A_i > A_c = \frac{N_p r_p}{2\,n^2}\,\frac{\pi R}{\sigma_y^2\left(1 + \frac{\sigma_x}{\sigma_y}\right)} \tag{11}$$

with R the radius of the machine, N_p the total number of particles, n the number of bunches, σ_x and σ_y respectively the rms horizontal and vertical beam sizes.

All ions with a mass-to-charge ratio less than the critical mass are unstable and cannot be trapped. Given that all possible ions lie usually in a mass-to-charge range of 1 to 44, then if the critical mass is less than 1, neutralization can reach high values, close to 1, similarly to coasting beams. This is the case of many small electron storage rings with large number of bunches.

Other machines which do not fulfill this condition, i.e. with critical masses higher than 1 or 2, or which have non-uniform bunch distributions, may also be plagued by some neutralization.

However, in principle, their ion density should be limited to the molecular density, yielding a neutralization of a fraction to a few percent, depending on the vacuum performance [3-12]. Finally, in the case of large colliders, for which the theory of ion trapping has also been well developed [11], their size prevents them from having neutralization problems.

4. TOLERABLE LIMITS IN NEUTRALIZATION

4.1 Incoherent effects

4.1.1 Incoherent tune shifts

In analogy to the radial (centripetal) force exerted by the beam space charge on an ion, a beam particle senses the focusing force from the ion-cloud space charge. Integration of this force around the ring yields a change of particle tunes ΔQ_{xy}. Assuming a perfect superposition of the ion and beam particle transverse distributions (elliptic bi-Gaussian) one finds for a particle in the beam center [13] :

$$\Delta Q_{x,y} = \frac{r_o \beta_{x,y}}{2\pi\beta\gamma} \frac{\eta N_p}{\sigma_{y,x}^2 \left(1 + \frac{\sigma_{x,y}}{\sigma_{y,x}}\right)} \tag{12}$$

with the classical proton (r_p) or electron (r_e) radius for respectively a proton (or antiproton) and an electron beam, $\beta_{x,y}$ the average β functions in the two planes x or y.

Expression (12) gives the maximum tune shift. Beam particles at the edge of the transverse momentum distribution or with large betatron amplitudes experience less focusing from the ion space charge (zero at the limit), such that $\Delta Q_{x,y}$ also represent approximately the ion-induced tune spreads. This is illustrated in figure 4, taken from the AA.

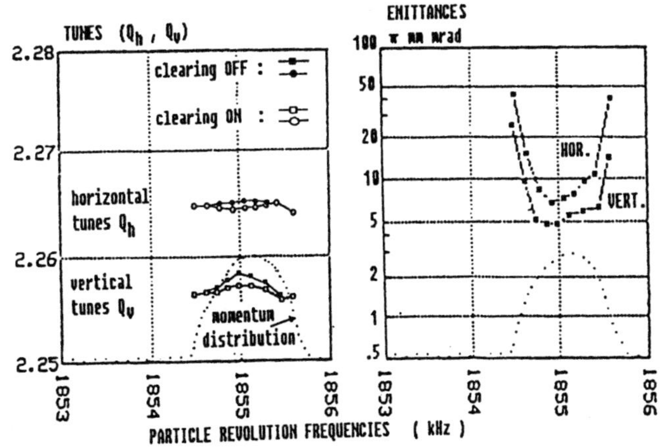

Figure 4 : Left : antiproton transverse tunes as function of their momentum, with full (no clearing) and little (clearing) neutralization. Right : particle emittances (independant of neutralization during the measurement).

The ion-induced incoherent tune shift (and spread) can take very large values for small machines (γ small). As examples, for EPA (E = 600 MeV, $\gamma = 1200$, $N_p = 6.10^{11}$ electrons) [14] :

$$\Delta Q_{ions} = 0.15\,\eta \text{ (i.e. } \sim 0.15 \text{ without clearing)}$$

and for the AA (p = 3.5 GeV/c, $N_{\bar{p}} = 10^{12}$ antiprotons) :

$$\Delta Q_{ions} = 0.01\,\eta \text{ (i.e. } \sim 0.01 \text{ with no clearing).}$$

On some small, low energy electron storage rings, the ion space-charge tune shift could even reach half integer values.

It is interesting to compare the neutralization (focusing) and direct beam space charge (defocusing) effects. The maximum total tune shifts produced by the neutralization and the beam space charge are given by :

$$\Delta Q_{x,y} = \frac{r_o\,\beta_{x,y}}{2\pi\beta\gamma B}\;\frac{N_p}{\sigma_{y,x}^2\left(1+\dfrac{\sigma_{x,y}}{\sigma_{y,x}}\right)}\left(\hat{\eta}-\frac{1}{\gamma^2}\right) \tag{13}$$

with B the bunching factor ($B = nl_B/2\pi R << 1$ for bunched beams) and $\hat{\eta} = B\eta$.

On small machines (which trap ions most easily), ions may dominate the beam space-charge tune shift. For the same conditions as above :

CERN AA : $\Delta Q_{ions} \lesssim 10\,\Delta Q_{sp.\,charge}$

CERN EPA : $\Delta Q_{ions} \lesssim 10^4\,\Delta Q_{sp.\,charge}$.

The incoherent space-charge tune shift expression (13) provides a means to measure the average neutralization in a ring [12]. This supposes the possibility of clearing the beam of its ions (clearing electrodes, beam shaking, asymmetric bunch filling) so that from measurements of the tunes Q, with and without ions, η can be obtained from (13) (note that the sign of the shift already tells us if the beam is space-charge or ion dominated). ΔQ is usually extracted from frequency spectral analysis of a transverse betatron side-band (the difference signal from a pair of electrodes around a beam mode frequency $(n \pm Q)f_{rev}$. Obtaining the maximum frequency shift experienced by a particle by comparison of two transverse spectra is only approximate (the measurement shown on figure 4 requires unfolding several spectra [15]). If one supposes identical initial bi-Gaussian transverse distributions for the beam particles and the ions, the maximum frequency shift ΔQ can be obtained by multiplying the shift of the peaks of the distributions by approximately 2.5 [16].

4.1.2 Resonances

Tune shifts due to ions cause particles to cross resonance lines in the tune diagram :

$$n\,Q_x + m\,Q_y = \text{integer}$$
$$n, m = 0, \pm 1, \pm 2...$$

resulting in transverse amplitude growth (emittance blow up) and eventually loss.

Very similarly to the excitation of non-linear resonances by the beam-beam interaction in colliding beam machines, the electrostatic field of ion clouds in neutralization pockets causes non-linear detuning, and its uneven distribution may excite very high-order resonances (up to 15th order in the AA, [9]). This effect, still present at low neutralization levels of a fraction of a percent, is very detrimental for antiproton accumulators which have low transverse emittance cooling rates (compared to electron storage rings). As the beams are unbunched, the neutralization can only be coped with by reducing the ion density through shaking of the beam (see section 5.3). Figure 5 shows a measurement of transverse heating rates due to the excitation of high-order resonances by ion pockets in the AA [9].

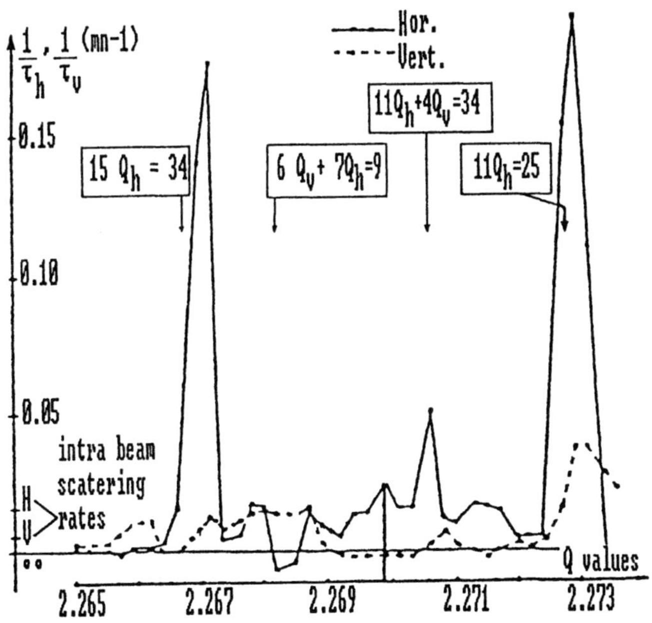

Figure 5 : Net resonance-induced emittance heating rates versus tune in the AA, due to residual neutralization pockets.

These heating rates occur at a neutralization level of a few percent, if the betatron frequencies of the particles concerned cross the resonance continuously, gaining amplitude at each crossing. This is explained in this case by a tune modulation caused by the ripple of the quadrupole power supplies [17] (the equivalent of the synchrotron tune modulation for the beam-beam effect in colliders).

In electron storage rings, the emission of synchrotron radiation by the circulating particles results in high damping rates for their transverse motion, so that it is unlikely (but not proven) that such high-order resonances can be harmful, since their strength decreases as their order increases. However due to the strong coupling introduced by the ion electrostatic force (which depends on both transverse coordinates x and y), and due to the fact that no other natural vertically heating mechanism exists in these machines (scattering on gas being negligible), even a small neutralization level may quickly manifest itself by an increased vertical beam size. Experience on modern machines shows that ion-induced tune shifts of a few 10^{-3} caused by neutralization levels of a few percent or less, already present annoying emittance effects.

One of these, common to some synchrotron radiation sources, is a pulsation of the vertical beam size. This occurs when the emittance growth rate due to an ion-induced resonance is larger, and the neutralization rate smaller, than the natural damping rate. As the beam inflates ions are chased away by the resonance, the beam then shrinks, neutralization builds up again and the process repeats itself in a sort of relaxation mechanism [18].

To summarise, experience on incoherent effects so far shows that tolerable neutralization levels should limit the maximum ion-induced tune shifts to $\sim 10^{-2}$ in electron storage rings with radiation damping, and to less than 10^{-3} in antiproton storage rings. The corresponding tolerable average neutralization may range from a fraction to a few percent.

4.2 Coherent instabilities

In addition to single-particle phenomena, transverse coherent instabilities, where collective oscillations of the beam center or the beam shape grow exponentially due to the interaction with the trapped species, have been identified on several machines (electron-proton in the CERN ISR and LBL Bevatron [10], ion-antiproton in the CERN and FNAL Antiproton Accumulators [9-20], ion-electron in the CERN PS [21]). These instabilities, which cause fast emittance blow-up of the beams and eventually loss of particles, can occur at lower neutralization levels than single-particle phenomena. Ion-driven dipole and quadrupole coherent instabilities have represented severe intensity-limiting mechanisms in the CERN AA, where they have been studied in detail [19].

In order to guess the tolerable neutralization with respect to these instabilities, we shall glance at the classical two-beam instability theory developed by Koshkarev and Zenkevich [22] and Lasslett, Sessler and Möhl [23], restricted to coasting beams.

In collective motion, the beam center or shape moves in a travelling wave pattern described by :

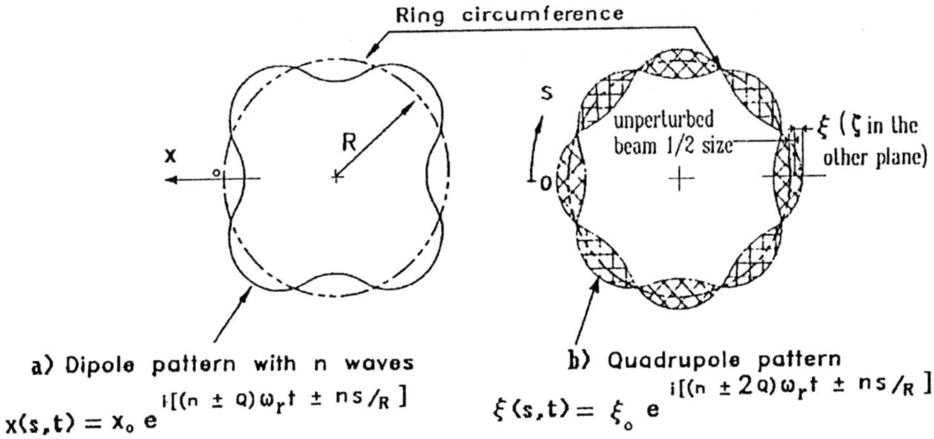

Figure 6 : Beam wave patterns for coherent modes. a) dipole mode - b) quadrupolar mode.

Here ω_r is the revolution frequency, R the ring radius, s the azimuth along the beam. The mode number n = 1, 2, 3... is equal to the number of nodes around the ring.

The mode frequency, as seen by a pick-up electrode at a fixed azimuth is

$$\omega_n^{\ell} = (n \pm \ell Q)\, \omega_r$$

where $\ell = 1$ for a dipole mode, 2 for a quadrupole mode, etc...

Due to the spread in betatron frequencies, the signal seen by the pick-up (Schottky band) has a certain width in frequency as sketched below :

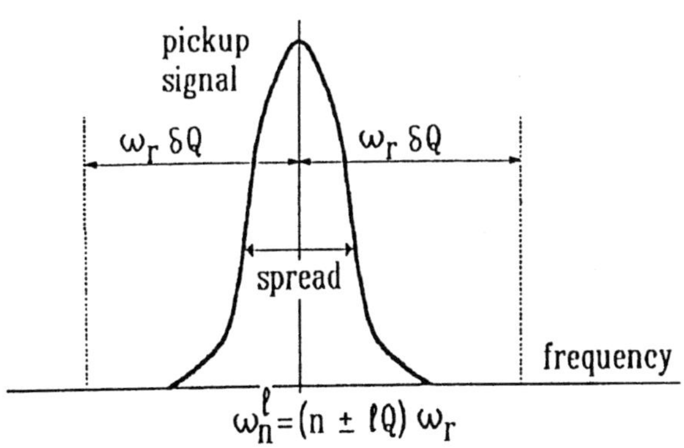

Figure 7 : Pick-up signal around a transverse beam mode frequency for a coasting beam (Schottky band). $\omega_r\ \delta Q$ is the maximum mode frequency shift induced by the ion force on the beam.

Transverse coherent ion-beam coupled oscillations leading to instability are possible when the ion-bounce frequency lies close to the natural beam mode, in a band $\omega_r\ \delta Q$. δQ reflects the force applied by ions on the beam and is obtained by solving the dispersion relation of the equation of motion without frequency spreads.

With the approximations of linear forces (i.e. uniform particle distributions superposed), neglecting image forces and averaging the lattice contributions, these equations are for a dipole mode [19] :

$$\text{particle} \quad \frac{1}{\omega_r^2}\ddot{y} + Q_o^2\, y - Q^2_{sc}(y - \bar{y}) + Q^2_c(y - \bar{y}_i) = 0$$

$$\text{ion} \quad \frac{1}{\omega_r^2}\ddot{y}_i\, q^2_{sc}(y_i - \bar{y}_i) + q^2_c(y_i - \bar{y}_i) = 0 \tag{14}$$

where the bar on the transverse coordinate represents the motion of the beam center, Q_o the particle unperturbed tune (only lattice dependent), and the other Q-q being the various space-charge forces:

Q_{sc} the particle-particle space charge :

$$Q^2_{sc} = \frac{2\,N_p\,r_o\,R}{2\,\pi\,\sigma_x\,(\sigma_x + \sigma_y)\,\beta^2\,\gamma^3}\ . \tag{15}$$

Q_c the ion space charge on particle :

$$Q^2_c = \frac{2\eta\,N_p\,r_o\,R}{2\,\pi\,\sigma_x\,(\sigma_x + \sigma_y)\,\beta^2\,\gamma} \tag{16}$$

(with r_o being r_e or r_p for an electron or hadron beam).
q_c the beam space charge on the ion of charge g e and atomic mass A (g = A = 1 for a trapped electron) :

$$q^2_c = \frac{2\,N_p\,r_{p,e}\,R\,g}{2\,\pi\,\sigma_x\,(\sigma_x + \sigma_y)\,\beta^2\,A}\ . \tag{17}$$

q_{sc} the ion-ion space charge : $q^2_{sc} = \eta\, z\, q^2_c$ and putting $Q^2 = Q_o^2 + Q_c^2 - Q_{sc}^2$; $q^2 = q^2_c - q^2_{sc}$.

Starting from Eq. (14), four similar linear equations can be derived for the beam transverse shape first-order variations in a quadrupole mode (variables ζ and ξ of figure 6).

In the approximation of round beams ($\sigma_x = \sigma_y$), it is found that the risk of transverse instabilities exists in a band given by :

$$\delta Q = p\,\frac{q_c\,Q_c}{\sqrt{q_c\,Q_o}} \tag{18}$$

where p = 1 for a dipole mode, p = 1/2, 1/4 respectively for a symmetric ($\zeta = \xi$) or antisymmetric ($\zeta = -\xi$) quadrupole mode. The width of this band, reflecting the risk of instability, is proportional to the strength of the force on the beam exerted by the ions Q_c, i.e. to the square root of the neutralization coefficient (Eq. (16)). It is smaller for quadrupole than for dipole modes.

For realistic frequency distributions the spreads in mode and ion frequencies Δn and Δi, i.e. the spreads of ω_n/ω_r and q respectively, introduce Landau damping in this band when three conditions are simultaneously satisfied :

$$\Delta n > p \left| \frac{Q^2_{sc}}{Q} \right| ; \Delta i > p \left| \frac{q^2_{sc}}{q} \right| ; \Delta n \, \Delta i > p^2 \left| \frac{Q_c^2 \, q^2}{qQ} \right| \tag{19}$$

with p=1, p = 1/2 or p = 1/4 as defined under Eq. (18).

Thus, the required spreads are smaller by p and the threshold neutralisation ($\eta \sim Q_c^2$) is higher by p^{-2} for quadrupole instabilities. In the AA, quadrupole instabilities appeared after dipole ones had been damped by transverse feedback, at 2 to 3 times higher beam intensities, causing large beam amplitudes and loss of particles. The application of the theory outlined above fitted observations when assuming a single ion pocket representing an average neutralization of 10^{-3} or less [19]. Expressions for these instability growth rates and maximum beam amplitudes can be derived. In general and especially for the dipole mode, the amplitudes of the trapped species are much larger than those of the main beam. Therefore, for beam transverse excursions of a fraction of a millimeter, the trapped particles are lost to the walls of the vacuum chamber, or simply detuned, causing a slight emittance blow-up of the main beam. Ions are cleared by these instabilities; the continuous creation of new ions from the residual gas may either feed a constant amplitude wave pattern of the beam, or provoke a relaxation mechanism similar to the incoherent effects [24].

Much work is still required to understand quantitatively the tolerable limits of neutralization to avoid coherent instabilities. Many necessary ingredients such as the types and the lifetimes of trapped ions and the location of neutralization pockets around a ring are usually unknown; if they were known, much refinement would have to be added to the theory to describe the experimental observations.

However, from the experience of existing machines, it is certain that to avoid coherent instabilities neutralization levels have to be extremely low, certainly well below 1% on the average. This level is difficult to achieve with the actual clearing means for small electron or antiproton machines [25].

5. CURES

At present there are four methods to fight against ion build up, maintenance of the best possible vacuum, uneven bunch filling, clearing electrodes and beam shaking.

5.1 Vacuum conditions

Good vacuum acts through the ion production rate R_p (thus reducing $\eta = R_p/R_c$) and ion type. UHV conditions, defined as $p < 10^{-9}$ mbar, imply mainly hydrogen as the residual gas. This has small ionization cross sections, and low z and mass (i.e. eventually untrappable or easy to remove because of high velocities). It is therefore essential that neutralization-prone machines have an extremely good vacuum. This means an efficient pumping system, together with a careful choice, preparation and in situ baking of all components in contact with the machine vacuum.

5.2 Uneven bunch filling

Many small electron storage rings partly solve the problem of ion trapping by introducing one or several gaps in the bunch train by not filling certain buckets at injection. This has the effect of introducing large stop-bands in the ion transverse frequency spectrum, or, in other words, bands of unstable mass-to-charge ratios in the ion mass spectrum [26]. Because an initially stable trapped ion can be further ionized, its mass-to-charge ratio decreases, and thus it has a chance to fall in an unstable band and to be cleared away. The neutralization is therefore reduced, but at the expense of the total beam intensity of the machine.

5.3 Clearing electrodes

Another common way to extract ions from the beam consists in installing a negatively polarized plate close to the beam, providing a transverse electric field which extracts beam-channelled ions. The potential U_c applied to the electrode must be such that it exceeds the maximum beam space-charge field in order to capture all the passing ions [3] :

$$\frac{U_c}{d} > E_{max} \sim \frac{I}{4\pi\,\varepsilon_o\,\beta c\,a} \tag{20}$$

for a round uniform beam of current I and radius a, d being the distance between the plate and the opposite vacuum chamber wall.

Because ions may have large velocities (transverse if born on the edge of the beam, longitudinal if born at the center) and electrodes are necessarily limited in size, the voltage to be applied for optimum clearing may represent several times this value [7].

A large number of electrodes is required to achieve a low neutralization. Strategic locations are straight sections close to fringe fields, dipole magnets, quadrupoles and pockets created by vacuum chamber cross-sectional changes, all places where ions are likely to accumulate or have low drift velocities. However there are also arguments against a large number of electrodes; they may represent a sizeable contribution to the machine transverse and longitudinal beam coupling inpedance unless great care is taken in their design. Figure 8 shows the distribution of electrodes, presenting negligible longitudinal coupling impedance, in the EPA [1]. The button electrodes are made of an alumina body, coated with a few 10 μm thick resistive layer (R_{square} ~ 30 kΩ), thus making it transparent to electromagnetic waves (R_{square} >> 377Ω).

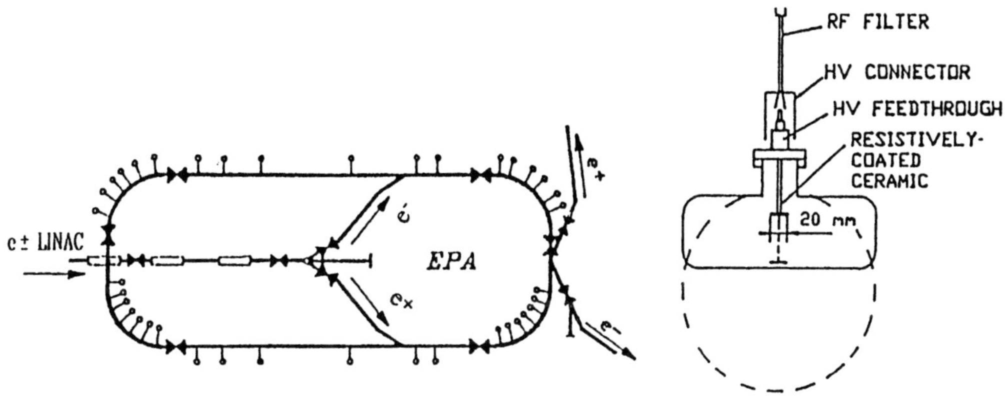

Figure 8 : Distribution of clearing electrodes around EPA and cross section of one electrode

5.4 Beam shaking

In practise it is extremely difficult to design a fully efficient clearing system for ions. Beam shaking has been successfully applied to reduce the ion concentration - and thus their action on the beam - in uncleared ion pockets [27]. It consists in using the beam, transversally driven in coherent motion by a pair of kicker electrodes fed by an RF voltage, to excite the trapped ions to large amplitudes. Experimentally [20], beam shaking has been shown to reduce neutralization effects and to have the following characteristics :

- It works best when applied vertically and when the frequency is just above an n+Q side band, and just below an n-Q band.
- It is efficient even with a low RF field (10 V/m) when the shaking frequency is close to a betatron side band (resonant mode shaking [4]) of the main beam.
- The shaking frequencies have to be carefully chosen to be close to the ion bounce frequencies (H_2^+, CO^+) in the potential of the circulating beam.
- The average neutralization has to be low (i.e. there must be other clearing mechanisms).

To understand some of these features, and in particular how ions can respond to a single harmonic excitation of the beam, the special features of resonance in non-linear oscillations (the ion in the non-linear beam space charge) will be briefly described [25].

The motion of the normalized vertical displacement v of an ion from a beam center oscillating harmonically at the shaking frequency ω :

$$y_c = y_{co} \sin(\omega t)$$

defined as

$$v = \frac{y - y_c}{2\,\sigma_y},$$

can be written as

$$\ddot{v} + \omega^2_{iy}\, v = f(v) + \omega^2 H \sin(\omega t) \tag{21}$$

where $H = y_{co}/2\,\sigma_y$ is the amplitude of the external forcing, and

$$\omega_{iy} = \left[\frac{g\,I}{4\,\pi\,\varepsilon_o\,\beta c\, m\, \sigma_y\,(\sigma_x + \sigma_y)} \right]^{1/2} \tag{22}$$

is the vertical ion bounce frequency. f (v) retains the non-linear terms resulting from a series expansion of the electric field (f (v) ~ o for small v). The small amplitude solutions of equation (21) at equilibrium, studied by a perturbative method, for shaking frequencies ω close to ω_i satisfy [25] :

$$\frac{\omega}{\omega_i} = \left[\frac{a\,G\,(a^2)}{a \pm H} \right]^{1/2} \tag{23}$$

where G (a^2) is a non-linear function of the system. Figure 9 shows the well-known curves giving the equilibrium amplitudes a as a function of ω/ω_i

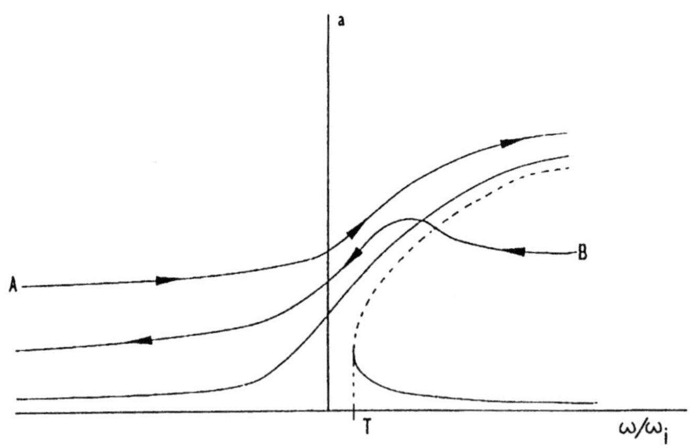

Figure 9 : Equilibrium ion amplitudes as a function of the shaking frequency ω. The dotted line corresponds to unstable solutions. For ω/ω_i above the transition point T, ions can have either small (lower curve) or large amplitudes, depending on initial conditions.

For highly neutralized beams (no clearing), ions have a broad spectrum of plasma frequencies, dependent on the neutralization by other species. In addition, due to the small beam-plus-ion space-charge potential, their drift velocities are very low such that their frequencies do not change rapidly. In this case shaking at one fixed frequency only affects a small fraction of them and is not efficient. This probably explains why shaking does not work when the clearing electrodes are turned off in AA and EPA.

When the clearing electrodes are on, only a low neutralization remains, concentrated in a few portions of a machine, thus reducing the spectrum of ion frequencies. In addition ions move faster along the beam due to higher potential gradients, and their frequencies scan this spectrum. Ions for which frequencies decrease follow the path A of figure 9 and provided this scanning of the resonance is not too fast, they "lock-on" to it and keep large amplitudes, thus reducing their density in the center of the beam. Ions following path B, that is those with increasing plasma frequencies, experience only a small jump at the crossing of the resonance and are not excited to large amplitude. This hysteresis phenomenon, jump bifurcation or so-called lock-on effect characteristic of non-linear resonances [27] explains the efficiency of shaking, and leads to an asymmetry in the ion behaviour. In particular, in conjunction with clearing current measurements, beam shaking has permitted the identification of ion neutralization pockets in the bending magnets of the CERN AA [7].

Figure 10 illustrates the effect of shaking the coasting antiproton beam vertically at 400 kHz (i.e. just below the fractional vertical tune frequency corresponding to the (o + q) band, where q is the non-integer part of the tune) in the FNAL Antiproton Accumulator. The decrease in beam emittances and the increased stacking rate are a consequence of reduced ion densities [28].

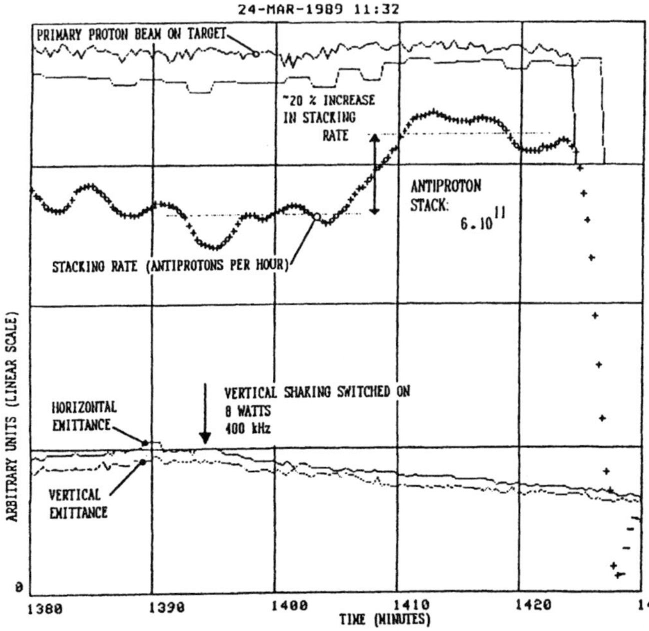

Figure 10 : Reduced beam emittance and increased stacking rate in the FNAL Antiproton Accumulator with beam shaking

6. CONCLUSION

Neutralization by ions from the residual gas in modern machines such as antiproton accumulators, synchrotron light sources and future B factories is still a challenge. The physics of neutralization and the ion dynamics are only partly understood, as are also the effects on the beam. Despite some progress made in minimizing their coupling with the beam, clearing electrodes are probably not sufficient to reduce the neutralization to acceptable levels in high intensity machines. Beam shaking has been shown to be an efficient additional antidote on some machines, but much progress has to be made both in its understanding and application. Ideas such as frequency modulated shaking, multi-frequency and cyclotron shaking have been proposed [25], which still need to be tested experimentally. All aim at improving further the efficiency of the process, to reduce the neutralization to acceptable levels even in the most difficult cases.

REFERENCES

[1] F. Caspers, J.P. Delahaye, J.C. Godot, K. Hübner, A. Poncet, *EPA Beam-Vacuum Interaction and Ion Clearing System*, CERN PS/88-37 (ML), 1988, EPAC, Rome, 1988.

[2] Y. Baconnier, *Neutralization of Accelerator Beams by Ionization of the Residual Gas*, CERN/PS/PSR/84-24, and CERN Accelerator School, Gif-sur-Yvette, France, 1984, CERN 85-19.

[3] A. Poncet, *Ion Trapping and Clearing*, CERN/MT 90-1 (ES), and CERN Accelerator School, Uppsala, Sweden, 1990, CERN 90-04, pp. 74-98.

[4] The PS Staff, *The CERN PS Complex as an Antiproton Source*, IEEE Trans. Nucl. Sci., Vol. NS-32, p. 2218, 1985.

[5] B. Rossi, *High Energy Particles*, Prentice Hall, New York, 1952.

[6] J. D. Jackson, *Classical Electrodynamics*, John Wiley & Sons, Inc., New York, 1962.

[7] F. Pedersen, A. Poncet, L. Søby, *The CERN Antiproton Accumulator Clearing System*, CERN PS/89-17 (ML), and Proc. Part. Acc. Conf., Chicago, 1989.

[8] Y. Miyamara, K. Takayama, G. Horikoshi, *Dynamical Analysis on the Longitudinal Motion of Trapped Ions in Electron Storage Rings*. Nucl. Inst. and Meth. in Ph. Res., A 270, p. 217, 1988.

[9] E. Jones, F. Pedersen, S. van der Meer, A. Poncet, *Transverse Instabilities Due to Beam Trapped Ions and Charge Matter in the CERN AA*, IEEE Trans. Nucl. Sci., Vol. NS-32, p. 2218, 1985.

[10] H. G. Hereward, *Coherent Instability due to Electrons in a Coasting Proton Beam,* CERN 71-15, 1971.

[11] G. Brianti, Y. Baconnier, O. Gröbner, E. Jones, D. Potaux, H. Schönauer, *Proceedings of the Workshop on p$\bar{p}$ in the SPS p$\bar{p}$. 1*, p. 121, 1980.

[12] M.E. Biagini, S. Guiducci, M. Preger, M. Serio, S. Tazzari, *Observation of Ion Trapping at ADONE, Proc.* 11th Int. Conf. on High Energy Accelerators, CERN, Geneva, 1980, p. 687.

[13] R.D. Kohaupt, *Mechanismus der Ionenabsaugung im Elektron-Positron Speicherring DORIS*, Interner Bericht DESY H1-71/2, 1971.

[14] A. Poncet, *Ion Clearing in EPA*, CERN PS/ML/Note 83-1.

[15] S. Van der Meer, *Measurement of Transverse Stack Emittances from Schottky Scans*, PS/AA/Note 84-11, 1984.

[16] W. Marsh, *How to Measure Beam Neutralization in the Accumulator,* Fermilab III Instabilities Workshop, Saint Charles, Illinois, June 1990 (unpublished).

[17] A. Dainelli, *Antiproton-Positive Ion Transverse Instabilities in the CERN AA. A Tune Modulated Direct Map Simulation*, CERN/PS/87-13 (AA), 1987.

[18] E. M. Rowe, *Trapped Ion Effects and Their Treatment in Electron Storage Rings,* American Vacuum Society Series 5, American Institute of Physics, Conf. Proc. 171, p.193, New York, 1988.

[19] R. Alves-Pires et al. *On the Theory of Coherent Instabilities Due to Coupling between a Dense Cooled Beam and Charged Particles from the Residual Gas*, CERN/PS/89-14 (AR), 1989, 1989 Part. Acc. Conf. Chicago, 1989.

[20] J. Marriner, D. Möhl, Y. Orlov, S. van der Meer, A. Poncet, *Experiments and Practice in Beam Shaking*, CERN/PS/89-48 (AR), 1989, 14th Int. Conf. on High Energy Accelerators, Tsukuba, Japan, 1989.

[21] R. Cappi and J.P. Riunaud, *Trapped Ion Effects with Electron Beams in the CERN PS*, EPAC 90, June 1990, Nice (F) and CERN/PS 90-17 (PA).

[22] D.G. Koshkarev and P.R. Zenkevich, *Resonance of Coupled Transverse Oscillations in Two Circular Beams*, Particle Accelerators, 1972, Vol. 3, pp. 1-9.

[23] L.J. Lasslett, A.M. Sessler, D. Möhl, *Transverse Two-Stream Instability in the Presence of Strong Species - Species and Image Forces*, Nucl. Instr. Meth., 121 (1974), p. 517.

[24] F. Pedersen, A. Poncet, *Proton-Antiproton Instability in the AA*, CERN PS/AA/ME Note 81, 1981.

[25] R. Alves Pires, J. Marriner, W. Marsh, A. Poncet, J. Rosenzweig, P. Zhou, *Ion Trapping*. Proceedings of the Fermilab III Instabilities Workshop, Saint Charles, Illinois, June, 1990.

[26] M.Q. Barton, *Ion Trapping with Asymmetric Bunch Filling of the NSLS VUV Ring*, Nucl. Inst. and Methods, A243, 1986, p.278.

[27] Y. Orlov, *The Suppression of Transverse Instabilities in the CERN AA by Shaking the $\bar{p}$ Beam*, CERN PS/89-01 (AR), 1989.

[28] J. Marriner and A. Poncet, *Neutralization Experiments with Protons and Antiproton Stacks - Ion shaking*, $\bar{p}$-Note 481, Fermilab Internal Note, 1989.

BEAM LOADING IN RF CAVITIES

R. Garoby
CERN PS - RF
1211 Geneva 23, Switzerland

1 INTRODUCTION

Beam loading is the conventional term for the effect of beam passage through an RF cavity installed for the control of some beam parameters such as energy, bunch length, energy spread, etc... The fields developed in such a structure and their effect on the beam can be analyzed as for all other discontinuities of the vacuum chamber. In particular, F. Sacherer's theory of impedances and beam instabilities [1] is perfectly relevant.

However, RF cavities differ largely from the other kind of impedances because they are deliberately tuned at (or near) a frequency where the beam current is large, and because they are equipped with an excitation port for connection to an RF amplifier. For various reasons, a cavity-amplifier system is generally optimized to provide the maximum acceleration per unit of length, with the minimum RF input power. Consequently the cavity impedance is large, and the current from the RF generator is minimized. When the beam current increases, there is then a limit where it becomes non-negligible and a beam loading situation is encountered. The following questions have then to be addressed:

- stability against small perturbations (case of a simple cavity-amplifier set-up, and generalization to the case of a full RF system with servo-loops),
- large transient situations,
- requirements for the RF power amplifier.

Review papers concerning these matters have already been published [2,3]. The present one reflects the personal experience of the writer, trying to describe and comment on the current knowledge about these issues, and the available cures.

2 CONVENTIONS

2.1 EQUIVALENT CIRCUIT

Figure 1 shows the equivalent circuit { terminology is the one used by F. Pedersen [4]}. The cavity is represented as a simple parallel R,L,C, resonator. The amplifier output impedance, transformed to the cavity gap, is included in R as well as the cavity losses. The beam and the RF amplifier are modelled as ideal current generators.

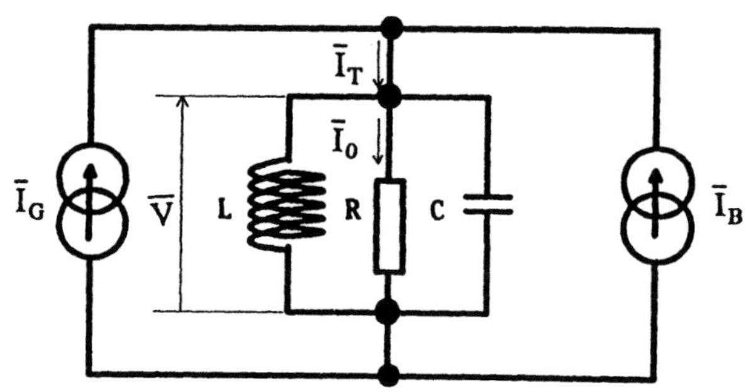

$\bar{V}$: voltage across the cavity gap
$\bar{I}_G$: generator current (transformed to the gap impedance)
$\bar{I}_B$: beam current
$\bar{I}_T$: total current in the cavity
$\bar{I}_0$: resistive current (current through R)
f_R : cavity resonant frequency
f_C: RF carrier frequency

Figure 1 : Equivalent circuit

2.2 STEADY STATE VECTOR DIAGRAM

Only the RF components (peak values) are considered. This is fully justified if the voltage induced at other frequencies is negligible, either because the currents have negligible spurious frequency components, or because the resonator impedance is very small at these other frequencies. As shown in the vector diagram in figure 2, the total current $\bar{I}_T$ flowing through the cavity impedance is the vector sum of the beam and generator currents ($\bar{I}_B$ and $\bar{I}_G$). $\bar{V}$ and $\bar{I}_T$ are linked by : $\bar{I}_T=\bar{V}/Z_{cav}=\bar{V}(1/R+jB)$ with $B=C\omega-1/(L\omega)$.
$\bar{I}_0=\bar{V}/R$ is the minimum current needed to develop a voltage $\bar{V}$ when the cavity is tuned. The tuning angle φ_Z is the argument of the complex impedance Z_{cav}. The loading angle φ_L is the equivalent impedance angle as observed by the RF amplifier (phase angle between generator current $\bar{I}_G$ and voltage $\bar{V}$).

For practical cases $\Delta f/f_R<<1$. Then : $\tan\varphi_Z \sim 2Q\Delta f/f_R$ where Q is the quality factor of the resonator and $\Delta f= f_R-f_C$ is the detuning with respect to the RF frequency f_C.

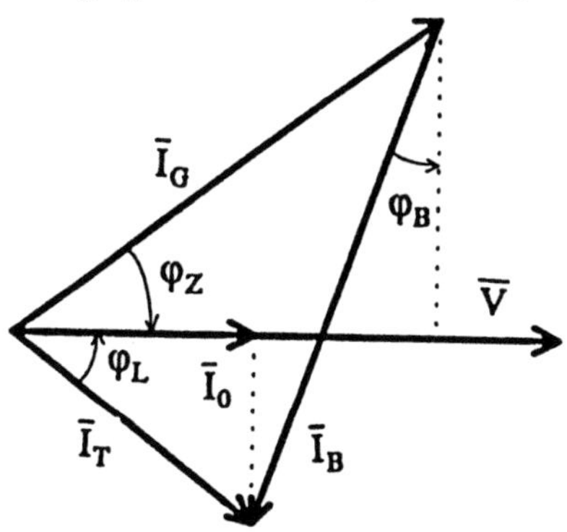

I_B : peak beam current at f_C . $I_B=2\,I_{DC}$ for short bunches, otherwise $I_B < 2\,I_{DC}$, I_{DC} being the DC beam current

φ_B : stable phase angle between RF components of beam current and cavity voltage

φ_S : stable phase angle for very short bunches $\{I_B \sin\varphi_B /2= I_{DC} \sin\varphi_S\}$

Y : by definition the relative beam loading Y is given by : $Y=I_B/I_0$

Figure 2 : Steady state vector diagram below transition energy : $\bar{I}_B$ downwards
and with acceleration (φ_B>0) : $\bar{I}_B$ leftwards

3 CONTINUOUS BEAM LOADING

3.1 ROBINSON INSTABILITY

The cavity tune (Z_{cav}) and the RF drive ($\bar{I}_G$) are assumed constant. Because of the vector addition of currents through the resonator (figure 2) any modulation of $\bar{I}_B$ induces changes in $\bar{I}_T$ and $\bar{V}$, which in turn modulate the phase φ_B of $\bar{I}_B$. The stability limits of this closed loop system are of two different types:

- an impedance limitation. It is intensity independent and corresponds to in-phase dipolar oscillations of all bunches (mode m=1 n=0) [1],
- an RF power limitation. It is attained when the RF focusing is annihilated by the action of the beam current.

K.W. Robinson [5] derived rigorously the analytical criteria, and another derivation of the impedance limitation appears in this book [6]. This last limit can also be obtained directly from Sacherer's theory [1] applied to mode m=1, n=0. The analysis given below puts the emphasis on the illustration of the physical processes using deliberately some simplifications. The rigorous demonstration can be found in the above references.

3.1.1 Impedance limitation

Let us assume the case of a beam energy below transition, and a cavity with a moderate Q, such that its filling time τ is small compared to the period T_S of synchrotron oscillation of the beam :

$$\tau = Q/(\pi f_R) << T_S = 1/f_S.$$

When a point charge makes a synchrotron oscillation around the synchronous particle, its frequency oscillates around the RF carrier frequency f_C driving the cavity. At each moment it loses energy proportionally to the real part of the cavity impedance (figure 3).

Qualitative approach

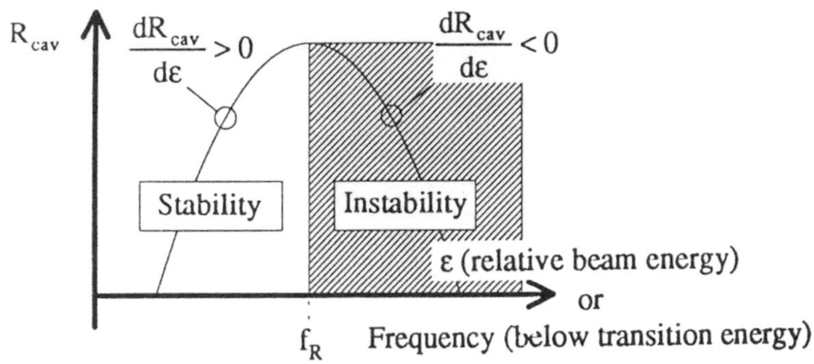

Figure 3 : Impedance limitation below transition energy

- If the resonator is tuned below the RF frequency ($f_R < f_C$), the resistive part of the impedance decreases when frequency or energy increases (below transition a higher frequency corresponds to a higher energy). When the beam is at a higher energy than the synchronous particle, it dissipates less energy in the cavity. Conversely, when the beam is at a lower energy than the synchronous particle it loses more energy. The synchrotron oscillation amplitude is growing (anti-damping) : there is instability.

- If the resonator is tuned above the RF frequency ($f_R > f_C$) the reverse is true. The synchrotron oscillation amplitude decreases because of energy dissipation in the cavity. The system is stable.

The criterion is then $\varphi_Z > 0$ below transition energy, and it can be demonstrated to be the reverse above the transition energy.

Quantitative approach

Let us consider the synchrotron oscillation of a very short bunch of N particles in the longitudinal phase plane with the coordinate system (φ,ε).

φ is the phase of the RF component of the beam current with respect to the RF voltage,

$\varepsilon = \Delta E/E$,

E is the synchronous particle energy, and ΔE the energy difference between beam and synchronous particle.

The voltage V in the cavity changes the relative beam energy per charge with respect to the synchronous particle at a rate given by

$$\dot{\varepsilon}_1 = \frac{I_{DC} V(\sin\varphi - \sin\varphi_S)}{NE}.$$

We shall not consider the effect of the reactive part of the cavity impedance, since it does not contribute to the damping or anti-damping of the oscillation. Dissipation in the real part R_{cav} of the impedance of the cavity leads to a rate of loss of relative energy

$$\dot{\varepsilon}_2 = -\frac{I_B^2}{2NE}\left(\frac{dR_{cav}}{d\varepsilon}\right)\varepsilon \approx -\frac{2I_{DC}^2}{NE}\left(\frac{dR_{cav}}{d\varepsilon}\right)\varepsilon.$$

The total rate of change of ε per charge is then

$$\dot{\varepsilon} = \frac{I_{DC} V(\sin\varphi - \sin\varphi_S)}{NE} - \frac{2I_{DC}^2}{NE}\left(\frac{dR_{cav}}{d\varepsilon}\right)\varepsilon. \tag{1}$$

In a synchrotron at a given B field, a relative frequency change $\Delta\omega/\omega$ is linked to a relative momentum difference $\Delta p/p$ by

$$\frac{\Delta\omega}{\omega} = -\eta\frac{\Delta p}{p} = -\frac{\eta}{\beta^2}\varepsilon \quad \text{using the relation :} \quad \frac{\Delta p}{p} = \frac{1}{\beta^2}\frac{\Delta E}{E} = \frac{1}{\beta^2}\varepsilon$$

where $\eta = \frac{1}{\gamma_T^2} - \frac{1}{\gamma^2}$, $\gamma = \frac{E}{E_0}$ and β are the usual relativistic parameters, γ_T being the value of γ at transition.

Consequently
$$\dot{\varphi} = -\Delta\omega = \frac{\omega_C\eta}{\beta^2}\varepsilon. \tag{2}$$

Differentiating (1) with respect to time, and using (2), the following second-order equation is derived :

$$\ddot{\varepsilon}+\frac{2I_{DC}^2}{NE}\left(\frac{dR_{cav}}{d\varepsilon}\right)\dot{\varepsilon}-\frac{I_{DC}V\cos\varphi_S}{NE}\frac{\omega_C\eta}{\beta^2}\varepsilon=0. \tag{3}$$

Searching for a solution of the form exp(jωt) we obtain $-\omega^2+j2\alpha\omega+\omega_{S0}^2=0$ (4)

using $\alpha=\frac{I_{DC}^2}{NE}\left(\frac{dR_{cav}}{d\varepsilon}\right)$

$$\omega_{S0}^2=-\frac{I_{DC}V\cos\varphi_S}{NE}\frac{\omega_C\eta}{\beta^2}. \tag{5}$$

Roots of (4) are of the form $\omega_i=j\alpha\pm\sqrt{\omega_{S0}^2-\alpha^2}$.

Solutions of (3) are linear combinations of $e^{-\alpha t}e^{+j\left(\sqrt{\omega_{S0}^2-\alpha^2}\right)t}$ and $e^{-\alpha t}e^{-j\left(\sqrt{\omega_{S0}^2-\alpha^2}\right)t}$

α is the damping rate of the synchrotron oscillation. Since α is of the sign of $\left(\frac{dR_{cav}}{d\varepsilon}\right)$ (see (5)), stability can only be preserved for $\left(\frac{dR_{cav}}{d\varepsilon}\right)>0$.

For the system to be stable against the Robinson instability, the resistive part of the cavity impedance must be a growing function of energy. Below transition for instance, this means that the cavity tune must be higher than the RF frequency, as predicted with the preceding qualitative analysis.

Remark : α can be derived from the rigorous expression [6] : $\alpha=\frac{\omega_{S0}I_{DC}(Z_R^+-Z_R^-)}{2V\cos\varphi_S}$

where Z_R^+, Z_R^- are the real part of the impedance at $\omega_C+\omega_{S0}$ and $\omega_C-\omega_{S0}$ respectively.

3.1.2 RF power limitation

The total RF power delivered to the bunch(es) sitting at the phase φ_B is

$$P_B=\frac{VI_B\sin\varphi_B}{2}=\frac{RI_B^2\sin\varphi_B}{2Y} \tag{6}$$

using $V = RI_0$ and $Y = \frac{I_B}{I_0}$.

Focusing of the bunch(es) towards the stable phase disappears when

$$\frac{dP_B}{d\varphi_B}(@I_G, \varphi_Z, I_B = \text{const.}) = 0 \tag{7}$$

$$\frac{dP_B}{d\varphi_B} = \frac{RI_B^2}{2}\left(\frac{\cos\varphi_B}{Y}d\varphi_B - \frac{\sin\varphi_B}{Y^2}dY\right). \tag{8}$$

From geometrical considerations in figure 2, we can write

$$I_G^2 = I_T^2 + I_B^2 - 2I_T I_B \sin(\varphi_Z - \varphi_B)$$

which we transform into :

$$\left(\frac{I_G}{I_B}\right)^2 = 1 + \frac{1}{Y^2\cos^2\varphi_Z} - \frac{2\sin(\varphi_Z - \varphi_B)}{Y\cos\varphi_Z}. \tag{9}$$

Differentiating (9) with the constraint I_G, φ_Z, I_B=constant leads to the relation

$$dY\left[Y\cos\varphi_Z \sin(\varphi_Z - \varphi_B) - 1\right] = d\varphi_B\left[-Y^2\cos\varphi_Z\cos(\varphi_Z - \varphi_B)\right]. \tag{10}$$

Bringing (10) into (8) and after some algebra we finally get

$$Y\sin 2\varphi_Z = 2\cos\varphi_B \tag{11}$$

which is precisely the second stability criterion from Robinson.

This second limit of stability is then clearly due to the limitation in the available RF power. When the threshold (11) is attained the instability grows exponentially, without oscillation, contrarily to the one due to the impedance limitation (section 3.1.1).

3.1.3 Complete Robinson criteria

Below transition energy the full Robinson criteria [5] are

$$0 < Y \sin 2\varphi_Z < 2\cos\varphi_B \,. \tag{12}$$

In the coordinate system (φ_Z,Y) they take the form illustrated in figure 4, for φ_B=0.

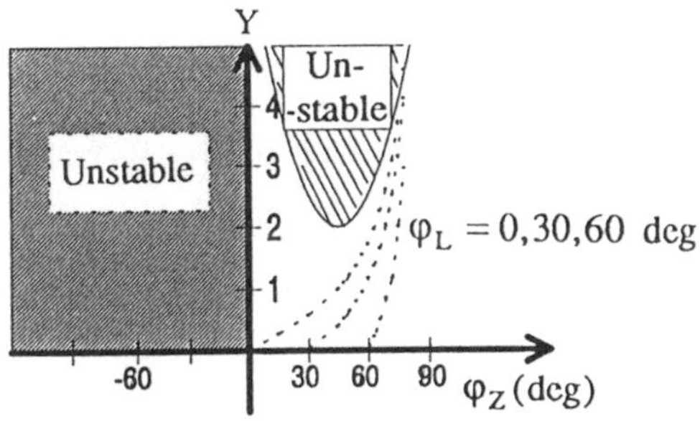

Figure 4 : Robinson stability limits for φ_B=0

According to the previous analysis the Y axis is the limit for an anti-damped oscillation around the stable particle ("Impedance limitation"), while the parabola in the first quadrant is the limit for an aperiodic type of movement away from the stable phase ("RF power limitation").

When a slow tuning system is active ("slow" meaning much slower than the period of synchrotron oscillation) which controls the reactive power required from the amplifier (φ_L=constant), then

$$\tan\varphi_Z = Y(\cos\varphi_B + \sin\varphi_B \tan\varphi_L) + \tan\varphi_L \,. \tag{13}$$

This relation is represented by the dotted lines in figure 4. It is clear that keeping φ_L slightly positive is sufficient to guarantee stability.

Remark : For a beam energy above transition, the sign of the inequalities in (12) is reversed. In the graphics of figure 4, the picture is symmetrical to the one below transition with respect to the Y axis.

3.2 MULTI-LOOP INSTABILITY

3.2.1 Modelling

Robinson criteria rarely apply to real RF systems because of the presence of various servo-loops for cavity tuning, gap voltage regulation, beam phase oscillation damping, etc... Neither $\overline{I}_G$ nor φ_Z are independent variables any more. Figure 5 shows these loops in the case of a typical RF system for a proton synchrotron.

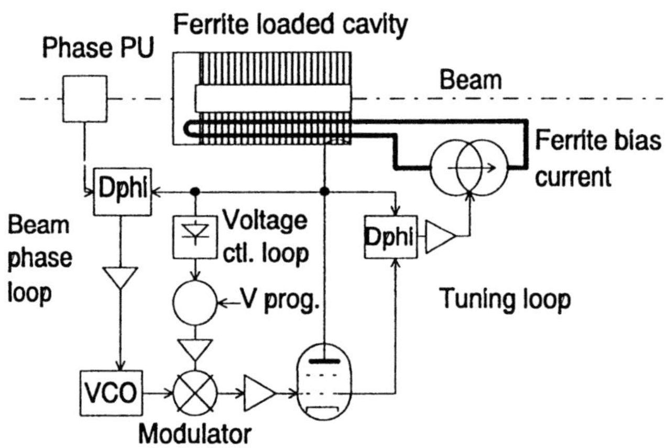

Figure 5 : Typical RF system for a proton synchrotron

Small beam intensity

The loops are coupled if the cavity is not tuned at the RF frequency because of Amplitude Modulation (AM) to Phase Modulation (PM) and PM to AM conversion in the detuned resonator. This is minimized by keeping the best possible tune and by choosing largely different cut-off frequencies. In such conditions each loop can be optimized for stability. In particular the beam phase loop damps the coherent oscillations of the bunch(es) (suppression of the "impedance limitation" leading to the instability in quadrant 2 of figure 4).

Large beam intensity

Because of vector addition of $\bar{I}_B$ with $\bar{I}_G$ to form $\bar{I}_T$ in the cavity, coupling is unavoidable between the loops. Stability analysis has to enclose all the loops and their coupling [4].

Small modulations of phase (p_x) and relative amplitude (a_x) of a variable x are defined as

$$x(t) = \mathrm{Re}\left[X(1+a_x(t))e^{j(\omega_c t+p_x(t))}\right] \qquad \text{with } a_x \ll 1 \text{ and } p_x \ll 1.$$

For such modulations we can derive the flow graph in figure 6 (from [4]), where the following transfer functions (Laplace transforms) are used :

C_p, C_a, C_t: Transmission through the low frequency electronics of the phase, amplitude and tuning loops

G^G_{pp} (G^G_{aa}): Transmission of phase and amplitude modulation of $\bar{I}_G$ into phase (amplitude) modulation of $\bar{V}$

G^G_{pa} (G^G_{ap}): Coupling of phase (amplitude) modulation of $\bar{I}_G$ into amplitude (phase) modulation of $\bar{V}$

G^B_{pp} (G^B_{pa}): Transmission of phase modulation of $\bar{I}_B$ into phase (amplitude) modulation of $\bar{V}$

G_{xp} (G_{xa}) : Transmission of tuning parameter modulation into phase (amplitude) modulation of $\bar{V}$

B : Transfer function of the phase of $\bar{V}$ with respect to the stable phase through the beam and into the phase of $\bar{I}_B$.

$$B = \frac{\omega_S^2}{s^2+\omega_S^2}. \tag{14}$$

The cavity impedance is

$$Z(s) = \frac{2\sigma Rs}{s^2+2\sigma s+\omega_R^2} \tag{15}$$

where $\sigma = \dfrac{\omega_R}{2Q}$ is the damping rate and ω_R the cavity resonant frequency (in rad/s).

$\omega_R, \omega_C, \sigma, \varphi_Z$ are linked by $\omega_R - \omega_C = \sigma \tan \varphi_Z$. (16)

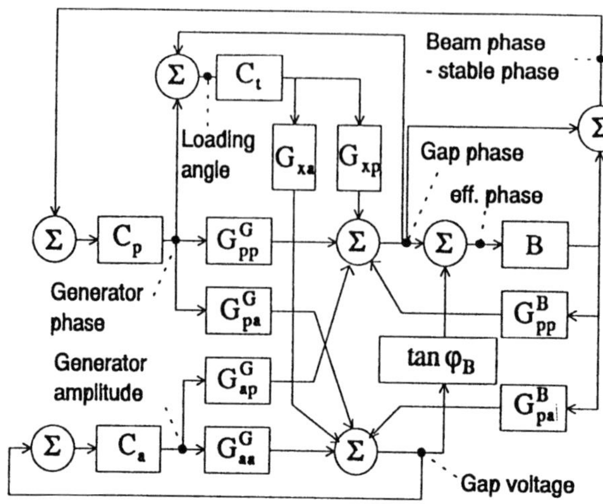

Figure 6 : Flow graph for a complete RF system (from F. Pedersen [4])

Transmission of small modulations of $\bar{I}_T$ through the cavity impedance into modulations of $\bar{V}$ are linear and given by

$$G_{pp} = G_{aa} = \frac{1}{2}\left\{\frac{Z(s+j\omega_C)}{Z(j\omega_C)} + \frac{Z(s-j\omega_C)}{Z(-j\omega_C)}\right\}$$

$$G_{pa} = -G_{ap} = \frac{1}{2}\left\{\frac{Z(s+j\omega_C)}{Z(j\omega_C)} - \frac{Z(s-j\omega_C)}{Z(-j\omega_C)}\right\}. \tag{17}$$

Using (15) and (17) we get

$$G_{pp} = G_{aa} = \frac{\sigma^2\left(1+\tan^2\varphi_Z\right)+\sigma s}{s^2+2\sigma s+\sigma^2\left(1+\tan^2\varphi_Z\right)}$$

$$G_{pa} = -G_{ap} = \frac{\sigma \tan\varphi_Z s}{s^2+2\sigma s+\sigma^2\left(1+\tan^2\varphi_Z\right)}. \tag{18}$$

$\bar{I}_T$ being the vector sum of $\bar{I}_G$ and $\bar{I}_B$, we need to project properly their modulations on I_T to get finally

$$G^G_{pp} = G^G_{aa} = \frac{\sigma^2\left(1+\tan^2\varphi_Z + Y(\sin\varphi_B - \tan\varphi_Z\cos\varphi_B)\right) + \sigma(1+Y\sin\varphi_B)s}{s^2+2\sigma s+\sigma^2\left(1+\tan^2\varphi_Z\right)}$$

$$G^G_{pa} = -G^G_{ap} = \frac{-\sigma^2 Y(\cos\varphi_B + \tan\varphi_Z\sin\varphi_B) + \sigma(\tan\varphi_Z - Y\cos\varphi_B)s}{s^2+2\sigma s+\sigma^2\left(1+\tan^2\varphi_Z\right)}$$

$$G^B_{pp} = \frac{Y\left(\sigma^2(\tan\varphi_Z\cos\varphi_B - \sin\varphi_B) - \sigma\sin\varphi_B s\right)}{s^2+2\sigma s+\sigma^2\left(1+\tan^2\varphi_Z\right)} \tag{19}$$

$$G^B_{pa} = -G^B_{ap} = \frac{Y\left(\sigma^2(\tan\varphi_Z\sin\varphi_B + \cos\varphi_B) + \sigma\cos\varphi_B s\right)}{s^2+2\sigma s+\sigma^2\left(1+\tan^2\varphi_Z\right)}.$$

Letting the tuning control parameter x be $x = \frac{\Delta\omega_R}{\sigma}$, we also get

$$G_{xp} = \frac{\sigma^2+\sigma s}{s^2+2\sigma s+\sigma^2\left(1+\tan^2\varphi_Z\right)}$$

$$G_{xa} = \frac{-\sigma^2\tan\varphi_Z}{s^2+2\sigma s+\sigma^2\left(1+\tan^2\varphi_Z\right)}. \tag{20}$$

3.2.2 Analysis

Even with simple integrators in the electronics of the loops ($C_p = \frac{\omega_p}{s}$, etc...) the characteristic equation of the system is already of degree 8 in s. The analytical treatment of the full system is very complicated, and the degrees of freedom are many. However numerical determination of the limits of stability is easy with a computer once the various parameters are known.

With the simplifying assumptions $\varphi_B=0$, $\varphi_L=0$, $\sigma=0$ and neglecting transmission through the beam ($B(s)=0$), the following criterion is obtained [4]:

$$Y < \sqrt{2+\frac{\omega_a}{\omega_T}+\frac{\omega_T}{\omega_a}+\frac{\omega_p}{\omega_T}+\frac{\omega_T}{\omega_p}+\frac{\omega_a}{\omega_p}+\frac{\omega_p}{\omega_a}}. \tag{21}$$

The stability limits in the case of a "typical" RF system for a proton synchrotron at $\varphi_B=0$ and no tuning loop is shown in figure 7 for comparison with the no-loop situation corresponding to the Robinson criteria (figure 4).

$\frac{\omega_S}{2\pi}=1.8$ kHz $\quad \frac{\sigma}{2\pi}=50$ kHz (1/2 cavity 3 db bandwidth)

$\frac{\omega_p}{2\pi}=18$ kHz $\quad \frac{\omega_a}{2\pi}=6$ kHz

$\frac{\omega_T}{2\pi}=0$ kHz (no tuning loop)

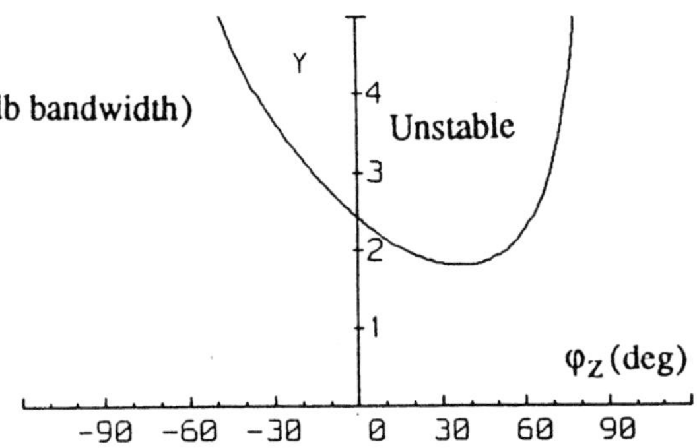

Figure 7 : Stability limits for a "typical" RF system with $\varphi_B=0$

Case of a very fast tuning loop

It is worthwhile to note that in the case of a very fast tuning loop maintaining $\varphi_L=0$, and with no other loop active, the instability limit is independent of φ_Z and is simply given by

$$Y \sin\varphi_B < 1 \tag{22}$$

which corresponds to an optimum transfer of power from the generator to the beam. It is an RF power limitation which can be derived with the procedure applied in section 3.1.2.

4 TRANSIENT BEAM LOADING

4.1 BASIC RELATIONS

In the analysis of section 3, low frequency (a few times f_S) modulations of small amplitude around the steady state have been considered. Consequently only rigid dipolar coherent oscillations (m=1, n=0) of the beam have been taken into account. The situation is different when the system is perturbed by fast transients in the beam current, with a repetition period which is larger than the cavity filling time $\tau=1/\sigma$. The spectral components of the beam current cover a wide frequency range where the cavity impedance is non-negligible. This is the case at injection or ejection, and also when the beam has gaps which are large with respect to τ.

Contrarily to the steady-state case where an ideal current generator is the best model for the beam effect, the transient analysis is often easier when calculating voltages [7]. The passage of a point charge across the cavity loads the capacitor C, which is equivalent to a step-like perturbation of V.

"Fundamental theorem of beam loading" [7]

When a point charge crosses an empty cavity, it leaves behind a voltage V_{b0}. The point charge itself is decelerated by $V_{b0}/2$.

An intuitive demonstration of this effect can be given as follows. Since the charge is point like, $|dI_B/dt|$ is infinite and no energy is left in the inductance L. The voltage V_{b0} developed across the capacitance C is related to the beam charge q by

$$V_{b0} = \frac{q}{C}. \tag{23}$$

The energy stored in C is $\Delta U_C = \frac{1}{2} C V_{b0}^2$.

Letting V_{DEC} be the decelerating voltage experienced by the charge q, the energy lost by the beam is

$$\Delta U_0 = qV_{DEC} = CV_{b0}V_{DEC} \cdot$$

Since the energy lost by the charge ΔU_0 has been transferred to the capacitance C as stored energy ΔU_C

$$\Delta U_0 = \Delta U_C \Rightarrow V_{DEC} = \frac{V_{b0}}{2} \qquad (24)$$

Transient response to a single beam passage

When a point charge crosses the cavity gap at time t=0s, the voltage V starts a damped oscillation at the cavity resonance frequency ω_R (in rad/s), with a damping rate $\sigma=1/\tau$ (in s^{-1}), the initial value being V_{b0}

$$\overline{V} = V_{b0}e^{(-\sigma+j\omega_R)t} . \qquad (25)$$

If a voltage V_g at frequency ω_C is forced across the gap by the RF amplifier before the beam passage, the result is simply the vector sum of both voltages.

4.2 NON-PERIODIC CURRENT TRANSIENTS

Such transients are encountered at beam transfer (injection/ejection), or while crossing transition energy in proton accelerators for instance. These disturbances being rare (a few per machine cycle) they correspond to step-like perturbations, and the usual servo-loops can be sufficient to get a properly damped response. The loop parameters must be such that the full RF system is stable under the analysis of section 3.2. On top of that the RF amplifier must be powerful enough to cope with the situation until the tuning loop has settled to the new optimum. This matter is treated in section 6.

Let us assume a periodic beam traversal with a period T_b (evenly spaced bunches of equal intensity), and a voltage $\bar{V}_g$ at frequency ω_C forced across the unperturbed cavity by the

RF amplifier. Taking the phase of $\bar{V}_g$ as reference, the voltage seen by the beam after many traversals is given by [7] :

$$\overline{V} = \overline{V}_g + V_{b0} e^{-j\left(\frac{\pi}{2}+\varphi_s\right)} \left(\frac{1}{2} + e^{(-\sigma + j\Delta\omega)T_b} + e^{2(-\sigma + j\Delta\omega)T_b} + \ldots \right) \tag{26}$$

where :

$$\Delta\omega = \omega_R - \omega_C . \tag{27}$$

This vector composition is shown in figure 8. Some algebraic transformations lead to :

$$\overline{V} = \overline{V}_g + \frac{V_{b0}}{2} e^{-j\left(\frac{\pi}{2}+\varphi_s\right)} \left(\frac{1 + e^{(-\sigma + j\Delta\omega)T_b}}{1 - e^{(-\sigma + j\Delta\omega)T_b}} \right) . \tag{28}$$

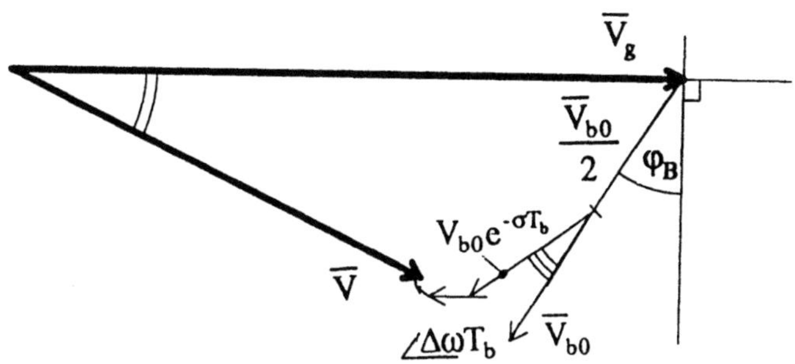

Figure 8 : Vector addition of voltages for repetitive bunch passages

After a transient period (a few times τ) following the beginning of the bunch train, a stationary situation is attained, and all bunches experience the same voltage V. Such a case is quasi-steady-state and can be handled as in section 3, although the gap voltage is clearly modulated at the bunch frequency $1/T_b$.

4.3 PERIODIC CURRENT TRANSIENTS

Such a situation occurs when the spacing between bunches is unequal, or when the spacing is regular but not the intensity of the bunches. We shall illustrate this with the example of a batch of five adjacent bunches filling one half of the circumference (h_{RF}=10). If $\tau << T_b$ all transients have decayed between the bunches and the situation is comparable to the non-periodic transient case (section 4.2). If not, and if $\tau<T_{REV}$ the system is in a periodic transient case. Assuming $\tau<<T_{REV}$ then no voltage is left from the previous beam passage when the first bunch of the batch arrives. The first bunch sees $\bar{V}_{b1}$, the second $\bar{V}_{b2}$, etc.. according to (26):

$$\overline{V}_{b1} = \overline{V}_g + V_{b0}e^{-j\left(\frac{\pi}{2}+\varphi_s\right)}\left(\frac{1}{2}\right)$$

$$\overline{V}_{b2} = \overline{V}_{b1} + V_{b0}e^{-j\left(\frac{\pi}{2}+\varphi_s\right)}e^{(-\sigma+j\Delta\omega)T_b} \qquad (29)$$

$$\overline{V}_{b3} = \overline{V}_{b2} + V_{b0}e^{-j\left(\frac{\pi}{2}+\varphi_s\right)}e^{2(-\sigma+j\Delta\omega)T_b}$$

etc....

Figure 9 shows the composition of vectors for the first two bunches.The successive bunches see voltages affected by phase and amplitude modulations.

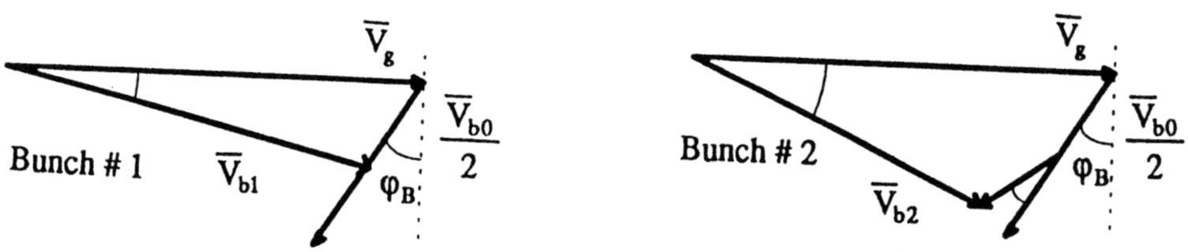

Figure 9 : Voltage seen by successive bunches

This result can also be derived from a frequency domain analysis, by complex addition of the voltages developed in the cavity impedance by the various spectral components of the beam current. Figure 10 shows the frequency domain situation alongside its time domain counterpart.

The important parameter is the cavity impedance at the various harmonics of the revolution frequency where large beam current components exist. The gap voltage V is modulated in amplitude and phase at a rate which is too fast and out of control for all the conventional loops envisaged in section 3.2. The amplifier works on a "pulsed" reactive load.

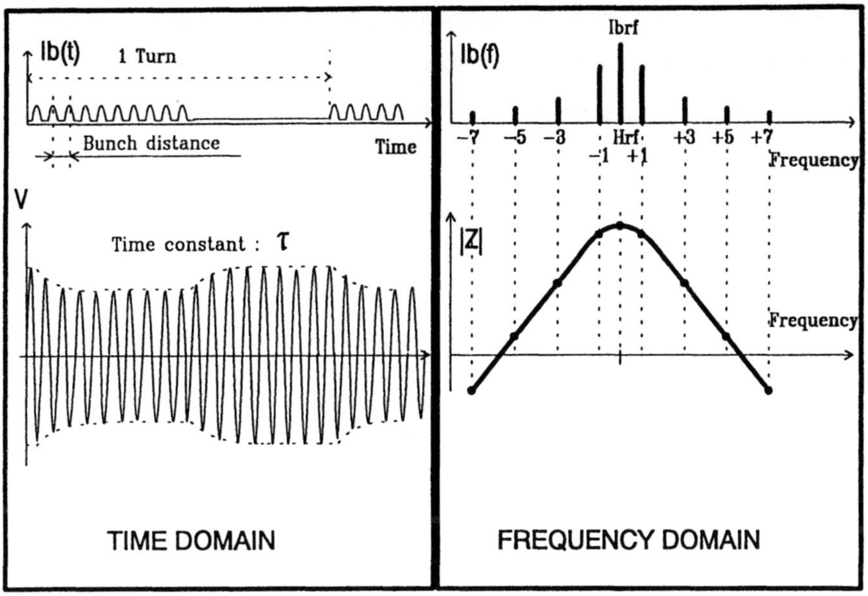

Figure 10 : Time and frequency domain analysis

The successive bunches of the batch experience different longitudinal focusing voltages. The excitation of non-coherent beam oscillations is favoured, leading to longitudinal emittance blow-up after filamentation.

5 CURES

5.1 PASSIVE DAMPING

The brute force method to decrease the relative beam loading $Y=I_B/I_0$ is to increase I_0. The immediate consequence is the requirement for more RF power from the amplifier, because of the higher dissipation in the resistance of the equivalent resonator. Various practical means can be applied for this purpose, such as :

- using a low output impedance amplifier (triode, tetrode with a pulsed bias [8], amplifier overcoupled to the cavity when driving from a transmission line...)
- putting a dummy load in parallel with the cavity.

The performance obtained is easily estimated in stability diagrams as in figures 4 and 7 by scaling down the Y-axis values proportionally to the increase of I_0.

5.2 FEEDFORWARD [9,10]

If the RF amplifier is driven to inject continuously a current $-\bar{I}_B$ in the cavity, the equivalent impedance seen by the beam is zero. This technique is illustrated in figure 11, where $-\bar{I}_B$ is added to the low level RF drive $\bar{I}_G'$ to get the generator current $\bar{I}_G$.

$$\left.\begin{array}{l}\bar{I}_G = \bar{I}'_G - \bar{I}_B \\ \bar{I}_T = \bar{I}_G + \bar{I}_B\end{array}\right\} \Rightarrow \bar{I}_T = \bar{I}'_G$$

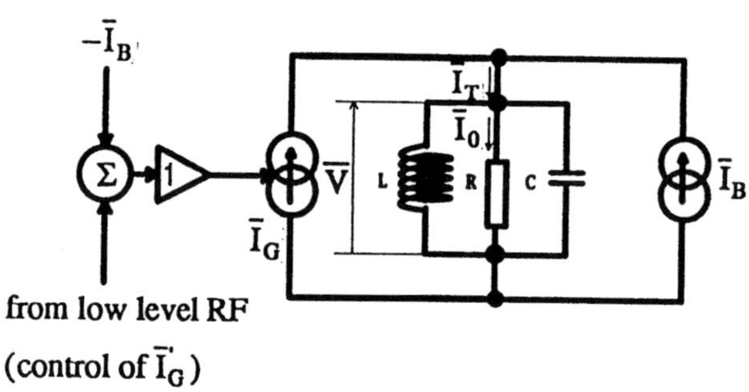

Figure 11 : Lay-out of a feedforward beam loading compensation

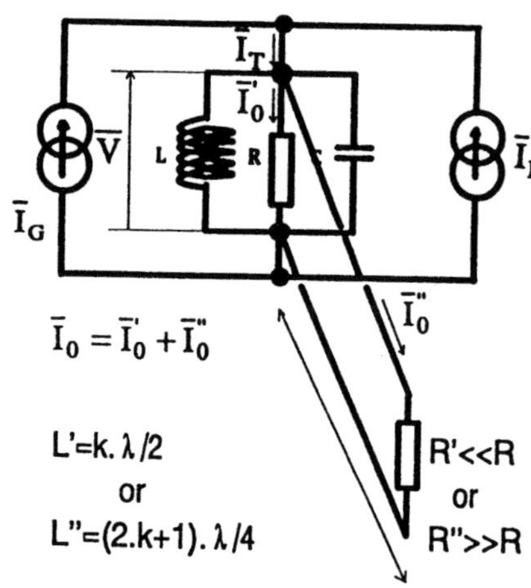

Figure 12 : Feedforward variant

Another implementation of the same principle is possible, making use of the total reflection of the beam induced signal on the output of a strongly mismatched amplifier (figure 12). The electrical length of the transmission line is adjusted so that there is a voltage node in the standing wave pattern at the input port of the cavity ($(2k+1)\lambda/4$ for an open circuit on the amplifier side).

The net result is that the low level RF system (Beam Control) controls directly $\bar{I}_T$, instead of $\bar{I}_G$ in the ordinary case. The cross-coupling between loops is considerably reduced and the global stability of the system is largely improved [9]. (The cross-coupling is not fully suppressed since modulations of $\bar{I}_B$ still induce modulation of the loading angle φ_L regulated by the tuning loop.) The requirements for the high power hardware are unchanged.

5.3 WIDE BAND FEEDBACK [11]

Enclosing the high power amplifier and the cavity in a closed loop with an important loop gain (see figure 13) brings in the usual benefit of reducing the sensitivity of any controlled parameter versus an external perturbation. In particular the equivalent impedance seen by the beam ($Z_{eq}=dV/dI_B$) is changed according to

$$Z_{eq} = \frac{Z_{cav}(j\omega)}{1+G_{OL}(j\omega)} \tag{30}$$

where $G_{OL}(j\omega)$ is the loop gain. In a first approximation Z_{cav} is the only frequency sensitive element in $G_{OL}(j\omega)$. The loop gain is then maximum and real at the cavity resonance ω_R, where we get

$$G_{OL}(j\omega_R) = AgRH$$

$$R_{eq} = \frac{R}{1+G_{OL}(j\omega_R)} << R. \tag{31}$$

The transfer function from the low level input v to the gap voltage V at ω_R is

$$\frac{V}{v} = \frac{AgR}{1+AgRH} \approx \frac{1}{H}. \tag{32}$$

This demonstrates one very interesting feature of this technique : the low level system directly controls the gap voltage V, which is the most relevant physical parameter for the beam.

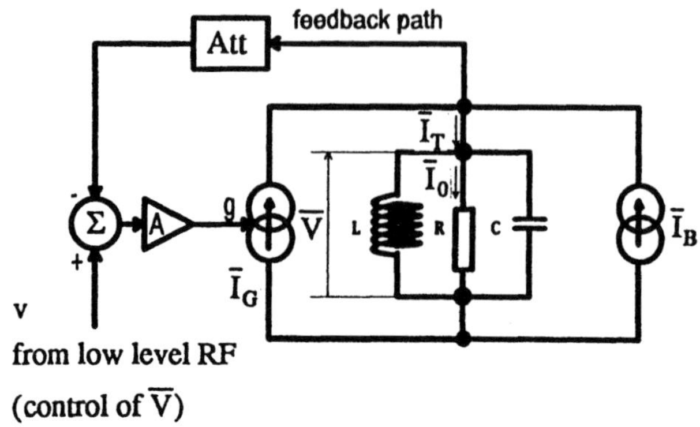

Figure 13 : Lay-out of wide-band feedback

However the group delay τ in the loop must be tightly controlled. Let us assume a fixed frequency cavity, where the loop gain has to be larger than 0 dB over a bandwidth B (in rad/s) :

$$B >> \omega_{3dB} = \frac{\omega_R}{Q} = 2\sigma. \tag{33}$$

The phase shift over this frequency range cannot exceed $3\pi/2$ rad if a phase margin of $\pi/4$ rad is requested at 0 dB gain. Since the resonator itself contributes already for π radians, then the phase shift due to τ must be smaller than 2 times $\pi/4$ rad over B rad/s : $B\tau < \frac{\pi}{2}$ rad. (34)

From the approximate expression of the cavity impedance, far from resonance,

$$Z_{cav}\left(j(\omega_R + \Delta\omega)\right) = \frac{R}{2jQ\frac{\Delta\omega}{\omega_R}}, \tag{35}$$

we deduce that the maximum loop gain at resonance is

$$G_{OLmax}(j\omega_R) = \frac{\pi Q}{2\omega_R \tau} \tag{36}$$

and the corresponding minimum impedance at resonance is

$$R_{eq\,min} = \frac{2\omega_R}{\pi}\frac{R}{Q}\tau. \tag{37}$$

It is worthwhile to notice that this result only depends on R/Q (cavity geometry) and τ.

If the cavity is tunable over a frequency range $B_{Tune} = \omega_{Rmax} - \omega_{Rmin}$, the requirement for the delay τ is even more stringent. With the same request for the phase margin one needs

$$(B_{Tune} + B)\tau < \frac{\pi}{2} \text{ rad}, \tag{38}$$

leading to a minimum impedance at resonance given by

$$R_{eq\,min} = \frac{\omega_R}{2\pi}\frac{R}{Q}\left(\frac{1}{\frac{1}{4\tau} - B_{Tune}}\right) \tag{39}$$

which is always less favourable than the fixed frequency optimum (35). A tunable low-Q resonator in the amplifier chain is a convenient means to ease the practical realization [12].

Implementing RF feedback is equivalent to increasing I_0 by the factor $(1+G_{OL}(j\omega_R))$, and consequently to reduce the relative beam loading Y in the same proportion. The effect is the same as for passive damping (section 5.1), but no more power is required from the amplifier.

Spurious effect of RF feedback

The longitudinal coupled bunch instabilities threshold [1] is degraded for two main reasons:

- the real part of Z_{eq}, which is the determining factor for the instabilities, is increased outside the loop bandwidth, because of the coefficient $\frac{1}{(1+G_{OL}(j\omega_R))}$ in the expression (30) of Z_{eq},
- if the phase shift in the amplifier chain is not ideally controlled to make $G_{OL}(j\omega_R)$ real, the maximum of Z_{eq} is shifted in frequency. The real part of Z_{eq} becomes asymmetrical with respect to the RF frequency, which combines with the first effect mentioned to help trigger beam instabilities.

However a conventional multipath damping system [13] can still be used for compensation, if one takes properly into account the modification of the low level to gap voltage transfer function [14].

5.4 LONG DELAY FEEDBACK [15,16]

Moderate bandwidths of a few times the synchrotron frequency are of interest when considering the voltage induced on revolution harmonics different from the RF frequency. A feedback with a long delay can be considered, as long as the stability requirements are met. This is the case when the feedback path contains a comb filter with gain maxima at the revolution harmonics, and when the total electrical delay is made equal to one revolution period in the machine (Figure 14).

The long delay makes it possible to make the necessary hardware at low power level, where digital electronics can be used, and to install it outside the radiation area in a convenient equipment room.

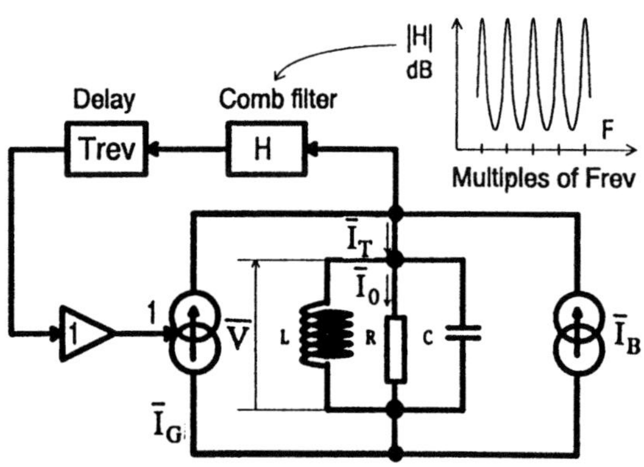

Figure 14 : Lay-out of "One-turn delay" feedback

A simple recursive filter gives the following comb-like transfer function

$$H(j\omega) = \frac{G_0}{1 - Ke^{-j\Delta\omega T_{REV}}} \tag{40}$$

where K and G_0 are constants, $0<K<1$, and $\Delta\omega = \omega - \omega_R$. Cascaded with the one turn delay ($e^{-j\Delta\omega T_{REV}}$) and the cavity impedance $Z_{cav}(j\omega)$, the overall open loop transfer function is

$$G_{OL}(j\omega) = \frac{G_0}{e^{j\Delta\omega T_{REV}} - K} Z_{cav}(j\omega). \tag{41}$$

With $Z_{cav}(j\omega)$ constant and real (case of a travelling wave cavity for instance [15]), it is represented in the complex plane by a circle centered on the real axis and crossing it at $\left[\frac{-G_0Z}{1+K},0\right]$ and $\left[\frac{G_0Z}{1-K},0\right]$ (figure 15).

Stability is preserved as long as $\frac{G_0Z}{1+K}<1$. The equivalent impedance for the beam is then

$$Z_{eq}(j\omega) = Z_{cav}(j\omega)\left(\frac{e^{j\Delta\omega T_{REV}} - K}{e^{j\Delta\omega T_{REV}} - K + G_0 Z_{cav}(j\omega)}\right). \tag{42}$$

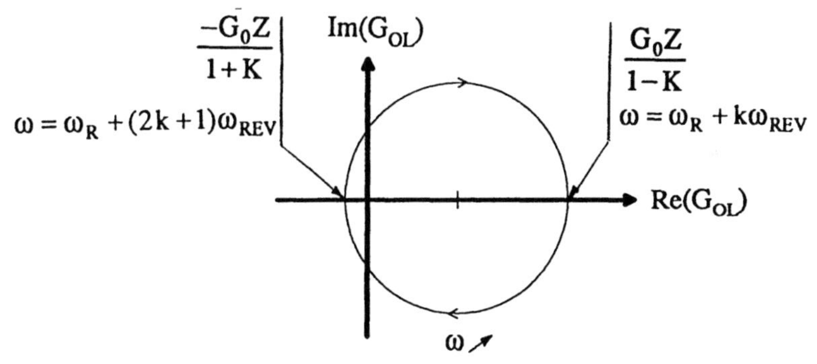

Figure 15 : Open loop transfer function

Minima of Z_{eq} occur at the revolution harmonics :

$$\omega = \omega_R + k\omega_{REV} \quad \rightarrow \quad Z_{eq\,min} = Z_{cav}\left(\frac{1-K}{1-K+G_0Z_{cav}}\right) \approx \frac{1-K}{G_0} << Z_{cav} \tag{43}$$

while impedance maxima are at

$$\omega = \omega_R + (2k+1)\omega_{REV} \quad \rightarrow \quad Z_{eq\,max} = Z_{cav}\left(\frac{1+K}{1+K-G_0 Z_{cav}}\right). \tag{44}$$

$$\text{Selecting} \quad Z_{eq\,max} = 2Z_{cav} \quad \Rightarrow \quad \frac{G_0 Z_{cav}}{1+K} = \frac{1}{2}$$

the minimum impedance attained (43) is $Z_{eqmin} \approx Z_{cav}(1-K)$ since $K \approx 1$.

For smaller cavity bandwidth (non-constant $Z_{cav}(j\omega)$) stability is still possible, and equivalent results are obtainable. A recent practical implementation of this principle has been made at the CERN PS, where the beam revolution frequency changes by 15% from injection to high energy, and the cavity is tunable from 4 to 9.5 MHz (harmonic number from 10 to 20) [16].

The long delay feedback technique reduces the cavity impedance on many revolution harmonics, which is very efficient against periodic transient beam loading and longitudinal coupled-bunch instabilities (section 4.3).

The high power amplifier has to be able to provide the current requested at these frequencies, where its load is strongly mismatched (in fact the load is like a short circuit when the system works properly, since current is injected and no voltage is developed).

5.5 COMPARISON OF CURES

The characteristics of the various cures are gathered in Table 1. The wide band feed-back technique is clearly the most interesting solution, as demonstrated by the number of developments in this direction over the recent years. A long delay feedback is a useful complement to reduce the impedance over a larger number of revolution harmonics than the wide band method alone can cover.

	Passive damping	Feedforward	Wide band feedback	Long delay feedback
References	8	9,10,20	11,12,17,18,19	15,16
Effect on continuous beam loading	Positive	Positive	Positive	Non-relevant
Effect on transient beam loading	Weak	Delicate and limited in performance	Positive but function of loop gain	Positive
Spurious effect on beam	None	Marginally positive	Degrades instab. threshold	Improves instab. threshold
Hardware complexity	Small	Small to medium	Large	Medium
Run-time adjustment	Easy (on bench)	Lengthy (with beam)	Easy (on bench)	Easy (on bench)
Stability against drift	Good	Bad (open loop)	Good (closed loop)	Good (closed loop)
Oper. range (β and h_{RF})	Unlimited	Lim. for good performance	Unlimited	Unlimited
Cost	Large	Small	Medium	Small
Oper. from low level RF	Easy	Complex	Easy	Easy

Table 1 : Comparison of cures

Remark : Comments on the feedforward method assume that the beam signal is fed at the low power end of the amplification chain (long delay). Better ratings can be given when a small overall electrical delay is used[19].

6 RF POWER REQUIREMENTS

The two kinds of regimes investigated previously have to be analyzed to derive what is expected from the amplifier-cavity system and their associated equipment. The steady-state situation helps define the coupling factor and the CW RF power. Analysis of the transient cases gives indications on the peak capabilities needed (RF power, generator current, mismatch,...) and on the speed of the servo-loops (tuning and voltage). These aspects are treated in the following paragraphs for the two most frequently used amplifier-cavity set-ups.

6.1 CURRENT GENERATOR NEAR TO THE CAVITY

This is the case of medium power systems (up to 100 kW), where the anode of a high power RF tube directly drives the cavity through a short connection. The vector diagram in figure 2 is used for the analysis.

Steady-state operation

When the tuning is not optimum ($\varphi_L \neq 0$) the current I_G is given by

$$I_G = \frac{I_0 + I_B \sin\varphi_B}{\cos\varphi_L}$$

under the voltage V, which means the amplifier delivers a real power : $P_{real} = \frac{VI_G \cos\varphi_L}{2}$ and a reactive power $P_{reactive} = \frac{VI_G \sin\varphi_L}{2}$.

With the optimum cavity tuning ($\varphi_L=0$) the current I_G and the real power P_{real} needed from the amplifier are minima. However, since $I_G=I_0+I_B\sin\varphi_B$, the apparent load impedance is a function of I_B when $\varphi_B \neq 0$.

Transient analysis

Let us assume that the voltage and beam phase loops are fast enough to correct the gap voltage in amplitude and phase in a fraction of a synchrotron period. For a step-like perturbation $\Delta\bar{I}_B$ in the beam current $\bar{I}_{B0}$ (case of section 4.2) the amplifier must "instantaneously" (faster than a synchrotron period) provide $-\Delta\bar{I}_B+\bar{I}_{G0}$ ($\bar{I}_{B0}$ and $\bar{I}_{G0}$ being the values before the perturbation). It lasts as long as the tuning loop has not settled again to $\varphi_L=0$: the faster the tuning loop, the shorter the requirement.

To cancel the effects of periodic transient beam loading (section 4.3) I_G has to be equal to the beam current at the various harmonics of the revolution frequency. The amplifier load at these frequencies is very low impedance when the system works properly (beam current properly compensated).

6.2 AMPLIFIER AND ISOLATOR SET-UP [21]

This is the case of very high power systems (~MW) used in large and very large synchrotrons, where RF is sent via a transmission line to the cavity, and tolerance to reflected power is limited. An isolator is inserted to guarantee a matched load to the amplifier (figure 16). RF power is then the only parameter to specify for the amplifier.

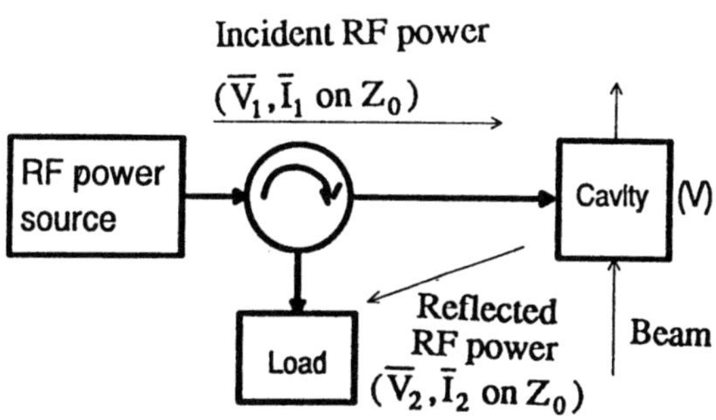

Figure 16 : Amplifier with isolator set-up

To get the required voltage $\bar{V}$ in the cavity, the incident RF power (wave ($\bar{V}_1,\bar{I}_1$) on the impedance Z_0) is a function of the beam loading which changes the apparent load at the end of the transmission line. The reflected power (wave ($\bar{V}_2,\bar{I}_2$) on the impedance Z_0) is dissipated in the load. The equivalent circuit is shown in figure 17, where the generator is represented as a current source matched to the line impedance Z_0 (current $2\bar{I}_1$ and shunt impedance Z_0) and driving the cavity impedance Z_{cav}. The beam current $\bar{I}_B$ is also injected into Z_{cav} and Z_0.

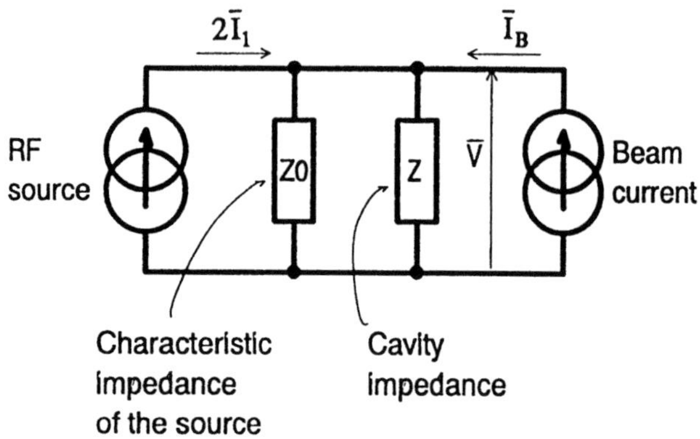

Figure 17 : Equivalent circuit

From figure 17 we derive $2\bar{I}_1 = \bar{V}\left(\frac{1}{Z_0} + \frac{1}{Z}\right) - \bar{I}_B$. (45)

By definition the voltage $\bar{V}_2$ of the reflected wave at the cavity is given by $\bar{V}_2 = \bar{V} - \bar{V}_1$.

Dividing by Z_0 and multiplying by 2 we get $2\bar{I}_2 = 2\frac{\bar{V}}{Z_0} - 2\bar{I}_1$. (46)

From (45) and (46) we deduce $2\bar{I}_2 = \bar{V}\left(\frac{1}{Z_0} - \frac{1}{Z}\right) + \bar{I}_B$. (47)

The reflected wave is generated at the cavity input port by a current source matched to the line impedance Z_0 (current $2I_2$ and shunt impedance Z_0), and is propagating backwards to the load through the isolator.

Steady-state operation

With an ideal tuning system $\bar{V}$ is kept in phase with $\bar{I}_1$. Consequently

$$2I_1 = V\left(\frac{1}{Z_0}+\frac{1}{R}\right)+I_B \sin\varphi_B$$

$$2I_2 = V\left(\frac{1}{Z_0}-\frac{1}{R}\right)-I_B \sin\varphi_B \cdot \tag{48}$$

Let us select the critical coupling for the cavity (Z_0=R) to minimize the RF power during storage (φ_B=0). The incident RF power being

$$P_{incident} = \frac{1}{2} Z_0 I_1^2 \tag{49}$$

we get
$$P_{incident} = \frac{1}{2}\frac{V^2}{R}+\frac{1}{2} V I_B \sin\varphi_B + \frac{1}{2} Z_0\left(\frac{I_B \sin\varphi_B}{2}\right)^2 . \tag{50}$$

The first two terms in (49) correspond to the minimum power required to compensate for the cavity losses and to accelerate the beam. The third term is due to the mismatch induced by the beam loading (even though the tuning is perfect). For supra-conducting cavities, where Z_0 is very large, that power loss becomes important so that the coupling coefficient (Z_0/R) is rather selected to cancel the reflected power during beam acceleration [21].

Transient analysis

For a step-like current perturbation $\Delta\bar{I}_B$ in the beam current $\bar{I}_B$ the situation is the same as in section 6.1. The amplifier must deliver $\bar{I}_{10}$-$\Delta\bar{I}_B$ in the cavity, as long as the tuning loop has not settled again at φ_L=0. Assuming that critical coupling is used (room temperature cavity : Z_0=R), the incident power required can be shown to be

$$\hat{P}_{incident} = P_0\left(1+\left(\frac{V\Delta I_B}{4P_0}\right)^2\right) \tag{51}$$

where P_0 is the power needed to get V when no acceleration takes place

$$P_0 = \frac{1}{2}\frac{V^2}{Z_0}. \tag{52}$$

In the case of periodic transient beam loading, with $\tau << T_{REV}$, $\Delta I_B = \hat{I}_B$ (local beam current). Consequently the peak RF power required is

$$\hat{P}_{incident} = P_0\left(1 + \left(\frac{V\hat{I}_B}{4P_0}\right)^2\right) \tag{53}$$

which is minimum for $$P_0 = \frac{V\hat{I}_B}{4}. \tag{54}$$

The minimum peak RF power is then $$\hat{P}_{incident} = \frac{V\hat{I}_B}{2}. \tag{55}$$

which depends on the local beam current $\hat{I}_B$.

REFERENCES

[1] F.J. Sacherer, "Bunch lengthening and microwave instability", IEEE Trans. Nuc. Sci., NS-24,1977, p. 1393

[2] D. Boussard, "Beam loading", 1985 CERN Accelerator School, Oxford (G.B.), CERN 87-03, and CERN SPS/86-10 (ARF)

[3] D. Boussard, "Control of cavities with high beam loading", IEEE Trans. Nuc. Sci., NS-32, 1985, p.1852

[4] F. Pedersen, "Beam loading effects in the CERN PS Booster", IEEE Trans. Nuc. Sci., NS-22, 1975, p.1906

[5] K.W. Robinson, "Stability of beams in Radiofrequency systems", CEAL-1010, Feb. 1964

[6] A. Hofmann, "Coherent beam instabilities", these proceedings

[7] P.B. Wilson, "Transient beam loading in electron-positron storage rings", PEP Note-276

[8] G. Gelato et al., "Evolution of the RF system of the CPS Booster since the beginning of its operation", IEEE Trans. Nuc. Sci., NS-22, 1975, p.1334

[9] D. Boussard, "Cavity compensation and beam loading instabilities", CERN/SPS/ARF Note 78-16

[10] H. Frischholz, W. Schnell, "Compensation of beam loading in the ISR RF cavities", IEEE Trans. Nuc. Sci., NS-24, 1977, p. 1683

[11] F. Ferger, W. Schnell, "The high power part of the RF system for the Intersecting Storage Rings (ISR)", 2nd Natl. Conf. on Part. Acc., Moscow, Sept. 1970

[12] R. Garoby et al., "RF system for high beam intensity acceleration in the CERN PS", Proc. of the 1989 IEEE Part. Acc. Conf., Chicago, March 20-23, p. 135

[13] F. Pedersen, F. Sacherer, "Theory and performance of the longitudinal active damping system for the CERN PS Booster", IEEE Trans. Nuc. Sci., NS-24,1977, p. 1396

[14] G. Schneider, "Group delay equalizer for improvement of longitudinal beam stability in the PSB", CERN PS/RF/Note 91-4

[15] D. Boussard, G. Lambert, "Reduction of the apparent impedance of wide band accelerating cavities by RF feedback", IEEE Trans. Nuc. Sci., NS-30, 1983, p. 2239

[16] F. Blas, R. Garoby, "Design and operational results of a one turn delay feedback for beam loading compensation on the CERN PS ferrite cavities", to be published in the Proc. of the 1991 Part. Acc. Conf., San Francisco, May 6-9

[17 J. M. Baillod et al., "A second harmonic RF system with feedback-reduced gap impedance", IEEE Trans. Nuc. Sci., NS-30, 1983, p. 3499

[18] D. Boussard et al., "RF feedback applied to a multicell superconducting cavity", Proc. of the European Part. Acc. Conf., Rome, June 7-11, 1988, p. 985

[19] L.W. Funk et al., "52 MHz RF systems for PETRA II and HERA proton synchrotrons", Proc. of the 14th intern. conf. on high energy acc., Particle Accelerators, 1990, Vol.29, p.71

[20] S. Ninomiya et al., "The RF power system with beam loading cancellation for the KEK-PS Booster", Proc. of the 14th intern. conf. on high energy acc., Particle Accelerators, 1990, Vol.32, p.69

[21] D. Boussard, "RF power estimates for a hadron collider", CERN SPS/ARF/Note 84-9

High-Brightness Injectors for Hadron Colliders

Thomas P. Wangler

Accelerator Technology Division, MS-H817
Los Alamos National Laboratory, Los Alamos, NM 87545 USA

1 Introduction

The counterrotating beams in collider rings consist of trains of beam bunches with N_B particles per bunch, spaced a distance S_B apart. When the bunches collide, the interaction rate is determined by the luminosity, which is defined as the interaction rate per unit cross section. For head-on collisions between cylindrical Gaussian beams moving at speed βc, the luminosity is given by

$$L = \frac{N_B^2 \beta c}{4\pi\sigma^2 S_B} \quad , \tag{1.1}$$

where σ is the rms beam size projected onto a transverse plane (the two transverse planes are assumed identical) at the interaction point. This beam size depends on the rms emittance of the beam and the focusing strength, and is given by

$$\sigma = \left[\frac{\varepsilon_n \tilde{\beta}^*}{\beta\gamma}\right]^{1/2} \quad , \tag{1.2}$$

where ε_n is the transverse normalized rms emittance, which is a measure of the 2-D phase-space area in each transverse plane, and is defined in terms of the second moments of the beam distribution. Our convention is to use the true rms normalized emittance, without factors of 4 or 6 that are sometimes used. The quantity $\tilde{\beta}^*$ is the Courant-Snyder betatron amplitude function at the interaction point, a characteristic of the focusing lattice, and γ is the relativistic Lorentz factor. Using this expression for σ, the luminosity becomes

$$L = \frac{N_B^2 \gamma \beta^2 c}{4\pi\tilde{\beta}^* \varepsilon_n S_B} \quad . \tag{1.3}$$

Achieving high luminosity at a given energy, and at practical values of $\tilde{\beta}^*$ and S_B, requires a large value for the ratio N_B^2/ε_n, which implies high intensity and small emittance. Thus, specification of the luminosity sets the requirements for beam intensity and emittance, and establishes the requirements on the performance of the injector to the collider ring. In general, for fixed N_B, the luminosity can be increased if ε_n can be reduced.

The minimum emittance of the collider is limited by the performance of the injector; consequently the design of the injector is of great importance for the ultimate performance of the collider. As a numerical example of luminosity that one hopes to achieve, we consider the SSC design.[1] Assuming $N_B = 0.75\times10^{10}$ protons per bunch, $\gamma = 21\times10^3$, $\beta = 1, \tilde{\beta}^* = 0.5$ m, $\varepsilon_n\pi = 1\times10^{-6}\pi$ m-rad, $S_B = 5$ m, we obtain $\sigma = 4.9\times10^{-6}$ m, and $L = 1.1\times10^{33}$ cm^{-2} s^{-1}.

1.1 Injector System Architecture

The injector system for a collider is itself a major high-energy accelerator facility. It consists of a dc injector, an rf linac, and a multistaged synchrotron accelerator. There are practical reasons for this sequence of accelerators, each of which is ideal within a given energy range, which allows the beam energy to be raised by orders of magnitude while maintaining high intensity and low beam emittances. As will be discribed later, the emittances can grow for a variety of reasons specific to each individual section, and for maximum luminosity the emittance growth in each of these sections must be carefully controlled.

The H^- beam for a p-p collider originates in the ion source, from which the beam is extracted and injected into the linac for further acceleration. H^- ions are chosen because, using multiturn injection through a stripper foil into a synchrotron ring,[2] a large circulating current of H^+ ions can be formed by the overlaying of many turns into the same phase-space area. Because many turns can be used to fill the ring, the procedure does not require a large current in the ion source and linac, and this reduces the linac space-charge problems. The output beam from the linac can be allowed to debunch, and subsequently be rebunched after injection into the ring at an rf frequency unrelated to the linac frequency. Alternatively the bunch structure from the linac can be maintained, and the bunches can be injected into preexisting buckets in a ring, whose rf frequency is synchronized with that of the linac. After acceleration to full energy in the lowest energy ring, the beam is extracted in a single turn, transported, and injected into the next ring, which has a larger circumference, consistent with it's higher output energy. The process is repeated, and beams can be stacked end-to-end until the second ring is full. Then acceleration of the H^+ beam in the second ring brings the beam to the full energy of the second ring, where it can be single-turn extracted and transferred to the next ring. This process of stacking and transferring beams continues until both collider rings are filled with beams from the injector system.

The advantages of using a multistage synchrotron for a high-brightness injector[3] include two technical reasons: 1) reduced rf frequency range for each ring, and 2) avoidance of low-field operation of the magnets, where the field quality is not as good. Also, an important fundamental reason for a multistage design is the space-charge force, which causes a spread in the transverse (betatron) single-particle oscillation frequencies (tunes) and can result in excitation of beam resonances, with consequent growth of emittance and loss of beam. The effect of the space-charge forces is to defocus the beam and consequently to reduce the net focusing, or cause a downward shift of the transverse tune. The expression for the space-charge-induced tune shift[4,5] is

$$\Delta\nu = \frac{-r_p N_T}{4\pi\varepsilon_n\gamma^2\beta}\frac{F\,G\,H}{\hat{B}} \quad , \tag{1.4}$$

where $r_p = e^2/4\pi\varepsilon_0 mc^2 (1.535\times10^{-18}\,\mathrm{m}$ for the proton), N_T is the total number of particles in the ring, F describes the image forces from the vacuum chamber and magnet poles ($F \approx 1$ when image terms can be neglected), G describes the transverse density distribution ($G = 1$ for uniform, 2 for parabolic, and 1.56 for Gaussian beams), and H is a correction for an elliptical beam ($H = 1/2$ for a circular beam). The quantity $\hat{B}$ is a bunching factor, the ratio of bunch length to bunch spacing (smaller $\hat{B}$ corresponds to more bunching). For further discussion we will assume a circular parabolic beam, whose image forces can be ignored; then $F\,G\,H = 1$. For a ring with circumference C, we may write $N_T = N_B C/S_B$.

Substituting this into Eq. (1.4) gives

$$\Delta\nu = \frac{-r_P N_B C}{4\pi\varepsilon_n\gamma^2\beta S_B}\frac{F\,G\,H}{\hat{B}} \quad . \tag{1.5}$$

For a given choice of the operating tune, there is a limit on $|\Delta\nu|$ to avoid serious resonances and loss of the beam. For a uniform beam the tune shift for all particles is the same, and the shift could in principle be canceled by increasing the external focusing. For a nonuniform beam distribution, the full tune shift is experienced only by the small-amplitude particles, and $|\Delta\nu|$ also approximately represents the space-charge tune spread of the beam. Then a simple cancellation is not obtained for all the particles and one must restrict $|\Delta\nu|$. The maximum admissible value for $|\Delta\nu|$ depends on the resonances near the beam operating points and their effective widths or stopbands. When a particle tune lies within a stopband, a large growth of the amplitude occurs, leading eventually to loss of the particles. A common rule-of-thumb for design to avoid dangerous resonances is to keep the tune spread below about 0.25. We see that $|\Delta\nu|$ is largest at the injection (lowest) energy, where $\beta\gamma^2$ is smallest. For the SSC low-energy booster, at 600 MeV where $\beta\gamma^2 = 2.13$, we assume $\hat{\beta} = 0.25, \varepsilon_n = 0.6\times10^{-6}$ m-rad, $C = 540\,\text{m}, S_B = 5\,\text{m}, N_B = 1.0\times10^{10}, G = 1.56$(Gaussian), $F = 1$, and $H = 1/2$. The tune shift calculated from Eq. (1.4) is $|\Delta\nu| = 0.32$, slightly larger than the conventional rule of thumb. Values of $|\Delta\nu| = 0.37$ at the Fermilab booster,[6] and $|\Delta\nu| = 0.6$ at the CERN booster[7] have been achieved. These results have shown that some compensation for the effects of the resonances can be obtained experimentally, allowing larger tune shifts and higher brightness performance in the rings.

For given values of N_B, S_B, and ε_n, we see from Eq. (1.5) that $|\Delta\nu|$ can be reduced by reducing the ring circumference C, or increasing the bunching factor $\hat{B}$, or increasing the injection energy. Increasing the injection energy helps for three reasons: 1) reduces the total self force (factor of γ^2), because of the larger magnetic self force at higher energies, 2) increases relativistic inertia (factor of γ), and 3) increases the revolution frequency corresponding to a stronger focusing force for a given tune ν, because tune is defined as the ratio of betatron frequency to revolution frequency. Larger $\hat{B}$ helps because it reduces the peak current. Reduced circumference helps because it increases the revolution frequency for a given ν.

The bunching factor is determined by the choice of synchronous phase, which also affects the acceleration rate. If $\hat{B}$ is fixed, there are two practical ways of reducing $|\Delta\nu|$, 1) increasing the injection energy, and 2) decreasing the ring circumference C. The latter choice would be at the expense of reduced final energy (for a given lattice and maximum dipole field). Thus, achieving an acceptable value for $|\Delta\nu|$ can limit both the injection and final energies, and provides a fundamental reason for a multistage synchrotron.

2 Emittance Concept and Definitions

Beams are composed of collections of particles, which can be described by six normalized phase-space coordinates $(x, \Delta P_x/mc, y, \Delta P_y/mc, z, \Delta P_z/mc)$. The longitudinal coordinates can be given in terms of either position and normalized momentum differences relative to a synchronous or reference particle $(z, \Delta P_z/mc)$, or time and kinetic energy differences $(\Delta t, \Delta W)$, or phase and kinetic energy differences $(\Delta\phi, \Delta W)$, with $\Delta\phi = 2\pi c\Delta t/\lambda$, where λ is the rf wavelength. In general we can talk about a 6-D phase-space distribution that describes the beam at any location in the accelerator.

As the beam is transported and accelerated, the distribution obeys Liouville's theorem,[8] provided that the proper conditions are satisfied. These are that 1) the continuity equation is valid in 6-D phase space, which means that no particles are created or lost, and 2) the forces are derivable from a potential function that depends only on the three spatial coordinates and time. Subject tô these requirements the theorem states that the density in 6-D phase space, measured along the trajectory of a particle, is invariant, or equivalently that the 6-D volume enclosed by any isodensity contour is invariant. Only when the motion is uncoupled, so that the three 2-D phase-space projections $(x, \Delta P_x/mc)$, $(y, \Delta P_y/mc)$, and $(z, \Delta P_z/mc)$ are independent, does Liouville's theorem apply separately in each of these projections.[9]

For the more general case where the beam particles interact with each other through their mutual Coulomb forces, the forces are not in principle derivable from a potential function that depends only on the three spatial coordinates. But, in most circumstances for practical beams, these interactions can be accurately described by an average collective potential, arising from the contributions of all the particles, and individual binary Coulomb collisions represent only a small effect. Under these circumstances, Liouville's theorem remains a good approximation. The condition for which the binary Coulomb collisions can be ignored and the self interactions treated as a collective space-charge force, is that the number of particles, N_D, within a Debye sphere is much greater than unity.[10] Thus

$$N_D = n\frac{4}{3}\pi\lambda_D^3 \gg 1 \tag{2.1}$$

where n is the nominal beam density, and nonrelativistically

$$\lambda_D = \sqrt{\varepsilon_0 kT/ne^2} \tag{2.2}$$

is the Debye length of the beam, where $kT = mc^2\varepsilon_n^2/a^2$. This criterion reflects the idea that the forces on a given particle from binary Coulomb collisions are important only for particles within approximately a sphere of radius equal to λ_D. The perturbing influence of individual particles further away than λ_D is small because of the shielding effect of the other charges. Thus, particles further than λ_D contribute only to the longer-range collective force. Within the Debye sphere the effect of individual collisions is also not important, when there are a lot of other particles in the sphere that can exert forces to approximately cancel the force of any single particle. For an equivalent uniform beam, whose total beam radius is twice the rms size of the beam, and for the case of a spherical bunch geometry, we find

$$N_D = \frac{4\pi}{3n^{1/2}}\left[\frac{\varepsilon_0 mc^2}{e^2}\right]^{3/2}\left[\frac{\varepsilon_n}{a}\right]^3 \tag{2.3}$$

where n is the density and a is the rms beam size in any projection. If the Debye length is larger than the beam radius, all particles in the bunch are contained within a Deybe sphere, and the criterion expressed by Eq. (2.1) is replaced by $N_B \gg 1$, where N_B is the number of particles in the bunch. For practical high-intensity beams, for example in low-energy linacs, we find that $\lambda_D \approx a, N_D \gg 1$, and Liouville's theorem is a good approximation. As an example, we consider a bunched 1-MeV, 25-mA proton beam in a 400-MHz linac, with an rms normalized emittance of $\varepsilon_n\pi = 0.2\pi$ mm-rad, and an rms beam size of $x_{rms} = 1$ mm. The number of particles per bunch is $N_B = I\lambda/ec = 3.9\times10^8$, $kT = mc^2(\varepsilon_n/a)^2 = 38$ eV, and the density is $n = 1.2\times10^{16}$ m^{-3}. This corresponds to

$\lambda_D = 0.42$ mm, and $N_D = 3.7\times10^6 \gg 1$, so that Liouville's theorem should be a very good approximation.

The total emittances for an idealized hard-edged uniform 2-D phase-space distribution in our convention are defined as $1/\pi$ times the phase-space area. This definition corresponds to the product of the semi-axes of an ellipse, and it is useful because the distributions of beams that are matched when only linear forces are present have concentric elliptical contours. For real beams, however, the phase-space density is usually nonuniform, and rarely are the phase-space boundaries sharp. Also the elliptical contours can become broadened or deformed by processes that violate Liouville's theorem, such as scattering or collimation, and can become deformed by forces that couple the degrees of freedom, by nonlinear forces arising from beam-optical aberrations, or by space-charge forces associated with nonuniform density distributions. For a single degree of freedom the deformation from nonlinear forces takes the appearance of a filamentation of an initial elliptical phase-space distribution, and is fundamentally a consequence of dependence of the particle oscillation frequency on variables such as amplitude, energy, etc., so that different particles rotate in phase space at different rates. Unless a beam is properly matched, which is not always possible, these filamentation-type distortions further increase some amplitudes, especially of the outermost particles in phase space, and result in an increase in the average beam size. As the filamentation process continues, the phase-space distribution can take a shape much like a spiral galaxy, as regions devoid of particles are captured within outer spiral arms. Because the average phase-space density including the empty regions decreases, the process is sometimes referred to as emittance dilution. For two and three degrees of freedom with coupled nonlinear forces, the projection of outer filaments onto 2-D phase-space plots can create a diffuse outer halo.

The resulting increase in rms beam size can be attributed to an increase in effective emittance of the beam. The magnitude of this effective emittance will depend both on processes that violate Liouville's theorem and on processes that cause distortions and filamentations in the distribution. A convenient definition of an effective emittance that has these properties and is applicable to all distributions, is the normalized rms emittance, which for the x-direction is defined, using the Sacherer definition,[11] in terms of the second moments of the distribution $\overline{x^2}$, $\overline{p_x^2}/(mc)^2$, and $\overline{xp_x}/(mc)$ as

$$\varepsilon_{n,x} = \left[\overline{x^2}\,\overline{p_x^2} - \overline{xp_x}^2\right]^{1/2} /(mc). \tag{2.4}$$

In the paraxial approximation for the transverse motion, the divergence x' is related to the transverse momentum component p_x by $p_x = mc\beta\gamma x'$. Using this result the rms normalized emittance for the transverse planes can be written in a more familiar form as

$$\varepsilon_{n,x} = \beta\gamma\left[\overline{x^2}\,\overline{x'^2} - \overline{xx'}^2\right]^{1/2} . \tag{2.5}$$

Another convention for emittance, first introduced by Lapostolle,[12] is frequently used, in which the right sides of Eqs. (2.4) and (2.5) are multiplied by 4; in this convention the emittance reduces to the total emittance for a beam with a uniform hard-edged distribution in the $x - x'$ projection. It has become more common recently to refer to the Lapostolle definition as the 4-rms emittance.

It is easily shown that a straight line in phase space, where $x \propto x'$, has zero rms emittance, whereas a nonlinear relation between x and x' gives a nonzero value, even though the thickness of the line is zero. The definitions in Eqs. (2.4) and (2.5) can be further motivated as follows. Consider an arbitrary phase-space distribution, whose second moments are known. Suppose we want to describe the distribution using an average ellipse, whose maximum projections on the displacement and momentum axes are chosen to equal the rms values along these axes. This allows us to relate the second moments of the distribution to the emittance ε_n of the ellipse and to the Courant-Snyder ellipse parameters $\tilde{\alpha}$, $\tilde{\beta}$, and $\tilde{\gamma}$. Thus we require

$$\overline{x^2} = \varepsilon_n \tilde{\beta}/\beta\gamma \ , \tag{2.6}$$

and

$$\overline{x'^2} = \varepsilon_n \tilde{\gamma}/\beta\gamma \ . \tag{2.7}$$

The other parameter $\tilde{\alpha}$ that describes the orientation of the ellipse is kinematically related to the rate of change of the amplitude parameter $\tilde{\beta}$ through the expression $\tilde{\alpha} = -(1/2)d\tilde{\beta}/ds$, where s is the beam direction. We also use the result $d\overline{x^2}/ds = 2\overline{xx'}$, to obtain

$$\tilde{\alpha} = -\overline{xx'}\beta\gamma/\varepsilon_n \ . \tag{2.8}$$

The Courant-Snyder parameters for the ellipse are related by $\tilde{\beta}\tilde{\gamma} - \tilde{\alpha}^2 = 1$, and when Eqs. (2.6), (2.7), and (2.8) are substituted into the result, we obtain the definition of rms emittance given by Eq. (2.5).

The usefulness of this rms emittance definition was further confirmed by Sacherer[11] and Lapostolle,[12] who showed that the rms beam size $a = (\overline{x^2})^{1/2}$ satisfies an envelope equation of motion, which is almost independent of the distribution. For a 2-D round beam in a uniform focusing channel with a wave number k_0, this becomes

$$a'' + k_0^2 a - \frac{\varepsilon^2}{\beta^2\gamma^2 a^3} - \frac{K}{4a} = 0 \ , \tag{2.9}$$

where K is an intensity parameter called the generalized perveance, to be defined later.

Some useful properties of rms emittance can be obtained by differentiating Eq. (2.5) with respect to the coordinate s along the beam direction. It can then[13] be shown that

$$\frac{d\varepsilon_n^2}{ds} = \frac{2\gamma^2}{mc^2}\left[\overline{x^2}\ \overline{x'F} - \overline{xx'}\ \overline{xF}\right] \ , \tag{2.10}$$

where F is the total force, the sum of the external plus the space-charge force. When the total force is linear in x, the two terms on the right side of Eq. (2.10) cancel, and there is no rms-emittance growth. Both nonlinear forces and non-Liouvillian processes can affect the averages $\overline{x'F}$ and $\overline{xF}$, to make the right side of Eq. (2.10) nonzero, and cause rms-emittance growth. However, nonlinear forces will not cause emittance growth if the two terms in brackets cancel. This is what happens when the beam reaches an equilibrium state with a nonuniform density distribution where the space-charge force is nonlinear.

For linear external forces, the external force contribution to Eq. (2.10) cancels, leaving only the space-charge force, which is nonlinear unless the spatial density distribution is uniform. Then, the right side of the equation can be expressed in a form proportional to the rate of change of space-charge field energy,[14] which depends on beam intensity and on the rate of change of the spatial-density distribution. This formulation has led to some useful analytic formulas for space-charge-induced emittance growth for beams injected into transport and accelerator channels, as will be discussed in more detail later.

Another effect associated with time-dependent phenomena can lead to apparent growth of rms emittance. When accelerator parameters vary with time (jitter), the beam centroids and ellipse parameters can fluctuate with time. If the beam variations are too fast for the measurement apparatus to resolve, this will appear as an enlarged phase-space area and an apparent increase in rms emittance as a result of the superposition of ellipse contours at different times. Examples include beam current variations caused by ion source fluctuation, mechanical vibration of rf cavity elements such as drift tubes, and excitation of higher-order cavity modes. The beam will be mismatched or missteered at certain times and, if nonlinear external forces act on the beam, these effects can result in real growth of rms emittance.

3 Emittance

3.1 Emittance in the DC Injector

The first accelerator subsystem in the H^- injector system is the dc injector, which must perform the functions of beam formation in an H^- ion source, followed by extraction, dc acceleration, and finally transport and matching into the first component of the linac, the RFQ.

Some useful scaling relations have been derived for ion sources.[15,16] If the extracted ions have a Maxwellian velocity distribution, and the beam is spatially uniform, the intrinsic rms-normalized emittance in each transverse plane can be written nonrelativistically for an assumed waist condition at the extraction aperture as

$$\varepsilon_n = 0.5r\sqrt{kT/mc^2} \ . \tag{3.1.1}$$

The quantity r is the extraction-aperture radius, kT is the ion temperature, and mc^2 is the ion rest energy. The space-charge-limited ion current is obtained from the Child-Langmuir law as

$$I(A) = 0.00173\pi V(kV)^{1.5}\left[\frac{r}{d}\right]^2\left[\frac{Z}{A}\right]^{0.5} \tag{3.1.2}$$

where V is the extraction voltage, d is the effective extraction gap width, and Z and A are the charge and mass numbers ($Z = 1$, and $A = 1$ for H^- ions). Keller assumes the maximum value of r/d is constrained by the extraction aberrations to about unity, and proposes an empirical electric breakdown limit of

$$d\,(\mathrm{cm}) = 0.0014V(\mathrm{kV})^{1.5} \ . \tag{3.13}$$

Reiser uses these assumptions to derive a maximum phase-space density (current divided by rms-normalized emittance), which is independent of extraction voltage, and using our emittance definitions is given by

$$\frac{I}{\varepsilon_n}\left[\frac{A}{\mathrm{mm}-\mathrm{mrad}}\right] = 3.0\sqrt{\frac{Z}{kT(\mathrm{eV})}} \ . \tag{3.1.4}$$

If kT represents the actual ion temperature, Eq. (3.1.4) can be considered as a theoretical upper limit for ion-source phase-space density. Generally, emittance is the quantity that is actually measured, and ion temperatures are often considered to be effective values that include emittance growth.

The emittance of the extracted beam[17] is determined by three effects: 1) the ion temperature kT, 2) beam-optics aberrations that produce nonlinear motion during the extraction process, and 3) time-dependent fluctuations in the plasma resulting in variations in current and other parameters. Which effects are dominant depends on the type of ion source and the specific design characteristics.

For a sufficiently stable and quiescient plasma, the emittance growth from fluctuations, especially current fluctuations, can be kept small. If this is not the case, the emittance will be enlarged as a result of the superposition of extracted emittances with different orientations at different currents. For surface-type ion sources, where the H^- are mainly produced on a cesium-coated cathode surface, bombarded by H^+ ions, large ion temperatures $kT > 10$ eV are obtained. For volume-type sources where the H^- ions are produced mainly by charge exchange of fast cathode-produced ions on H atoms, the ion temperatures can be kT <0.5 eV. For the surface-type sources, if the extraction optics can be designed to minimize the effects of aberrations, the lowest measured emittance values are consistent with those calculated from the H^- ion temperature at emission. For volume-type sources, where the intrinsic temperatures are low, the emittance is usually determined by the aberrations at extraction. From Eq. (3.1.1) we obtain estimates of phase-space density of 0.95 A/mm-mrad for the surface-type source (assuming $kT = 10$ eV), and 4.2 A/mm-mrad (assuming $kT = 0.5$ eV) for the volume-type source.

As a numerical example of an operating source, we consider the circular aperture magnetron source (surface-type source), designed by the Brookhaven group[18] for injection into an RFQ. The reported values of x- and y-normalized 90% emittances are $\varepsilon_n(90\%)\pi = 0.94$ and 1.0π mm-mrad at both 70 mA and 36 mA. The fact that emittance is independent of current is consistent with the assumption that for this source the emittances are not determined by the intrinsic ion temperature but by other effects such as aberrations. Assuming the rms-normalized emittance equals $\varepsilon_n(90\%)/4.6$ (valid for a Gaussian phase-space density), and averaging over the x- and y-planes, gives $\varepsilon\pi = 0.21\pi$ mm-mrad, and at 70 mA a brightness $I/\varepsilon_n = 0.33$ A/mm-mrad, which is about a factor of three lower than the theoretical limit of 0.95 A/mm-mrad estimated above. Eq. (3.1.1) gives an effective temperature of $kT = 41$ eV, compared with the estimated value of $kT > 10$ eV, consistent with the hypothesis that aberrations or other effects have increased the emittance.

After the beam is extracted from the source, it is transported in the low-energy beam transport (LEBT) section to the RFQ entrance, where a small-diameter, converging beam must be produced for a proper match.[19] In general two design approaches have been considered: 1) magnetic focusing and space-charge neutralized (compensated) transport in a plasma formed by beam-induced ionization of the residual gas, and 2) electric focusing (which removes the background plasma) and transport with no neutralization. Both approaches have advantages and disadvantages, and most of the operational experience has been with space-charge neutralized transport.

In space-charge neutralized transport, the H^- beam ionizes the residual gas, creating both electrons and positive ions. Initially, the space-charge potential of the beam drives the electrons out and attracts the positive ions. If the gas density is sufficiently high, the positive-ion charge can build up until the space-charge forces of the beam are canceled, with a characteristic time constant $\tau = 1/n\sigma v$ of typically about 50 μsec. A steady state can result, where the positive-ion density slightly exceeds the beam density, leading to net focusing and a pinch effect. Although the space-charge forces are more than canceled, external focusing is still required to counteract the natural tendency of the beam to diverge because of its finite emittance. This focusing is now provided by magnetic lenses, which are compatible with the presence of the neutralizing plasma because they do not sweep away the positive ions.

The advantage of space-charge neutralized transport for control of the space-charge defocusing forces can be seen from the Kapchinskii-Vladimirskii (K-V) envelope equation for transport of a round beam,

$$a'' + k_0^2 a - \frac{\varepsilon^2}{\beta^2\gamma^2 a^3} - \frac{K}{4a} = 0 \tag{3.1.5}$$

where a is the rms beam size, k_0 is a measure of the average focusing strength, and the dimensionless generalized perveance describing the space-charge force is

$$K = \frac{qI}{2\pi\varepsilon_0 mc^3(\beta\gamma)^3} \quad . \tag{3.1.6}$$

The ratio of the space-charge to the emittance term is

$$\rho = K\left[\frac{\beta\gamma a}{2\varepsilon_n}\right]^2 \quad . \tag{3.1.7}$$

The matched beam size in a uniform focusing channel is determined by the solution of Eq. (3.1.5) when $a'' = 0$, and is determined by the space-charge term when $\rho > 1$ (space-charge dominated) and by the emittance term when $\rho < 1$ (emittance dominated).

As an example, consider 50 keV and 50 mA of H^- beam with an rms-normalized emittance of $\varepsilon_n\pi = 0.2\pi$ mm-mrad. Assuming solenoid focusing in a smooth approximation with a magnetic field of 1 T and solenoids that occupy half the space, we assume that

$$k_0 = \frac{eB}{2\sqrt{2}mc\beta\gamma} = 11.3 \text{ m}^{-1} \tag{3.1.8}$$

where $1/\sqrt{2}$ accounts for the filling factor of the solenoids. With no neutralization the full space-charge force is present, and we obtain $K = 0.0032$, a matched beam size of $a = 2.6$ mm, and the ratio of the space-charge to emittance term is $\rho = 13.5$. The beam is space-charge dominated and it will be more difficult to focus it to a small diameter for matching into an RFQ. An interesting quantity is the space-charge potential between the center and the edge of the beam, which is

$$\Delta\phi = \frac{I}{4\pi\varepsilon_0 c\beta} = 150 \text{ V} \quad . \tag{3.1.9}$$

For space-charge neutralized transport the steady-state space-charge potential is not zero but is estimated as $\Delta_\phi = kT_e/e$. The electron temperature can be estimated[20] from the relation $kT_e/e = 0.67\ [V_b Vi/(M_b/M_e)]^{1/2}$, where V_b is the beam energy, V_i is the ionization potential of the background gas, and M_b and M_e are the masses of the beam ions and electrons. Using the ionization potential of molecular hydrogen of $V_i = 15.4$ eV, we obtain $kT_e = 13.6$ GV. From Eq. (3.1.9) the residual or effective current of 4.6 mA is obtained, with a corresponding effective perveance of $K = 0.00029$. The rms beam size becomes $a = 1.34$ mm, and the ratio $\rho = 0.33$, ie. an emittance-dominated beam that is easier to focus into the RFQ.

Space-charge neutralization is easy to implement, by simply increasing the gas pressure, either by control of the pumping rate or by deliberately introducing a background gas. But there are significant disadvantages such as 1) difficulty in providing constant and uniform gas density, 2) stripping loss of the H^- beam, and 3) instability and emittance growth from the associated plasma fluctuations.[21]

The dominant emittance growth mechanism for space-charge neutralized transport, identified by Allison,[22] is the variation of the output ellipse orientation caused by fluctuations in either the current from the ion source or the fraction of neutralization. As a result, the time-averaged phase-space area is increased, and this corresponds to an apparent growth of emittance. Allison quotes a frequency range of about 1 to 1000 kHz for these fluctuations.

In addition to the fast current fluctuations, the amount of space-charge neutralization, the corresponding effective current, and the output phase-space ellipse parameters vary with time for a pulsed ion source. When the pulse length is small compared with the space-charge neutralization time constant τ (typically about 50 μsec), the output ellipse parameters correspond approximately to unneutralized (full space-charge) transport. For pulse lengths large compared to τ, the average ellipse is dominated by the steady state (space-charge neutralized transport). For pulse lengths near τ, one can observe strong variations of the ellipse parameters throughout the pulse as the beam approaches equilibrium. These time variations as the neutralization varies from 0 to near 100% have been measured.[23] If the time resolution of the emittance measurement exceeds the pulse length, these variations result in apparent emittance growth. Nevertheless this effect causes a time-dependent beam mismatch at the entrance to the RFQ, which generally would result in time-dependent emittance growth within the RFQ.

The main emittance growth concern for the unneutralized transport with electrostatic focusing is from aberrations.[24] The main problem area may be the final match into the RFQ, which requires a large beam within the final lens to obtain a small beam diameter inside the RFQ. A second source of potential emittance growth also occurs when the beam is expanded before final matching into the RFQ, and space-charge field energy is converted to thermal energy in the beam.[25,14]

3.2 Emittance in the RF Linac

The conventional high-current rf ion-linac configuration begins with a dc injector and is followed by the radio-frequency quadrupole (RFQ) linac[26,27] used to bunch and accelerate the beam from 100 keV or below to a few MeV for protons. The RFQ uses rf electric-quadrupole fields to provide strong focusing for low-velocity particles. To maintain the acceleration efficiency and transverse focusing at higher energies, the drift-tube linac (DTL) with magnetic quadrupole lenses within the drift tubes becomes a better choice. The RFQ-to-DTL transition can occur at a few MeV for a proton or H^- beam. At much higher energies, the rf-power efficiency of the DTL decreases below that of the coupled-cavity linac (CCL) structures.[28] The transition from DTL to CCL becomes attractive at about 100 MeV for a proton linac.

The transverse emittance growth mechanisms in a high-intensity linac can be separated into three general categories: 1) effects in a perfect linac with aligned and rms-matched beams, 2) effects for misaligned and rms-mismatched beams, and 3) effects associated with misalignments and errors in the linac structure.

When comparing emittance-growth effects, one should be cautious about making general conclusions as to which mechanisms are most important, because different factors can dominate in different accelerators. The primary sources of transverse emittance growth in a perfect linac with no misalignments and no field errors are generally from self forces and rf forces. Both forces are nonlinear, couple the degrees of freedom, and are the cause of emittance growth because the transverse oscillation frequencies depend on amplitude (nonlinearities), and on the other coordinates (coupling). A resulting decoherence of transverse oscillations for different particles causes distortion of the initial phase space, unless the distribution is already internally matched to these forces.

For high-current linacs we will be concerned with emittance growth resulting from self forces as the dominant effect. In general the self forces include both electric and magnetic forces. The latter are usually much smaller in a proton linac, so we will refer to these self forces just as space-charge forces. We begin by discussing space-charge-induced emittance growth with a qualitative discussion of beam matching into an accelerator channel. We distinguish between what we will call internal matching and rms matching. Internal matching constrains the 6-D phase-space distribution so that the isodensity contours coincide with the particle phase-space trajectories. For an internally matched beam the distribution will be in equilibrium in the accelerator channel, and no emittance growth will occur, even though nonlinear forces may act on the beam. Such an equilibrium distribution is independent of time (stationary) if the focusing is uniform along the accelerator, or is periodic for a periodic-focusing channel. Examples of equilibrium distributions have been studied for 2-D transport channels.[29] The most frequently studied is the K-V distribution, which unfortunately is physically unrealistic because the beam is distributed on the surface of a hyperellipsoid in 4-D phase space, resulting in no particles in the central core of the 4-D space. This distribution results in uniform ellipses for all 2-D projections.[30]

Given a beam that is not internally matched, one would like to be able to transform it into an internally matched, equilibrium distribution for the given accelerator channel, without increasing the rms emittance in the process. We do not know whether it is even possible in principle to perform such a transformation, without accompanying emittance growth. Nevertheless, it is feasible to match the rms beam sizes in each degree of freedom. This is accomplished by providing a beam-optics transformation, so that the rms sizes are exactly right to balance the sum of the focusing, space-charge, and emittance terms in the rms-envelope equations. In a uniform focusing channel the matched rms sizes are then constant. In a periodic channel the rms sizes will undergo a periodic flutter about their average values. An rms-matched beam is not necessarily internally matched. If it is not, the beam distribution is not in equilibrium, and the beam will evolve with the possibility of irreversible emittance growth. Nevertheless, rms matching is an important characteristic of an internally matched equilibrium distribution, and injection of an rms-matched beam can be considered a first approximation to providing the desired internally matched beam.

Numerical simulations of nonequilibrium linac beams show that such beams do evolve to quasi-equilibrium distributions, which change only slowly as the beam is accelerated. The evolution of the beams is usually accompanied by rms-emittance change as a result of both nonlinearity and coupling between degrees of freedom. Experience has shown that the velocity distributions of the final beams are Maxwellian-like and, when focusing is linear, the spatial distribution of a space-charge-dominated beam consists of an approximately uniform charge-density core of density n. The density decreases to zero over a finite distance approximately equal to the Debye length λ_D given nonrelativistically by Eq. (2.2). For space-charge-dominated beams $\lambda_D \ll a$, and the equilibrium spatial distribution is approximately uniform with a sharp falloff at the edges. For emittance-dominated beams

$\lambda_D \gg a$, and the Debye tail occupies essentially the entire spatial extent of the beam, resulting in a peaked Gaussian-like charge density. Among 2-D continuous equilibrium beams, the K-V distribution is anomalous because it is always uniform in its projection to real space, regardless of the relative importance of emittance and space charge. However, this distribution does not correspond to the final equilibrium state of beams observed in numerical simulation studies.

It is further observed in numerical simulation that the emittance growth of beams that evolve to a final equilibrium distribution is associated mostly with a halo of low-density particles in phase space. This halo is especially undesirable for high-duty-factor linacs because it results in particle losses on the accelerator walls and radioactivation of the accelerator.

In early emittance growth studies[31,32] of error-free, high-current proton drift-tube linacs, it was concluded that emittance growth caused by the transverse rf forces was small. Nonlinear space-charge forces, associated with longitudinal to transverse coupling, were identified as the primary source of the observed transverse emittance growth. Chasman also found that this emittance-growth mechanism leads to a lower limit for the output emittance as input emittance is decreased at fixed beam current. Later studies[33] showed that emittance growth from longitudinal to transverse coupling could be physically correlated with the dependence of the transverse oscillation frequency on the longitudinal position of the particles in the bunch. Furthermore, the results were not very sensitive to the input particle distribution.

An observation of longitudinal emittance decrease associated with the transverse emittance increase led to the suggestion of equipartioning.[34] In this picture the emittance changes are the result of the evolution of a high-current beam towards an equilibrium distribution in which approximate kinetic-energy balance in the center of mass, or equipartioning, would be established. This suggests that if the input beam could be equipartioned in the accelerator in addition to being rms matched, there would result a better approximation to the ideal internally matched beam, and therefore space-charge-induced emittance growth would be minimized.

Equipartitioning would mean that the mean-square center-of-mass velocities in each degree of freedom would be equal. This corresponds to the condition that

$$\sigma_x \varepsilon_x = \sigma_y \varepsilon_y = \sigma_z \varepsilon_z \tag{3.2.1}$$

where σ_x, σ_y, and σ_z are phase advances per focusing period, associated with both the focusing and the space-charge forces. For a given set of input beam emittances, $\varepsilon_x, \varepsilon_y$, and ε_z, and a given beam intensity, the condition given by Eq. (3.2.1) imposes a constraint on the relative focusing forces in the three degrees of freedom of the accelerator channel. Thus, while rms matching is achieved by providing a suitable beam-optical matching section before the beam is injected into the linac, the equipartitioning condition depends both on the input beam, through the beam current and the emittances, and the accelerator design, which determines the focusing and the phase advances.

Additional understanding of the equipartitioning dynamics was obtained from the work of Hofmann,[35] who identified the longitudinal-to-transverse space-charge coupling effects with coherent coupling instabilities associated with anisotropy in the beam. Underlying this approach is a particular point of view for the description of the emittance growth.[36] The idea is that the emittance growth is the result of the excitation of unstable collective modes of oscillation of the beam. The beam is characterized by an infinite number of eigenmodes of collective oscillation. For example, a dipole mode corresponds to the displacement of the centroid and coherent oscillations of the whole beam about the axis as a result of the external focusing forces. Also there are higher-order modes, some of which are internal, involving radial and azimuthal (quadrupole, sextupole, etc.) density oscillations of the beam. The first reported study of such modes for a 2-D round K-V distribution in a uniform focusing channel was made by Gluckstern,[37] who identified many modes and derived their stability characteristics. Studies for the K-V beam in a quadrupole channel have also been made.[38] Some of the modes were found to be unstable at a certain threshold values of intensity. Numerical simulation is always required to determine whether an unstable mode produces emittance growth before the beam stabilizes.

The existence of unstable collective modes with different intensity thresholds suggests that it is predominantly through these instabilities involving the internal rearrangement of the beam distribution that the nonlinear space-charge forces cause emittance growth.[36] This picture is useful if it can be determined which modes are easily excited and accompanied by significant emittance growth, and which are relatively harmless and lead to redistribution of charge without much growth.

With this background in mind, Hofmann studied the K-V distribution with an asymmetry between the x- and y-planes and derived the instability thresholds for the different modes. Although the studies correspond to continuous beams in an x-y geometry, Hofmann found that the same instability thresholds were approximately valid for the r-z geometry of a bunched beam. It was found that equipartitioned beams were stable with respect to these instabilities, and that generally the requirement for avoiding emittance growth even allowed some relaxation of exact equipartitioning. The predictions of Hofmann's model were further tested for high-current beams in drift-tube linacs with constant accelerating fields by Jameson,[39] who confirmed that equipartitioned input beams produced the minimum emittance growth, while non-equipartitioned beams could produce a significant transfer of energy and emittance between the longitudinal and transverse planes. Jameson showed from simulation how the parameters of the non-equipartitioned accelerating beam can change in the space defined by the mode-stability plots derived by Hofmann.[35] This behavior can be complicated and makes it difficult to derive simple design guidelines for avoidance of emittance growth. Perhaps the simplest design approach is to require exact equipartitioning as defined by Eq. (3.2.1). A less restrictive guideline is suggested by Hofmann, whose conservative criterion is that energy anisotropy is generally tolerable when the phase advance ratio $\sigma_l/\sigma_t < 1$ where σ_l and σ_t are the phase advances for longitudinal and transverse motion respectively. The growth times from numerical simulation were typically about 1 to 2 transverse oscillation periods.

Later, a second space-charge-induced emittance-growth mechanism was discovered[25,14] that even affects beams that are both rms matched and symmetric or equipartitioned, but are internally mismatched. This mechanism has been called charge redistribution. When a beam is injected into a transport or accelerator channel, the charged beam particles, behaving like a plasma, adjust their positions to shield the external fields from the interior of the beam. For linear external fields in the extreme space-charge (zero-emittance) limit, this implies a charge rearrangement to a uniform density to produce a linear space-charge field for exact shielding. Finite emittance beams are observed in numerical simulation to evolve to an internally matched charge density with a central uniform core and a finite thickness boundary, whose width is about equal to the Debye length. The rms-emittance growth arises from the nonlinear space-charge fields while the beam has nonuniform density and is undergoing internal plasma oscillations. The emittance growth can also be described as the result of the decoherence of the plasma oscillation phases for particles with different amplitudes, resulting in distortion of the phase-space area. This mechanism of emittance growth has the smallest known growth time; the full emittance growth occurs during only one-quarter of a plasma period, followed by damped oscillations of the emittance for typically ten or so additional plasma periods. In a high-current linac the full growth can occur within a single cell. This mechanism can become important when beams that have been internally matched to very strong focusing channels are injected after rms matching into weaker focusing channels. In the strong focusing channel where the matched beam size is small, the emittance term in the envelope equation dominates over the space-charge term, and the equilibrium distribution has a strongly peaked spatial density, generally having the appearance of a Gaussian distribution. In the weak focusing channel the rms beam size is large, the space-charge term dominates over the emittance term, and the corresponding equilibrium distribution is nearly uniform.

If the rms-matched input beam has the peaked spatial profile characteristic of the strong focusing channel, the beam density will change from peaked to nearly uniform in the weak focusing channel, and the change in shape of the distribution can be used to calculate the emittance growth. This results from the fact that, for a fixed rms beam size, the space-charge field energy is minimum for a uniform beam and increases as beams become more nonuniform. The evolution of the beam from peaked to uniform is accompanied by conversion of space-charge field energy to center-of-mass kinetic energy, which causes an increase in temperature and emittance. The emittance growth for a spherical bunch containing N particles, each with charge q, is obtained from the expression[40]

$$\frac{\varepsilon_{n,f}}{\varepsilon_{n,i}} = \left[1 + q^2 N a U_{ni}/60\sqrt{5}\pi\varepsilon_0\gamma^3 mc^2\varepsilon_{n,i}^2\right]^{1/2} \tag{3.2.2}$$

where a is the rms beam size and U_{ni} is the initial nonlinear field-energy parameter, a function only of the shape of the initial distribution. Emittance growth from charge redistribution is sensitive to the initial spatial charge density. For an initial Gaussian profile, $U_{ni} = 0.308$, whereas for a uniform density $U_{ni} = 0.0$.

To avoid emittance growth from charge redistribution it is necessary either to avoid transitions to accelerator channels with weaker focusing, or always to provide input beams with spatial profiles that are as uniform as possible. Other guidelines for minimizing emittance growth from charge redistribution can be inferred from Eq. (3.2.2). For a given beam current I, defined as the average value over an rf period, the number N of particles per bunch is given by $N = I\lambda/qc$, where λ is the rf wavelength. Equation (3.2.2) predicts that the emittance growth is less at high frequencies, a result that appears because a high frequency linac has less charge per bunch for a given current. This condition was first reported[33] in studies of emittance growth that probably included equipartitioning effects, and is more generally valid than for the charge-redistribution effect alone. Equation (3.2.2) also shows that larger beams are more susceptible to increased emittance growth from nonlinear space-charge forces.

Numerical simulation studies of transverse emittance growth in an RFQ linac have also been reported.[41] The main features are: 1) the emittance growth is predominantly caused by space-charge forces, 2) most of the growth occurs while bunching the beam and is a strong function of the longitudinal beam size, 3) the growth is weakly dependent on the beam current, 4) above a certain current the growth is almost independent of the initial distribution, and 5) as for the drift-tube linac the final emittance approaches a lower limit as the initial emittance approaches zero at fixed beam current. The emittance growth in the RFQ bunching section may be a combination of the equipartitioning effect and charge redistribution as the bunching forces increase the peak value of the beam current and drive the beam into a more space-charge-dominated regime. A semi-empirical emittance growth formula was obtained [41] based on Eq. (3.2.2), which is in good agreement with the numerical simulation results. This formula shows the advantage of high frequency and strong focusing for control of space-charge-induced emittance growth in the RFQ.

Next we discuss the additional emittance growth experienced by beams that are misaligned or rms mismatched. Misalignment results in a beam which undergoes coherent oscillations about the axis. These oscillations produce emittance growth if there are external forces that are nonlinear or couple the degrees of freedom. Then the misalignments put the beam into regions where the nonlinearities and coupling terms have a greater effect on the motion. Both nonlinearity and coupling are characteristic properties of the transverse rf-defocusing forces in the rf gaps, and therefore in practice additional emittance growth does result from beam misalignment. Consequently, particles at different transverse distances from the axis, and with different values of the longitudinal coordinate or phase, rotate in phase space at different frequencies. An initial ellipse eventually becomes distorted, resulting in an increase in rms emittance. Space-charge forces do not contribute to additional emittance growth for misaligned beams. Forces from image charges induced on the walls by the beam are nonlinear and could produce emittance growth. These effects are generally small for aligned beams in proton linacs, but for misaligned beams can lead to growth.[42]

If the beam is only rms mismatched, it is centered on the beam axis, but the initial shape and orientation of the ellipse are not the same as for the matched beam at low intensity. The rms beam envelope oscillates as the mismatched ellipse rotates in phase space. For a space-charge-dominated beam the mismatch oscillations are plasma oscillations. Additional potential energy becomes available for emittance growth of mismatched beams, which means that rms matching of beams is very important for the control of emittance growth.[43,44]

A significant source of emittance growth from beam instability is associated with a beam in a periodic channel. This is caused by the envelope instability,[38,45] which occurs as a result of excitation of coherent modes by the periodic focusing structure. Some of the modes have nonuniform density distribution, whose nonlinear fields can produce significant emittance growth. The safest course to avoid the resonances is to design the accelerator channel at a zero current phase advance per focusing period, no larger than $\sigma_0 = \pi/2$.

The final general category of emittance growth effects is the effect of linac errors. Errors of transverse misalignment are often of most concern. They cause steering of the beam from the nominal beam axis, and result in the same kind of emittance growth effects as for a misaligned beam. The real cause of emittance growth is not the misalignments but the increased nonlinearity in the transverse rf fields experienced by the misaligned beams. In longitudinal space the corresponding effect is tank rf-phasing errors, which can increase the phase excursions of the beam, and contribute to transverse emittance growth because of the longitudinal-to-transverse coupling. Quadrupole gradient errors produce rms mismatch in the beam, which increases the particle amplitudes. RF phase and amplitude errors will produce a corresponding longitudinal mismatch, which likewise can affect the transverse emittance because of the coupling.

Another source of error-induced emittance growth comes from quadrupole rotation about the longitudinal axis, which causes coupling of x- and y-motions. Other effects, such as higher (nonlinear) magnetic multipoles, some of which are magnet-fringe fields, are usually smaller in linacs.

3.3 Emittance in Synchrotron Injector Rings

Some important emittance growth mechanisms in the injector rings include the following.[46]

a) Mismatch of either the centroid or rms amplitude at injection into the ring, which results in larger excursions of the beam into regions of nonlinear fields.
b) Multiple Coulomb scattering in the stripper foil during the injection process of the H^- beam.
c) Coherent transverse instabilities, which if uncorrected result in larger beam displacements and nonlinear fields.
d) Increase of emittance from momentum dispersion because of the dependence of the ellipse orientation on momentum.
e) Nonlinear space-charge-induced tune spread into resonance stop bands, which causes the tunes of beam particles repeatedly to cross stopbands of strong resonances as the particles undergo transverse and longitudinal oscillations about the center of the bunch. The resulting growth of betatron amplitudes exposes the beam particles to nonlinear fields, which causes the emittance to grow.

The space-charge tune-spread mechanism is important at high intensity and is unique to circular accelerators. The resonances and nonlinear fields are caused by imperfections in the dipole and quadrupole magnets, which form the focusing lattice. These imperfections can result from field errors and misalignments, and the real field distribution can be described as an infinite series of multipoles (dipole, quadrupole, sextupole, octupole, etc.). Multipole terms higher than quadrupole describe the nonlinear fields and contribute mostly at the larger displacements from the design orbit. Each of the multipole terms can be further decomposed into Fourier components in the azimuthal coordinate around the ring.[47,48] Each Fourier component of a given multipole can drive the beam particles into resonance at certain values of the particle tunes, resulting in growth of the amplitudes, which can eventually result in loss of the beam on the apertures. Dipole or first-order resonances drive the beam at integer tunes, quadrupole or second-order resonances at half-integer tunes, sextupole or third-order resonances at third-integer tunes, octupole or fourth-order resonances at quarter-integer tunes, etc. The particle tune does not have to lie exactly at the resonance value to be under the influence of the resonance; there is a stopband width within which a beam particle becomes locked into the resonance and will be driven to larger amplitudes.

In the Fermilab booster ring the space-charge tune-shift mechanism is believed to be the main source of intensity-dependent growth.[6] Two competing effects occur after the full beam has been injected. The beam is adiabatically bunched before acceleration to achieve high capture efficiency, resulting in increased tune shift and tune spread because of the reduced bunching factor in the tune-shift formula Eq. (1.4). Transverse emittance growth occurs mostly during this time of maximum tune spread, and particle losses can also result because of the growth in amplitudes driven by resonances whose stopbands lie within the increased range of particle tunes. A second and beneficial effect is acceleration, which increases the $\beta\gamma^2$ factor in the tune-shift formula and, as the energy increases, the tune spread is eventually reduced below a level where significant resonances are excited.

The transverse emittance growth occurs because of the nonlinear forces that act on the particles, especially when their amplitudes are large. The resonant forces drive the particle amplitudes into the regions of stronger nonlinear fields. As can be seen from the tune-shift formula, the emittance growth reduces the tune spread, and emittance growth continues until either the particle tunes are safely outside the resonant stopbands or sufficient beam is lost on the apertures. For a given accelerator at a given choice of transverse zero-current tunes, there will be a characteristic maximum acceptable tune spread, determined by the operating point and strengths of the nearby stopbands. In the Fermilab booster the experimental relationship between emittance and intensity is linear (in accordance with the tune-shift formula) with a slope that corresponds to a limiting space-charge tune shift of $\delta\nu = -0.38$.

Inspection of the tune-shift formula suggests several ways of reducing the emittance growth. One way is to reduce the tune spread by increasing the injection energy, and this approach is the main motivation for the project to upgrade the Fermilab linac from 200 to 400 MeV.[49] The tune-spread limit can be increased by reducing the strengths of the nearby resonances. This can be accomplished by introducing correction magnets of the proper multipole content and azimuthal Fourier component to compensate for the strength of the given stopband. This approach has been successfully implemented for the CERN synchrotron booster,[7] and a very large tune-shift value of $\Delta\nu = -0.60$ was achieved after compensating for one second-order and three third-order stopbands. Correction dipoles are routinely used to center the orbit, which reduces the strength of first-order stop-bands and also keeps the beam in the more linear-field region of the accelerator.

If a certain tune-spread limit is in effect for a given accelerator at a given operating point, it is important to know what will happen to an injected beam with too small an emittance (an emittance which results in too large a tune spread). The experience with the Fermilab booster, described above, suggests that the emittance will grow and the tune spread will decrease until the uncompensated stopbands are outside the tune spread of the beam. The emittance growth within the ring appears to provide a natural stabilizing mechanism, so external methods of spoiling the emittance before injection are unnecessary. However, the increased emittance is not necessarily desirable, because it results in reduced luminosity for the collider.

4 Performance of High Brightness Injectors

Having reviewed transverse emittance growth mechanisms in dc injectors, linacs, and synchrotrons, we now follow the evolution of transverse emittance of the proton beam through the world's highest energy accelerator system, at Fermilab. We recommend an earlier article by Finley[50] as an excellent source of material on this subject for the Fermilab $\bar{p}p$ Tevatron collider facility. Table I shows the results of transverse emittance measurements on the proton beam at different locations and energies at Fermilab.[50]

Table I. Summary of Measured Transverse RMS-Normalized Emittances in the Fermilab Accelerator Chain Given as $\pi\varepsilon_n$ (mm·mrad)

Location	Energy (GeV)	Horizontal	Vertical
Linac In	0.00075	0.46π	0.57π
Linac Out	0.2	1.1π	1.1π
Booster In	0.2	2.5π	
Booster Out	8	2.5π	
8-GeV Line	8	1.7π to 2.5π	1.7π
8-GeV Main Ring	8	1.7π to $2.5\ \pi$	1.3π to 2.0π
Main Ring Accelerated	150	2.8π	
Main Ring Coalesced	150	4.5π	
Tevatron Collider	800	5π to $10\ \pi$	3.3π to 5.0π

The emittances in the rings quoted in Finley's article are the 6-rms values, and for the linac are 90% values. In Table I we have divided the 6-rms values by 6 and the 90% values by 4.6 to convert them to rms values. The 4.6 factor assumes Gaussian distributions. The overall emittance growth factor from the linac input to full energy ranges from 10 to 20 in the horizontal plane and ranges from 6 to 9 in the vertical plane.

Table II shows the design rms-normalized emittances for the SSC accelerator chain. It shows an overall growth factor of only 5, and a final emittance lower by a factor of 5 to 10 than the Fermilab value in the horizontal plane and lower by a factor of 3 to 5 in the vertical plane. This improvement in beam quality is a significant challenge which will require careful alignment and matching throughout the accelerator chain, and sufficient diagnostics to understand the beam characteristics at every stage.

Table II. Summary of Design Transverse RMS-Normalized Emittances in the SSC Accelerator Chain.

Location	Energy (GeV)	$\pi\varepsilon_{n,rms}$ (mm·mmrad)
Linac In	0.000035	0.2
Linac Out	0.6	0.4
LEB Out	11	0.6
MEB Out	200	0.7
HEB Out	2000	0.8
Collider	20000	1.0

Acknowledgements

I wish to thank Martin Reiser for valuable suggestions and comments on this material.

References

1. Helen T. Edwards, "SSC Design Status," Part. Accel. **26** (1990) 151.
2. C. Ankenbrandt, C. Curtis, C. Hojvat, R. R. Johnson, C. Owen, C. Schmidt, L. Teng, and R. C. Webber, "H^- Charge Exchange Systems," Proc. 11th Int. Conf. on High-Energy Accelerators, Geneva, July 1990, p. 260.
3. M. K. Craddock, "High-Intensity Circular Proton Accelerators," High-Brightness Accelerators, NATO ASI Series B: Physics **178** (1986) 43.
4. L. J. Laslett, "On Intensity Limitations Imposed by Transverse Space-Charge Effects in Circular Particle Accelerators," Proc. 1963 Summer Study on Storage Rings, Accelerators, and Experimentation at Super-High Energies, BNL-7534, p. 325.
5. K. H. Reich, K. Schindl, and H. Schonauer, "An Approach to the Design of Space-Charge Limited High-Intensity Synchrotrons," Proc. 12th Int. Conf. on High-Energy Accelerators, Fermilab, 1982, p. 438.
6. C. Ankenbrandt and S. D. Holmes, "Limits in the Transverse Phase-Space Density in the Fermilab Booster," Proc. 1987 IEEE Part. Accel. Conf., Washington, DC, March, 1987, p. 1066.
7. G. Gelato, L. Magnani, N. Rasmussen, K. Schindl, and H. Schonauer, "Progress in Space-Charge Limited Machines: Four Times the Design Intensity in the CERN Proton Synchrotron Booster," Ibid., p. 1298.

8. J. D. Lawson, "The Physics of Charged Particle Beams," Clarendon Press, Oxford, 1977, p. 124.
9. Claude Lejeune and Jean Aubert, "Emittance and Brightness: Definitions and Measurements", Adv. Electronics and Electron Physics, Suppl. 13A, Applied Charged Particle Optics (1980) 184.
10. J. D. Lawson, "The Physics of Charged Particle Beams," Clarendon Press, Oxford 1977, p. 207.
11. F. J. Sacherer, "RMS Envelope Equations with Space Charge," IEEE Trans. Nucl. Sci. **18** (1971) 1105.
12. P. M. Lapostolle, "Possible Emittance Increase Through Filamentation Due to Space Charge In Continuous Beams," IEEE Trans. Nucl. Sci. **18** (1971) 1101.
13. T. P. Wangler, "Energy and Emittance Relations from RMS Equations of Motion for a Continuous Beam with Azimuthal Symmetry," Los Alamos Group AT-1 Memo AT-1:84-323, Oct. 1984.
14. T. P. Wangler, K. R. Crandall, R. S. Mills, and M. Reiser, "Relationship Between Field Energy and RMS Emittance in Intense Particle Beams," IEEE Trans. Nucl. Sci. **32** (1985) 2196.
15. M. Reiser, "Current and Phase-Space Density Limits of Ion Sources," Los Alamos Group AT-1 Memo AT-1:86-45, Feb. 1986.
16. R. Keller, "Brightness Limits for Ion Sources," High-Brightness Accelerators, NATO ASI Series B: Physics **178** (1986) 395.
17. P. Allison, "Some Comments on Emittance of H^- Ion Beams," Proc. 4th Int. Symp. on Production and Neutralization of Negative Ions and Beams, AIP Conf. Proc. **158** (1986) 465.
18. J. G. Alessi, "A Circular Aperture Magnetron for Injection into an RFQ," Proc. 4th Int. Symp. on Production and Neutralization of Negative Ions and Beams, AIP Conf. Proc. **158** (1986) 419.
19. K. R. Crandall, R. H. Stokes, and T. P. Wangler, "RF Quadrupole Beam Dynamics Design Studies," Proc. 1979 Linear Accelerator Conf., BNL 51134, Sept. 1979, p. 205.
20. M. E. Rudd, Phys. Rev. A20 (1979), 787.
21. P. Allison, A. J. T. Holmes, A. F. Lietzki, J. H. Whealton, and L. Wright, "Panel Discussion: Transport of Negative Ion Beams," Proc. 4th Int. Symp. on Production and Neutralization of Negative Ions and Beams, AIP Conf. Proc. **158** (1986) 520.
22. P. Allison, "Emittance Growth Caused by Current Variations in a Beam Transport Channel," IEEE Trans. Nucl. Sci. **NS-32** (1985) 2556.
23. D. S. Barton, "Negative Ion Source Tests for H^- Injection at the Brookhaven AGS," IEEE Trans. Nucl. Sci. **NS-28** (1981) 2681.
24. O. A. Anderson, C. F. Chan, W. S. Cooper, W. B. Kunkel, J. W. Kwan, A. F. Lietzke, C. A. Matuk, P. Purgolis, and L. Soroka, "Design of a 200 mA DC H^- Injector for an RFQ," Proc. 1987 IEEE Part. Accel. Conf., Washington, DC, 1987, p. 289.
25. J. Struckmeier, J.Klabunde, and M. Reiser, Part. Accel. **15** (1984) 47.
26. I. M. Kapchinskij and V. A. Tepliakov, "Linear Ion Accelerator with Spatially Homogeneous Strong Focusing," Prib. Tekh. Eksp. **2** (1970) 19.
27. Richard H. Stokes and Thomas P. Wangler, "Radio-Frequency Quadrupole Accelerators and Their Applications," Ann. Rev. of Nucl. Sci. **38** (1988) 97.
28. S. O. Schriber, "High-Beta Linac Structures," Proc. 10th Linear Accelerator Conf., BNL 51134, 1980, p. 164.
29. I. M. Kapchinskij, "Theory of Resonance Linear Accelerators," Harwood Academic Publishers, 1985, p. 273.

30. I. M. Kapchinskij and V. V. Vladimirskij, "Limitations of Proton Beam Current in a Strong Focusing Linear Accelerator Associated with the Beam Space Charge," Proc. Int. Conf. on High-Energy Accelerators and Instrumentation, CERN, Geneva, 1959, p. 274.
31. R. Chasman, "Numerical Calculations of the Effects of Space Charge on Six Dimensional Beam Dynamics in Proton Linear Accelerators," Proc. 1968 Proton Linear Accelerator Conf., BNL 50120, 1968, p. 372.
32. R. Chasman, "Numerical Calculations on Transverse Emittance Growth in Bright Linac Beams," IEEE Trans. Nucl. Sci. **NS-16** (1969) 202.
33. J. W. Staples and R. A. Jameson, "Possible Lower Limit to Linac Emittance", IEEE Trans. Nucl. Sci. **NS-26** (1979) 3698.
34. P. M. Lapostolle, "Round Table Discussion of Space Charge and Related Effects," Proc. 1968 Proton Linear Accelerator Conf., BNL 50120, 1968, p. 437.
35. I. Hofmann and I. Bozsik, "Computer Simulation of Longitudinal-Transverse Space Charge Effects in Bunched Beams," Proc. 1981 Linear Accelerator Conf., Los Alamos, LA-9234-C, 1982, p. 116.
36. I. Hofmann, "Emittance Growth of Ion Beams with Space Charge," Nucl. Instrum. Meth. **187** (1981) 281.
37. R. L. Gluckstern, "Oscillation Modes in Two Dimensional Beams," Proc. 1970 Linear Accelerator Conf., Fermilab, 1970, p. 811.
38. I. Hofmann, L. J. Laslett, L. Smith, and I. Haber, "Stability of the Kapchinskij-Vladimirsky (K-V) Distributions in Long Periodic Transport Systems," Part. Accel. **13** (1983) 145.
39. R. A. Jameson, "Equipartitioning in Linear Accelerators," Proc. 1981 Linear Accelerator Conf., Los Alamos, LA-9234-C, 1982, p. 125.
40. T. P. Wangler, K. R. Crandall, and R. S. Mills, "Emittance Growth from Charge Density Changes in High-Current Beams," Proc. Int. Symp. on Heavy Ion Fusion, Washington, DC, AIP Conf. Proc. **152** (1986) 166.
41. T. P. Wangler, R. S. Mills, and K. R. Crandall, "Emittance Growth in Intense Beams," Proc. 1987 Part. Accel. Conf., IEEE Cat. No. 87CH2387-9, 1987, p. 1006.
42. C. M. Celeta, "The Effect of Nonlinear Forces on Coherently Oscillating Space-Charge Dominated Beams," Ibid., p. 996.
43. M. Reiser, "Emittance Growth in Mismatched Space-Charge Dominated Beams," Submitted to 1991 IEEE Part. Accel. Conf., San Francisco.
44. A. Cucchetti, M. Reiser, and T. Wangler, "Simulation Studies of Emittance Growth in RMS Mismatched Beams," Ibid.
45. J. Struckmeier and M. Reiser, "Theoretical Studies of Envelope Oscillations and Instabilities of Mismatched Intense Charged Particle Beams in Periodic Focusing Channels and Particle Accelerators," Part. Accel. **14** (1984) 227.
46. "Site-Specific Conceptual Design of the Superconducting Super Collider," SSC Lab. Report SSC-SR-1051, June 1990, p. 254.
47. E. J. N. Wilson, "Circular Accelerators—Transverse," Physics of Particle Accelerators, Vol. 1, AIP Conf. Proc. **153** (1987) 4.
48. E. J. N. Wilson, "The Importance of Non-Linearities in Modern Proton Synchrotrons," Theoretical Aspects of the Behavior of Beams in Accelerators and Storage Rings, CERN 77-13, July 1977, p. 111.
49. D. E. Young and R. J. Noble, "400-MeV Upgrade for the Fermilab Linac," Proc. 14th Int. Conf. on High-Energy Accelerators, Part 1, Part. Accel. **26** (1990) 205.
50. D. Finley, "Emittance Through the Fermilab Accelerator Chain," Physics of Particle Accelerators, Vol. 2., AIP Conf. Proc. **184** (1989) 1910.

THE NEXT LINEAR COLLIDER*

Ronald D. Ruth
Stanford Linear Accelerator Center
Stanford University, Stanford, California 94309

1 Introduction

There is now broad agreement in the high energy physics community that to continue exploring the energy frontier in e^+e^- interactions, we will have to abandon circular colliders and adopt linear colliders. This realization has led to active research throughout the world towards the next generation of linear colliders. The past few years have seen great strides in our understanding of both the accelerator physics and the technology of linear colliders. We are now at the point where we can discuss in fair detail the design of such a "Next Linear Collider," or NLC.[1-5]

The two key design parameters of the NLC are its energy and luminosity. A broad consensus has emerged over the past couple of years that the energy should be 0.5 TeV (total electron plus positron energy), upgradable to at least 1.0 TeV. One reason for this choice of energy range is the great potential of such a collider for significant high-energy physics research in the era of the SSC. Another is that this energy range is a natural next step; it is a factor of 5 to 10 beyond that of the present Stanford Linear Collider (SLC). In order to obtain a sufficient event rate to perform detailed measurements, the luminosity of the collider should increase with the square of its energy. For an NLC in the TeV energy range, a luminosity of $10^{33} - 10^{34}$ $\mathrm{cm^{-2}s^{-1}}$ is required.

A factor of 5–10 energy increase can be obtained in two ways: by increasing the collider length to 10–20 times that of the SLC (3 km), or by raising its accelerating field to 5–10 times the SLC gradient (20 MV/m). The present consensus is that we should first increase the accelerating field by about a factor of three to five—up to about 50 to 100 MV/m. To limit the RF power required, this field should be provided by structures similar to those used in the SLC but at a higher RF frequency of 10–30 GHz. At SLAC, the frequency choice for the NLC is 11.4 GHz, or four times the present SLC frequency. Of course, the ultimate tradeoff between length and accelerating field is governed by the overall cost and the upgradability. A broad optimum occurs at the point where the linear costs (accelerating structure, magnets, tunnel, etc.) equal the cost of the RF power source.

The choice of luminosity range also greatly influences the design of the linear collider. In principle, one could increase the luminosity simply by raising the repetition rate of the accelerator, but the wall-plug power increases in direct proportion. In a reasonable design, the wall-plug power should fall in the range 100–200 MW. Given this constraint, the best way to increase the luminosity is to shrink the beam size at the interaction point (IP). In addition, the beam cross section must be kept flat at the IP in order to minimize the amount of "beamstrahlung" radiation emitted as energetic electrons or positrons interact with the electromagnetic field of the opposing bunch.

The luminosity can be further enhanced by accelerating several bunches on each machine cycle. A single bunch of particles can, in practice, extract only a few percent of the energy available in the accelerating structure. With additional bunches, we get both greater luminosity and higher efficiency of energy transfer to the beam. The number of particles in each bunch, another factor that directly affects the luminosity, is limited by the RF energy that can be stored in the accelerating structure and by the amount of beamstrahlung radiation that can be tolerated. The obvious solution is to generate trains of successive bunches, each with a fairly moderate number of particles.

* Work supported by the Department of Energy, contract DE–AC03–76SF00515.

Given these goals and constraints, we can now sketch a rough design of a linear collider able to achieve both the desired energy and luminosity. A possible layout is shown in Fig. 1. There are two complete linear accelerators, one for electrons and the other for positrons. Each linac is supplied with particle beams by a damping ring followed by a preacceleration section consisting of two bunch compressors and a 16-GeV linac. After passing through the main linacs and final focus system, the beams collide at a small crossing angle inside a large particle detector similar to the SLD detector at SLAC.

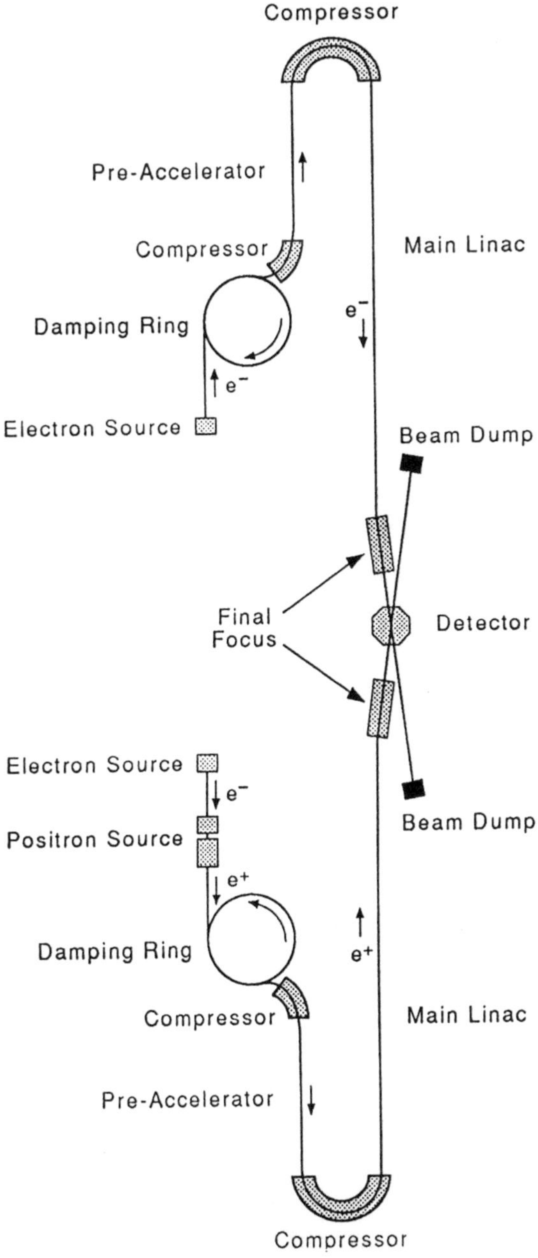

Fig. 1. Schematic diagram of the NLC.

To illustrate the basic features of the NLC operation, let's follow some electron bunches through the collider. A sequence of 10 bunches or so is created at the source and accelerated up to about 1.8 GeV in a preaccelerator. This "batch" of bunches is then injected into a damping ring that serves to reduce the transverse and longitudinal phase space occupied by the electrons in each bunch. At the proper moment, these bunches are extracted from the ring and then compressed along their direction of motion by a bunch compressor, after which they are accelerated up to about 16 GeV and compressed a second time just prior to injection into the main, high-gradient, linac. The entire batch is carefully steered and focused as the electrons are accelerated up to full energy in the linac. Precision magnets in the final focus system squeeze the bunches down by about a factor of 300 just before they collide at the IP with similar bunches of positrons. Except for the fact that they were created differently, from the shower of particles that occurs when a bunch of electrons hits a metal target, these high energy positron bunches have followed a similar evolution. After the beams collide, their debris is channeled out of the detector area and into shielded dumps.

1.1 NLC Parameter Options

The parameters for an NLC are not definite yet; however, over the past few years we have narrowed down the range of possibilities considerably. At SLAC, we have recently reviewed several options for an NLC which has an initial energy of 0.5 TeV in the CM and an upgraded energy of 1.0–1.5 TeV. Table 1 lists three parameter options: the first two columns are for 0.5 TeV in the CM, while the final column is for 1.0 TeV. In Option 1, a short linear collider is constructed with the full acceleration gradient of 100 MV/m. This can be upgraded to Option 3 by doubling the length of the linac while keeping the injection system fixed. In Option 2, a long linear collider is constructed with a reduced acceleration gradient of 50 MV/m. This can be upgraded to Option 3 by the addition of power sources to the linac. In both upgrade paths, the final focus must be modified somewhat.

Option 1 is quite short and may be less expensive than Option 2, but we are required to face all the problems of the high acceleration gradient and the required high peak power RF sources. In Option 2, we relax the requirements for RF power by a factor of four and begin with a reduced acceleration gradient. The price is an initially longer accelerator with the increased conventional construction.

We have found in the design process that it is very important to realize that the intensity and emittance at the final focus are quite different from those in the damping ring. To model this, the intensity has been allowed to decrease as shown in Table 1. In addition, the emittance at the final focus is assumed to be diluted by about 65%.

To discuss the NLC in more detail, we divide the problem into the two primary parameters: the energy and the luminosity. In the next section we discuss how to obtain the energy in an NLC.

Table 1. NLC Parameter Options.

Option		1	2	3
Energy		$\frac{1}{4}+\frac{1}{4}$ TeV	$\frac{1}{4}+\frac{1}{4}$ TeV	$\frac{1}{2}+\frac{1}{2}$ TeV
Luminosity		2×10^{33}	2×10^{33}	1×10^{34}
Linac Length		7 km	14 km	14 km
Accel. Gradient		100 MV/m	50 MV/m	100 MV/m
RF Frequency		11.4 GHz	11.4 GHz	11.4 GHz
# Particles/bunch:	DR	2×10^{10}	1×10^{10}	2×10^{10}
	Linac	1.8×10^{10}	9×10^{9}	1.8×10^{10}
	FF	1.5×10^{10}	7×10^{9}	1.5×10^{10}
# Bunches, n_b		10	10	10
Repetition Freq.		120 Hz	180 Hz	180 Hz
Wall-Plug Power		66 MW	50 MW	200 MW
IP Beam Size:	σ_y	4 nm	4 nm	2.5 nm
	σ_x	320 nm	200 nm	220 nm
	σ_z	100 μm	100 μm	100 μm

2 The Energy

As discussed in the Introduction, the energy of the linear collider is obtained through a combination of length and acceleration gradient,

$$E = \mathcal{E}_z L \ , \tag{1}$$

where $\mathcal{E}_z$ is the acceleration gradient and L is the length of the linac. This relation is over-simplified, for reasons which we discuss later, in that the average acceleration gradient may differ from the peak and the length may include space for focusing magnets, etc.

The acceleration is obtained with the use of radio-frequency (RF) structures as shown in Fig. 2. The structure shown is a travelling-wave structure. It is basically a long copper cylinder periodically interrupted by disks with holes along the center line. Every so often (every 1.5 m or so), the structure is interrupted by a feed for fresh RF power and a load to remove the depleted upstream power.

The RF power is provided by the RF source and is transported to the accelerating structure in a copper waveguide. The structure is designed as a travelling-wave structure and, as such, has a characteristic phase velocity and group velocity. The phase velocity is designed to be the speed of light. In this way, if a relativistic electron enters the structure at the correct phase for acceleration, then it will be continually accelerated as it traverses the entire structure. The group velocity v_g is the rate at which the structure is filled with energy; it is the velocity of the envelope of the RF pulse as it traverses the travelling-wave structure. If L_s is the length of the structure, then the time to fill the structure T_f is given by

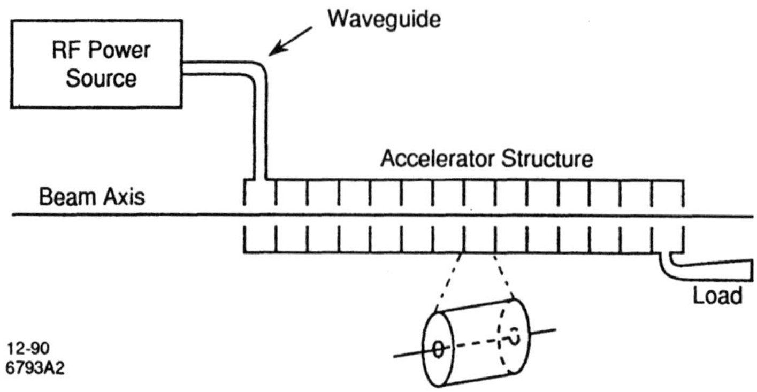

Fig. 2. Diagram of a travelling-wave RF structure.

$$T_f = \frac{L_s}{v_g} \; . \tag{2}$$

For cases of interest, the group velocity is somewhat less than one tenth of the velocity of light.

2.1 The Extraction of Energy

The energy gain of a test particle at the head of a bunch traversing a structure of length L_s is

$$\Delta E = \mathcal{E}_z L_s \cos\varphi \; , \tag{3}$$

where φ is the phase on the RF wave. The trailing particles see not only the field supplied by external sources but also the field induced by the bunch itself, the longitudinal wakefield. The bunch charge induces fields in all the synchronous modes of oscillation of the accelerating structure. The field induced in the fundamental accelerating mode is

$$\mathcal{E}_{wake} = -2k\, q \cos(\omega z/c) e^{-\lambda z} \; , \tag{4}$$

where k is called the loss parameter, q is the charge in a bunch which is short compared to the wavelength of the RF, ω is the RF frequency, z is the distance behind the point-like bunch, and λ is the decay constant due to losses in the structure walls. The wakefield ahead of a speed-of-light bunch vanishes due to causality. The total wakefield induced consists of a sum over all the synchronous modes of the structure; the dominant term is given in Eq. (4).

This field induced by the particle bunch causes problems which must be dealt with. The particles at the head of the bunch feel the full acceleration, while those at the tail feel an accelerating field reduced by $2kq$. If the particles are being accelerated on the crest of the RF, this causes a reduction of the average energy gained by the bunch, and it also causes an energy spread:

$$\Delta E_{ave} = (\mathcal{E}_z - kq)L_s$$
$$(\Delta E)_{spread} = \pm kqL_s \; . \qquad (5)$$

These effects are due to the extraction of energy from the RF wave. The field induced by the bunch reduces the electric field in the structure an amount which corresponds to the energy extracted by the bunch of particles. The fraction of energy extracted from a full structure by a bunch at the crest of the RF is

$$\eta_0 = 1 - \frac{(\mathcal{E}_z - 2kq)^2}{\mathcal{E}_z^2}$$
$$\simeq \frac{4kq}{\mathcal{E}_z} \qquad \text{for small } \eta_0 \; . \qquad (6)$$

The reduction of the average value of the energy gain can be compensated either by increasing the accelerating field $\mathcal{E}_z$ or by adding more accelerator sections to make up the lost energy. The spread of energy in a bunch can be compensated for by shifting the phase of the bunch on the RF wave. In this way, with very little loss of acceleration, it is possible to obtain a slope sufficient to cancel the variation induced by the bunch wakefield. For a uniform particle distribution with a full width $\Delta\varphi$, the phase offset is

$$\sin\varphi_0 = \frac{2kq}{\Delta\varphi\mathcal{E}_z} \; . \qquad (7)$$

Provided that the phase offset is small, this compensation technique works well. In order to achieve a small phase offset, the bunch can be lengthened or the intensity reduced. With very long bunches, the curvature of the RF must be taken into account and can, in fact, be used to provide compensation of the nonlinear variation of energy along the bunch. With short bunches, only the linear variation can be cancelled, which leaves a residual nonlinear energy variation. This residual must be kept smaller than the energy acceptance of the final focus system.

2.2 Multi-Bunch Energy Compensation[6]

In the Introduction, we discussed the acceleration of a short train of bunches in order to improve the luminosity by extracting more energy from the RF structure. From the analysis of the previous section, the second bunch must have an energy which is lower by

$$\Delta E_2 = -2kqL_s \; . \qquad (8)$$

Once again, this is simply due to the extraction of energy from the RF wave.

This problem can be solved by changing the effective structure length for the two bunches. If the structure is partially filled when the first bunch passes through, and if the additional energy entering the structure prior to the passage of the second bunch matches the energy extracted by the first bunch, then the second bunch will have the same energy as the first. This is illustrated in Fig. 3; the cross-hatched areas of the external field and wakefield must match in order to compensate the energy difference. This technique is used at the SLC to adjust the relative energy of the positrons and electrons.

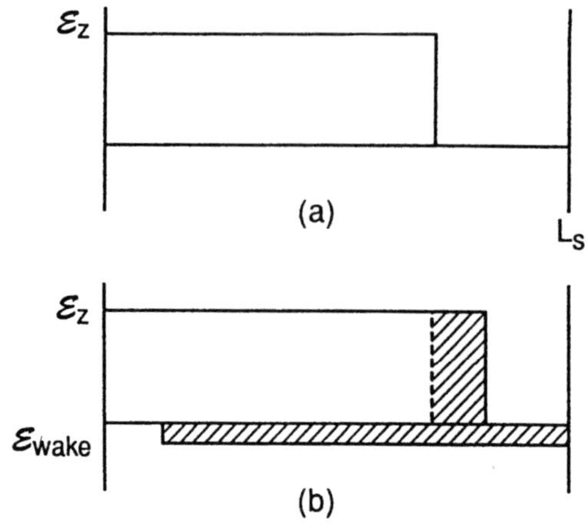

Fig. 3. The electric field profile in the structure (a) just before passage of bunch one and (b) just before passage of bunch two.

2.3 Multibunch Beam Breakup

Let us assume that we can match the energy of a short train of bunches as described in the previous section. There are still other problems caused by the wakefield induced by a bunch of particles. If the bunch is offset in the structure, then it induces a deflecting force behind it which is proportional to the offset of the bunch. This transverse wakefield is similar in form to the longitudinal and consists of the sum of many modes which cause deflection,

$$W_{\perp}(z) = \sum_{n} W_n \sin(\omega_n z/c) e^{-\lambda_n z} \, , \tag{9}$$

where W_n is the strength of a particular mode, z is the distance behind a short bunch, ω_n is the mode frequency and λ_n is the decay constant for mode n. The transverse wakefield differs from the longitudinal in that the transverse is sine-like while the longitudinal is cosine-like.

The transverse wakefield can cause an instability known as beam breakup.[7,8] It is caused by the combination of the coupling from bunch to bunch and also by a resonance effect. To see this, consider just two bunches. If these two are offset coming into the structure, the magnet focusing causes them to oscillate with a wavelength $2\pi\beta$. The first bunch oscillates freely down the linac according to

$$\frac{d^2 x_1}{ds^2} + \frac{x_1}{\beta^2} = 0 \, . \tag{10}$$

The second bunch is also focused but, in addition, is deflected by the leading bunch's wakefield

$$\frac{d^2 x_2}{ds^2} + \frac{x_2}{\beta^2} = \frac{N e^2 W_{\perp}(l)}{E} \, x_1 \, , \tag{11}$$

where N is the number of particles in bunch one and E is the energy of the bunches. The deflecting force is proportional to the position of bunch one. Because bunch one oscillates in the focusing system, the force on the right-hand side of Eq. (11) oscillates and bunch two is driven at resonance. Therefore, the amplitude of the second bunch grows linearly down the accelerator.

The effect is similar for many bunches: bunch three is driven on resonance by bunch one and bunch two and so on. The result is that the bunches at the end of the train can reach large amplitudes unless something is done to ameliorate the problem. The solution is to eliminate, to the extent possible, the force coupling the bunches together. This can be done by a special design of the RF structure and is the subject of the next section.

2.4 Accelerating Structures

As discussed earlier, the job of the RF structure is to accelerate the beam. As such, it is usually optimized to provide the greatest acceleration for the lowest RF power. In addition, the design can have a large impact on the stability of a single bunch (to be discussed in Section 3.6.2) and on the stability of a train of bunches. To assure the stability of a train of bunches, we would like to reduce the deflecting field induced by a bunch as much as possible before the passage of the next bunch.

This can be accomplished in two ways (see Fig. 4). In the first method, shown in Fig. 4(a), the cavity design is altered so that the deflecting fields are strongly coupled to external waveguides. After a bunch passage, the fields in the cavity die out quickly as they are propagated out the waveguide into a matched load. The design shown in Fig. 4(a) shows radial waveguides coupled via slots cut in the irises of the accelerator.[9]

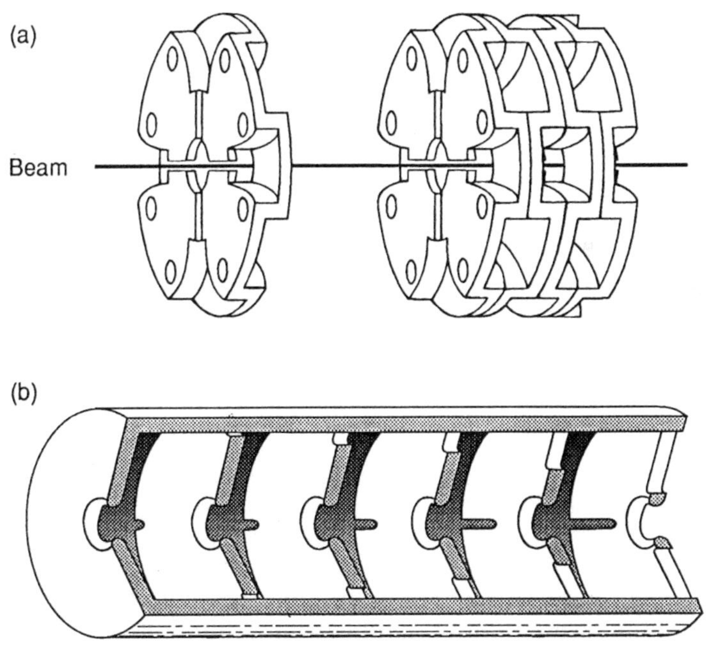

Fig. 4. Two methods of wakefield damping: (a) radial waveguides transmit the energy out of the structure; (b) variation of cell construction causes decoherence and effective damping.

The second technique, shown in Fig. 4(b) relies on the cancellation of the deflections from cell to cell. If the cells in a single short structure are designed so that the deflecting modes oscillate at different frequencies, then the average deflection over the structure effectively damps due to the decoherence of the various cell wakefields. The initial decoherence time is just the inverse of the spread in frequency. This technique is illustrated in Fig. 4(b), where the change in frequency is accomplished with three radial slots of varying depth cut into the irises of the structure.

Damped structures similar to that shown in Fig. 4(a) have been constructed at SLAC and have achieved a quality factor Q ~ 8 for the dominant higher-order mode.[10,11] This damping is completely sufficient to eliminate the beam breakup discussed in the previous section.[7,8]

The second technique of detuning is an alternative, and possibly simpler, technique which is presently under investigation at SLAC.

2.5 RF Power Sources

To achieve the desired acceleration gradient for the Next Linear Collider, RF power sources must be provided which give the required peak power and pulse length at the desired frequency. In the designs in Table 1 we find acceleration gradients of 50–100 MV/m at an RF frequency of 11.4 GHz. To achieve the larger gradient, it is necessary to provide about 350 MW of peak RF power in a pulse of about 100 ns in length to be fed into a structure about 1.5 m in length.

In the SLC, the acceleration is accomplished with 2.8-GHz accelerating structures 3 m in length. Each of these is fed by a 40-MW pulse about 1 μs long which yields an acceleration gradient of about 20 MV/m. In order to increase the acceleration gradient in such a structure to 100 MV/m, it would be necessary to increase the RF peak power and the stored energy by a factor of 25.

At the higher frequency of 11.4 GHZ, the energy density must once again increase by a factor of 25; however, the cross-sectional area drops by a factor of 1/16. Thus, the energy per unit length in the RF structure only changes by a factor of two, provided the higher frequency is utilized. Although the energy is changed very little, the structure of the RF pulse is very different from that at the SLC. The necessary RF pulses are higher in peak power but shorter in duration and feed a shorter accelerator structure. The primary challenge for the RF power system is to provide a source with the characteristics described above.

There are basically two approaches to this problem as outlined in Fig. 5. The first approach, shown in Fig. 5(a), uses RF pulse compression. With this technique, a modulated power pulse of ~ 1 μs is provided by a conventional pulsed power transformer, a modulator. This pulse is converted to an RF pulse of the same duration by some device such as a klystron. After the RF is created, it can be compressed by RF pulse compression to the desired pulse length with a correspondingly higher peak power.

The second technique, magnetic compression, begins with the same modulated power pulse and then compresses this pulse using a technique called magnetic pulse compression; the time structure is achieved before the creation of RF. After this stage, RF can be created by a device such as a relativistic klystron or by an array of multiple power sources driven in parallel. This second technique has been under experimental investigation by a collaboration of SLAC, LLNL, and LBL. The relativistic klystron achieved a power output of 330 MW with a 20-ns pulse.[12] This technique, however, is presently not considered a candidate for the power source due to inefficiency and cost. The remainder of this section is devoted to the first alternative.

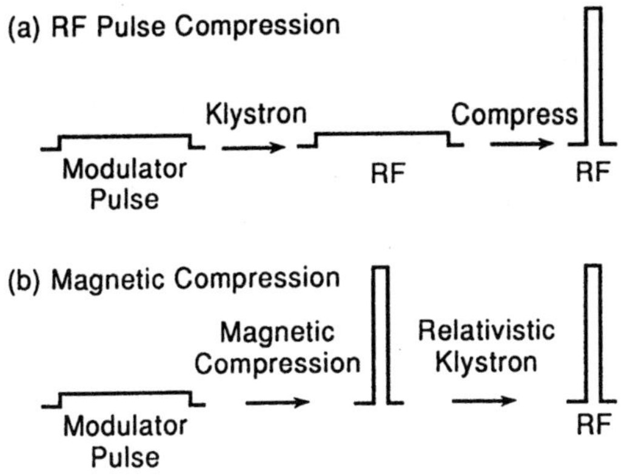

Fig. 5. Two methods of producing short high-peak-power RF pulses.

2.5.1 The Klystron

A block diagram of the RF power system with RF pulse compression is shown in Fig. 6. The modulator power supply is conventional and is similar in most respects to those used at SLAC for the SLC. Therefore, we will begin the discussion with the klystron.

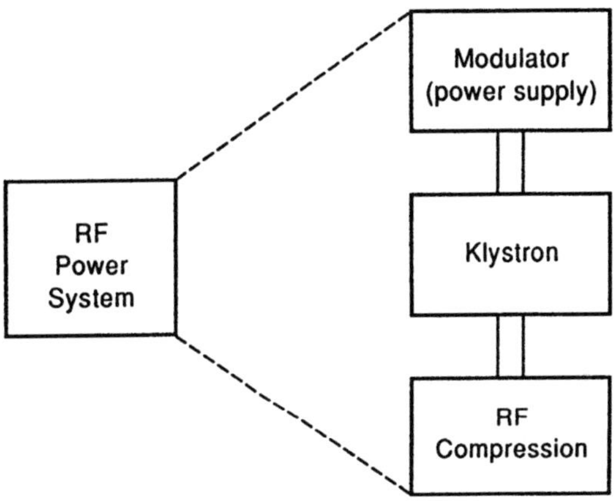

Fig. 6. Block diagram of RF power system with RF pulse compression.

A schematic diagram of a klystron is shown in Fig. 7. Put very simply, a klystron is a narrow-band, high-gain radio-frequency amplifier. To achieve this amplification, an electron beam is created by the voltage induced by the modulator across the cathode and anode. The electrons are accelerated to an energy of about 400 kV with a current of about 500 A; they are transported down a narrow tube with a solenoid magnet providing the focusing. A small amount of RF power applied to the input cavity ($\lesssim 1$ kW) modulates the beam energy at the RF frequency. Due to the induced velocity difference, the faster electrons catch up to those that were decelerated. This creates a small density modulation of the beam as it enters the first gain cavity.

This cavity is resonantly excited by the RF electric field of the modulated beam to a field of approximately 10 times that in the input cavity. This field acting back on the beam provides much

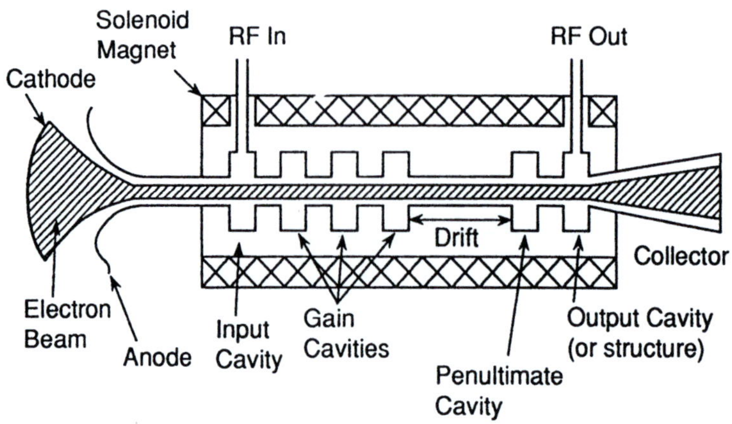

Fig. 7. Schematic diagram of a klystron.

deeper bunching by the time the beam reaches the second gain cavity. This process continues until the final gain cavity, where the energy of the beam is modulated by as much as 30% of its value. This modulated beam bunches strongly in the final drift region, is further compressed by the penultimate cavity and then enters the output structure. This may be one or more standing-wave cavities, or it may be a travelling-wave structure. The beam induces a field in the output structure; however, this structure is designed so that the phase of the RF field is such as to decelerate the sequence of bunches entering it. This deceleration extracts the RF energy in the bunches and transfers it to the cavity fields which are coupled to an external waveguide. The RF power flows out the waveguide and can be transported for further use. The beam is deposited in a water-cooled collector with approximately one half of its energy removed by this process. With the parameters given above, the klystron produces about 100 MW of RF power.

Klystrons similar in all respects to the one just described have been built at SLAC and have achieved 75 MW in short pulses and 50 MW in long pulses as of this writing.[13] The design goal is to achieve a 100-MW klystron at 11.4 GHz with a pulse length of about 1 μs. This pulse length is much too long to be used directly in the acceleration process; we need RF pulse compression.

2.5.2 RF Pulse Compression

The object of RF pulse compression is to convert a long RF pulse of moderate power into a short RF pulse with high power. Ideally, a factor of five decrease in pulse length could yield a factor of five increase in peak power. Due to inefficiencies, the factor is always somewhat less. The RF pulse compression system SLED (SLAC Energy Doubler) is presently used at SLAC to boost the klystron power by about a factor of three before powering the SLAC linac. This system uses storage cavities to allow the RF to build up. A phase switch from the klystron effectively releases the energy. Unfortunately, this system gives a pulse shape which is sharply spiked due to the exponential decay of the fields in the storage cavities. For an NLC it is useful to have a flat-top pulse to control multibunch energy spread.

This flat-top pulse can be obtained by two different methods. The first method, called binary pulse compression (BPC), uses delay lines to delay the leading portion of an RF pulse so that it is coincident in time with the trailing portion. This yields an RF pulse which is one half as long, but with nearly twice the power. This process can be repeated in a sequence to achieve more and more multiplication. Due to losses in components and waveguides, the method is limited to about three compressions.

Figure 8 shows a schematic diagram of a two-stage BPC system which was constructed at SLAC.[14–16] The 3db hybrid shown in Fig. 8 is a four-port device which combines two power inputs into one or another output port depending upon relative phase. In this way phase shifts can be used as high power RF switches. A three-stage system of analogous design has been constructed at SLAC and has achieved a multiplication factor of 5.5 while reducing the pulse length by a factor of eight.[17] This system, together with two 100-MW klystrons, would produce RF power sufficient for a 5-m-long accelerator with an acceleration gradient of about 100 MV/m. High power tests of this three-stage system are continuing.

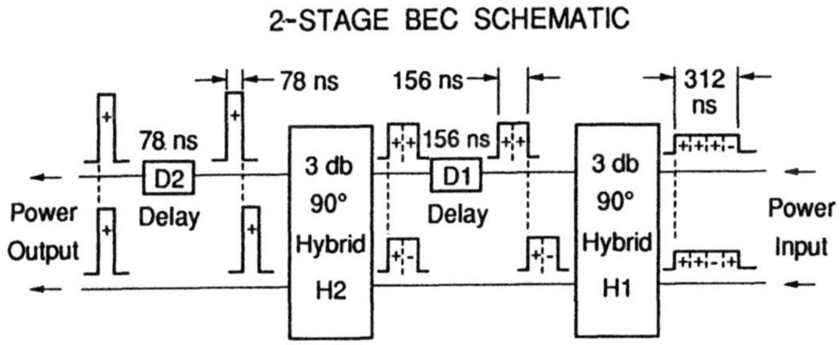

Fig. 8. Schematic diagram of a BPC system.

One disadvantage of the BPC method of pulse compression is that it uses rather long delay lines. The waveguides which are used have a group velocity very close to the speed of light, and they are only used once as transmissive delay lines. This problem has led to the development of a new pulse compression scheme called SLED II.[18] The system as shown in Fig. 9 is similar to the SLED system at SLAC except that the cavities for storing the RF are replaced by resonant delay lines. Each of these delay lines has a round-trip delay time equal to the output pulse length. A resonant buildup of energy stored in the lines takes place during an input pulse length which is an integral number of delay periods, typically in the range of four to eight. A phase reversal of the input pulse effectively releases the stored energy to produce a flat-top output pulse during the final delay period. An example of a SLED II pulse compression by a factor of four is shown in Fig. 10. Measurements from a low power SLED II system with a power gain of four have shown excellent agreement with theory.

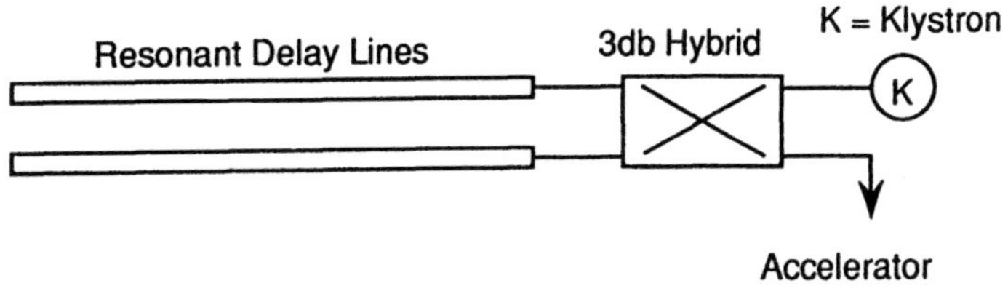

Fig. 9. A block diagram of SLED II.

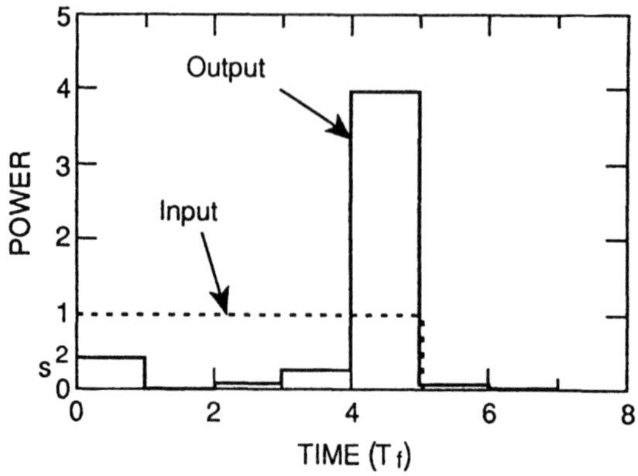

Fig. 10. SLED II pulse compression.

Comparing SLED II with BPC, the amount of waveguide delay line to achieve a similar compression is reduced by more than a factor of five. This is due to the reflective nature of the scheme; the delay lines are used repeatedly as the RF wave builds up. In addition, this method can be staged by placing the SLED II systems in series to provide even larger compressions if necessary. A high power SLED II pulse compression system will be constructed at SLAC in 1991 to investigate this promising technique further.

3 The Luminosity

In the first two sections, we discussed the basic layout and how to obtain the energy in a linear collider; in this section, we discuss how to obtain the luminosity. Although the luminosity depends upon beam properties at the interaction point, those properties depend upon beam dynamics throughout the entire linear collider; therefore, we must trace this influence throughout the collider. Before doing that, however, let's examine the luminosity formula. Luminosity for a linear collider is the same as for a circular collider except that there is an additional term, an enhancement factor due to the mutual pinching of the beams. The luminosity is given by

$$\mathcal{L} = \frac{N_+ N_- f_{rep} n_b}{4\pi\sigma_x\sigma_y} H_D \ , \tag{12}$$

where $N_\pm$ is the number of positrons/electrons per bunch, f_{rep} is the repetition frequency, n_b is the number of bunches accelerated on each cycle of the accelerator, H_D is the pinch enhancement factor, and finally, σ_x and σ_y are the rms beam size of the Gaussian spot at the interaction point. Each bunch is assumed to collide with only one other bunch in the opposing bunch train.

The object is to increase the luminosity to 10^{33}–10^{34} cm^{-2}s^{-1}, for the energy range $\frac{1}{2}$ to 1 TeV. To do that we must increase the numerator of Eq. (12) and decrease the denominator as much as possible. For the numerator, we have at our disposal the number of particles per bunch, the repetition rate, and the number of bunches on each cycle, but we must satisfy the constraint that the wall-plug power is in the range of 100–200 MW. For the denominator, we can decrease the cross-sectional area by decreasing σ_x and σ_y, but to do this we must keep the beam flat to control beamstrahlung.

In the next few sections we discuss each term in the luminosity formula. The discussion of beam size is subdivided into several sections. In the next section we begin with the numerator of Eq. (12).

3.1 Intensity and Repetition Rate

First let's discuss the single-bunch intensity $N_\pm$ and the repetition rate f_{rep}. From conservation of energy, we must have

$$eNE_{cm}f_{rep} = \eta_{rf}\eta_b P_{wall} \ , \tag{13}$$

where η_{rf} is the efficiency for converting wall-plug power to RF power, η_b is the fraction of the energy extracted by a single bunch, and P_{wall} is the total wall-plug power supplied to the linacs. The wall-plug-to-RF efficiency, η_{rf}, is about 20% for the projected RF system. This is a fairly realistic estimate including all of the factors in the power system which were discussed in the first section. There are new ideas which could raise this to perhaps 30–40%; however, with the system shown in Section 2.5, η_{rf} is about 20%.

For somewhat different reasons, the single-bunch extraction efficiency is limited to about 2%. In Section 2.1, we discussed the single-bunch energy spread which is induced by longitudinal wakefields. Although the linear part can be compensated by shifting the RF phase to obtain a linear slope, the higher-order effects are difficult to compensate. This limits single-bunch energy extraction to a few percent.

For the purpose of this discussion, let's select a wall-plug power of 150 MW for an E_{cm} = 1 TeV.

Because the required bunches have a very small transverse dimension, it is necessary to control their offset pulse-to-pulse with a feedback system. In order for this feedback system to work efficiently, the sample rate must be at least six times the rate at which the beam centroid is changing. Because ground motion is an important source of bunch motion, and because the spectrum drops off rapidly above 10 Hz, the repetition rate of the accelerator must be greater than 60 Hz. In order to have it sufficiently greater, we set the repetition frequency to 180 Hz. It could be dropped as low as 120 Hz; however, 60 Hz is probably too low. Substituting the previous parameter values in Eq. (13), we find that the maximum number of particles per bunch is $N_\pm \simeq 2 \times 10^{10}$.

3.2 The Number of Bunches

As discussed in the Introduction, the designs for the NLC include the acceleration of many bunches on each cycle of the collider. The purpose of this is, of course, to increase the luminosity linearly with increasing number of bunches. If there were no constraints, the largest luminosity would be obtained by placing all the charge in the bunch train into one bunch because in this case there is quadratic gain with increasing intensity. As discussed in the previous section, the single-bunch intensity is limited by the amount of energy it can extract while retaining a small relative energy spread. It turns out that this intensity is also consistent with transverse stability (Section 3.6.2) and with beam-beam effects (Section 3.7.1). Thus, the quadratic gain is stopped by these bounds; however, since there is about 98% of the energy left in the structure, it is possible to continue to gain linearly by increasing the number of bunches.

A large number of bunches brings along a host of other complications. Some of these were discussed earlier. The bunches must be stable transversely which means that the structure must be designed in a special way (Sections 2.3 and 2.4). The energy spread bunch-to-bunch must be controlled. Although the solution presented in Section 2.2 does keep the energy spread small, only about 20% of the energy can be extracted without introducing more complicated compensation techniques. This limits the number of bunches to about 10; although the single-bunch intensity can be traded off somewhat with the number of bunches.

The RF pulse must be of rather high quality. Systematic phase and amplitude variations over the bunch train must be less than about 2% (such tolerances are not unrealistic with the power sources discussed). Because a significant fraction of the fields felt by the trailing bunches are due to the leading bunches, the intensity of the bunch train must be controlled with a precision less than 2%. The damping rings which produce these trains of bunches must be able to accelerate them without instability. If small position or energy changes occur, a compensation system must be developed to assure that the bunches enter the final focus system on the same trajectory and with the same energy. The final focus system must be designed so that the distant crossings of bunches do not disrupt the primary collisions at the interaction point.

Although the addition of many bunches appears to be "free" in that we simply use energy that would normally be wasted, it introduces complexity into every subsystem of the entire collider. The benefit is an order of magnitude increase in the luminosity.

3.3 The Beam Size

The transverse size of a beam in an accelerator is determined by two basic parameters: the emittance ϵ and the beta function β,

$$\sigma = \sqrt{\epsilon\beta} \,. \tag{14}$$

The emittance is a parameter that is proportional to the area occupied by the beam distribution in transverse phase space (x, p_x). It is defined by

$$\epsilon_x = \frac{1}{p_0}[< x^2 >< p_x^2 > - < xp_x >^2]^{\frac{1}{2}} \,, \tag{15}$$

where x is the transverse position, p_x is the corresponding transverse momentum, and p_0 is the central momentum of the bunch of particles. The angle brackets in Eq. (15) indicate an average over the distribution of particles in a bunch. Because the quantity in the square brackets is an adiabatic invariant (in the absence of synchrotron radiation), the emittance decreases inversely with the momentum of the bunch in a linear accelerator.

The longitudinal emittance is defined in a similar way,

$$\epsilon_z = \frac{1}{p_0}[< z^2 >< \Delta p^2 > - < z\Delta p >^2]^{\frac{1}{2}} \,, \tag{16}$$

where z is the longitudinal deviation from a central position within the bunch, and Δp is the deviation of the particle momentum from a central momentum. Once again, the quantity in the square brackets is an adiabatic invariant, which causes ϵ_z to decrease inversely with the beam momentum in a linear accelerator. In the special case of a high-energy electron linac, the longitudinal distribution and the bunch length are fixed because the particles all travel at essentially the speed of light. In this case, the fractional momentum spread varies inversely with the beam momentum.

The beta function β was first introduced by Courant and Snyder in their description of the alternating gradient focusing of particle beams.[19] The parameter not only determines the particle beam size through Eq. (14), it also determines the instantaneous wavelength of the oscillations of particles within the beam envelope as they traverse the focusing magnets (wavelength = $2\pi\beta$).

The beta function also plays an important role at the interaction point. In a magnet-free region, it has the particularly simple form

$$\beta(s) = \beta^* + \frac{(s - s_0)^2}{\beta^*} , \tag{17}$$

where β^* is the minimum value of $\beta(s)$ and s_0 is the location of that minimum, the IP in this case. According to Eq. (14), the beam size near the interaction point is therefore

$$\sigma^2(s) = \epsilon\beta^* + \frac{\epsilon}{\beta^*}(s - s_0)^2 . \tag{18}$$

From this form, it is obvious that β^* is the depth of focus because the beam size increases by $\sqrt{2}$ when $s - s_0 = \beta^*$. Thus, the beta function plays two important roles at the IP—it determines both the spot size and depth of focus.

3.4 The Damping Ring[20,21]

The damping ring serves to reduce the emittance of the bunches of particles in all three degrees of freedom. It is an electron storage ring similar in all essential features to the storage rings used for colliding beams or synchrotron light production. The particles in an electron storage ring radiate a substantial fraction of their energy on each turn—energy that is restored by RF accelerating cavities. In the process of radiation, the particles lose energy from all three degrees of freedom, but it is restored only along one, the direction of motion; the proper amount is supplied at a single RF phase for a particle with the design energy, which leads to damping in all three dimensions. The fact that radiation is emitted as discrete quanta, however, introduces stochastic noise that causes diffusion of particle trajectories.

The competition between these damping and diffusion effects leads to an equilibrium value for the emittance of an electron storage ring. Damping rings are designed to enhance the damping effects using strong magnetic fields (such as those in wiggler magnets), while limiting the diffusion by the special design of the transverse focusing in the ring. In addition, there is a unique feature of electron storage rings that can be used to advantage. Due to the lack of vertical bending, the vertical emittance of the beam is much smaller than the horizontal—typically two orders of magnitude smaller. Such naturally flat beams are a key feature of many NLC designs.

One possible design for a future damping ring is about a factor of five larger and operates at an energy 50 percent higher than that of the SLC damping rings (see Fig. 11). The final emittance of the beam is more than an order of magnitude smaller than that of the SLC beams, which leads to much smaller sizes. In fact, the vertical extent of a beam emerging from this damping ring would be a few microns, or about equal to the final spot size at the SLC interaction point.

Another key difference is the simultaneous damping of many batches of bunches. In the SLC, at most two bunches are damped simultaneously, whereas this NLC ring will damp 10 batches of 10 bunches all at once. This feature allows a longer damping time for any given bunch, because we can extract the "oldest" batch and inject a new "young" batch while leaving those in their "adolescence" to continue damping undisturbed.

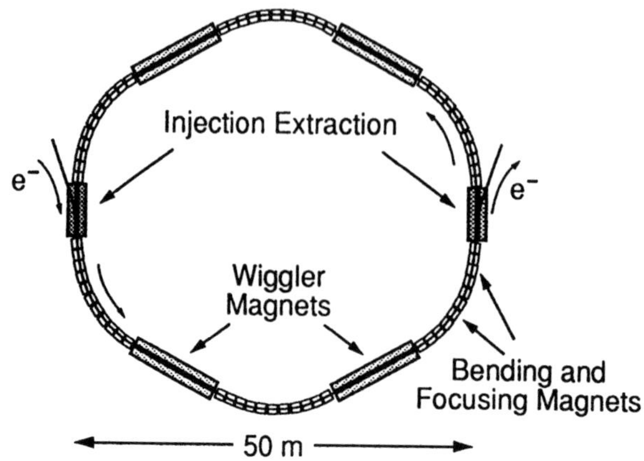

Fig. 11. A design of an NLC damping ring.

Because the bunches forget their origins in the damping ring, their conditions upon emerging are entirely determined by their behavior in the damping ring. This places special emphasis on the stability of the magnets in the damping ring and extraction system.

3.5 Bunch Compression/Pre-Acceleration[22,23]

Although the longitudinal emittance obtained in the damping ring is small enough, the bunch is still much too long for acceleration in a linac. In the SLC and NLC, this problem is solved by a technique called bunch compression, which shortens the bunch while increasing its energy spread. Each bunch passes through an RF accelerating structure phased so that the trailing particles emerge with lower energy than the leading particles. Then the bunch passes through a sequence of magnets that disperses the beam so that particles of different momenta travel on different paths. Particles with higher momentum (at the head of the bunch) travel a longer path than those of lower momentum (at the tail). The tail of the bunch can therefore catch up with the head, producing a shorter bunch—but at the cost of a greater energy spread.

This type of bunch compression has been used routinely in the SLC, where bunches 5 mm long are compressed to 0.5 mm for acceleration in the linac. Much shorter bunches will be required in the NLC. Short bunches will suffer less from transverse wakefields in the linac, and they permit a smaller depth of focus at the IP (about 100 microns for the NLC). In principle, another order of magnitude in compression could be obtained in a single stage; in practice, however, this approach would lead to other deleterious effects due to the large energy spread that would be induced in the beam. For this reason, the extra compression is provided by a second bunch compressor operating at a higher energy.

In the NLC, the bunch is first compressed as in the SLC to 0.5 mm in length, after which the beam is accelerated to about 16 GeV. The longitudinal spread of the beam is unaffected by this acceleration, but the relative energy spread decreases linearly with energy. The compression is then repeated, resulting in a bunch length as low as 50 microns. By separating the compression process into two discrete steps, we can keep the relative energy spread small throughout.

3.6 Linac Emittance Preservation[24]

The linac is the heart of the linear collider. As the beam is almost continuously accelerated, it is also focused transversely. During this process various effects conspire to dilute the emittance unless special care is taken. Because the linac is so vital and the potential for emittance dilution and beam size increase is so great, we will discuss various contributing factors in the next few subsections.

3.6.1 Injection Errors

After the bunch is compressed in length and as it enters the high-gradient linac, the bunch is about 2 μm high, 20 μm wide and 100 μm long. To obtain the necessary luminosity, the beam must be demagnified to the size shown in Table 1, $\sigma_y \times \sigma_x = 4$ nm $\times$ 320 nm. All of the offsets or angular kicks of the beam which occur upstream of the final focus system, however, get demagnified right along with the beam size. This means that the local beam size sets the scale for any offset and the local beam divergence sets the scale for any angular kick. If we examine the beam at some location along the accelerator, and if the beam motion from pulse to pulse is large compared to the beam size, then the beams will miss at the interaction point. In order to avoid this problem, these pulse-to-pulse offsets must be small compared to the local beam size. Equivalently, if a particular magnetic component has a varying amplitude, the variation of the angular kick must be small compared to the beam divergence at that point.

The emittance can also be destroyed by initial errors in beam size at the entrance to the linac. The beam size in an accelerator was discussed in Section 3.3. If there is bending or if the beam is offset in quadrupoles, the beam is dispersed with different momenta occupying different positions. In this more general case, the beam size is

$$\sigma^2 = \epsilon\beta + D^2\delta^2 , \tag{19}$$

where D is called the dispersion function and δ is the momentum variation in the beam prior to the bending field. For example, at the end of the compression section, $\delta \simeq 0.01$. At the entrance to the linac D should vanish. If not, this error in beam size results in emittance dilution in the acceleration process. For typical flat beam parameters the tolerance on dispersion D is given by

$$\begin{aligned} D_y &< 0.2 \text{ mm} \\ D_x &< 2 \text{ mm} . \end{aligned} \tag{20}$$

The dilution caused by residual dispersion is additive. There are also multiplicative effects due to the mismatch of the beta function of the magnetic focusing lattice. If the beam were mono-energetic, these mismatches would not filament and could be compensated at any point along the linac. Since there is a significant energy spread, this mismatch must be avoided. For a small error in β at the entrance to the linac, and provided the filamentation is complete, the emittance dilution is given by

$$\frac{\Delta\epsilon}{\epsilon_o} \simeq \frac{1}{2}\left(\frac{\Delta\beta}{\beta}\right)^2 . \tag{21}$$

For incomplete filamentation, the emittance dilution will be somewhat less.

3.6.2 *Wakefields and BNS Damping*

Wakefields are a key problem not only for linear colliders but for all accelerators and storage rings. The standard solution to this problem is to first reduce the wakefield forces until they are small compared to the applied external fields. Then compensation can be used, either feedback or modification of beam parameters,or we can simply live within the limits by keeping the number of particles in the bunch sufficiently small.

In Sections 2.2 and 2.3 we discussed the effects of the long-range wakefield. The multibunch beam breakup can be controlled by damping the undesirable modes in the RF structure; this reduces the long-range wake at the second bunch but has little effect within the first bunch. Now we examine the effect of the short-range wakefield on the stability of a single bunch.

The short-range wakefield can be expressed again as a sum of modes; however, in this case it is necessary to include modes at very high frequency. A typical short-range wake is shown in Fig. 12. It rises from zero, has a large peak and then oscillates with a frequency determined by the dominant mode. The bunches which will be in an NLC are so short that they fall on the initial rise of the wakefield. This is sometimes approximated as a linear rise (shown as the dotted line in Fig. 12).

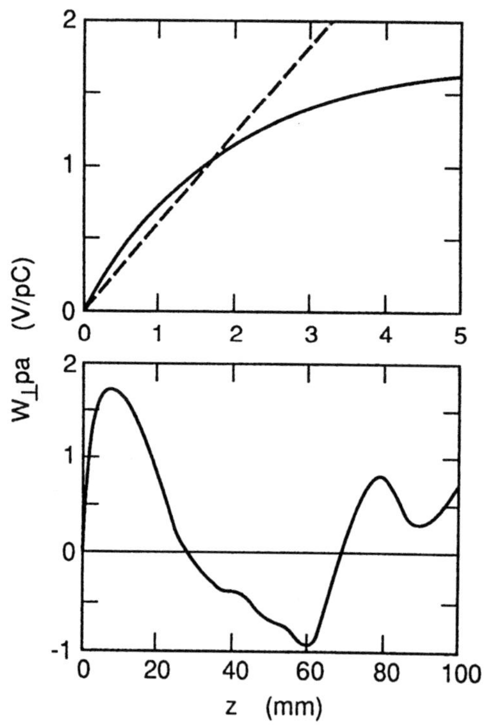

Fig. 12. The short-range transverse wakefield at the SLC. The upper graph shows a detail of the behavior 5 mm behind a point bunch.

The transverse wakefield increases rapidly with increasing frequency. If all dimensions are scaled, then

$$W_\perp(z) = \left(\frac{\lambda_o}{\lambda}\right)^3 W_\perp^o(z\lambda_o/\lambda) \, , \tag{22}$$

where λ is the scaled wavelength and λ_o is a reference wavelength. The initial slope varies inversely with the fourth power of the wavelength. Most of this variation comes only from the proximity of the iris hole to the beam. (By causality the short-range wakefield must be independent of the distance to the outer wall of a structure.) It is, therefore, possible to reduce the short-range wakefield by increasing the iris hole size relative to the wavelength. This reduces the short-range transverse wakes, but it also decreases the effectiveness of the accelerating structure. Therefore, one must balance the transverse benefit of increasing the iris size with the increased RF power necessary to achieve a given acceleration field with the larger iris size.

Even with the reduced wakefield within the bunch, there is still an instability induced within a bunch due to the coupling of the head and tail by the wakefield. The situation is completely analagous to that for multibunch instability discussed in Section 2.3; the same two-particle model suffices. In this case the head of the bunch, particle one, drives the tail of the bunch, particle two, on resonance. The growth is initially linear with distance but becomes exponential as the simple model breaks down.

Fortunately, there is a technique, called BNS damping, which can be used to compensate the instability.[25] The problem and solution are illustrated in Fig. 13 where a two-particle model is shown. If the two particles are offset to one side of the structure, the wakefield force deflects the tail particle away from the axis. We add to this the external fields due to the focusing magnets; on the average there is a focusing force in the opposite direction. If we reduce the energy of tail of the bunch by inducing an energy correlation along the bunch (this occurs naturally and is controlled by the phase offset discussed earlier), then the tail particle experiences a stronger force than the head particle. Finally, if the additional force can be adjusted to cancel the wakefield force, then the two particles, the head and tail, move coherently together, and the growth is completely eliminated. The BNS correlated energy spread is given by

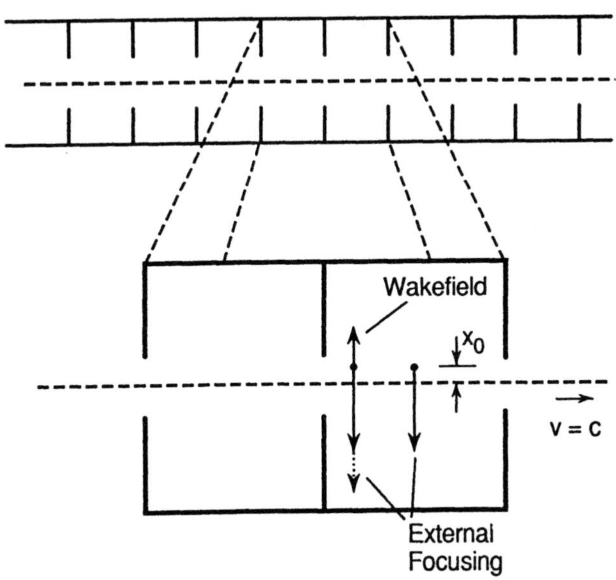

Fig. 13. Illustration of BNS damping. The additional focusing force on the lower energy trailing particle (dotted line) exactly cancels the wakefield force of opposite sign.

$$\left(\frac{\Delta E}{E}\right)_{\mathrm{BNS}} \equiv \delta_{\mathrm{BNS}} = \frac{e^2 N W_{\perp}(\sigma_z)\beta_o^2}{4E_o} \,, \tag{23}$$

where N is the number of particles in a bunch, $W_{\perp}(\sigma_z)$ is the transverse wakefield evaluated at σ_z, and β_o is the β-function at energy E_o.

If the wakefield is large, then one can still satisfy Eq. (23) by using RF focusing rather than energy variation to vary the focusing strength. In this case, however, trajectories can filament rapidly. To avoid emittance dilution with strong wakes, the alignment and trajectory tolerances are less than the beam size. This leads to 1 μm alignment tolerances.[26,27] As we shall see in the next sections, these tiny tolerances can be avoided by keeping the wakefields weak. With weak wakefields, tolerances are dominated by chromatic effects.

In the weak wakefield regime, BNS damping has been tested at the SLC linac.[28] In this case the tail growth due to a coherent oscillation was reduced by an order of magnitude. BNS damping has since been adopted as the normal running configuration for SLC.

3.6.3 *Chromatic Effects*

Upon injection into the linac, the compressed bunch has about a 1% uncorrelated energy spread. As the beam is accelerated, this relative spread decreases inversely with energy. At the same time a correlation between energy and bunch position is introduced due to the longitudinal wake and the curvature of the RF. Thus, the distribution in phase space becomes a wavy line which, when projected on the energy axis, yields an effective energy spread. At any location along the accelerator, the overall energy spread is a combination of the damping injected energy spread and the variation of energy along the bunch. After the bunch emittance is sufficiently damped, the relative energy spread remains constant unless deliberately increased by phase changes. For this reason it is useful to consider two models; one with constant energy spread and one with damping energy spread. In all cases considered below, we give not only the formula but also the value for the first design from Table 1 with energy 0.5 TeV in the center of mass.

The first chromatic effect to consider is that of a coherent betatron oscillation. If the variation of the phase advance with momentum (chromatic phase advance) is much greater than unity, the oscillation filaments. In this case the oscillation amplitude must be less than the beam size to avoid emittance dilution. If the chromatic phase advance is small ($\delta\psi_{\mathrm{tot}} < 1$), then the tolerance on a coherent oscillation of size $\hat{x}_o$ is

$$\begin{aligned} \hat{x}_o &< \frac{\sigma_\beta}{\delta_o \psi_{\mathrm{tot}}} = \frac{\sigma_\beta}{\delta_o \psi_{\mathrm{cell}}} \frac{2}{N_q} \,, \\ \hat{x}_o &< 2\sigma_\beta \,, \end{aligned} \tag{24}$$

where $\delta_o = 2 \times 10^{-3}$ is the constant relative momentum, ψ_{cell} and ψ_{tot} are the phase advance per cell and total phase advance respectively, and N_q is the number of quadrupoles. For the case of a damping energy spread with initial value $\delta_i = 0.01$, the tolerance is

$$\begin{aligned} \hat{x}_o &< \frac{\sigma_\beta}{\delta_i \psi_{\mathrm{cell}}} \frac{2}{N_q} \left(\frac{\gamma_f}{\gamma_i}\right) , \\ \hat{x}_o &< 5\sigma_\beta \,. \end{aligned} \tag{25}$$

For the case of a corrected trajectory let us consider the model of a sequence of random trajectory bumps. In this case the tolerance on the trajectory or alignment is

$$(\Delta x)_{\rm rms} < \frac{\sigma_\beta}{\delta_o \psi_{\rm cell}} \left(\frac{3}{N_q}\right)^{1/2} , \tag{26}$$
$$(\Delta x)_{\rm rms} < 30\mu{\rm m}$$

for a constant energy spread δ_o. For an initial damped energy spread δ_i, we have

$$(\Delta x)_{\rm rms} < \frac{\sigma_\beta}{\delta_i \psi_{\rm cell}} \left(\frac{1}{N_q}\right)^{1/2} \left(\frac{\gamma_f}{\gamma_i}\right)^{3/4} , \tag{27}$$
$$(\Delta x)_{\rm rms} < 30\mu{\rm m} \; .$$

3.6.4 *Misaligned Accelerator Sections*

BNS damping only cures the growth and filamentation of coherent oscillations in the linac; it is an average compensation rather than a local one. In an actual linac, the wakefield kicks are not cancelled locally by adjacent quadrupoles. This leads to an incoherent growth of wakefield tails due to the random sequence of misalignments between the trajectory and the accelerator structure. If we parameterize the strength of the wakefield kick by $\delta_{\rm BNS}$ as defined in Eq. (23), the tolerance on random accelerator misalignments is given by

$$(\Delta x_{\rm structure})_{\rm rms} < \frac{\sigma_\beta}{\delta_{\rm BNS} \psi_{\rm cell}} \left(\frac{3}{N_q}\right)^{1/2} , \tag{28}$$
$$(\Delta x_{\rm structure})_{\rm rms} < 25\mu{\rm m} \; ,$$

for $\delta_{\rm BNS} = 2.5 \times 10^{-3}$. From Eqs. (26) and (28) above, we see that the structure tolerances and quadrupole alignment tolerances are comparable provided that $\delta_{\rm BNS} \sim \delta_0$, that is, provided that the energy correlation needed for BNS damping is equal to the minimum energy spread in the linac.

3.6.5 *Compensation of Chromatic/Wakefield Effects*

The alignment tolerances shown above assume that the trajectory is a random sequence of bumps. There is no particular reason that it has to be random. Let us for the moment neglect wakefields. Then it is possible to measure the trajectories for particles of *different* energy and choose a trajectory which yields a small difference. Such a difference trajectory can be generated by scaling all the magnetic fields in the linac by a small amount so that the entire beam has an effective energy which is changed. By choosing the corrector sequence to minimize this difference trajectory (as well as the actual trajectory), the dispersion generated by misalignments is cancelled locally.

This technique is called dispersion-free correction. Provided that the beam position monitors have a *precision* the order of 1 μm, it is possible to essentially decouple the quadrupole misalignments from the dispersive effects.[29,30] This increases the tolerances given in Eqs. (26) and (27) by an order of magnitude.

When we include wakefields, the coherent motion is BNS damped and the incoherent motion gives rise to a random tail growth which can be controlled by tight tolerances. All that really matters for this effect is the value of the offset of the bunch within the structure. The offsets can be caused by two effects: misalignments of structures and trajectory offsets in structures. The trajectory is under our control; therefore, it is possible to use a trajectory which cancels the wakefield effects locally. Recently, it has been shown that by modifying the dispersion-free trajection technique, we can obtain a trajectory which cancels both the wakefield effects and the energy variation of the trajectory.[31]

Finally, we are left with the misalignments of accelerating structures. The most straightforward technique is to simply align the structure to the beam by using a BPM which is geometrically linked to the structure center. Such a BPM could consist of simply measuring the transverse wakefields induced by the beam.[32] One can use this information to either move the structure or move the trajectory to minimize the wakefield effects. Alternatively, for weak wakes, it is possible to deliberately move the beam or the structure to add a wakefield which cancels the effect of the rest of the accelerator.[33]

3.6.6 Beam Tilt

If there are RF kicks due to construction errors in the accelerator sections, the tail of the beam receives a different kick than the head. This can give a tilt to the beam. If we assume a random uncorrelated sequence of RF kicks, and compensate the center of the bunch with dipole correctors, the tilt tolerance is given by

$$(\Theta_{\rm rms})\left\{\beta_o < \sin\phi_o >_{\rm rms} \frac{2\pi}{\lambda_{\rm rf}} \sqrt{N_s} \left(\frac{\gamma_o}{\gamma_f}\right)^{1/2}\right\} < \frac{\sigma_y}{\sigma_z} , \tag{29}$$

where $\Theta_{\rm rms}$ is the rms RF kick angle for a beam with energy γ_o, N_s is the number of accelerator sections, ϕ_o is the phase angle of the transverse kick relative to the bunch, and σ_z is the bunch length. For the parameters for collider number one in Table 1, we have

$$\Theta_{\rm rms} < 2\mu{\rm rad} . \tag{30}$$

If such a kick is caused entirely by the systematic tilting of irises in an accelerating section (the bookshelf effect), then the tolerance on the systematic tilt angle of all the irises is given by

$$\Theta_{\rm iris} < 0.3 \text{ mrad} . \tag{31}$$

3.6.7 Jitter and Vibration: Motion Pulse to Pulse

Feedback is essential to handle the "slow" drift of x, x', y, y', E. In practical cases it is possible to feedback at $f \lesssim \frac{f_{rep}}{6}$. This sets the scale for what we consider slow. Time variation has many sources in linear colliders, for example: damping ring kicker jitter, power supply variations and ground motion. The jitter of the kicker in the damping ring must be kept small compared to the natural divergence of the beam at the kicker. Tolerances in power supply variations are also set in many cases by the beam divergence. The effects of ground motion depend upon the design and assumptions for the motion. If the wakes are weak and chromatic effects are kept small, there is no filamentation, and the beam moves coherently from pulse to pulse. If wakes are strong, and there is a large spread of betatron wave number, there is filamentation so that the beam size varies from pulse to pulse with a smaller centroid motion.

If we assume coherent motion, then for random magnet-to-magnet jitter the tolerance is

$$(\Delta x_{\text{jitter}})_{\text{rms}} < \frac{\sigma_\beta F}{\beta}\left(\frac{3}{N_q}\right)^{1/2} , \qquad (32)$$
$$(\Delta x_{\text{jitter}})_{\text{rms}} < (0.04)\sigma_\beta ,$$

where F is the focal length of a lens. If, on the other hand, there is magnet-to-magnet correlated motion of amplitude Δx_λ, then the dominant effect occurs when the wavelength λ is equal to the betatron wavelength. However, since in the planned designs the betatron wavelength changes $\propto \gamma^{1/2}$, the resonance is only temporary. If $2\pi\beta_i < \lambda < 2\pi\beta_f$, then the tolerance is given by

$$\Delta x_\lambda < \sigma_\beta \frac{2}{(\pi\psi_{\text{cell}})^{1/2}}\left(\frac{\gamma_f}{\gamma}\right)^{1/2}\left(\frac{2}{N_q}\right)^{1/2} , \qquad (33)$$
$$\Delta x_\lambda < (0.1 \text{ to } 0.4)\sigma_\beta ,$$

where γ is the energy at which $2\pi\beta = \lambda$.

3.7 THE FINAL FOCUS[34–37]

At the end of each linear accelerator is a final focus system whose purpose is to compress the tiny bunches to sub-micron dimensions. To obtain the luminosity desired, the cross-sectional area of each bunch must be only a few hundred square nanometers. In addition, we must focus it to the shape of a flat ribbon (rather than a string) in order to minimize the radiation emitted as the particles in the bunch encounter the intense electromagnetic field of the opposing bunch. These goals are accomplished by the use of a complex magnetic focusing system analogous (in reverse) to an optical telescope used to magnify distant objects. This system uses quadrupole magnets as focusing elements in a unique combination that provides a very large demagnification.

A major problem is the so-called "chromatic" effect of the final quadrupole magnets. Two parallel electron beams with different momenta entering a perfect quadrupole magnet are brought to a focus at slightly different longitudinal positions because the lower energy beam is bent slightly more than the higher energy beam by the magnetic field (see Fig. 14). For it not to affect the spot size, this shift of focal point must be smaller than β^*, the depth of focus of the beam. Due to the requirement of flat or ribbon-shaped beams, this depth of focus is about 100 microns in the vertical dimension.

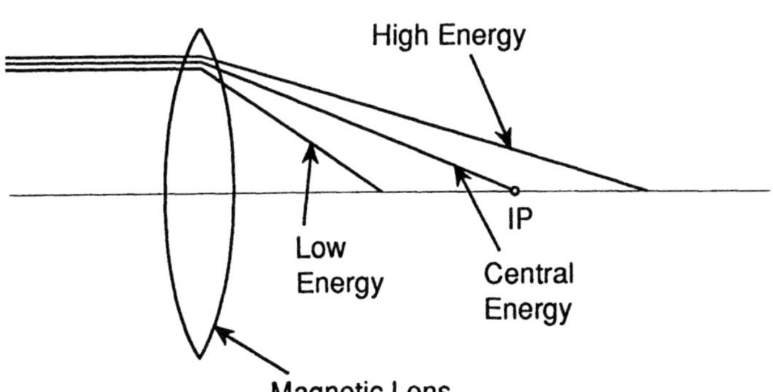

Fig. 14. The "chromatic" effect of the final quadrupole focusing magnet. Particles of differing energy are focused to different locations.

Such a small depth of focus makes the chromatic effects particularly serious. The chromatic correction of the final quadrupoles is in fact the key to the final focus. Upstream of these quadrupoles, a combination of bending magnets that disperse the beam combined with nonlinear sextupole magnets ensures that higher energy particles get a bit *more* focusing than lower energy particles. When a bunch arrives at the last quadrupole, the chromatic effect of the magnetic field upon it is exactly canceled.

The basic principles of the chromatic correction for particle beams have been known for about 30 years. Their first application in a linear collider was in the SLC, where the beams are demagnified by about a factor of 30, yielding spot sizes of about three microns. Because the demagnification necessary in the NLC is about a factor of 300, however, the design of its final focus system will be substantially different from that of the SLC.

In order to test such a next-generation final focus experimentally, an international collaboration including SLAC, INP, KEK, Orsay and DESY has been formed to design and construct a Final Focus Test Beam (FFTB) at SLAC.[38] This facility will use the SLC beam emerging straight ahead from the linac as its source of electron bunches.

Figure 15 shows a schematic of the location and layout of the FFTB. It is a scaled version of an NLC final focus, and as such, is qualitatively similar to NLC designs. A special feature of the design is that the chromaticity-correcting sextupoles are grouped in separate pairs, one for the horizontal dimension and one for the vertical. This pair of magnets is arranged so that the nonlinear aberrations introduced are cancelled, while the chromatic effects add. The bends shown in Fig. 15 horizontally disperse the different momenta in the beam so that the sextupoles give somewhat more focusing to the higher energy particles. This additional focusing is arranged so as to cancel the lack of focusing of the higher energy particles in the final quadrupoles.

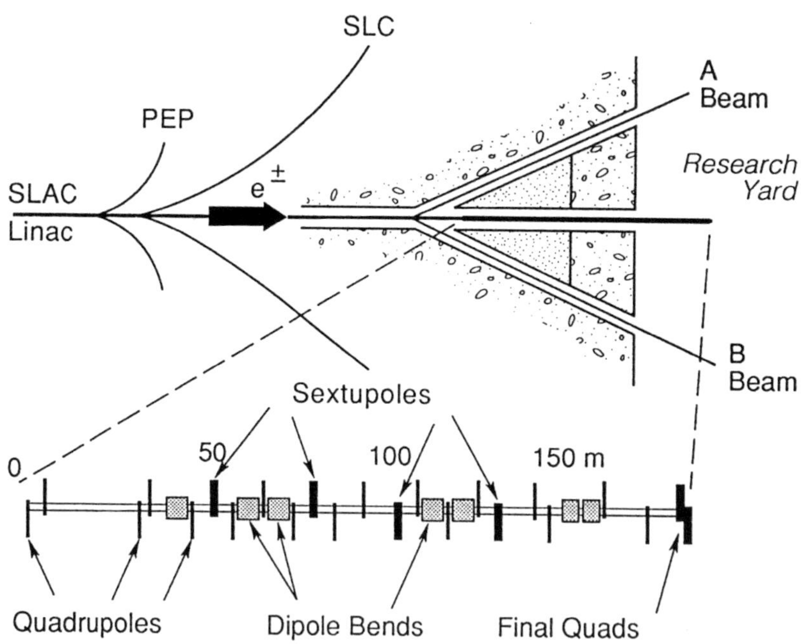

Fig. 15. The location and schematic layout of the Final Focus Test Beam.

The goal of the FFTB is to produce bunches with transverse dimensions of 60 nanometers high by 1 micron wide. Figure 16 shows the vertical beam size plotted versus the vertical β^* at the IP. In an ideal linear system, as discussed in Section 3.3, the beam size is just proportional to the square root of β^*. This is shown as the dotted line in Fig. 16. If the bunch has finite energy spread and with no correction, this linear decrease is modified by chromatic aberrations so that the beam size reaches a minimum of about 1 μm (the solid line in Fig. 16). Finally, if the chromatic-correction sextupoles are powered and if the system is properly tuned and adjusted, the vertical beam size follows the linear optics down to a size of about 60 nm before other high-order effects spoil the compensation (the dashed line in Fig. 16).

The FFTB will not achieve the beam size necessary for NLC due to the lack of a suitable low-emittance source. In fact, to achieve such low emittance, we need the NLC damping ring and linac. The FFTB *will*, however, test the optical demagnification necessary for an NLC. In fact, the design β^* for FFTB is identical to that for NLC. In addition to this primary goal, the collaboration will use this facility to test the alignment, stability and instrumentation requirements needed to achieve such small spots. The Final Focus Test Beam is a key component of the worldwide research effort towards the NLC.

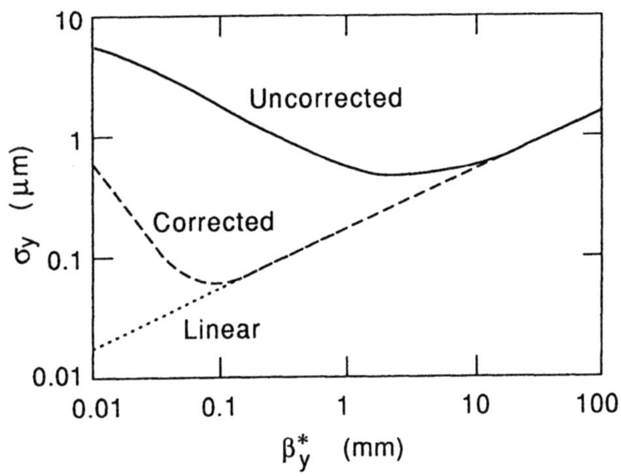

Fig. 16. Beam size versus optical tuning for uncorrected optics (solid), corrected optics (dashed) and linear optics (dotted).

3.7.1 *Beam-Beam Effects*[39–44]

When two oppositely-charged bunches collide at the IP, the intense electromagnetic field generated by the bunches tends to mutually focus them. This leads to disruption of the bunch and to a pinch enhancement of the luminosity. The enhancement factor H_D was given in Eq. (12) for the luminosity. For round bunches, this enhancement can be quite large ($\gtrsim 5$); for flat bunches, however, it is considerably reduced ($\lesssim 2$) because the pinch only occurs in one dimension. If, in addition, the bunches are misaligned relative to each other, the centroids are attracted during the bunch passage. This leads to a two-stream instability which for moderate disruption actually helps the collision process; if the bunches are misaligned, they bend toward each other and collide partially anyway.

The combination of very high electromagnetic fields and high particle energy yields substantial amounts of synchrotron radiation known as beamstrahlung. The average energy loss due to beamstrahlung ranges from 1 to 30 percent in various NLC designs. In extreme cases, many of these photons can subsequently generate electron-positron pairs in the intense electromagnetic fields present. The radiated photons or charged particles can strike detector components, causing undesirable backgrounds.

The train of bunches on each cycle also presents a problem at the final focus. In order to have a separate channel for the outgoing disrupted bunch, collisions take place at a small angle. As the bunches approach the collision point, they feel the field from those bunches which are exiting and have already collided. This sequence of bunches can induce a multibunch kink instability which can cause trailing bunches to miss the interaction point. This effect can be controlled by the charge per bunch or by the crossing angle.

In practice, these beam-beam effects are what impose the ultimate limits on the possible charge per bunch—and thus on the luminosity. In the design described above, the luminosity limit is bypassed by using a short train of bunches, each with moderate total charge. This approach allows us to maintain the desired luminosity while keeping beam-beam effects under control.

4 Summary and Concluding Remarks

In the previous sections we have outlined the basic issues important in the design of a Next Linear Collider. The energy can be obtained by essentially conventional means, with the use of RF accelerating structures combined with high peak power RF sources—klystrons—which are similar to those used presently in the SLC. The key difference is the change of frequency by a factor of 4. For the structures this change of frequency presents no problem. Structures at 11.4 GHz have been constructed; damped and detuned structures have been built or are being designed. The power source is very close to realization. The klystron discussed in Section 2.5.1 could easily provide enough power for the lower gradient option (option 2) in Table 1. The RF pulse compression necessary to achieve the proper pulse length has been tested and has behaved as theory would indicate.

The luminosity of the Next Linear Collider is perhaps the more difficult problem. To reach the desired levels of $10^{33} - 10^{34}$ cm^{-2}sec^{-1}, it is necessary to compress the beam spot to a few hundred square nanometers. This situation is forced upon us by conservation of energy; the wall-plug power must be kept within reasonable bounds. In spite of the small size required, many of the tolerances can be brought to conventional values when compensation techniques are applied. Many of the issues of producing small spots will be addressed by the Final Focus Test Beam.

The second major component to the luminosity increase is the acceleration of many bunches on each machine cycle. This increases the efficiency of the collider but also introduces many complications throughout all the subsystems. Experience has been gained at the SLC which accelerates three bunches on each cycle, and also at all long-pulse linacs which accelerate sometimes thousands of bunches on each cycle. Thus far, no fundamental problems have been discovered which would preclude the acceleration of trains of bunches.

To conclude, the outlook for obtaining both the energy and luminosity of a Next Generation Linear Collider is bright. Provided that the engineering effort on the power source is successful, an NLC design could become a reality by the mid 1990s.

Acknowledgements

I would like to thank Tanya Boysen for help in preparing the manuscript and Perry Wilson for his careful reading and good suggestions. Finally, I would like to thank the members of the Accelerator Theory and Special Projects Department and the Accelerator Department at SLAC for their continuing work on SLC and NLC; without their effort this paper could not have been written.

References

1. Proceedings of the Workshop on Physics of Linear Colliders, Eds. L. Palumbo, S. Tazzari and V.G. Vaccaro, Capri, Italy (1988), available from INFN Frascati, Italy.
2. Proceedings of the Summer Study on High Energy Physics in the 1990's, Snowmass, Colorado, July 1988, World Scientific, Singapore (1989).
3. Proceedings of the International Workshop on Next Generation Linear Colliders, SLAC, Stanford, CA, Dec. 1988, SLAC-Report-335.
4. Linear Collider Working Group Reports From Snowmass '88, Ed. R. D. Ruth, SLAC-Report-334.
5. Proceedings of the 2nd International Workshop on Next-Generation Linear Colliders, Eds. S. Kurokawa, H. Nakayama and M. Yoshioka, KEK, Tsukuba, Japan, March 1990, KEK Internal 90–22.
6. R.D. Ruth, "Multibunch Energy Compensation," SLAC–PUB–4541 (1989), and in Ref. 1.
7. K.A. Thompson and R.D. Ruth, "Controlling Transverse Multibunch Instabilities in Linacs of High Energy Linear Colliders," Phys. Rev. D 41, p. 964 (1990), and in SLAC–PUB–4801 (1989).
8. K.A. Thompson and R.D. Ruth, "Multibunch Instabilities in Subsystems of 0.5 and 1.0 TeV Linear Colliders," SLAC–PUB–4800 (1988), and in Refs. 2 and 4.
9. R.B. Palmer, "Damped Accelerator Cavities," SLAC–PUB–4542 (1988), and in Refs. 2 and 4.
10. H. Deruyter *et al.*, "Damped Accelerator Structures," Proceedings of the 2nd European Particle Accelerator Conference, Nice, France (1990), and in SLAC–PUB–5263.
11. N.M. Kroll and D.U.L. Yu, "Computer Determination of the External Q and Resonant Frequency of Waveguide Loaded Cavities," Particle Accel. 34, 231 (1990), and in SLAC–PUB–5171.
12. M.A. Allen *et al.*, "High Gradient Electron Accelerator Powered by a Relativistic Klystron," Phys. Rev. Lett. 63, p. 2472 (1989).
13. M.A. Allen *et al.*, "RF Power Sources for Linear Colliders," Proceedings of the 2nd European Particle Accelerator Conference, Nice, France (1990), and in SLAC–PUB–5274.
14. Z.D. Farkas, "Binary Peak Power Multiplier and its Application to Linear Accelerator Design," IEEE Transcripts on Microwave Theory and Techniques, MTT–34, No. 10, p. 1036 (1986), and SLAC–PUB–3694.
15. P.B. Wilson, "RF Pulse Compression and Alternative RF Sources," SLAC–PUB–4803 (1988), and in Refs. 2 and 4.

16. Z.D. Farkas, G. Spalek and P.B. Wilson, "RF Pulse Compression Experiment at SLAC," Proceedings of the 1989 IEEE Particle Accelerator Conference, Chicago, IL (1989), and in SLAC–PUB–4911.

17. T.L. Lavine *et al.*, "Binary RF Pulse Compression Experiment at SLAC," Proceedings of the 2nd European Particle Accelerator Conference, Nice, France (1990), and in SLAC–PUB–5277.

18. P.B. Wilson, Z.D. Farkas and R.D. Ruth, "SLED-II: A New Method of RF Pulse Compression," Proceedings of the Linear Accelerator Conference, Albuquerque, NM (1990), and in SLAC–PUB–5330.

19. E.D. Courant and H.S. Snyder, "Theory of the Alternating-Gradient Synchrotron," Ann. of Phys. 3, 1 (1958).

20. T.O. Raubenheimer, L.Z. Rivkin and R.D. Ruth, "Damping Ring Designs for a TeV Linear Collider" SLAC–PUB–4808 (1988) and in Refs. 2 and 4.

21. T.O. Raubenheimer *et al.*, "A Damping Ring Design for Future Linear Colliders," Proc. of 1989 IEEE Part. Acc. Conf., Chicago, IL, p. 1316, and in SLAC–PUB–4912.

22. S.A. Kheifets, R.D. Ruth, J.J. Murray and T.H. Fieguth, "Bunch Compression for the TLC, Preliminary Design," SLAC–PUB–4802 (1988), and in Refs. 2 and 4.

23. S.A. Kheifets, R.D. Ruth and T.H. Fieguth, "Bunch Compression for the TLC," Proc. of Int. Conf. on High Energy Acc., Tsukuba, Japan (1989) and in SLAC–PUB–5034.

24. R.D. Ruth, "Beam Dynamics in Linear Colliders," Proceedings of the Linear Accelerator Conference, Albuquerque, NM (1990), and in SLAC–PUB–5360.

25. V. Balakin, A. Novokhatsky and V. Smirnov, Proc. of the 12th Int. Conf. on High Energy Accelerators, Fermilab, p. 119 (1983).

26. J. Tuckmantel, "Beam Tracking With RF-Focussing in CLIC," CLIC-Note 87 (1989).

27. G. Guignard, "Status of CERN Linear Collider Studies," Proceedings of the 1990 Linear Accelerator Conference, Albuquerque, NM.

28. J. Seeman et al., SLAC-PUB-4968, to be published.

29. T.O. Raubenheimer and R.D. Ruth, "A New Trajectory Correction Technique for Linacs," Proceedings of the 2nd European Particle Accelerator Conference, Nice, France (1990), and in SLAC–PUB–5279.

30. T.O. Raubenheimer and R.D. Ruth, "A Dispersion-Free Trajectory Correction Technique for Linear Colliders," Nucl. Inst. & Meth. A302, 191–208, and in SLAC–PUB–5222 (1990).

31. T.O. Raubenheimer, "A New Technique for Correcting Emittance Dilutions in Linear Accelerators," SLAC–PUB–5355 (1990), to be published in Nucl. Inst. & Meth. A.

32. This technique was suggested by W. Schnell and V. Balakin.

33. J. Seeman, "New Control of Transverse Wakefield in a Linac by Displacing Accelerating Structures," SLAC-PUB-5337 (1990).

34. K. Oide, "Final Focus System for TLC," SLAC–PUB–4806 (1988), and in Refs. 2 and 4.

35. J. Irwin, "The Applications of Lie Algebra Techniques to Beam Transport Design," Nucl. Inst. & Meth. A298, 460–472, and in SLAC–PUB–5315 (1990).

36. J.J. Murray, K.L. Brown and T.H. Fieguth, "The Completed Design of the SLC Final Focus System," Proceedings of the 1989 Particle Accelerator Conference, p. 331, and SLAC–PUB–4219 (1987).

37. K. Oide, "Synchrotron Radiation Limit on the Focusing of Electron Beams," Phys. Rev. Lett. 61, 1713 (1988).

38. J. Buon, "Final Focus Test Beam for the Next Linear Collider," Proc. of the 2nd European Part. Acc. Conf., Nice, France, 1990.

39. P. Chen, "Disruption, Beamstrahlung, and Beamstrahlung Pair Creation," SLAC-PUB-4822, 1988, and Refs. 2 and 4.

40. R. Blankenbecler, S.D. Drell and N. Kroll, "Pair Production From Photon Pulse Collisions," Phys. Rev. D 40, p. 2462, 1989, and SLAC-PUB-4954.

41. P. Chen and V.I. Telnov, "Coherent Pair Creation in Linear Colliders," Phys. Rev. Lett. 63, p. 1796, 1989, and SLAC-PUB-4923.

42. M. Jacob and T.T. Wu, "Pair Production in Bunch Crossing," Phys. Lett. B 221, p. 203, 1989.

43. V.N. Baier, V.M. Katkov and V.M. Strakhovenko, Proc. 14th International Conf. on High Energy Particle Accelerators, Tsukuba, Japan, 1989.

44. M.S. Zolotarev, E.A. Kuraev and V.G. Serbo, "Estimates of Electromagnetic Background Processes for the VLEPP Project," Inst. Yadernoi Fiziki, Preprint 81–63, 1981; English Translation SLAC TRANS–0227, 1987.

HADRON COLLIDER LUMINOSITY LIMITATIONS

L. Evans

CERN, Geneva, Switzerland

1. INTRODUCTION

The era of high energy hadron colliders began with the construction of the Intersecting Storage Rings (ISR) at CERN in the years 1966-1970. The ISR was the first and only proton-proton colliding beam device built to date. Two separate rings stored counter-rotating coasting beams of protons intersecting at eight collision points at an angle of about 15 degrees. It operated mostly at the fixed energy of its injector (26 GeV) although in its latter years the beams could be taken to the maximum energy of 31 GeV using a technique called phase displacement acceleration. The ISR was a pioneering device on which accelerator physicists learned how to handle stored hadron beams and particle physicists learned how to build detectors and do experiments. It operated for physics between 1971 and 1983 during which time a great deal was learned about hadron storage rings and new techniques were developed, in particular that of stochastic cooling, which were to play a major role in the development of the only two other hadron colliders ever operated.

These two machines are quite different from the ISR in many respects. The SPS proton-antiproton collider at CERN and the Tevatron collider at FNAL were not built as dedicated colliding beam storage rings. Both machines are single-ring devices, the SPS capable of accelerating protons to 450 GeV/c and the Tevatron, about the same size but the first large machine using superconducting magnets, accelerating to 900 GeV/c. The development of stochastic cooling techniques at the ISR in the early 1970's and subsequent development at CERN made available for the first time substantial quantities of antiprotons and opened the door to the possibility of emulating the much lower energy electron-positron colliders, already in operation for a number of years. Counter-rotating beams of protons and antiprotons circulate in the common ring, colliding at the locations of the experimental detectors. Unlike the ISR, these beams must be kept bunched in order to make the highest possible luminosity with the available number of particles as well as to limit the strength of the beam-beam interaction, the perturbation of one beam by the electromagnetic field of the other, to the minimum possible value.

In the coming years the high energy frontier of particle physics will be addressed by the two hadron colliders now under active study. The Superconducting Super Collider, to be built near Dallas will be a proton-proton collider with a centre-of-mass energy of 40 TeV. The Large Hadron Collider (LHC) planned to be installed in the LEP tunnel at CERN will also be a two-ring proton-proton collider with a centre-of-mass energy around 15 TeV.

Both the SSC and the LHC will be superconducting machines with the two beams normally crossing at an angle, much smaller than in the ISR, of a few hundred microradians. Unlike the ISR, the beams must be kept bunched in order to minimise the total charge needed for a given luminosity, compensate for the non-negligible synchrotron radiation loss at these high energies, and to reduce the longitudinal size of the collision length to a minimum. A hole must also be kept in the azimuthal beam distribution in order for the beam abort kickers to rise without quenching the machine due to beam losses in the cryogenic part of the magnet system. Both these machines will operate at much higher luminosities than the other hadron colliders and will pose a real challenge for the design of the detectors. As far as the machine designs are concerned, much of the information about the luminosity limitations gleaned from the other machines is of relevance. Some of these limitations can be removed by proper choice of parameters and by careful engineering. Other limitations, in particular that of the beam-beam interaction are more fundamental in nature.

TABLE 1

EXISTING AND FUTURE HADRON COLLIDERS

Machine	Years of operation	Energy GeV/c	Luminosity $cm^{-2}\ s^{-1}$
SR	1971-1983	26-31	$1.3*10^{32}$
SPPS	1981-1990	315	$5.9*10^{30}$
TEV	1986-	900	$2.0*10^{30}$
LHC	?	7700	10^{33}-10^{34}
SSC	"	20000	"

One unique machine worth mentioning is the HERA lepton-proton collider under construction at DESY in Hamburg. This is a 2-ring device consisting of a superconducting machine which accelerates protons from 40 GeV to 820 GeV, colliding with electrons at 26 GeV in the other ring. Although it does not fall into the category of hadron colliders, many of the problems encountered in other proton storage rings will manifest themselves in the HERA proton ring.

2. THE INTERSECTING STORAGE RINGS

In the ISR[1], [2], the beams were stacked in longitudinal phase space. The maximum current was about 60 Amps, giving a stored energy of 5 MJ and a beam power of 1.5 TW. The intensity decay rate under good conditions corresponded to a beam lifetime of around 2 years. As well as protons, the ISR has stored alpha particles, deuterons and antiprotons. More than ten years of machine studies yielded an enormous amount of information on how both bunched and debunched beams behave in storage rings.

Three main effects limited the maximum achievable current.

1. Ions produced by beam-gas collisions were accelerated to the walls by the space-charge potential of the beam, where they produced desorption of gas molecules leading to pressure runaway. This is typically a problem associated with coasting beams and is unlikely to be a major limitation in future bunched-beam colliders since the ion cloud generally has time to disperse between bunches.

The quality of the machine vacuum was one of the most important parameters limiting the ISR performance throughout its lifetime. Although ion-induced desorption will not play a role in the next generation of hadron colliders, this does not mean to say that the quality of the vacuum will not be a problem. In fact, most of the pumping is obtained from the cold bore of the vacuum chamber, around 4.2 K for the SSC and 1.8 K for the LHC. Because of the high vapour pressure of hydrogen at 4.2 K, (of the order of 10^{-6} torr), after some time of operation a monolayer of hydrogen will build up on the surface of the chamber and the vacuum may degrade to the point where the machine would have to be warmed up to get rid of the hydrogen. This problem is under detailed investigation.

2. Above a certain current threshold, the longitudinal phase-space density was diluted by a fast blowup of the longitudinal emittance of the beam during stacking. This limited the amount of current which could be stored inside a given momentum acceptance. This instability (the microwave instability) is caused by the interaction of the bunched beam with the high-frequency part of the machine impedance causing self-bunching and a blowup of the momentum spread. It has been observed in several other machines and in fact limits the single-bunch intensity in the SPS collider.

3. Transverse instabilities in the debunched beam were controlled by introducing a betatron frequency spread through finite chromaticity (Landau damping). The acceptable frequency spread was limited by the space available between nonlinear resonances. These resonances can be driven either by imperfections in the guide fields or by the strong nonlinearity of the beam-beam force. In the ISR the working area straddled resonances of order 7 and above so the allowable tune spread was quite large.

Both the Tevatron and the SPS proton-antiproton collider suffer from much stronger beam-beam forces than the ISR. These induce both nonlinear resonance excitation and amplitude-dependent tune shifts causing a spread in betatron frequencies which must be accommodated between resonances of much higher order. The total tune spread is limited to about one fifth of that achieved in the ISR.

3. THE SPS PROTON-ANTIPROTON COLLIDER

The first proton-antiproton collisions were recorded in the SPS in the summer of 1981. The first physics run took place at the end of 1981 when 0.2 nb^{-1} of integrated luminosity was produced. In the second run, at the end of 1982, the peak luminosity was increased to $5.3*10^{28}$ cm^{-2} s^{-1} and a total integrated luminosity of 28 nb^{-1} was produced, enough to reveal the first W candidates. In the 1990 collider run, the peak luminosity was $6.0*10^{30}$ cm^{-2} s^{-1} and the highest daily integrated luminosity was 260 nb^{-1}. This shows the steady progress in mastering a highly complex operation together with an improvement in the understanding of the basic limitations of the chain of accelerators from the antiproton accumulator to the SPS[3].

In the collider itself the main parameters of importance are the peak luminosity and the luminosity lifetime. The peak luminosity is governed by the current per bunch and the number of bunches per beam as well as the transverse emittance and the values of the horizontal and vertical beta functions at the interaction points. The bunch current is limited by incoherent space charge at the 26 GeV/c injection energy and by the microwave instability. The luminosity lifetime is mainly limited by intrabeam scattering, multiple Coulomb scattering between particles in the same bunch causing a slow blowup of both radial and longitudinal emittance.

The microwave instability in the SPS manifests itself as a variation of the bunch length with current in a similar way that it did in the ISR. At low intensity the bunch lengthening is simply a result of the loss of focussing voltage due to the interaction of the beam with the low-frequency inductive part of the machine impedance, the "inductive wall effect". Above a threshold intensity of around $9*10^{10}$ protons per bunch the microwave instability sets in, causing a longitudinal emittance blowup until the bunch finally stabilises due to Landau damping. Above about $1.2*10^{11}$ protons per bunch the emittance becomes too large to be contained in the available bucket area during acceleration so part of the beam is lost.

Both the ISR and SPS were constructed in the days when much less was known about beam stability than it is now. In particular, although it was fully realised that high-Q parasitic resonators should be avoided, the influence of the broad-band impedance was less well appreciated. Modern machines are constructed with much stricter control of the impedance budget. For example, the longitudinal impedance of LEP is only about a quarter of an ohm instead of 20 ohms or so in the SPS. For the future machines it is expected to be even smaller so the microwave instability should no longer be important.

Low-beta insertions are vital for the attainment of useful luminosity in both the SPS and Tevatron. Separately powered quadrupoles strongly perturb the lattice functions locally in order to produce very small values of the horizontal and vertical beta functions inside the experimental detectors. The SPS uses doublets to reduce the beta functions from their normal values of about 50 m in each plane to 0.6 m horizontal and 0.15 m vertical. In the Tevatron, triplets are used to produce a beta value of about 0.5 m in each plane. The low-beta insertions produce strong chromatic aberrations which need sophisticated sextupole schemes for their correction.

In machines operating above transition energy, intrabeam scattering causes a continuous blowup of one or more of the phase-space areas depending on the relative emittance distribution between the three phase planes. In the SPS both radial and longitudinal emittances increase as a function of time resulting in a reduction of the luminosity lifetime. The measured growth rates agree quite well with the theory of intrabeam scattering. The blowup rates are proportional to the density in 6-dimensional phase space. Since only the transverse blowup contributes to the luminosity lifetime to first order, intrabeam scattering can be controlled in future machines by having a sufficiently large longitudinal emittance and designing the radiofrequency system appropriately.

4. THE TEVATRON

The Tevatron at Fermilab started operation as a proton-antiproton collider in 1986[4]. Bunches are injected at 150 GeV/c after a sophisticated coalescing procedure in the Fermilab main ring and slowly ramped to the storage energy of 900 GeV/c. In spite of its rather short life, the peak luminosity in the Tevatron has already exceeded $2*10^{30}$ cm^{-2} s^{-1}.

Due to its high injection energy and low coupling impedance the Tevatron does not suffer from either space-charge or instability problems. In addition the longitudinal emittance is naturally blown up by the bunch coalescing gymnastics needed to form dense single bunches, so the emittance growth rate due to intrabeam scattering is low. Consequently the luminosity lifetime is somewhat longer than in the SPS. Of course, being the first machine of its type, the Tevatron has reserved one or two surprises of its own. In particular, the discovery of time-dependent persistent currents is of great consequence for future superconducting machines.

Persistent current effects in superconducting magnets produce strong multipoles in the magnetic field. The phenomenon is strongest at low fields and is highly nonlinear in field and temperature. The multipole fields also exhibit a slow ln(t) time dependance.

The first indication of this phenomenon in the Tevatron was the observation that the machine chromaticity slowly drifted with time on the 150 GeV/c injection platform[5]. This was traced to the slow drift of the sextupole harmonic in the Tevatron dipoles . The magnitude of this variation corresponds to a change in machine chromaticity of 60 units and must be dynamically compensated.

The other field harmonics demonstrate a similar behaviour. The slope and intercept of this curve depend on the history of the magnetic cycle. As well as this slow drift there is a rapid change at the beginning of acceleration where the multipoles effectively recover from the prior decay.

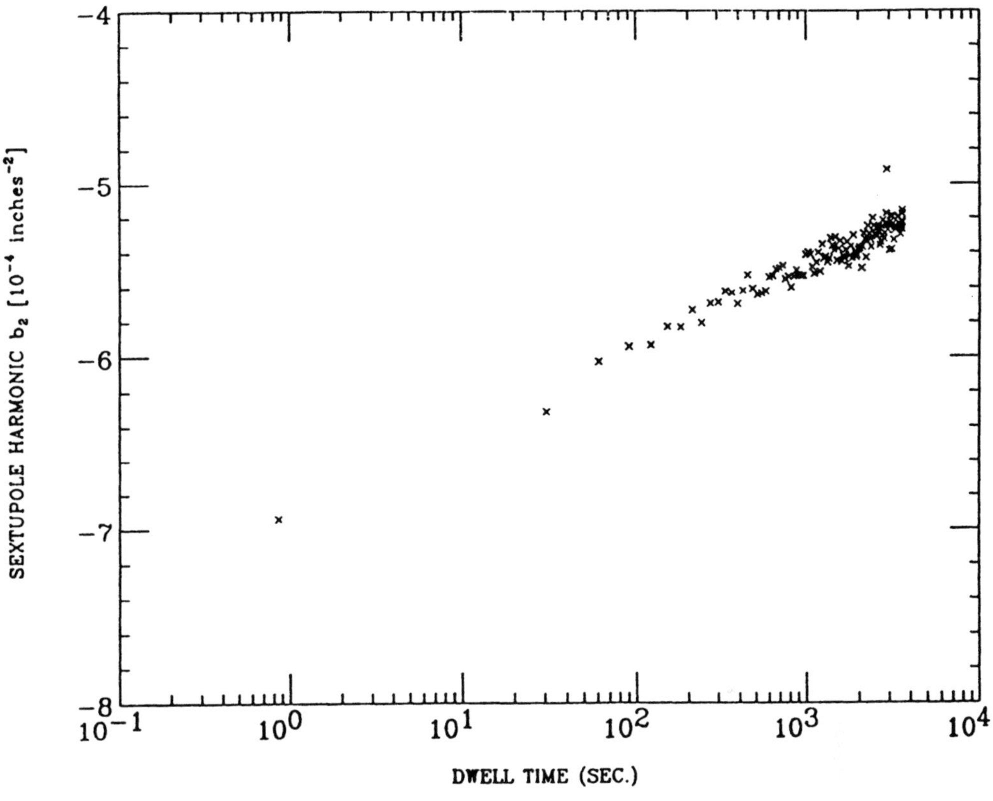

Fig. 1 Slow drift of sextupole harmonic with time in the Tevatron dipole magnets

The injection energy of the Tevatron is rather high compared with the next generation of colliders where this effect would prove to be catastrophic if it could not be attenuated substantially. In particular the low injection energy into the HERA proton ring (40 GeV) will make the phenomenon very important. Studies are underway of methods of dynamic correction of the chromaticity drift and the sharp change normally observed at the start of acceleration. In addition, significant progress in the understanding of its dependance on various parameters has been made in the past few years. In particular the size of the superconducting filaments plays a very important role. A major effort in technological development is underway to produce much finer filaments, and initial results are encouraging.

5. THE EFFECT OF NONLINEAR FIELDS

Most of the effects mentioned above can be eliminated or attenuated to a tolerable level by careful design. Coherent instabilities can be brought under control by carefully controlling the impedance of equipment surrounding the beam. Intrabeam scattering can be greatly reduced by correctly choosing beam and radiofrequency parameters. Persistent current effects can be reduced by improved technology of superconducting filament production and by sophisticated correction techniques.

However, the very purpose of colliders is to make collisions between the constituent particles of the two beams inside the experimental detectors. Together with these desired interactions a very undesirable phenomenon occurs which can strongly affect the stability of the circulating beams themselves. This is caused by the interaction of the particles in one beam with the strongly nonlinear global electromagnetic field of the other beam. The first effect is the excitation of nonlinear resonances; for certain rational combinations of horizontal and vertical tune values the particle orbits become unstable. The remedy is to stay clear of these forbidden tune values. However a second effect of the beam-beam interaction is the creation of a tune spread, a variation of tune depending on particle amplitude. This makes it more and more difficult to find resonance-free regions as the strength of the interaction increases.

The observations from both the SPS and the Tevatron[6] indicate that the tunes must be kept away from nonlinear resonances of rather high order. In the SPS the working point sits between 3rd and 10th order resonances, clear of resonances up to 16th order. For the Tevatron, the working point sits between 5th and 7th order resonances, clear up to 12th order. In the SPS, resonances of order above 4 are mainly excited by the beam-beam interaction whereas in the Tevatron the influence of the multipoles of the superconducting magnets is felt up to order 7, i.e. even with only one beam in the machine the effect of these resonances can be observed. The resonances limit the total tune shift, i.e. the sum of the linear tune shifts over all of the interaction points, to a value of between 0.02 and 0.03 for both the SPS and Tevatron. In the Tevatron a maximum tune shift of 0.0025 per interaction point has been reached for 12 interaction points; in the SPS the number of interaction points is limited to 3 by separating the beams outside the experiments using electrostatic separators. The maximum tune shift achieved per interaction point is about 0.007.

For many years the beam-beam interaction has been the subject of intense theoretical study. As a result some success in at least a qualitative understanding of the phenomenon has been achieved. The loss of stability of particle orbits is due to stochastic motion of particles due to the strongly nonlinear fields. The threshold for the onset of stochasticity is strongly reduced by the fact that the beam is bunched. Individual resonances are split into a set of closely spaced synchrotron sidebands which can overlap and cause stochastic behavior. The problem of the beam-beam interaction forms part of a much more general class of nonlinear problems involving chaotic systems.

In the absence of any sound quantitative theory, criteria based on the experimental data from the SPS and Tevatron are being used for the design of LHC and SSC. For the SSC the total allowable tune spread is assumed to be 0.024. For the LHC, a slightly more conservative spread of 0.02 is assumed. This includes the tune spread due to the head-on beam-beam interactions together with that coming from long-range effects due to the small crossing angle, and the tune spread due to magnetic field imperfections

Magnetic field imperfections can produce single-beam effects similar to those produced by the beam-beam interaction. Whereas the magnetic field uniformity in the older machines using classical magnets can be made very good, superconducting magnets pose quite another problem. The field quality is determined by the size of the bore and by the precision in the location of the superconductor. Both the Tevatron and the HERA proton rings have dipole magnet bores (3 inch and 75 mm diameter respectively) that are quite a bit larger than those foreseen for SSC and LHC (50 mm). Even then, in HERA high-order correctors are foreseen to correct systematic multipoles at the rather low injection field. For the LHC and SSC considerable effort is being put into a thorough investigation, mainly by long-time tracking, of the influence of both systematic and random multipole errors on the dynamic aperture of the machines.

6. CONCLUSIONS

The three hadron colliders operated to date have taught us a great deal about the behaviour of both bunched and debunched beams in storage rings. The main luminosity limitations are now well enough understood that most of them can be strongly attenuated or eliminated by appropriate design precautions. Experience with the beam-beam interaction in both the SPS and the Tevatron allow us to predict the performance of the new generation of colliders with some degree of confidence. One of the main challenges that the accelerator physicist faces is the problem of the dynamic aperture limitations due to the lower field quality expected, imposed by economic and other constraints.

References

1] E. Keil, CERN 72-14 (1972).

2] K. Hübner, CERN 77-15 (1977).

3] L. Evans, Proc. 1987 IEEE Particle Accelerator Conference, Washington D C, p. 12 (1987).

4] R. Johnson, Proc. 1987 IEEE Particle Accelerator Conference, Washington D C, p. 8 (1987).

5] D. Herrup et al., Proc. 1989 IEEE Particle Accelerator Conference, Chicago, p. 518 (1989).

6] M. Harrison, R. Schmidt, Proc. 2nd European Particle Accelerator Conference, Nice, p. 55 (1990).

PHYSICS AND TECHNOLOGY CHALLENGES OF $B\overline{B}$ FACTORIES*†

Michael S. Zisman
Exploratory Studies Group
Accelerator & Fusion Research Division
Lawrence Berkeley Laboratory
Berkeley, CA 94720

Abstract

An e^+e^- collider designed to serve as a B factory requires a luminosity of 3×10^{33} cm^{-2} s^{-1}—a factor of 15 beyond that of the best present collider (the CESR ring)—and thus presents a considerable challenge to the accelerator builder. To optimize the experiment, it is necessary that the $B\overline{B}$ system have a moving center-of-mass, which implies different energies for the two beams (hence an "asymmetric" collider). This feature dictates that a two-ring configuration be used. Accelerator physics issues that arise in such a design are related to the need to tightly focus the beams to a vertical beta function on the order of 1 cm, to bring the beams from two different rings into collision and then cleanly separate them again, and to mask the detector region sufficiently to permit measurements with very large beam currents passing through the interaction region. In addition, the process of optimizing the luminosity for asymmetric collisions breaks new ground.

Because the luminosity is limited by the beam-beam interaction, any large improvement must come from considerably increasing both the beam current and the number of bunches in the ring. These choices place many demands on accelerator technology as well as accelerator physics. Vacuum systems must be designed to handle the thermal load from a multi-ampere beam of 8–9 GeV and to maintain an adequate running pressure (below 10 nTorr) in the face of a large gas load from synchrotron radiation induced photodesorption. An RF system capable of supporting the high beam currents must be developed. To reduce the growth of potentially strong multibunch instabilities, the cavity higher-order modes (HOMs) must be highly damped to $Q \leq 70$. Even with a well-optimized RF system, the high beam currents typically mean that wideband multibunch feedback systems (both longitudinal and transverse) are needed to maintain beam stability. Effective approaches to deal with these issues have been identified by the various B factory design groups, and representative examples will be mentioned.

I. INTRODUCTION

There has been growing interest in the past several years in the design of a high-luminosity e^+e^- collider, operating at the Υ(4S) resonance, to serve as a "B factory." The primary physics motivation for

* This work was supported by the Director, Office of Energy Research, Office of High Energy and Nuclear Physics, High Energy Physics Division, U.S. Department of Energy, under Contract No. DE-AC03-76SF00098.
† This paper is based on an invited talk presented at the 1991 Particle Accelerator Conference, San Francisco, May 6–9, 1991 and will appear in the IEEE Conference Proceedings.

such a facility is to determine the origins of *CP* violation. This phenomenon is expected to be easily observable in the B system, and determining its origins will provide a stringent test of the Standard Model. *CP*-violation studies benefit considerably from having a moving center of mass for the $B\overline{B}$ system, so an asymmetric collider is preferred. The physics capability of such a facility is not restricted solely to *CP*-violation studies; rich programs in rare B decays, Υ spectroscopy, charm and tau physics, and two-photon physics will also be available.

Although both dual-storage-ring [1–6] and linac-plus-storage-ring [7] designs have been studied, the focus here will be on the former configuration, storage-ring-based systems. All presently active proposals have chosen this design approach.

II. REQUIREMENTS

To study *CP* violation at the $\Upsilon(4S)$ resonance with an asymmetric collider, a peak luminosity of 3×10^{33} cm^{-2} s^{-1} is needed [8]. The actual figure-of-merit for the collider, however, is not the peak but the *integrated* luminosity. This is because the physics measurements require the study of an abundant sample of B decays to obtain statistically significant results. It is in this sense that we refer to the collider as a "factory."

The luminosity can be expressed in terms of the appropriate collider parameters as [5]

$$\mathcal{L}\left[\text{cm}^{-2}\text{s}^{-1}\right] = 2.17 \times 10^{34} \xi (1+r) \left(\frac{I \cdot E}{\beta_y^*}\right)_{+,-} \tag{1}$$

where I is the total beam current (A), β_y^* is the vertical beta function at the interaction point (cm), r is the beam aspect ratio (σ_y^*/σ_x^*, i.e., 0 for flat, 1 for round beams), E is the beam energy (GeV), and ξ is the beam-beam tune shift parameter. The subscript on the rightmost factor in Eq. (1) signifies that it can be evaluated using the parameters from either the electron (–) or positron (+) ring. The beam-beam tune shift parameter is not really under our control, and the beam energy is constrained by the need to run at the $\Upsilon(4S)$ resonance, requiring that $E_+ \cdot E_- = 28$ GeV^2.

It is clear from inspection of Eq. (1) that a fifteen-fold increase in luminosity compared with existing colliders requires high beam currents and small beta functions at the interaction point (IP). The requirement for low beta functions leads to some practical difficulties. For example, low beta functions are produced by strong quadrupoles, and these make the chromaticity correction difficult. Moreover, to take advantage of the low beta functions, there is a concomitant need for short bunches, such that $\sigma_\ell \leq \beta^*$. To produce the short bunches takes a high RF voltage, and thus considerable RF hardware. Taken together, these considerations imply a practical limit corresponding to $\beta_y^* = 1–2$ cm.

Because of the limitation from the beam-beam interaction, that is, the limit on the maximum value of ξ, a large increase in beam current implies the use of many more bunches than is typical of today's colliders. (Clearly it is possible to put high current in fewer bunches, but the single-bunch intensity is limited by the transverse mode-coupling instability, and the beam-beam limit pushes the design towards an unreasonably large emittance.) Given little maneuvering room, it is reasonable for the designer to choose the number of bunches to be sufficiently large that *the parameters of a single bunch remain relatively standard*. This is the approach followed by essentially all B factory design groups. Typical parameter ranges for the designs considered here appear in Table 1.

Table 1
Typical B Factory Parameter Ranges

Parameter	Value
Total current, I (A)	1–3
Single-bunch current, I_b (mA)	1–5
No. of bunches, k_B	100–2000
Horizontal emittance, ε_x (nm·rad)	100
Bunch length, σ_ℓ (cm)	1
Energy, E_-/E_+ (GeV)	8/3.5 or 9/3.1
Luminosity, $\mathcal{L}$ (cm^{-2} s^{-1})	$1\text{–}3 \times 10^{33}$

III. Physics Challenges

The design of a high-luminosity asymmetric B factory leads to physics challenges primarily in the areas of lattice design and the beam-beam interaction. In the first area, the issues are related to the production of low β_y^* values, the separation of the two beams, and the design of the masking system. In the second area, the physics issues are centered around the techniques for optimizing the luminosity for the new parameter regime of asymmetric collisions.

Lattice Design

Low beta function. To provide the required luminosity, it is necessary to produce low β_y^* values, on the order of 1 cm, without introducing excessive chromaticity into the lattice. To accomplish this, the low-beta quadrupoles must be located as close as possible to the IP, as shown in Fig. 1 for the SLAC/LBL/LLNL (PEP-II) design [5]. Although the permissible chromaticity can only be determined by actual particle tracking simulations, a good rule to apply is that β/s_Q should be less than 100, where s_Q is the distance of the quadrupole from the IP and β is the beta function at the quadrupole location. As can be seen in Fig. 1, the low-energy beam (LEB) focusing does not present a problem, but the high-energy beam (HEB) is more difficult. To locate the HEB quadrupole closer to the IP, it is designed as a superconducting Panofsky-style septum quadrupole. Equivalent design approaches with conventional magnets have been followed by other groups [2, 3, 4].

Because the LEB focusing quadrupoles are close to the IP, they lie within the solenoidal field of the detector. This restricts the choice of technology to either permanent magnets or superconducting magnets. Solutions using one or both of these technologies have been adopted by various designers. An example of an interaction region layout based on superconducting magnets is shown in Fig. 2, taken from Ref. [4]. In this case, the solenoid field is compensated by means of "anti-solenoid" windings to avoid coupling the horizontal and vertical beam motions. When using permanent magnets, as in Fig. 1, coupling is compensated with skew quadrupoles located outside the detector region. The placement and dimensions of the low-beta quadrupoles are restricted by the "detector stay-clear" area, usually defined as a 300 mrad cone.

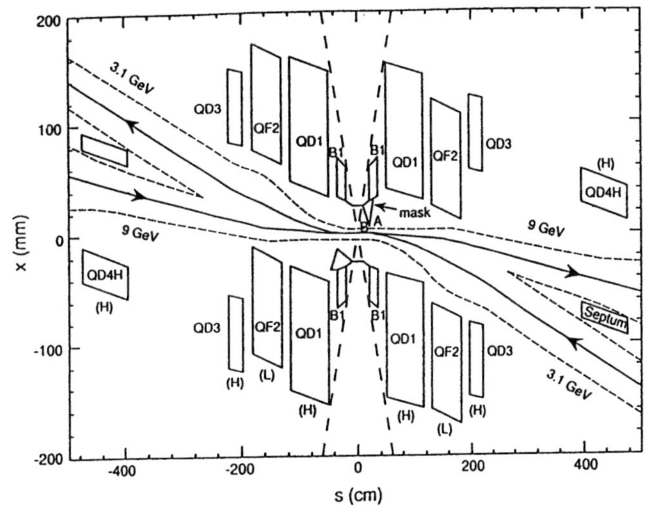

Figure 1. Anamorphic plan view of a B Factory interaction region for head-on collisions [5].

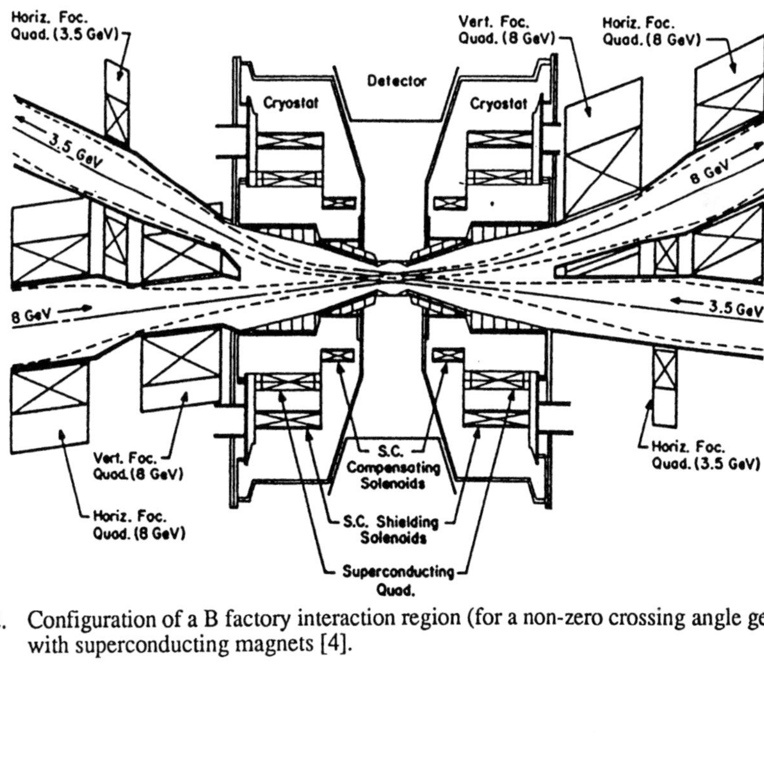

Figure 2. Configuration of a B factory interaction region (for a non-zero crossing angle geometry) with superconducting magnets [4].

Beam separation. The technique used for beam separation in an asymmetric B factory depends in large measure on the design approach. For the commonly adopted head-on collision case, the separation is accomplished by means of dipoles located close to the IP followed by offset quadrupoles. The separation dipoles could either be run in a symmetric or an asymmetric configuration; the latter case, illustrated in Fig. 1, is referred to as an "S-bend" geometry. The advantages of the S-bend geometry are that it decouples the masking solutions for the two rings, and it permits the synchrotron radiation fans generated by the separation magnets to exit the interaction region without creating severe background problems. It is worth noting here that an S-bend layout of the type shown in Fig. 1 lends itself well to being converted into a non-zero crossing angle scheme (cf. Fig. 2) without major hardware rearrangements.

Masking. A successful masking scheme must take into account all sources of backgrounds, including synchrotron radiation from the separation magnets and offset quadrupoles, lost particles from beam-gas and beam-beam interactions, and lost particles during injection [9]. It is also important that the solution adopted be insensitive to the details of the beam tail distribution and to small displacements of masks, magnets, and beam orbits. In general, backgrounds are never completely understood, so it is desirable to aim for safety margins of more like a factor of ten than a factor of two. It is good practice for designers of high luminosity accelerators to have close involvement with the detector users. The machine-detector interface is one of the most crucial aspects of the machine design, and the effort and care spent on it are evident in the various design reports that are now available [1–6].

Beam-Beam Interaction

Choice of Tune Shift. The beam-beam tune shift in the case of an asymmetric collider has not been studied experimentally. In the absence of such data, most design groups have taken guidance from the existing body of data on symmetric collisions [10]. It can be seen from such data that the beam-beam tune shift parameter ξ lies in the range from 0.02 to 0.06 for present colliders. Because most machines have reached $\xi = 0.03$, this value has generally been adopted by B factory design groups as a prudent target figure. (The KEK group [2] has adopted a larger tune shift value of 0.05, based on their choice to use very short bunches, 0.5 cm.) Note that this value is *not* intended to represent a beam-beam limit, it is merely a design parameter. To stay closer to the existing body of knowledge, head-on collisions are the initial design choice of all but one group [4]. For both the head-on and non-zero crossing angle cases, beam-beam simulations are being carried out to demonstrate that the design choice is a realistic one. Thus far, it is fair to say that no new physics issues have arisen that are related to the asymmetry itself.

Energy Transparency. At present, most designers have adopted some set of conditions intended to make the asymmetric beam-beam collisions behave similarly to the well-studied symmetric case. The so-called "energy transparency" conditions postulated by Chin [11] require equality of beam-beam parameters, beam sizes, tune modulation from synchrotron oscillations at the IP, and damping decrements, $\lambda = T_0/\tau_{SR}$. Thus, we choose parameters such that

$$\xi_{x,+} = \xi_{x,-} \text{ and } \xi_{y,+} = \xi_{y,-}$$

$$\sigma_{x,+} = \sigma_{x,-} \text{ and } \sigma_{y,+} = \sigma_{y,-}$$

$$\left(\frac{\sigma_\ell \, \nu_s}{\beta^*_{x,y}}\right)_+ = \left(\frac{\sigma_\ell \, \nu_s}{\beta^*_{x,y}}\right)_-$$

$$\lambda_+ = \lambda_-$$

Further constraints have been put forth by Krishnagopal and Siemann [12] and these "equal tuneprint" conditions have been adopted in some designs [2,4]. The present view is that such symmetrization attempts are convenient (in the sense of restricting the parameter space available), but may not be entirely necessary. It is also unclear whether the restricted parameters corresponding to the symmetry conditions guarantee the optimum luminosity. It has been shown in one case [5] that the effects of parasitic collisions intrinsically tend to break the symmetry between the two beams anyway. This aspect of the parameter optimization needs further work.

Crab Crossing. To permit a non-zero crossing angle while avoiding the excitation of synchrobetatron resonances, it is attractive to consider the possibility of crab crossing. This scheme [13] involves the use of a transverse deflecting mode of crab RF cavities, located at a phase difference of

$$\Delta\phi = (n \pm 1/4)\, 2\pi$$

from the IP, to rotate the head and tail of the bunches such that they collide head-on at the IP, but in a transversely moving reference frame.

The voltage required to perform the rotation is given by

$$V_c = \frac{(E/e)\phi\lambda_c}{2\pi\sqrt{\beta_{xc}\beta^*_x}} \quad . \tag{2}$$

For typical parameters, V_c is about 2 MV. Simulations done to date [4,5] suggest that voltage and phase tolerances are reasonable, so the technique should be viable. Nonetheless, prudence dictates that a small crab angle, on the order of 10 mrad, is the best choice. Such an angle is sufficiently small that it does not obviate the need for common quadrupoles for the two beams, as shown in Fig. 2. It is clear that crab crossing is a promising technique, though it has not yet been tested. Because of the absence of separation dipoles, the synchrotron radiation liberated near the IP is reduced with the crab crossing scheme compared with the head-on case; this should be of benefit in terms of detector backgrounds.

IV. TECHNOLOGY CHALLENGES

The physics issues discussed in Section III make certain implicit assumptions about the hardware capabilities in a B factory. For example, beam lifetime estimates assume that the average pressure in the storage rings will remain below about 10 nTorr (N_2 equivalent) despite the high gas loads associated with possibly several amperes of circulating beam. Similarly, luminosity estimates assume that these high beam currents can be supported without melting anything. The assessment of growth times for coupled-bunch instabilities is based on the ability to damp the dangerous HOMs of the RF cavities to $Q \leq 70$. Perhaps most importantly, we assume that the integrated luminosity can be maintained, that is, that the reliability of the components is such that the collider does not "spend all of its time in the shop."

In this section we discuss the technology areas where the main challenges arise. These include the vacuum system, the RF system, and the feedback system. It is worth commenting here that some other items, such as the separation magnets indicated in Figs. 1 and 2 (and the equivalent components in each of the other interaction region designs), are nontrivial design tasks as well.

Vacuum System

There are two main challenges for a B factory vacuum system:

- withstanding the high thermal flux from the synchrotron radiation power
- maintaining a low pressure in the face of considerable synchrotron radiation induced gas desorption

The average linear power density for the chamber is given by

$$P_L = \frac{P_{SR}}{2\pi\rho} \propto \frac{E^4 I}{\rho^2} . \tag{3}$$

This quantity varies widely among the various designs, as it depends on both the beam current requirement and the bend radius of the ring magnets. The lowest power density is that of the KEK design [3], 1.5 kW/m; the highest value, 25 kW/m comes from the "hard-bend" region of the Cornell design [4]. In terms of thermal management, an equally important quantity is the areal density. The height of the synchrotron radiation fan at the chamber wall is typically about 0.4 mm, in which case the areal power densities range from 0.4 to 5.6 kW/cm^2.

The photodesorption gas load in the B factory rings can be written as

$$Q_{gas} = 2.42 \times 10^{-2}\, E_{[\mathrm{GeV}]}\, I_{[\mathrm{mA}]}\, \eta_F \quad [\mathrm{Torr{\cdot}L/s}] \tag{4}$$

where the desorption coefficient, η_F, corresponds to the number of molecules produced per incident photon. The desorption coefficient depends on the chamber material, its history, and the photon dose to which the material has been exposed. After exposure to about one hundred ampere-hours of beam, values of low-to-mid 10^{-6} are expected for a copper chamber.

The two approaches that can be adopted for a B factory are a standard chamber shape, with a pumping channel on the inner radius, or an antechamber design in which the synchrotron radiation photons exit through a slot in the wall into an external pumping chamber. For cases where the design pressure can be achieved with a pumping speed of $S \approx 100$ L/s/m, no antechamber is needed. For cases where $S \geq 500$ L/s/m is required, standard distributed ion pumps will not suffice. Then the system of choice is to use non-evaporable getter (NEG) or titanium sublimation pumps (TSPs). In a difficult case, such as the hard-bend region of the Cornell design, where the photon flux is high and where the pressure has been held to 1 nTorr to reduce detector backgrounds, both types of pumps are used with an antechamber configuration (see Fig. 3) to give a total pumping speed of about 2500 L/s/m.

Most designers favor a chamber made from copper or a copper alloy, similar to the chamber installed in the electron ring at HERA. In addition to the low desorption coefficient mentioned above, copper has good thermal properties and is self-shielding for the synchrotron radiation emitted by the beams (thus obviating the need for a lead liner on the outside of the chamber).

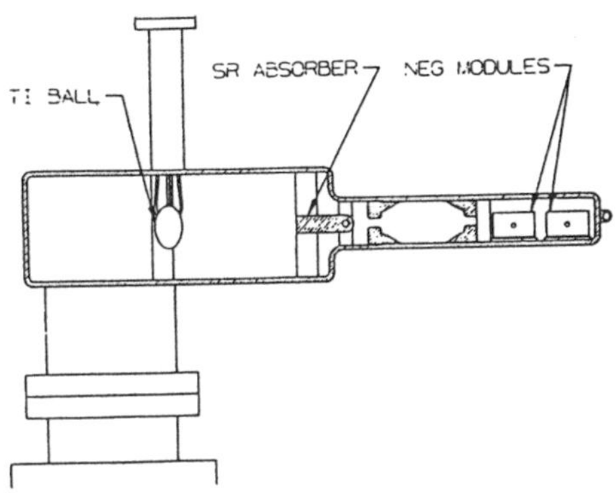

Figure 3. Vacuum chamber for the transition region of CESR-B [4] where the bend radius is only 45 m and the pressure must be held to less than 1 nTorr of CO and CO_2.

RF System

The main challenges for the RF system include:

- replacing the large synchrotron radiation power loss
- minimizing the HOM impedance per cell

The synchrotron radiation losses for an 8 or 9 GeV beam in the high-energy ring of a B factory could be 5 MW at a design luminosity of 3×10^{33} cm^{-2} s^{-1}. The issue is not the power per se, however, but is related to the need for controlling the HOM impedance by reducing the number of cavities. This results in a requirement for high input power through the cavity window—up to 500 kW for a room-temperature system. (To put this value in context, it is only half of the power transmitted through the output window of a modern klystron.) Special windows are being designed to handle this power level. It is also important to minimize the HOM impedance of an individual cavity by damping techniques in order to ensure practical parameters for the feedback system or, possibly, to eliminate the need for feedback entirely.

Both room-temperature [1–3,5] and superconducting [4] cavity designs are being actively developed for B factory use. In the room-temperature case, single- [1,3,5] or two-cell [2] cavities are being considered. Waveguides or slots in the cavity body are used to couple out the dangerous HOMs. With this technique, damping to a Q of about 30 has been demonstrated (at low power) in a pillbox cavity [5]. It is not possible to use the waveguide technique with superconducting cavities, but in this case it is not necessary to optimize the shunt impedance of the cavity and a large beam aperture is acceptable. In the Cornell approach, the aperture is sufficiently large that the HOMs propagate to a room-temperature ferrite load on the inner surface of the beam tube. Calculated damping to the level of $Q \approx 70$ is obtained [4].

The choice of superconducting technology will minimize the number of RF cells required. However, in the heavily beam loaded regime of a B factory, the advantage is only about 30% (assuming the same limitation on cavity window power as in the room-temperature case). In designs involving crab cavities, the use of superconducting technology is likely to be preferred. For this application the requirements are high voltage and low power, which match well with the strengths of superconducting RF. To serve as a crab cavity, the cell must be driven at a transverse deflecting mode (TM110) rather than at the fundamental.

Feedback System

The requirement here is to control the growth of potentially strong coupled-bunch instabilities driven by the HOMs of the RF system. Due to the high beam current and large number of bunches, the instabilities can grow rapidly (≈ 1 ms), and the bandwidth requirements can be high (≈ 100 MHz). It is worth noting that the response of the feedback system to injection transients may dominate the power requirements. This issue favors an injection system that is phase-locked to the ring RF systems. It also helps to inject the beam in many small portions rather than large amounts of charge all at once.

A promising approach is to use a bunch-by-bunch system operating in the time domain [14]. An advantage of this choice is that the system can damp dipole motion from any source, including injection transients and beam-beam disturbances as well as coupled-bunch instabilities.

V. Summary and Outlook

The construction of a high-luminosity asymmetric B factory provides excellent scientific opportunities, combining first-rate particle physics incentives (to study the origins of *CP* violation) with equally exciting challenges in both the accelerator physics and accelerator technology areas. Challenges in accelerator physics include:

- development of lattices to collide and then cleanly separate two unequal energy beams
- achieving high luminosity in asymmetric beam-beam collisions
- designing effective masking techniques to protect the detector

Challenges in accelerator technology include:

- designing vacuum systems capable of handling large thermal loads, providing adequate pumping speed, and having acceptable impedance characteristics
- designing RF systems capable of handling high beam power and providing greatly reduced HOM impedance
- designing wideband bunch-by-bunch feedback systems

Effective approaches to all of these challenges have been identified and R&D activities are being vigorously pursued at many laboratories to optimize designs and finalize design choices. Extensive simulation studies of accelerator physics issues are also being carried out to better understand the beam-beam interaction and beam instabilities.

It is recognized by the various B factory design groups that making a large jump in luminosity will not be an easy task. Perhaps the most important ingredient in ensuring the success of a B factory will be to constantly remember to treat these challenges with proper respect.

Acknowledgments

I would like to thank my many "co-conspirators" in B factory design, not only at LBL, SLAC, Caltech, and LLNL, but at CERN, Cornell, DESY, KEK, Novosibirsk, and PSI, for numerous profitable interactions. I would especially like to acknowledge my colleague Swapan Chattopadhyay, who dragged me kicking and screaming into this exciting endeavor.

REFERENCES

[1] K. Balewski, et al., "Study of an Asymmetric B-Factory," *Beam Dynamics Issues of High-Luminosity Asymmetric Collider Rings*, in AIP Conf. Proc. 214, Berkeley, CA, August 1990, p. 565-574.

[2] "Accelerator Design of a KEK B Factory," ed. by S. Kurokawa et al., KEK Report 90-24, March 1991.

[3] "Feasibility Study for a B-Meson Factory in the CERN ISR Tunnel," ed. by T. Nakada, CERN 90-02, PSI PR-90-08, March 1990.

[4] K. Berkelman, et al., "CESR-B: Conceptual Design for a B Factory Based on CESR," CLNS 91-1050, January 1991.

[5] "An Asymmetric B Factory Based on PEP," Conceptual Design Report, LBL PUB-5303, SLAC-372, CALT-68-1715, UCRL-ID-106426, UC-IIRPA-91-01, February 1991.

[6] A. Zholents, "Novosibirsk B-Factory," *Beam Dynamics Issues of High-Luminosity Asymmetric Collider Rings*, in AIP Conf. Proc. 214, Berkeley, CA, August 1990, pp. 592-593.

[7] P. Grosse Wiesmann, et al., "A Linac-on-Ring Collider B-Factory Study," *Beam Dynamics Issues of High-Luminosity Asymmetric Collider Rings*, in AIP Conf. Proc. 214, Berkeley, CA, August 1990, p. 602-615.

[8] "The Physics Program of a High-Luminosity Asymmetric B Factory at SLAC," ed. by D. Hitlin, SLAC-353, LBL-27856, CALT-68-1588, October 1989.

[9] H. DeStaebler, "Interaction Region Considerations," *Beam Dynamics Issues of High-Luminosity Asymmetric Collider Rings*, in AIP Conf. Proc. 214, Berkeley, CA August 1990, p. 59-80.

[10] D. Rice, "Beam-Beam Interaction: Experimental," *Beam Dynamics Issues of High-Luminosity Asymmetric Collider Rings*, in AIP Conf. Proc. 214, Berkeley, CA August 1990, p. 219-234.

[11] Y.H. Chin, "Symmetrization of the Beam-Beam Interaction in an Asymmetric Collider," *Beam Dynamics Issues of High-Luminosity Asymmetric Collider Rings*, in AIP Conf. Proc. 214, Berkeley, CA August 1990, p. 424-433.

[12] S. Krishnagopal and R. Siemann, Phys. Rev. D **41**, 1741 (1990).

[13] K. Oide and K. Yokoya, Phys. Rev. A **40**, 315 (1989).

[14] F. Pedersen, "Feedback Systems," *Beam Dynamics Issues of High-Luminosity Asymmetric Collider Rings*, in AIP Conf. Proc. 214, Berkeley, CA, August 1990, p. 246-269.

List of Participants

Baartman, Rick	TRIUMF	Vancouver		Canada
Banerjee, Sreeparna	University of Virginia	Charlottesville	VA	USA
Bogacz, S. Alex	Fermilab	Batavia	IL	USA
Bohn, Courtlandt	ANL	Argonne	IL	USA
Brennan, Mike	BNL	Upton	NY	USA
Brown, Kevin	BNL	Upton	NY	USA
Bryant, Philip	CERN	Geneva		Switzerland
Byrd, John	Cornell University	Ithaca	NY	USA
Caspers, Friedhelm	CERN	Geneva		Switzerland
Chan, Kwok-Chi	LANL	Los Alamos	NM	USA
Chang, Chu Rui	SSC Laboratory	Dallas	TX	USA
Chang, Tzu-Hsien	U. of Massachusetts	Amherst	MA	USA
Chao, Alexander	SSC Laboratory	Dallas	TX	USA
Chao, Y.-C.	SLAC	Stanford	CA	USA
Chattopadhyay, S.	LBL	Berkeley	CA	USA
Chen, Pisin	SLAC	Stanford	CA	USA
Chen, Yu-Jiuan	LLNL	Livermore	CA	USA
Colestock, Patrick	Fermilab	Batavia	IL	USA
Corbett, Jeff	SLAC	Stanford	CA	USA
Craig, George	LLNL	Livermore	CA	USA
Curtin, Mark S.	Rockwell International	Canoga Park	CA	USA
Curtin, Michael	Booz-Allen & Hamilton	Arlington	VA	USA
Dainelli, Antonio	INFN-Laboratori	Padova		Italy
Dekkers, Daniel	CERN	Geneva		Switzerland
Delayen, Jean	ANL	Argonne	IL	USA
Dylla, Fred	CEBAF	Newport News	VA	USA
Edwards, Donald	SSC Laboratory	Dallas	TX	USA
Emma, Paul	SLAC	Stanford	CA	USA
Evans, John	CERN	Geneva		Switzerland
Evans, Lyndon	CERN	Geneva		Switzerland
Fauchet, Anne-Marie	BNL	Upton	NY	USA
Faugier, Andre	CERN	Geneva		Switzerland
Fink, Charles	ANL	Argonne	IL	USA
Fitzgerald, Dan	LANL	Los Alamos	NM	USA
Forster, Martin	University of Dortmund	Dortmund		Germany
Funk, L. Warren	SSC Laboratory	Dallas	TX	USA
Gangopadhyaya, Asim	Loyola University	Chicago	IL	USA
Gareyte, Jacques	CERN	Geneva		Switzerland
Garnett, Robert	LANL	Los Alamos	NM	USA

Garoby, Roland	CERN	Geneva		Switzerland
Ghigo, Andrea	INFN-Frascati	Frascati		Italy
Goderre, Glenn	Fermilab	Batavia	IL	USA
Gounder, K.	U. of So. Carolina	Columbia	SC	USA
Graham, William	SLAC	Menlo Park	CA	USA
Greenwald, Shlomo	Cornell University	Ithaca	NY	USA
Greenwald, Zipi	Cornell University	Ithaca	NY	USA
Grobner, Oswald	CERN	Geneva		Switzerland
Grunder, Hermann	CEBAF	Newport News	VA	USA
Guerra, Eduardo	SLAC	Menlo Park	CA	USA
Hahn, Alan	Fermilab	Batavia	IL	USA
Hanna, Samy Maurice	BNL	Upton	NY	USA
Hartman, Spencer	UCLA	Los Angeles	CA	USA
Harwood, Leigh	CEBAF	Newport News	VA	USA
Herr, Werner	CERN	Geneva		Switzerland
Ho, Ching-Hung	UCLA	Argonne	IL	USA
Hofmann, Albert	CERN	Geneva		Switzerland
Holmes, Stephen	Fermilab	Batavia	IL	USA
Howe, David	U.S. State Department	APO	NY	USA
Humphries, Jr., Stanley	U. of New Mexico	Albuquerque	NM	USA
Hutson, Richard	LANL	Los Alamos	NM	USA
Jackson, Gerald	Fermilab	Batavia	IL	USA
Jacobs, Ken	MIT Bates Linac	Middleton	MA	USA
Jason, Andrew	LANL	Los Alamos	NM	USA
Johnson, Kenneth F.	LANL	Los Alamos	NM	USA
Jones, Kevin	LANL	Los Alamos	NM	USA
Jowett, John M.	CERN	Geneva		Switzerland
Kim, Sun Kee	Rutgers University	Piscataway	NJ	USA
Ko, In Soo	Pohang Inst. of Sci.	Kyungbuk		Korea
Koscielniak, Shane	TRIUMF	Vancouver		Canada
Kraus, Jr., R.H.	LANL	Los Alamos	NM	USA
Krejcik, Patrick	SLAC	Stanford	CA	USA
Kuske, Peter	BESSY	Berlin		Germany
Leemann, Christoph	CEBAF	Newport News	VA	USA
Leifeste, Gordon Travis	SSC Laboratory	Dallas	TX	USA
Li, Rui	CEBAF	Newport News	VA	USA
Liger, Philippe	CEBAF	Newport News	VA	USA
Limberg, Torsten	SLAC	Stanford	CA	USA
Lofnes, Tor	Uppsala University	Uppsala		Sweden

Lombardi, Augusto	INFN-Legnaro	Stony Brook	NY	USA
Lu, Xianping	University of Colorado	Boulder	CO	USA
MacLachlan, James A.	Fermilab	Batavia	IL	USA
Mahoney, Kelly	CEBAF	Newport News	VA	USA
Mais, Helmut	DESY	Hamburg		Germany
Mane, Sateesh	BNL	Upton	NY	USA
Martin, Philip	Fermilab	Batavia	IL	USA
Mazarakis, Michael	Sandia Nat. Lab.	Albuquerque	NM	USA
Meisner, Keith	Fermilab	Batavia	IL	USA
Merminga, Lia	SLAC	Stanford	CA	USA
Miller, R. Bruce	Titan Technologies	Albuquerque	NM	USA
Milton, Stephen	Paul Scherrer Institute	Villigen		Switzerland
Minty, Michiko	Indiana University	Bloomington	IN	USA
Month, Mel	BNL	Upton	NY	USA
Moore, Craig	Fermilab	Batavia	IL	USA
Nakai, Hirotaka	KEK	Tsukuba		Japan
Nakamura, Norio	KEK	Tsukuba		Japan
Nam, Kwanghee	Pohang Accel. Lab.	Kyungbuk		South Korea
Ng, King-Yuen	Fermilab	Batavia	IL	USA
Peggs, Steve	Fermilab	Batavia	IL	USA
Peschardt, Ernst	CERN	Geneva		Switzerland
Placidi, Massimo	CERN	Geneva		Switzerland
Plum, Michael	LANL	Los Alamos	NM	USA
Poncet, Alain	CERN	Geneva		Switzerland
Potier, Jean Pierre	CERN	Geneva		Switzerland
Rangarajan, Govindan	LBL	Berkeley	CA	USA
Raparia, Deepak	SSC Laboratory	Dallas	TX	USA
Raubenheimer, Tor	Stanford University	Stanford	CA	USA
Rees, Grahame	Rutherford Appleton	Chilton, Didcot		England
Riche, Alain Joe	CERN	Geneva		Switzerland
Rinolfi, Louis	CERN	Geneva		Switzerland
Robin, David Seth	UCLA	Los Angeles	CA	USA
Rosenzweig, James	Fermilab	Batavia	IL	USA
Ruth, Ronald	SLAC	Stanford	CA	USA
Sagalovsky, Leonid	ANL	Argonne	IL	USA
Sakanaka, Shyogo	KEK	Tsukuba		Japan
Saritepe, Selcuk	Fermilab	Batavia	IL	USA
Seeman, John T.	SLAC	Stanford	CA	USA
Serio, Mario	INFN-LNF	Frascati		Italy

Shaposhnikova, Elena	Inst. for Nuc. Res.	Moscow		USSR
Shea, Thomas	BNL	Upton	NY	USA
Sorensen, Allan	University of Aarhus	Aarhus		Denmark
Spjeldvik, Walther	Weber State College	Ogden	UT	USA
Stampke, Stuart	SSC Laboratory	Dallas	TX	USA
Stockhorst, Hans	KFA	Juelich		Germany
Sutter, David F.	USDOE	Washington	DC	USA
Syphers, Michael J.	SSC Laboratory	Dallas	TX	USA
Tennyson, Jeff	SLAC	Stanford	CA	USA
Thompson, Kathleen	SLAC	Stanford	CA	USA
Turner, Stuart	CERN	Geneva		Switzerland
Turner, William	LLNL	Livermore	CA	USA
von Holtey, Georg	CERN	Geneva		Switzerland
Walker, Nicholas	SLAC	Stanford	CA	USA
Wang, Haipeng	SUNY at Stony Brook	Stony Brook	NY	USA
Wang, Xijie	UCLA	Los Angeles	CA	USA
Wangler, Thomas	LANL	Los Alamos	NM	USA
Wanzenberg, R.	DESY	Hamburg		Germany
Weiland, Thomas	DESY	Hamburg		Germany
Weng, Wu-Tsung	BNL	Upton	NY	USA
Wildman, David	Fermilab	Batavia	IL	USA
Wilkinson, Carol	LANL	Los Alamos	NM	USA
Yamamoto, Hitoshi	University of Chicago	Chicago	IL	USA
Yokoya, Kaoru	KEK	Tsukuba		Japan
Yu, Xiao-Tong	MIT	Cambridge	MA	USA
Zholents, Alexander	Inst. of Nuclear Phys.	Novosibirsk		USSR
Zisman, Michael S.	LBL	Berkeley	CA	USA
Zolfaghari, Abbi	MIT	Middleton	MA	USA

MIX
Papier aus verantwortungsvollen Quellen
Paper from responsible sources
FSC® C105338

If you have any concerns about our products,
you can contact us on
ProductSafety@springernature.com

In case Publisher is established outside the EU,
the EU authorized representative is:
Springer Nature Customer Service Center GmbH
Europaplatz 3, 69115 Heidelberg, Germany

Printed by Libri Plureos GmbH
in Hamburg, Germany